Photofunctional Nanomaterials for Biomedical Applications

Photofunctional Nanomaterials for Biomedical Applications

Edited by
Chunxia Li and Jun Lin

WILEY-VCH

Editors

Prof. Chunxia Li
Shandong University
72 Binhai Road
266237 Qingdao
China

Prof. Jun Lin
Changchun Institute of Applied
Chemistry, Chinese Academy of Sciences,
5625 Renmin Street
130022 Changchun
China

Cover Image: © LAGUNA DESIGN/Getty Images

Library of Congress Card No.: applied for

British Library Cataloguing-in-Publication Data
A catalogue record for this book is available from the British Library.

Bibliographic information published by the Deutsche Nationalbibliothek
The Deutsche Nationalbibliothek lists this publication in the Deutsche Nationalbibliografie; detailed bibliographic data are available on the Internet at <http://dnb.d-nb.de>.

Print ISBN: 978-3-527-35333-0
ePDF ISBN: 978-3-527-84532-3
ePub ISBN: 978-3-527-84533-0
oBook ISBN: 978-3-527-84534-7

Typesetting Straive, Chennai, India

Contents

Foreword

According to the data by the International Agency for Research on Cancer of the World Health Organization, the global incidence rate of cancer can potentially escalate to 29.5 million by 2040. Even in the present era, cancer continues to persist as a prominent contributor to the deaths of humans worldwide. Tumor metastasis, as the leading factor, accounts for approximately 90% of cancer-related fatalities. In fact, metastasis may occur even at the time of cancer diagnosis for a number of patients. Importantly, plenty of individuals afflicted with lung cancer exhibit metastatic disease, as evidenced by an over 80% diagnosis rate. Furthermore, recent research have revealed the presence of diverse bacteria within the majority of solid tumors, which has been evidenced to be responsible for tumor metastasis and colonization. The current prevailing treatment modalities, such as chemotherapy, hormonal therapy, and radiation therapy, have achieved significant progress in treating metastatic cancer and prolonging the survival of patients to certain extents; unfortunately, the overall therapeutic efficacy remains unsatisfactory.

The advancements in nanotechnology have aroused significant progress or even breakthroughs in the biomedical fields. Photofunctional nanomaterials, being a novel class of light-responsive materials, have gained extensive attention and utilization. For example, they as carriers can achieve targeted drug delivery and controlled release, resulting in the enhanced efficacy of medications. After ingenious modifications with functional molecules, such as photosensitizers, photothermal agents, chemotherapeutic drugs, and gene composites, photofunctional nanomaterials exhibit great potential for advancing various therapeutic modalities such as photodynamic therapy, photothermal therapy, chemotherapy, antibacterial therapy, gene therapy, and biosensing. Moreover, these nanomaterials also enable simultaneous image-guided multi-modal therapy. Therefore, photofunctional nanomaterials hold immense promise in the biomedical fields.

Based on this, in this book, Professor Chunxia Li, Professor Jun Lin, and their collaborators have reviewed and summarized recent research progress on photofunctional nanomaterials for biomedical fields in a comprehensive and well-organized manner. Impressively, the authors have elucidated the underlying functioning mechanisms of these nanomaterials in depth and extensively discussed the challenges encountered by photofunctional nanomaterials along with their future development prospects. This book offers researchers in the nanomedicine field a highly methodical and professional compendium and reference summarizing the application of photofunctional nanomaterials in a broad spectrum.

Finally, it is anticipated that this publication will greatly benefit further advancements in the research and applications of photofunctional nanomaterials and establish a robust groundwork for the future clinical translation.

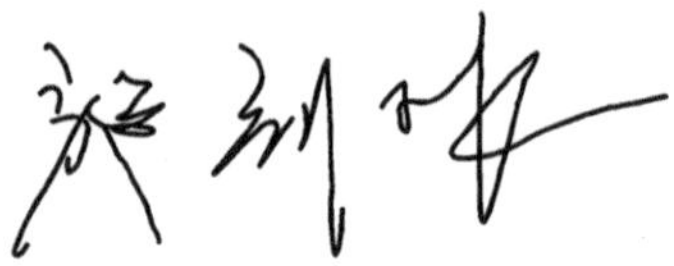

January 30, 2024

Prof. Jianlin Shi
State Key Laboratory of High Performance
Ceramics and Superfine Microstructure
Shanghai Institute of Ceramics
Chinese Academy of Sciences
Shanghai, China

Preface

In recent years, morbidity and mortality of cancer have obviously increased; therefore, cancer has become a serious threat to human health. Thus, how to provide new clues to prewarning and early diagnosis and therapy of cancer is highly demanded. To achieve this goal, the new material chemistry is essential because the nanoparticles have to be uniform and tunable in terms of physical properties such as size, shape, and surface. Photofunctional nanomaterials are light-driven functional materials that can effectively convert and utilize light energy, an inexhaustible source of clean and renewable energy. Especially, near-infrared (NIR) light is a highly accurate non-physical and non-invasive external stimulation method that has the unique advantages of high penetration depth and low fluorescence background interference, achieving synchronous control of time and space. Therefore, the light can be utilized to achieve controllable chemotherapy or photodynamic therapy (PDT). Photothermal nanomaterials can convert light energy into heat energy, resulting in local overheating and tumor ablation (photothermal therapy, PTT). Therefore, the design and applications of photofunctional nanocomposites are an important branch of biomedical functional materials and have become the frontier research direction in interdisciplinary fields.

This book will focus on providing researchers and graduate/undergraduate students in interdisciplinary fields including chemistry, materials science, and biomedicine, highlighting their full potential in biomedical applications including tumor diagnosis and therapy, photogenetically engineered bacteria, optical information storage, and biosensing.

Chapter 1 demonstrates a basic introduction and classification of photofunctional nanomaterials as well as their applications in biomedical fields.

From Chapters 2–7, the rare earth luminescence nanomaterials are systematically introduced, mainly shedding light on their luminescence mechanism, multiform luminescence modulation, construction of composites, and the corresponding applications in gene delivery as well as the biosafety evaluation.

Chapter 8 provides an overview of photosensitizers for PDT. The basic principles of PDT, classifications of various photosensitizers, and the mechanisms during treatment are outlined.

Chapter 9 introduces the persistent luminescent materials for optical information storage applications. And the design of ternary quantum dots and their application in tumor-related marker detection, imaging, and therapy are mentioned in Chapter 10.

Chapter 11 summarizes nanomaterial-induced pyroptosis and immunotherapy including pyroptosis pathways and the potential for immunotherapy, especially in activating effector T cells and promoting dendritic cell maturation.

Chapter 12 mainly highlights the latest developments of inorganic nanomaterials in PTT. The mechanism and application of nanomaterial-based PTT against cancer by photothermal immunotherapy are also expounded.

Chapter 13 begins by presenting an overview of the antibacterial mechanisms inherent in photofunctional antibacterial nanomaterials, focusing on the approaches developed to overcome resistance in multidrug-resistant (MDR) pathogens and the prominent challenges in the field of photofunctional antibacterial nanomaterials.

Chapter 14 focuses on the applications of photoelectrochemicals nanomaterials in biosensing.

Chapter 15 describes the concept of X-ray-activated PDT, outlines the interaction mechanisms between X-rays and nanosystems, and introduces the application of X-ray-activated nanosystems in photodynamic therapy for deep-seated tumors.

And, the last chapter, Chapter 16, introduces the perspectives on the opportunities and potential benefits of photofunctional nanomaterials for biomedicine.

Finally, we greatly acknowledge the substantial contribution from Prof. Hongjie Zhang, Prof. Rongjun Xie, Prof. Yulei Chang, Prof. Ruichan Lv, Prof. Xiaoji Xie, Prof. Renren Deng, Prof. Tao Zhang, Prof. Guanying Chen, Prof. Jing Feng, Prof. Yixi Zhuang, Prof. Ziyong Cheng, Prof. Bingbing Ding, Prof. Ping'an Ma, Prof. Zhiyao Hou, Prof. Biao Dong, Prof. Lin Wang, Prof. Piaoping Yang, Prof. Jinliang Liu, and Dr. Qianqian Sun to this book.

26 January 2024
Qingdao, China

Professor Chunxia Li
Shandong University, Qingdao, China
Professor Jun Lin
Changchun Institute of Applied Chemistry,
Chinese Academy of Sciences,
Changchun, China

Acknowledgments

We greatly acknowledge the substantial contribution from Prof. Hongjie Zhang, Prof. Rongjun Xie, Prof. Yulei Chang, Prof. Ruichan Lv, Prof. Xiaoji Xie, Prof. Renren Deng, Prof. Tao Zhang, Prof. Guanying Chen, Prof. Jing Feng, Prof. Yixi Zhuang, Prof. Ziyong Cheng, Prof. Ping'an Ma, Prof. Zhiyao Hou, Prof. Biao Dong, Prof. Lin Wang, Prof. Piaoping Yang, Prof. Jinliang Liu, and Dr. Qianqian Sun to this book.

1

General Introduction and Background of Photofunctional Nanomaterials in Biomedical Applications

Chunxia Li[1] *and Jun Lin*[2]

[1]*Shandong University, School of Chemistry and Chemical Engineering, Institute of Frontier Chemistry, Binhai Road, Qingdao 266237, P.R. China*
[2]*Chinese Academy of Sciences, Changchun Institute of Applied Chemistry, State Key Laboratory of Rare Earth Resource Utilization, Renmin Avenue, Changchun 130022, P.R. China*

1.1 Introduction to Nanomaterials

From the perspective of human history, the history of human knowledge of the world is the history of the development of scale and materials. With the continuous development of science and technology, human cognition of the world has long exceeded the single macroscopic world, and the emergence of nanotechnology in 1984 has directly brought human beings into an infinite and mysterious "small size, big world." Therefore, as an important tool for human beings to explore the microscopic world, the research and application of nanomaterials have been widely valued by many disciplines in recent years [1].

Nanomaterials are materials that have at least one dimension in three-dimensional space in the nanoscale range (1–100 nm, 1 nm = 10^{-9} m) or are composed of them as basic units [2, 3] and are known as "the most promising materials of the 21st century." As research progresses, scientists are gradually discovering that materials with dimensions at the nanoscale can exhibit unique properties that are superior to those of conventional materials in terms of physics, chemistry, optics, thermodynamics, and magnetism [4–6]. This is because the ratio of the number of atoms on the surface of nanomaterials to the total number of atoms increases dramatically as the particle size decreases. The special properties of nanomaterials are as follows.

1.1.1 Surface and Interfacial Effects

When the size of a material is reduced to the nanoscale, the number of surface atoms, the surface area, and the surface energy increase dramatically, and at the same time, a large number of unsaturated bonds, dangling bonds, and active centers

Photofunctional Nanomaterials for Biomedical Applications, First Edition.
Edited by Chunxia Li and Jun Lin.

appear in nanomaterials, and the surface defects of the material also increase. These defects introduce many surface states in the energy barrier band gap, which become traps for electrons or holes, seriously affecting the optical, photoelectrochemical, and nonlinear optical properties of the materials [7–10]. Therefore, many new properties of nanomaterials are inextricably linked to their surface and interfacial effects.

1.1.2 Small Size Effect

When the size of the material is comparable to or smaller than physical quantities, such as the wavelength of light waves (less than 100 nm), the radius of exciton bands of De Broglie wavelengths (1–10 nm), or the coherence length of superconductivity, the periodic boundary conditions of the internal crystals will be disrupted, and the density of atoms near the surface layer of the particles of amorphous nanoparticles will be reduced. This will lead to significant changes in the macroscopic physical and chemical properties (such as sound, light, electricity, magnetism, heat, and mechanics) of the nanomaterials [11–14].

1.1.3 Quantum Size Effect

When the size of the material is reduced to the nanometer scale, the electronic energy levels near the metallic Fermi energy level change from continuous to discrete, and the continuous energy band, valence band, and conduction band of semiconductors become discrete energy level structures, and the bandgap broadening phenomenon is called the quantum size effect [15–18]. When the energy level spacing is greater than the thermal, magnetic, electrostatic, photonic, or superconducting condensation energy, nanomaterials will exhibit a range of properties that are very different from those of bulk materials.

1.1.4 Macroscopic Quantum Tunneling Effects

When the scale of a material enters the nanometer range, certain macroscopic quantities of nanoparticles (such as particle magnetization intensity, magnetic flux in quantum coherent devices, and electric charge) exhibit tunneling effects that can cross the potential barriers of the macroscopic system and produce changes, known as macroscopic quantum tunneling effect [19–22]. Macroscopic quantum tunneling is the theoretical basis for future microelectronic and optoelectronic devices.

Because of the special properties mentioned above, nanomaterials have many more excellent physicochemical properties than macroscopic materials. Physically, nanomaterials have good electrical conductivity, dielectricity, magnetism, and mechanical properties. From a chemical point of view, nanomaterials are highly active on the surface and are particularly prone to adsorbing other atoms or chemically reacting with other atoms, which greatly improves the catalytic ability of the reaction. As a result, nanomaterials have broad application prospects in many fields such as optoelectronics, environmental science, and biomedicine.

1.2 Introduction and Classification of Photofunctional Nanomaterials

The development and application of materials are a sign of the progress of time and human civilization. The history of materials is as long as the history of mankind. Mankind has gone through a long period of Stone Age, Bronze Age, and Iron Age. Nowadays, with the continuous improvement of nanomaterial synthesis technology, research on nanomaterials has gradually shifted to refinement and functionalization. At the same time, with the growing demand for material functions in science and technology and living standards, as well as the cross-fertilization of the frontiers of various disciplines, a wide range of functionalized nanomaterials have emerged.

Functionalized nanomaterials are diverse and wide-ranging, and there are many ways to classify them. Functional nanomaterials can be classified into electrical nanomaterials, magnetic nanomaterials, optical nanomaterials, thermal nanomaterials, acoustic nanomaterials, chemical nanomaterials, invisible nanomaterials, and so on according to their performance, which is spectacular, and a large number of new functional nanomaterials have been introduced every year. Among them, a series of nanomaterials with unique optical properties (light-functional nanomaterials) have been successfully prepared [23–28], which has become one of the hotspots in the field of functionalized nanomaterials in recent years and has been widely applied in many fields, such as bioanalysis and sensing, illumination and display, environmental monitoring and purification, energy conversion and storage, biomedicine, anticounterfeiting, and information.

In fact, the understanding of optical phenomena in nature has a long history, and it can be said that human beings have progressed along with the knowledge of optical phenomena in nature. For example, fire is man's first perception of light, without which human society would not have survived. The propagation and absorption of light profoundly affect all aspects of nature, and without the phenomena of selective absorption, scattering, transmission, and reflection of light by matter, there would be no colors, no heat, no energy transformed by light, and no production of all biomass in nature. Photofunctional nanomaterials are light-driven functional materials that can effectively convert and utilize light energy, an inexhaustible source of clean and renewable energy. To date, the applications of photofunctional nanomaterials are very diverse, but essentially, the same physical and chemical processes take place and follow similar laws, including photon capture, photon absorption and utilization (conversion), and physicochemical processes at the surface interface [29].

1.2.1 Capture of Photons

Light has a fluctuating and particle duality (wave-particle duality), and when considering the energy conversion between light and electrons, light is treated as a particle called a photon. The trapping of photons is the first step in the process of light conversion and utilization by photofunctional nanomaterials, and the more photons that are trapped, the greater the chance that they will be absorbed. It can also be assumed that the photon trapping ability of a material determines the upper limit of its light conversion efficiency. In general, the photon trapping capability of photofunctional nanomaterials can be increased in three directions: broadening the absorption spectral range of the material, reducing the loss of light after it has passed through the material, and increasing the optical range of light in the material.

1.2.2 Absorption and Conversion of Photons

In addition to trapping as many photons as possible, photofunctional nanomaterials must have the ability to absorb and convert the trapped photons into phonon vibrations, photogenerated electron-hole pairs, or other energies. When light is shone on a material, various physicochemical effects such as photothermal effect, photoluminescence, photoelectric effect, and photochemical effect, are produced due to electromagnetic vibrations of electromagnetic waves or inelastic collisions of photons.

Among them, the photothermal effect refers to the fact that photothermal materials, after absorbing the energy of light radiation, do not directly cause a change in the internal electronic state but convert the absorbed light energy into the vibration of the crystal point structure (which means the production of phonons), thus causing a temperature rise and the generation of thermal energy [30]. As the research progressed, it was found that the photothermal effect of photofunctional nanomaterials depends not only on the incident light but also on the absorption spectrum of the material itself. Materials are dense systems consisting of a large number of atoms and molecules, and thermal energy is the average kinetic energy of the irregular motion of these particles. In other words, the accelerated motion of atoms and molecules at the microscopic level corresponds exactly to the increased temperature of the material at the macroscopic level (known as heating). The characteristic frequencies of the continuous relative vibrations and rotations of the atoms and molecules in the material are similar to those of infrared light, so they can resonate with external infrared light. Therefore, when a material is irradiated with infrared light, the motion of the atoms and molecules of the material will be enhanced, and a large amount of heat will be emitted. In contrast, the material absorbs less blue-violet light, resulting in less heat generation and a poorer thermal effect.

Photoluminescence is the process by which a material absorbs photons and then re-radiates them [31]. When the material is irradiated with light of a certain wavelength, the electrons in the ground state of the material (mainly π electrons and f and d electrons) are excited to a high-energy state, and when the external light stops, the electrons in the excited state will jump back to the ground state. In the process of the electron jump, some of the energy is emitted in the form of photons to accomplish the purpose of the light. Photoluminescence can be divided into two categories according to the delay time: fluorescence and phosphorescence [32]. Fluorescence is the emission of photons immediately after the substance is excited, and the luminescence time is $\leq 10^{-8}$ seconds. Phosphorescence can continue to emit light for a long time, and usually, the luminescence time is $\geq 10^{-8}$ seconds. According to light excitation and emission, photoluminescent nanomaterials can be classified into upconversion luminescence or downconversion luminescence nanomaterials according to the nature of their light emission. Among them, upconversion luminescence is a photoluminescence phenomenon that violates Stokes' law, which is manifested by the conversion of several low-energy (long-wavelength) photons into one high-energy (short-wavelength) photon [33]. Downconversion luminescence is a photoluminescence phenomenon that obeys Stokes' law and is manifested by the conversion of a high-energy photon into one or more low-energy photons [34]. When one low-energy photon is emitted, it is commonly referred to as luminescence, and when two or more low-energy photons are emitted, it is referred to as quantum clipping of luminescence.

Photoelectric effect, which is the electrical effect of light, refers to the material in the light of the phenomenon of emission of electrons, the essence of photon excitation material to produce electrons and hole pairs; electrons migrate into the external circuit to do work, manifested as electrical energy [35]. Photochemical effect is the chemical effect of light, which refers to the material used in the photon excitation to produce electrons and hole pairs. They were with the reactant redox reaction, stored as chemical energy phenomenon [36].

1.2.3 Physical-chemical Processes at the Surface Interface

The physical processes at the surface interface of photofunctional nanomaterials mainly involve thermal radiation dissipation and relaxation quenching of photogenerated photons. In photothermal materials, by controlling the composition and structure of the nanomaterials, the radiation angle coefficients of the materials can be effectively reduced, thereby reducing the dissipation of thermal energy and improving the utilization of photothermal conversion energy. In luminescent materials, photons generated by photoluminescence can be recaptured by the material and achieve relaxation by nonradiative bonding, thus reducing the luminescence effect. In semiconductor nanomaterials, photoelectrons migrate to the surface of the material and return to the interior of the material to recombine with holes after work is done by the external circuit, achieving the conversion of light energy into electrical energy.

Surface interface chemical reactions of photofunctional nanomaterials usually occur during photocatalysis of the materials. Photogenerated electrons and holes migrate to the surface or interface of the material and combine with electron acceptors or donors adsorbed on the surface of the photocatalytic material to undergo a redox reaction, storing the energy in the chemical bonds of the products and realizing the conversion of light energy into chemical energy. The detailed process of the chemical reaction at the surface interface roughly involves the following processes. First, the reactants must diffuse around the material and pass through the Helmholtz layer to be adsorbed onto the active sites of the material. Next, the reactants undergo structural rearrangement at the active site and undergo redox reactions with photogenerated electron holes. Finally, the reaction products desorb from the surface of the material and return to the reaction solution. In the whole process, a series of elementary reactions such as adsorption of reactants, charge exchange, and desorption of products together constitute the whole interfacial reaction process.

1.3 Introduction to Nanobiomedicine

In recent years, nanomaterials have been gradually integrated with biology and medicine in the development process, gradually forming a new discipline, namely "nanobiomedicine" (Figure 1.1) [38, 39]. Some people call the cooperation between nanomaterials and biomedicine "a great change" because it not only opens the door for nanobiology and nanomedicine but also brings new opportunities for biomedical research and clinical application. Governments around the world have increased the investment of funds and personnel in nanomedicine research, and nanobiomedicine has rapidly become a frontier and hot topic in the development of the biotechnology field in various countries, attracting more and more attention and expectations.

To be more precise, nanobiomedicine is an emerging discipline that applies nanomaterials to biomedical field, which involves materials science, physics, chemistry, biology, medicine, quantum science, and many other fields, and has very distinctive multidisciplinary cross characteristics. The research directions of nanobiomedicine include but are not limited to drug delivery and release [40–42], bioimaging [43], diagnosis and treatment of diseases (especially tumors) [44, 45], biosensing [46, 47], tissue engineering, and so on [48–50].

1.3.1 Nano-drug Delivery Systems

Through the design and preparation of nanocarriers such as nanoparticles, nanocapsules, nanofibers, and drugs are encapsulated in nanoscale structures to achieve precise drug delivery and release. Such nano-drug delivery systems can improve drug bioavailability, reduce side effects, and enable targeted therapy.

1.3.2 Nano-imaging Technology

High-resolution nano-imaging technology has been developed by exploiting the special optical, magnetic, and acoustic properties of nanomaterials. For example, nanoparticles can be used as contrast agents for fluorescence imaging, magnetic resonance imaging (MRI), and photoacoustic imaging (PAI) to achieve accurate imaging of biological tissues and lesions.

1.3.3 Nano-diagnostic Technologies

By exploiting the special properties of nanomaterials, nano-diagnostic technologies have been developed for biomarker detection, early diagnosis of diseases, and prognostic assessment. For example, nanoprobes can be used to detect the presence and activity state of specific molecules, enabling highly sensitive and selective diagnosis.

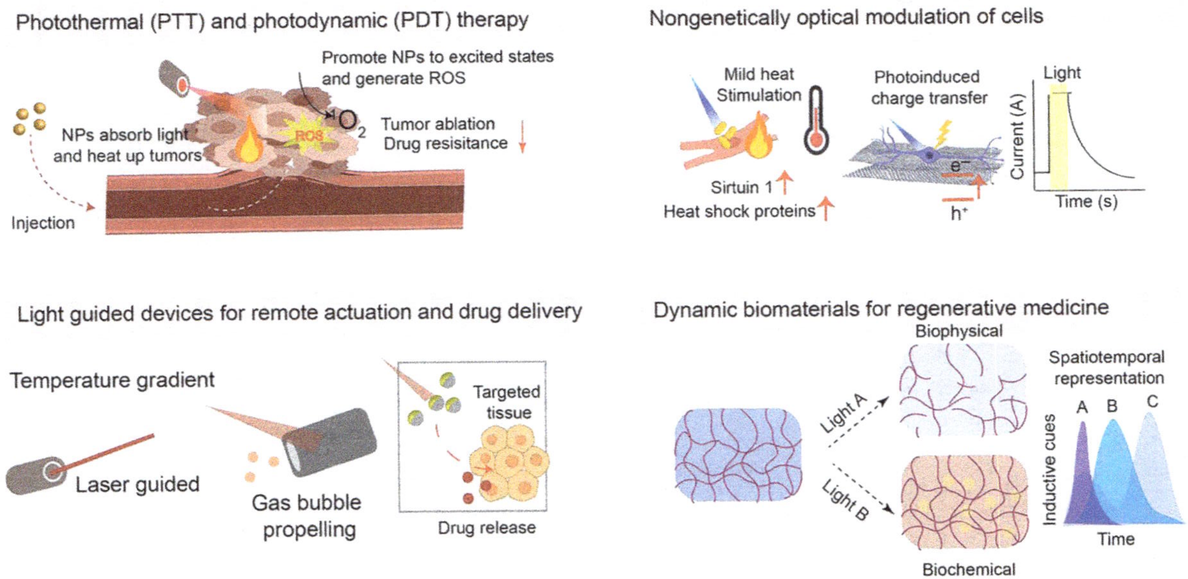

Figure 1.1 Inorganic light-responsive nanomaterials have been widely applied to four main research fields of light-responsive biomaterials, which consist of PTT and PDT, light-guided devices for remote actuation and drug delivery, nongenetically optical modulation of cells, and dynamic biomaterials for regenerative medicine. Source: [51]/John Wiley & Sons/CC by 4.0.

1.3.4 Nanotherapeutic Technology

Using the special properties of nanomaterials, therapeutic methods have been developed for thermotherapy, phototherapy, drug release, and other therapeutic methods. For example, thermal therapy for tumors can be achieved by absorbing light energy from nanomaterials to produce thermal effects or by using nanoparticles with specific responsiveness to achieve targeted drug release.

1.3.5 Nano-biosensors

Using the special properties of nanomaterials and biorecognition molecules, nano-biosensors have been developed to monitor biomolecules, cellular activities, and biological processes. These sensors can monitor physiological and pathological changes in organisms in real time and provide timely diagnostic and therapeutic feedback.

1.3.6 Tissue Engineering

Nanotechnology can be used to create materials such as scaffolds, nanoparticles, and nanofibers with nanoscale structures to support and guide cell growth and differentiation. These nanomaterials can mimic the microstructure of human tissues and provide a suitable physical and chemical environment to promote cell adhesion, proliferation, and differentiation. In addition, nanotechnology can be used to prepare nanocarriers with controlled drug release to enhance the therapeutic efficacy of tissue-engineered constructs.

At present, nanobiomedicine has become an important direction in the development of nanotechnology, and its booming momentum will continue to provide new technologies and methods for modern biomedical research, open up new horizons for important biomedical problems at the nanoscale, and reveal the relevant new principles and possible practical applications.

In the biomedical field, the choice of light has unique requirements. As a common external stimulus with the advantages of noninvasiveness, high spatial and temporal resolution, and spatial and temporal controllability (including controllable light intensity and wavelength), light sources play an important role in biology and medicine [37, 51] and can induce organisms to perform or regulate many specific biological processes at a given site, such as gene transfection, cell function, signaling, ion channel opening, protein activity, molecular isolation, and tissue regeneration (Figure 1.2) [52]. However, most current photosensitive components respond only to ultraviolet or visible light, so the use of light as an excitation source for biomedical applications is subject to certain dilemmas that jeopardize its potential applications; for example, ultraviolet or visible light is readily absorbed and scattered in living tissues and has a very shallow tissue penetration depth [53, 54]. In addition, ultraviolet is phototoxic and is likely to damage biomolecules such as nucleic acids, proteins, and lipids. These issues need to be addressed from two perspectives. On the one hand, since biological tissues cannot efficiently absorb near-infrared (NIR) light

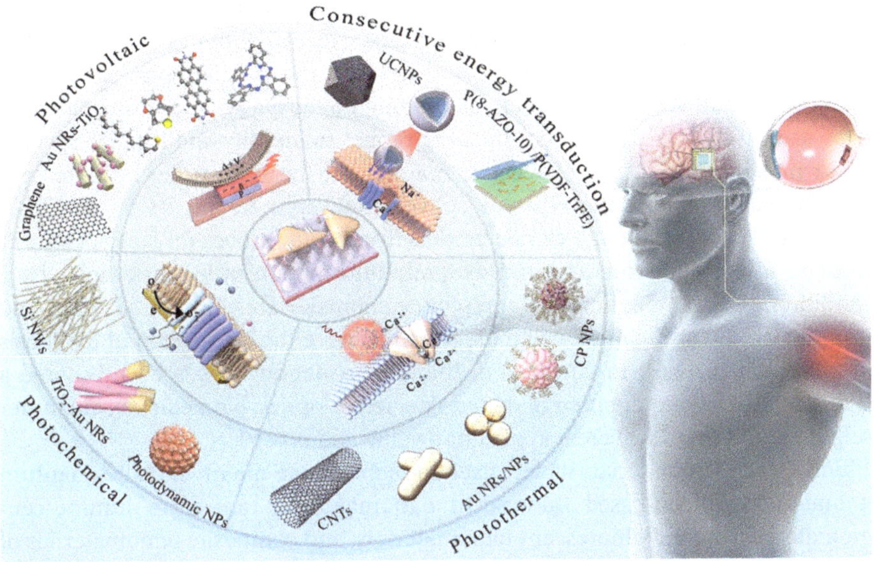

Figure 1.2 Summary of photoactive nanomaterials for regulation of biomolecules, neurons, and nerve tissues. Source: Reproduced with permission from Ref. [52]; © 2022/American Chemical Society.

themselves, NIR light sources with lower tissue absorption, less light scattering, and deeper tissue penetration should be used to replace ultraviolet and visible light for biomedical research and applications [55]. On the other hand, there is a need to actively research and develop photofunctional nanomaterials or devices that can absorb NIR light and convert it into the desired signal or energy.

1.4 Classification of Photofunctional Nanomaterials

At this point, it should be stated in advance that due to the wide range of photofunctional nanomaterials, it is impossible to study them in detail one by one. Therefore, in connection with the theme of this book-"Photofunctional nanomaterials for biomedical applications," we will discuss the fluorescent, photothermal, photodynamic, photoelectrochemical, and photoacoustic parts of photofunctional nanomaterials and their research in biomedical fields such as biodetection [56], optogenetics [57], antibacterial [46], tumor diagnosis and treatment [45, 58], and other biomedical fields.

1.4.1 Fluorescent Nanomaterials

Fluorescent nanomaterials are nanoscale materials with special optical properties that absorb light energy at specific wavelengths and emit fluorescence at specific wavelengths. Compared with traditional organic dye fluorescent molecules, fluorescent nanomaterials have the advantages of better optical stability, higher fluorescence intensity, and easy regulation of shape, size, and function [59, 60]. As a result, fluorescent nanomaterials have attracted great attention and have undergone rapid development in the biomedical field. At the cellular level, small-sized fluorescent nanomaterials can enter cells through endocytosis and infiltration and can be used as cellular fluorescent imaging probes to identify and localize specific cells or molecules in a highly selective manner. At the in vivo level, fluorescent nanomaterials can be passively and actively delivered to the lesion site after entering the bloodstream and can be used as fluorescent probes for early diagnosis of disease to provide structural and dynamic information of living samples [61]. In addition, the combination of fluorescent nanomaterials with antibacterial and antitumor drugs through nanotechnology can achieve the integration of diagnosis and treatment [62–65]. In conclusion, fluorescent nanomaterials have become a key link in the cross-integration of materials science, optics, biomedicine, and other disciplines, and they have great application value and prospects.

Fluorescent nanomaterials widely used in biomedicine mainly include quantum dot materials, silicon-based fluorescent nanomaterials, rare earth luminescent nanomaterials, organic fluorescent nanomaterials, and composite nanomaterials of different fluorescent materials.

1.4.1.1 Quantum Dots

Quantum dots, also known as artificial atoms, are generally spherical or spheroidal in shape, and their diameters are often between 2-20 nm, with high crystallinity and micro-size being their typical features. When the size of quantum dots is less than or equal to the exciton Bohr radius, the electrons and holes inside them are quantum limited, and the continuous energy band structure becomes a discrete energy level structure with molecular properties, so the quantum dots produce strong light absorption and luminescence effects (Figure 1.3) [61]. As the size of the quantum dots decreases, the energy difference between the highest valence band and the lowest conduction band increases, which means that the quantum dots require more energy to excite, and at the same time, more energy is released when the crystal returns to the ground state. At this point, the wavelengths at which the quantum dots absorb and emit light are shifted toward shorter wavelengths of higher energy, known as the blue shift. This allows researchers to tune the optical properties of quantum dots according to their size for better applications in biology, medicine, and other fields. In addition to size tuning, the development of new methods for synthesizing quantum dots, such as core-shell structures, alloying, doping, surface ion modification, and voltage tuning, is likely to further facilitate the application of quantum dots in biomedical fields.

Quantum dots can be divided into semiconductor quantum dots (such as II–VI and III–V elements) [56], IV (carbon and silicon) quantum dots, and lead halide chalcogenide (perovskite) quantum dots according to the different constituent elements. Quantum dot materials used in biomedical applications do not contain toxic heavy metal elements such as cadmium or lead, and the most commonly used are carbon quantum dots, zinc-based quantum dots (ZnS, ZnO, and ZnSe), and Ag-based quantum dots (Ag_2S and Ag_2Se) [65–68]. In addition, most of the current quantum

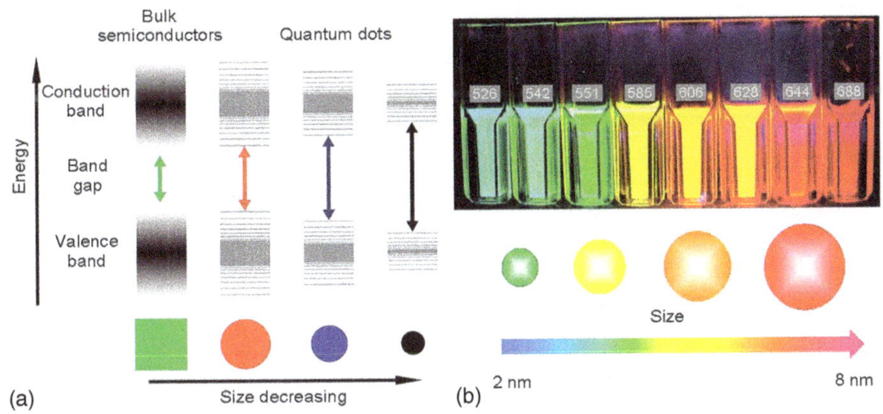

Figure 1.3 (a) Energy diagrams for bulk semiconductor and QDs. (b) Example of cysteine-capped CdTe/ZnTe core/shell QDs synthesized in aqueous media, with a size-dependent luminescence. Upper panel: photographic images of CdTe/ZnTe QDs dispersed in water under UV irradiation. The emitted peak wavelength is marked in white numbers on the top of the image. Lower panel provides an illustration of the corresponding core/shell nanoparticle size from 2 to 8 nm. These core/shell QDs were prepared at the Institute for Lasers, Photonics, and Biophotonics, University at Buffalo. Source: Reproduced with permission from Ref. [39]; © 2016/American Chemical Society.

dots are mainly synthesized by the most classical high-temperature pyrolysis method and therefore their surfaces are often covered with alkyl chains. Achieving phase transfer (oil to water) of quantum dots is an important first step in their entry into biological applications. In addition, some quantum dots (like carbon quantum dots) can be synthesized directly in the aqueous phase, making their use in biological environments more convenient [69].

Since the first publication of research using quantum dots as bioprobes for fluorescence imaging of living cells in science in 1998, the exploration of quantum dots in the field of biomedical imaging has mushroomed [70]. Carbon quantum dots are widely used in cellular/bacterial imaging and in vivo imaging due to their high fluorescence stability, tunable optical properties, wide source of synthetic raw materials, low cost, and good biocompatibility. Depending on the variation of their fluorescence intensity, they can also be used to detect a wide range of anions and cations (K^+, Al^{3+}, Fe^{3+}, $ONOO^-$) [71], drug molecules (antibiotics, tetracycline, methotrexate), biomolecules (DNA, proteins, cysteine, hemoglobin, and cholesterol), bacteria, pH, and fingerprints. Tan and coworkers have used carbon dots encapsulated in organosilicon shells to achieve targeted specific capture of bacteria (Figure 1.4) [72]. The carbon dots@organosilicon surface was modified with anti-*Staphylococcus aureus* antibodies, which could selectively capture the target bacteria. At the same time, the bacteria were separated using antibody-modified magnetic nanoparticles. The organosilicon shell is then destroyed by reducing the sodium boron hydride ($NaBH_4$), releasing the carbon dots, and causing them to fluoresce. The fluorescence signal was significantly enhanced by the hundreds of carbon dots encapsulated in each organosilicon particle. The fluorescence signal of this

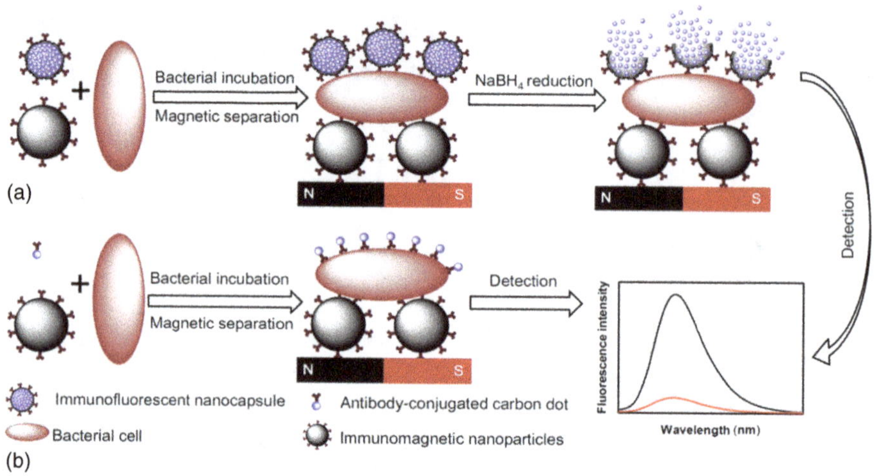

Figure 1.4 (a) Illustration of the detection of pathogen bacteria with the proposed method and (b) conventional method. Source: Reproduced with permission from Ref. [72]; © 2018/American Chemical Society.

method is enhanced by a factor of 108 compared to the conventional assay using only carbon dots as markers. Ag_2Se quantum dots with a narrow direct band gap (0.15 eV) are a new type of fluorescent probe in the NIR-II region that has received great attention in recent years [68]. To overcome the problem of low fluorescence quantum, yield due to a large number of cation vacancies and crystal defects caused by the high mobility of Ag ions, Wang and coworkers synthesized silver-gold-selenium (AgAuSe) quantum dots with tunable fluorescence emission spectra in the range of 820–1170 nm by alloying for the first time, which is an important impetus for the research of in vivo fluorescence imaging as well as therapeutic tracing [73].

1.4.1.2 Silicon-Based Fluorescent Nanomaterials

Silicon is a fundamental building block of the human body. Silicon-based fluorescent nanomaterials exhibit good biocompatibility, attractive optical properties, and easy surface modification and have attracted widespread attention and research in the fields of biosensing, disease detection, and imaging. Fluorescence imaging based on silicon-based fluorescent nanomaterials enables long-range real-time imaging for the analysis of cancer, bacterial-related diseases, and ophthalmic diseases. In 2016, researchers in Japan developed a silicon-based fluorescent nanomaterial with a core-double-shell structure [74]. In this structure, crystalline silicon nanoparticles serve as the core, while hydrocarbon groups and surfactants are encapsulated on the surface. Two-photon excitation fluorescence imaging showed that the silicon nanoparticles would efficiently emit light when absorbing NIR light, while the hydrocarbon groups in the coating would increase the luminescence quantum yield. This work enabled the first successful bioimaging of silicon-based fluorescent nanomaterials under 650–1000 nm light irradiation, known as the "bio-optical window." As the research progressed, the scientists found that conventional fluorescent probes suffered from the need to reintroduce probe molecules when changing labels and the difficulty of assessing the total concentration of probes at a target site, so color-switchable probes were developed to provide more reliable information at the cell or tissue level. Yang and coworkers recently reported a silicon-based color-switchable fluorescent nanomaterial that can successfully suppress oxidation-induced photoluminescence loss through a two-step surface passivation strategy to achieve a fluorescence jump from intrinsic NIR light on the microsecond scale to blue light on the nanosecond scale [75]. Specifically, red-emitting silicon nanocrystals are first passivated with acrolein and then interact with amines via a Schiff base reaction mechanism to achieve an in situ red-to-blue fluorescence transition in living cells. Taking the presence of abundant amino acids in the biological environment as an example, the blue fluorescence intensity displayed by the cells is dependent on the dosage of amino acids, suggesting that the silicon-based fluorescent nanomaterials can reflect the distribution and concentration of amino acids in different regions of living cells in real time and differentiate between living and dead cells. The silicon-based color-switchable fluorescent nanomaterials provide in situ multi-color imaging at the cell or tissue level for many fundamental life activities involving amino acids, proteins and nucleic acids, and have a wide range of development and application.

1.4.1.3 Rare Earth Luminescent Nanomaterials

Rare earth luminescent nanomaterials refer to a kind of materials doped with rare earth ions or containing rare earth elements in the matrix. Rare earth elements are known as "vitamins of modern industry" and "luminescent treasures" (Figure 1.5). The luminescent properties of rare earth luminescent nanomaterials arise from the abundant electronic energy levels of rare earth elements and the unique 4f electron jump properties. Rare earth elements have unfilled 4f electron shell layer, and the arrangement of electrons on this electron layer will produce different energy levels. These electrons in the different energy levels between the jump will produce a large number of absorption and emission [57]. Rare earth luminescence can be divided into two categories: sharp line spectra (f–f transitions) and band spectra (f–d transitions). Rare earth luminescent nanomaterials are an important class of optical materials with the advantages of narrow emission spectral bands, high color purity, strong light absorption, high luminescence intensity, long fluorescence lifetime, and stable physicochemical properties [76]. By adjusting the type and proportion of doped rare earth ions, the full band of luminescence from ultraviolet to infrared can be realized under the excitation of a single excitation light source. Meanwhile, the low toxicity of rare earth luminescent nanomaterials has great potential for application in biofluorescent labeling and medical imaging and is now attracting the attention of researchers in the field of biomedicine.

Rare earth luminescent nanomaterials are generally composed of two parts: the matrix (the main body of the luminescent materials, mainly various compounds of Y, Gd, and Lu) and the activator (which is the dopant ion used as the luminescent centers, such as Er^{3+}, Tm^{3+}, Ho^{3+}, Tb^{3+}, Sm^{3+}, and Eu^{3+}) [33, 77]. The matrix provides a suitable crystalline environment for the activator ions, which

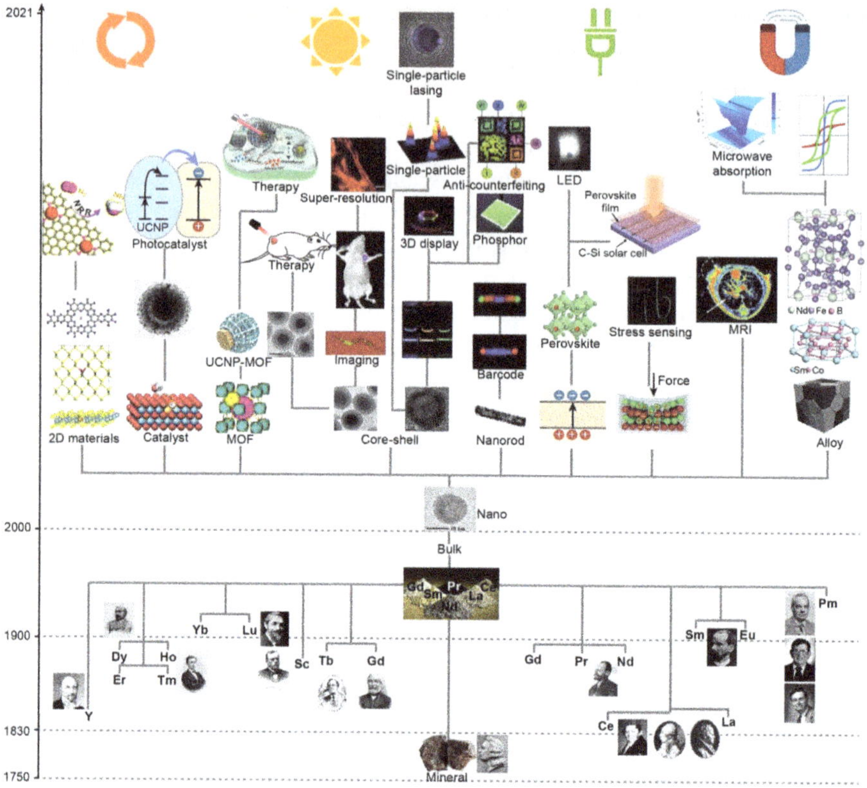

Figure 1.5 Development and application of rare earth-based nanomaterials. The history of rare-earths begins with the discovery of rare earth minerals and their separation and purification by scientists. Since the advent of nanotechnology, doping of rare-earths in nanomaterials of different morphology and composition has enabled many technological applications. Source: Reproduced with permission from Ref. [57]; © 2022/American Chemical Society.

themselves usually constitute only a few luminescent energy levels. The activator acts as a luminescence inducer for the luminescent material. In addition, some sensitizers (such as Yb^{3+} and Nd^{3+}) can be added to the material, which can be transferred to the neighboring luminescent centers by Foster resonance energy transfer (FRET), thus achieving efficient absorption of the excitation light and increasing the luminescence efficiency of the upconversion nanomaterials [78]. Based on the different luminescence mechanisms, rare earth luminescent nanomaterials can be classified into rare earth upconversion fluorescent nanomaterials and rare earth downconversion fluorescent nanomaterials. Downconversion luminescence follows Stokes' law, which refers to the process by which luminescent materials emit low-energy, long-wavelength light (e.g. visible light) after being excited by high-energy, short-wavelength light (e.g. ultraviolet light). Upconversion luminescence, also known as anti-Stokes luminescence, refers to the process by which a luminescent material absorbs low-energy long-wavelength light and emits high-energy short-wavelength light [76].

Rare earth luminescent nanomaterials have low toxicity and do not disrupt the normal physiological state of living organisms when imaged at the cellular, tissue, and in vivo levels. Rare earth luminescent nanomaterials have good photostability and chemical stability and can continue to emit light when excited by a light source, and their luminescence properties remain undisturbed even in acidic or alkaline chemical environments [39]. Rare earth luminescent nanomaterials used in biomedical fields are mostly excited by NIR light, which is highly penetrating and can be used for deep tissue imaging. At the same time, biological autofluorescence is not excited by NIR light (organisms basically do not produce fluorescence in the NIR region), which reduces the detection of background fluorescence and achieves a higher signal-to-noise ratio. In addition, the lower excitation energy of NIR light results in less photodamage to the organism and better safety, which is a unique advantage in biological applications [79]. As the study progressed, the researchers divided NIR light into NIR-I (650–950 nm) and NIR-II (1000–1700 nm). Compared to the visible range and NIR-I, NIR-II light, which is in the optically 'transparent' window of biological tissue, suffers from greatly reduced absorption and scattering effects as it propagates through biological tissues such as skin, fat, and bone, resulting in deeper tissue penetration and less energy attenuation in the delivery process [80]. At the same time, the autofluorescence of biological tissues is much lower in this band [81]. In particular, the imaging window at wavelengths of 1500–1700 nm is recognized as a promising window for noninvasive imaging of deep tissues, as the 'discount rate' caused by tissues is further reduced and autofluorescence from various pigments in the organism is almost eliminated, resulting in a further improvement in imaging depth and spatiotemporal resolution (Figure 1.6).

To date, a variety of imaging techniques based on rare earth luminescent nanomaterials have been successively developed, including upconversion luminescence imaging [82], downshifted luminescence imaging [83], time-resolved imaging [84], sustained luminescence imaging and multifunctional imaging [57], which have great potential for application in the fields of gene assay, immunoassay, fluorescent labeling, cellular imaging, biodetection, and biosensing. As shown in Figure 1.7, Lin and coworkers used the FRET phenomenon between $NaYF_4$:Yb/Er and gold nanoparticles to detect *Escherichia coli* [85]. The gold nanoparticles were modified with an aptamer of the target strain ATCC8739, and a cDNA nucleotide strand complementary to the target strain aptamer was modified on $NaYF_4$:Yb/Er. In the absence of *E. coli*, the complementary pairing of the cDNA nucleotide strand with the *E. coli* aptamer causes $NaYF_4$:Yb/Er to be in close proximity to the gold nanoparticles, and the fluorescence emitted by $NaYF_4$:Yb/Er is quenched. When *E. coli* is present, the aptamer dissociates from the cDNA and binds to *E. coli*, resulting in the dissociation of $NaYF_4$:Yb/Er from the gold nanoparticles and the recovery of upconversion fluorescence. Chen and coworkers synthesized a four-shell layer of upconversion nanoparticles co-doped with Yb^{3+} and Tm^{3+}, achieved controllable regulation of the fluorescence lifetime by adjusting the thickness of the ion migrating layer and the concentration of rare earth ions doped in the upconversion energy transfer layer, and developed a time-resolved upconversion system based on

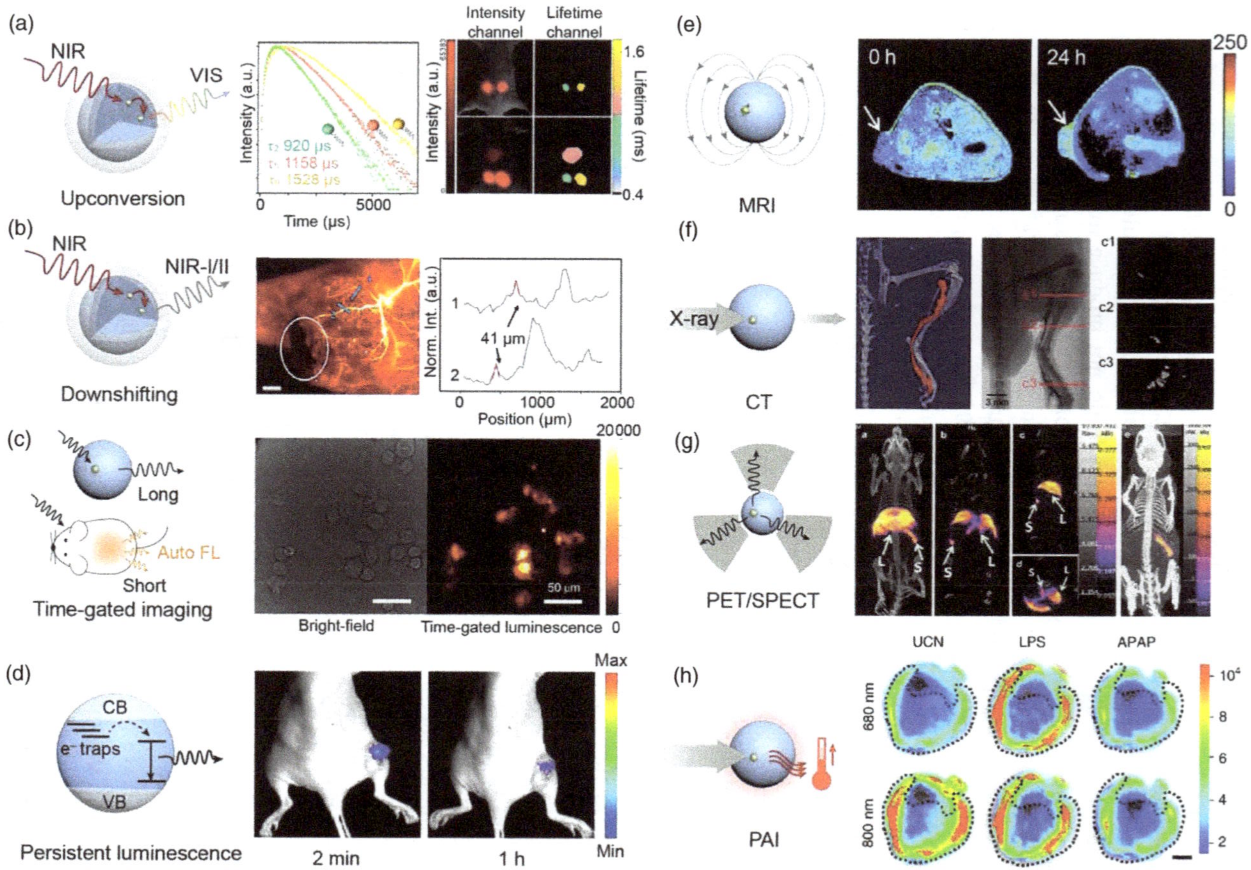

Figure 1.6 Schematic illustration of the principles of imaging modes and corresponding (a) visible fluorescence imaging, (b) near-infrared fluorescence imaging (scale bar, 2 mm), (c) time-gated imaging (scale bars, 50 μm), (d) persistent luminescence imaging, (e) magnetic resonance imaging (MRI), (f) computed tomography (CT scale bar, 3 mm), (g) positron emission computed tomography/single-photon emission computed tomography (PET/SPECT), and (h) photoacoustic imaging (PAI) of rare earth doped nanomaterials. Source: Reproduced with permission from Ref. [57]; © 2022/American Chemical Society.

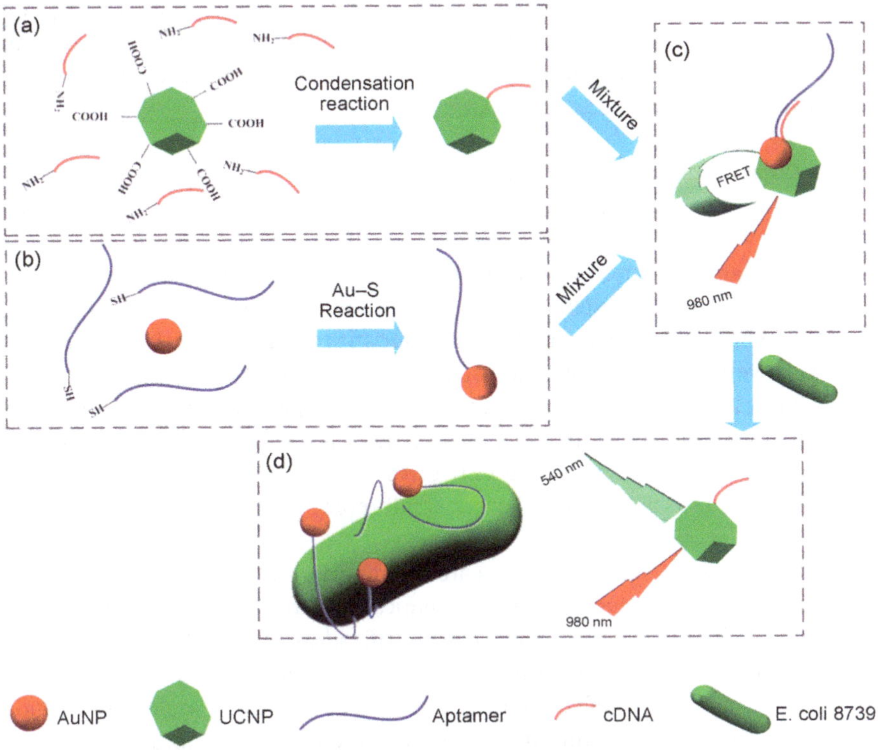

Figure 1.7 Schematic illustration of upconversion nanoparticles based FRET aptasensor for rapid and ultrasensitive bacteria detection. (a) The amino-modified complementary DNA of the aptamer is attached to the carboxyl-functionalized UCNPs by condensation reaction (b) Conjugating thiol-modified aptamers to the AuNPs through Au-S chemistry. (c) The FRET system is established between a donor-acceptor pair: UCNPs-cDNA hybridized with AuNPs-aptamers. (d) By introducing target bacteria into the FRET system, aptamers preferentially bind to target bacteria, resulting in the dissociation of cDNA. Thereby, aptamers-DNA pairs are destroyed and the green fluorescence recovers. Source: Reproduced with permission from Ref. [85]; © 2016/Elsevier.

this, which was successfully applied to in vivo multiplexed pathway imaging [86]. Zhang and coworkers developed rare earth-alkali metal fluoride nanofluorescent probes based on Tm^{3+}, Er^{3+}, and Ho^{3+}-doped cubic crystal phases, which achieved different degrees of downshifted luminescence enhancement at 1632, 1530, and 1180 nm, respectively, with the excitation of 980 nm NIR light, further enriching the fluorescence emission of rare earth elements in the NIR-II and realizing real-time dynamic multiple in vivo imaging [87]. Meanwhile, rare earth doped nanomaterials can also be used as contrast agents; for example, Gd^{3+} with seven unpaired electrons is suitable as a positive (T_1) contrast agent for MRI, and Dy^{3+} and Ho^{3+} are suitable as negative (T_2) contrast agents due to their high magnetic moments (~10.6 μB) and short electronic relaxation times (~0.5 ps) [88]. In addition, rare earth luminescent nanomaterials can also be used as contrast agents for X-ray computed tomography (CT) imaging due to the high atomic number and lanthanide contraction. Among the lanthanides, lutetium (Lu) has the highest atomic number, and $NaLuF_4$ and $LiLuF_4$ have been shown to be effective CT contrast agents [89, 90]. Several research teams have also used rare earth-doped nanoparticles as contrast agents for PAI [91, 92].

1.4.1.4 Organic Fluorescent Nanomaterials

Organic fluorescent nanomaterials are organic nanomaterials that can emit low-energy light waves after absorbing high-energy light waves. Through reasonable structural design, the luminescence performance of organic fluorescent nanomaterials can be effectively adjusted by changing the conjugation length and increasing the Stokes shift, which shows the characteristics of the material's structural diversity as well as its functional adjustability and predictability [93]. Meanwhile, organic fluorescent nanomaterials have good biocompatibility, which is of great concern to researchers at the intersection of materials, chemistry, and biomedicine. According to the different molecular structures, organic fluorescent nanomaterials can be classified into organic fluorescent small molecules and organic fluorescent polymers.

Organic fluorescent small molecules are an important class of fluorescent probe materials with the advantages of strong chemical modification, low synthesis cost, wide range of fluorescence emission, easy purification, and no heavy metal contamination. The structure of organic fluorescent small molecules is easy to design, and there are many types of them. Common fluorescent small molecules include fluorescein, rhodamine, coumarin, porphyrin, anthocyanin, fluoroborodipyrrole (BODIPY), and its derivatives. Recently, Lei and coworkers developed Chrodol-3, a hydroxyl-containing anthocyanin dye with NIR-II fluorescence with high quantum yield (0.34%) and low pK_a (6.40), and prepared PN910, a NIR-II fluorescent probe for monitoring reactive oxygen species (ROS) and reactive nitrogen species, based on Chrodol-3 [94]. In vivo fluorescence imaging showed that PN910 could be used to accurately monitor cystitis and colitis. Smith and coworkers reported a modular synthetic approach to synthesize a series of high-performance ionic heptamethine cyanine dyes, which greatly enhanced the fluorescence emission and chemical stability of the dyes [95]. The produced NIR-I and NIR-II dyes can specifically bind to proteins and target specific sites in mice, enabling efficient in vivo tumor fluorescence imaging.

It is well-known that the application of most organic fluorescent molecules is limited by the aggregation-induced quenching (ACQ) phenomenon, which is manifested by the fact that they exhibit good fluorescence emission properties in dilute solutions [96, 97]. However, in the solid state or when the molecules are aggregated, their fluorescence emission intensity decreases or even disappears due to strong π–π interactions. It was only in 2001 that Tang and coworkers discovered a class of organic small-molecule fluorescent materials with special fluorescence properties, which are basically nonfluorescent in the solution state, while the fluorescence is significantly enhanced when they are prepared into nanoparticles or thin films, a phenomenon known as aggregation-induced luminescence (AIE) phenomenon (Figure 1.8) [98, 99]. The discovery of the AIE phenomenon opens up a new

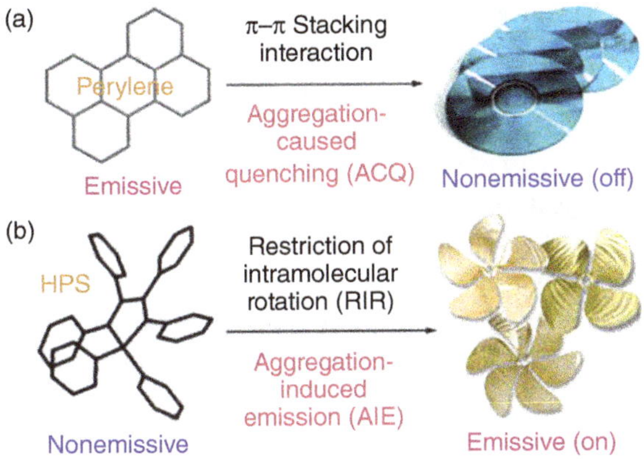

Figure 1.8 (a) Planar luminophoric molecules such as perylene tend to aggregate as discs pile up due to the strong p–p stacking interactions between the aromatic rings, which commonly turns "off" light emission. (b) Nonplanar luminogenic molecules such as hexaphenylsilole (HPS) behave oppositely, with their light emissions turned "on" by aggregate formation due to the restriction of the intramolecular rotation (RIR) of the multiple phenyl rotors against the silole stator in the aggregate state. Source: Reproduced with permission from Ref. [98]; © 2011/The Royal Society of Chemistry.

avenue for applied research of organic functional optical materials. In 2016, AIE nanoparticles were listed in Nature as one of the four nanomaterials that will support the upcoming "nanolight revolution" [100]. Over the past two decades, research on AIE-emitting nanomaterials has made great progress, and many AIE molecules and probes have been designed and prepared for applications in cell- or organelle-specific imaging, single- or multiphoton imaging of blood vessels, disease sensing, in vivo drug distribution monitoring, long-term tracking, and fluorescence imaging-guided surgery [101–104]. For example, in the field of organelle-specific imaging, researchers have designed and synthesized DTPAP-P cationic polymers [105]. Due to anion–π^+ interactions, DTPAP-P exhibits excellent optical properties in the solid state and can be targeted to mitochondria or nuclei depending on the cellular state, enabling organelle-specific imaging and dynamic tracking at the nanoscale. In the field of in vivo tumor imaging, Liu and coworkers reported for the first time an NIR-II AIE nanoparticle for targeted NIR-II fluorescence and NIR-I PAI of brain tumors and evaluated in detail the penetration depth, spatial resolution, sensitivity, and imaging effect through in situ brain tumor models with intact scalp and skull, providing a powerful technical tool for monitoring and visualizing cerebrovascular and brain tissue abnormalities [106].

Organic fluorescent polymers can be divided into conjugated and nonconjugated fluorescent polymers according to their structural properties. The luminescence mechanisms of the two types of fluorescent polymers are different. Conjugated fluorescent polymers have a π–π conjugated structure with benzene rings or aromatic heterocycles as the main chain luminescent units, which have optical properties similar to those of semiconductor materials, and there are various photophysical processes between the ground state and the excited state. Nonconjugated fluorescent polymers have more luminescence mechanisms, including luminescence induced by polymer side chain chromophores and luminescence induced by cochromophores such as aliphatic amines, ester groups, and carbonyl groups under certain conditions. In recent years, organic fluorescent polymers have shown great potential for application in a variety of fields such as sensor detection, drug screening, bioimaging, and pathological analysis, attracting the attention of more and more researchers. Tian and coworkers have synthesized a new type of organic luminescent "spherical nucleic acid," which can enhance the signal transduction of nucleic acid detection by exploiting the "light trapping antenna effect" of fluorescent π-conjugated polymers, thus realizing the detection of nucleic acids at ultra-low concentrations [107].

1.4.2 Photothermal Nanomaterials

Photothermal nanomaterials, also known as photothermal conversion nanomaterials, are a class of materials that can convert light energy into heat energy (Figure 1.9) [109]. When the size of the material reaches the nanoscale, the absorption of light changes significantly, and these nanomaterials can convert the absorbed light energy into thermal energy. By selecting suitable photothermal nanomaterials, surface modification, and functionalization to improve photothermal conversion efficiency, stability, and biocompatibility, they can be used in photothermal therapies for cancer treatment, antimicrobial, tissue repair, and so on. In photothermal therapy (PTT), photothermal reagents are used to increase the heating effect of the lesion area [110]. Depending on the degree of warming, it can be subdivided into irreversible damage therapy (above 48 °C), hyperthermia (41–48 °C, which is the PTT discussed in this chapter) and hypothermia (below 41 °C) [37, 111]. High temperatures can cause thermal damage or thermal lethality of the cells in the lesion to treat the disease. With the advantages of noninvasiveness, high selectivity, and localized thermal effect, PTT can effectively treat a wide range of diseases, including cancer, bacterial infections, and inflammation [112].

With the gradual deepening of research, the types of photothermal nanomaterials are becoming richer and richer, and at present, the more researched ones are mainly metal photothermal nanomaterials, semiconductor photothermal nanomaterials, organic photothermal nanomaterials, carbon matrix photothermal nanomaterials, certain 2D nanomaterials, and biomass photothermal nanomaterials, as well as the composite materials of these materials, and so on [108, 113–115].

1.4.2.1 Metallic Photothermal Nanomaterials

The photothermal conversion process of metallic photothermal nanomaterials is highly dependent on the spatial motion of free electrons. When metal nanoparticles (such as gold, silver, copper, platinum, and palladium) are irradiated with light of a certain wavelength, the free electrons in the particles interact with the photons, absorb the energy of the photons, and are excited to form a periodic electron density oscillation [116]. This oscillation resonates with the electromagnetic field

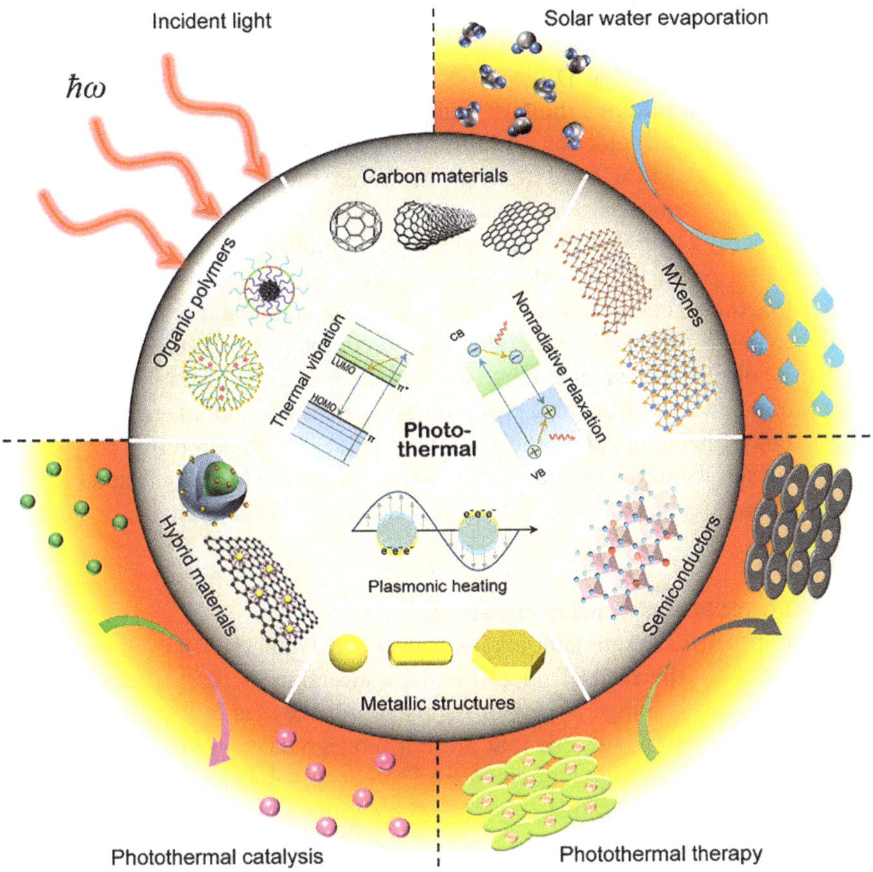

Figure 1.9 Overview of the mechanisms, categories, and applications of photothermal nanomaterials. Source: [108]/American Chemical Society/CC BY 4.0.

of the incident light, giving rise to the localized surface plasmon resonance (LSPR) effect [117]. In this resonance process, the excited free electrons convert energy into heat by colliding with other electrons, which is a key step in the photothermal conversion process of metal nanoparticles. Furthermore, as the temperature of the metal nanoparticles increases, the heat is transferred to the surroundings through the interaction of electrons and metal lattice vibrations [108, 118]. In particular, the photothermal conversion process of metallic photothermal nanomaterials depends on factors such as the size and shape of the particles and the properties of the surrounding medium. Therefore, by finely controlling these parameters, it is possible to regulate the motion mode and energy level structure of free electrons, which in turn affects the strength and efficiency of the surface plasmon resonance effect, electron–electron, electron–photon, and electron–phonon interactions and achieves the regulation and optimization of the photothermal conversion performance.

1.4.2.2 Semiconductor Photothermal Nanomaterials

Semiconductor nanomaterials with a fixed energy band gap, such as copper sulfide [119, 120], copper selenide [121, 122], and molybdenum disulfide [123], have excellent light absorption in the NIR region. Their energy band gap refers to the energy required for an electron to jump from the valence band (full electronic state) to the conduction band (empty electronic state). When semiconductor photothermal nanomaterials are irradiated by light, photons with energy greater than their energy band gap can excite electrons to jump from the valence band to the conduction band, forming electron-hole pairs. In the absence of an external electric field, the electrons in the excited state are unstable and will jump back to the valence band from the conduction band to compound with the holes. In this process, the electron releases energy equal to the energy band gap. The released energy is transferred to the crystal lattice in the form of vibrational energy (thermal energy), thus realizing the photothermal conversion. The photothermal conversion properties of semiconductor materials can be optimized by changing factors such as the type of material, crystal structure, defect state, particle size, shape, or external environment.

1.4.2.3 Organic Photothermal Nanomaterials

Organic photothermal nanomaterials are a new type of materials that have attracted much attention in the field of photothermal conversion in recent years because their light absorption range is easy to regulate, and they can overcome the characteristics of inorganic materials that are not biodegradable, and they have been developed rapidly. Currently, common organic photothermal nanomaterials include small-molecule dyes, conjugated polymers, and organometallic complexes.

Small-molecule dyes with low quantum yields tend to have good photothermal conversion effects because low quantum yields mean that after absorbing a photon, only a small amount of the dye's energy is re-radiated in the form of fluorescence or phosphorescence, while more energy is released in a nonradiative way (e.g. thermal vibration). At present, small-molecule dyes that have been relatively well studied include indocyanine green (ICG), Prussian blue, and so on [124, 125]. In the case of ICG, for example, the presence of a large π-conjugated system in its molecule makes the absorption peak of this dye mainly in the range of 600–850 nm. When irradiated with laser light, the electron energy in the dye molecule will increase, and then the electrons will jump from the ground state to the excited state. When the electrons return to the ground state, the energy is released in the form of light and heat, allowing this dye to have the dual ability of fluorescence and photothermal conversion, giving it a wide range of applications in PTT, bioimaging, and biomarkers.

Conjugated polymers have received increasing attention from researchers due to their better photostability, biocompatibility, and temperature rise stability, as well as higher light absorption properties and photothermal conversion efficiencies compared to small-molecule dyes [126]. These conjugated polymers are typically derived from organic small-molecule monomers through polymerization reactions. Typical polymer-based photothermal agents mainly include

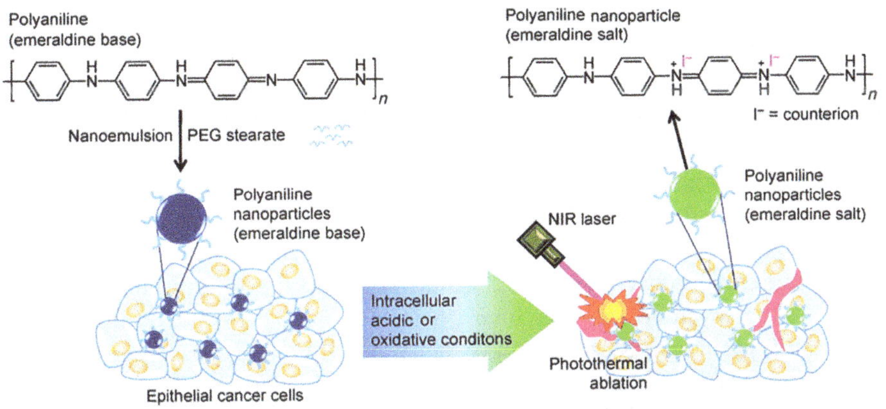

Figure 1.10 Schematic illustration of the preparation of organic photothermal agents based on polyaniline nanoparticles and their application for the photothermal ablation of epithelial cancer cells by NIR laser irradiation. Source: Reproduced with permission from Ref. [127]; © 2011/John Wiley & Sons.

polyaniline, polypyrrole, polythiophene, and polydopamine, which have different photothermal conversion mechanisms due to their large structural differences [108]. Among them, polyaniline is a conductive polymer material with good biocompatibility, which has important applications in biomedical fields such as cell and tissue engineering due to its structural tunability, excellent electrical properties, and adaptability to the environment. In 2011, Haam and coworkers reported a novel organic photothermal agent based on polyaniline that induces hyperthermia for the treatment of epithelial cancer (Figure 1.10) [127]. Under the effect of protonation doping or the oxidative environment in cancer cells, a new bandgap state is created between the valence and conduction bands of polyaniline, inducing electron shifts and lowering the excited state energy level. At this point, the imine group on polyaniline is converted to an imine salt, and its absorption peak is significantly red-shifted, which can effectively ablate epithelial cancer cells under NIR light. Another representative class of polymer-based photothermal agents is polydopamine, a multifunctional biomimetic polymer material obtained by self-polymerization of dopamine monomer in a weakly alkaline environment [128–130]. On the one hand, polydopamine has broadband light-absorbing properties similar to those of natural melanin (broad light absorption throughout the ultraviolet, visible, and NIR regions), and on the other hand, it possesses nonselective and strong adhesion properties to the surfaces of a wide range of solid materials, mimicking the function of adsorbing proteins secreted by mollusks.

Organometallic complexes, such as metal phthalocyanine complexes and metal carbazole complexes, are compounds consisting of one or more metal ions and organic ligands. In such compounds, the metal ion usually acts as the central atom, while the organic ligand forms a chemical bond with the metal ion through an electron-donating group. The organometallic complexes have higher electron cloud density, better light absorption, and higher photothermal conversion efficiency, and their light absorption properties can be tuned by changing the type and structure of the ligand, which has important application prospects in the field of photothermal conversion. Phthalocyanine is a planar macrocyclic conjugated system composed of four isoindole units, and there is a cavity with a diameter of about 2.70×10^{-10} m in the center of its structure, and more than 70 metal elements can be chelated with phthalocyanine at the position of the cavity through two covalent bonds and two coordination bonds to form a highly stable metal phthalocyanine [131–133]. According to the aggregation leading to bursting effect, Zhang and coworkers synthesized special zinc phthalocyanine aggregates with better photothermal stability and photothermal conversion effect, which inhibited the fluorescence emission of zinc phthalocyanine and the inter-system jumping path and promoted the nonradiative jumping process to achieve higher photothermal conversion efficiency [134]. Zhang and coworkers synthesized a series of complexes with different rare earth centers (Ln = Gd, Yb, Er) based on NIR-absorbing carbazole porphyrin ligands [135]. These complexes have strong absorption in the phototherapy window of 650–850 nm.

Meanwhile, the fluorescence of the complexes was significantly burst due to the heavy atom effect and exhibited high photothermal conversion efficiency as well as singlet oxygen ($^1O_2$) generation capacity associated with the center metals.

1.4.2.4 Carbon-Based Photothermal Nanomaterials

The photothermal conversion performance of carbon-based nanomaterials (e.g. carbon dots, fullerenes, carbon nanotubes [wires], graphene, and carbon nitride) is mainly due to the unique electronic structure and optical properties [123, 136–139]. Carbon-based nanomaterials have dense and sparse π-electron clouds with very close energy levels and generally exhibit broadband absorption, meaning that the vast majority of photon energy from the ultraviolet to the NIR can be absorbed by the electrons, placing them in an excited state. Meanwhile, thanks to the high thermal conductivity, the excited electrons in carbon-based nanomaterials fall back to the ground state and release heat, which can be quickly transferred to the surrounding materials and environment to avoid their own temperature being too high, thus improving the efficiency of light and heat conversion. In addition, carbon-based nanomaterials have good chemical stability, thermal stability, and photothermal tunability. By forming carbon nanostructures with different morphologies or by compositing carbon materials with other materials, it is possible to effectively maintain stable properties and modulate their photothermal properties [108, 140]. Carbon-based nanomaterials have been widely used in thermal therapy, imaging, and sensing platforms.

As shown in Figure 1.11, Tang and coworkers reported a graphene-functionalized magnetic microbead that can rapidly and sensitively detect tumor cells by using the magnetic effect of magnetic microbeads and the photothermal effect of graphene oxide, greatly solving the traditional problem of the difficulty of counting cells under the microscope [141]. Specifically, graphene oxide and magnetic beads are modified with specific antibodies and then combined to form a graphene-magnetic bead complex through antigen-antibody interaction. To identify cancer cells, the complex was incubated with the sample and then subjected to a membrane filtration

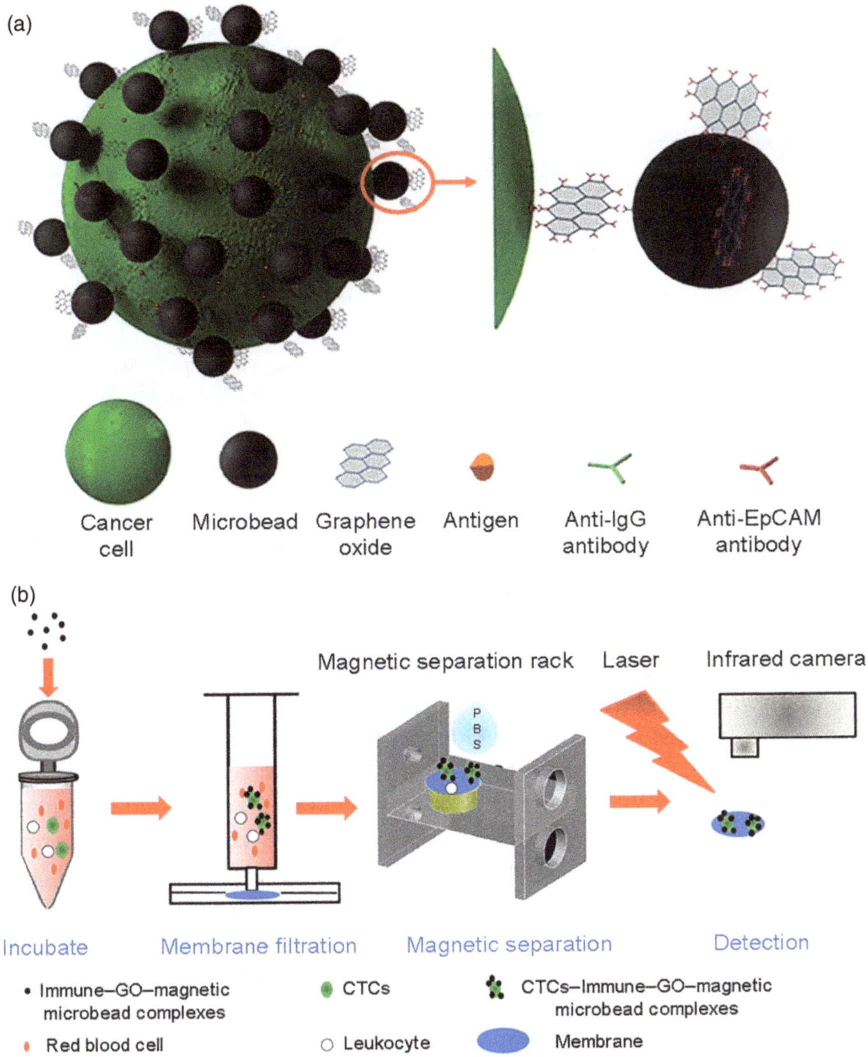

Figure 1.11 (a) Schematic view of the cancer cells recognized by the immune-GO-magnetic microbead complexes. (b) Procedure of cell capture and thermal contrast detection based on the immune-GO-magnetic microbead complexes. Source: Reproduced with permission from Ref. [141]; © 2016/American Chemical Society.

and magnetic separation process to retain the cells attached to the graphene oxide-magnetic microbead complex and remove other contaminants. Finally, when a laser is used to irradiate graphene oxide on the cell-retaining membrane, graphene oxide, which has a photothermal effect, generates heat and causes a temperature change. A standard curve of temperature change versus number of cancer cells can therefore be established, and the number of cancer cells can then be determined by infrared thermography.

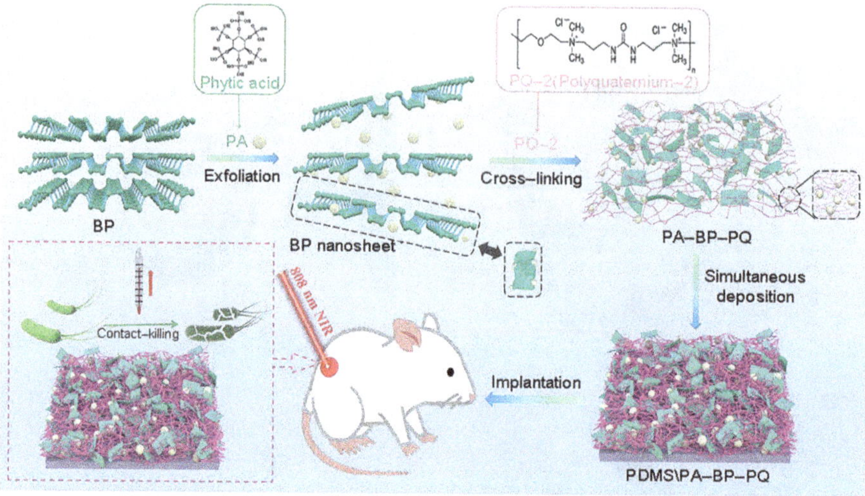

Figure 1.12 Schematic representation of the preparation of PDMS-PA-BP-PQ nano-coating and evaluation of its contact-killing and antibacterial photothermal therapy (aPTT) effect. Source: Reproduced with permission from Ref. [150]; © 2023/John Wiley & Sons.

1.4.2.5 Certain Two-Dimensional (2D) Nanomaterials

Black phosphorus and metal-carbon nitride/nitride materials are among the emerging two-dimensional nanomaterials that have rapidly attracted considerable attention from researchers since their discovery [142, 143]. As the most thermodynamically stable allotropic isomer of phosphorus [144], 2D semiconductor black phosphorus not only has a high specific surface area and excellent photothermal properties but also can be completely degraded in vivo to nontoxic phosphates, phosphites, and other phosphorus oxides [145, 146], which has excellent biocompatibility and safety and has a wide range of applications in the fields of drug carriers, antibacterial, and oncological therapy. The photothermal conversion efficiency of black phosphorus is higher than that of traditional photothermal agents, mainly due to its broadband light absorption properties, unique energy band structure, high thermal conductivity, and adjustable layer thickness. These properties enable black phosphorus to more effectively absorb light energy, convert it into heat energy and conduct it into the heat carrier, thereby improving the photothermal conversion efficiency [147, 148]. Wei et al. prepared a photothermal antibacterial hyaluronic acid-based nanocomposite injectable hydrogel for infected diabetic wound repair [149]. In this hydrogel, it is the functionalized black phosphorus nanosheets that exert the photothermal properties. Kang and coworkers prepared black phosphorus nanosheets by phytate-assisted ultrasonic stripping of black phosphorus powder, which were further deposited on the surface of polydimethylsiloxane with the participation of cationic polymers to form a stable functionalized coating (PA-BP-PQ) (Figure 1.12) [150]. The coating was found to have good contact bactericidal and photothermal antimicrobial activity against *S. aureus* and *E. coli* in anti-infection experiments.

2D transition metal carbon/nitride (MXene) is typically composed of $n+1$ layers of transition metal atoms (M) and n layers of carbon or nitrogen atoms (X) [151, 152]. Since it was first reported in 2011 [153], it has become a research hotspot due to its excellent properties such as broad wavelength absorption, high electrical conductivity, porous structure, excellent mechanical properties, and tunable chemical stability. In terms of photothermal conversion, MXene has good electromagnetic wave absorption capability and can absorb light in a wide range of wavelengths, including the visible and NIR spectra. Meanwhile, the high specific surface area and abundant free radical distribution are one of the important factors for its excellent photothermal conversion performance. In addition, the high density of surface charge carriers of MXene also plays an important role in their photothermal activities [154]. When MXene absorbs light energy, the photons excite the charge carriers on its surface, and the recombination of these charge carriers generates heat. Therefore, the excellent electromagnetic wave absorption capability, high specific surface area, abundant radical distribution and high density of surface carriers of MXene work together to give them strong light energy absorption capability and excellent photothermal conversion performance. The Ti_3C_2 MXene prepared by Hong and coworkers can quickly and efficiently kill a variety of drug-resistant bacteria and bacterial membrane structures within 20 minutes of NIR light irradiation [155]. Guo and coworkers developed a NIR light-activated photothermal-chemotherapeutic synergistic therapeutic platform by combining the nano-photothermal agent Ti_3C_2TX MXene, the chemotherapeutic agent doxorubicin (DOX), and a thermoresponsive DNA hydrogel [156]. The heat generated by Ti_3C_2TX in the presence of NIR light causes the double-stranded DNA cross-linking structure to unfold and release DOX, enabling the controlled release of photothermal and chemotherapeutic agents for the effective treatment of localized cancers.

1.4.2.6 Biomass Photothermal Nanomaterials

Melanin and hemoglobin are two well-known biomass photothermal nanomaterials that are naturally produced by living organisms as biomolecules with chromophores that can absorb long-wavelength photons and convert them into heat [157–159]. Melanin, a biopolymer widely found in the skin, hair, feathers, and eyes of animals, is produced by melanocytes in living organisms. In general, melanin can be classified into eumelanin, pheomelanin, neuromelanin, allomelanin, and pyomelanin based on the chemical precursors in biosynthesis [160]. Polydopamine in conjugated polymers is a typical class of artificial melanin materials. In particular, the light absorption of melanin extends into the NIR region. As shown in Figure 1.13a, Yu et al. fabricated a hyaluronic acid-based microneedle that exploits the antioxidant and photothermal functions of natural melanin nanoparticles extracted from cuttlefish ink to scavenge ROS and enable PTT [161]. In addition, the amorphous silica shell formed by biomineralization on the surface of melanin nanoparticles continuously degrades to release SiO_4^-, which can control the inflammatory environment and stimulate the proliferation and differentiation of endothelial cells, thus promoting skin tissue regeneration. Therefore, this microneedle has multiple functions of tumor eradication and wound healing at the same time and

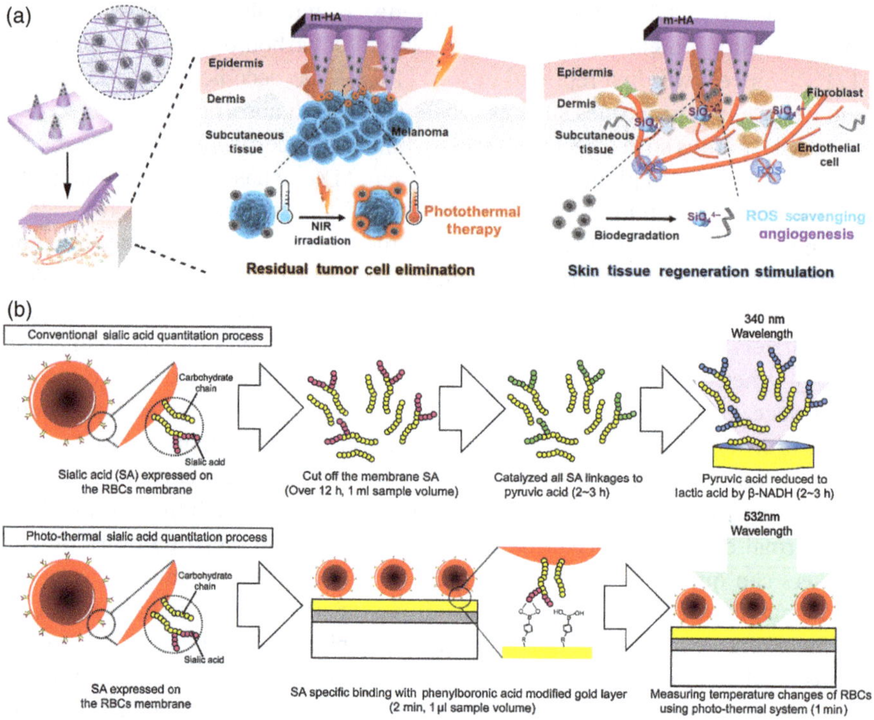

Figure 1.13 (a) Schematic therapeutic process of the biomineralized melanin nanoparticles-loaded microneedle patches for the subcutaneous melanoma postoperative wound. Source: Reproduced with permission from Ref. [161]; © 2022, Wiley-VCH GmbH. (b) Comparison of SA analysis using a photothermal sensing system with a conventional method. Source: Reproduced with permission from Ref. [162]; © 2012/Elsevier.

is expected to be an effective adjunct to the postoperative treatment of skin tumor excision.

There are few reports on sensors based on biomass photothermal nanomaterials, which combine the good biocompatibility of biomass materials, the high precision and sensitivity of nanotechnology, and the special properties of the photothermal effect. A photothermal sensing system based on hemoglobin molecules in erythrocytes is presented here. Hemoglobin is a protein in erythrocytes that is mainly responsible for carrying and delivering oxygen [157]. Hemoglobin consists of four polypeptide chains, each containing a porphyrin ring structure with an iron ion in the center. The ability of hemoglobin to convert light energy into heat energy is primarily related to the porphyrin ring structure and its interaction with the iron ion. When light hits hemoglobin, the porphyrin ring can absorb some of the light energy, which excites electrons to jump from the ground state to the excited state. These electrons then release energy in a nonradiative way, known as heat, which returns the electrons to their ground state. Using this property of the hemoglobin molecule, Jung et al. developed a photothermal biosensor that can analyze the hemoglobin content of whole blood through temperature changes, making the diagnosis of anemia easier and faster [163]. Building on this, they developed a new technique based on monitoring temperature changes in red blood cells to assess the concentration of sialic acid on their membranes (Figure 1.13b) [162]. A self-assembled monolayer of phenylboronic acid-modified thermoresponsive material was used to modify the sensor surface to detect red blood cells that overexpress sialic acid. Hemoglobin in the erythrocytes generates heat when irradiated with a 532 nm laser, which is used to determine the amount of sialic acid based on temperature changes on the sensor surface. The increase in erythrocyte temperature was found to be proportional to the level of sialic acid expression and laser intensity. The technique can directly measure the level of sialic acid in blood without complicated preprocessing steps (e.g. enzyme treatment and labeling), is easy to use, and can achieve sensitive detection of sialic acid, providing a new way of thinking about disease diagnosis.

1.4.3 Photodynamic Nanomaterials

Photodynamic nanomaterials are nanomaterials with specific structures and components that can act as carriers for photosensitizing small molecules, or that generate specific light responses and activities when irradiated with light, generating ROS for the realization of photodynamic therapy (PDT) [164, 165]. As a light-triggered therapy, the photodynamic nanomaterials are first injected or targeted to the diseased area and then irradiated by an external light, which activates the nanomaterials to generate a local chemical reaction, resulting in the destruction of tumor cells or disease-related tissues [166]. The treatment is noninvasive, highly selective, and localized and can reduce damage to normal tissues; it has been widely studied and applied in the treatment of tumors, bacterial infections, and some other chronic diseases.

As the core component of PDT, photodynamic nanomaterials are usually composed of nanoparticles or nanostructures whose surface or internal chemical components and structures are designed to have the ability to transport photosensitizers or to absorb, convert, and transmit light energy (Figure 1.14). Photodynamic nanomaterials can be divided into the following categories.

1.4.3.1 Photosensitizer-Loaded Nanomaterials

At the current stage of research on small-molecule organic photosensitizers, most photosensitizers have poor stability and water solubility in vivo and are easily degraded or deactivated by agglomeration [167–169]. The use of nanocarrier materials can improve photosensitizer stability, biodistribution, and lifetime in vivo. In addition, the targeting of photosensitizers is poor, and it is difficult to achieve effective accumulation at lesion sites (such as tumors) [170]. By modifying ligands or antibodies on the surface of the vector materials, targeted delivery to the lesion site can be achieved, improving therapeutic efficacy and reducing damage to normal tissues. Carrier materials can also be used as storage and release platforms for photosensitizers, such as the orderly release of photosensitizers in response to in vivo environment (e.g. pH and temperature) or external stimuli (e.g. light and

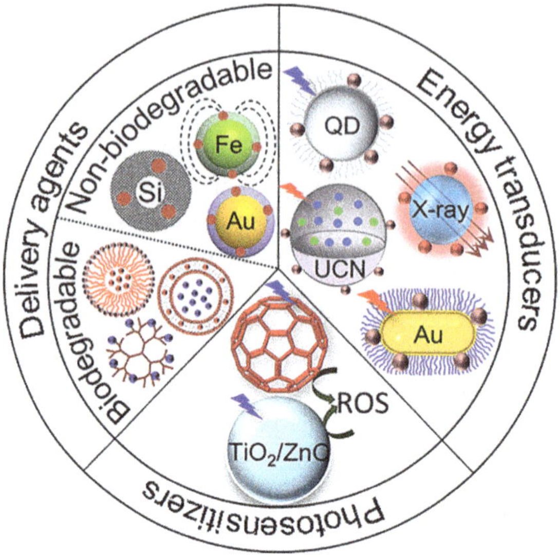

Figure 1.14 Nanoparticles in photodynamic therapy. Source: Reproduced with permission from Ref. [164]; © 2015/American Chemical Society.

magnetic field). Nanomaterials used for photosensitizer loading are usually divided into two categories.

Biodegradable Nanomaterials Biodegradable nanomaterials are made of natural biomaterials or synthetic polymeric nanomaterials with good biocompatibility and degradation properties that, after transporting the photosensitizer to the lesion site, undergo an enzymatic or hydrolytic reaction and release the photosensitizer, thereby improving the bioavailability and therapeutic efficacy of the photosensitizer and reducing the risk to normal tissues. Some common biodegradable nanocarrier materials are listed below.

Liposomes Liposomes are small nanoscale vesicles composed of phospholipids, cholesterol, etc., which can well encapsulate hydrophilic and hydrophobic drugs in their central lumen and membrane layer structures, respectively [171–174]. Liposomes can provide a stable environment to protect photosensitizers from the in vivo environment. Due to the presence of many active sites on the liposome membrane, their properties can be modified by attaching various functional ligands (such as antibodies, peptides, glycoprotein coupling, and folic acid) to enhance targeting to the lesion site. The metabolic properties and stability of liposomes can also be improved by tailoring their composition and structure. For example, the rate of drug release and stability can be controlled by adjusting the lipid fraction and the thickness of the membrane layer of liposomes, thereby improving the bioavailability and therapeutic efficacy of photosensitizers and making them more desirable carriers for photosensitizers.

Polymeric Nanoparticles Poly(lactic-*co*-glycolic acid) (PLGA) and polyethylene glycol (PEG) are both highly biocompatible and biodegradable polymers that can be prepared as photosensitizer-loaded nanoparticles [175–178]. The surface of polymeric nanoparticles can also be chemically modified to improve their targeting and stability. It is worth noting that both PLGA and PEG have been approved by the US Food and Drug Administration (FDA). For PLGA, the FDA has approved it for use in a variety of areas including drug delivery (such as the preparation of nanoparticles and microspheres), tissue engineering, and medical devices. PEG is approved by the FDA as a pharmaceutical excipient for use as an inactive ingredient in drug formulations to improve drug solubility, stability and bioavailability [179].

Nanogels Nanogels are highly cross-linked polymers with a network structure and have excellent biocompatibility and biodegradability [180, 181]. This highly cross-linked network structure provides good stability with a large pore structure, allowing drugs (including photosensitizers) to be evenly distributed within the nanogel, preventing premature degradation or inactivation in vivo [182]. Another advantage of nanogels is that they can respond rapidly to changes in the in vivo environment and provide controlled drug release [183]. For example, when the nanogel is exposed to the low pH environment of tumor tissue, the cross-linked structure of the gel can be altered, leading to drug release [184, 185]. In addition, the temperature-sensitive properties of nanogels can be used to achieve temperature-sensitive drug release [186]. In conclusion, nanogels have promising applications in areas such as drug delivery and photosensitive therapy.

Natural Macromolecules Proteins, chitosan, DNA, gelatin, etc., which have good biocompatibility and biodegradability, can be loaded with photosensitizers in various ways (such as wrapping, adsorption, and chemical bonding) [187–190]. Among them, albumin has many advantages as a photosensitizer carrier [191–194]. Firstly, as a naturally occurring protein in living organisms, albumin can reduce adverse reactions to tissues and cells and reduce the toxicity and side effects of photosensitizer. Secondly, albumin has a long-circulating life and high stability, which allows it to carry and deliver photosensitizers stably in the body. Thirdly, albumin as a carrier can improve the stability of photosensitizers. Many photosensitizers are susceptible to inactivation by light and oxygen in the free state, and encapsulating them in albumin can provide protection and prolong the life of photosensitizers. In addition, albumin can modulate the activation state of photosensitizers by interacting with them, further improving their stability and efficacy. Fourthly, albumin has good targeting for drug delivery. Albumin can achieve targeted delivery to specific tissues or cells by binding to specific receptors or molecules. For example, by modifying the ligands on the surface of albumin, selective recognition and localization of tumor cells can be achieved, thereby improving the accumulation of photosensitizer in tumor tissues and the therapeutic effect. Fifthly, albumin as a carrier can also provide controlled drug release. By adjusting the structure of albumin or complexing it with other materials, it is possible to achieve a slow release of the photosensitizers or a controlled release in response to changes in the specific in vivo environment.

Biodegradable Mesoporous Silica Mesoporous silica nanoparticles are nanomaterials with regular pores and high specific surface area for efficient encapsulation of small-molecule photosensitizers. By introducing metal elements (e.g. elemental iron, manganese) or metal oxides (e.g. calcium oxide) as well as organic compounds (e.g. compounds containing disulfide or selenium bonds) into the –Si–O–Si– backbone of mesoporous silica [195–199], the degree of shrinkage of mesoporous silica nanoparticles' backbones can be reduced and porosity can be increased to enhance the dissolution rate of the mesoporous silica in the in vivo environment and to facilitate the release of the loaded substances. There are two main methods to regulate the degradation rate of mesoporous silica by doping with organic compounds, one of which is to bind noncovalently doped organic molecules (e.g. methylene blue) into the mesoporous silica backbone [200, 201], causing oligomerization of the silica nanoscale, thereby accelerating its hydrolysis/degradation process. The second is through the hydrolysis reaction of organosilanes [202, 203], which allows organic groups that can be cleaved in the physiological environment to participate in the formation of the silica skeleton.

Nondegradable Nanomaterials Nondegradable nanomaterials, such as mesoporous silicon [204–208], have higher stability and longer lifetime than biodegradable nanomaterials and are less sensitive to pH and temperature. This enables them to stably carry and release photosensitizers in vivo, providing longer-lasting therapeutic effects. In addition, the stability of nondegradable nanomaterials reduces the loss and degradation of photosensitizers and increases their utilization. Another advantage of nondegradable nanomaterials is their high drug-loading capacity and more flexible surface modification capability. They typically have a large specific surface area and tunable pore structure, which can effectively encapsulate photosensitizers and chemically modify the surface for targeted delivery and controlled release of drugs. Therefore, nondegradable nanomaterials also have unique advantages in the field of photosensitizer carriers.

Overall, the choice of nondegradable or biodegradable nanomaterials for photosensitizers loading requires a comprehensive consideration of the specific application needs and objectives. If long-term stable drug release and high drug loading capacity are required, nondegradable nanomaterials may be more appropriate. If gradual drug release is required to reduce toxicity and side effects, or to achieve drug release at a specific time, biodegradable nanomaterials may be more competent. In addition, factors such as the cost of producing nanomaterials, scalability, and production safety also need to be considered.

1.4.3.2 Nanomaterials with Intrinsic Photodynamic Effects

Some of the nanomaterials themselves can be used as downhconversion photosensitive reagents [119, 145, 164, 209–212]; for example, some nonmetallic semiconductor nanomaterials (e.g. fullerene nanoparticles and carbon nitride), metal oxide semiconductor nanomaterials (e.g. titanium dioxide nanoparticles and zinc oxide), metal sulfide semiconductor nanomaterials (e.g. copper sulfide), and metal nanoparticles (e.g. gold, platinum, and silver), which can be excited by light and transfer the energy to the surrounding oxygen, exciting the triplet oxygen into $^1O_2$ (a type of ROS). Electron-hole pairs are also generated, and the electrons can reduce the oxygen to produce superoxide anion, which is also a ROS. Typically, metallic semiconductor nanomaterials generate ROS mainly through the superoxide anion generation mechanism, whereas nonmetallic semiconductor nanomaterials may generate ROS mainly through the monoclinic oxygen generation mechanism.

Among them, functionalized fullerenes have made promising progress in the field of antimicrobial photodynamic inactivation [213]. The amphiphilic C60 is obtained by introducing hydrophilic cationic groups through chemical substitution reaction, ion exchange reaction, photochemical reaction, or electrochemical method, etc. Since C60 itself weakly absorbs visible light, its ability to absorb visible light can be improved by introducing a light-trapping antenna on its core. C60 has good photostability and is not easily photobleached, and when excited by light, the light-trapping antenna absorbs the visible light and transfers the energy to the C60 molecule through a resonance energy transfer mechanism. Therefore, this process can more effectively excite C60 to participate in photochemical reactions and promote the occurrence of type I and type II photoreactions, which can form $^1O_2$ by energy transfer and superoxide anion radicals by electron transfer, thus exerting photodynamic effects. Photodynamic inactivation experiments have shown that the cationic C60 is an efficient photosensitizer and can be used as a broad-spectrum antimicrobial agent. In these structures, the hydrophobic nature of C60 enhances membrane permeability, while the presence of a positive charge increases the binding of fullerene derivatives to microbial cells, which greatly satisfies the antimicrobial requirement.

1.4.3.3 Energy Conversion Nanomaterials for Photosensitizers

These nanomaterials do not have photosensitivity per se, but they can act as carrier materials and energy converters for photosensitizers and can transfer energy to photosensitizers after being activated by external energy, thus indirectly activating photosensitizers [164, 214]. In other words, this process causes the photosensitizer to produce ROS by changing the nature of the light (such as wavelength and intensity). The main advantage of these nanomaterials is that they change the UV or visible light absorption range of traditional photosensitizers and increase the light absorption of photosensitizers in the NIR range. They mainly include X-ray activated nanomaterials, luciferase quantum dot nanomaterials, two-photon absorption nanomaterials, and rare earth upconversion nanomaterials.

X-Ray-activated Nanomaterials X-ray-activated nanomaterials generally have strong X-ray absorption and can convert it into UV or visible light. The photodynamic process is performed by activating surface-loaded photosensitizer molecules through energy transfer. The advantages of using this type of nanomaterial for PDT include the following [215–220]: X-rays have a strong penetrating ability compared to other photodynamic methods and have a better therapeutic effect on deep tissue lesions; X-rays work synergistically with PDT to produce a better therapeutic effect; X-ray-activated nanomaterials emit fluorescent light, which enables them to achieve precise localization of the lesion and irradiated area to avoid unnecessary damage to normal tissues.

Chen et al. have achieved important research results in X-ray-induced PDT (Figure 1.15a) [221]. They successfully prepared mZGGOs long afterglow nanoparticles with uniform size and monodispersed spherical morphology using mesoporous silica template method. It was found that under low-dose X-ray irradiation, the mZGGOs could emit obvious NIR persistent luminescence. This fluorescence has a long lifetime and can penetrate 1 cm of muscle tissue, which can achieve an effective imaging function in mice. In addition, by loading silicon phthalocyanine (Si–Pc) photosensitizer, the X-ray excited persistent luminescence can effectively stimulate Si–Pc to generate ROS and kill cancer cells. Furthermore, in vivo luminescence imaging of in situ hepatocellular carcinoma showed that these nanoparticles accumulated in liver cancer tissues through a passive targeting mechanism. X-ray-activated PDT can effectively inhibit the growth of in situ liver tumors. Using organic phosphorescent nanoscintillators with dual roles as scintillators and photosensitizers, the deep-seated tumors can be effectively eliminated at a low dose of 0.4 Gy with almost no toxic side effects on normal tissues and cells (Figure 1.15b) [222]. The above study provides an important reference for the further development and application of X-ray-induced PDT in clinical applications.

Luciferase Quantum Dot Nanomaterials The biofluorescent resonance energy transfer system based on luciferase quantum dot nanomaterials is a new type of energy converter that combines biotechnology and nanotechnology. The light produced by the reaction between luciferin and luciferase is called "molecular light" or "autofluorescence" [220, 223–225]. Although the total photon flux of molecular light is much lower than that of an external light source (for example, a laser), molecular light can operate within a few nanometers of the target and achieve relatively high photosensitivity efficiency at much lower photon fluxes through near-field energy transfer. Lai et al. synthesized a Renilla luciferase-immobilized quantum dot 655 (QD-RLuc8) for the realization of bioluminescence resonance energy transfer-mediated PDT [226]. Upon addition of coelenterazine, a substrate for the action of luciferase, QD-RLuc8 emitted autofluorescence at a wavelength of 655 nm, which in turn activated the photosensitizer Foscan to generate ROS.

Yun et al. have performed theoretical calculations of the photosensitization efficiency of molecular light using Chlorin e6 (Ce6) as a photosensitizer and bioluminescence resonance energy transfer from quantum dots as molecular light [227]. The calculations show that an irradiation intensity of up to $1\,\mathrm{mM\,cm^{-2}}$ per Ce6 molecule is required in the minimum diffraction-limited region when irradiating with conventional physical light sources such as LED or laser. However, using bioluminescence resonance energy transfer-triggered photosensitization, one photon from a luciferin/luciferase reaction is sufficient to activate one Ce6 molecule if the bioluminescence resonance energy transfer efficiency is 100%. When the bioluminescence resonance energy transfer efficiency is about 50%, two photons can activate one Ce6 molecule. In comparison, the efficiency of activating Ce6 using bioluminescence resonance energy transfer is about 6830 times higher

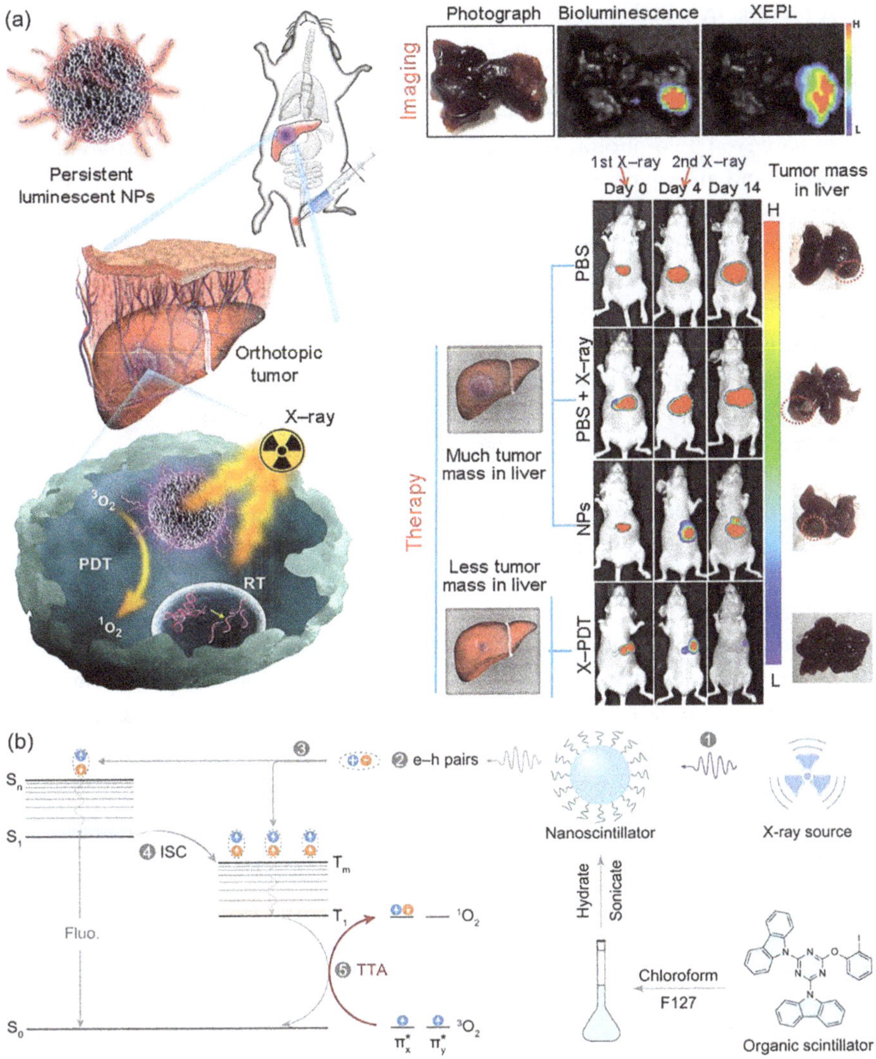

Figure 1.15 (a) Schematic illustration demonstrates that X-ray-induced persistent luminescence promotes ultrasensitive imaging and effective inhibition of orthotopic hepatic tumors. Source: Reproduced with permission from Ref. [221]; © 2020/John Wiley & Sons. (b) Schematic illustration demonstrates X-ray-induced persistent luminescence promotes ultrasensitive imaging and effective inhibition of orthotopic hepatic tumors. Source: [222]/Springer Nature/CC BY 4.0.

than that of conventional LED or laser. Thus, the work provides good theoretical and computational evidence that molecular light can effectively activate photoreactions and in vivo photosensitization by near-field energy transfer.

Two-photon Absorption Nanomaterials Two-photon absorbing nanomaterials are nanoscale materials that absorb two photons. Two-photon absorption is a nonlinear optical process in which two photons are absorbed simultaneously by an atom or molecule, causing the atom or molecule to jump from an initial energy level to a higher energy level [228, 229]. This process requires the total energy of the two photons to be equal to or greater than the jump energy level difference. Photosensitizers based on two-photon excited nanoparticles, which combine two-photon absorption and nanotechnology to facilitate deep light penetration and reduce photobleaching of photosensitizers, have received much attention from researchers. A large number of photosensitizers have been developed, including mainly those based on two-photon excited quantum dots, two-photon excited carbon-based nanomaterials, two-photon excited silica nanomaterials, two-photon excited gold nanomaterials, and two-photon excited polymer nanomaterials. Regarding the related research progress in this field, Chen et al. have published a review entitled "Two-photon excitation nanoparticles for photodynamic therapy" in *Chemical Society Reviews* [230]. We will not repeat the contents here.

Rare Earth Upconversion Nanoparticles (UCNPs) As mentioned above, UCNPs have the ability to absorb two or more low-energy photons and then "upconvert" them into a single high-energy photon, converting near-infrared light into visible or even ultraviolet light [231]. This is an inverse-Stokes-law process realized by a sequence of multiphoton absorption and energy transfer, and this unique property of UCNPs could not be better exploited to improve the efficiency of photosensitizers [76, 232, 233]. This is because conventional photosensitizers, such as dyes and organic compounds, typically absorb only a portion of the visible light spectrum, whereas UCNPs can absorb infrared or NIR light and convert it to visible or ultraviolet light, greatly expanding the spectral range of photosensitizer applications. Therefore, this unique optical property gives UCNPs a significant advantage in activating photosensitizers by overcoming the problems of short absorption of conventional photosensitizers in terms of wavelength and insufficient penetration depth of visible light, which can improve the efficiency of photosensitizers while reducing the potential damage to biological tissues from high-energy photons [234–237]. In addition, UCNPs are more stable to light and heat than conventional photosensitizers, making them ideal light energy converters, especially in biomedical applications. Overall, UCNPs offer a novel and efficient way of converting and harnessing light energy with a wide range of applications.

Ju et al. synthesized UCNPs with energy absorbing ($NaYF_4$:Nd,Yb), energy emitting ($NaYF_4$:Yb,Er), and inert ($NaYF_4$:Gd) core-shell structures, limiting energy absorption and transfer to a local space of 5 nm thickness on a single nanoparticle, and binding a NIR dye (800CW) to the surface of the $NaYF_4$:Nd,Yb layer for further sensitization (Figure 1.16a) [238]. Under NIR light irradiation, 800CW absorbs a large number of photons and then transfers the energy to Nd^{3+} and Er^{3+} sequentially via resonance energy transfer, which, coupled with the effect of inert nuclei blocking energy transfer to the interior of the nanomaterials, results in a 3600-fold increase in the emitted light intensity of Er^{3+} compared to that of conventionally structured UCNPs. When a receptor molecule (such as the dye N719) is attached to the surface of UCNPs, the target to be measured substance, Hg^{2+}, reacts with N719,

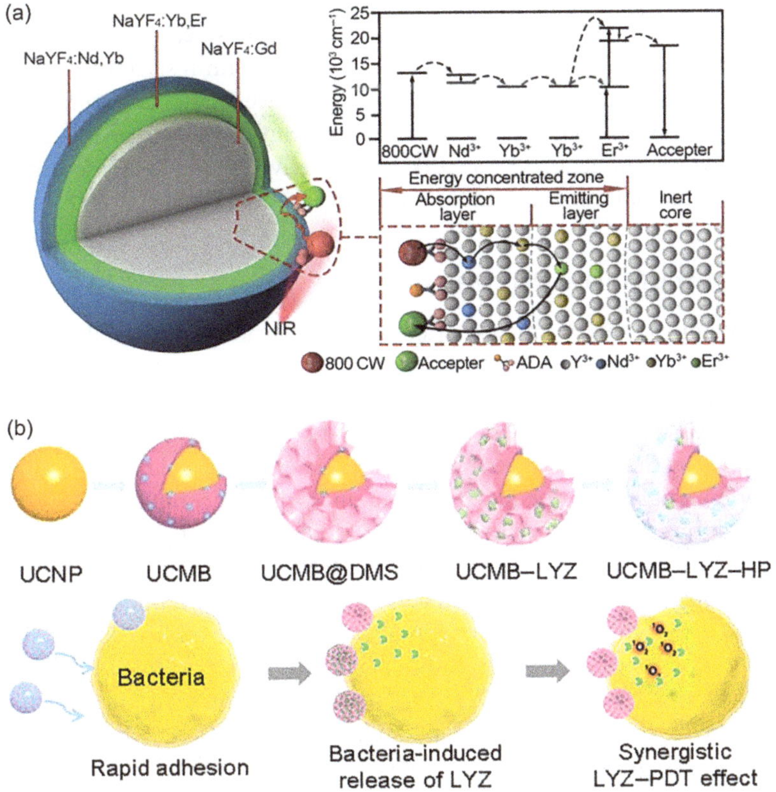

Figure 1.16 (a) UCNPs structured with an energy-concentrating zone to boost energy transfer efficiency and a simplified energy level diagram. Source: Reproduced with permission from Ref. [238]; © 2019/John Wiley & Sons. (b) Illustration of the synthesis and mechanism of action of antibacterial nanohybrid UCMB-LYZ-HP. Source: Reproduced with permission from Ref. [239]; © 2021/John Wiley & Sons.

which in turn induces the recovery of Er^{3+}-emitting fluorescence, significantly improving the sensitivity of the bioanalysis and lowering the detection limit by three orders of magnitude. When the photosensitizer Rose Bengal was used as an energy acceptor molecule modified on the surface of the UCNPs, the ROS yield was also significantly improved, resulting in better tumor therapeutic outcomes even under anoxic conditions. This study provides an important option for the development of highly sensitive biochemical analysis methods and cancer therapeutic protocols.

Chen et al. reported a bioinorganic hybrid nanomaterial that presents a combined photodynamic and lysozyme antimicrobial therapy based on UCNPs, which successfully fights drug-resistant bacterial infections in deep tissues (Figure 1.16b) [239]. The key to this therapy is that they used a type of UCNP encapsulated by an oversized mesoporous silica layer that can be loaded with photosensitizers and lysozyme. They then modified the surface of the material by self-assembly of hyaluronic acid and polylysine to form a "valve" that responds to bacterial hyaluronidase, allowing intelligent release of lysozyme from the silica pore channels. Antimicrobial experiments demonstrated the ability of this hybrid material to exert synergistic antimicrobial effects through bacterial cell wall disruption by releasing lysozyme and PDT, with significant disinfectant effects against methicillin-resistant *S. aureus*. The excellent therapeutic efficacy of the material against deep tissue infections with drug-resistant bacteria, without toxic side effects, was confirmed by experiments in mice models of wound infection. This work is instructive for the development of nonantibiotic antimicrobial agents and for mitigating the resistance crisis.

1.4.4 Photoelectrochemical Nanomaterials

Photoelectric nanomaterials are a class of energy-transforming functional materials that can convert light energy into electrical, chemical, or other energy [240–242]. There are various classification criteria for such materials, which can be divided into photoelectric conversion materials and photocatalytic materials in terms of their applications, and organic photoelectric functional materials, inorganic photoelectric functional materials, and organic–inorganic photoelectric complexes in terms of their compositions. Due to their unique physical and chemical properties, photoelectric nanomaterials have a wide range of applications in various fields such as photocatalysis, solar cells, light detection, optical communication, optical storage, lighting, and display technologies. Undoubtedly, photoelectric nanomaterials also play an important role in biomedical fields, such as biosensing and analytical detection, disease diagnosis, drug analysis, biomarkers and tracking, photothermal and photodynamic therapies, life analysis, and pathogenic microbial research [243, 244]. In this section, we will focus on the applications of photoelectrochemical nanomaterials in biosensing and analytical detection and nothing else.

Photoelectrochemical analysis is a new type of analytical method based on the photoelectric conversion properties of nanomaterials for detection, which has been established in recent years following the mature development of optical, photochemical, and electrochemical methods. This method is mainly based on the photoelectric conversion of photoelectrochemically active materials on the electrode surface and the recognition process of the target to be measured and is characterized by high sensitivity, simple equipment, low cost, fast detection speed, and easy miniaturization [245, 246]. As shown in Figure 1.17a, the photoelectrochemical sensors used in photoelectrochemical detection generally include, among other components, an excitation light source (laser, LED, xenon lamp, etc.), a monochromator, an electrolytic cell and an electrochemical detection device [247]. In the presence of the light source and the electron donor/acceptor, the photoelectrically active material on the working electrode undergoes charge jump and transfer, resulting in the formation of a stable photocurrent in the external circuit. Quantitative photoelectrochemical analysis is based on the relationship between the change in photocurrent or photovoltage and the concentration of the substance to be measured before and after the direct or indirect biorecognition interaction process. Since the photoelectrochemical process couples the photoexcitation process with electrochemical detection, the excitation and detection signals belong to different energy forms, which can significantly reduce the background noise compared

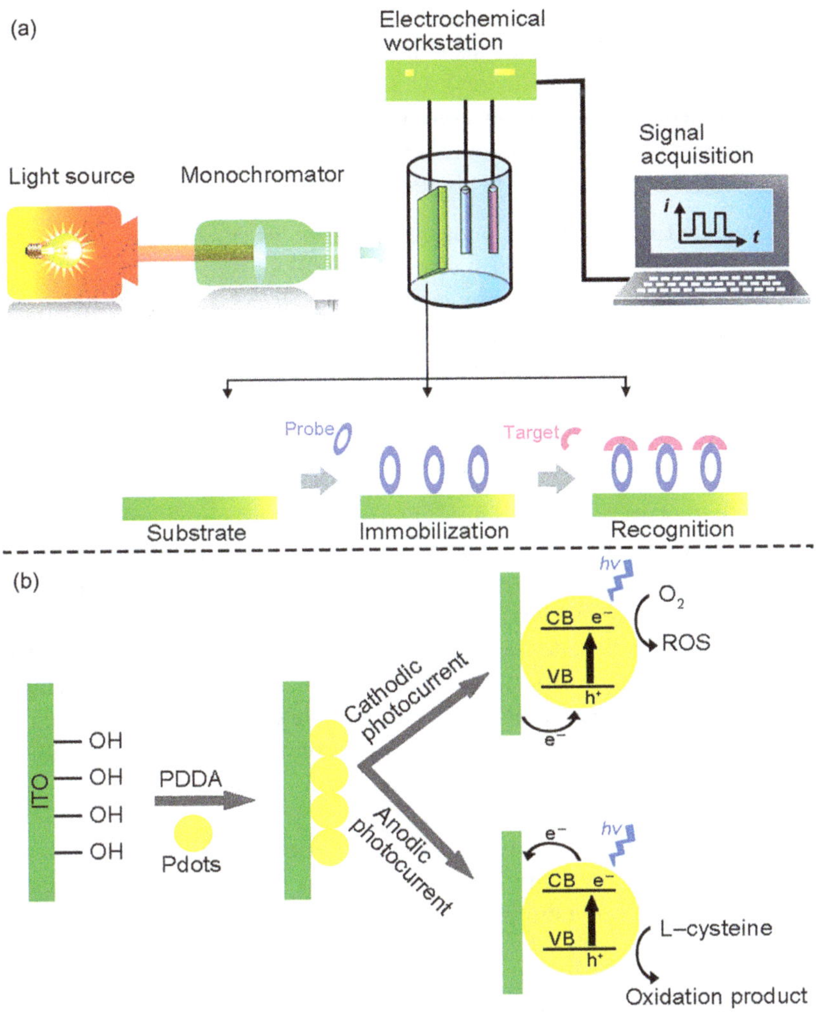

Figure 1.17 (a) Schematic of the general instrumentation and working principle of PEC bioanalysis. Source: Reproduced with permission from Ref. [247]; © 2015/The Royal Society of Chemistry. (b) Schematic illustration for the preparation of TPP-doped PFBT Pdots and its application for the PEC detection of L-cysteine. Source: Reproduced with permission from Ref. [248]; © 2017/American Chemical Society.

to traditional detection methods (such as chromatography, enzyme labeling, and atomic absorption spectroscopy). At the same time, the lower signal-to-noise ratio helps to further reduce the detection limit [249].

1.4.4.1 Photocurrent Signal Generation Mechanism

Polymer dots (Pdots) are used as an example to elucidate the mechanism by which the sensor exerts photoelectrochemical action (Figure 1.17b) [248]. When the energy of an incident photon ($h\nu$) is greater than the forbidden bandwidth of a photoelectrochemical material, electrons in the valence band of the material absorb the photon energy and jump to the conduction band to form photogenerated electrons. At the same time, photogenerated holes are also formed in the valence band, resulting in electron-hole pairs (two types of carriers). The carriers have short lifetimes and are in an unstable excited state, so they must release energy to return to their steady state. In the case of excited photoelectrochemical materials, there are four main pathways for the de-activation energy process of photogenerated electron-hole pairs:

(I) Photogenerated electrons and photogenerated holes complex as they move through the material, known as in vivo complexation;
(II) Complexation occurs when photogenerated electrons and photogenerated holes migrate to the surface of the material, namely surface complexation;
(III) Photogenerated electrons migrate to the surface of the material and then react with electron acceptors in solution, that is, oxidized;
(IV) Photogenerated holes migrate to the surface of the material and then react with the electron donor in solution, which is reduced.

Of these four pathways, complexation of photogenerated electrons and photogenerated holes leads to bursting of the photocurrent signal, which is detrimental to the photocurrent signal and is therefore considered a side reaction. To inhibit the complexation of these two carriers, electron donors or electron acceptors can be added to the solution to deplete the photogenerated electrons or photogenerated holes accordingly, which is an effective method of achieving a continuous and stable photocurrent. When the electron donor in the electrolyte solution is consumed in large quantities, the photogenerated electrons become the main carriers, and an anodic photocurrent is generated in the external circuit. If the electron acceptor in the electrolyte solution is consumed in large quantities, the photogenerated holes become the main carriers, and a corresponding cathodic photocurrent is generated in the external circuit. It is not difficult to see that the presence of electron donors or electron acceptors can provide a stable and reliable signal output for carrying out sensing of living matter.

1.4.4.2 Core Elements of Photoelectrochemical Biosensors

Photoelectrochemically active materials and biorecognition elements are the two core elements for the construction of photoelectrochemical biosensors, and both are indispensable [247]. They work together to enable the sensor to achieve high sensitivity, selectivity, and specificity for detecting target biomolecules. Photoelectrochemically active materials are materials that are capable of generating signals such as currents or potentials through photoexcitation or photocatalytic reactions and play the role of signal conversion in photoelectrochemical sensors [250]. So far, a variety of photoelectric conversion materials have been developed for the formation of photoelectrochemical biosensors, mainly in the following categories: inorganic photoelectric materials represented by TiO_x, ZnO, SnO_2, Cu_2O, Bi_2S_3, Ag_2S, CdTe, CdS, and so on; organic photoelectric materials, including organic small molecules, organometallic compatibilizers, polymer polymers, and part of COFs and MOFs; various types of composite photoelectric materials; and other photoelectric materials such as all-carbon materials, certain fluorescent proteins, and so on. These materials have good photoelectrochemical properties and are able to convert light energy into electrical signals or chemical reactions, thereby achieving signal amplification and detection of target biomolecules. For reasons of space, this is not dealt with here. See Chapter 14 for details. Biometric elements are biomolecules or biomaterials capable of interacting specifically with a target substance [251].

1.4.4.3 Types of Photoelectrochemical Biosensors

According to different classification methods, photoelectrochemical biosensors can be divided into different types. For example, according to different application fields, photoelectrochemical biosensors are mainly classified into photoelectrochemical small-molecule sensors, photoelectrochemical DNA sensors, photoelectrochemical immunosensors, photoelectrochemical enzyme sensors, and photoelectrochemical cell analysis. Based on the binding method of the target, photoelectrochemical biosensors can be classified into photoelectrochemical biosensors detected by direct method, photoelectrochemical biosensors detected by sandwich method, and photoelectrochemical biosensors detected by competition method. Based on the difference in the interaction between the biorecognition element and the target, photoelectrochemical biosensors can be further classified into two main categories: biocatalytic and biophilic. According to the type of photoelectrically active materials, photoelectrochemical sensors can be classified into photoelectrochemical sensors based on inorganic semiconductor nanomaterials, photoelectrochemical sensors based on organic semiconductor nanomaterials, and photoelectrochemical sensors based on composite semiconductor nanomaterials. According to the photocurrent signal change pattern of photoelectrochemical biosensors caused by the target, they can be classified into the following three types: signal enhancement type, signal burst type, and signal flip type. The classifications are as follows.

Signal-enhanced Photoelectrochemical Biosensors Such sensors use the specific interaction of biomolecules with the target analyte to detect the presence of the target analyte by means of a current or potential signal generated by a photoelectrochemical reaction. Signal enhancement can be achieved in a variety of ways, including the use of enzymes, nanoparticles, multilayer structures, or signal-enhancing molecules. This approach can dramatically improve the sensor's ability to detect target molecules at low concentrations, thereby increasing the sensitivity and lower limit of detection. Xu et al. first proposed a strategy based on the combined enhancement of carrier directional mobility by piezoelectric and plasma excitons and successfully constructed a portable photoelectrochemical biosensor for the detection of low-abundance cancer markers in human plasma [252]. They simplified the basic building blocks of the photoelectrochemical biosensor, such as using a handheld UV flashlight instead of a physical excitation light source and a digital multimeter instead of an electrochemical workstation (Figure 1.18). The sensor performs the capture and immunoincubation process of the target PSA on a separate microtiter plate, where H_2O_2 produced by the enzyme immunoassay

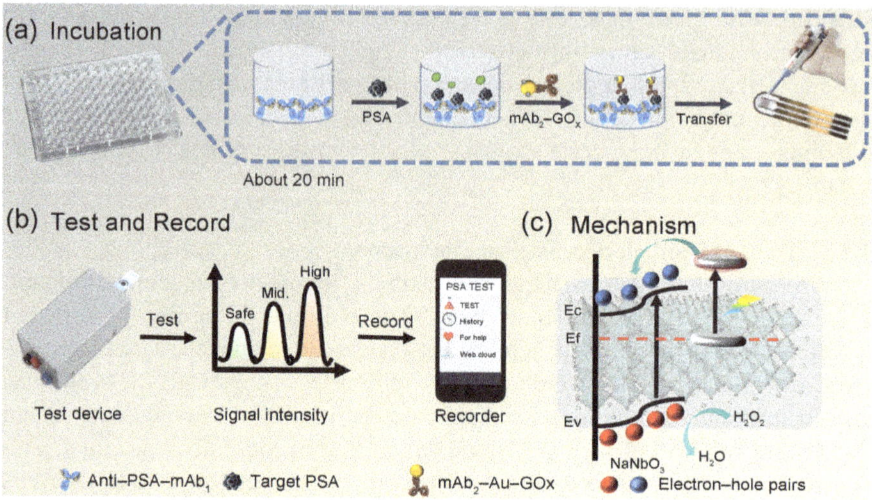

Figure 1.18 Schematic diagram of PEC bioassay based on local surface plasmon resonance and piezoelectric co-enhancement. (a) Schematic diagram of the incubation process of target PSA in a microtiter plate. (b) Schematic diagram of the PEC recording platform for 3D printed micro devices, with results displayed on a smartphone. (c) Schematic diagram of the energy level structure and photoelectron transfer of the Ag/$NaNbO_3$ composites. Source: Reproduced with permission from Ref. [252]; © 2022/Elsevier.

induces the directional separation of electrons and holes under light excitation, thereby amplifying the photocurrent signal. A digital multimeter was used to monitor the photocurrent in real time. In addition, the effect of Ag nanoparticle deposition on the surface electronic states of the piezoelectric chalcogenide $NaNbO_3$ was investigated by density flood theory calculations. The results show that the photoelectrochemical biosensor exhibits an ultra-wide linear range and an ultra-low detection limit for the target PSA under optimized conditions. The performance is comparable to commercial ELISA kits at the 95% confidence level. This work provides new ideas for the preparation of improved photoelectrochemical biosensors for the rapid and accurate detection of cancer-related proteins.

Signal-burst Photoelectrochemical Biosensors Unlike signal-enhanced sensors, signal-burst sensors cause a reduction in the detected signal when a specific biometric event occurs. This typically involves a biomolecule interacting with the sensor surface, resulting in a reduced photocurrent or current. Signal bursting can be achieved by physical or chemical mechanisms such as energy transfer, electron transfer, or competitive binding. Signal-bursting mechanisms are often used in detection strategies based on competition or site occupancy, and Yuan and coworkers have successfully prepared a signal-bursting photoelectrochemical biosensor for ultrasensitive DNA determination by using SnO_2/BiOBr complexes with ideal photoelectrochemical conversion efficiencies as the photoelectrically active substrate and SiO_2 as an effective signal quencher [253]. As shown in Figure 1.19, the SnO_2/BiOBr heterojunction was coated on the electrode surface

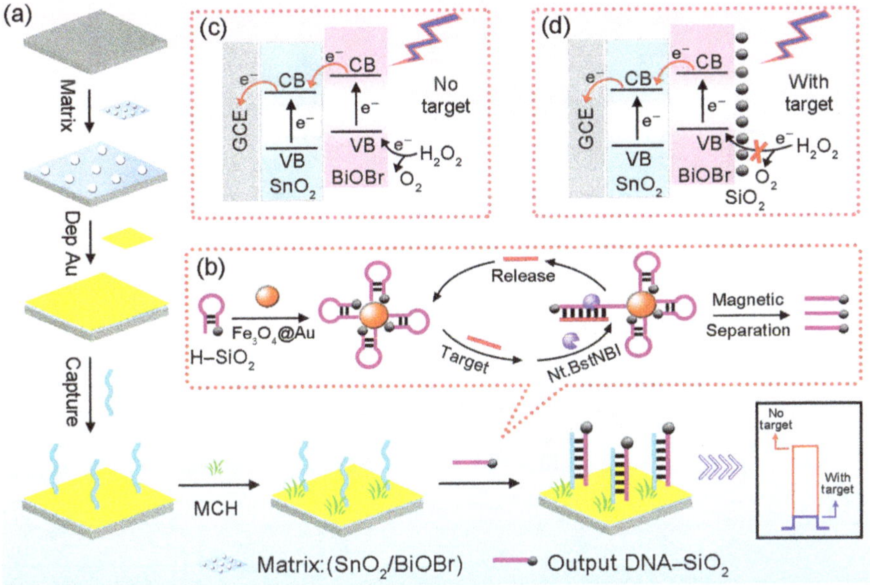

Figure 1.19 Schematic illustration of this PEC assay for DNA detection based on the SnO_2/BiOBr heterostructure. (a) Biosensor assembly procedure. (b) Enzyme-assisted signal amplification process. (c,d) Photocurrent generation mechanism. Source: Reproduced with permission from Ref. [253]; © 2021/American Chemical Society.

to form a large built-in electric field, which not only inhibited the electron-hole recombination but also improved the light absorption ability in the UV-visible region and significantly enhanced the intensity of the initial photocurrent signal generation. Meanwhile, a gold nanoparticle coating was deposited on the surface of SnO_2/BiOBr to immobilize and capture DNA. The target DNA was then converted into a large amount of SiO_2-modified single-stranded output DNA through an enzymatic cyclic amplification process. Due to the strong steric hindrance effect of SiO_2 on electron transport, the initial photocurrent generated by the SnO_2/BiOBr heterojunction can be effectively burst, thus achieving highly sensitive detection of the target. Overall, this signal-bursting photoelectrochemical biosensor opens up a new way to monitor various biomarkers.

Signal-flipping Photochemical Biosensors Signal flip-flop sensors are sensors in which the signal change triggered by a biometric event is reversed, i.e. flipped from one state to another. For example, a charge transfer process may occur after a photosensitive material or photoelectrode surface modification layer in a sensor interacts with a target analyte, resulting in a change in the intensity or polarity of the photoelectrochemical signal. This type of sensor may rely on complex biochemical or physical processes to translate a biometric event into a change in the electrical signal, a change that may be due to a change in conductivity, the opening or closing of ion channels, or the production, or consumption of electrochemically active substances. Dai and coworkers designed a signal-switchable photoelectrochemical

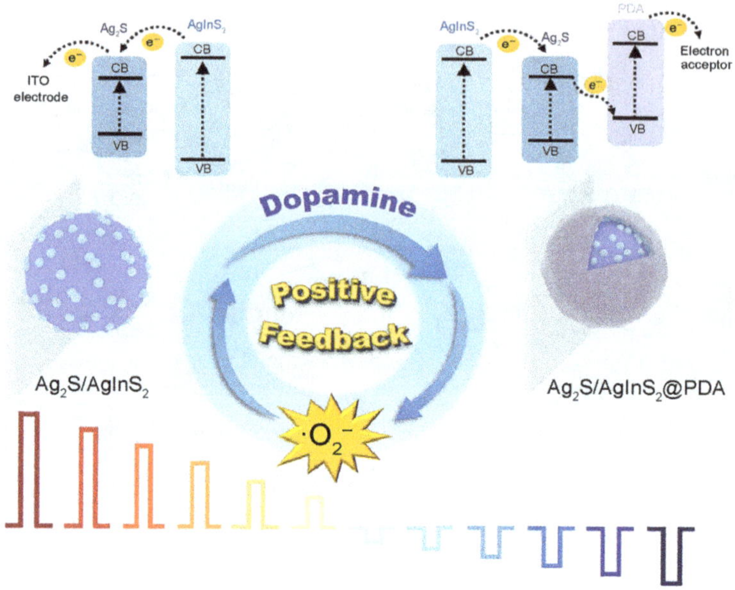

Figure 1.20 Illustration of a signal-switchable photoelectrochemical biosensor for the detection of dopamine. Source: Reproduced with permission from Ref. [254]; © 2023/American Chemical Society.

biosensor for the detection of dopamine (Figure 1.20) [254]. The polarity switching of the $Ag_2S/AgInS_2$ heterojunction was regulated by free radical-induced positive feedback polydopamine adhesion. In particular, the $Ag_2S/AgInS_2$ complex exhibited extremely stable anodic photoelectrochemical activity. In the presence of light, it is capable of generating superoxide radical ($\cdot O_2^-$) and hydroxyl radical ($\cdot OH$) intermediates that lead to the oxidative polymerization of dopamine and the subsequent attachment of the generated polydopamine to the $Ag_2S/AgInS_2$ heterojunction. In this way, the presence of polydopamine creates a new electron transfer pathway that causes the polarity of $Ag_2S/AgInS_2$ to switch and the photoelectrochemical response to change from anodic to cathodic photocurrent.

In addition, polydopamine with stronger photoreductive activity promotes more radical intermediates, creating a positive feedback mechanism for polydopamine generation and greatly improving the sensitivity of photoelectrochemical biosensing. The applicability of the photocurrent polarity shift achieved by the sensor in real sample assays (e.g. determination of polydopamine in fetal bovine serum) was determined by comparison with commercial ELISA methods.

Different types of photoelectrochemical biosensors have their own advantages and applicability in different application scenarios. The choice of sensor usually depends on factors such as the nature of the target analyte, the desired detection sensitivity, and the complexity of the experimental design. With advances in materials science and biotechnology, the design of these sensors is becoming increasingly sophisticated, enabling higher detection performance and application flexibility.

1.4.5 Photoacoustic Nanomaterials

Photoacoustic nanomaterials are materials that can convert light energy into acoustic energy and emit it in the form of ultrasonic waves. Figure 1.21 shows that photoacoustic nanomaterials undergo three processes to produce a photoacoustic signal under the action of a laser: light absorption, photothermal conversion, and thermoacoustic conversion [256, 257]. Specifically, irradiating a material with a pulsed laser or a continuous beam of light that has undergone periodic intensity modulation creates periodic temperature changes within the material, causing that part of the material and its adjacent medium to thermally expand and contract to produce periodic changes in stress (or pressure) and thus generate an acoustic signal, known as a photoacoustic signal [255, 258]. The frequency of the photoacoustic signal is the same as the optical modulation frequency, and its intensity and phase depend on the optical, thermal, elastic, and geometrical properties of the material. This photoacoustic effect has important applications in biomedical imaging, detection, and analysis.

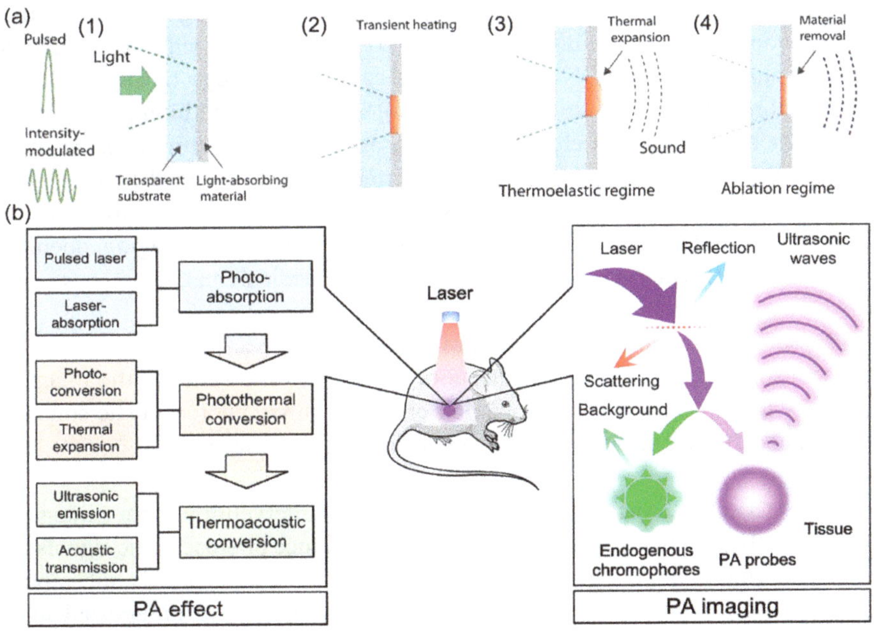

Figure 1.21 (a) Photoacoustic generation. (1) Light absorption in a light-absorbing material. Light is illuminated in the form of pulsed or intensity-modulated waves. (2) Transient heating of the material by light absorption. Sound generation through two different regimes: (3) thermoelastic regime and (4) ablation regime. In the thermoelastic regime, sound is generated by thermal expansion of the heated material. In the ablation regime, the parts of the material are removed through phase change (invasive), producing much stronger sound intensity. Source: Reproduced with permission from Ref. [255]; © 2018/John Wiley & Sons. (b) Schematic diagram showing the three processes of PA generation and the interaction of laser and biological tissue in PA imaging. Source: Reproduced with permission from Ref. [256]; © 2022/American Chemical Society.

1.4.5.1 Introduction to Photoacoustic Imaging

Photoacoustic imaging (PAI) is a noninvasive imaging technique developed based on the photoacoustic phenomenon, which uses photoacoustic detection to capture acoustic signals generated by biological tissues and then processes these signals through a computer to obtain an image of the internal structure of the biological tissues [259]. In PAI, photoacoustic detection, as an essential part of the process, is achieved by capturing and analyzing changes in the intensity of ultrasonic signals generated by laser energy absorbed by endogenous or exogenous photoacoustic probes to achieve the detection of the target analyte.

In PAI, not only can photoacoustic probes generate photoacoustic signals, but some endogenous pigments can also generate background signals [260–262]. At the same time, as the laser light passes through biological tissue, its intensity in the tissue is significantly reduced due to reflection and scattering at the surface and within the tissue, which can also affect the optical signal [260, 263]. However, the signal intensity in PAI is not represented by the optical signal but by ultrasound, and the scattering of ultrasound in biological tissue is much less than the optical signal. Therefore, although there is attenuation of the optical signal in biological tissue, this does not reduce the resolution of PAI [264]. PAI combines the high sensitivity and contrast of optical imaging with the deep penetration of ultrasound imaging and is therefore capable of imaging tissue up to several centimeters deep while maintaining high spatial resolution [265, 266]. This makes PAI promising for biomedical applications. For example, it can clearly depict the anatomical structure (e.g. melanin and lipid distribution) and physiological function (e.g. blood, blood flow, oxygen saturation, and metabolic rate) of tissues [256, 267–269]. PAI is therefore able to overcome the limitations of purely optical and purely ultrasonic imaging, opening up new possibilities in the field of biosensing and biomedical imaging.

1.4.5.2 Selection of Photoacoustic Contrast Agents

Photoacoustic contrast agents are substances used in PAI to enhance the imaging contrast of tissues and organs. Depending on their source, photoacoustic contrast agents can be divided into endogenous and exogenous contrast agents.

Endogenous Contrast Agents In the absence of external intervention, some biological tissues have the ability to absorb light energy and produce photoacoustic signals on their own, which are endogenous contrast agents. Some endogenous substances (such as water, hemoglobin, melanin, fat, collagen, etc.) have differential light absorption and can be used as endogenous contrast agents in PAI [267, 270, 271]. By modeling the propagation and detection of suitable photoacoustic signals, it is possible to image the structure and function of tissues and organs such as blood, skin, and the eye.

Endogenous contrast agents are substances that occur naturally in living organisms (namely nonspecific photoacoustic contrast agents) and do not require any additional introduction or special treatment, and are therefore highly biocompatible and less likely to cause immune or other adverse reactions. Certain endogenous contrast agents are present only in specific tissues or organs and can provide specific imaging and contrast that can aid in the study and diagnosis of relevant tissues and lesions. For example, the aforementioned hemoglobin is a protein found only in red blood cells and can be used to image blood, blood vessels and the heart [272]. Melanin is a pigment found only in tissues such as skin, hair, and eyes and is helpful in imaging skin lesions, melanoma and the fundus of the eye [273]. Bilirubin is a yellow pigment metabolized by the liver and found mainly in bile, and is useful for imaging the gallbladder, bile ducts and gallstones [274]. Calcitonin is a calcium salt found in bones and calcified lesions and can be used to image bones, calcified tumors and atherosclerosis [275–277].

However, an unsatisfactory aspect of actual imaging is that certain diseases do not produce endogenous photoacoustic signals, or produce weak photoacoustic signals with relatively low contrast, and may therefore be difficult to detect by PAI in some cases [278]. In addition, the nature and distribution of endogenous contrast agents cannot usually be regulated, optimized, or tuned by external means. Therefore, the use of exogenous contrast agents is often necessary to enhance the contrast of PAI for the study and diagnosis of various diseases.

Exogenous Contrast Agents Unlike endogenous contrast agents, exogenous contrast agents are the result of artificial intervention. These contrast agents are substances introduced into a living organism by injection or other means, which are then calibrated against specific biomarkers or diseased tissue to enhance the contrast of PAI. Exogenous contrast agents are usually light-absorbing nanomaterials such as nanoparticles and nano-dyes. These nanomaterials are capable of absorbing laser energy and generating photoacoustic signals, resulting in more pronounced contrast in imaging.

From an application point of view, tuning the absorption of exogenous contrast agents to the NIR region is key to enhancing image contrast and can improve the resolution and image output intensity of PAI. This is because, on the one hand, NIR light has a greater penetration depth and can better penetrate biological tissues to achieve a deeper imaging depth, enabling imaging of deep tissues and organs [257]. On the other hand, the wavelength range of NIR light is one of the least absorbing regions of biological tissues, and biological tissues absorb and scatter relatively less in this wavelength range, which can reduce the loss of light energy during the imaging process [53, 80]. In addition, the use of NIR light can avoid the damage and photothermal effect on biological organisms and reduce the adverse effects on biological organisms [53]. Therefore, adjusting the absorption of exogenous contrast agents to the NIR region can lead to more efficient generation and propagation of the photoacoustic signal, thereby increasing the intensity of the photoacoustic signal and improving image contrast.

In summary, exogenous photoacoustic contrast agents generally have the following advantages [279]: firstly, they provide a stronger photoacoustic signal, which enhances the contrast to tissues and organs, making subtle structures and functions more clearly visible. Secondly, their properties can be adapted and optimized to meet specific imaging requirements by changing their composition, size, surface modifications, etc. Thirdly, they can be appropriately modified and functionalized to achieve targeted imaging of specific tissues or cells, improving the accuracy and specificity of imaging.

Introduction to Photoacoustic Imaging Probes So far, many contrast agents (which can also be called PAI probes), for example, carbon-based nanomaterials, metal nanomaterials, transition metal sulfides, organic small molecules and polymers, and ionic liquids, have been widely used in PAI, which has greatly contributed to the development of this imaging technique [280]. In addition to the advancement of a part of PAI probes with strong light-absorbing properties, smart PAI probes triggered by the microenvironment at the lesion (e.g. tumor, Alzheimer's disease, and atherosclerosis) have also been gradually developed. The recent developments of these two types of PAI probes are summarized here.

Photoacoustic Imaging Probes with Strong Light Absorption Properties In general, photoacoustic contrast agents and photothermal converters share almost the same physicochemical properties, which means that nanomaterials used for PTT are usually effective for PAI [281, 282]. Such imaging probes can be appropriately modified and functionalized to allow targeted imaging of specific tissues or cells. Listed below are some PAI probes with strong light absorption properties.

A) *Gold-based nanoprobes*: Due to the LSPR effect, gold-based nanomaterials have high light-absorbing properties in the visible and NIR light range. In addition, the shape, size, and structure of gold nanoparticles can be tuned to optimize their light-absorbing properties and biocompatibility. Yao He and coworkers investigated the active endocytosis of glucose polymer-modified gold nanoparticles by different bacteria via the ATP-binding cassette transport pathway, and the small-sized gold nanoparticles after endocytosis could be aggregated under laser induction (Figure 1.22) [283]. The photoacoustic signal of the aggregates was enhanced about 15.2-fold compared to unaggregated gold nanoparticles, overcoming the difficulty of inaccessibility of large near-nanoparticles to bacteria and enabling high-resolution PAI of bacteria in deep tissues. Emelianov and coworkers synthesized hyperbranched gold nanoconstructs capable of acting as efficient photoacoustic contrast agents [284]. It was found that hyperbranched gold nanoconstructs exhibit superior optical properties, including strong NIR absorption, high absorption efficiency at different polarization angles, and excellent photostability, compared to conventional contrast agents based on equipartitioned excitations, such as gold nanorods and gold nanostars. In vitro and in vivo experiments confirm that hyperbranched gold nanostructures are well-suited for high-contrast imaging and stable PAI.
B) *Semiconductor nanoprobes*: These materials include metal sulfides, metal selenides, and metal carbides [285], which have important applications in the field of photoacoustic nanoprobes due to their unique physical and chemical properties such as high electrical conductivity, optoelectronic properties, and thermal stability. Tan and coworkers synthesized an ultrasmall single-layer of MoS_2 nanodots with metallic phase by *n*-butyllithium intercalation-assisted

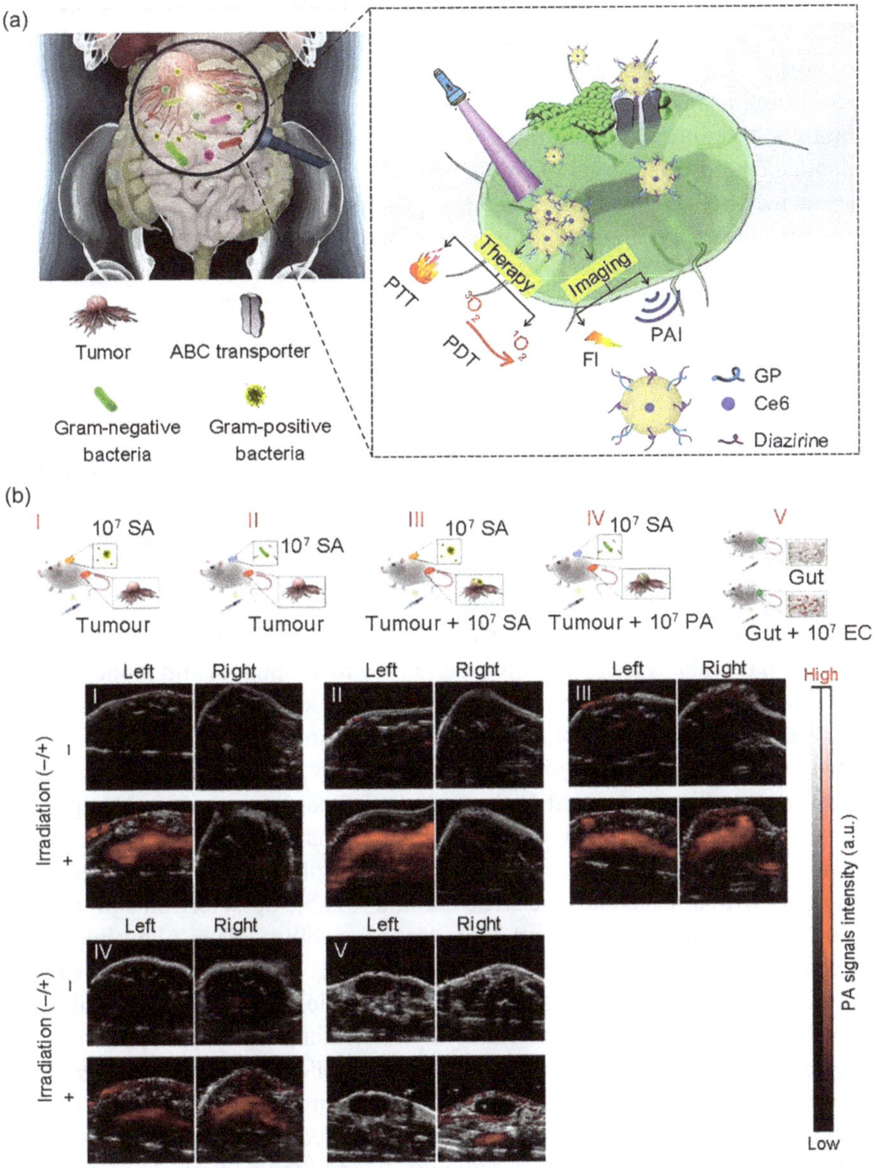

Figure 1.22 (a) Schematic showing bacteria eating gold nanoparticles for aggregation-enhanced imaging and killing bacteria. The imaging models include fluorescence imaging (FI) and photoacoustic imaging (PAI). The therapeutic methods include PTT and PDT. (b) Schematic illustrating the mice with different treatments and corresponding photoacoustic imaging of photoacoustic intensity of different infected sites with or without the irradiation of 405 nm laser. Source: [283]/Springer Nature/CC by 4.0.

liquid-phase sonication [286]. The nanodots were found to have strong absorption and PAI signals in the NIR-II region and exhibit high extinction coefficients and photothermal conversion efficiency at 1064 nm, which can be used as PAI-mediated photothermal nanoformulations in the NIR-II region. Jiang and coworkers proposed a strategy for activatable PAI using in situ cation exchange of ultrathin zinc selenide (ZnSe) nanoplates to monitor copper levels in the brain of Alzheimer's disease mice [287]. They attached ultrathin ZnSe probes to peptide ligands that target the blood-brain barrier, allowing them to effectively cross it. Once in the brain, the nanoplatelet probe rapidly exchanges with endogenous copper ions, enabling activatable PAI of copper levels in the brain. Wu et al. obtained MnO_x/Ta_4C_3 nanocomposites by growing manganese oxide nanoparticles in situ on the surface of Ta_4C_3 MXene [288]. Ta_4C_3 has good photothermal conversion properties, making the synthesized material an ideal contrast agent for PAI both in vitro and in vivo.

C) *Carbon-based nanoprobes*: Carbon-based nanomaterials (e.g. carbon dots, carbon nanotubes, and graphene) serve as effective photoacoustic probes for biomedical imaging, such as tumor detection and vascular imaging, due to their high light absorption properties in the ultraviolet to NIR range and good biocompatibility [139, 289, 290]. Smith and coworkers have prepared ultra-selective Ly-6C^{hi} carbon nanotubes for imaging and identification of inflammatory atherosclerotic plaques [291]. Since inflammatory cells such as macrophages and monocytes tend to accumulate on inflammatory patches, this specific carbon nanotube is preferentially taken up by macrophages and monocytes. In this technique, the vascular region of interest is irradiated with light, causing the carbon nanotubes to vibrate and emit an ultrasound signal. This photoacoustic effect makes it possible to visualize inflammatory plaques and identify which unstable patches are present in the body. This imaging technique is important for the early detection and localization of inflammatory atherosclerotic plaques.

D) *Black phosphorus-based nanoprobes*: Black phosphorus is a two-dimensional material with a special layered structure and tunable optoelectronic properties. Due to its wide band gap and high carrier mobility, black phosphorus has strong light absorption ability, which can effectively convert light into heat and produce a strong photoacoustic signal. Meanwhile, black phosphorus also has good biocompatibility and degradability. Therefore, black phosphorus-based nanoprobes have been widely studied and applied in the field of PAI and have potential applications in biomedical imaging and drug delivery disease diagnosis. Yu and coworkers successfully prepared a novel black phosphorus PAI contrast agent known as titanium ligand-modified black phosphorus quantum dots [292]. The quantum dots have excellent photoacoustic conversion properties and good stability in the physiological environment. Cell and animal in vivo tests have shown that the quantum dots can efficiently accumulate in tumor cells and absorb light pulses, resulting in detectable ultrasound signals. By analyzing these signals, high-resolution and high-contrast 3D tumor images can be generated. This study provides a new method for PAI of tumor tissue, which enables accurate imaging and localization of tumors and is expected to provide more accurate and reliable imaging guidance for clinical diagnosis and treatment.

E) *Organic photothermal molecule-based nanoprobes*: There are many common dyes, such as pyrrolidine dyes, acridine dyes, phthalocyanine dyes, indocyanine, and rhodamine, which have strong light-absorbing properties, especially in the NIR region, and are used as PAI probes [293–295]. These organic photothermal molecules have high light absorption cross-sections and high photothermal conversion efficiencies, which can improve the generation and detection sensitivity of photoacoustic signals. In addition, these molecules can often be chemically modified or conjugated to other nanomaterials (e.g. gold nanoparticles or silicon nanoparticles) to improve their stability and targeting in vivo. Lin and coworkers have developed a new class of rhodamine-like photoacoustic dye called GX, which has controllable photoacoustic properties and function in the NIR-II region (Figure 1.23) [296].

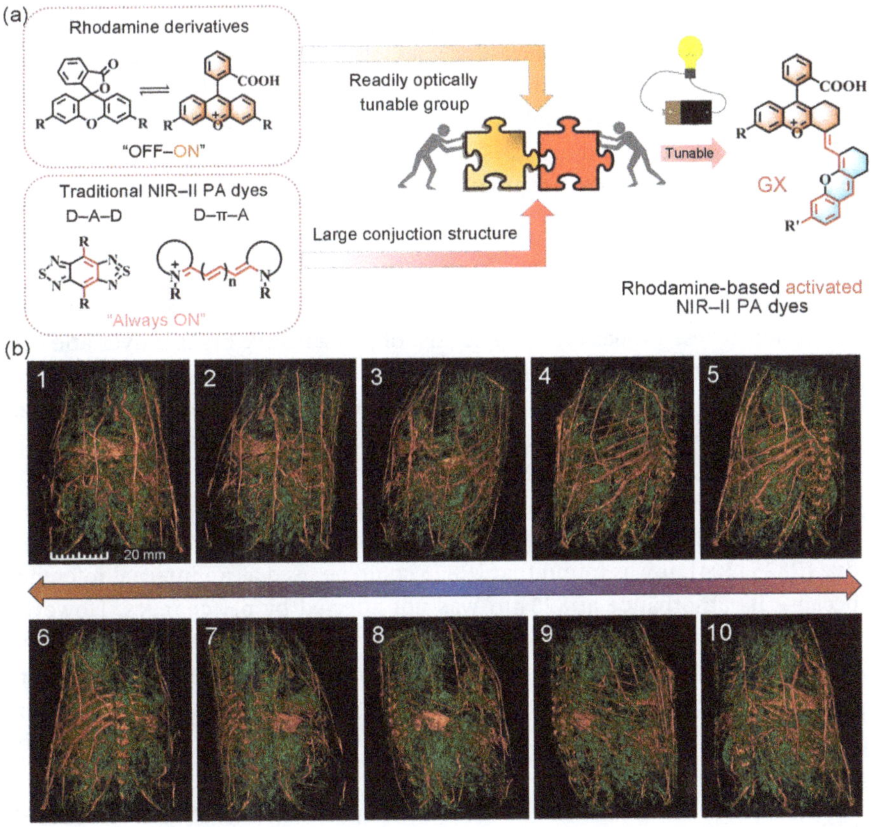

Figure 1.23 (a) Rational design of dyes GX as a novel class-activated NIR-II PA platform. (b) Multi-angle NIR-II PA 3D imaging of mice vasculature by dye GX-5. Source: Reproduced with permission from Ref. [296]; © 2023/John Wiley & Sons.

These novel NIR-II photoacoustic dyes possess a robust optical switching mechanism similar to that of conventional rhodamine and have absorption and emission wavelengths suitable for the NIR-II region. Among them, the dye GX-5 has absorption and emission wavelengths up to 1082 and 1360 nm, respectively. They performed PAI experiments using GX-5 dye and successfully visualized the distribution and structure of the mouse vasculature in a three-dimensional perspective. In addition, using GX-5 as an activatable dye platform, they constructed the first NIR-II probe, GX-5-CO, for the specific detection of carbon monoxide. The probe has been shown to reveal carbon monoxide levels in hypertensive mice using NIR-II PAI, demonstrating the potential value of this novel NIR-II photoacoustic dye, which is expected to be a powerful tool for biomedical imaging and disease diagnosis, among other applications.

Intelligent Photoacoustic Imaging Probes Triggered by the Lesion Microenvironment The lesion microenvironment refers to the specific environmental conditions surrounding the lesion area in diseases such as tumors, Alzheimer's disease, and atherosclerosis [297–302]. Lesion-triggered smart PAI probes are probes that respond and translate into photoacoustic signals based on the characteristics of the lesion microenvironment. These probes can provide a better understanding of the mechanisms of disease and the characteristics of the lesion area, and offer guidance and clues for the diagnosis and treatment of disease. The following are some common microenvironment-triggered smart PAI probes at lesions:

A) *pH-sensitive probes*: Diseases such as tumors and atherosclerosis are often associated with an acidic microenvironment [275, 303], so pH-sensitive probes are a common type of microenvironment-triggered photoacoustic probe at the site of a lesion. These probes typically consist of pH-sensitive organic dyes and carriers. In an acidic environment, these organic dyes undergo structural changes that alter their optical and photoacoustic properties and generate specific photoacoustic signals. pH-sensitive probes can achieve highly sensitive imaging and molecular diagnosis of the acidic microenvironment at the lesion site. Liu and coworkers reported a pH-responsive ratiometric photoacoustic probe obtained by self-assembly based on albumin, benzo[*a*]phenoxazine (BPOx), and the NIR dye IR825 [304]. It was found that IR825 could be used as an internal reference because its absorbance at 825 nm was not affected by pH changes. However, the photoacoustic signal at 680 nm of BPOx was significantly enhanced with decreasing pH. Therefore, the ratio of the intensity of the 680 nm to the 825 nm photoacoustic signal can be used to measure the pH of the tumor site. Using this pH-dependent ratio probe, immediate pH changes within the tumor can be detected.
B) *GSH-sensitive probes*: GSH is one of the major intracellular reductants involved in regulating intracellular redox reactions and maintaining intracellular redox homeostasis [305, 306]. However, the level of intracellular GSH is altered in some disease states, leading to dysregulation of redox homeostasis. In cancer, intracellular GSH levels are usually elevated due to factors such as high metabolic

activity, increased oxidative stress, and drug resistance in cancer cells [307, 308]. High levels of GSH help cancer cells resist oxidative stress and the toxicity of chemotherapeutic drugs, thereby promoting tumor growth and metastasis. In contrast, in some cardiovascular diseases, neurodegenerative disorders, and the aging process, GSH levels can decrease, making cells more susceptible to damage from oxidative stress [309–311]. Shi's team designed an intelligent pH/GSH dual-stimulation-responsive NIR probe, Cy-1, which can be activated by the synergistic effect of acidic pH and reducing GSH, followed by a bioorthogonal condensation reaction between the cyano group of 2-cyanobenzothiazole and 1,2-amino thiols to form amphiphilic cyclic dimers, which further self-assemble into nanoparticles with larger sizes [312]. The formed nanoparticles significantly enhanced the accumulation and retention of the probe in the tumor, enabling highly sensitive NIR/PAI.

C) *Hypoxia-sensitive probes*: Diseases such as tumors, atherosclerosis and Alzheimer's disease are often hypoxic [51, 313, 314], so hypoxia-sensitive probes are also a common microenvironment-triggered photoacoustic probe. These probes are usually nanoparticles or molecules containing specific fluorescent dyes or fluorescent proteins that undergo chemical or physical changes when exposed to an oxygen-deficient environment. This change can alter the way the probe absorbs light, which in turn affects the resulting acoustic signal. Hypoxia-sensitive probes provide information on the spatial distribution and temporal changes of oxygen deprivation, helping to understand disease mechanisms and guide therapeutic strategies. Nie and coworkers developed a hypoxia-sensitive molecular probe, HyP-650, for PAI [315]. The probe has an absorption peak at 650 nm, which is red-shifted in a hypoxic environment and converted to a substance red-HyP-650 with an absorption peak of 740 nm. Therefore, by simultaneously detecting the signal changes at both 650 and 740 nm, the conversion ratio of HyP-650 can be calculated, and the oxygen content in the tissue can be further inferred. This study overcomes some of the shortcomings of conventional methods of detecting tissue oxygen content and enables simultaneous imaging of tissue and blood oxygen content. This provides new ideas for hypoxia-related pathological analysis and disease detection. Using the HyP-650 probe, researchers can noninvasively detect hypoxia in tissues and obtain information about the extent and distribution of hypoxia. This is important for understanding the role of hypoxia in disease development, assessing disease severity, and guiding treatment.

D) *Enzyme-sensitive probes*: Diseases such as tumors and atherosclerosis usually exhibit high enzymatic activity [51, 275, 316]. Therefore, enzyme-sensitive photoacoustic probes can be used to accurately and sensitively detect the activity and distribution of specific enzymes in organisms by exploiting their molecular cleavage or structural rearrangement under the catalytic action of enzymes, which in turn changes the optical properties of the probes and affects the nature of the photoacoustic signals generated. Gao and coworkers have integrated a NIR fluorescent photoacoustic dye, a bone-targeting motif, and a histone K-responsive peptide sequence to prepare a histone K-activatable

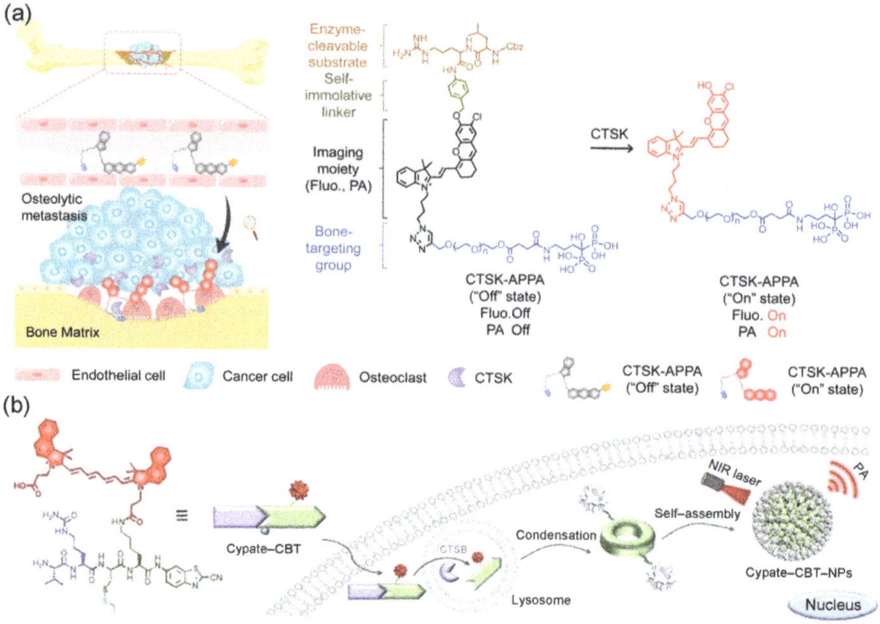

Figure 1.24 (a) Schematic diagram showing the mechanism of CTSK-APPA for imaging osteolytic metastasis. Source: Reproduced with permission from Ref. [317]; © 2023, The Authors. Advanced Science published by Wiley-VCH GmbH. (b) CTSB-triggered self-assembly of Cypate-CBT-NPs for PAI of CTSB activity in vitro. Source: Reproduced with permission from Ref. [318]; © 2021/John Wiley & Sons.

bone-targeting fluorescent photoacoustic probe (Figure 1.24a) [317]. The probe could be targeted to bone tissue and activated by histone kinase K in the microenvironment of osteolytic bone metastatic tumors, and high-contrast and specific dual-modality photoacoustic and fluorescence imaging could be achieved under laser irradiation. Liang and coworkers developed a smart photoacoustic probe, Cypate-CBT, which responds to abundant GSH and histone B in tumor cells to self-assemble into nanoparticles containing the NIR fluorescent dye cypate, which in turn can be used for tumor-specific PAI of histone B activity (Figure 1.24b) [318].

E) *Temperature-sensitive probes*: Malignant tumors, Alzheimer's disease, and atherosclerosis, among others, usually exhibit higher temperatures. Therefore, the design of temperature-sensitive probes can provide information on temperature distribution and changes, which can help to understand temperature regulation mechanisms in biological processes and play an important role in medical diagnosis and treatment. By introducing about 4% tungsten into vanadium dioxide nanoparticles, a thermochromic material, Xing and coworkers not only reduced the phase-transition temperature of vanadium dioxide but also enhanced its absorption capability in the NIR region, which improved the performance of PAI [319]. Furthermore, the above materials were encapsulated with PEG to prepare body temperature-triggered phase-transition photoacoustic nanoprobes (W-VO_2@PEG), which have good convertible optical absorption properties in the NIR region close to body temperature. This study provides a novel temperature-sensitive PAI nanoprobe with low phase-transition temperature and enhanced NIR absorption for high-resolution, high-contrast in vivo imaging.

F) *Metal ion-sensitive probes*: Metal elements play an important role in life processes, and they are involved in many key cellular processes such as enzyme catalysis, signal transduction, and regulation of gene expression. However, imbalances in metal ion homeostasis can severely disrupt these vital processes and lead to a variety of diseases [320]. Metal ion-sensitive probes allow the imaging of lesions through the binding of metal ions. These probes typically consist of metal ion-sensitive organic dyes and support that undergo structural changes when the metal ions are bound, thereby altering their optical properties and photoacoustic performance. He and coworkers developed a NIR absorbing photoacoustic probe (named HD-Zn) based on a merocyanine dye [321]. When the photoacoustic probe is not bound to Zn^{2+}, the lone pair of electrons on the amino group can freely flow into the dye, producing a PA_{750} signal. In contrast, when Zn^{2+} is present at the lesion, HD-Zn can sense Zn^{2+} and produce a distinct blue-shifted PA_{680} signal, together with a reduced absorbance at 750 nm. In in vivo experiments, the PA_{680}/PA_{750} ratio was increased approximately 3.5-fold in mice injected with HD-Zn, demonstrating the successful implementation of Zn^{2+}-activated PAI. Zhang and coworkers reported a series of small-molecule probes, all with molecular weights below 440, that highly selectively bind Cu^{2+} to form free radicals and exhibit strong NIR photoacoustic signals [322]. Among them, the screened RPS1 probe not only has high radical stability and photoacoustic intensity but also can easily cross the blood-brain barrier to achieve PAI of Cu^{2+} in the brain. This study provides an effective tool for the study of Alzheimer's disease at the molecular level, drug screening, and evaluation of therapeutic effects.

1.5 Conclusion

Photofunctional nanomaterials have unique optical properties and can be used in a variety of scenarios in the biomedical field, such as drug delivery, bioimaging, optical information storage, cancer diagnosis, and therapy, as well as biosensing. In this book, we will introduce the basic background of photofunctional nanomaterials and their recent progress in various biomedical fields. Finally, the challenges and future opportunities for photofunctional nanomaterials are discussed.

References

1 Rivera-Gil, P., Jimenez De Aberasturi, D., Wulf, V. et al. (2012). The challenge to relate the physicochemical properties of colloidal nanoparticles to their cytotoxicity. *Accounts of Chemical Research* 46: 743–749.

2 Moriarty, P. (2001). Nanostructured materials. *Reports on Progress in Physics* 64: 297–381.

3 Rosei, F. (2004). Nanostructured surfaces: challenges and frontiers in nanotechnology. *Journal of Physics: Condensed Matter* 16: S1373–S1436.

4 Martin, C.R. (1994). Nanomaterials: a membrane-based synthetic approach. *Science* 266: 1961–1966.

5 Wang, Z.L. (2000). Probing the properties of individual nanostructures by novel techniques. *Progress in Natural Science: Materials International* 10: 481–496.

6 Yang, D. and Sun, W. (2003). Structural characters and special properties of nanomaterials. *Materials Review* 17: 7–10.

7 Bai, H.Y., Luo, J.L., Jin, D., and Sun, J.R. (1996). Particle size and interfacial effect on the specific heat of nanocrystalline Fe. *Journal of Applied Physics* 79: 361–364.

8 Chen, S., Tang, M., Zhang, Z. et al. (2013). Interfacial effect on the ferromagnetic damping of CoFeB thin films with different under-layers. *Applied Physics Letters* 103: 032402.

9 Nagashima, K., Yanagida, T., Klamchuen, A. et al. (2010). Interfacial effect on metal/oxide nanowire junctions. *Applied Physics Letters* 96: 073110.

10 Ni, M., Leung, M., Leung, D., and Sumathy, K. (2006). Theoretical modeling of TiO_2/TCO interfacial effect on dye-sensitized solar cell performance. *Solar Energy Materials and Solar Cells* 90: 2000–2009.

11 Buffat, P. and Borel, J.P. (1976). Size effect on the melting temperature of gold particles. *Physical Review A* 13: 2287–2298.

12 Huang, W.C. and Lue, J.T. (1994). Quantum size effect on the optical properties of small metallic particles. *Physical Review B* 49: 17279–17285.

13 Solliard, C. and Flueli, M. (1985). Surface stress and size effect on the lattice parameter in small particles of gold and platinum. *Surface Science* 156: 487–494.

14 Tounsi, A., Heireche, H., Berrabah, H.M. et al. (2008). Effect of small size on wave propagation in double-walled carbon nanotubes under temperature field. *Journal of Applied Physics* 104: 104301.

15 Ciraci, S. and Batra, I.P. (1986). Theory of the quantum size effect in simple metals. *Physical Review B* 33: 4294–4297.

16 Mikhailov, G.M., Malikov, I.V., and Chernykh, A.V. (1997). Influence of the quantum size effect for grazing electrons on the electronic conductivity of metal films. *JETP Letters* 66: 725–731.

17 Ohno, T., Numakura, K., Itoh, H. et al. (2009). Control of the quantum size effect of TiO_2–SiO_2 hybrid particles. *Materials Letters* 63: 1737–1739.

18 Xiong, Y., Yu, K.N., and Xiong, C. (1994). Photoacoustic investigation of the quantum size effect and thermal properties in ZrO_2 nanoclusters. *Physical Review B* 49: 5607–5610.

19 Hendry, P.C., Lawson, N.S., McClintock, P.V.E. et al. (1988). Macroscopic quantum tunneling of vortices in He II. *Physical Review Letters* 60: 604–607.

20 Ivanov, B.A. and Kireev, V.E. (2004). Macroscopic quantum tunneling in small antiferromagnetic particles: effects of a strong magnetic field. *Physical Review B* 70: 214430.

21 Schwartz, D.B., Sen, B., Archie, C.N., and Lukens, J.E. (1985). Quantitative study of the effect of the environment on macroscopic quantum tunneling. *Physical Review Letters* 55: 1547–1550.

22 Tejada, J., Ziolo, R.F., and Zhang, X.X. (1996). Quantum tunneling of magnetization in nanostructured materials. *Chemistry of Materials* 8: 1784–1792.

23 West, J.L. and Halas, N.J. (2003). Engineered nanomaterials for biophotonics applications: improving sensing, imaging, and therapeutics. *Annual Review of Biomedical Engineering* 5: 285–292.

24 Wang, J., Zhang, L., and Li, Z. (2021). Aggregation-induced emission luminogens with photoresponsive behaviors for biomedical applications. *Advanced Healthcare Materials* 10: 2101169.

25 Ren, Y., Liu, H., Liu, X. et al. (2020). Photoresponsive materials for antibacterial applications. *Cell Reports Physical Science* 1: 100245.

26 Li, J., Li, G., Lu, X. et al. (2023). Magnetically responsive optical modulation: from anisotropic nanostructures to emerging applications. *Advanced Functional Materials* 34: 2308293.

27 Wang, C., Wang, H., Xu, B., and Liu, H. (2020). Photo-responsive nanozymes: mechanism, activity regulation, and biomedical applications. *View* 2: 20200045.

28 Li, Z. and Yin, Y. (2019). Stimuli-responsive optical nanomaterials. *Advanced Materials* 31: 1807061.

29 Wei, Y., Wang, Z., and Ranbo, Y. (2019). New media for efficient light energy conversion: hollow multi-shelled structures (HoMSs) (in chinese). *Chinese Science Bulletin-Chinese* 64: 3577–3593.

30 Jung, Y., Kim, M., Kim, T. et al. (2023). Functional materials and innovative strategies for wearable thermal management applications. *Nano-Micro Letters* 15: 160.

31 He, X.-P., Hu, X.-L., James, T.D. et al. (2017). Multiplexed photoluminescent sensors: towards improved disease diagnostics. *Chemical Society Reviews* 46: 6687–6696.

32 Zhang, Z.-Y. and Liu, Y. (2019). Ultralong room-temperature phosphorescence of a solid-state supramolecule between phenylmethylpyridinium and cucurbit[6]uril. *Chemical Science* 10: 7773–7778.

33 Zheng, K., Loh, K.Y., Wang, Y. et al. (2019). Recent advances in upconversion nanocrystals: expanding the kaleidoscopic toolbox for emerging applications. *Nano Today* 29: 100797.

34 Yu, D., Yu, T., Lin, H. et al. (2022). Recent advances in luminescent downconversion: new materials, techniques, and applications in solar cells. *Advanced Optical Materials* 10: 2200014.

35 Li, F., Shen, T., Wang, C. et al. (2020). Recent advances in strain-induced piezoelectric and piezoresistive effect-engineered 2d semiconductors for adaptive electronics and optoelectronics. *Nano-Micro Letters* 12: 106.

36 Arroyo, P.C., David, G., Alpert, P.A. et al. (2022). Amplification of light within aerosol particles accelerates in-particle photochemistry. *Science* 376: 293–296.

37 Lee, H.P. and Gaharwar, A.K. (2020). Light-responsive inorganic biomaterials for biomedical applications. *Advanced Science* 7: 2000863.

38 Wang, Y., Xia, K., Wang, L. et al. (2021). Peptide-engineered fluorescent nanomaterials: structure design, function tailoring, and biomedical applications. *Small* 17: 2005578.

39 Chen, G., Roy, I., Yang, C., and Prasad, P.N. (2016). Nanochemistry and nanomedicine for nanoparticle-based diagnostics and therapy. *Chemical Reviews* 116: 2826–2885.

40 Kong, X., Qi, Y., Wang, X. et al. (2023). Nanoparticle drug delivery systems and their applications as targeted therapies for triple negative breast cancer. *Progress in Materials Science* 134: 101070.

41 Wang, J., Ni, Q., Wang, Y. et al. (2021). Nanoscale drug delivery systems for controllable drug behaviors by multi-stage barrier penetration. *Journal of Controlled Release* 331: 282–295.

42 Jiang, Q. and Zhang, S. (2023). Stimulus-responsive drug delivery nanoplatforms for osteoarthritis therapy. *Small* 19: 2206929.

43 Mei, J. and Tian, H. (2021). Most recent advances on enzyme-activatable optical probes for bioimaging. *Aggregate* 2: e32.

44 Fan, W., Yung, B., Huang, P., and Chen, X. (2017). Nanotechnology for multimodal synergistic cancer therapy. *Chemical Reviews* 117: 13566–13638.

45 He, M., Chen, F., Shao, D. et al. (2021). Photoresponsive metallopolymer nanoparticles for cancer theranostics. *Biomaterials* 275: 120915.

46 Huang, X., Li, L., Chen, Z. et al. (2023). Nanomedicine for the detection and treatment of ocular bacterial infections. *Advanced Materials* 35: 2302431.

47 Lin, X., Zhao, M., Peng, T. et al. (2023). Detection and discrimination of pathogenic bacteria with nanomaterials-based optical biosensors: a review. *Food Chemistry* 426: 136578.

48 Zhang, X., Liu, F., and Gu, Z. (2023). Tissue engineering in neuroscience: applications and perspectives. *BME Frontiers* 4: 0007.

49 Bao, L., Cui, X., Mortimer, M. et al. (2023). The renaissance of one-dimensional carbon nanotubes in tissue engineering. *Nano Today* 49: 101784.

50 Pardo, A., Gómez-Florit, M., Barbosa, S. et al. (2021). Magnetic nanocomposite hydrogels for tissue engineering: design concepts and remote actuation strategies to control cell fate. *ACS Nano* 15: 175–209.

51 Sun, Q., Wang, Z., Liu, B. et al. (2022). Recent advances on endogenous/exogenous stimuli-triggered nanoplatforms for enhanced chemodynamic therapy. *Coordination Chemistry Reviews* 451: 214267.

52 Zhao, D., Huang, R., Gan, J.-M., and Shen, Q.-D. (2022). Photoactive nanomaterials for wireless neural biomimetics, stimulation, and regeneration. *ACS Nano* 16: 19892–19912.

53 Deng, K., Li, C., Huang, S. et al. (2017). Recent progress in near infrared light triggered photodynamic therapy. *Small* 13: 1702299.

54 Liu, H., Zhao, J., Xue, Y. et al. (2023). X-ray-induced drug release for cancer therapy. *Angewandte Chemie International Edition* 62: e202306100.

55 Zhang, X., Wang, S., Cheng, G. et al. (2022). Light-responsive nanomaterials for cancer therapy. *Engineering* 13: 18–30.

56 Zheng, X., Wu, Y., Zuo, H. et al. (2023). Metal nanoparticles as novel agents for lung cancer diagnosis and therapy. *Small* 19: 2206624.

57 Zheng, B., Fan, J., Chen, B. et al. (2022). Rare-earth doping in nanostructured inorganic materials. *Chemical Reviews* 122: 5519–5603.

58 Zhao, W., Zhao, Y., Wang, Q. et al. (2019). Remote light-responsive nanocarriers for controlled drug delivery: advances and perspectives. *Small* 15: 1903060.

59 Wu, G., Qiu, H., Liu, X. et al. (2023). Nanomaterials-based fluorescent assays for pathogenic bacteria in food-related matrices. *Trends in Food Science and Technology* 142: 104214.

60 Zhao, D., Wang, C., Su, L., and Zhang, X. (2020). Application of fluorescence nanomaterials in pathogenic bacteria detection. *Progress in Chemistry* 33: 1482–1495.

61 Pu, Y., Wang, D., Qian, J., and Chen, J. (2017). Fluorescent nanomaterials and their applications in bioimaging. *Materials China* 36: 103–111.

62 Yang, M., Guo, X., Mou, F., and Guan, J. (2022). Lighting up micro-/nanorobots with fluorescence. *Chemical Reviews* 123: 3944–3975.

63 Huang, J., He, B., Zhang, Z. et al. (2020). Aggregation-induced emission luminogens married to 2d black phosphorus nanosheets for highly efficient multimodal theranostics. *Advanced Materials* 32: 2003382.

64 Han, H., Jin, Q., Wang, H. et al. (2016). Intracellular dual fluorescent lightup bioprobes for image-guided photodynamic cancer therapy. *Small* 12: 3870–3878.

65 Li, C., Ma, M., Zhang, B. et al. (2022). A self-assembled nanoplatform based on Ag_2S quantum dots and tellurium nanorods for combined chemo-photothermal therapy guided by H_2O_2-activated near-infrared-ii fluorescence imaging. *Acta Biomaterialia* 140: 547–560.

66 Ding, C., Huang, Y., Shen, Z., and Chen, X. (2021). Synthesis and bioapplications of Ag_2S quantum dots with near-infrared fluorescence. *Advanced Materials* 33: 2007768.

67 Zhao, J., Chen, G., Gu, Y. et al. (2016). Ultrasmall magnetically engineered Ag_2Se quantum dots for instant efficient labeling and whole-body high-resolution multimodal real-time tracking of cell-derived microvesicles. *Journal of the American Chemical Society* 138: 1893–1903.

68 Yu, M., Yang, X., Zhang, Y. et al. (2021). Pb-doped Ag_2Se quantum dots with enhanced photoluminescence in the NIR-II window. *Small* 17: 2006111.

69 Chen, B., Liu, M., Gao, Y. et al. (2022). Design and applications of carbon dots-based ratiometric fluorescent probes: a review. *Nano Research* 16: 1064–1083.

70 Gao, X., Cui, Y., Levenson, R.M. et al. (2004). In vivo cancer targeting and imaging with semiconductor quantum dots. *Nature Biotechnology* 22: 969–976.

71 Yang, H., Bai, L., Geng, Z. et al. (2023). Carbon quantum dots: preparation, optical properties, and biomedical applications. *Materials Today Advances* 18: 100376.

72 Yang, L., Deng, W., Cheng, C. et al. (2018). Fluorescent immunoassay for the detection of pathogenic bacteria at the single-cell level using carbon dots-encapsulated breakable organosilica nanocapsule as labels. *ACS Applied Materials & Interfaces* 10: 3441–3448.

73 Yang, H., Li, R., Zhang, Y. et al. (2021). Colloidal alloyed quantum dots with enhanced photoluminescence quantum yield in the NIR-II window. *Journal of the American Chemical Society* 143: 2601–2607.

74 Chandra, S., Ghosh, B., Beaune, G. et al. (2016). Functional double-shelled silicon nanocrystals for two-photon fluorescence cell imaging: spectral evolution and tuning. *Nanoscale* 8: 9009–9019.

75 Chen, H., Xu, J., Wang, Y. et al. (2022). Color-switchable nanosilicon fluorescent probes. *ACS Nano* 16: 15450–15459.

76 Gai, S., Li, C., Yang, P., and Lin, J. (2014). Recent progress in rare earth micro/nanocrystals: soft chemical synthesis, luminescent properties, and biomedical applications. *Chemical Reviews* 114: 2343–2389.

77 Zhou, J., Liu, Q., Feng, W. et al. (2015). Upconversion luminescent materials: advances and applications. *Chemical Reviews* 115: 395–465.

78 Charron, D.M. and Zheng, G. (2018). Nanomedicine development guided by fret imaging. *Nano Today* 18: 124–136.

79 Dong, H., Du, S.R., Zheng, X.Y. et al. (2015). Lanthanide nanoparticles: from design toward bioimaging and therapy. *Chemical Reviews* 115: 10725–10815.

80 Xiang, H. and Chen, Y. (2020). Materdicine: interdiscipline of materials and medicine. *View* 1: 20200016.

81 Xu, J., Fu, M., Ji, C. et al. (2023). Plasmonic-enhanced NIR-II downconversion fluorescence beyond 1500 nm from core–shell–shell lanthanide nanoparticles. *Advanced Optical Materials* 11: 2300477.

82 Sun, G., Xie, Y., Wang, Y. et al. (2023). Cooperative sensitization upconversion in solution dispersions of co-crystal assemblies of mononuclear Yb^{3+} and Eu^{3+} complexes. *Angewandte Chemie International Edition* 62: e202304591.

83 Xu, R., Liu, J., Cao, H. et al. (2023). In vivo high-contrast biomedical imaging in the second near-infrared window using ultrabright rare-earth nanoparticles. *Nano Letters* 23: 11203–11210.

84 Wang, X., Han, X.-J., and Chen, G.-Y. (2020). Time-resolved imaging using lanthanide-doped nanomaterials. *Chinese Journal of Luminescence* 41: 1045–1057.

85 Jin, B., Wang, S., Lin, M. et al. (2017). Upconversion nanoparticles based fret aptasensor for rapid and ultrasenstive bacteria detection. *Biosensors and Bioelectronics* 90: 525–533.

86 Li, H., Tan, M., Wang, X. et al. (2020). Temporal multiplexed in vivo upconversion imaging. *Journal of the American Chemical Society* 142: 2023–2030.

87 Yang, Y., Chen, Y., Pei, P. et al. (2023). Fluorescence-amplified nanocrystals in the second near-infrared window for in vivo real-time dynamic multiplexed imaging. *Nature Nanotechnology* 18: 1195–1204.

88 Li, Y., Luo, Y., Wu, Z. et al. (2023). Research progress of T_1-T_2 dual-modal nano-magnetic resonance contrast agents. *Chinese Medicinal Biotechnology* 18: 516–522.

89 Li, Y., Younis, M.H., Wang, H. et al. (2022). Spectral computed tomography with inorganic nanomaterials: state-of-the-art. *Advanced Drug Delivery Reviews* 189: 114524.
90 Jin, Y., Ni, D., Gao, L. et al. (2018). Harness the power of upconversion nanoparticles for spectral computed tomography diagnosis of osteosarcoma. *Advanced Functional Materials* 28: 1802656.
91 Ai, X., Wang, Z., Cheong, H. et al. (2019). Multispectral optoacoustic imaging of dynamic redox correlation and pathophysiological progression utilizing upconversion nanoprobes. *Nature Communications* 10: 1087.
92 Wang, Z. and Xing, B. (2020). Near-infrared multipurpose lanthanide-imaging nanoprobes. *Chemistry--An Asian Journal* 15: 2076–2091.
93 Ashoka, A.H., Aparin, I.O., Reisch, A., and Klymchenko, A.S. (2023). Brightness of fluorescent organic nanomaterials. *Chemical Society Reviews* 52: 4525–4548.
94 Zhang, X., Chen, Y., He, H. et al. (2021). ROS/RNS and base dual activatable merocyanine-based NIR-II fluorescent molecular probe for in vivo biosensing. *Angewandte Chemie International Edition* 60: 26337–26341.
95 Li, D., Gamage, R.S., Oliver, A.G. et al. (2023). Doubly strapped zwitterionic NIR-I and NIR-II heptamethine cyanine dyes for bioconjugation and fluorescence imaging. *Angewandte Chemie International Edition* 62: 202305062.
96 Cheng, H., Li, Y., Tang, B., and Yoon, J. (2020). Assembly strategies of organic-based imaging agents for fluorescence and photoacoustic bioimaging applications. *Chemical Society Reviews* 49: 21–31.
97 Shi, Y., Zhu, D., Wang, D. et al. (2022). Recent advances of smart aiegens for photoacoustic imaging and phototherapy. *Coordination Chemistry Reviews* 471: 214725.
98 Hong, Y., Lam, J.W.Y., and Tang, B. (2011). Aggregation-induced emission. *Chemical Society Reviews* 40: 5361–5388.
99 Luo, J.D., Xie, Z.L., Lam, J.W.Y. et al. (2001). Aggregation-induced emission of 1-methyl-1,2,3,4,5-pentaphenylsilole. *Chemical Communications* 1740–1741.
100 Lim, X. (2016). The nanolight revolution is coming. *Nature* 531: 26.
101 Cai, X. and Liu, B. (2020). Aggregation-induced emission: recent advances in materials and biomedical applications. *Angewandte Chemie International Edition* 59: 9868–9886.
102 Mao, D. and Liu, B. (2021). Biology-oriented design strategies of AIE theranostic probes. *Matter* 4: 350–376.
103 Wang, D., Su, H., Kwok, R.T.K. et al. (2018). Rational design of a water-soluble NIR aiegen, and its application in ultrafast wash-free cellular imaging and photodynamic cancer cell ablation. *Chemical Science* 9: 3685–3693.
104 Zhang, J., He, B., Hu, Y. et al. (2021). Stimuli-responsive aiegens. *Advanced Materials* 33: 2008071.
105 Xu, Y., Dang, D., Zhang, N. et al. (2022). Aggregation-induced emission (AIE) in super-resolution imaging: cationic AIE luminogens (AIEgens) for tunable organelle-specific imaging and dynamic tracking in nanometer scale. *ACS Nano* 16: 5932–5942.

106 Sheng, Z., Guo, B., Hu, D. et al. (2018). Bright aggregation-induced-emission dots for targeted synergetic NIR-II fluorescence and NIR-I photoacoustic imaging of orthotopic brain tumors. *Advanced Materials* 30: 1800766.

107 Xiao, F., Fang, X., Li, H. et al. (2022). Light-harvesting fluorescent spherical nucleic acids self-assembled from a DNA-grafted conjugated polymer for amplified detection of nucleic acids. *Angewandte Chemie International Edition* 61: 202115812.

108 Cui, X., Ruan, Q., Zhuo, X. et al. (2023). Photothermal nanomaterials: a powerful light-to-heat converter. *Chemical Reviews* 123: 6891–6952.

109 Chen, Y., Gao, Y., Chen, Y. et al. (2020). Nanomaterials-based photothermal therapy and its potentials in antibacterial treatment. *Journal of Controlled Release* 328: 251–262.

110 Han, H.S. and Choi, K.Y. (2021). Advances in nanomaterial-mediated photothermal cancer therapies: toward clinical applications. *Biomedicine* 9: 305.

111 Chang, M., Wang, M., Hou, Z., and Lin, J. (2022). Problems and solutions of nanomaterials in antitumor photothermal therapy. *Chinese Journal of Luminescence* 43: 995–1013.

112 Xiong, Y., Rao, Y., Hu, J. et al. (2023). Nanoparticle-based photothermal therapy for breast cancer noninvasive treatment. *Advanced Materials* 33: 2305140.

113 Cheng, L., Wang, X., Gong, F. et al. (2019). 2d nanomaterials for cancer theranostic applications. *Advanced Materials* 32: 1902333.

114 Chen, J., Ning, C., Zhou, Z. et al. (2019). Nanomaterials as photothermal therapeutic agents. *Progress in Materials Science* 99: 1–26.

115 Wang, Z., Wang, M., Wang, X. et al. (2023). Photothermal-based nanomaterials and photothermal-sensing: an overview. *Biosensors and Bioelectronics* 220: 114883.

116 Zheng, L., Luo, Y., and Xu, W. (2018). Applications of optical and photothermal nanomaterials in biosensing, drug-targeted delivery and bioimaging. *Biotechnology Bulletin* 34: 79–89.

117 Huang, X.H., El-Sayed, I.H., Qian, W., and El-Sayed, M.A. (2006). Cancer cell imaging and photothermal therapy in the near-infrared region by using gold nanorods. *Journal of the American Chemical Society* 128: 2115–2120.

118 Yang, H., Zhang, Z., Liu, C., and Keming, J. (2022). Research progress of photothermal micro-nano materials. *Laser Journal* 43: 1–7.

119 Wang, S., Riedinger, A., Li, H. et al. (2015). Plasmonic copper sulfide nanocrystals exhibiting near-infrared photothermal and photodynamic therapeutic effects. *ACS Nano* 9: 1788–1800.

120 Ma, J., Li, N., Wang, J. et al. (2023). In vivo synergistic tumor therapies based on copper sulfide photothermal therapeutic nanoplatforms. *Exploration* 3: 20220161.

121 Huang, Q., Zhang, S., Zhang, H. et al. (2019). Boosting the radiosensitizing and photothermal performance of $Cu_{2-x}Se$ nanocrystals for synergetic radiophotothermal therapy of orthotopic breast cancer. *ACS Nano* 13: 1342–1353.

122 Chan, L., Liu, Y., Chen, M. et al. (2023). Cuproptosis-driven enhancement of thermotherapy by sequentially response $Cu_{2-x}Se$ via copper chemical transition. *Advanced Functional Materials* 33: 2302054.
123 Chen, Y., Wu, Y., Sun, B. et al. (2017). Two-dimensional nanomaterials for cancer nanotheranostics. *Small* 13: 1603446.
124 Wang, X. and Cheng, L. (2020). Multifunctional prussian blue-based nanomaterials: preparation, modification, and theranostic applications. *Coordination Chemistry Reviews* 419: 213393.
125 Xue, P., Hou, M., Sun, L. et al. (2018). Calcium-carbonate packaging magnetic polydopamine nanoparticles loaded with indocyanine green for near-infrared induced photothermal/photodynamic therapy. *Acta Biomaterialia* 81: 242–255.
126 Cheng, Y.J., Yang, S.H., and Hsu, C.S. (2009). Synthesis of conjugated polymers for organic solar cell applications. *Chemical Reviews* 109: 5868–5923.
127 Yang, J., Choi, J., Bang, D. et al. (2010). Convertible organic nanoparticles for near-infrared photothermal ablation of cancer cells. *Angewandte Chemie International Edition* 50: 441–444.
128 Wang, Z., Zou, Y., Li, Y., and Cheng, Y. (2020). Metal-containing polydopamine nanomaterials: catalysis, energy, and theranostics. *Small* 16: e1907042.
129 Zheng, Y., Cao, T., Han, X. et al. (2022). Structurally diverse polydopamine-based nanomedicines for cancer therapy. *Acta Materia Medica* 1: 427–444.
130 Ma, Z.-Y., Li, D.-Y., Jia, X. et al. (2023). Recent advances in bio-inspired versatile polydopamine platforms for "smart" cancer photothermal therapy. *Chinese Journal of Polymer Science* 41: 699–712.
131 Yang, S., Yu, Y., Gao, X. et al. (2021). Recent advances in electrocatalysis with phthalocyanines. *Chemical Society Reviews* 50: 12985–13011.
132 Nash, G.T., Luo, T., Lan, G. et al. (2021). Nanoscale metal–organic layer isolates phthalocyanines for efficient mitochondria-targeted photodynamic therapy. *Journal of the American Chemical Society* 143: 2194–2199.
133 Gao, D., Wong, R.C.H., Wang, Y. et al. (2020). Shifting the absorption to the near-infrared region and inducing a strong photothermal effect by encapsulating Zinc(II) phthalocyanine in poly(lactic-co-glycolic acid)-hyaluronic acid nanoparticles. *Acta Biomaterialia* 116: 329–343.
134 Gao, D., Chen, T., Chen, S. et al. (2021). Targeting hypoxic tumors with hybrid nanobullets for oxygen-independent synergistic photothermal and thermodynamic therapy. *Nano-Micro Letters* 13: 99.
135 Zhu, M., Zhang, H., Ran, G. et al. (2021). Metal modulation: an easy-to-implement tactic for tuning lanthanide phototheranostics. *Journal of the American Chemical Society* 143: 7541–7552.
136 Tian, Y., Zhao, D., Huang, X. et al. (2022). Extended π-conjugative carbon nitride for single 1064 nm laser-activated photodynamic/photothermal synergistic therapy and photoacoustic imaging. *ACS Applied Materials & Interfaces* 14: 7626–7635.

137 Yin, P.T., Shah, S., Chhowalla, M., and Lee, K.B. (2015). Design, synthesis, and characterization of graphene-nanoparticle hybrid materials for bioapplications. *Chemical Reviews* 115: 2483–2531.

138 Fusco, L., Gazzi, A., Peng, G. et al. (2020). Graphene and other 2D materials: a multidisciplinary analysis to uncover the hidden potential as cancer theranostics. *Theranostics* 10: 5435–5488.

139 Wang, B., Cai, H., Waterhouse, G.I.N. et al. (2022). Carbon dots in bioimaging, biosensing and therapeutics: a comprehensive review. *Small Science* 2: 2200012.

140 Xu, P. and Liang, F. (2020). Nanomaterial-based tumor photothermal immunotherapy. *International Journal of Nanomedicine* 15: 9159–9180.

141 Zhang, H., Zhang, Z., Wang, Y. et al. (2016). Rapid and sensitive detection of cancer cells based on the photothermal effect of graphene functionalized magnetic microbeads. *ACS Applied Materials & Interfaces* 8: 29933–29938.

142 Qu, G., Xia, T., Zhou, W. et al. (2020). Property–activity relationship of black phosphorus at the nano–bio interface: from molecules to organisms. *Chemical Reviews* 120: 2288–2346.

143 Liu, S., Pan, X., and Liu, H. (2020). Two-dimensional nanomaterials for photothermal therapy. *Angewandte Chemie International Edition* 59: 5890–5900.

144 Lei, W., Liu, G., Zhang, J., and Liu, M. (2017). Black phosphorus nanostructures: recent advances in hybridization, doping and functionalization. *Chemical Society Reviews* 46: 3492–3509.

145 Yang, B., Chen, Y., and Shi, J. (2018). Material chemistry of two-dimensional inorganic nanosheets in cancer theranostics. *Chem* 4: 1284–1313.

146 Wang, H., Zhong, L., Liu, Y. et al. (2018). A black phosphorus nanosheet-based sirna delivery system for synergistic photothermal and gene therapy. *Chemical Communications* 54: 3142–3145.

147 Sun, Z., Xie, H., Tang, S. et al. (2015). Ultrasmall black phosphorus quantum dots: synthesis and use as photothermal agents. *Angewandte Chemie International Edition* 54: 11526–11530.

148 Shao, J., Xie, H., Huang, H. et al. (2016). Biodegradable black phosphorus-based nanospheres for in vivo photothermal cancer therapy. *Nature Communications* 7: 12967.

149 Yang, X., He, S., Wang, J. et al. (2023). Hyaluronic acid-based injectable nanocomposite hydrogels with photo-thermal antibacterial properties for infected chronic diabetic wound healing. *International Journal of Biological Macromolecules* 242: 124872.

150 Wu, Z., Huang, T., Sathishkumar, G. et al. (2023). Phytic acid-promoted exfoliation of black phosphorus nanosheets for the fabrication of photothermal antibacterial coatings. *Advanced Healthcare Materials* 2302058.

151 Zhang, Y.Z., El-Demellawi, J.K., Jiang, Q. et al. (2020). Mxene hydrogels: fundamentals and applications. *Chemical Society Reviews* 49: 7229–7251.

152 Lukatskaya, M.R., Kota, S., Lin, Z. et al. (2017). Ultra-high-rate pseudocapacitive energy storage in two-dimensional transition metal carbides. *Nature Energy* 2: 17105.

153 Naguib, M., Mashtalir, O., Carle, J. et al. (2012). Two-dimensional transition metal carbides. *ACS Nano* 6: 1322–1331.

154 Hao, S., Han, H., Yang, Z. et al. (2022). Recent advancements on photothermal conversion and antibacterial applications over mxenes-based materials. *Nano-Micro Letters* 14: 178.

155 Wu, F., Zheng, H., Wang, W. et al. (2020). Rapid eradication of antibiotic-resistant bacteria and biofilms by mxene and near-infrared light through photothermal ablation. *Science China Materials* 64: 748–758.

156 He, P.P., Du, X., Cheng, Y. et al. (2022). Thermal-responsive mxene-DNA hydrogel for near-infrared light triggered localized photothermal-chemo synergistic cancer therapy. *Small* 18: 2200263.

157 Zou, Q., Abbas, M., Zhao, L. et al. (2017). Biological photothermal nanodots based on self-assembly of peptide–porphyrin conjugates for antitumor therapy. *Journal of the American Chemical Society* 139: 1921–1927.

158 Guo, K., Jiao, Z., Zhao, X. et al. (2023). Melanin-based immunoregulatory nanohybrids enhance antitumor immune responses in breast cancer mouse model. *ACS Nano* 17: 10792–10805.

159 Zhang, Y., Wang, Q., Ji, Y. et al. (2022). Mitochondrial targeted melanin@mSiO_2 yolk-shell nanostructures for NIR-II-driven photo-thermal-dynamic/immunotherapy. *Chemical Engineering Journal* 435: 134869.

160 Cao, W., Zhou, X., McCallum, N.C. et al. (2021). Unraveling the structure and function of melanin through synthesis. *Journal of the American Chemical Society* 143: 2622–2637.

161 Lei, Q., He, D., Ding, L. et al. (2022). Microneedle patches integrated with biomineralized melanin nanoparticles for simultaneous skin tumor photothermal therapy and wound healing. *Advanced Functional Materials* 32: 2113269.

162 Kwak, B.S., Kim, H.O., Kim, J.H. et al. (2012). Quantitative analysis of sialic acid on erythrocyte membranes using a photothermal biosensor. *Biosensors and Bioelectronics* 35: 484–488.

163 Kwak, B.S., Kim, H.J., Kim, H.O., and Jung, H.-I. (2010). An integrated photo-thermal sensing system for rapid and direct diagnosis of anemia. *Biosensors and Bioelectronics* 26: 1679–1683.

164 Lucky, S.S., Soo, K.C., and Zhang, Y. (2015). Nanoparticles in photodynamic therapy. *Chemical Reviews* 115: 1990–2042.

165 Chen, J., Fan, T., Xie, Z. et al. (2020). Advances in nanomaterials for photodynamic therapy applications: status and challenges. *Biomaterials* 237: 119827.

166 Xie, Z., Fan, T., An, J. et al. (2020). Emerging combination strategies with phototherapy in cancer nanomedicine. *Chemical Society Reviews* 49: 8065–8087.

167 Ke, L., Wei, F., Xie, L. et al. (2022). A biodegradable iridium(III) coordination polymer for enhanced two-photon photodynamic therapy using an apoptosis–ferroptosis hybrid pathway. *Angewandte Chemie International Edition* 61: 202205429.

168 Xiao, Q., Lin, H., Wu, J. et al. (2020). Pyridine-embedded phenothiazinium dyes as lysosome-targeted photosensitizers for highly efficient photodynamic antitumor therapy. *Journal of Medicinal Chemistry* 63: 4896–4907.

169 Lin, W., Liu, Y., Wang, J. et al. (2023). Engineered bacteria labeled with iridium(III) photosensitizers for enhanced photodynamic immunotherapy of solid tumors. *Angewandte Chemie International Edition* 135: 202310158.

170 Cui, M., Tang, D., Wang, B. et al. (2023). Bioorthogonal guided activation of cGAS-STING by AIE photosensitizer nanoparticles for targeted tumor therapy and imaging. *Advanced Materials* 35: 2305668.

171 Yang, Z., Wang, Z., and Wei, P. (2020). Research advances on stimuli-responsive liposomes in the targeted drug delivery system. *Chinese Journal of Biochemistry and Molecular Biology* 36: 1395–1403.

172 Liu, P., Chen, G., and Zhang, J. (2022). A review of liposomes as a drug delivery system: current status of approved products, regulatory environments, and future perspectives. *Molecules* 27: 1372.

173 Beltrán-Gracia, E., López-Camacho, A., Higuera-Ciapara, I. et al. (2019). Nanomedicine review: clinical developments in liposomal applications. *Cancer Nanotechnology* 10: 11.

174 Almeida, B., Nag, O.K., Rogers, K.E., and Delehanty, J.B. (2020). Recent progress in bioconjugation strategies for liposome-mediated drug delivery. *Molecules* 25: 5672.

175 Hao, L., Li, J., Wang, P. et al. (2021). Spatiotemporal magnetocaloric microenvironment for guiding the fate of biodegradable polymer implants. *Advanced Functional Materials* 31: 2009661.

176 Fan, J., Qin, Y., Xiao, C. et al. (2022). Biomimetic PLGA-based nanocomplexes for improved tumor penetration to enhance chemo-photodynamic therapy against metastasis of TNBC. *Materials Today Advances* 16: 100289.

177 Xin, J., Deng, C., Zheng, M., and An, F. (2023). Amphiphilic photosensitizer polymer as a nanocarrier of cytotoxic molecule for carrier-free combination therapy. *MedComm - Biomaterials and Applications* 2: 28.

178 Li, L., Shao, C., Liu, T. et al. (2020). An NIR-II-emissive photosensitizer for hypoxia-tolerant photodynamic theranostics. *Advanced Materials* 32: 2003471.

179 Chen, W., Zhou, S., Ge, L. et al. (2018). Translatable high drug loading drug delivery systems based on biocompatible polymer nanocarriers. *Biomacromolecules* 19: 1732–1745.

180 Qureshi, M.A. and Khatoon, F. (2019). Different types of smart nanogel for targeted delivery. *Journal of Science: Advanced Materials and Devices* 4: 201–212.

181 Hajebi, S., Rabiee, N., Bagherzadeh, M. et al. (2019). Stimulus-responsive polymeric nanogels as smart drug delivery systems. *Acta Biomaterialia* 92: 1–18.

182 Neamtu, I., Rusu, A.G., Diaconu, A. et al. (2017). Basic concepts and recent advances in nanogels as carriers for medical applications. *Drug Delivery* 24: 539–557.

183 Preman, N.K., Barki, R.R., Vijayan, A. et al. (2020). Recent developments in stimuli-responsive polymer nanogels for drug delivery and diagnostics: a review. *European Journal of Pharmaceutics and Biopharmaceutics* 157: 121–153.

184 Ma, H., Yu, G., Cheng, J. et al. (2023). Design of an injectable magnetic hydrogel based on the tumor microenvironment for multimodal synergistic cancer therapy. *Biomacromolecules* 24: 868–885.

185 Zheng, Y., Lv, X., Xu, Y. et al. (2019). pH-sensitive and pluronic-modified pullulan nanogels for greatly improved antitumor in vivo. *International Journal of Biological Macromolecules* 139: 277–289.

186 Zavgorodnya, O., Carmona-Moran, C.A., Kozlovskaya, V. et al. (2017). Temperature-responsive nanogel multilayers of poly(n-vinylcaprolactam) for topical drug delivery. *Journal of Colloid and Interface Science* 506: 589–602.

187 He, J., Sun, Y., Gao, Q. et al. (2023). Gelatin methacryloyl hydrogel, from standardization, performance, to biomedical application. *Advanced Healthcare Materials* 12: 2300395.

188 Tao, L., Huo, M., and Xu, W. (2020). Advances in the research of protein nanocarrier materials. *Journal of China Pharmaceutical University* 51: 121–129.

189 Hu, Q., Li, H., Wang, L. et al. (2018). DNA nanotechnology-enabled drug delivery systems. *Chemical Reviews* 119: 6459–6506.

190 Hejazi, R. and Amiji, M. (2003). Chitosan-based gastrointestinal delivery systems. *Journal of Controlled Release* 89: 151–165.

191 An, F.F. and Zhang, X.H. (2017). Strategies for preparing albumin-based nanoparticles for multifunctional bioimaging and drug delivery. *Theranostics* 7: 3667–3689.

192 Tao, H.-Y., Wang, R. -q., Sheng, W.-J., and Zhen, Y.-S. (2021). The development of human serum albumin-based drugs and relevant fusion proteins for cancer therapy. *International Journal of Biological Macromolecules* 187: 24–34.

193 Bhushan, B., Khanadeev, V., Khlebtsov, B. et al. (2017). Impact of albumin based approaches in nanomedicine: imaging, targeting and drug delivery. *Advances in Colloid and Interface Science* 246: 13–39.

194 Bern, M., Sand, K.M.K., Nilsen, J. et al. (2015). The role of albumin receptors in regulation of albumin homeostasis: implications for drug delivery. *Journal of Controlled Release* 211: 144–162.

195 Wu, J., Williams, G.R., Niu, S. et al. (2019). A multifunctional biodegradable nanocomposite for cancer theranostics. *Advanced Science* 6: 1802001.

196 Yang, B., Yao, H., Tian, H. et al. (2021). Intratumoral synthesis of nanometalchelate for tumor catalytic therapy by ligand field-enhanced coordination. *Nature Communications* 12: 3393.

197 Yang, B. and Shi, J. (2020). Ascorbate tumor chemotherapy by an iron-engineered nanomedicine-catalyzed tumor-specific pro-oxidation. *Journal of the American Chemical Society* 142: 21775–21785.

198 Yang, B., Guo, Y., Wang, Y. et al. (2022). Nanomedicine-leveraged intratumoral coordination and redox reactions of dopamine for tumor-specific chemotherapy. *CCS Chemistry* 4: 1499–1509.

199 Li, X., Zhang, L., Dong, X. et al. (2007). Preparation of mesoporous calcium doped silica spheres with narrow size dispersion and their drug loading and degradation behavior. *Microporous and Mesoporous Materials* 102: 151–158.

200 Zhao, S., Zhang, S., Ma, J. et al. (2015). Double loaded self-decomposable SiO_2 nanoparticles for sustained drug release. *Nanoscale* 7: 16389–16398.

201 Zhang, S., Chu, Z., Yin, C. et al. (2013). Controllable drug release and simultaneously carrier decomposition of SiO_2-drug composite nanoparticles. *Journal of the American Chemical Society* 135: 5709–5716.

202 Du, X., Kleitz, F., Li, X. et al. (2018). Disulfide-bridged organosilica frameworks: designed, synthesis, redox-triggered biodegradation, and nanobiomedical applications. *Advanced Functional Materials* 28: 1707325.

203 Yang, Y., Chen, F., Xu, N. et al. (2022). Red-light-triggered self-destructive mesoporous silica nanoparticles for cascade-amplifying chemo-photodynamic therapy favoring antitumor immune responses. *Biomaterials* 281: 121368.

204 Vallet-Regí, M., Schüth, F., Lozano, D. et al. (2022). Engineering mesoporous silica nanoparticles for drug delivery: where are we after two decades? *Chemical Society Reviews* 51: 5365–5451.

205 Kankala, R.K., Han, Y.H., Na, J. et al. (2020). Nanoarchitectured structure and surface biofunctionality of mesoporous silica nanoparticles. *Advanced Materials* 32: 1907035.

206 Cheng, Y.J., Hu, J.J., Qin, S.Y. et al. (2020). Recent advances in functional mesoporous silica-based nanoplatforms for combinational photo-chemotherapy of cancer. *Biomaterials* 232: 119738.

207 Yang, Y., Zhang, M., Song, H., and Yu, C. (2020). Silica-based nanoparticles for biomedical applications: from nanocarriers to biomodulators. *Accounts of Chemical Research* 53: 1545–1556.

208 Lin, F.-C., Xie, Y., Deng, T., and Zink, J.I. (2021). Magnetism, ultrasound, and light-stimulated mesoporous silica nanocarriers for theranostics and beyond. *Journal of the American Chemical Society* 143: 6025–6036.

209 Younis, M.R., He, G., Qu, J. et al. (2021). Inorganic nanomaterials with intrinsic singlet oxygen generation for photodynamic therapy. *Advanced Science* 8: 202102587.

210 Jin, L., Shen, S., Huang, Y. et al. (2021). Corn-like Au/Ag nanorod-mediated NIR-II photothermal/photodynamic therapy potentiates immune checkpoint antibody efficacy by reprogramming the cold tumor microenvironment. *Biomaterials* 268: 120582.

211 Han, R., Zhao, M., Wang, Z. et al. (2019). Super-efficient in vivo two-photon photodynamic therapy with a gold nanocluster as a type I photosensitizer. *ACS Nano* 14: 9532–9544.

212 Fatima, H., Jin, Z.Y., Shao, Z., and Chen, X.J. (2022). Recent advances in ZnO-based photosensitizers: synthesis, modification, and applications in photodynamic cancer therapy. *Journal of Colloid and Interface Science* 621: 440–463.

213 Heredia, D.A., Durantini, A.M., Durantini, J.E., and Durantini, E.N. (2022). Fullerene C60 derivatives as antimicrobial photodynamic agents. *Journal of Photochemistry and Photobiology C: Photochemistry Reviews* 51: 100471.

214 Chatterjee, D.K., Fong, L., and Zhang, Y. (2008). Nanoparticles in photodynamic therapy: an emerging paradigm. *Advanced Drug Delivery Reviews* 60: 1627–1637.

215 Wang, Y., Zhang, H., Liu, Y. et al. (2022). Catalytic radiosensitization: insights from materials physicochemistry. *Materials Today* 57: 262–278.

216 Xie, J., Wang, Y., Choi, W. et al. (2021). Overcoming barriers in photodynamic therapy harnessing nano-formulation strategies. *Chemical Society Reviews* 50: 9152–9201.

217 Chen, X., Song, J., Chen, X., and Yang, H. (2019). X-ray-activated nanosystems for theranostic applications. *Chemical Society Reviews* 48: 3073–3101.

218 He, L., Yu, X., and Li, W. (2022). Recent progress and trends in X-ray-induced photodynamic therapy with low radiation doses. *ACS Nano* 16: 19691–19721.

219 Jiang, M., Deng, Z., Zeng, S., and Hao, J. (2021). Recent progress on lanthanide scintillators for soft X-ray-triggered bioimaging and deep-tissue theranostics. *View* 2: 20200122.

220 Hu, J., Tang, Y., Elmenoufy, A.H. et al. (2015). Nanocomposite-based photodynamic therapy strategies for deep tumor treatment. *Small* 11: 5860–5887.

221 Shi, T., Sun, W., Qin, R. et al. (2020). X-ray-induced persistent luminescence promotes ultrasensitive imaging and effective inhibition of orthotopic hepatic tumors. *Advanced Functional Materials* 30: 2001166.

222 Wang, X., Sun, W., Shi, H. et al. (2022). Organic phosphorescent nanoscintillator for low-dose X-ray-induced photodynamic therapy. *Nature Communications* 13: 5091.

223 Cui, X., Li, X., Peng, C. et al. (2023). Beyond external light: on-spot light generation or light delivery for highly penetrated photodynamic therapy. *ACS Nano* 17: 20776–20803.

224 Ran, C. and Pu, K. (2023). Molecularly generated light and its biomedical applications. *Angewandte Chemie International Edition* 202314468.

225 Zhang, Y., Hao, Y., Chen, S., and Xu, M. (2020). Photodynamic therapy of cancers with internal light sources: chemiluminescence, bioluminescence, and cerenkov radiation. *Frontiers in Chemistry* 8: 770.

226 Hsu, C.-Y., Chen, C.-W., Yu, H.-P. et al. (2013). Bioluminescence resonance energy transfer using luciferase-immobilized quantum dots for self-illuminated photodynamic therapy. *Biomaterials* 34: 1204–1212.

227 Kim, Y.R., Kim, S., Choi, J.W. et al. (2015). Bioluminescence-activated deep-tissue photodynamic therapy of cancer. *Theranostics* 5: 805–817.

228 Hu, X., Wang, Z., Su, Y. et al. (2021). Metal–organic layers with an enhanced two-photon absorption cross-section and up-converted emission. *Chemistry of Materials* 33: 1618–1624.

229 Su, Y., Dai, Y., Zeng, Y. et al. (2023). Interpretable machine learning of two-photon absorption. *Advanced Science* 10: 2204902.

230 Shen, Y., Shuhendler, A.J., Ye, D. et al. (2016). Two-photon excitation nanoparticles for photodynamic therapy. *Chemical Society Reviews* 45: 6725–6741.

231 Wang, D., Zhu, L., Chen, J.F., and Dai, L. (2016). Liquid marbles based on magnetic upconversion nanoparticles as magnetically and optically responsive miniature reactors for photocatalysis and photodynamic therapy. *Angewandte Chemie International Edition* 55: 10795–10799.

232 Fan, Y., Liu, L., and Zhang, F. (2019). Exploiting lanthanide-doped upconversion nanoparticles with core/shell structures. *Nano Today* 25: 68–84.

233 Liu, S., Yan, L., Huang, J. et al. (2022). Controlling upconversion in emerging multilayer core–shell nanostructures: from fundamentals to frontier applications. *Chemical Society Reviews* 51: 1729–1765.

234 Ryu, K.A., Kaszuba, C.M., Bissonnette, N.B. et al. (2021). Interrogating biological systems using visible-light-powered catalysis. *Nature Reviews Chemistry* 5: 322–337.

235 Zhang, C., Gu, Z., and Yan, L. (2019). Intelligent response-type medication system based on up-conversion fluorescent nanomaterials for the application of cancer therapy. *Scientia Sinica Chimica* 49: 1179–1191.

236 Fan, W., Huang, P., and Chen, X. (2016). Overcoming the achilles' heel of photodynamic therapy. *Chemical Society Reviews* 45: 6488–6519.

237 Wang, F., Banerjee, D., Liu, Y.S. et al. (2010). Upconversion nanoparticles in biological labeling, imaging, and therapy. *Analyst* 135: 1839–1854.

238 Zhang, X., Chen, W., Xie, X. et al. (2019). Boosting luminance energy transfer efficiency in upconversion nanoparticles with an energy-concentrating zone. *Angewandte Chemie International Edition* 58: 12117–12122.

239 Li, Z., Lu, S., Liu, W. et al. (2021). Synergistic lysozyme-photodynamic therapy against resistant bacteria based on an intelligent upconversion nanoplatform. *Angewandte Chemie International Edition* 60: 19201–19206.

240 Ma, X., Kang, J., Wu, Y. et al. (2022). Recent advances in metal/covalent organic framework-based materials for photoelectrochemical sensing applications. *TrAC Trends in Analytical Chemistry* 157: 116793.

241 Lv, J., Xie, J., Mohamed, A.G.A. et al. (2022). Photoelectrochemical energy storage materials: design principles and functional devices towards direct solar to electrochemical energy storage. *Chemical Society Reviews* 51: 1511–1528.

242 Wang, S., Li, S., Wang, W. et al. (2019). A non-enzymatic photoelectrochemical glucose sensor based on bivo4 electrode under visible light. *Sensors and Actuators B: Chemical* 291: 34–41.

243 Wang, Y., Rong, Y., Ma, T. et al. (2023). Photoelectrochemical sensors based on paper and their emerging applications in point-of-care testing. *Biosensors and Bioelectronics* 236: 115400.

244 Devadoss, A., Sudhagar, P., Terashima, C. et al. (2015). Photoelectrochemical biosensors: new insights into promising photoelectrodes and signal amplification strategies. *Journal of Photochemistry and Photobiology C: Photochemistry Reviews* 24: 43–63.

245 Zhang, J., Shang, M., Gao, Y. et al. (2020). High-performance VS2 QDs-based type II heterostructured photoanode for ultrasensitive aptasensing of lysozyme. *Sensors and Actuators B: Chemical* 304: 127411.

246 Shang, M., Gao, Y., Zhang, J. et al. (2020). Signal-on cathodic photoelectrochemical aptasensing of insulin: plasmonic Au activated amorphous MoS_x photocathode coupled with target-induced sensitization effect. *Biosensors and Bioelectronics* 165: 112359.

247 Zhao, W.-W., Xu, J.-J., and Chen, H.-Y. (2015). Photoelectrochemical bioanalysis: the state of the art. *Chemical Society Reviews* 44: 729–741.

248 Li, Y., Zhang, N., Zhao, W.-W. et al. (2017). Polymer dots for photoelectrochemical bioanalysis. *Analytical Chemistry* 89: 4945–4950.

249 Zhao, C.-Q. and Ding, S.-N. (2019). Perspective on signal amplification strategies and sensing protocols in photoelectrochemical immunoassay. *Coordination Chemistry Reviews* 391: 1–14.

250 Liu, T., Ouyang, X., Cheng, X. et al. (2023). Applications of photoelectrochemical sensors for the detection of emerging contaminants. *Energy Environmental Protection* 37: 12–19.

251 Zhang, Z., Zhang, Y., Yu, H. et al. (2020). Types of biosensor recognition elements and their research progress in clinical testing. *Chinese Journal of Clinical Laboratory Science* 38: 767–771.

252 Xu, M., Lin, L., Jin, G. et al. (2022). Two-in-one: portable piezoelectric and plasmonic exciton effect-based co-enhanced photoelectrochemical biosensor for point-of-care testing of low-abundance cancer markers. *Biosensors and Bioelectronics* 211: 114413.

253 Long, D., Tu, Y., Chai, Y., and Yuan, R. (2021). Photoelectrochemical assay based on SnO_2/BiOBr p–n heterojunction for ultrasensitive DNA detection. *Analytical Chemistry* 93: 12995–13000.

254 Li, Z., Lu, J., Wu, F. et al. (2023). Polarity conversion of the $Ag_2S/AgInS_2$ heterojunction by radical-induced positive feedback polydopamine adhesion for signal-switchable photoelectrochemical biosensing. *Analytical Chemistry* 95: 15008–15016.

255 Lee, T., Baac, H.W., Li, Q., and Guo, L.J. (2018). Efficient photoacoustic conversion in optical nanomaterials and composites. *Advanced Optical Materials* 6: 1800491.

256 Liu, Y., Teng, L., Yin, B. et al. (2022). Chemical design of activatable photoacoustic probes for precise biomedical applications. *Chemical Reviews* 122: 6850–6918.

257 Fu, Q., Zhu, R., Song, J. et al. (2018). Photoacoustic imaging: contrast agents and their biomedical applications. *Advanced Materials* 31: 1805875.

258 Wang, J., Xie, K., Ren, W. et al. (2022). Material composition and application progress of photoacoustic transducer. *Journal of Sensor Technology and Application* 10: 528–537.

259 Park, B., Park, S., Kim, J., and Kim, C. (2022). Listening to drug delivery and responses via photoacoustic imaging. *Advanced Drug Delivery Reviews* 184: 114235.

260 Hong, G., Antaris, A.L., and Dai, H. (2017). Near-infrared fluorophores for biomedical imaging. *Nature Biomedical Engineering* 1: 0010.

261 Yang, Q., Ma, Z., Wang, H. et al. (2017). Rational design of molecular fluorophores for biological imaging in the NIR-II window. *Advanced Materials* 29: 1605497.

262 Antaris, A.L., Chen, H., Cheng, K. et al. (2015). A small-molecule dye for NIR-II imaging. *Nature Materials* 15: 235–242.

263 Lyu, Y. and Pu, K. (2017). Recent advances of activatable molecular probes based on semiconducting polymer nanoparticles in sensing and imaging. *Advanced Science* 4: 1600481.

264 Upputuri, P.K. and Pramanik, M. (2020). Recent advances in photoacoustic contrast agents for in vivo imaging. *WIREs Nanomedicine and Nanobiotechnology* 12: 1618.

265 Wang, L.V. and Hu, S. (2012). Photoacoustic tomography: in vivo imaging from organelles to organs. *Science* 335: 1458–1462.

266 Liu, X., Duan, Y., and Liu, B. (2021). Nanoparticles as contrast agents for photoacoustic brain imaging. *Aggregate* 2: 4–19.

267 Lin, L. and Wang, L.V. (2022). The emerging role of photoacoustic imaging in clinical oncology. *Nature Reviews Clinical Oncology* 19: 365–384.

268 Park, B., Oh, D., Kim, J., and Kim, C. (2023). Functional photoacoustic imaging: from nano-and micro- to macro-scale. *Nano Convergence* 10: 29.

269 Zhang, J., Qiao, Z., Yang, P. et al. (2014). Recent advances in near-infrared absorption nanomaterials as photoacoustic contrast agents for biomedical imaging. *Chinese Journal of Chemistry* 33: 35–52.

270 Shan, T., Zhao, Y., Jiang, S., and Jiang, H. (2020). In-vivo hemodynamic imaging of acute prenatal ethanol exposure in fetal brain by photoacoustic tomography. *Journal of Biophotonics* 13: e201960161.

271 Moreno, M.J., Ling, B., and Stanimirovic, D.B. (2020). In vivo near-infrared fluorescent optical imaging for CNS drug discovery. *Expert Opinion on Drug Discovery* 15: 903–915.

272 Mukaddim, R.A., Rodgers, A., Hacker, T.A. et al. (2018). Real-time in vivo photoacoustic imaging in the assessment of myocardial dynamics in murine model of myocardial ischemia. *Ultrasound in Medicine & Biology* 44: 2155–2164.

273 He, H., Schönmann, C., Schwarz, M. et al. (2022). Fast raster-scan optoacoustic mesoscopy enables assessment of human melanoma microvasculature in vivo. *Nature Communications* 13: 2803.

274 Li, Y.-Y., Hu, Z.-B., Sun, H.-T., and Sun, Z.-R. (2020). Density functional theory studies on the excited-state properties of bilirubin molecule. *Acta Physica Sinica* 69: 163101.

275 Björkegren, J.L.M. and Lusis and A.J. (2022). Atherosclerosis: recent developments. *Cell* 185: 1630–1645.

276 Bai, S., Lan, Y., Fu, S. et al. (2022). Connecting calcium-based nanomaterials and cancer: from diagnosis to therapy. *Nano-Micro Letters* 14: 145.

277 Thiabaud, G.D., Schwalm, M., Sen, S. et al. (2023). Texaphyrin-based calcium sensor for multimodal imaging. *ACS Sensors* 8: 3855–3861.

278 Zhang, X., Wu, Y., Chen, L. et al. (2023). Optical and photoacoustic imaging in vivo: opportunities and challenges. *Chemical and Biomedical Imaging* 1: 99–109.

279 Kruger, R.A., Liu, P., Fang, Y.R., and Appledorn, C.R. (1998). Photoacoustic ultrasound (PAUS)-reconstruction tomography. *Medical Physics* 22: 1605–1609.

280 Wu, Y., Zeng, F., Zhao, Y., and Wu, S. (2021). Emerging contrast agents for multispectral optoacoustic imaging and their biomedical applications. *Chemical Society Reviews* 50: 7924–7940.

281 Ouyang, J., Xie, A., Zhou, J. et al. (2022). Minimally invasive nanomedicine: nanotechnology in photo-/ultrasound-/radiation-/magnetism-mediated therapy and imaging. *Chemical Society Reviews* 51: 4996–5041.

282 Liu, Y., Yu, W., Wang, J. et al. (2021). Application of bismuth-based nanomaterials in imaging diagnosis and therapy for cancer. *Chinese Journal of Inorganic Chemistry* 37: 1–15.

283 Yang, Y., Chu, B., Cheng, J. et al. (2022). Bacteria eat nanoprobes for aggregation-enhanced imaging and killing diverse microorganisms. *Nature Communications* 13: 1255.

284 Kim, M., VanderLaan, D., Lee, J. et al. (2023). Hyper-branched gold nanoconstructs for photoacoustic imaging in the near-infrared optical window. *Nano Letters* 23: 9257–9265.

285 Gong, F., Cheng, L., and Liu, Z. (2020). Application of nanoprobes in photoacoustic cancer imaging. *Laser and Optoelectronics Progress* 57: 180004.

286 Zhou, Z., Li, B., Shen, C. et al. (2020). Metallic 1t phase enabling MoS_2 nanodots as an efficient agent for photoacoustic imaging guided photothermal therapy in the near-infrared-II window. *Small* 16: 2004173.

287 Han, Y., Yi, H., Wang, Y. et al. (2022). Ultrathin zinc selenide nanoplatelets boosting photoacoustic imaging of in situ copper exchange in Alzheimer's disease mice. *ACS Nano* 16: 19053–19066.

288 Dai, C., Chen, Y., Jing, X. et al. (2017). Two-dimensional tantalum carbide (MXenes) composite nanosheets for multiple imaging-guided photothermal tumor ablation. *ACS Nano* 11: 12696–12712.

289 Li, J., Lu, X., Fu, Y. et al. (2021). Research progress of novel graphene-like two-dimensional nanomaterials applied in photoacoustic imaging of tumor. *Chinese Bulletin of Life Sciences* 33: 479–489.

290 Yang, Z., Xu, T., Li, H. et al. (2023). Zero-dimensional carbon nanomaterials for fluorescent sensing and imaging. *Chemical Reviews* 123: 11047–11136.

291 Gifani, M., Eddins, D.J., Kosuge, H. et al. (2021). Ultraselective carbon nanotubes for photoacoustic imaging of inflamed atherosclerotic plaques. *Advanced Functional Materials* 31: 2101005.

292 Sun, Z., Zhao, Y., Li, Z. et al. (2017). TiL_4-coordinated black phosphorus quantum dots as an efficient contrast agent for in vivo photoacoustic imaging of cancer. *Small* 13: 1602896.

293 Zeng, Y., Dou, T., Ma, L., and Ma, J. (2022). Biomedical photoacoustic imaging for molecular detection and disease diagnosis: "always-on" and "turn-on" probes. *Advanced Science* 9: 2202384.

294 Liu, Y., Teng, L., Liu, H.-W. et al. (2019). Recent advances in organic-dye-based photoacoustic probes for biosensing and bioimaging. *Science China Chemistry* 62: 1275–1285.

295 Jiang, Y. and Pu, K. (2017). Advanced photoacoustic imaging applications of near-infrared absorbing organic nanoparticles. *Small* 13: 1700710.

296 Li, J., Wang, J., Xu, L. et al. (2023). A class of activatable NIR-II photoacoustic dyes for high-contrast bioimaging. *Angewandte Chemie International Edition* 202312632.

297 Ahmadi, S., Rabiee, N., Bagherzadeh, M. et al. (2020). Stimulus-responsive sequential release systems for drug and gene delivery. *Nano Today* 34: 100914.

298 Jack, C.R. (2022). Advances in Alzheimer's disease research over the past two decades. *The Lancet Neurology* 21: 866–869.

299 Song, Y., Jing, H., Vong, L.B. et al. (2022). Recent advances in targeted stimuli-responsive nano-based drug delivery systems combating atherosclerosis. *Chinese Chemical Letters* 33: 1705–1717.

300 Overchuk, M. and Zheng, G. (2018). Overcoming obstacles in the tumor microenvironment: recent advancements in nanoparticle delivery for cancer theranostics. *Biomaterials* 156: 217–237.

301 Khattab, M., Mansour, H., Fawzy, H., and El-Khatib, A. (2022). Repurposed anti-cancer epidermal growth factor receptor inhibitors: mechanisms of neuroprotective effects in Alzheimer's disease. *Neural Regeneration Research* 17: 1913.

302 Zhao, L., Meng, F., Li, Y. et al. (2023). Multivalent nanobody conjugate with rigid, reactive oxygen species scavenging scaffold for multi-target therapy of Alzheimer's disease. *Advanced Materials* 35: 2210879.

303 Peng, S., Xiao, F., Chen, M., and Gao, H. (2021). Tumor-microenvironment-responsive nanomedicine for enhanced cancer immunotherapy. *Advanced Science* 9: 2103836.

304 Chen, Q., Liu, X., Chen, J. et al. (2015). A self-assembled albumin-based nanoprobe for in vivo ratiometric photoacoustic pH imaging. *Advanced Materials* 27: 6820–6827.

305 Wang, X., Wang, X., Jin, S. et al. (2019). Stimuli-responsive therapeutic metallodrugs. *Chemical Reviews* 119: 1138–1192.

306 Jiao, Y.T., Kang, Y.R., Wen, M.Y. et al. (2023). Fast antioxidation kinetics of glutathione intracellularly monitored by a dual-wire nanosensor. *Angewandte Chemie International Edition* 62: e202313612.

307 Niu, B., Liao, K., Zhou, Y. et al. (2021). Application of glutathione depletion in cancer therapy: enhanced ROS-based therapy, ferroptosis, and chemotherapy. *Biomaterials* 277: 121110.

308 Hecht, F., Zocchi, M., Alimohammadi, F., and Harris, I.S. (2023). Regulation of antioxidants in cancer. *Molecular Cell* 11: 001.

309 Matuz-Mares, D., Riveros-Rosas, H., Vilchis-Landeros, M.M., and Vázquez-Meza, H. (2021). Glutathione participation in the prevention of cardiovascular diseases. *Antioxidants* 10: 1220.

310 Verkhratsky, A., Butt, A., Li, B. et al. (2023). Astrocytes in human central nervous system diseases: a frontier for new therapies. *Signal Transduction and Targeted Therapy* 8: 396.

311 Park, S., Kim, Y.A., Lee, J. et al. (2023). 4-Hydroxycinnamic acid attenuates neuronal cell death by inducing expression of plasma membrane redox enzymes and improving mitochondrial functions. *Food Science and Human Wellness* 12: 1287–1299.

312 Wang, A., Mao, Q., Zhao, M. et al. (2020). pH/reduction dual stimuli-triggered self-assembly of NIR theranostic probes for enhanced dual-modal imaging and photothermal therapy of tumors. *Analytical Chemistry* 92: 16113–16121.

313 Somaa, F. (2021). A review of the application of hyperbaric oxygen therapy in Alzheimer's disease. *Journal of Alzheimer's Disease* 81: 1361–1367.

314 Liang, X., Li, H., Li, X. et al. (2023). Highly sensitive H_2O_2-scavenging nano-bionic system for precise treatment of atherosclerosis. *Acta Pharmaceutica Sinica B* 13: 372–389.

315 Chen, M., Knox, H.J., Tang, Y. et al. (2019). Simultaneous photoacoustic imaging of intravascular and tissue oxygenation. *Optics Letters* 44: 3773.

316 de Visser, K.E. and Joyce, J.A. (2023). The evolving tumor microenvironment: from cancer initiation to metastatic outgrowth. *Cancer Cell* 41: 374–403.

317 Song, Z., Miao, J., Miao, M. et al. (2023). Cathepsin K-activated probe for fluoro-photoacoustic imaging of early osteolytic metastasis. *Advanced Science* 10: 2300217.

318 Wang, C., Du, W., Wu, C. et al. (2021). Cathespin B-initiated cypate nanoparticle formation for tumor photoacoustic imaging. *Angewandte Chemie International Edition* 61: 202114766.

319 Li, L., Chen, H., Shi, Y., and Xing, D. (2022). Human-body-temperature triggerable phase transition of W-VO_2@PEG nanoprobes with strong and switchable NIR-II absorption for deep and contrast-enhanced photoacoustic imaging. *ACS Nano* 16: 2066–2076.

320 Swartchick, C.B. and Chan, J. (2023). Leveraging coordination chemistry to visualize metal ions via photoacoustic imaging. *Current Opinion in Chemical Biology* 74: 102312.

321 Fang, H., Wang, C., Chen, Y. et al. (2021). A photoacoustic Zn^{2+} sensor based on a merocyanine/xanthene-6-ol hybrid chromophore and its ratiometric imaging in mice. *Inorganic Chemistry Frontiers* 8: 3402–3410.

322 Wang, S., Sheng, Z., Yang, Z. et al. (2019). Activatable small-molecule photoacoustic probes that cross the blood–brain barrier for visualization of copper(II) in mice with Alzheimer's disease. *Angewandte Chemie International Edition* 58: 12415–12419.

2

Mechanism in Rare-Earth-Doped Luminescence Nanomaterials

Yulei Chang

Chinese Academy of Sciences, Changchun Institute of Optics, Fine Mechanics and Physics, State Key Laboratory of Luminescence Science and Technology, Dongnanhu road 3888, Changchun 130033, China

2.1 Introduction

Rare-earth (RE) elements refer to the 15 elements from lanthanum (La) to lutetium (Lu) in the periodic table, along with scandium (Sc) and yttrium (Y) of the same group (IIIB), for a total of 17 elements. The studies on RE elements can be traced back to the eighteenth century. With more than two centuries of advancements in RE chemistry, these elements have rich applications in the military, chemical, agricultural, and biomedical fields. RE-doped photoluminescence (PL) nanomaterials are fascinating materials that find applications in various fields, including optics, electronics, and biomedicine. The luminescent properties of RE ions are attributed to the particularity of their electronic configurations. Emission and excitation mainly originate from the electronic transition between 4f or between 5d → 4f energy levels (ELs). In other words, the study of RE luminescent materials is actually the study of materials related to the physical properties of 4f electrons, which go through three main stages of absorption, energy transfer (ET), and light emission, and with mainly three luminescence processes involved, including downshifting (DS), upconversion (UC), and downconversion/quantum cutting (QC).

2.2 Composition of RE-Doped Luminescence Nanomaterials: Substrate (Host), Activator, and Sensitizer

RE-doped luminescent materials mainly comprise host matrix materials and dopants. For hosts, RE-doped nanomaterials typically comprise an inorganic matrix hosting lanthanide (Ln^{3+}) ion, but some complexes also show photoluminescence, which is not to be emphasized in this book. Although the host generally does not emit luminescence, it can create an appropriate crystal field for the activator to generate tailored emissions.

Photofunctional Nanomaterials for Biomedical Applications, First Edition.
Edited by Chunxia Li and Jun Lin.

Take upconversion processes as an example (the detailed luminescence mechanism will be introduced in Section 2.3), which have been extensively explored within diverse nanoscale host matrices, encompassing halides such as fluorides, chlorides, bromides, and iodides, along with oxides, sulfur oxides, phosphates, vanadates, and more. Ideal matrix materials ought to possess minimal lattice phonon energies to mitigate nonradiative losses while enhancing radiative transitions. This is due to the necessity of phonon assistance within the host lattice to facilitate nonradiative energy dissipation. Heavy halide compounds, including chlorides, bromides, and iodides, manifest low phonon energies, often below 300 cm^{-1}. Nevertheless, these halides tend to be hygroscopic and have restricted usage scenarios. Despite their phonon energies commonly reaching above 500 cm^{-1}, which is relatively high, oxides demonstrate elevated chemical stability. Conversely, fluorides commonly demonstrate diminished phonon energies (approximately 350 cm^{-1}) and heightened photochemical stability, frequently serving as preferred host materials for photon upconversion processes. Generally, as the phonon energy decreases, the quantity of phonons required for the nonradiative process increases, resulting in a reduction in the nonradiative rate and an improvement in upconversion efficiency. Hence, it is advantageous to integrate lanthanide ions into dielectric matrices characterized by extremely low-frequency phonons, as it enhances emission efficiency by mitigating nonradiative decay rates.

In addition to the host material type, the crystal field (CF) significantly impacts the photoluminescence efficacy of lanthanide activators since the local crystal symmetry surrounding the lanthanide ions impacts their optical properties. Crystal phases with lower symmetry are typically advantageous for upconversion luminescence (UCL) efficiency because the interaction between the 4f configurations is enhanced, leading to a more pronounced mixing of energy levels. For example, hexagonal Yb^{3+}/Er^{3+} co-doped microcrystals show visible UCL that is 4.4-fold greater than those of cubic microcrystals [1], whereas monoclinic ZrO_2 nanoparticles exhibit enhanced upconversion emission compared to tetragonal phase nanoparticles [2]. However, under the influence of other factors, the situation will change. For example, with the participation of phonons, the downshifting luminescence of cubic phase one is stronger than the luminescence of its hexagonal phase counterpart under a comparable condition [3].

Furthermore, dopants comprise activators (luminescent centers), co-activators, and sensitizers. The doped ions Er^{3+}, Tm^{3+}, and Ho^{3+}, distinguished by their stepped energy level (EL) arrangements, are frequently selected as emitters for generating photoluminescence emission under continuous wavelength laser pumping. Typically, the doping level remains relatively low (<2 mol%) to mitigate energy loss due to cross-relaxation (CR). Compared to typical lanthanide emitters, the absorption cross-section of Yb^{3+} is much larger. Upon excitation by a 980-nm laser, Yb^{3+} ions are pumped into the $^2F_{5/2}$ state, which matches well with various 4f-4f transition levels in Er^{3+}, Tm^{3+}, and Ho^{3+}, enabling effective ET from Yb^{3+} to these emitter ions. Hence, Yb^{3+} is frequently co-doped alongside Er^{3+}, Tm^{3+}, or Ho^{3+} as a sensitizer to augment UCL.

For Yb/Er co-doped upconversion nanoparticles (UCNPs), the green UCL originates from transitions between $^2H_{11/2}$ and $^4S_{3/2} \rightarrow {}^4I_{15/2}$, yielding peaks at 525 and 540 nm, respectively. Additionally, red UCL peaked at 660 nm is attributed to the transition of $^4F_{9/2} \rightarrow {}^4I_{15/2}$. Furthermore, weak blue emission (445–505 nm) attributed to the $^4F_{5/2} \rightarrow {}^4I_{15/2}$, $^2P_{3/2} \rightarrow {}^4I_{11/2}$, and $^4F_{7/2} \rightarrow {}^4I_{15/2}$ transitions, respectively. For Yb^{3+}/Tm^{3+} co-doped UCNPs, the UCL emissions at 450, 480, 645, and 800 nm are attributed to the $^1D_2 \rightarrow {}^3F_4$, $^1G_4 \rightarrow {}^3H_6$, $^1G_4 \rightarrow {}^3F_4$, and $^3H_4 \rightarrow {}^3H_6$ transitions, respectively. Moreover, UV UCL centered at 345 and 365 nm is ascribed to the $^3P_6 \rightarrow {}^3F_4$ and $^1D_2 \rightarrow {}^3H_6$ transitions. And in Yb^{3+}/Ho^{3+} co-doped UCNPs, blue UCL peaked at 486 nm, resulting from the $^4F_3 \rightarrow {}^5I_8$ transition, along with green emission at 510–570 nm attributed to the $^5F_4/^5S_2 \rightarrow {}^5I_8$ transitions. Red UCL of 647 nm is also detected during the transition from the $^5F_5 \rightarrow {}^5I_8$. In addition, the NIR UCL emission at 751 nm is attributed to the $^5F_4$, $^5S_2 \rightarrow {}^5I_7$ transition.

2.3 Mechanism of RE-Doped Luminescence Nanomaterials

2.3.1 Luminescence: Downshifting, Upconversion, and Downconversion

2.3.1.1 Downshifting Luminescence

Ln^{3+} ions are considered attractive luminescent centers because they have rich electronic levels and transitions. Moreover, because of the shielding effect of the $5s^2 5p^6$ subshells, their characteristic spectra can be less influenced by their ionic surroundings, resulting in weak absorption and sharp-band emissions in the 4f-4f transitions [4]. And the unique 4f energy levels produce exceptional photostability, a large Stokes shift, and a long luminescence decay. The downshifting luminescence (DSL) of Ln^{3+} ions refers to the Stokes shift emission, mainly involving 4f-4f and 5d-4f, two transition processes.

4f-4f Transition Because Ln^{3+} ions have similar electron configurations ($[Xe]4f^{n-1} 5d^{0-1} 6s^2$), they have analogous physical and chemical characteristics [5, 6]. Besides, they have large quantum numbers ($n = 4$, $l = 3$) and abundant electronic states derived from the 4f electron configuration (Figure 2.1). Due to the rule of electron transition, the 4f-4f intraconfigurational transition is parity forbidden in freely existing Ln^{3+} ions. When the Ln^{3+} ions are doped in a crystal lattice, the parity-forbidden principle can be partially alleviated due to the intervention of an odd-parity crystal field, resulting in some allowable transitions. Thus, photoluminescence can occur in Ln^{3+} ions when transitioning from a higher-energy state to another. If the energy gap is large enough and luminescence wavelength ranges from UV (e.g. Pr^{3+} and Gd^{3+}) to visible (e.g. Eu^{3+}, Tb^{3+}) and to NIR (e.g. Er^{3+}, Tm^{3+}, and Yb^{3+}) [7–9]. Due to the parity-forbidden 4f-4f transitions for trivalent Ln^{3+} ions, their luminescence characteristic generally has narrow emission bands and long luminescent decay time curves up to tens of milliseconds. For instance, in $KGdF_4$:Eu^{3+} [10],

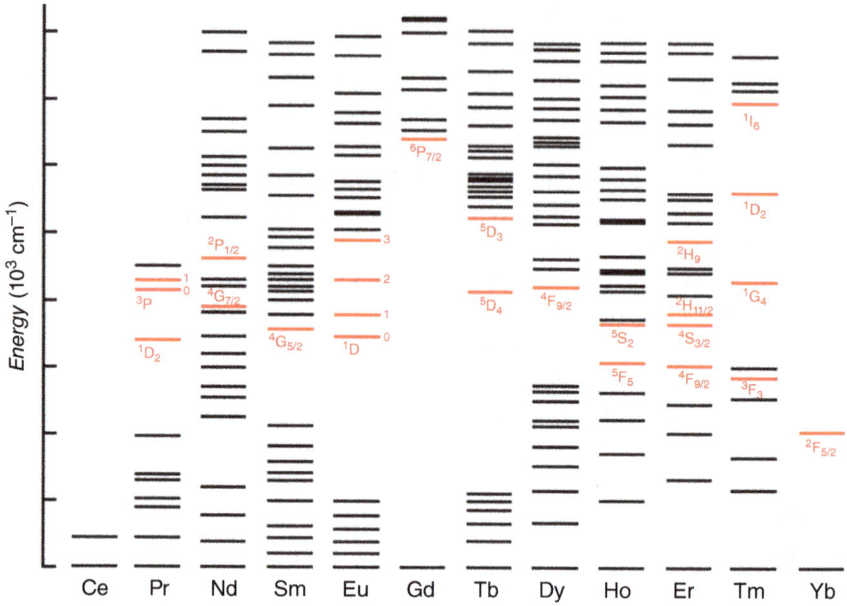

Figure 2.1 Energy level diagrams of trivalent Ln^{3+} ions in CaF_2. Typical emissive excited states are emphasized in red lines. Source: Reproduced with permission from Liu et al. [7]/The Royal Society of Chemistry.

$NaGdF_4:Eu^{3+}$ [11], and $Gd_2O_3:Eu^{3+}$ [12] nanomaterials, the corresponding lifetimes of Eu^{3+} were measured to be 10.34, 5.7, and 2.2 ms, respectively. Such prolonged luminescence of Ln^{3+} ions is highly desirable in background-free, time-resolved bioimaging applications.

When the Ln^{3+} ions are doped in a host material, the degenerate levels are split into several sub-levels by the CF effects, called Stark splitting. The 4f-4f transitions are mainly induced by the odd-parity electric dipole (ED) operator, the even-parity magnetic dipole (MD), and electric quadrupole (EQ) operators. According to Judd-Ofelt (JO) theory, the probabilities of the induced ED 4f-4f transitions can be calculated from the dipole strength, which is expressed by [13]:

$$P_{\text{ED}} = e^2 \cdot \sum_{\lambda=2,4,6} \Omega_\lambda |\langle \Psi \| U^\lambda \| \Psi' \rangle|^2$$

Here e is the electron charge, Ω_λ are three phenomenological parameters, and the squared bracketed expressions $|\langle \Psi \| U^\lambda \| \Psi' \rangle|^2$ are dimensionless doubly reduced matrix elements. Some induced ED transitions have been experimentally validated and show special sensitive to variations in the surrounding metal-ion environment. These transitions are known as hypersensitive transitions or pseudo-quadrupolar transitions. For MD operators, the transition intensity is expressed by:

$$P_{\text{MD}} = \chi_{\text{MD}} \cdot P'$$

where χ_{MD} is refractive index factor and P' is oscillator strength parameters and tabulated.

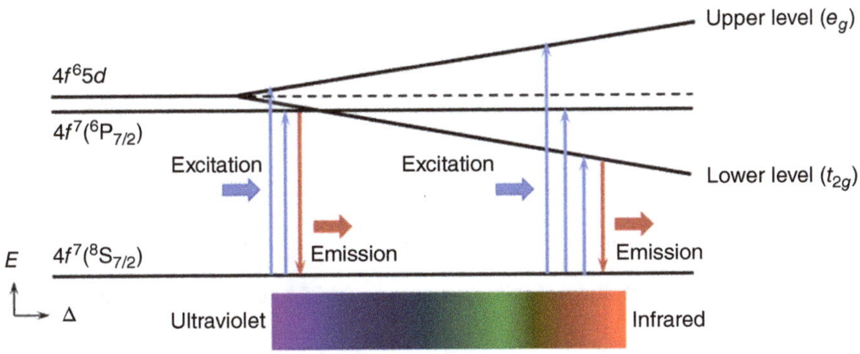

Figure 2.2 Scheme for the CF splitting of the electronic ELs of Eu^{2+} ions in phosphors. Source: Reproduced with permission from Adachi [14]/Electrochemical Society, Inc.

For Ln^{3+} ions, probabilities of ED transitions have the same order of magnitude as that of magnetic dipole (MD) transitions, and both can be observed in the luminescence spectra. However, EQ transitions are much weaker than ED, and corresponding peaks are usually unrealized. Thus, the experimental intensities of transition strength are determined by:

$$P_{EXP} = P_{\mathrm{ED}} + P_{\mathrm{MD}}$$

5d-4f Transition The transition of a 5d electron into the 4f configurations of lanthanide ions is parity allowed. Therefore, the corresponding luminescence intensity is greatly stronger than that of 4f-4f transitions. The spin-allowed 5d-4f transitions of trivalent Ln^{3+} ions have large energies ($E > 50\,000\ \mathrm{cm^{-1}}$, $\lambda > 200$ nm), excluding Ce^{3+}, Pr^{3+}, and Tb^{3+}, so they are rarely observed in coordination compounds. However, some 5d-4f transitions have been observed in divalent lanthanide ions, such as Eu^{2+}, Sm^{2+}, Yb^{2+}, and Tm^{2+}, due to their lower transition energies ($E < 40\,000\ \mathrm{cm^{-1}}$, $\lambda < 250$ nm). Besides, due to the outermost position of 5d electrons, their energy gaps greatly depend on the surroundings of metal ions, and the emission of 5d-4f transition tends to band spectrum, just like the emissions in transition metal ions.

In the 4f-5d transition of divalent lanthanide ions, the emission of Eu^{2+} is a general example and has attracted more attention [14]. The luminescence wavelengths span from the near-UV (~310 nm) to the near-infrared (~850 nm) region and strongly depend on the host material. The diagram of electronic energy levels for Eu^{2+} is shown in Figure 2.2. The wavelength dependence of emission of Eu^{2+}-activated phosphors originates from the CF splitting Δ of the excited state 5d level, which means the energy difference between the e_g and t_{2g} orbitals. With the value increase, the emission band redshifts. For instance, the emission peaks of Eu^{2+} in $Sr_{1-x}Ba_xAl_{12}O_{19}$: Eu^{2+} phosphor is redshifted with decreasing Ba^{2+} concentration [15].

2.3.1.2 Upconversion Luminescence (UCL)

Photon upconversion is a nonlinear anti-Stokes phenomenon where multiple low-energy photons combine to produce a single higher-energy photon. Ln^{3+} ions

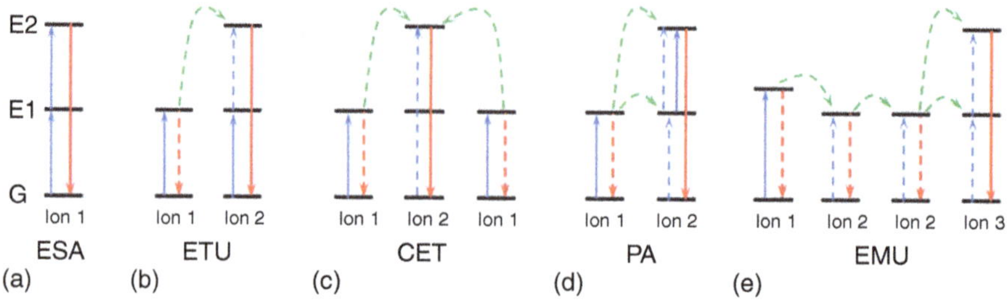

Figure 2.3 Schematic illustration of UCL mechanisms: (a) ESA, (b) ETU, (c) CET, (d) PA, (e) EMU.

have abundant f configurations and d configurations of the metastable energy states, giving the Ln-doped materials tremendous possibility for upconversion processes, which are the distinct categories outlined as excited state absorption (ESA), energy transfer upconversion (ETU), cooperative energy transfer upconversion (CET), photon avalanche (PA), and energy migration upconversion (EMU) as displayed in Figure 2.3 [16, 17].

In the ESA process, Ln^{3+} ion continuously absorbs two low-energy photons, and the electron transitions from the ground state to the intermediate state E1 and finally to the excited state E2. When the energy of the excited state is relaxed, UCL occurs, as shown in Figure 2.3a.

In the ETU process, in general, two different lanthanide ions are involved, including the sensitizer (ion 1) and the activator (ion 2) (Figure 2.3b). After absorbing photons, the sensitizers are excited from the ground state G to a higher-lying E1 excited state. Then, the sensitizer transfers energy to the activator nonradiatively, and the activator is excited to E2. Upconversion photons are emitted when the activator transitions from E2 to G [18, 19].

The CET mechanism engages two essential ion species: a cooperative sensitizer, labeled as ion 1, and an activator, designated as ion 2. Distinguishing it from ETU, CET is characterized by the absence of extended lifetime intermediate ELs within the activators. In CET, ET from two sensitizers to one activator simultaneously occurred, leading to upconversion from E2. However, its UC emission efficiency is extremely weak and much lower than that of ETU [20, 21].

In the PA process, three necessary conditions are required: weak G absorption, strong E2 state absorption, and efficient CR. The excited Ln^{3+} ion 2 first transits from G to metastable E1 via GSA and then transits to E2 via ETU between ions 1 and 2. When the CR occurs between the activator ion and a neighboring ion at G, these two ions transit to E1. It means that one excitation photon and one ion 1 at the metastable E1 state can encourage two ion 2s at E1 state. As this sequence recurs, two ion 2s at the E1 state generate four; four yield eight, resulting in a PA effect that amplifies the population of ion 2 in the E1 state. In the meantime, the PA upconversion from the E2 state occurs [22].

The EMU process exists in UCNPs that feature a core–shell (CS) structure, and three sorts of Ln^{3+} ions, called sensitizer, migrator, and activator, are doped into

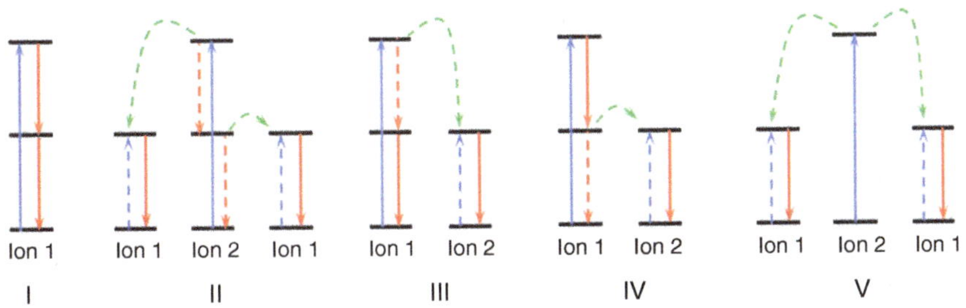

Figure 2.4 Schematic illustration of typical QC mechanisms.

different spatial regions. Under the excitation, excited energy (EE) transfers from the sensitizer to the migrator and finally to the activator, generating UCL [21].

2.3.1.3 Downconversion/Quantum Cutting (QC)

QC materials can absorb a single photon with high energy and transform it into two or more lower-energy photons, achieving an internal quantum efficiency exceeding 100%. This process is also called downconversion. Wegh et al. first found the quantum-cutting process in a $LiGdF_4:Eu^{3+}$ phosphor, which theoretically demonstrated a quantum efficiency of 190% in 1999 [23]. There are several mechanisms in the QC process, which are demonstrated in Figure 2.4. The first one is based on one activator with three ELs (Figure 2.4I). After absorbing one high-energy photon, the lanthanide ion produces two photons with the participation of the intermediate energy level. This process has been reported in some single ions, such as Ho^{3+}, Tm^{3+}, or Er^{3+} [24–28]. The QC mechanisms, as shown in Figure 2.4II–V, require two or three kinds of ions and ET between two close ions, and the population of an intermediate energy level of ion 1 achieves the energy splitting. In order to perform an efficient QC process, the combined energy of the emission from the two acceptors must match the energy of the donor absorption [16].

2.3.2 Nonradiative Transition: Energy Transfer and Migration

2.3.2.1 Energy Transfer (ET)

ET is a phenomenon involving energy transfer or exchange between two or more dopant ions. Regarding RE-doped luminescence nanomaterials, ET mainly focuses on the electron–electron ET between the excited Ln^{3+} (Figure 2.3). And the prerequisite for realizing the ET process is that there is a good energy-matching relationship between the EL pairs involved. Generally, there are two types of ET in RE-doped luminescence nanomaterials: radiative and nonradiative processes. The former is also called radiation reabsorption ET because it involves the reabsorption of radiation energy. There is little research on this type of ET in practical applications because of its low efficiency and the harsh conditions for rare-earth ions to achieve. The latter usually requires the assistance of a sensitizer in the system, and the sensitizer does not emit any radiation during this ET process, leading to a higher ET efficiency. Nonradiative ET is the leading way of ET among RE ions, and resonance energy transfer (RET), CR, and phonon-assisted energy transfer (PAET) are its main forms. The ELs of the sensitizer and the activator must be in the same position for RET to occur, resulting in the reversibility of the process. And this resonance transfer works only when the effective transition probability of the activator is less than the transfer probability of the sensitizer.

Moreover, the specifics of CR and PAET will be discussed in subsequent chapters. In a word, ET is the basic mechanism to realize the photoluminescence of RE-doped luminescence nanomaterials, and the intrinsic ET of Ln^{3+} significantly influences the corresponding luminescence efficiency and color outputs. As a matter of course, engineering the ET process is crucial to related nanomaterials to achieve the wanted PL for applications. Concerning optical properties, the modulation of ET can be categorized into three distinct groups: multicolor tuning, enhanced luminescence intensity, and new luminescence excitations triggered by ET.

2.3.2.2 Energy Migration (EM)

EM is structurally comparable to long-range ET, to put it briefly. Sensitizer, accumulator, migrator, and activator are the main luminescent centers that an EM process inserts into distinct layers at specific concentrations (Figure 2.3). The process involves harvesting pump photons with a sensitizer ion and then promoting an adjacent accumulator ion to excited states. Once the activator randomly bounces over a migrator ion, it obtains the excited energy from the high-lying ESs of the accumulator and traps the energy. In distinct layers of the CS nanostructure, the luminescent centers are spatially confined to control the energy exchange. Preventing detrimental CR is essential (see infra). Migrators within the CS interface are essential to facilitate the ET, thereby enabling an effective EMU process. A pivotal feature of the EMU design is its capacity for one-step ET to the activator while concurrently accumulating EE through successive ET, a critical component.

In tandem with the advantage of ETU's intrinsic high conversion efficiency for low-energy excitation photons, rigorous selection requirements for activators may be substantially waived. While reducing luminescence quenching, the CS construction enables facile modulation of activator composition and concentration. Therefore, tunable UCL will probably be attained for luminescent centers without long-lived intermediate ES [29]. Liu and coworkers employed a $NaGdF_4$@$NaGdF_4$ CS NP in which the activator Tb^{3+}, Eu^{3+}, Dy^{3+}, or Sm^{3+} as well as sensitizer/accumulator (Yb/Tm) was added to the shell and core layers of the nanoparticle, respectively [30]. The host material, β-$NaGdF_4$, was selected owing to its high UC efficiency and the advantage of utilizing the Gd sublattice for rapid energy movement. Interestingly, the lowest EL of Gd^{3+} is located in the UV spectral region, where most Ln^{3+} shows overlapping absorption bands. Another benefit is a relatively wide energy gap between the ground state and the $^6P_{7/2}$ state of Gd^{3+}, which minimizes energy loss from CR and multiphonon emission. The excitation photons were accumulated using the Yb/Tm sensitizer/accumulator array. It is well known that as a result of the effective ET between these two ions, five photons can combine to achieve the population of the target EL. In a stoichiometric Yb^{3+} lattice, they also demonstrated a five-photon UCL of Tm^{3+}, revealing that confining EM is a broad and adaptable technique for controlling multiphoton UCL while avoiding the detrimental effect of concentration quenching. High UCL intensity is clearly demonstrated by the realization of a room-temperature laser at roughly 310 nm. Two orders of magnitude more optical harvest were obtained than expected from a conventional Yb/Tm system [31].

Moreover, the QC may also be accomplished via the EM process [23, 32–34]. A high-energy vacuum ultraviolet photon is absorbed by a Gd^{3+} ion, which then delivers the energy to the two Eu^{3+} ions, releasing two red photons. The research team reported that the DC system in $LiGdF_4$ was composed of Er^{3+} and Tb^{3+}. Visible light has a quantum efficiency of 110% [35]. Then, Gd^{3+}-Eu^{3+} was also investigated by other research teams. With the Gd^{3+}-Tb^{3+} fluoride combination, effective DCL was obtained.

2.4 Luminescence Modulation

2.4.1 Crystal Field (CF) Regulation

The intricate properties of Ln^{3+} can be attributed to the distinctive characteristics of their [Xe]$4f^N$ configurations, where N ranges from 0 to 14. These configurations are governed by three quantum numbers: S, L, and J [36]. As mentioned above, the shielding effect of filled $5s^2 5p^6$ subshells on the 4f orbitals allows the ELs of Ln^{3+} to be highly defined, making the ions almost insensitive to the ionic surrounding environment. As a result, the inner-shell 4f-4f transitions of Ln^{3+}in NPs are distinct and clearly identifiable, similar to the spectra of corresponding free ions. Upon doping an Ln^{3+} ion into a dielectric crystal, there is a consequent division in ELs. The highly confined states of the f-electrons are typically modeled using the theoretical framework of CF theory [37]. The energy level fitting (ELF) methodology, which involves the parametrization of an effective operator Hamiltonian that includes both free ions and CF interactions, is frequently utilized to determine the electronic ELs of Ln^{3+} and their corresponding CF parameters within a specific host. The effective operator Hamiltonian that is widely utilized has been described by Wybourne et al., Carnall, and Crosswhite [38]. The f-shell empirical programs facilitate a comprehensive diagonalization without truncating the $4f^N$ wave functions. This tool can be utilized for ELF once a series of experimental CF levels of Ln^{3+} occupy a doping site within a particular host derived from high-resolution PL and/or absorption spectra [7]. The computational tools have been widely used and distributed in the Ln^{3+} spectroscopy community for many years.

Ln^{3+}, when incorporated into NPs with numerous sites, has a distinct PL and luminescence decay. This is attributed to the diverse CF environments they encounter. Optical characteristics can be employed to discern the local structures, or site symmetry (SS), of these Ln^{3+}. For example, the Eu^{3+} is recognized as a viable probe among all Ln^{3+}[4, 39, 40]. While its $^5D_0 - {}^7F_1$ transition exhibits an MD character and demonstrates insensitivity to SS, this transition is markedly hypersensitive to its local SS. Consequently, this suggests that SS information can be gleaned from the intensity ratio of Eu^{3+} emission [41, 42]. The transition for $^5D_0 - {}^7F_J$ (where $J = 0$–6) with

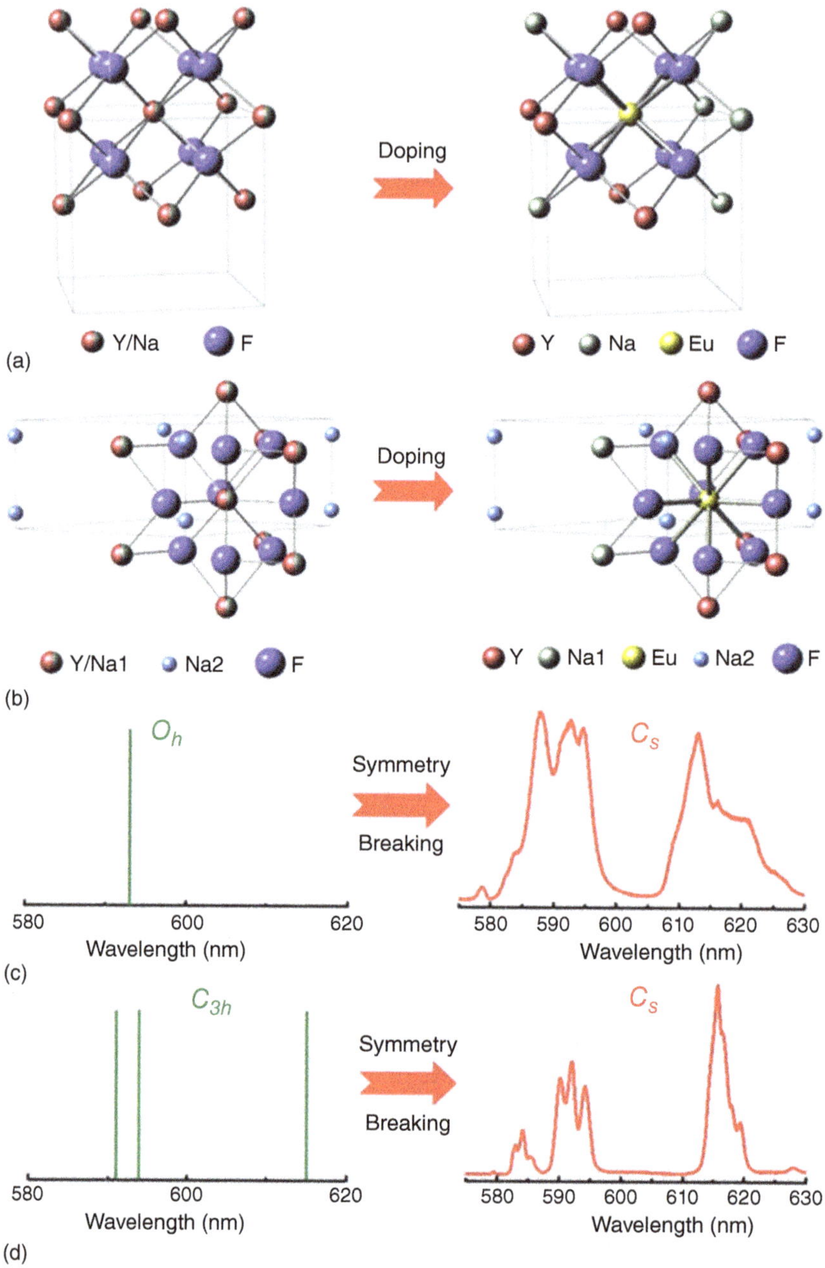

Figure 2.5 Schematic representation illustrates the disruption of crystallographic SS in Eu^{3+} within (a) α- and (b) β- $NaYF_4$ crystals. Introduction of Ln^{3+} triggers a transformation in the disordered Y/Na site, consequently diminishing the crystallographic SS of O_h in (a) to C_s and C_{3h} in (b) to C_s. This is substantiated by the significant alteration observed in the spectral features of Eu^{3+} when doped in (c) α- and (d) β- $NaYF_4$ crystals, compared to their theoretically predicted counterparts. Source: Reproduced with permission from Tu et al. [44]/John Wiley & Sons.

Eu^{3+} at different space groups has also been determined [43]. Therefore, multiple luminescent sites of Eu^{3+} inside various CF surrounds can be identified in NPs by referring to the CF splitting numbers of the above transitions.

In the study of Ln^{3+}-doping disordered crystals, Chen et al. demonstrated a pervasive disruption of crystallographic SS (Figure 2.5a,b). Eu^{3+} was employed as the structural probe to facilitate this investigation [1, 44, 45]. As depicted in Figure 2.5c,d, the most intense SSs were observed. The relative fitting further validated this observation. Comparable symmetry collapse was observed in α-$KLaF_4$ and α-$KGdF_4$ [10, 46].

Simultaneously, an extensive collection of CF parameters and ELs for Er^{3+} in both cubic- and hexagonal-$NaYF_4$ was determined in an independent investigation. The creation of highly luminous $NaYF_4$-based UC crystals has long been searching for these crucial properties. Similar research was done for Er^{3+} in semiconductor NPs made of anatase TiO_2 [47]. A stable group of CF parameters and Er^{3+} ELs, assuming a C_{2v} SS, was found by applying fitting. Larger CF strength often presents more intense CF interaction that Ln^{3+} are experiencing. Furthermore, a significant increase in CF strength is frequently the consequence of an impoverished crystallographic SS that Ln^{3+} fills typically in the host.

Tuning the local CF helps enlarge intra-configurational transitions due to a breakdown in the local symmetry. Using optically inert non-Ln^{3+} as compensation is an effective technique. It is standard practice to alter the local CF to incorporate various alkali and transition metal ions. Due to the reduced ionic radii, Li^+ is designed to disperse into the lattice. In Yb^{3+} and Er^{3+} co-doped UCNPs, Li^+ was added to Y_2O_3 by Zhang and coworkers [48]. Compared to Y_2O_3: Yb, Er UCNPs, the UCL of Y_2O_3: Li^+, Yb^{3+}, and Er^{3+} UCNPs was boosted by two orders of magnitude. Subsequently, Wang et al. observed a 30-times surge in UCL for $NaYF_4$: Yb, Er UCNPs following an 80% Li^+ doping [49]. Cai and coworkers studied the Li^+-induced luminescence enhancement of UCNPs [50]. Some transition metal ions, including Bi and Fe, have been used to alter the surrounding CF. This suggests that the modification of the local CF is not solely reliant on doping with Ln^{3+} [51, 52].

Hao and coworkers have elucidated a fascinating method for amplifying UCL using low voltages to BTO: Yb, Er films [53]. They posited that this amplification is attributable to the heightened radiative probability resulting from the diminished symmetry of Er^{3+} sites.

2.4.2 Surface Defects Passivation

Surface defects are significant contributors affecting the luminescence of RE-doped nanoparticles, and the CS architecture is a fundamental method in passivation surface defects. The CS structure involves encapsulating an RE-doped core with different material shells, such as a homogeneous or heterogeneous core/shell structure. Such design offers several advantages, including enhanced luminescence, adjusting excitation wavelength, stability, and biocompatibility.

A homogeneous CS structure is defined as a CS configuration wherein both the core and shell are composed of identical host material. This CS structure type is straightforward yet highly efficient, as it facilitates epitaxial growth and establishes a uniform interface between the core and shell materials. For example, Yan and coworkers found that overlaying a layer of $NaYF_4$ through epitaxial growth onto $NaYF_4$:Yb^{3+}/Er^{3+} nanoparticles effectively mitigates the impacts of surface defects and surface ligands and thus reduces the associated nonradiative decay [54]. Frank and coworkers demonstrated that the absolute quantum yield of $NaYF_4$:Yb^{3+}/Er^{3+}@$NaYF_4$ homogeneous CS nanoparticles is three-fold brighter than that of $NaYF_4$:Yb^{3+}/Er^{3+} core nanoparticles under identical conditions, which is about 0.3% under 980 nm excitation [55].

Through kinetic analysis of UCL, Kong and coworkers showed that the luminescent centers of $NaYF_4$:Yb^{3+}/Er^{3+} core nanoparticles alone produce severe nonradiative decay, whereas the $NaYF_4$ homogeneous coatings can effectively reduce the nonradiative decay and enhance the UCL [56]. The core/shell/shell-structured $NaGdF_4$:Er^{3+}@$NaGdF_4$:Ho^{3+}@$NaGdF_4$ nanoparticles designed by Wang and coworkers showed broad excitations in the NIR range. Moreover, UC emission was significantly enhanced after coating the outermost $NaGdF_4$ homogeneous shell layer [57]. It is evident that the homogeneous CS structure successfully enhanced the UCL of the RE-doped nanoparticles. Such structures can physically separate the core from the surrounding defects (ligands, solvents, etc.), thus effectively suppressing the surface-related quenching. The creation of CS structures can usually be explained by increased size distribution, improved luminescence efficiency, and prolonged UCL decay.

In addition to homogeneous CS structures, heterogeneous CS structures with different core and shell main matrix compositions have been investigated to improve the luminescence efficiency of core nanoparticles. Heterostructured CS nanoparticles, as multicomponent nanomaterials, enable surface defect passivation while providing opportunities for single-component utilization and demonstrating bifunctional or multipurpose characteristics. This expansion enhances their applications in biological imaging, detection, and phototherapy.

The benefit of employing CS nanoparticles lies in integrating multiple imaging modalities into a single entity. For instance, Zhang et al. studied a set of $NaYF_4$:Yb^{3+}, Er^{3+}@$NaGdF_4$ nanoparticles with thin shell thickness in the 0–2.5 nm range and confirmed that the $NaGdF_4$ heterogeneous CS structure could effectively passivate the surface defects to improve the UCL performance of the nanoparticles [58]. Liu and coworkers also demonstrated that the $NaGdF_4$ heterogeneous shell coating could effectively inhibit the surface loss of EE [59]. CaF_2 coating is more efficient at mitigating quenching in aqueous environments to maintain UC emission and suppressing lanthanide leakage than the commonly used heterogeneous shells of $NaGdF_4$. Yan and coworkers found that RE-doped $NaYF_4$ nanoparticles coated with a CaF_2 heterogeneous shell layer increased the UC of the nanoparticles by about 300 times [60]. Chen et al. coated heterogeneous shells of CaF_2 on the surface of $NaYbF_4$:Tm^{3+} nanoparticle, which resulted in a 35-fold enhancement in the NIR to NIR UCL intensity and a quantum yield of 0.6% (0.3 $W\,cm^{-2}$) [61]. Silicon dioxide coatings are also very common for heterogeneous shells. Coating a SiO_2 layer on the surface of $NaYF_4$:Yb^{3+}/Tm^{3+} or Yb^{3+}/Er^{3+} nanoparticles can efficiently counteract surface defect-induced quenching, thereby improving its UC emission [62]. From the chemical point of view, besides enhancing UCL by passivating surface defects and integrating multimodal states, another unique advantage of heterogeneous CS structures lies in their capability to characterize the morphology of the CS comprehensively. This task remains challenging for homogeneous CS structures.

2.4.3 ET Regulation

2.4.3.1 Multicolor Tuning (MCT) of UCL

The main effective emitters in an ETU system are Er^{3+}, Tm^{3+}, and Ho^{3+}. These few chosen parameters constrain the restricted emission color outputs. Take Yb^{3+}/Er^{3+} co-doped UCNPs as an example; they can emit red and green emissions. Similarly, emissions with similar wavelengths are produced by Yb^{3+} and Ho^{3+} co-doped UCNPs, which severely restricts the use of UCNPs for multiplexed analyte detection and multicolor encoding. As such, MCT research is quite important. Many methods have been suggested, such as inducing more ELs, screening the host materials, and varying the Ln^{3+} doping level [29].

Regulating the Concentration of RE Ions Doping The number of activators and their distribution within the host materials are contingent upon the level of Ln^{3+} doping. Alterations in the doping concentration of the luminescent centers typically result in the novel CR and BET emergence.

Introducing Extraneous Energy Levels Doping different RE activators may produce multicolor emissions, and abundant ELs offer Ln^{3+} multiple paths for ET. Moreover, the CR induced by different RE activators may alter the initial emissions in the activator co-doped system, resulting in fresh emission bands or different branch ratios.

Fine-tuning VIS UCL in UCNPs was accomplished by Wang et al. [63, 64]. They successfully adjusted the luminescence from blue to white by incrementally increasing the doping level of Er^{3+} [65, 66]. With different emission outputs and spectral fingerprints, Wang et al. created a variety of Yb-based UCNPs [67, 68].

Zhang and coworkers effectively tuned UCL in Yb/Ho/Ce tri-doped UCNPs [69]. As the concentration of Ce^{3+} increased, a distinct shift in color output was observed. This color change was attributed to multiphoton CR [70]. Qin and coworkers have shown that the modulation of UCL in $NaYF_4$:Er, Tm UCNPs, and multiphoton CR results in an increase in the red-to-green ratio of Er^{3+} with the doping of Tm^{3+} [71, 72]. An investigation into the effect of co-doping with additional Ln^{3+} on the UCL was conducted by Chan et al. The results demonstrated that UCNPs may generate nearly pure green or red UCL [71, 73]. Near monochromatic red emission was caused via CR, but monochromatic green emission was suggested to be caused by selective quenching of red emission.

Several transition metal ions have been introduced as transitional states for UC. By utilizing the $^4T_1$ state of Mn^{2+}, it is possible to tailor the UCL of Ln^{3+}. Li and coworkers first unveiled the effect of Mn^{2+} on the UCL of Er^{3+} in Mn^{2+}-based UCNPs

[73]. Spectral analysis revealed that this method generated pure red UCL spectrally. The ET from Yb^{3+} to Er^{3+} to Mn^{2+} to Er^{3+} increased red UCL while reducing the intensity of green UCL. Subsequently, Liu and coworkers verified the red UCL of $KMnF_3$:Yb^{3+} and Er^{3+} UCNPs [74]. Furthermore, the researchers detected spectrally pure NIR and red emission in $KMnF_3$:Yb^{3+}, Tm^{3+} and $KMnF_3$:Yb^{3+}, Ho^{3+} UCNPs, respectively. Both of these observations facilitated ET through the $^4T_1$ EL of Mn^{2+}.

Excitation Wavelength Modulation Multicolor tunable emission also benefits from varied excitation light sources for Ln^{3+} with rich EL configurations. By manipulating the pulse width and the wavelength of infrared laser beams, Liu and coworkers present a practical and adaptable method for fine-tuning UCL throughout the entire color spectrum [75].

2.4.3.2 Energy Transfer–Triggered Novel Upconversion Excitation

Traditional ETU-mechanized UCNPs have a sharp excitation band around 980 nm because of the absorption of Yb^{3+} sensitizers. The limited variety of excitations nonetheless restricts the versatility of further UCNP applications. For instance, a large fraction of the 980-nm excitation coincides with water absorption, which is essential to all living things.

To address the above problem, Nd^{3+}, Tm^{3+}, and Er^{3+}-sensitized UCNPs were developed [75–77]. For example, Nd^{3+} were excited by the NIR light at 730, 808, and 865 nm, and the energy was then transferred to adjoining Yb^{3+} by a DS way. Activators then captured the energy and released UCL. Nd^{3+} and Er^{3+} were doped into different layers to prevent mutual quenching. It was found that the excitation efficiencies were comparable. Additionally, the excitation of 808 nm significantly reduced the heating impact brought on by 980 nm [78–80]. Sustaining a low concentration of Nd^{3+} is essential to reduce multiphoton CR between Er^{3+} and Nd^{3+}.

2.4.4 Cross-Relaxation (CR) Regulation

The CR process requires the sensitizer and activator to have one or more EL pairs with different positions but a matchable energy gap, which is effective and irreversible. The two kinds of CR can be initiated by the EL self-matching of the same ion and the EL pair matching of the different ions, respectively [81]. Notably, the CR process relies highly on the doping concentration of luminescent centers. Once the doping concentration of active ions is high, severe CR occurs between the doped ions. Especially when the EL of sensitizer ions exhibits a multi-step distribution, self-quenching brought on by increased doping level counteracts the emission improvement and may ultimately attenuate the total luminescence intensity. Recently, numerous methods have been developed due to developments in NP research to lessen energy loss at high doping levels and modulate associated CR processes.

2.4.4.1 Alleviating Concentration Quenching (CQ) for Highly Doped UCNPs

CQ affects the interaction between active ions, including sensitizers and activators in highly doped UCNPs. EL structures of each kind of dopant ion vary, and several mechanisms predominate throughout the quenching processes. A specific strategy is often necessary to minimize the quenching of ESs in Ln^{3+}.

Confining Excitation Energy The primary cause of CQ in UCNPs is attributed to EM across the dopant sublattice. This phenomenon predominantly occurs during the depopulation of sensitizer ions, which typically possess relatively straightforward EF structures. EE can easily be transported to quenching defects or impurities via long-distance energy migration (LDEM), resulting in nonradiative relaxation loss. In UCNPs, it is simple to spatially confine EE by preventing energy sinks from dissipating.

Introducing Ln^{3+} into a CS structure is an effective technique for containing EE in UCNPs. Most quenching sites (QSs) in nanoscale materials are found on the NP surface and in this manner, the protective shell may effectively separate the surface quenchers and dopant ions. The CS arrangement minimizes the quenching of migratory energy reserved within the dopant sublattice [82, 83]. Several groups have successfully enhanced the UCL by doping feasible Yb^{3+} or Er^{3+} as high as possible and surface passivation by inert shells [61, 84, 85]. Since 2017, we and other researchers have reported $NaErF_4@NaYF_4(NaLuF_4)$ NPs, which significantly advances our comprehension of the luminescence mechanism in heavily doped UCNPs [83, 86, 87].

In addition, introducing appropriate trapping centers can effectively obstruct LDEM within the nanocrystal lattice. Then, doping Tm^{3+} into the $NaErF_4$ lattice may cause a transient energy trap, which aids in directing the EE away from defects, as reported by Liu and coworkers [88]. The UCL of $NaErF_4@NaYF_4$ UCNPs was enhanced three times under 980-nm excitation by doping Tm^{3+}. Yang and coworkers utilized other Ln^{3+} as additional trapping centers to selectively improve UCL bands in Er@Y UCNPs to reduce energy losses at defective lattice locations [89]. One helpful method for enhancing the UCL of highly doped UCNPs is the ease with which trapping center doping might be achieved during NP manufacturing.

Controlling EM patterns is another way to achieve the confinement of EE. Liu and coworkers used KYb_2F_7 NPs to present the Yb^{3+} dispersed in arrangements of tetrad clusters [90]. Since the ionic distance in the cluster is much smaller than the inter-cluster distance, the EE is essentially preserved inside the sublattice region. As a result, even without a protective shell, EE migration to QSs is reduced. Additionally, the spatial concentration of EE makes it easier for Er^{3+} activators to absorb the energy, generating very potent violet UCL.

Using hierarchical nanostructures to regulate the spatial distribution of Ln^{3+} provides a more adaptable method of intervening with EM. For restricting EM in Yb^{3+}, Wang and coworkers designed a sandwich nanostructure employing the substances $NaYF_4@NaYbF_4@NaYF_4$ [31]. The researchers demonstrated logical tuning over the average EM distance and the local EE density by varying the thickness of the $NaYbF_4$ layer. They successfully obtained robust 290-nm emission at an optimal shell thickness, almost 45 times more intense than the UCNPs doped with 19 mol% of Yb^{3+}. Utilizing the identical nanostructure, they successfully achieved efficient UCL of Er^{3+} and Ho^{3+} [91].

High-Density Excitation Er^{3+} and Tm^{3+} are typical photon UC activators with numerous ESs. CR and PAET are confined mechanisms that are easily used to depopulate the high-lying ESs when the concentration of the activator increases. These processes are unlikely to be dealt with through spatial energy confinement [92–95]. A high excitation power density (PD) is required to stop the depopulation processes in that scenario.

Benz et al. developed a rate formula that imitates the UCL intensity of Ln^{3+} ions as a function of excitation rate and doping level to avoid CQ in GaN: Er^{3+} [96, 97]. According to their findings, powerful irradiance could efficiently minimize the odds of CR by reducing the population density. Under high-power stimulation, they hypothesized that the UCL intensity would increase in correlation with the concentration.

Notably, the low absorbance of the materials severely restricts the increase in UCL intensity at low doping levels with increasing excitation PD [98]. Cohen and coworkers developed a population balance model in an independent study to elucidate the relationship between the excitation PD and UCL intensity [99]. When exposed to low excitation PD, the UCL emission of Yb^{3+}/Er^{3+} co-doped UCNPs was most prominent with a 0.5 mol% doping of Er^{3+}. However, the optimal doping level dramatically rose to ~20 mol% at high PD. Optical testing of a single UCNP verified that traditional UCNPs showed fast brightness saturation as the excitation PD rose. Remarkably, the highly doped UCNPs finally beat conventional ones in brightness at increased excitation PD.

Using a high excitation PD through a micro-structured optical fiber and substantially doping Tm^{3+} activators in UCNPs, strong UCL was obtained as opposed to Er^{3+} [100]. By increasing the doping level of $NaYF_4$ UCNPs, it was possible to detect a 70 times boost in UCL of Tm^{3+} under 980-nm excitation. Upon high PD excitation, enough brightness of the UCNPs is quite significant. Being willing to compromise for safe operating conditions is critical for some fields that require low excitation PD.

An integrated optical waveguide circuit platform was recently presented by Wang and coworkers to deliver a high PD of excitation at low input power and to magnify the excitation field. The researchers achieved a high PD by employing a microresonator to cycle the excited photons and low input power [101]. By excitation of 1550 nm, this platform exhibited an intense UCL of 380 nm for Er@Y UCNPs. The study revealed that the total power conversion efficiency exceeded 5%. Surface passivation and high PD irradiation were combined by EM and CR to reduce CQ, an essential component in the efficient multiphoton UC process.

2.4.4.2 NIR Downshifting Modulation by CR

As mentioned above, the CR process highly depends on the doping level of sensitizer and activator ions in the NPs. As the concentration of activator ions is artificially increased, the CR process among the activator ions will intensify. Thus, due to the copious EL structure of Ln^{3+}, the related downshifting process of NIR light will be amplified when stimulated by the excitation light source. Moreover, a typical case is the employment of Tm^{3+}-based NPs for downshifting modulation by our group [76]. By continuously increasing the doping concentration of Tm^{3+} in the core, the noticeable emission enhancement of 1600–2100 nm can be observed using $LiTmF_4$@$LiYF_4$ NPs. This emission band mainly originates from the EL transition from $^3F_4$ to $^3H_6$ of Tm^{3+}.

Moreover, the population enhancement of the $^3F_4$ state is mainly ascribed to the CR process. The DSL enhancement is primarily due to the population enhancement of the $^3F_4$ EL caused by the CR process of $^3H_4 + {}^3H_6 \rightarrow 2\,{}^3F_4$. Afterward, the same phenomenon can be observed in Zhang and coworker's two reports [102, 103]; however, it is due to the introduction of different concentrations of Yb^{3+}in Zhang's study that the optimal concentration is 8% and 5%, rather than 100% in the non-doped system. In addition, in the Ho^{3+}-based system, the 2030 and 2842 nm DSL can also be modulated by tuning the CR process of $^5F_4/^5S_2 + {}^5I_8 \rightarrow {}^5F_5 + {}^5I_4$ and $^5F_4/^5S_2 + {}^5I_7 \rightarrow {}^5F_5 + {}^5I_6$, respectively [103–105]. And the CR process of $^5I_6 + {}^2F_{5/2} \rightarrow {}^5I_7 + {}^2F_{7/2}$ and $^5F_4/^5S_2 + {}^2F_{5/2} \rightarrow {}^5F_5 + {}^2F_{7/2}$ between Ho^{3+} and Ce^{3+} can also be used to modulate the 2030 and 2842 nm DSL of Ho^{3+}-based system [106]. As for the Er^{3+}-based system, upon an 800-nm excitation, 1525-nm downshifting could be modulated based on the CR process of $2\,{}^4I_{9/2} \rightarrow {}^4S_{3/2} + {}^4I_{13/2}$ [77, 83]. And the CR process of $^4I_{11/2} + {}^2F_{5/2} \rightarrow {}^4I_{13/2} + {}^2F_{7/2}$ between Er^{3+} and Ce^{3+} can also be used to modulate the 1525-nm DSL of Er^{3+}-based system [107].

2.4.5 Phonon-Assisted Energy Transfer (PAET)

In the PAET process, the energy between the EL pairs do not match well, and the energy difference is close to the phonon energy of the system. Therefore, with the help of phonons, the energy matching between the EL pairs is realized by absorbing or releasing one or more phonons during the electron-phonon transition process.

In most cases, PAET will lead to the quenching of the ES, and finally, heat will be released as a result of the quenched energy, and the thermal field may readily influence this process [101]. In the Er^{3+}-rich $NaErF_4$@$NaYF_4$ CS nanosystem, the UCL of Er^{3+} is extremely quenched by the PAET process. With the help of multiphonon assistant ET, the population of higher levels ($^2H_{11/2}/^4S_{3/2}$) is extremely quenched by the lower levels, such as $^4I_{9/2}$ and $^4I_{15/2}$ (Figure 2.6a). Therefore, the cryogenic field can suppress the PAET and increase the UCL of Er^{3+} due to the decrease of phonon in the crystal matrix, resulting in a drastically ~2000-fold enhancement for green emission (Figure 2.6b,c) [108]. The PAET-induced thermal quenching is also found in other Ln^{3+} ions, such as Tm^{3+}, Ho^{3+}, and Eu^{3+} [109].

However, the thermal activation of UCL in NPs has recently opened up new possibilities in bioscience and nano-photonics. In general, changing the ambient temperature has become an essential means to modulate the PL properties of RE-doped luminescence nanomaterials by the PAET process [110].

Temperature measurement is essential in various fields, such as automotive, biomedicine, and fluid dynamics. Traditional thermometers are unsuitable for

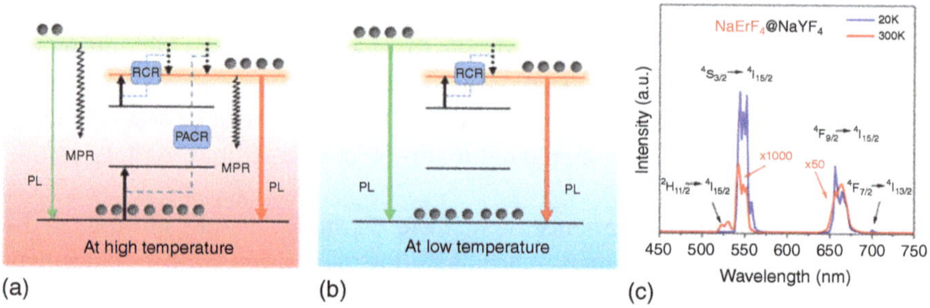

Figure 2.6 (a,b) The PAET process in $NaErF_4$@$NaYF_4$ NPs occurs at high temperatures and is suppressed at low temperatures. (c) The UCL spectra of $NaErF_4$@$NaYF_4$ NPs at 20 and 300 K. Source: Reproduced with permission from Li et al. [108]/American Chemical Society.

accurate measurements in small apparatuses and are affected by electromagnetic interference [109, 110]. However, UCNPs can accurately determine temperature spectroscopically without needing physical contact, utilizing temperature-dependent PL intensity or intensity ratios. PL nanomaterials, characterized by their stable, sharp emission bands and temperature-dependent luminescence, are driving significant progress in non-contact thermometers.

Temperature is often determined by utilizing a specific luminescent characteristic, such as the delay in luminescence or the change in the emission spectrum. These concepts are discussed in recent literature. Undesirable effects from factors like variations in the excitation source and absorption by the material can be eliminated by using the intensity ratio. For example, Er^{3+} has thermally linked ELs of $^2H_{11/2}$ and $^4S_{3/2}$, resulting in these two levels generating thermosensitive green emissions. Similar to peasecods, hollow $LiZrF_4$:Yb, Er NPs demonstrate a relative sensitivity (Sr) of 5.65% K^{-1} [111].

Exploring other Ln^{3+} ions, such as Nd^{3+} and Eu^{3+}, for potential utilization can be considered. By manipulating the UC process through nano-structural design, it is possible to enhance the sensitivity of nano-thermometers that utilize Nd^{3+} [112, 113]. Luminescence nanothermometry (NT) based on phase transition has been demonstrated in $LiYO_2$: Eu with S_r up to 12.5% K^{-1} [112]. Additionally, ratio-metric temperature sensing may be achieved by a combination of multiple luminescent ions. The incorporation of Eu^{2+} in co-doped Sc_2O_3 NPs with Eu^{3+} enables superior performance in NT due to the intrinsic calibration luminescence of Eu^{2+}. Furthermore, temperature-independent Tb^{3+} UCL and thermally quenched Mn^{4+} UCL exhibited a sensitivity of 2.8% °C^{-1} [114]. Dual-excitation decoding is feasible using hybrids made of UCNPs and quantum dots, with a minimal systematic error of about 2.3 °C [115]. A higher sensitivity than basic thermally coupled systems can be achieved by integrating thermally enhancement (Yb/Nd) and quenching (Yb/Er) systems into a nanostructure [116]. Incorporating temperature-independent intensities derived from Eu^{3+} or Ho^{3+} facilitates non-contact temperature monitoring and 2D mapping across a broad temperature spectrum while offering the dual benefits of quantitative modeling and chemically stable Ln^+ emission [117, 118]. UC-based thermometers can precisely detect the temperature of biological tissues [119], which has been investigated extensively in various standard operating procedures for luminous NT use in biological contexts [117, 120–124]. When combined with therapies based on microscopic temperature manipulations, chemotherapy (CT) can significantly increase the effectiveness of therapeutic interventions [125]. Exceptionally sensitive temperature gradients were observed within HeLa cancer cells through 3D optical manipulation using thermal-sensing NPs within micro-sized systems [126]. Despite concerns about reliability due to tissue-induced spectrum distortions resulting from the temperature-dependent nature of light absorption and scattering in tissues [127], Ag_2S NPs have shown promise for amalgamating multiple regression with multi-parametric sensing to enhance temperature sensitivity and resolution tenfold [128]. This advancement enables precise detection of even minute fluctuations in live tissues [128], making Ag_2S-based NT potentially valuable for providing accurate absolute thermal readings for diagnosis and treatment purposes related to liver health [129]. The lifetime stability under varying conditions makes it suitable for implementing real-time monitoring applications [130], such as $NaYF_4$@$NaYF_4$:Yb, Nd@CaF_2-based nanothermometer exhibiting consistent detection capability at depths up to 4 mm within living tissues over a range from 10 to 64 °C at sensitivities ranging from 1.4% to 1.1% $°C^{-1}$ [131]. PAET couples involving Yb/Nd could also facilitate NIR lifelong NT [132], while further research suggests that reducing decay duration Tm^{3+} state could augment initial biological window NT capabilities [133].

2.4.6 Dye Sensitization

Many promising applications are based on NPs, such as small animal bioimaging, photo-triggered drug release, and photodynamic therapy. However, the light absorption cross-section of RE-doped NP is extremely weak ($\sim 10^{-20}$ cm^2) and within a very narrow spectral band, resulting in limited light brightness of NPs, which restricts their practice application. Dye sensitization is an effective strategy for improving the light-harvesting ability and spectrally broadened absorption of NPs since the absorption cross-section of dye is 1000–10 000 times larger than that of NPs. Dye sensitization is usually realized by two composite configurations: (I) dye-sensitized pure core and (II) dye-sensitized CS nanoparticle (Figure 2.7). Both structures relate to ET from NIR dyes to Ln^{3+} ions at the organic–inorganic interface.

Additionally, this organic–inorganic hybrid structure can enlarge the regulation of molecular absorption band gap and other complementary molecules. Notably, the dye sensitization problem is mainly concentrated in the aggregation-caused quenching (ACQ) of fluorescent dyes, hindering the increase in the number of coupling dye molecules. Thus, aggregation-induced emission (AIE) fluorescent dyes are a more promising demand for dye sensitization strategy.

2.4.6.1 Dye-Sensitized Core Nanoparticles

Most organic dyes used to sensitize NPs are small NIR organic molecules with emission wavelengths of 750–1150 nm, which overlap with the absorption of active Ln^{3+} ions, including sensitizers and activators. This provides the possibility of allowing ET from organic dyes to Ln^{3+} ions.

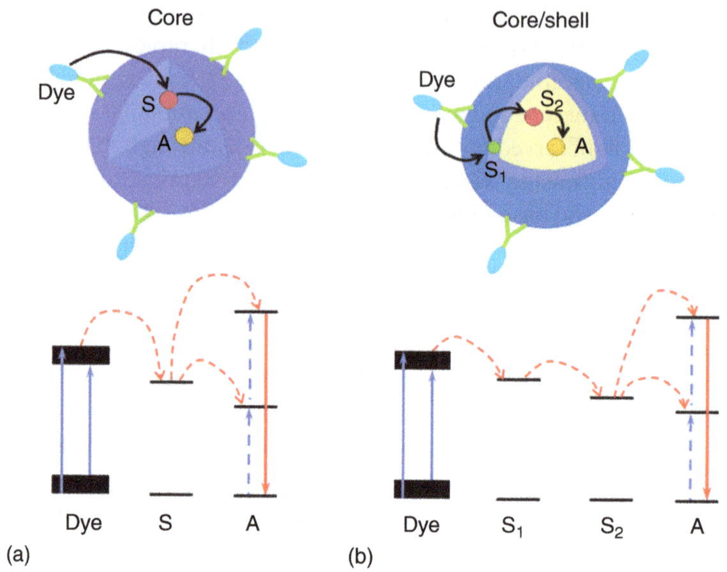

Figure 2.7 Schematic illustrations of dye-sensitized enhanced luminescence of NPs in (a) core and (b) core–shell structure.

For dye-sensitized core NPs, dye molecules transfer energy to sensitizer S (usually S = Yb^{3+}, Nd^{3+}, and Er^{3+}) and activator ions A (A = Er^{3+}, Ho^{3+}, and Tm^{3+}) through Förster resonance energy transfer (FRET). The advantage of this structure is that the distance of energy transfer from organic dyes to S is short, leading to high FRET efficiency; the disadvantage is that S and A ions are close to surface quenchers, which reduces the luminescence efficiency. In general, the total UC luminescence of dye-sensitized $NaYF_4$:Yb, Er nanoparticles was significantly enhanced (approximately 3300 times) due to the increase in injected energy [134]. Additionally, with doping 30% Gd^{3+} in the $NaYF_4$ matrix, Garfield et al. accelerated the intersystem crossing (ISC) rate of IR806 molecules on the surface of nanoparticles through the heavy atom. Transferring energy to Yb^{3+} ions triplet–triplet state energy transfer enhanced the intensity of UCL by 33 000 times, and the QY was increased by 100 times. In recent years, dye-sensitized DSL has also been realized. By combining indocyanine green (ICG) dye molecules with inorganic $NaYF_4$:Er (20 mol%) nanoparticles, dye sensitization was achieved, which increased the DSL (1520 nm) and UCL (540 nm) by 5-fold and 32-fold, respectively [135]. Therefore, dye-sensitized lanthanide-doped nanoparticles have opened up huge opportunities for many applications. However, the photobleaching problem remains, which needs to be improved.

2.4.6.2 Dye-Sensitized Core–Shell Nanoparticles

In order to solve the problems of (i) mismatch between the emission of NIR dyes and the absorption of sensitizers, (ii) the strong surface quenching in dye-sensitized pure core NPs, the NIR dye-sensitized CS structured NPs were introduced (Figure 2.7).

For dye-sensitized CS structure, the ET occurs from dye molecules to the sensitizer (S_1) in the shell; the excitation energy transfers to the sensitizer S_2 and then to the activator in the core, which is excited via the ETU mechanism to produce luminescence emission. Due to the structural design of the CS, which isolates the core from the spatial environment, the shell significantly suppresses the quenching of surface-related defects in the core. In addition, the absorption peak of fluorescent dyes, which cannot exhibit obvious spectral overlap with the absorption of usually used lanthanide sensitizers, can achieve strong spectral overlap through cascade ET of different sensitizers. For example, a design of CS $NaYbF_4$:0.5%Tm^{3+}@$NaYF_4$:Nd^{3+} nanoparticles sensitized with NIR dye (IR808) was reported to prove this theory. This efficient ET between dye molecules and Nd^{3+} leads to a 14 times enhancement in luminescence intensity to $NaYbF_4$:0.5%Tm^{3+} pure core [136]. Therefore, the dye-sensitized CS nanostructures technology integrates the merits of CS nanostructures and the "antenna" effect of dyes, generating ultra-bright luminescence.

2.4.7 Combined Excitation Regulation

In recent years, the double/multi-wavelength combined excitation technology has rapidly developed in PL of RE-doped NPs and its applications. Its main mechanism can be categorized as excited state absorption (ESA), stimulated emission depletion (STED), ground state depletion (GSD), and STED-inspired methods. All these luminescence processes refer to the unstable intermediate state EL, where energy can be stimulated to the intermediate metastable state by absorbing specific energy under the first excitation beam. Afterward, another laser with a different wavelength from the first beam controls the electron-filling state of the metastable energy level, resulting in different photoluminescence models, including luminescence enhancement, luminescence loss, and new wavelength emission. In the dual-wavelength excitation effect, the most common research mechanism in RE-doped materials is ESA, and the excitation beams need to be strictly matched with the energy levels of the doped ions. Compared with single wavelength stimulation, ETU and ESA can improve UC efficiency by reducing the excitation energy loss caused by energy mismatch. It is worth noting that dual-wavelength excitation cannot only induce enhancement but also produce attenuation in the UCL. For instance, the STED mechanism can deplete the electron-filling state of the metastable energy level, leading to luminescence loss. Recently, according to the STED-like mechanism, rare-earth doped super-resolution probes have emerged, breaking through the limitations of the traditional STED probes depending on dyes and protein luminescent materials.

2.4.7.1 ESA

In the dual-wavelength excitation photoluminescence process, the ESA mechanism mainly has three functions: (I) UC emission, (II) sublevel luminescence emission, and (III) suppression of luminescence emission. Due to its rich energy levels, lanthanide ions exhibit multiple ESA processes within a wide spectrum. So far, the existing ESA given the dual-wavelength photoluminescence effect consists of ESA for UCL (Figure 2.8a), ESA for energy transfer of UCL (Figure 2.8b), cyclic

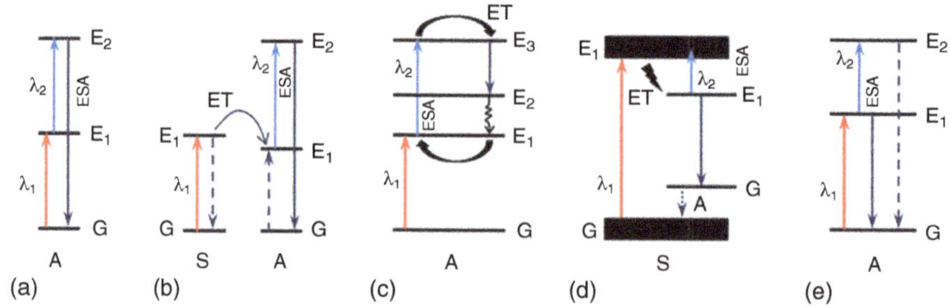

Figure 2.8 Schematic of different ESA mechanisms upon dual-wavelength excitation: (a) ESA for UCL; (b) ESA for UCL through energy transfer; (c) recycle ESA; (d,e) ESA to upper dark state: (d) GSA → ET → ESA; and (e) GSA → ESA.

ESA (Figure 2.8c), and ESA for luminescence depletion (Figure 2.8d,e). Qiu et al. observed effective broadband excitation of NIR to visible UCL in LaF_3: Ho^{3+}. When excited simultaneously at 970 and 1150 nm, the UCL intensity is significantly boosted, owing to the ETU mechanism [137].

Subsequently, they raised a method to enhance the upconversion efficiency of $NaYF_4$:1%Er nanoparticles through multi-wavelength excitation. Compared with the overall luminescence intensities under 790 and 1520 nm excitation, the green and red emission intensities of $NaYF_4$:1%Er nanoparticles enhanced by 3.05 and 1.53 folds, respectively, upon co-excitation at both wavelengths [138].

2.4.7.2 STED

In addition to enhancing the UC luminescence of the target wavelength, two-wavelength excitation also yields luminescence depletion, which is generally used in super-resolution STED microscopes. In a classic report (Figure 2.9), Jin and coworkers utilized strong cross-relaxation in Yb:1%Tm systems to increase the population of $^3H_4$ levels, using 980 nm (0.66 MW cm^{-2}) as a stable excitation beam and 800 nm (9.75 MW cm^{-2}) as a depleted excitation beam, leading to suppress the upconversion process [139]. Furthermore, Zhan et al. make complete utilization of the concentration-dependent CR between dopants in the highly doped Tm system (18% Yb: 10% Tm), achieving almost entirely emission weakening by conducting EE of the emission state with a second laser beam (975 nm at 700 kW cm^{-2}, 810 nm at 17.7 MW cm^{-2}) [140]. These characteristics have achieved a low-power super-resolution STED microscope and a nanomicroscope. The luminescence intensity and saturation of the "STED beam" influence the imaging resolution. In theory, resolution can be improved by increasing beam intensity or decreasing saturation.

2.4.8 External Field Modulation

2.4.8.1 Magnetic Field Modulation

The so-called magnetoluminescence refers to the phenomenon of photoluminescence or electroluminescence changes observed in transition metal-doped

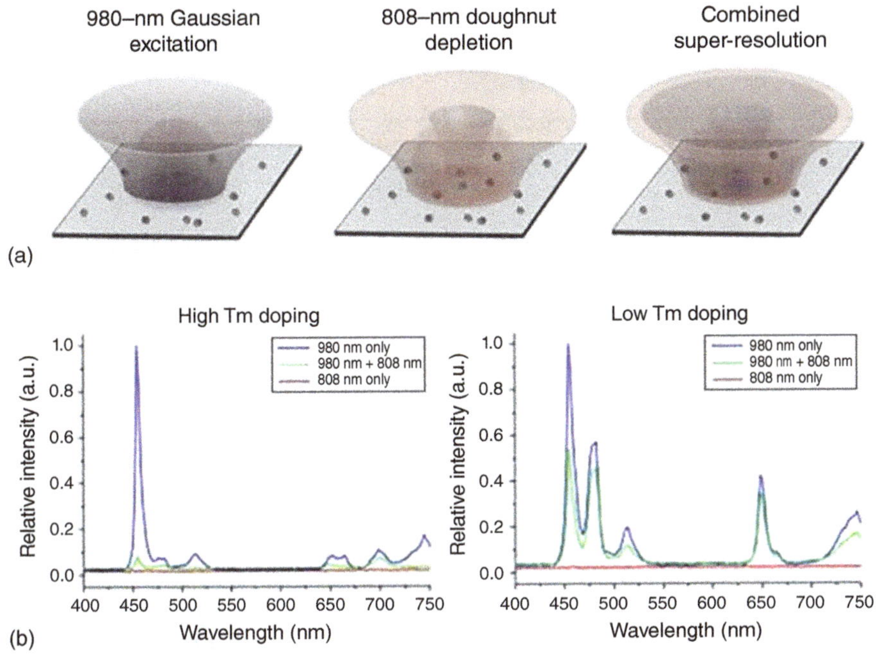

Figure 2.9 (a) Schematic diagram of STED microscope. (b) Low-doped Tm (1%) and high-doped Tm (8%) luminescence emission spectra. Source: Reproduced with permission from Liu et al. [139]/Springer Nature.

phosphors, 2D electron gases, organic compounds, and so on in the presence of an applied magnetic field (MF), usually at high MF and low temperatures. For example, Köche et al. observed magnetoluminescence in electron gases within a helium cryostat under an MF of 15 T and a temperature of 290 mK [141]. Magnetoluminescence in InP/GaP quantum dots was measured at MF up to 60 T and temperatures below 100 K by Dewitz et al. [142]. Magnetization is often observed when an MF is applied in UCL materials with magnetic elements. The magnetic response arises from the non-interacting localized properties of the magnetic moments in the compounds, and the magnetization and relaxation of UCL materials containing magnetic elements have been developed for diverse applications, including bioseparation and magnetic resonance imaging (MRI).

However, magnetoluminescence in UCL materials has been reported only in a few papers [143, 144]. The mechanism associated with the magnetoluminescence effect in UCL materials can be understood as the Zeeman splitting of the energy levels of Ln^{3+} emitter (e.g. Er^{3+} ions). Figure 2.10 depicts a schematic energy level diagram of Er^{3+} with only $^4S_{3/2}$ and $^4I_{15/2}$ states. Under an applied MF, the $^4S_{3/2}$ state of the Er^{3+} ion splits into four states, $|+3/2\rangle$, $|-3/2\rangle$, $|+1/2\rangle$, and $|-1/2\rangle$, via the Zeeman effect. The splitting between $|-3/2\rangle$ and $|-1/2\rangle$ increases as the MF increases. This leads to a double-state splitting of $|\pm3/2\rangle$, while the probability of radiation from $|+3/2\rangle$ to $|-3/2\rangle$ is negligible. Simultaneously, the predominant portion of UCL emission emanates from the lowest $|-3/2\rangle$ energy level within the $^4S_{3/2}$

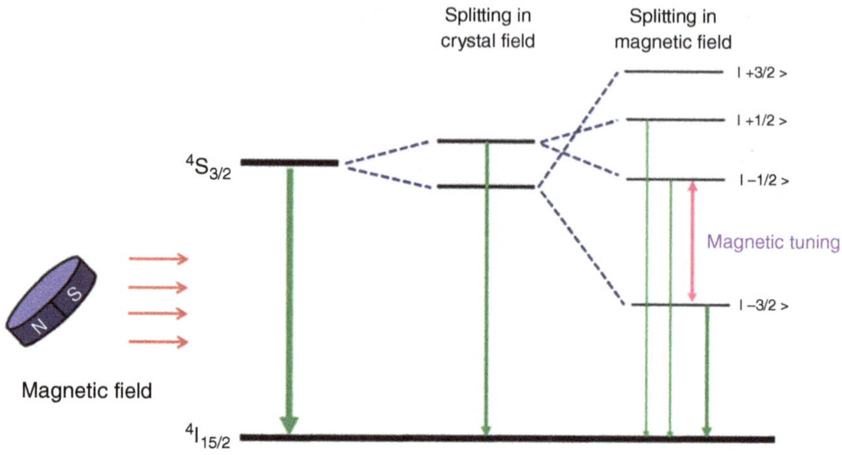

Figure 2.10 A simplified energy level diagram illustrating the splitting of energy level $^4S_{3/2}$ of the Er^{3+} ion where the gap between the Zeeman levels |−1/2> and |−3/2> increases with an external magnetic field. Source: Reproduced with permission from Yin et al. [145]/The Royal Society of Chemistry.

quartet of states. Consequently, applying an MF diminishes the intensity of green UCL emissions.

Li et al. have found that Nd^{3+}-doped $NaGdF_4$:Nd^{3+}/Yb^{3+}/Er^{3+} BFSL nanoplates with 3 mol% doping levels have significant magnetic response effects at 300 and 10 K (Figure 2.11a,b). At room and low temperatures, the UCL intensity experienced a reduction of approximately 60% and 15% relative to the initial UCL value. Furthermore, the adjustable magnetic response effect was accomplished by changing the concentration of Nd^{3+}, and the best one was 4% at 20 mol% Yb^{3+} and 2 mol% Er^{3+} for the greatest magnetic response effect (Figure 2.11c,d).

2.4.8.2 Electric Field Modulation

In low-symmetry media, the CF around the activator of Ln^{3+} ions exerts an inhomogeneous influence on the optical properties of UCL materials. This principle is the most common reason for the UCL enhancement of hexagonal $NaYF_4$ relative to its cubic phase. Altering the symmetry of the host and the CF of the UCL material is usually achieved chemically, that is by changing the doping element and crystal phase. In contrast, the electric field (EF) responsive action of UCL materials employs a physical approach that dynamically adjusts the symmetry of the lattice position surrounding activator ions within the same host in situ and in real time. The resulting symmetry alterations lead to varying transition probabilities, resulting in changes in the UCL.

On the basis of classical J–O theory, the spontaneous emission probability A_{ed} corresponds to an electric dipole (ED) transition from the initial J manifold $|[S, L]J\rangle$ to a final J manifold $|[S', L']J'\rangle$ as represented in Eq. (2.1),

$$A_{ed} = \frac{64\pi^4 e^2}{3h(2J+1)\lambda^3}\left[\frac{n(n^2+2)^2}{9}\right]S_{ed} \tag{2.1}$$

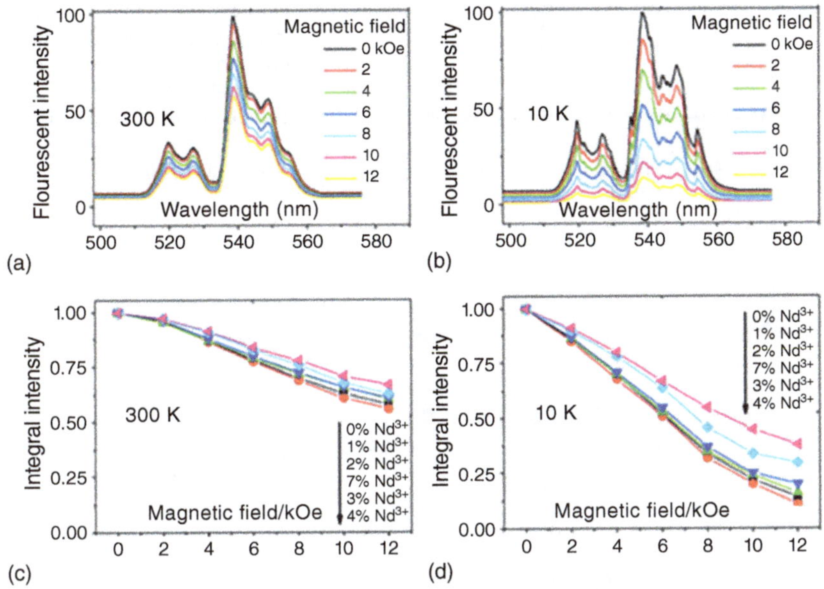

Figure 2.11 Effect of external magnetic field on the green UCL of $NaGdF_4$: Nd, Yb, Er BFSL at (a) 300 K and (b) 10 K. Dependence of the integral luminescence intensity of $NaGdF_4$: Nd, Yb, Er BFSL with different doping concentration of Nd^{3+} on the applied magnetic field at (c) 300 K and (d) 10 K. Source: Reproduced with permission from Tikhomirov et al. [144]/American Chemical Society.

where e represents the electron charge, λ denotes the average wavelength of the transition, n represents the refractive index at the wavelength of the transition, and h stands for the Planck constant. The line strength S_{ed} for the electric dipole (ED) transition is expressed as shown in Eq. (2.2),

$$S_{ed} = \sum_{t=2,4,6} \Omega_t |\langle 4f^n [S, L]J \| U^{(t)} \| 4f^n \left[S', L'\right] J' \rangle|^2 \tag{2.2}$$

where three terms $\langle \| U^{(t)} \| \rangle$ are reduced matrix elements of the unit tensor operators, and three J–O intensity parameters Ω_t ($t = 2, 4, 6$), encapsulating the influence of the CF and radial electron integrals, among other factors. Amid these intricate components, the Ω_2 is particularly prominent in facilitating what is termed hypersensitive transition. Notably, this parameter is recognized for its strong correlation with the degree of asymmetry at the sites occupied by lanthanide ions.

To investigate the effect of the EF on the UCL spectra, it is essential to devise a suitable combination of activator and host, which should have the asymmetry of the activator ionic sites varying with the EF. The following is an example of BTO:Yb/Er (Figure 2.12) to illustrate the EF response.

Barium titanate (BTO) was chosen as the ferroelectric host for UCL due to its reputation as a paradigmatic system in understanding crystal structure transformations induced by mechanical stress, EF, and temperature. The lanthanide-activated Er^{3+} ion consists of a partially filled 4f inner layer and two outer layers. At room temperature, the BTO lattice is noncentrosymmetric in the tetragonal structure

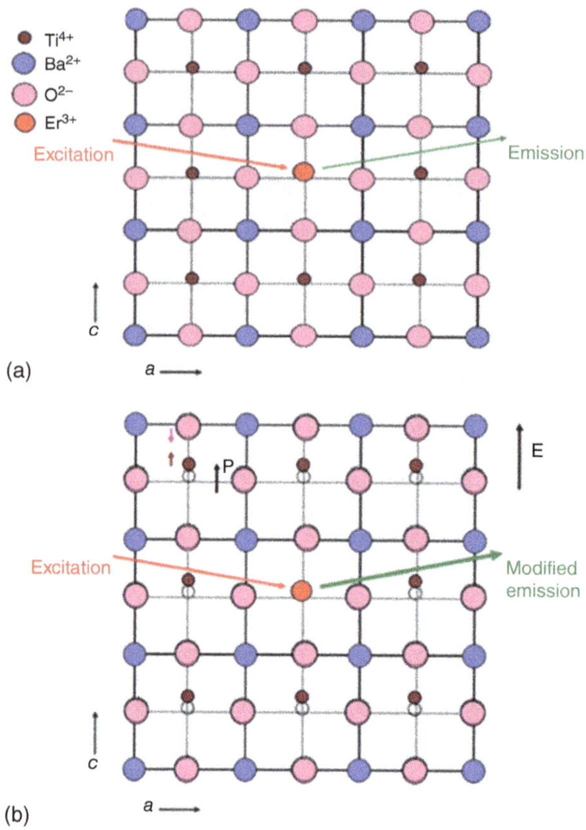

Figure 2.12 The tetragonal lattice of Er-doped BTO films (a) without and (b) with an external electric field E. Source: Reproduced with permission from Liu et al. [143]/John Wiley & Sons.

belonging to a point group of 4 mm (C_{4v}), and Ti is changed in association with negatively charged oxygen atoms (Figure 2.12a), producing polarization. Previous studies have shown that doped Er^{3+} ions can replace Ti^{4+} ions in the BTO [146], and thus, the Er^{3+} ions in the lattice are noncentrosymmetric even in the absence of an EF. Upon applying an EF to the system, the lattice of BTO undergoes distortion, resulting in a reduction in crystal symmetry around the Er^{3+} ion (Figure 2.12b). Essentially, the lower symmetry around the Ln^{3+} implies that the more inhomogeneous CF can mix opposite parity into the 4f conformational energy levels, thereby increasing the electric dipole transition probability of the doped ion. As a result, the hypersensitivity of the Er^{3+} ion transition leads to the EF response of the BTO.

UCL enhancement can be notably augmented by elevating the DC bias voltage, as depicted in Figure 2.13a. The figure illustrates that the enhancement factor for the green emission band reaches up to 2.7, while the red UCL at 656 nm ($^4F_{9/2} \rightarrow {}^4I_{15/2}$) exhibits minimal variation despite an increase in voltage. This suggests that the amplification of emission is strongly reliant on the wavelength.

Given that the enhancement factor for the green emission band surpasses that of the red emission band, as depicted in Figure 2.13b, the photoluminescence intensity ratio between green and red emission experiences significant augmentation with increasing DC bias voltage. The discrepancy in enhancement factors between the

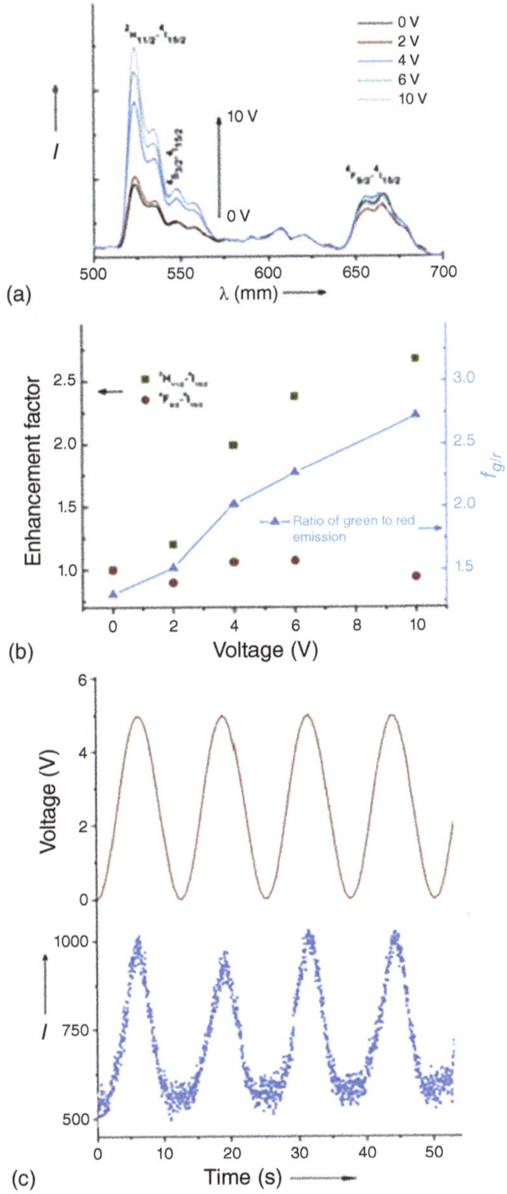

Figure 2.13 (a) The UCL spectra of the BTO:Yb/Er film under different DC bias voltages (0–10 V). (b) The enhancement factors (left axis) for green and red UCL and their PL intensity ratio (right axis) as a function of the applied bias voltages were also measured under 980-nm excitation. (c) Sinusoidal AC electric voltage was applied to the BTO:Yb/Er film (upper in red), and UCL emission at 523 nm (bottom in blue) was recorded as a function of time. Source: Reproduced with permission from Liu et al. [143]/John Wiley & Sons.

green and red UCL arises from the variation in the impact of local symmetry on these distinct emission bands, which is contingent upon the respective 4f internal transitions. Crucially, the switchable symmetry of the ferroelectric material provides the necessary conditions for achieving reversible and dynamic adjustment of photoluminescence emission via external EF modulation. As depicted in Figure 2.13c, the photoluminescence intensity of BTO:Yb/Er thin-film exhibits capability for frequency-matched luminescence control under sinusoidal AC bias conditions.

In addition to ferroelectric materials, materials obtained by integrating RE-doped nanoparticles with electrosensitive photonic crystals (PPCs) WO_{3-x} can also dynamically respond to changes in EFs [147]. In the WO_{3-x} PPCs/UCNPs hybrid materials, the emission enhancement of UCNPs reaches 5–26 times by external electrical stimulation at +1.6 and −1.6 V. The UCNPs can also respond dynamically to changes in the EF. This phenomenon arises from reversible alterations in the refractive index and concentration of oxygen vacancies within WO_{3-x}, enabling dynamic manipulation of the photonic band gap and localized surface plasmon resonance of WO_{3-x} PPCs. By employing programmable logic gate arrays that combine WO_{3-x} PPCs with UCNPs, which convert information-encrypted electrical signals into visible modes, novel technologies for display and optical storage devices can be realized.

2.4.8.3 Plasma Resonance Enhancement

The enhancement of UCL primarily stems from the reduction in the nonradiative rate/cross-section relaxation rate, changes in the CF surroundings, and a decrease in the symmetry of Ln^{3+} ions. In contrast, local electromagnetic field modulation of UC augmentation is attributed to alterations in the excitation rate and/or radiative rate, achieving a 10 000-fold enhancement. Plasma enhancement systems can be broadly categorized into three types: metal plasmas (including assembled metal/upconverters, metal/UCNPs CS composites, and plasma arrays/cavities), photonic crystal modulation materials, and semiconductor LSPR to boost UC intensity.

In analogy to metal plasma enhancement, an increase in UCL was monitored in Mn^{2+} doped UCNPs/Ag composites [148]. The Mn^{2+} doping led to a gradual color shift from green to red in $NaYF_4$:Yb^{3+}, Er^{3+} UCNPs, albeit with an overall intensity decrease with increasing Mn^{2+} doping concentration. The results show that the UCL of the hybrid material is enhanced by about a factor of two, and the intensity ratios of I_{520}/I_{540} and I_{650}/I_{540} vary more with the increase of excitation power, which is due to the greater sustainability of the synergistic energy transfer. Yin et al. recently reported significant UC enhancement in $Ag@SiO_2@Lu_2O_3$:Gd,Yb,Er CS composites, with their localized surface plasmon resonance (LSPR) widths ranging from 300 nm to over 900 nm, centered at approximately 520 nm [149]. They achieved a 30-fold luminescence enhancement compared to a control sample without silver, with optimizations for a 75 nm silver core and a 30 nm thick SiO_2 spacer layer. The green UCL near the center of the LSPR exhibits a more significant enhancement factor (40-fold) than the red one due to the emission enhancement by plasma coupling. Wang and coworkers reported a notable enhancement of UCL from Tm^{3+} doped nanoparticles utilizing a laser cavity [150]. Through precise control of laser parameters, the population of intermediate excited states of Tm^{3+} can be constrained to the desired level above the excitation threshold. The findings reveal that the upconversion efficiency in the 300~620 nm emission range under 976 nm excitation is augmented by an order of magnitude compared to scenarios lacking a laser cavity. This enhancement is ascribed to the suppression of intrinsic complexation processes in the intermediate excited state, with excess excitation power directly contributing to multiphoton upconversion enhancement.

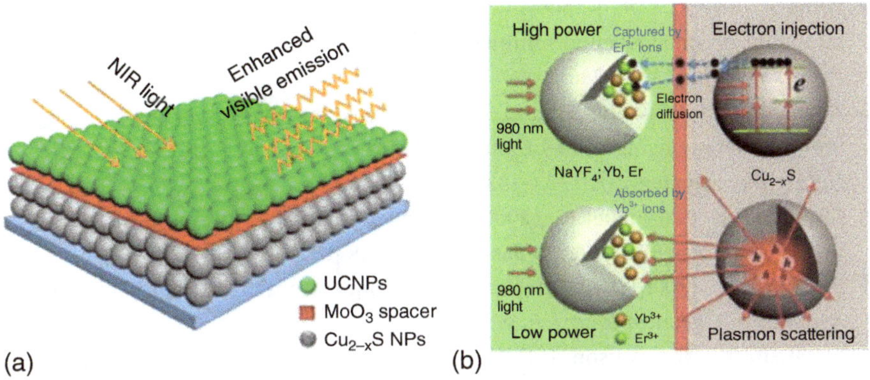

Figure 2.14 (a) Diagrammatic representation of $Cu_{2-x}S$-MoO_3-$NaYF_4$:Yb^{3+}, Er^{3+} composite film. (b) Illustrative scheme explaining the UCL enhancement in the $Cu_{2-x}S$-MoO_3-$NaYF_4$: Yb^{3+}, Er^{3+} hybrids at high and low excitation power. Source: Reproduced with permission from Zhou et al. [152]/American Chemical Society.

In terms of enhanced emission from photonic crystals, Song and coworkers first realized the modulation of UCL by photonic crystals in $NaYF_4$:Yb^{3+}, Tm^{3+} UCNP /PMMA OPCs [145]. In this study, $NaYF_4$:Yb^{3+}, Tm^{3+} upconversion nanoparticles (UCNPs) were loosely self-organized within the voids of PMMA optical photonic crystals (OPCs) or directly deposited onto glass substrates. The overall UCL intensity of the PMMA/$NaYF_4$:Yb^{3+},Tm^{3+} film exhibited over a 32-fold increase compared to that of the $NaYF_4$:Yb^{3+}, Tm^{3+} film alone, primarily due to effective photon blocking at the excitation wavelength (980 nm). Through adjustment of the photon-blocking band position and excitation angle, it was observed that the enhancement degree and luminescence lifetime of the photonic crystals were strongly influenced by the size of the photon-blocking band.

Finally, PMMA-based OPCs were also effectively utilized to boost the luminescence intensity of cell imaging.

The plasmonic behavior of semiconductor nanoparticles is generally associated with the collective oscillations of excess free carriers in the lattice, linked to structural vacancies or ionization-doped impurities, resulting in strong extinction bands at near-infrared wavelengths [151]. Song and coworkers designed and prepared $Cu_{2-x}S$-MoO_3-$NaYF_4$:Yb^{3+}, Er^{3+} hybridized films, as shown in Figure 2.14, in which the $Cu_{2-x}S$ NPs exhibit tunable LSPR in the NIR region, allowing coupling with NIR excitation to visible UCL [152]. When incorporated with an appropriate amount of MoO_3, the UCL of Er^{3+} in the $Cu_{2-x}S$-MoO_3-$NaYF_4$: Yb^{3+}, Er^{3+} hybridized film is increased compared with the corresponding $NaYF_4$: Yb^{3+}, Er^{3+} film on glass at optimal spacer layer thickness (MoO_3: ~8 nm).

References

1 Krämer, K.W., Biner, D., Frei, G. et al. (2004). Hexagonal sodium yttrium fluoride based green and blue emitting upconversion phosphors. *Chemistry of Materials* 16: 1244–1251.

2 Patra, A., Friend, C.S., Kapoor, R., and Prasad, P.N. (2003). Effect of crystal nature on upconversion luminescence in Er^{3+}:ZrO_2 nanocrystals. *Applied Physics Letters* 83: 284–286.
3 Zhong, Y., Ma, Z., Wang, F. et al. (2019). In vivo molecular imaging for immunotherapy using ultra-bright near-infrared-IIb rare-earth nanoparticles. *Nature Biotechnology* 37: 1322–1331.
4 Bünzli, J.C.G. and Piguet, C. (2005). Taking advantage of luminescent lanthanide ions. *Chemical Society Reviews* 34: 1048–1077.
5 Bünzli, J.C.G. (2006). Benefiting from the unique properties of lanthanide ions. *Accounts of Chemical Research* 39: 53–61.
6 Freeman, A.J. and Watson, R.E. (1962). Theoretical Investigation of some magnetic and spectroscopic properties of rare-earth ions. *Physical Review* 127: 2058–2075.
7 Liu, Y.S., Tu, D.T., Zhu, H.M., and Chen, X. (2013). Lanthanide-doped luminescent nanoprobes: controlled synthesis, optical spectroscopy, and bioapplications. *Chemical Society Reviews* 42: 6924–6958.
8 Bünzli, J.C.G. (2009). Lanthanide luminescent bioprobes (LLBs). *Chemistry Letters* 38: 104–109.
9 Ju, Q., Liu, Y., Tu, D. et al. (2011). Lanthanide-doped multicolor GdF_3 nanocrystals for time-resolved photoluminescent biodetection. *Chemistry – A European Journal* 17: 8549–8554.
10 Ju, Q., Tu, D., Liu, Y. et al. (2011). Amine-Functionalized lanthanide-doped $KGdF_4$ nanocrystals as potential optical/magnetic multimodal bioprobes. *Journal of the American Chemical Society* 134: 1323–1330.
11 Liu, Y., Tu, D., Zhu, H. et al. (2010). A strategy to achieve efficient dual-mode luminescence of Eu^{3+} in lanthanides doped multifunctional $NaGdF_4$ nanocrystals. *Advanced Materials* 22: 3266–3271.
12 Liu, L. and Chen, X. (2007). Energy levels, fluorescence lifetime and judd–ofelt parameters of Eu^{3+} in Gd_2O_3nanocrystals. *Nanotechnology* 18: 255704.
13 Eliseeva, S.V. and Bünzli, J.C.G. (2010). Lanthanide luminescence for functional materials and bio-sciences. *Chemical Society Reviews* 39: 189–227.
14 Adachi, S. (2023). Review–Photoluminescence spectroscopy of Eu^{2+}-activated phosphors: from near-uv to deep red luminescence. *Ecs Journal of Solid State Science and Technology* 12: 016002.
15 Wu, Q., Liu, Z., and Jiao, H. (2009). Luminescent properties of stabled hexagonal phase $Sr_{1-x}Ba_xAl_2O_4$:Eu^{2+} ($x = 0.37$–0.70). *Physica B: Condensed Matter* 404: 2499–2502.
16 Huang, X.Y., Han, S.Y., Huang, W., and Liu, X. (2013). Enhancing solar cell efficiency: the search for luminescent materials as spectral converters. *Chemical Society Reviews* 42: 173–201.
17 Chen, G.Y., Qiu, H.L., Prasad, P.N., and Chen, X. (2014). Upconversion nanoparticles: design, nanochemistry, and applications in theranostics. *Chemical Reviews* 114: 5161–5214.

18 Zheng, X., Kankala, R.K., Liu, C.G. et al. (2021). Lanthanides-doped near-infrared active upconversion nanocrystals: upconversion mechanisms and synthesis. *Coordination Chemistry Review* 438: 213870.

19 Li, Z., Liang, T., Wang, Q.R., and Liu, Z.H. (2020). Strategies for constructing upconversion luminescence nanoprobes to improve signal contrast. *Small* 16: 1905084.

20 Strek, W., Deren, P., and Bednarkiewicz, A. (2000). Cooperative processes in $KYb(WO_4)_2$ crystal doped with Eu^{3+} and Tb^{3+} ions. *Journal of Luminescence* 87–89: 999–1001.

21 Zhou, B., Shi, B.Y., Jin, D.Y., and Liu, X.G. (2015). Controlling upconversion nanocrystals for emerging applications. *Nature Nanotechnology* 10: 924–936.

22 Bell Gatch, D., Holmstrom, S.A., and Yen, W.M. (1999). Photon avalanche in Pr^{3+}:$LaCl_3$. *Journal of Luminescence* 83–84: 55–59.

23 Wegh, R.T., Donker, H., Oskam, K.D., and Meijerink, A. (1999). Visible quantum cutting in $LiGdF_4$:Eu^{3+} through downconversion. *Science* 283: 663–666.

24 Yu, D.C., Huang, X.Y., Ye, S. et al. (2011). Three-photon near-infrared quantum splitting in β-$NaYF_4$:Ho^{3+}. *Applied Physics Letters* 99: 161904.

25 Yu, D.C., Huang, X.Y., Ye, S. et al. (2011). A sequential two-step near-infrared quantum splitting in Ho^{3+} singly doped $NaYF_4$. *AIP Advances* 1: 042161.

26 Chen, X., Wu, J., Xu, X. et al. (2009). Three-photon infrared quantum cutting from single species of rare-earth Er^{3+} ions in $Er_{0.3}Gd_{0.7}VO_4$ crystalline. *Optics Letters* 34: 887–889.

27 Miritello, M., Lo Savio, R., Cardile, P., and Priolo, F. (2010). Enhanced down conversion of photons emitted by photoexcited Er_xY_2–xSi_2O_7 films grown on silicon. *Physical Review B* 81: 041411.

28 Yu, D.C., Ye, S., Peng, M.Y. et al. (2012). Sequential three-step three-photon near-infrared quantum splitting in β-$NaYF_4$:Tm^{3+}. *Applied Physics Letters* 100: 191911.

29 Sun, L.-.D., Dong, H., and Zhang, P.-Z. and Yan C.-H. (2015). Upconversion of rare earth nanomaterials. *Annual Review of Physical Chemistry* 66: 619–642.

30 Wang, F., Deng, R., Wang, J. et al. (2011). Tuning upconversion through energy migration in core-shell nanoparticles. *Nature Material* 10: 968–973.

31 Chen, X., Jin, L., Kong, W. et al. (2016). Confining energy migration in upconversion nanoparticles towards deep ultraviolet lasing. *Nature Communications* 7: 10304.

32 Oskam, K.D., Wegh, R.T., Donker, H. et al. (2000). Downconversion: a new route to visible quantum cutting. *Journal of Alloys and Compounds* 300–301: 421–425.

33 Wegh, R.T., Donker, H., Oskam, K.D., and Meijerink, A. (1999). Visible quantum cutting in Eu^{3+}-doped gadolinium fluorides via downconversion. *Journal of Luminescence* 82: 93–104.

34 Wegh, R.T., Donker, H., van Loef, E.V.D., and Meijerink, A. (2000). Quantum cutting through downconversion in rare-earth compounds. *Journal of Luminescence* 87–89: 1017–1019.

35 Wegh, R.T., van Loef, E.V.D., and Meijerink, A. (2000). Visible quantum cutting via downconversion in $LiGdF_4$:Er^{3+}, Tb^{3+} upon Er^{3+} $4f^{11}\rightarrow4f^{10}5d$ excitation. *Journal of Luminescence* 90: 111–122.

36 Spector, N. (1965). Spectroscopic properties of rare earths. Brian G. Wybourne. Interscience (Wiley), New York, 1965. *Science* 148: 1082.

37 Wybourne, B.G. and Meggers, W.F. (1965). Spectroscopic properties of rare earths. *Physics Today* 18: 70–72.

38 Carnall, W.T., Goodman, G.L., Rajnak, K., and Rana, R.S. (1989). A systematic analysis of the spectra of the lanthanides doped into single crystal LaF_3. *The Journal of Chemical Physics* 90: 3443–3457.

39 Chen, X.Y. and Liu, G.K. (2005). The standard and anomalous crystal–field spectra of Eu^{3+}. *Journal of Solid State Chemistry* 178: 419–428.

40 Tanner, P.A. (2013). Some misconceptions concerning the electronic spectra of tri–positive europium and cerium. *Chemical Society Reviews* 42: 5090–5101.

41 Judd, B.R. (1962). Optical absorption intensities of rare–earth ions. *Physical Review* 127: 750–761.

42 Ofelt, G.S. (1962). Intensities of crystal spectra of rare–earth ions. *The Journal of Chemical Physics* 37: 511–520.

43 Mason, J. (1990). Lanthanide probes in life, chemical and earth sciences: theory and practice. *Elsevier, Amsterdam* 28: 737–738.

44 Tu, D., Liu, Y., Zhu, H. et al. (2012). Breakdown of crystallographic site symmetry in lanthanide-doped $NaYF_4$ crystals. *Angewandte Chemie International Edition* 52: 1128–1133.

45 Aebischer, A., Hostettler, M., Hauser, J. et al. (2006). Structural and spectroscopic characterization of active sites in a family of light-emitting sodium lanthanide tetrafluorides. *Angewandte Chemie International Edition* 45: 2802–2806.

46 Liu, R., Tu, D., Liu, Y. et al. (2012). Controlled synthesis and optical spectroscopy of lanthanide–doped $KLaF_4$ nanocrystals. *Nanoscale* 4: 4485–4491.

47 Luo, W., Fu, C., Li, R. et al. (2011). Er^{3+}-Doped anatase TiO_2 nanocrystals: crystal-field levels, excited-state dynamics, upconversion, and defect luminescence. *Small* 7: 3046–3056.

48 Chen, G., Liu, H., Liang, H. et al. (2008). Upconversion emission enhancement in Yb^{3+}/Er^{3+}–codoped Y_2O_3 nanocrystals by tridoping with Li^+ ions. *The Journal of Physical Chemistry C* 112: 12030–12036.

49 Wang, H.-Q. and Nann, T. (2009). Monodisperse upconverting nanocrystals by microwave–assisted synthesis. *ACS Nano* 3: 3804–3808.

50 Cheng, Q., Sui, J., and Cai, W. (2012). Enhanced upconversion emission in $Yb^{3+}$and Er^{3+}codoped $NaGdF_4$ nanocrystals by introducing Li^+ ions. *Nanoscale* 4: 779–784.

51 Jiang, L., Xiao, S., Yang, X. et al. (2012). Enhancement of up–conversion luminescence in Zn_2SiO_4:Yb^{3+}, Er^{3+} by co–doping with Li^+ or Bi^{3+}. *Applied Physics B* 107: 477–481.

52 Ramasamy, P., Chandra, P., Rhee, S.W., and Kim, J. (2013). Enhanced upconversion luminescence in $NaGdF_4$:Yb, Er nanocrystals by Fe^{3+} doping and their application in bioimaging. *Nanoscale* 5: 8711–8717.

53 Hao, J., Zhang, Y., and Wei, X. (2011). Electric-Induced enhancement and modulation of upconversion photoluminescence in epitaxial $BaTiO_3$: Yb/Er thin films. *Angewandte Chemie International Edition* 50: 6876–6880.

54 Mai, H.-X., Zhang, Y.-W., Sun, L.-D., and Yan, C.-H. (2007). Highly efficient multicolor up-conversion emissions and their mechanisms of monodisperse $NaYF_4$: Yb, Er core and core/shell-structured nanocrystals. *The Journal of Physical Chemistry C* 111: 13721–13729.

55 Boyer, J.-C. and van Veggel, F.C.J.M. (2010). Absolute quantum yield measurements of colloidal $NaYF_4$: Er^{3+}, Yb^{3+} upconverting nanoparticles. *Nanoscale* 2: 1417–1419.

56 Wang, Y., Tu, L., Zhao, J. et al. (2009). Upconversion luminescence of β–$NaYF_4$: Yb^{3+}, Er^{3+}@β–$NaYF_4$ core/shell nanoparticles: excitation power density and surface dependence. *The Journal of Physical Chemistry C* 113: 7164–7169.

57 Chen, D., Lei, L., Yang, A. et al. (2012). Ultra–broadband near–infrared excitable upconversion core/shell nanocrystals. *Chemical Communications* 48: 5898–5900.

58 Zhang, F., Che, R., Li, X. et al. (2012). Direct imaging the upconversion nanocrystal core/shell structure at the subnanometer level: shell thickness dependence in upconverting optical properties. *Nano Letters* 12: 2852–2858.

59 Su, Q., Han, S., Xie, X. et al. (2012). The effect of surface coating on energy migration–mediated upconversion. *Journal of the American Chemical Society* 34: 20849–20857.

60 Wang, Y.-F., Sun, L.-D., Xiao, J.-W. et al. (2012). Rare–Earth nanoparticles with enhanced upconversion emission and suppressed rare–earth–ion leakage. *Chemistry – A European Journal* 18: 5558–5564.

61 Chen, G., Shen, J., Ohulchanskyy, T.Y. et al. (2012). (α–$NaYbF_4$:Tm^{3+})/CaF_2 core/shell nanoparticles with efficient near–infrared to near–infrared upconversion for high–contrast deep tissue bioimaging. *ACS Nano* 6: 8280–8287.

62 Li, Z., Guo, H., Qian, H., and Hu, Y. (2010). Facile microemulsion route to coat carbonized glucose on upconversion nanocrystals as high luminescence and biocompatible cell–imaging probes. *Nanotechnology* 21: 315105.

63 Vetrone, F., Boyer, J.-C., Capobianco, J.A. et al. (2004). Significance of Yb^{3+} concentration on the upconversion mechanisms in codoped Y_2O_3:Er^{3+}, Yb^{3+} nanocrystals. *Journal of Applied Physics* 96: 661–667.

64 Wang, F. and Liu, X. (2008). Upconversion multicolor fine–tuning: visible to near–infrared emission from lanthanide–doped $NaYF_4$ nanoparticles. *Journal of the American Chemical Society* 130: 5642–5643.

65 Yin, A., Zhang, Y., Sun, L., and Yan, C. (2010). Colloidal synthesis and blue based multicolor upconversion emissions of size and composition controlled monodisperse hexagonal $NaYF_4$: Yb,Tm nanocrystals. *Nanoscale* 2: 953–959.

66 Qiu, H., Chen, G., Fan, R. et al. (2014). Intense ultraviolet upconversion emission from water-dispersed colloidal YF_3:Yb^{3+}/Tm^{3+}rhombic nanodisks. *Nanoscale* 6: 753–757.

67 Shen, J., Chen, G., Ohulchanskyy, T.Y. et al. (2013). Tunable near infrared to ultraviolet upconversion luminescence enhancement in (α-$NaYF_4$:Yb,Tm)/CaF_2 core/shell nanoparticles for in situ real-time recorded biocompatible photoactivation. *Small* 9: 3213–3217.

68 Wang, M., Mi, C., Zhang, Y. et al. (2009). NIR–responsive silica–coated $NaYbF_4$: Er/Tm/Ho upconversion fluorescent nanoparticles with tunable emission colors and their applications in immunolabeling and fluorescent imaging of cancer cells. *The Journal of Physical Chemistry C* 113: 19021–19027.

69 Chen, G., Liu, H., Somesfalean, G. et al. (2009). Upconversion emission tuning from green to red in Yb^{3+}/Ho^{3+}–codoped $NaYF_4$ nanocrystals by tridoping with Ce^{3+} ions. *Nanotechnology* 20: 385704.

70 Wang, G., Peng, Q., and Li, Y. (2010). Luminescence tuning of upconversion nanocrystals. *Chemistry – A European Journal* 16: 4923–4931.

71 Chan, E.M., Han, G., Goldberg, J.D. et al. (2012). Combinatorial discovery of lanthanide–doped nanocrystals with spectrally pure upconverted emission. *Nano Letters* 12: 3839–3845.

72 Huang, L., Wang, L., Xue, X. et al. (2011). Enhanced red upconversion luminescence in Er–Tm codoped $NaYF_4$ phosphor. *Journal of Nanoscience and Nanotechnology* 11: 9498–9501.

73 Zeng, J.H., Xie, T., Li, Z.H., and Li., Y. (2007). Monodispersed nanocrystalline fluoroperovskite up–conversion phosphors. *Crystal Growth and Design* 7: 2774–2777.

74 Wang, J., Wang, F., Wang, C. et al. (2011). Single-Band upconversion emission in lanthanide-doped $KMnF_3$ nanocrystals. *Angewandte Chemie International Edition* 50: 10369–10372.

75 Deng, R., Qin, F., Chen, R. et al. (2015). Temporal full–colour tuning through non–steady–state upconversion. *Nature Nanotechnology* 10: 237–242.

76 Chang, Y., Chen, H., Xie, X. et al. (2023). Bright Tm^{3+}–based downshifting luminescence nanoprobe operating around 1800 nm for NIR–IIb and c bioimaging. *Nature Communications* 14: 1079.

77 Wang, W., Feng, Z., Li, B. et al. (2021). Er^{3+} self–sensitized nanoprobes with enhanced 1525 nm downshifting emission for NIR–IIb in vivo bio–imaging. *Journal of Materials Chemistry B* 9: 2899–2908.

78 Shen, J., Chen, G., Vu, A.-M. et al. (2013). Engineering the upconversion nanoparticle excitation wavelength: cascade sensitization of tri-doped upconversion colloidal nanoparticles at 800 nm. *Advanced Optical Materials* 1: 644–650.

79 Wang, Y.-F., Liu, G.-Y., Sun, L.-D. et al. (2013). Nd^{3+}–Sensitized upconversion nanophosphors: efficient in vivo bioimaging probes with minimized heating effect. *ACS Nano* 7: 7200–7206.

80 Wen, H., Zhu, H., Chen, X. et al. (2013). Upconverting near-infrared light through energy management in core–shell–shell nanoparticles. *Angewandte Chemie International Edition* 52: 13419–13423.

81 Pei, S., Ge, X., and Sun, L. (2020). Metal Ions doping for boosting luminescence of lanthanide-doped nanocrystals. *Frontiers in Chemistry* 8: 610481.

82 Cheng, T., Marin, R., Skripka, A., and Vetrone, F. (2018). Small and bright lithium-based upconverting nanoparticles. *Journal of the American Chemical Society* 140: 12890–12899.

83 Zuo, J., Li, Q., Xue, B. et al. (2017). Employing shells to eliminate concentration quenching in photonic upconversion nanostructure. *Nanoscale* 9: 7941–7946.

84 Chen, B. and Wang, F. (2018). $NaYbF_4$@CaF_2 core–satellite upconversion nanoparticles: one-pot synthesis and sensitive detection of glutathione. *Nanoscale* 10: 19898–19905.

85 Shen, B., Cheng, S., Gu, Y. et al. (2017). Revisiting the optimized doping ratio in core/shell nanostructured upconversion particles. *Nanoscale* 9: 1964–1971.

86 Li, Q., Li, X., Zhang, L. et al. (2018). An 800 nm driven $NaErF_4$@$NaLuF_4$ upconversion platform for multimodality imaging and photodynamic therapy. *Nanoscale* 10: 12356–12363.

87 Johnson, N.J., He, S., Diao, S. et al. (2017). Direct evidence for coupled surface and concentration quenching dynamics in lanthanide-doped nanocrystals. *Journal of the American Chemical Society* 139: 3275–3282.

88 Chen, Q., Xie, X., Huang, B. et al. (2017). Confining excitation energy in Er^{3+}-sensitized upconversion nanocrystals through Tm^{3+}-mediated transient energy trapping. *Angewandte Chemie International Edition* 56: 7605–7609.

89 Shang, Y., Hao, S., Lv, W. et al. (2018). Confining excitation energy of Er^{3+}–sensitized upconversion nanoparticles through introducing various energy trapping centers. *Journal of Materials Chemistry C* 6: 3869–3875.

90 Wang, J., Deng, R., MacDonald, M.A. et al. (2013). Enhancing multiphoton upconversion through energy clustering at sublattice level. *Nature Materials* 13: 157–162.

91 Chen, B., Liu, Y., Xiao, Y. et al. (2016). Amplifying excitation–power sensitivity of photon upconversion in a $NaYbF_4$: Ho nanostructure for direct visualization of electromagnetic hotspots. *The Journal of Physical Chemistry Letters* 7: 4916–4921.

92 Fu, X., Fu, S., Lu, Q. et al. (2022). Excitation energy mediated cross-relaxation for tunable upconversion luminescence from a single lanthanide ion. *Nature Communications* 13: 4741.

93 Liu, H., Huang, K., Valiev, R.R. et al. (2017). Photon upconversion kinetic nanosystems and their optical response. *Laser and Photonics Reviews* 12: 1700144.

94 Tu, L., Liu, X., Wu, F., and Zhang, H. (2015). Excitation energy migration dynamics in upconversion nanomaterials. *Chemical Society Reviews* 44: 1331–1345.

95 Wang, Z. and Meijerink, A. (2018). Concentration quenching in upconversion nanocrystals. *The Journal of Physical Chemistry C* 122: 26298–26306.

96 Benz, F. and Strunk, H.P. (2012). Rare earth luminescence: a way to overcome concentration quenching. *AIP Advances* 2: 042115.

97 Chen, S., Dierre, B., Lee, W. et al. (2010). Suppression of concentration quenching of Er–related luminescence in Er–doped GaN. *Applied Physics Letters* 96: 181901.

98 Tian, B., Fernandez-Bravo, A., Najafiaghdam, H. et al. (2018). Low irradiance multiphoton imaging with alloyed lanthanide nanocrystals. *Nature Communications* 9: 3082.

99 Gargas, D.J., Chan, E.M., Ostrowski, A.D. et al. (2014). Engineering bright sub–10–nm upconverting nanocrystals for single–molecule imaging. *Nature Nanotechnology* 9: 300–305.

100 Zhao, J., Jin, D., Schartner, E.P. et al. (2013). Single–nanocrystal sensitivity achieved by enhanced upconversion luminescence. *Nature Nanotechnology* 8: 729–734.

101 Sun, T., Li, Y., Ho, W.L. et al. (2019). Integrating temporal and spatial control of electronic transitions for bright multiphoton upconversion. *Nature Communications* 10: 1811.

102 Chen, Z.H., Wang, X., Yang, M. et al. (2023). An extended NIR-II superior imaging window from 1500 to 1900 nm for high-resolution in vivo multiplexed imaging based on lanthanide nanocrystals. *Angewandte Chemie International Edition* 62: e2023118.

103 Yang, Y., Chen, Y., Pei, P. et al. (2023). Fluorescence–amplified nanocrystals in the second near–infrared window for in vivo real–time dynamic multiplexed imaging. *Nature Nanotechnology* 18: 1195–1204.

104 Ding, M., Dong, B., Lu, Y. et al. (2020). Energy manipulation in lanthanide-doped core–shell nanoparticles for tunable dual-mode luminescence toward advanced anti-counterfeiting. *Advanced Materials* 32: 2002121.

105 Huang, R., Liu, S., Huang, J. et al. (2021). Tunable upconversion of holmium sublattice through interfacial energy transfer for anti–counterfeiting. *Nanoscale* 13: 4812–4820.

106 Chen, D., Liu, L., Huang, P. et al. (2015). Nd^{3+}–Sensitized Ho^{3+} single–band red upconversion luminescence in core–shell nanoarchitecture. *The Journal of Physical Chemistry Letters* 6: 2833–2840.

107 Zhong, Y., Ma, Z., Zhu, S. et al. (2017). Boosting the down–shifting luminescence of rare–earth nanocrystals for biological imaging beyond 1500 nm. *Nature Communications* 8: 737.

108 Li, Q., Xie, X., Wu, H. et al. (2023). Superenhancement photon upconversion nanoparticles for photoactivated nanocryometer. *Nano Letters* 23: 3444–3450.

109 Allison, S.W. and Gillies, G.T. (1997). Remote thermometry with thermographic phosphors: Instrumentation and applications. *Review of Scientific Instruments* 68: 2615–2650.

110 Yan, L., Huang, J., An, Z. et al. (2022). Activating ultrahigh thermoresponsive upconversion in an erbium sublattice for nanothermometry and information security. *Nano Letters* 22: 7042–7048.

111 Fu, H., Liu, C., Peng, P. et al. (2020). Peasecod-Like hollow upconversion nanocrystals with excellent optical thermometric performance. *Advanced Science* 7: 2000731.

112 Balabhadra, S., Debasu, M.L., Brites, C.D.S. et al. (2015). Boosting the sensitivity of Nd^{3+}–based luminescent nanothermometers. *Nanoscale* 7: 17261–17267.

113 Wawrzynczyk, D., Bednarkiewicz, A., Nyk, M. et al. (2012). Neodymium(iii) doped fluoride nanoparticles as non–contact optical temperature sensors. *Nanoscale* 4: 6959–6961.

114 Piotrowski, W., Trejgis, K., Maciejewska, K. et al. (2020). Thermochromic luminescent nanomaterials based on Mn^{4+}/Tb^{3+} codoping for temperature imaging with digital cameras. *ACS Applied Materials and Interfaces* 12: 44039–44048.

115 Yu, S., Xu, J., Shang, X. et al. (2020). A dual-excitation decoding strategy based on NIR hybrid nanocomposites for high-accuracy thermal sensing. *Advanced Science* 7: 2001589.

116 Mi, C., Zhou, J., Wang, F. et al. (2019). Ultrasensitive ratiometric nanothermometer with large dynamic range and photostability. *Chemistry of Materials* 31: 9480–9487.

117 Trejgis, K., Bednarkiewicz, A., and Marciniak, L. (2020). Engineering excited state absorption based nanothermometry for temperature sensing and imaging. *Nanoscale* 12: 4667–4675.

118 van Swieten, T.P., van Omme, T., van den Heuvel, D.J. et al. (2021). Mapping elevated temperatures with a micrometer resolution using the luminescence of chemically stable upconversion nanoparticles. *ACS Applied Nano Materials* 4: 4208–4215.

119 Vetrone, F., Naccache, R., Zamarrón, A., and de la Fuente, A.J. (2010). Temperature sensing using fluorescent nanothermometers. *ACS Nano* 4: 3254–3258.

120 Di, X., Wang, D., Zhou, J. et al. (2021). Quantitatively monitoring in situ mitochondrial thermal dynamics by upconversion nanoparticles. *Nano Letters* 21: 1651–1658.

121 Gong, Z., Wu, T., Chen, X. et al. (2021). Upconversion nanoparticle decorated spider silks as single–cell thermometers. *Nano Letters* 21: 1469–1476.

122 Kaczmarek, A.M., Suta, M., Rijckaert, H. et al. (2020). Visible and NIR upconverting Er^{3+}–Yb^{3+} luminescent nanorattles and other hybrid PMO-inorganic structures for In vivo nanothermometr. *Advanced Functional Materials* 30: 2003101.

123 Ximendes, E.C., Rocha, U., Sales, T.O. et al. (2017). In vivo subcutaneous thermal video recording by supersensitive infrared nanothermometers. *Advanced Functional Materials* 27: 1702249.

124 Zhu, X., Feng, W., Chang, J. et al. (2016). Temperature–feedback upconversion nanocomposite for accurate photothermal therapy at facile temperature. *Nature Communications* 7: 10437.

125 Zhu, X., Li, J., Qiu, X. et al. (2018). Upconversion nanocomposite for programming combination cancer therapy by precise control of microscopic temperature. *Nature Communications* 9: 2176.

126 Rodríguez-Sevilla, P., Zhang, Y., Haro-González, P. et al. (2016). Thermal scanning at the cellular level by an optically trapped upconverting fluorescent particle. *Advanced Materials* 28: 2421–2426.

127 Shen, Y., Lifante, J., Fernández, N. et al. (2020). In vivo spectral distortions of infrared luminescent nanothermometers compromise their reliability. *ACS Nano* 14: 4122–4133.

128 Maturi, F.E., Brites, C.D.S., Ximendes, E.C. et al. (2021). Going above and beyond: a tenfold gain in the performance of luminescence thermometers joining multiparametric sensing and multiple regression. *Laser and Photonics Reviews* 15: 2100301.

129 Lifante, J., Shen, Y., Zabala Gutierrez, I. et al. (2021). Reaching deeper: absolute in vivo thermal reading of liver by combining superbright Ag_2S nanothermometers and in silico simulations. *Advanced Science* 8: 2003838.

130 Brites, C.D.S., Balabhadra, S., and Carlos, L.D. (2018). Lanthanide-based thermometers: at the cutting-edge of luminescence thermometry. *Advanced Optical Materials* 7: 1801239.

131 Tan, M., Li, F., Cao, N. et al. (2020). Accurate in vivo nanothermometry through NIR-II lanthanide luminescence lifetime. *Small* 16: 2004118.

132 Maciejewska, K., Bednarkiewicz, A., and Marciniak, L. (2021). NIR luminescence lifetime nanothermometry based on phonon assisted Yb^{3+}–Nd^{3+}energy transfer. *Nanoscale Advances* 3: 4918–4925.

133 Raab, M.E., Maurizio, S.L., Capobianco, J.A., and Prasad, P.N. (2021). Lifetime of the 3H4 electronic state in Tm^{3+}–doped upconverting nanoparticles for NIR nanothermometry. *The Journal of Physical Chemistry B* 125: 13132–13136.

134 Zou, W., Visser, C., Maduro, J.A. et al. (2012). Broadband dye–sensitized upconversion of near–infrared light. *Nature Photonics* 6: 560–564.

135 Wang, D., Wang, D., Kuzmin, A. et al. (2018). ICG-Sensitized $NaYF_4$: Er nanostructure for theranostics. *Advanced Optical Materials* 6: 1701142.

136 Chen, G., Damasco, J., Qiu, H. et al. (2015). Energy–cascaded upconversion in an organic dye–sensitized core/shell fluoride nanocrystal. *Nano Letters* 15: 7400–7407.

137 Zhou, J., Deng, J., Zhu, H. et al. (2013). Up–conversion luminescence in LaF_3:Ho^{3+}via two–wavelength excitation for use in solar cells. *Journal of Materials Chemistry C* 1: 7.

138 Chen, P., Yu, S., Xu, B. et al. (2014). Enhanced upconversion luminescence in $NaYF_4$: Er nanoparticles with multi–wavelength excitation. *Materials Letters* 128: 299–302.

139 Liu, Y., Lu, Y., Yang, X. et al. (2017). Amplified stimulated emission in upconversion nanoparticles for super–resolution nanoscopy. *Nature* 543: 229–233.

140 Zhan, Q., Liu, H., Wang, B. et al. (2017). Achieving high–efficiency emission depletion nanoscopy by employing cross relaxation in upconversion nanoparticles. *Nature Communications* 8: 1058.

141 Köche, N., Plentz, F., Rodrigues, W.N. et al. (2001). Magnetoluminescence of InAs self–assembled dots embedded in a two–dimensional electron gas. *Physica B: Condensed Matter* 298: 295–301.

142 Dewitz, C.V., Hatami, F., Millot, M. et al. (2009). Evidence of type–I direct recombination in InP/GaP quantum dots via magnetoluminescence. *Applied Physics Letters* 95: 151105.

143 Liu, Y., Wang, D., Shi, J. et al. (2013). Magnetic tuning of upconversion luminescence in lanthanide-doped bifunctional nanocrystals. *Angewandte Chemie International Edition* 52: 4366–4369.

144 Tikhomirov, V.K., Chibotaru, L.F., Saurel, D. et al. (2009). Er^{3+}-doped nanoparticles for optical detection of magnetic Field. *Nano Letters* 9: 721–724.

145 Yin, Z., Zhu, Y., Xu, W. et al. (2013). Remarkable enhancement of upconversion fluorescence and confocal imaging of PMMA Opal/$NaYF_4$:Yb^{3+}, Tm^{3+}/Er^{3+} nanocrystals. *Chemical Communications* 49: 3781–3783.

146 Garrido-Hernández, A., García-Murillo, A., de Carrillo-Romo, F. et al. (2014). Structural studies of $BaTiO_3$:Er^{3+} and $BaTiO_3$:Yb^{3+} powders synthesized by hydrothermal method. *Journal of Rare Earths* 32: 1016–1021.

147 Xia, H., Li, D., Shang, J. et al. (2023). Electrically sensitive plasmonic photonic crystals for dynamic upconversion manipulation. *Advanced Functional Materials* 33: 2304045.

148 Wang, Y., Nan, F., Cheng, Z. et al. (2015). Strong tunability of cooperative energy transfer in Mn^{2+}-doped (Yb^{3+}, Er^{3+})/$NaYF_4$ nanocrystals by coupling with silver nanorod array. *Nano Research* 8: 2970–2977.

149 Yin, D., Wang, C., Ouyang, J. et al. (2014). Synthesis of a novel core–shell nanocomposite Ag@SiO_2@Lu_2O_3: Gd/Yb/Er for large enhancing upconversion luminescence and bioimaging. *ACS Applied Materials and Interfaces* 6: 18480–18488.

150 Jin, L.M., Chen, X., Siu, C.K. et al. (2017). Enhancing multiphoton upconversion from $NaYF_4$: Yb/Tm@$NaYF_4$ core–shell nanoparticles via the use of laser cavity. *ACS Nano* 11: 843–849.

151 Xie, Y., Carbone, L., Nobile, C. et al. (2013). Metallic-like stoichiometric copper sulfide nanocrystals: phase- and shape-selective synthesis, near-infrared surface plasmon resonance properties, and their modeling. *ACS Nano* 7: 7352–7369.

152 Zhou, D., Liu, D., Xu, W. et al. (2016). Observation of considerable upconversion enhancement induced by $Cu_{2-x}S$ plasmon nanoparticles. *ACS Nano* 10: 5169–5179.

3

Upconversion and NIR-II Luminescence Modulation of Rare-Earth Composites Using Material Informatics

Wenjing Li and Ruichan Lv

Xidian University, Laboratory of Electromechanical Integrated Manufacturing of High-performance Electronic Equipments, Interdisciplinary Research Center of Smart Sensor, School of Life Science and Technology, Xihoo, Shaanxi, 710071, P.R. China

3.1 Introduction

Upconversion luminescence (UCL) is a luminescence phenomenon caused by the anti-Stokes effect. The luminous substance in UCL absorbs low-energy long-wave radiation while emitting high-energy short-wave radiation. As a result, at UCL, the excitation wavelength is longer than the emission wavelength. UCL's mechanism is a two-photon or multiphoton process. The luminous center absorbs two or more lower-energy photons, transitions to the ground state, and releases a high-energy photon after absorbing two or more lower-energy photons. The metastable luminous center must have a long energy level lifetime in order to achieve an efficient two-photon or multiphoton effect. Because of the luminescence mechanism of UCL and the properties of rare-earth ions, most UCL materials are compounds doped with rare-earth ions.

Rare-earth elements feature significant Stokes shifts, sharp emission peaks, extended fluorescence lifetimes, strong chemical or photochemical stability, minimal toxicity, and weak light fading due to their unique structure. Rare-earth elements outperform other elements in terms of light, electricity, magnetic properties, and thermal conductivity. At the same time, because the transition between rare-earth ion levels is a forbidden f-f transition, it has a long lifetime and can aid in achieving a practical two-photon or multiphoton effect. The capacity of UCL to emit UV photons into the visible spectrum via absorption of numerous NIR photons makes it appealing. This feature has resulted in a wide range of applications. Its anti-Stokes feature has been exploited to create a number of applications.

To achieve synergistic effects, upconversion nanoparticles (UCNPs) are mixed with other functional materials to form nanocomposites. Based on the precise design and components, the resulting nanocomposites have considerable promise in a variety of applications. This chapter discusses the UCL luminescence mechanism, various synthesis and optimization methodologies for UCNPs, and emerging applications in bioimaging and cancer therapy.

Photofunctional Nanomaterials for Biomedical Applications, First Edition.
Edited by Chunxia Li and Jun Lin.

3.2 Typical Processes of Upconversion Luminescence

The energy level transition of rare-earth ions is the primary focus of the UCL process research. Generally, according to the luminescence process and mechanism, it can be divided into excited state absorption (ESA), photon avalanche (PA), energy transfer (ET), cross-relaxation (CR), cooperative upconversion (CU), and second harmonic generation (SHG) [1–4]. Figure 3.1 illustrates the several pathways that result in UCL [5].

3.2.1 Excited State Absorption

ESA is the most basic process of UCL. It is based on the principle that the same activator ion reaches the higher energy excited state energy level from the ground state energy level through continuous two-photon or multiphoton absorption. Then radiation transition occurs to return to the ground state to produce UCL. In ESA, when an excited electron in the metastable state E1 absorbs a photon, it transitions from E1 to E2, which is the excited electron's new state. In general, the electron's extended lifespan at the E1 energy level and near resonance between the G → E1 and E1 → E2 transitions are necessary for the ESA of rare-earth ions. Rare-earth ions can continually absorb photons because of their extended life span [6]. Prior to the electron decaying into G, ESA may happen more than once while releasing photons that are more energetic than the incident photon.

As shown in Figure 3.2, a family of lanthanide-doped imaging probes was created by Levy et al. that resonant with the ESA transition and stimulate excitation at

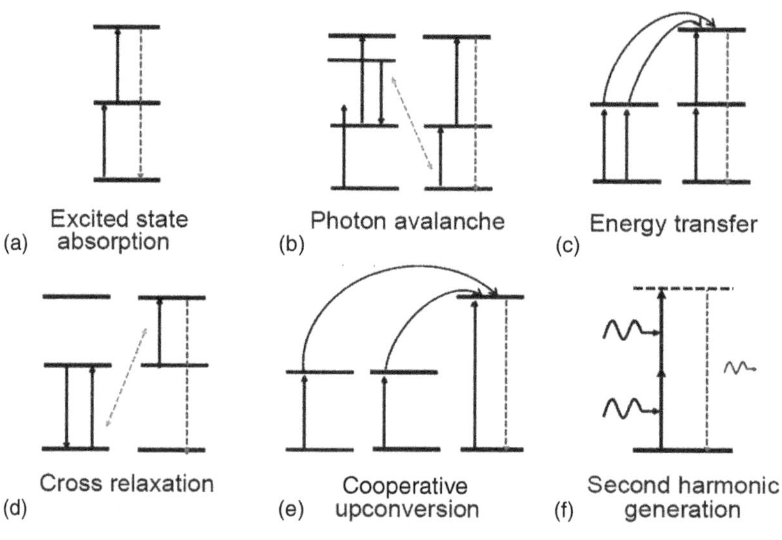

Figure 3.1 UCL mechanisms of (a) excited state absorption (ESA), (b) photon avalanche (PA), (c) energy transfer (ET), (d) cross-relaxation (CR), (e) cooperative upconversion (CU), and (f) second harmonic generation (SHG) shown by energy level. Source: Reproduced with permission from Lv et al. [5]/Elsevier.

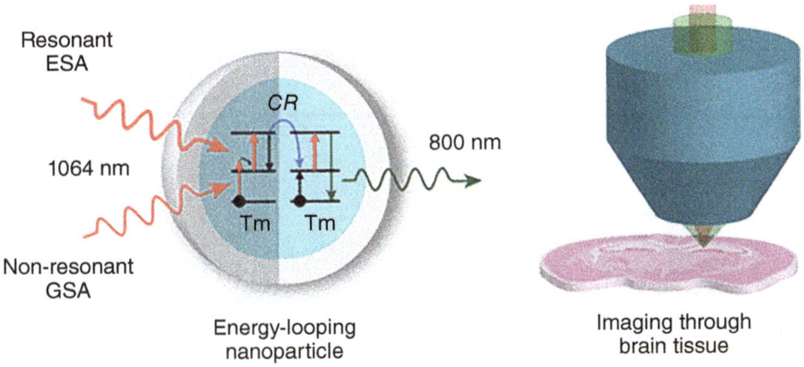

Figure 3.2 Lanthanide-doped imaging probe based on ESA. Source: Reproduced with permission from Levy et al. [7]/American Chemical Society.

NIR-II wavelengths via the energy loop mechanism. Tm^{3+}-doped RNEPs were stimulated at 1064 nm and emitted emissions at 800 nm, indicating the UCL process of Tm^{3+}-doped RNEPs, according to their research [7].

3.2.2 Photon Avalanche

PA combines the ET and ESA processes. The pump light energy corresponds to the E2 and E3 energies of ions. An ion on the E2 is excited to E3 after absorbing energy. The E3 level and E1 level have a CR process. The ions are accumulated at the E2 level, resulting in an avalanche increase of particles on the E2 level, called the PA process. The PA path depends on the accumulation of particle numbers in the excited state, so the obvious PA process will occur only when the rare-earth ion doping concentration is high enough. The basis of this mechanism is that particles at one energy level produce pumping effects with quantum efficiency greater than one at another energy level through CR. Increasing excitation light intensity will shorten the time to establish equilibrium, and the intensity of equilibrium absorption will increase, which may form a very effective upconversion.

3.2.3 Energy Transfer

The sensitizer in ET absorbs the excitation energy and transfers it to the activator. Thus, one electron in the activator is raised in energy. For ET to occur, the sensitizer and activator must be sufficiently close to one another, and the changes involved typically need to resonate for this process to occur. For instance, at 980 and 800 nm, Yb^{3+} and Nd^{3+} have enormous absorption cross sections. Therefore, they are often utilized as sensitizers. Tm^{3+} and Er^{3+} are frequent UCL activators [8–11]. Specifically, visible UCL is enabled when Yb^{3+} or Nd^{3+} is doped with Er^{3+} [12]. Emory achieved effective UCL by using concentrated and recovered energy from NaYF4:Er^{3+}, Tm^{3+} [13]. Er^{3+} can also transfer energy to the activator, acting as a sensitizer to produce UCL.

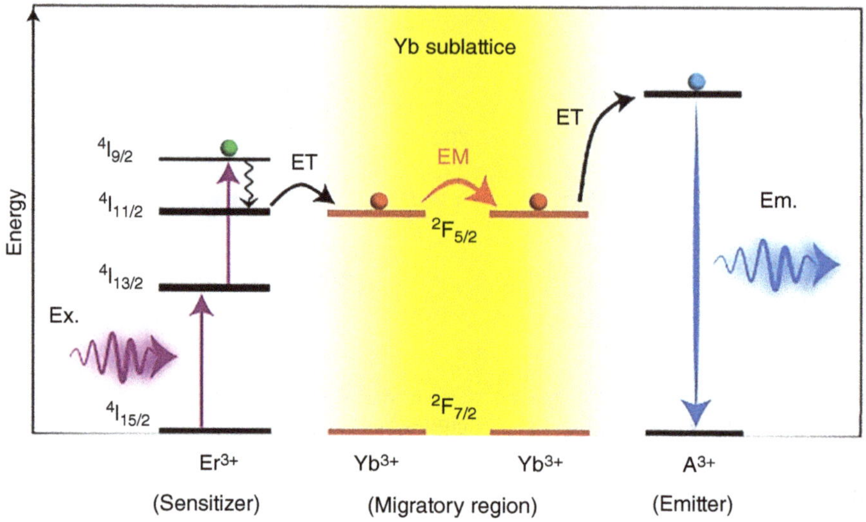

Figure 3.3 Multilayer core–shell nanostructure based on ET. Source: Reproduced with permission from Zhou et al. [14]/Springer Nature.

As shown in Figure 3.3, Zhou et al. utilized Er^{3+} as a sensitizer to excite the activator by gathering 1530 nm radiant energy. Afterward, UCL was enhanced because of the absorption transformation at ${}^4I_{13/2} \leftarrow {}^4I_{15/2}$ [14].

3.2.4 Cross-Relaxation

CR may take place within ions of same type or distinct types with similar properties. When enough ions are excited to the intermediate state, two physically close excited ions may be coupled by a nonradiative transition, with one returning to the ground or lower intermediate energy state and the other jumping to the upper excited energy level and then producing a radiative transition. CR can provide a luminescence mechanism for upconversion, but sometimes this phenomenon leads to fluorescence quenching. It should be noted that CR is a critical process for quenching high-energy emissions. For example, when the doping concentration of Tb^{3+} and Tm^{3+} is too high, it is possible to cause the quenching of the high-level emission due to the CR process and enhance the descending NIR-II emission. The CR process depends only on the interaction between the two centers, so CR only plays a role when the concentration of the luminous center exceeds a specific value [15].

3.2.5 Cooperative Upconversion

CU occurs when two or more ions transfer energy to another ion, thereby causing the ion to transition from the ground state to the excited state [16]. CU occurs between ions of the same type simultaneously located in the excited state and is an interaction between multiple ions. The emitter ion in CU does not interact with the excited state of the sensitizer ion through an intermediate metastable energy level. The most common types of doping in CU are Yb^{3+}/Tb^{3+} and Yb^{3+}/Eu^{3+} ions [17–19].

3.2.6 Second Harmonic Generation

Wong and Phillips reported the synthesis of Tb3+ compounds with adjustable second/third harmonic and explained the mechanism of indirect multiphoton upconversion-sensitized luminescence [20]. When the organic ligand sensitizes the Tb3+ center and it possesses the correct symmetry center, the phenomenon happens. Organic lanthanide complexes as probe molecules may be further developed thanks to this novel simultaneous multiphoton method. In addition, they offer a different potential mechanism that might be investigated further in order to explore the possibility of probe and imaging connectors, especially in biological systems where the metal core is attached to both the antenna and the target [21]. SHG circumvents bleaching impact of fluorescence thanks to its exceptionally high resolution and great imaging depth of nonlinear optics.

3.3 Synthesis Methods of Upconversion Nanoparticles

3.3.1 Thermal Decomposition Methods

The synthesis of lanthanide UCNPs is achieved by pyrolysis of metal precursors in organic solvents in an anaerobic environment by thermal decomposition. The ratio of organic precursor to solvent is critical to the success of the reaction. In this method, organometallic compounds are usually used as precursors, which are dissolved in high-boiling organic solvents with the help of lipophilic surfactants, and then high temperatures are used to decompose and nucleate the precursors rapidly, and the nanoparticles are prepared by growth. Among them, rare-earth organic acid salts (such as trifluoroacetate, oleate, and acetate) are usually used as organometallic precursors, and octadecene is used as a high-boiling organic solvent. Oleic acid (OA) and oleylamine (ODE), often used as surfactants, can also act as coordinating solvents, and their adsorption on nanoparticles can help to control the size and shape of the nanoparticles.

The thermal decomposition method requires an experimental environment free of water and oxygen. This requirement is usually achieved by inert gas protection or vacuum. In addition, the temperature, heating rate, and concentration of precursors affect the size and homogeneity of the nanoparticles, which can be adjusted to control the crystal growth process during the experiment. UCNPs prepared by the thermal decomposition method have been widely used in research due to their good monodispersity, high particle crystallinity, regular morphology, and small nanocrystalline particle size. Due to the surface being covered with a layer of surfactant molecules, the products obtained by this method are straightforward to disperse in organic phase solvents and are also less prone to agglomeration.

LaF_3 crystals were originally produced by Yan and colleagues utilizing a single-source precursor and thermal decomposition [22]. Subsequently, they pioneered the usage of multisource precursors to create superior nanocrystals via thermal decomposition, which significantly aided in the advancement of UCNPs [23]. Through an examination of the impact of reaction duration and the proportion of sodium to lanthanide precursors, Murray and collaborators obtained a series of beautiful rare-earth-based nanocrystals [24]. Lv and coworkers applied $LnCl_3$ ($Ln = Y^{3+}$ and Er^{3+}) as precursors in an OA and ODE mixture at a high temperature. After full response, adding NH_4F and NaOH produces nanoparticles of $NaYF_4:Er^{3+}$ as core [25].

3.3.2 Hydrothermal/Solvothermal Method

The hydrothermal/solvothermal method accomplishes the synthesis of UCNPs in a closed system with an aqueous or organic solution as the medium. When the mixture in the closed vessel is heated, a high-pressure environment is formed, which will help to increase the solubility and reactivity of the solid reactants in solution, accelerate the rate of migration and diffusion between ions, and promote the chemical reactions of the substances in the system to produce new phases or new substances. The method provides a high-pressure environment through heating, enabling reactions that cannot be carried out at atmospheric pressure.

Based on the reaction temperature, the hydrothermal/solvent thermal method can be classified into low-temperature hydrothermal/solvent thermal method (operating temperature below 100 °C), medium-temperature hydrothermal/solvent thermal method (operating temperature between 100 and 300 °C), and high-temperature hydrothermal/solvent thermal method (operating temperature can be as high as 1000 °C). In most cases, the preparation of UCNPs is done at low and medium temperatures, and the resulting products possess the benefits of excellent dispersion, high purity, high crystallinity, regular morphology, and uniform particle size.

Usually, the preparation of UCNPs using the hydrothermal/solvothermal approach is completed in a particular reaction apparatus. This apparatus is hermetically sealed to provide a high-pressure reaction environment for the reaction system. In addition, due to the advantages of material and design, the device is resistant to acids and alkalis, has good corrosion resistance, and is airtight, which is safer and more environmentally friendly as it favors the improvement of the purity of the product and at the same time avoids the spillage of hazardous substances. Thanks to the unique reaction environment, there are fewer constraints when using the hydrothermal/solvothermal method. Except for some substances that can decompose or release a large amount of gas and lead to the explosion of the sealed reactor, there are no other constraints on the available reactants and solvents, and the range of choices is extensive. Hydrothermal/solvothermal method is often used in the synthesis of UCNPs [26–29]. In preparing UCNPs, selecting suitable reactant precursors, reaction solvents, and organic additives is necessary. Water, ethanol, glycol, and other basic hydrophilic inorganic/organic solvents are the most often used solvents [30]. Li and associates presented OA to the field of solvothermal synthesis. It expanded the spectrum of materials and structures that can be created using this process because it is an oleophilic organic solvent [26].

3.3.3 Co-precipitation Method

In order to prepare ultrafine powders of composite oxides containing two or more metal elements, co-precipitation is necessary. When a homogeneous phase of two or more metal cations is present in solution, different metal ions are precipitated as precipitates simultaneously after adding precipitating agents, and homogeneous precipitates of various compositions can be obtained. The co-precipitation method has unique advantages in preparing rare-earth ion-doped nanoparticles. Co-precipitation is one of the older and more traditional methods for preparing lanthanide ion-doped nano/microcrystals, and it is currently the most widely used method for synthesizing ultra-micron powdered materials in the laboratory and industry. Due to the mild reaction conditions, the crystallinity of the products obtained by this method is relatively low, resulting in low fluorescence intensity. Hence, the precipitation products usually require calcination and other treatments. The co-precipitation reaction is simple and can be controlled by controlling the concentration of the solution particles, the reaction temperature, and so on, to obtain chemically homogeneous products with good morphology and appropriate particle size. pH is critical to the precipitation process, as high pH may lead to the re-dissolution of some precipitates, while low pH may lead to incomplete precipitation. Therefore, special attention should be paid to the choice of precipitant and pH control for the precipitation of substances with significant differences in solubility. In addition, for the preparation of complex multi-component systems by the multi-ion co-precipitation process, it is required that each component has the same or similar hydrolysis or precipitation conditions so that the different metal ions can generate precipitates at the same time as far as possible, in order to ensure the uniformity of the chemical composition of the product. Co-precipitation method is widely researched in biomedical [31–34]. Jin and associates developed a near-infrared fluorescence "turn-on" system that utilized gold and rare-earth nanoparticles [34]. A straightforward, uniform, and extremely selective HIV-1-based DNA detection technology was developed as a result of the research.

3.3.4 Sol–Gel Method

The sol–gel method uses compounds containing highly chemically active components as precursors. It uniformly mixes these raw materials in a liquid-phase environment. It undergoes hydrolysis, polymerization, condensation, gelation, gelling, drying, and pyrolysis, forming a stable and transparent sol–gel system in solution. Afterward, the gel particles in the sol slowly cross-link and aggregate into a gel with a three-dimensional spatial network structure; the gel is dried, sintered, and cured to prepare molecular and even nano-substructured materials.

The sol–gel method requires that the raw material itself should have enough reactivity to participate in the gel formation process and must be able to dissolve in the reaction medium. The most commonly used feedstocks as precursors are metal–alcohol salts; hydroxides, complexes, and certain salts can also be used. Depending on the raw materials used, the sol–gel method can be divided into two categories: the alcohol salt sol–gel method and the aqueous solution sol–gel method. Among them, the alcohol salt sol–gel method uses metal-alcohol salts as the primary raw material, and the reaction process can be controlled by adjusting pH during the reaction. The method mainly consists of four steps. (i) Sol formation: metal-alcohol salts and other raw materials through hydrolysis and polycondensation reaction to form sol. (ii) Gel formation: sol through aging, coating, fiber forming, shaping, and other processes to form a wet gel. (iii) Gel drying: wet gel is further dried to get the dry gel. (iv) Calcination: dry gel after heat treatment to get the final product. Although the alcohol salt sol–gel method is technically mature, the raw materials are expensive and highly toxic. At the same time, the process has low production efficiency, so it takes work to mass production. The aqueous solution sol–gel method, to a certain extent, can compensate for these shortcomings.

Aqueous solution sol–gel method to inorganic salt solution as raw materials, the main steps and alcohol salt sol–gel method are similar, including sol formation, gel formation, gel drying, and calcination four steps. (i) Sol formation: through the polycondensation (condensation) or dispersion to form a sol. (ii) Gel formation: by raising the pH of the solution or heating dehydration to achieve the conversion of sol to gel. (iii) Gel drying: drying to remove the solvent in the wet gel. (iv) Calcination: Eliminate the air holes in the dry gel to obtain solid luminescent materials.

Nanoparticles prepared by the sol–gel method usually have irregular spherical shapes, a broad size distribution range, and a certain degree of agglomeration. Moreover, the raw materials used in this method are expensive and have specific toxic side effects, and the process is cumbersome and difficult to control. However, it can be combined with soft lithography and electrostatic spinning, so it is suitable for preparing unique rare-earth fluorescent materials, such as thin films, patterns, and fiber phosphors. In addition, due to the inclusiveness of the network structure of inorganic sol–gel, chemical stability, optical transparency, elasticity, as well as rigidity and plasticity, it is also possible to use gel impregnation, in-situ synthesis. and pre-doping method, such as various rare-earth complexes, rare-earth ions doped into it, constituting optically active hybrid polymer materials. Lin and coworkers developed a novel aqueous sol–gel method called Pechini-type sol–gel process and prepared a large variety of micro/nano-luminescent powders and various core–shell structured phosphors and dyes by this method [35–39]. Compared with the traditional method, this method has the advantages of lower toxicity, cheaper cost, simple reaction, and easy control.

3.3.5 Other Methods

In addition to the above methods, there are many new synthesis methods for UCNPs, such as the microemulsion method, microwave synthesis method, and ionic liquid method. These methods have their characteristics and advantages and have been gradually improved and developed in scientific research.

(1) *Microemulsion method*: A microemulsion system is an isotropic, transparent/semitransparent, thermodynamically stable, and uniformly dispersed system

formed by two immiscible liquids, generally composed of the oil phase, water phase, surfactant, and co-surfactant, usually, water droplets in oil (water in oil, W/O) or oil droplets in water (oil in water, O/W) to form a monodisperse system. The microstructure is a monodisperse system formed by dropping water in oil (W/O) or oil in water (O/W), and its particle size is 5–70 nm. It can be divided into normal phase (O/W), reversed phase (W/O), and bicontinuous phase microemulsion. These are the ordered combinations of surfactant molecules at the interface of oil/water.

The microemulsion method obtains UCNPs by completing a chemical reaction in a microemulsion droplet. In this process, the UCNP's nucleation and growth can be affected by the volume of water in the droplet and the concentration of the reactants, and nanoparticles with different particle sizes can be obtained by controlling these factors. In addition, the particle surface can be modified, and the particle size can be controlled by selecting different surfactant molecules. The product's surface obtained by the microemulsion method will be coated with a layer of surfactant molecules, resulting in a high inter-particle dispersion and less likely to agglomerate. It is worth noting that if the surfactant is not removed, it can be uniformly dispersed in various organic solvents to form a dispersed system to facilitate the study of its optical properties and the effect of surfactants and other media.

(2) *Microwave method*: This method will have to directly absorb microwaves to produce high-temperature compounds and monomers as raw materials in the microwave oven and quickly synthesize the substrate or light-emitting materials. If the raw material reactants themselves do not have the nature of absorbing microwaves to generate high temperatures, amorphous carbon powder and graphite powder can be pressed into pellets and placed among the reactants. The microwave effect generates high temperatures, indirectly heating the reactants, and prompts the reaction to occur. The preparation of UCNPs by microwave heating has the advantages of simple equipment, rapid reaction, easy operation, mild conditions, low production costs, and relatively small product size and not easy to agglomerate. Because the heat source is distributed inside the mixture, the method can be uniformly heated, the reaction warming speed is fast, and the thermal efficiency is high, which enables the rapid preparation of many luminescent materials and shortens the reaction time from dozens or tens of hours to dozens or minutes to complete. At the same time, many performance indicators of luminescent materials can be improved.

(3) *Ionic liquid method*: An ionic liquid is a liquid composed entirely of anions and cations. Because many inorganic salts and organic substances have good solubility and stabilization ability, ionic liquids in inorganic synthesis can be used as a reaction solvent, capping agent, or surfactant. Compared with conventional solvents, ionic liquids have a wider electrochemical window, extended hydrogen bonding, templating agent, and capping agent functions.

They are primarily used as green and novel solvents, often combined with other methods, such as solvothermal and microwave heating, to obtain new products, properties, and morphologies [40].

3.4 Material Informatics in UCL

Researchers use experimental experience in traditional methods to randomly find the luminescent materials with the highest quantum yield [41, 42]. This approach takes so much time with low efficiency [43–48]. The doping optimization of rare-earth luminescent powders is a combinatorial optimization problem. Thanks to the development of information technology, it is possible for materials informatics to find the optimized composition and structure [49–51].

Heuristic algorithms (HAs) is an intuitive or empirically constructed algorithm that gives a feasible solution to the problem at an acceptable cost (computational time and space) [52–55]. Compared with traditional techniques, HAs compromise optimality, accuracy, precision, or completeness for speed in order to achieve rapid optimization, which is targeted to address fitness problems. Many researchers studied the doping of luminous nanoparticles by using the HAs. Sohn Kee-Sun et al. proposed a HAs combinatorial chemical strategy based on traditional calcined phosphors to create promising red phosphors for three-band white LED applications [56]. Lv et al. used HAs to adjust luminescence intensity according to the amount of rare-earth element [57]. They also predicted and searched for optimum materials using classification methods (support vector machine, regression algorithm, etc.) [57, 58]. There are three kinds of well-studied heuristics: genetic algorithm (GA), particle swarm optimization (PSO), and simulated annealing (SA) [59–61]. Interestingly, the several algorithms are all inspired by nature and life.

3.4.1 Genetic Algorithm

Darwin's theory of evolution inspired GA. In 1975, Holland et al. abstracted Darwinian evolution into a mathematical model and proposed a GA. The GA is a global random search optimization algorithm that uses the "survival of the fittest" mechanism to derive the relationship between the material properties and the molar percentage of each ingredient. The proposed algorithm makes essential contributions to guiding the optimization problem of chemical synthesis [57, 62]. In order to discover new magnetic compounds with improved functionality, Chelikowsky illustrated the efficacy of an adaptive GA based on datasets comprising thermodynamic parameters like formation energy, magnetic property, and crystallography [63]. As shown in Figure 3.4, GA was used to optimize the luminescence intensity of the UCNPs [57]. In another research, the luminescence intensity of UCL and NIR-II in individual fluorogens was optimized using classical heuristic genetic algorithms, and the final results show that this GA can be used to simultaneously optimize bimodality imaging [46]. Sun et al. used regression equations to explore the relationship between luminescence and doping concentration. The GA obtained the doping concentration corresponding to the maximum luminescence intensity [63]. Zhao and coworkers used the GA to modulate the UCL and finally obtained the effective red light emission [64, 65]. Lv and coworkers optimized the UCL and NIR-II luminescence intensity in the single phosphor using the traditional heuristic GA [66]. The ultimate

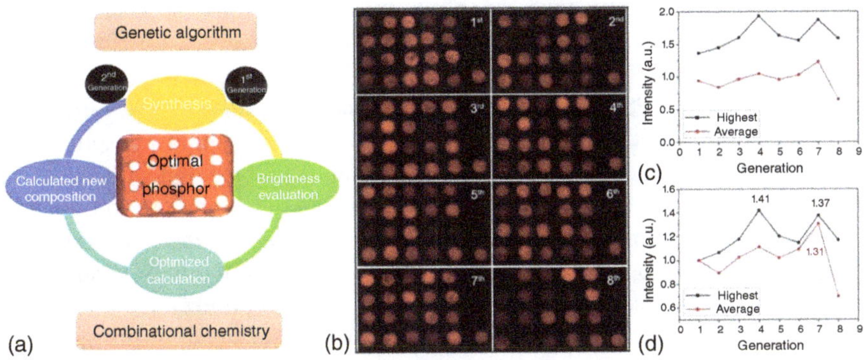

Figure 3.4 The highest intensity, spectra, and optimization results of each generation of the optimal luminescent powder by GA. (a) Schematic diagram of optimal phosphor prediction using experiments combined with GA algorithm. (b) Photographs of phosphors in eight generations under 365 nm excitation. The average and highest intensity (c) and normalized intensity (d) of each generation. Source: Reproduced with permission from Lv et al. [57]/American Chemical Society.

findings indicate that the dual-modal imaging can be simultaneously improved by the use of GA.

3.4.2 Particle Swarm Optimization

The flock activity inspires PSO. In flocks, birds choose flight direction only by a certain number of neighbors. Nevertheless, the whole flock forms an orderly group around a center. It follows that the complex global behavior of an individual is only composed of simple rules. The particle swarm algorithm is proposed based on this idea. PSO is a swarm intelligence technique that takes into account a D-dimensional continuous solution space with a swarm of p particles. The algorithm is easy to understand and program and has a global search capability. It has now been applied in the field of combinatorial chemistry.

3.4.3 Simulated Annealing

Salt-making inspires SA. When the salt solution is heated, the water evaporates, and the solution concentration increases to saturation. After stopping the heating, the table salt crystallized to form crystals. In this process, the system temperature decreases with time, and the energy decreases. The process is called annealing [67]. The annealing process can find the lowest internal energy in the system. A mathematical model of the SA is proposed based on this. However, in the chemical optimization synthesis problem, mature applications have yet to use the annealing algorithm.

Lv and coworkers used four HAs to optimize the luminescence intensity of the phosphors, including SA, improved annealing with a harmony search algorithm (HSA), PSO, and GA [58]. According to the results, HSA showed better than SA.

Traditional SA has a slow convergence rate with poor optimization results and high cost. The effect is not ideal even if iterations were repeated many times. And, HSA has a much better optimization result by using the code information of the powders. When using PSO to optimize the phosphors, the luminescence intensity gradually enhanced with the increasing generation. The optimal phosphor was shown in the final fifth generation (the fifth generation). Compared with PSO, GA could find the best local phosphors in shorter time. When using GA to optimize the phosphors, the optimal phosphor was shown in the third generation. However, the highest brightness optimized by GA was lower than that of the PSO, indicating that the optimization speed of GA is faster while the optimization process of PSO is more stable. Furthermore, the study explored the relationship between the luminescence intensity and activator/sensitizer concentration. As shown in Figure 3.5a,b, the intensity of 340 luminescent powders doping with Ce^{3+} and Tb^{3+} was explored. Then the powders were divided into eight groups by concentrations (Figure 3.5c,d). It is clearly obvious that when the Ce^{3+} concentration is in the range of 0.02–0.03 and the Tb^{3+} at 0.21–0.26, the powders have the optimal brightness. When using PSO, the variance of all concentrations decreases gradually, indicating its highest local searching ability (Figure 3.5e). The variance of the four methods is decreasing, indicating that the optimization direction is correct (Figure 3.5f).

3.4.4 Other Methods

Support vector machine (SVM) algorithm can be used to predict the effect of doping elements and analyze the luminescence stability of UCNPs. Yang et al. used GA and SVM to find the optimal phosphor with enhanced red UCL emission using multielement $K^+/Li^+/Mn^{2+}$ metal modulation [68]. After optimizing by GA, the best phosphors' fluorescence intensity is 4.91 fold than the first generation. Furthermore, the relative quantum yield and the tissue penetration depth of the optimized phosphors are also better than before.

For UCNPs, the conditions of preparation and synthesis will affect the crystallinity, phase, morphology, and UCL properties. Accordingly, a large scale of experiments is needed to obtain ideal products. Aimed at the problem, Qin et al. used the automatic nanomaterial synthesizer (ANS) to explore the growth regularity of complex upconversion nanocrystals (UCNCs) [69]. With the help of ANS, the controllability, repeatability, and success rate of the experiment were significantly enhanced. According to the results, the nanocrystals' size can be changed by controlling the concentration of Y^{3+}, which will help researchers obtain materials for different uses. As shown in Figure 3.6, when the concentration of Y^{3+} increased from 0 to 0.50, the UCNCs showed uniform spherical morphology and the sizes decreased from 42.53 to 14.94 nm. The shape of UCNCs turned to irregular spheres or ellipsoids when the concentration of Yb^{3+} increased from 0.60 to 0.795.

Regression algorithm is another favorable method to clarify the stability of optimization. Regression analyses include principal component analysis, logistics analysis, ridge analysis, linear regression (least squares method), nonlinear regression (nonlinear least squares method), and partial least squares analysis. As the most commonly used statistical method, regression analysis is usually used to

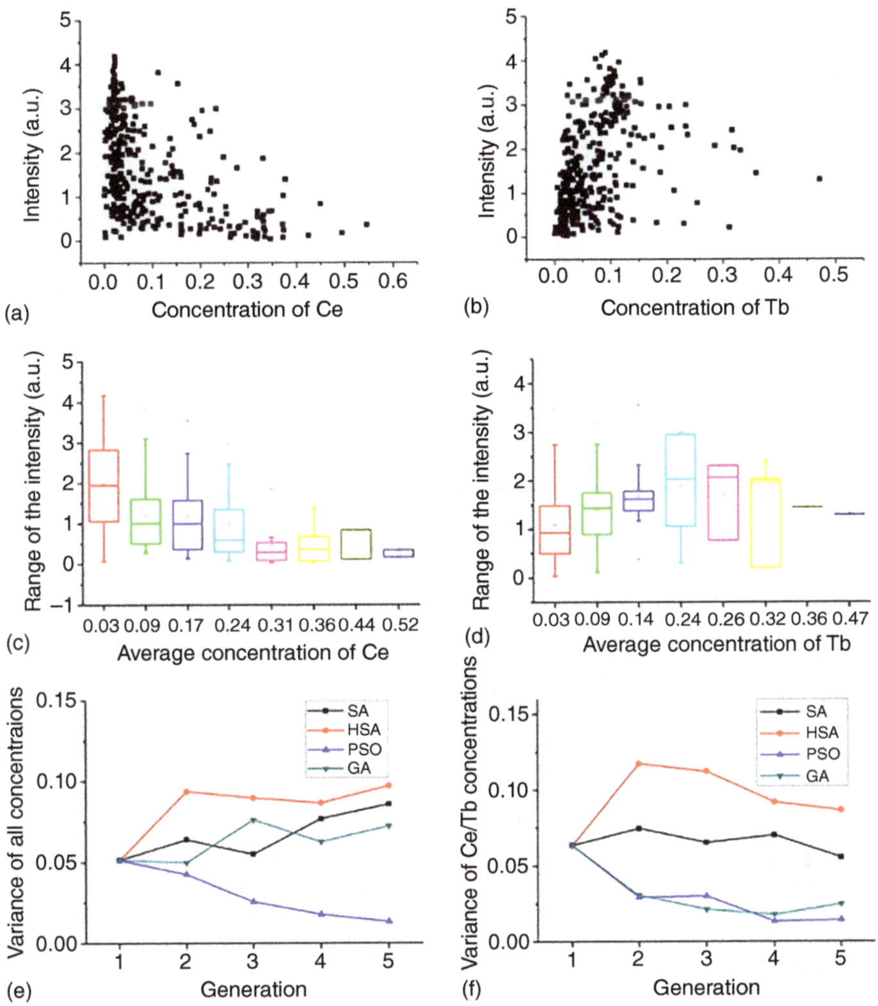

Figure 3.5 Intensities of 340 phosphors doping with (a) Ce^{3+} and (b) Tb^{3+}. Divided the powders into eight groups by (c) Ce^{3+} and (d) Tb^{3+} concentrations. Variance of (e) all and (f) Ce^{3+} /Tb^{3+} concentrations in different generations. Source: Reproduced with permission from Lv et al. [58]/American Chemical Society.

establish a quantifiable connection between the variables and the anticipated feature. In the optimization problem of the luminous powder, a regression algorithm can be used to explore the relationship between phosphors' brightness and doping concentrations. When the concentration increases, the intensity first enhances and then decreases. The scatter point distribution law indicates that the variation law of the one-dimensional quadratic equation is better suited to the functions of UCL emission intensity to the codopant concentration. Using the statistical product and service solutions (SPSS) software, we fit the scatter points in each research using all of the data in multiple generations with quadratic curves, and the table of variance analysis was generated as a result. And for curve fitting, R square and F values are

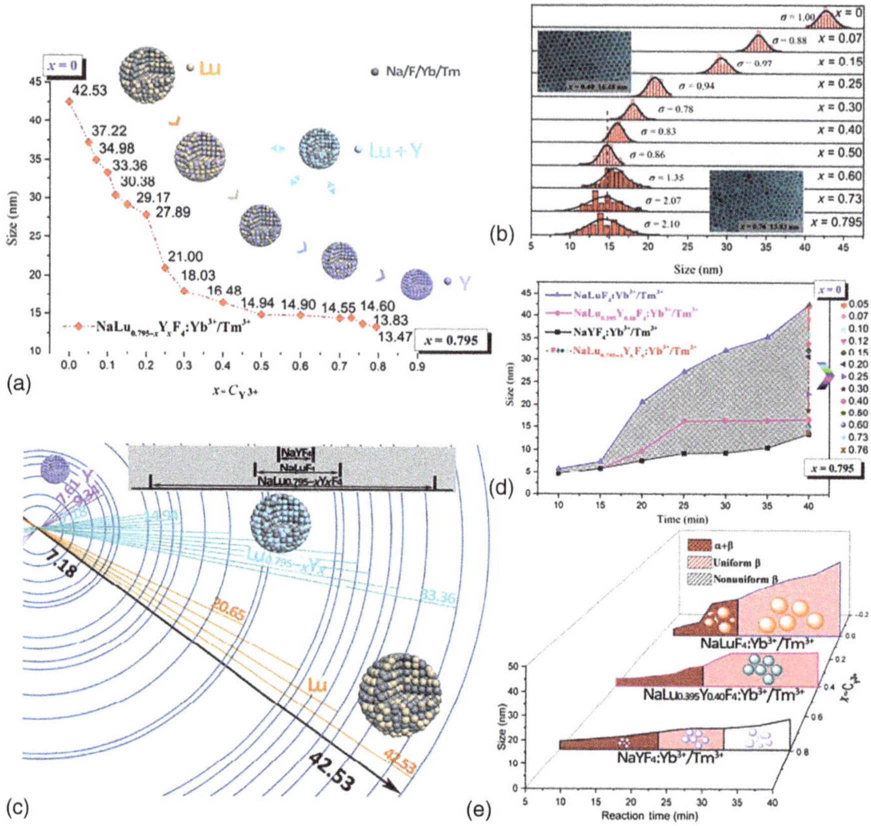

Figure 3.6 (a) Size variation against the concentration of Y^{3+}. (b) Size distributions and representative TEM images. (c) Schematic illustration of the sizes of the uniform UCNCs. (d) Size variation against the reaction time. (e) C-t-P-S growth phase diagram of $NaLu_{0.795-x}Y_xF_4$:Yb^{3+}/Tm^{3+} UCNCs. Source: Reproduced with permission from Zhang et al. [69]/Springer Nature.

computed. R square, which is denoted as the determinant coefficient, is the ratio of regression square sum to total square sum. The F value is one test statistic that comes from variance analysis. Whether the entire regression equation is useful is determined by the overall test. When the regression equation's Sig value for the F value is less than 0.05, it may be deemed beneficial. The F value and R square value of the final model can be used to show how accurate the optimization was.

The data derived by SPSS show that, with the exception of the Li element, the other elements of Er, Mn, Yb, and K have good predictions to the UCL intensity in the alkali metal optimization research.

3.5 Cancer Therapy Based on UCNPs

The advancement of nanochemistry allows for quick improvements in early detection and treatment, encourages successful cancer prevention, and enhances patient survival therapy [70–76]. The main areas of the role of UCNPs in cancer research are imaging and phototherapy. In the field of phototherapy of cancer, UCNPs can be effective in living organisms stimulated by near-infrared light. Photodynamic therapy (PDT) and photothermal therapy (PTT) properties based on UCNPs have been widely demonstrated, and phototherapeutic efficiency has been demonstrated in cells, in vitro [17, 77–87].

3.5.1 Photodynamic Therapy

PDT is a cancer therapy technique that kills diseased cells by producing cytotoxic reactive oxygen species (ROS) from the high-energy light emitted by a photosensitizer [88, 89]. The first photosensitizer for PDT is porphyrin [90]. Commonly used photosensitizers include zinc phthalocyanine (ZnPc), tetraphenyl porphyrin (TPP), methylene blue (MB), rose-bengal (RB), and chlorin dihydrogen E6 (Ce6) [91–97]. At present, UCNPs have been widely used in the study of photodynamics [89, 91–104]. The researchers used a conventional drug carrier based on nanoparticles to load a hydrophobic photosensitizer that would produce singlet oxygen when exposed to light [105]. PDT effects have been characterized using a variety of techniques, including drug loading/release properties, in vitro assessment of PDT effects, and nanochemical ways to create micelle cavities and confirm singlet oxygen formation [105, 106]. Zhang has proposed employing UCNPs as remotely controlled nano-transducers for in vivo PDT, in which the upconverted light emission from UCNP can more efficiently activate photosensitive agents like ZnPc and MC540. Thus, a 980 nm laser's in vivo anti-cancer efficacy was demonstrated [105].

Lv et al. developed a nanoplatform to achieve mutually enhanced dual PDT of tumors [107]. At 808 nm light, UCNP can generate UV emission to trigger CeO_2 and ICG to produce ROS (Figure 3.7).

3.5.2 Photothermal Therapy

UCNPs with blue/green luminescence can be used as a PTT material for ET [108, 109]. When it comes to PTT, UCNPs have to work with organic or inorganic PTT agents, like carbon, manganese dioxide, copper sulfide, and polydopamine. In the resulting composite nanomaterials, the PTT effect is mainly derived from the absorption of NIR and, in part, from the upconversion ET of UCL light upon NIR laser excitation. The temperature of the synthetic reaction greatly influences the biostability of composite nanomaterials. At a higher temperature, the synthesis of the compound is more stable. Since UCNPs@MnO_2 has the highest biostability and can adapt to a micro acid environment, it is the most promising nanocomposite for cancer diagnostics and treatment among the other options. The composite is quenched by the luminescence of the manganese dioxide coating to generate heat for PTT tumor cells. As manganese dioxide is reduced to a soluble colorless Mn^{2+} ion, the visible light emission is gradually restored, thus dissolving its coating and illuminating the treated cancer cells [110].

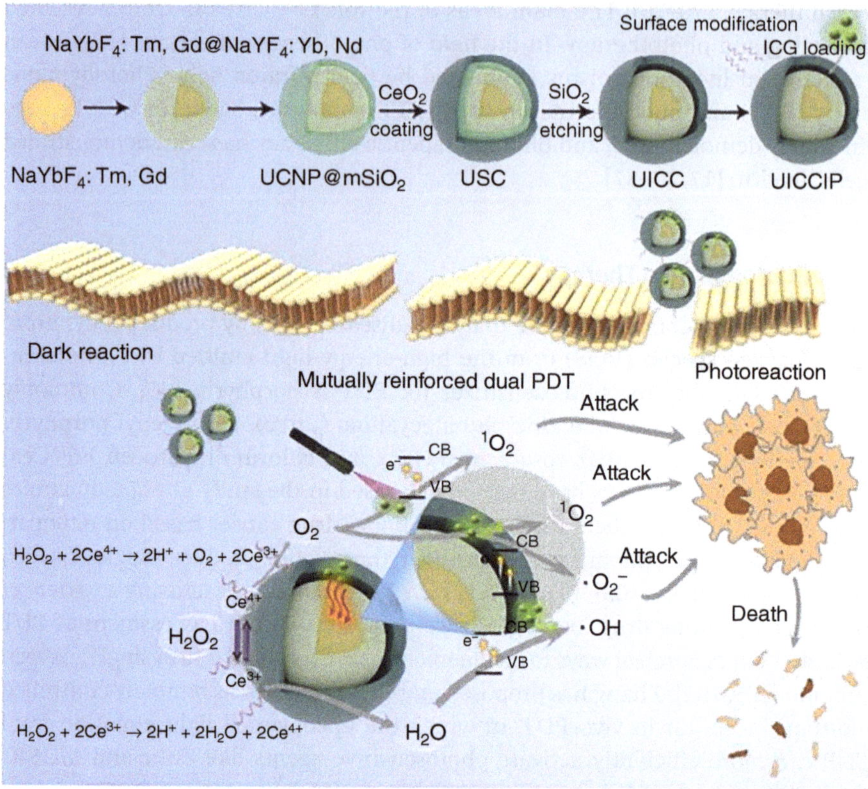

Figure 3.7 Schematic illustration of the preparation and functional diagram for mutually reinforcing dual photodynamic therapy. Source: Reproduced with permission from Lv et al. [107]/Elsevier.

Heat production during photothermal treatment has side effects on normal tissues and cells adjacent to the lesion [111–115]. This problem can be improved by improving the efficiency of phototherapy at the disease site using various strategies. These tactics include improved identification of highly expressed proteins in cancers, accurate targeting of the tumor, and stimulation response to the tumor microenvironment. By attaching particular reactive enzymes to UCNPs, Xing presented a microenvironment-sensitive method to cause the build-up of phototherapeutic medications in tumors [116]. The photothermal conversion efficiency of UCNPs and black phosphorus reaches 30.84% at 980 nm laser, which is much better than that of conventional gold nanorods (23.33%) and nanostars (22.63%) [29]. For guided PTT imaging, various photothermal agents including $MnFe_2O_4$ and carbon composites have been suggested in conjunction with UCNPs [117, 118].

Some studies have combined PTT with PDT to form a multifunctional nanoplatform [119–123]. The efficacy of cancer ablation can be considerably increased by this combined therapy. Nanomotors (UCNPs@mSiO_2-Au-Cys) were developed by Liu et al. for PTT/PDT combination therapy and tumor multimode imaging [124]. In this work, UCNPs were loaded with photosensitizer (Cy-S-Ph-NH2) and

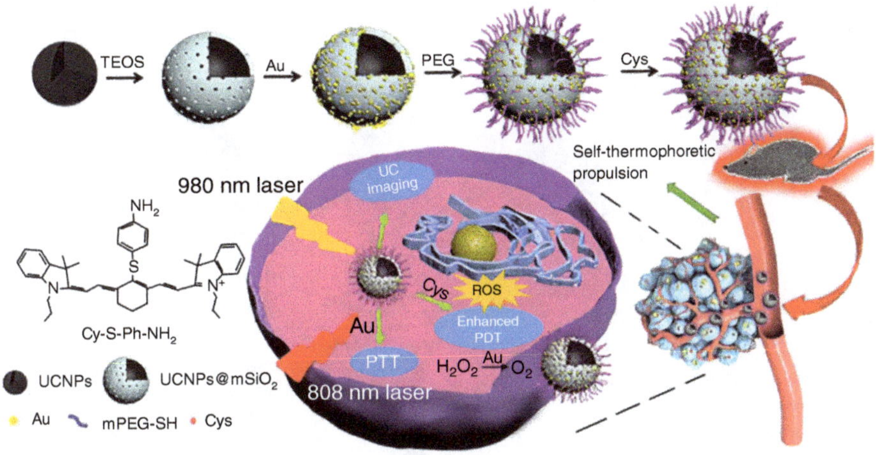

Figure 3.8 Synthesis and application of UCNPs@mSiO_2-Au-Cys nanomotors for nanotheranostics. Source: Reproduced with permission from Liu et al. [124]/Elsevier.

photothermal agent (Au NPs) (Figure 3.8). The nanocatalytic enzyme activity of Au NPs can boost PDT's effectiveness. Besides, the Au NPs can furthermore offer a photothermal conversion. In the meantime, Cys can boost the photothermal effect and accomplish PDT. Furthermore, the tumor tissue permeability significantly increases under 808 nm laser irradiation thanks to the autonomous motion generated by nanomotors.

3.5.3 Photo-Immunotherapy

Immunotherapy's exceptional targeting precision and safety profile have garnered significant interest. External stimuli are not required for tumor immunotherapy to work.

However, it activates immune responses to culture immune cells and attack cancer cells [125]. Phototherapy synergistic immunotherapy combining UCNPs and immunomodulation is a novel approach to tumor treatment [15, 98, 126–132]. Typically, local treatment releases signals from tumor-associated antigens and damage-associated molecular patterns [133]. These compounds have the ability to strengthen immunity. The therapeutic efficacy based on UCNPs can be improved after combining phototherapy and immunotherapy. A core–shell structure consisting of PDA as the core and $NaGdF_4$:Yb/Er as the shell was created by Liu et al. and then modified by Ce6. The PDA component functions as a photothermal agent and Ce6 as a photodynamic therapeutic agent when exposed to a 980 nm laser [134]. In the study, the immunological response induced by phototherapy activated T lymphocytes and T memory cells, which in turn inhibited tumor spread and recurrence. The investigators used anti-CTLA-4 as an immune checkpoint blockade to enhance PDT and promote the activation of immune responses to kill cancer cells [135]. Li and colleagues used PCpG and UCNPs in combination for immunotherapy. The core–shell nanoparticles $NaGdF_4$:70%Yb,1%Tm@$NaGdF_4$ emitted at 313, 363,

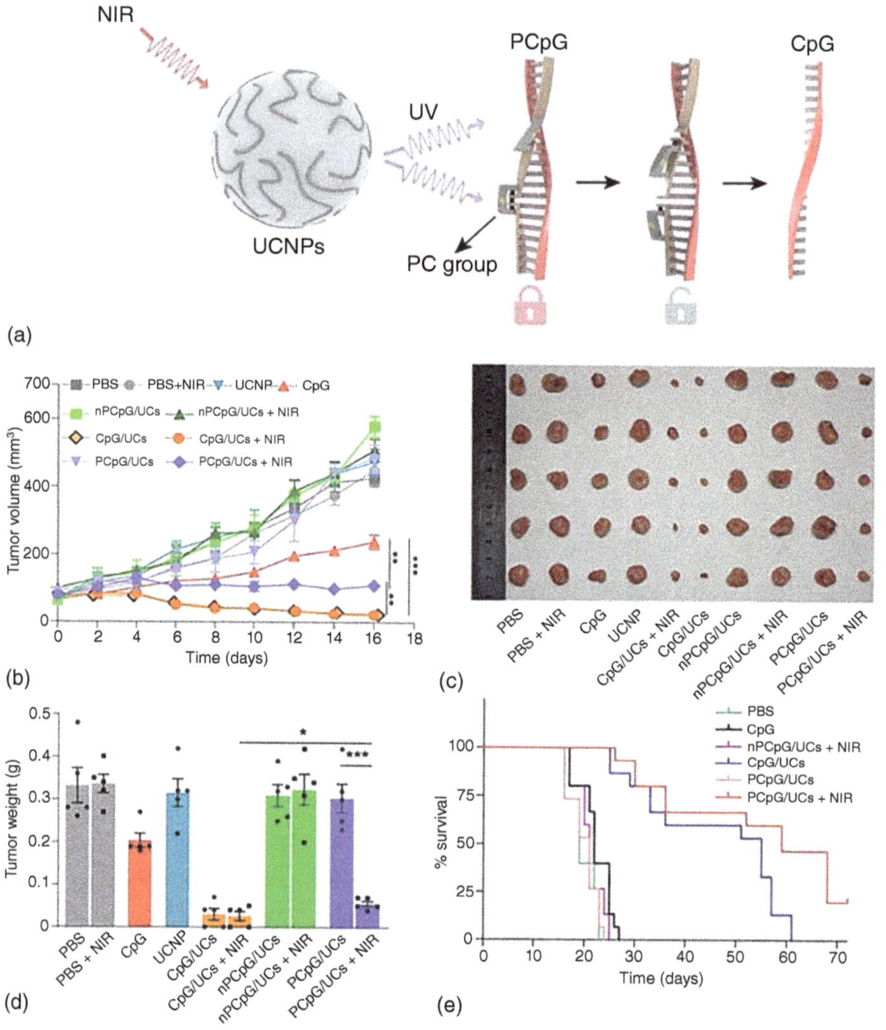

Figure 3.9 Photoactivatable immunodevice through the integration of UCNPs with the UV light-responsive PCpG. (a) Schematic of the design of photoactivatable immunodevice through the integration of UCNPs with the UV light-responsive PCpG. (b) Tumor growth curves up to day 16. (c) Representative pictures of the tumors. (d) Average tumor weights of mice at day 16 in the different treatment groups. (e) Kaplan–Meier survival curves of mice in the different treatment groups. Source: Reproduced with permission from Chu et al. [136]/Springer Nature/CC BY 4.0.

453, and 478 nm upon activation by a 980 nm laser, as illustrated in Figure 3.9. The immunotherapy agent CpG was released by PCpG upon activation by UV radiation. In the in vivo trials, the CpG/UCs combination proved to be more effective than PCpG/UCs + NIR. However, the former produced significant systemic toxicity, underscoring the usefulness of UCNPs in this procedure [136].

Tumor-associated macrophages (TAMs) are a significant immunoadjuvant treatment target because they are found in substantial tumor populations and stimulate tumor growth. Chemotherapy and radiation have little effect on TAMs. Through repolarized TAMs and transfer of M2 macrophages to M1 macrophages under X-ray irradiation, researchers have employed Bi-based UCNP loaded with DOX to trigger the immunogenic tumor death, which can release pro-inflammatory cytokines and kill tumor cells [137].

3.5.4 Photo-Gene Therapy

Gene therapy is the transduction or knockout of a specific gene, such as DNA or small interfering RNA (siRNA) molecules. This approach has already played a crucial role in alternative drug treatments to treat or prevent disease [138]. UCNPs can be combined with GT, significantly improving the efficiency and feasibility of gene therapy. A fraction of UCNPs showed excellent gene delivery capacity in UCNPs-based gene therapy. The other part plays an essential role in light activation.

In gene therapy, gene delivery is a crucial link [139]. The face-modified UCNP is a modern approach to increase the flexibility of GT because of its unique advantages [139, 140]. Jayakumar et al. have combined PDT-gene therapy with UCNPs and performed surface modification to achieve a synergistic treatment of PDT and GT [141]. In that study, on the one hand, UCNPs could help to achieve the efficient delivery of therapeutic genes, thus leading to the silencing of specific proteins. On the other hand, PDT, formed by UCNPs, can cooperate with PDT and gene therapy. When exogenous genes are transported to the cell by the vector, endosomal vesicle sequestration may result in material degradation or transit outside the cell [142]. Zhang and associates improved the therapeutic efficacy by increasing the delivered genes' endosomal escape by photochemical internalization (PCI) techniques [141]. In the research, UCNPs are bound to TPPS2a and photo-morpholino. In the presence of the NIR, TPPS2a is triggered by UCNPs to produce ROS, which locally disrupts the endosomal vesicle wall and triggers the release of the nanoparticles. Meanwhile, the UV light emitted by UCNPs cleaves the photo-morpholino sense strand and releases the antisense morpholine, further inhibiting STAT 3 gene expression. Efficient gene knockdown was achieved against the controls without promoting nuclear endosomal escape, indicating that the platform has good potential for clinical application. Kuang et al. designed polyprodrug/siRNA-loaded UCNPs (PUCNP@Pt@siPlk1) to achieve a combination of chemotherapy and GT [143]. This strategy can achieve simultaneous multimodal imaging and synergistic therapy. The nanoparticles enabled on-demand delivery of drugs and siRNA, showing excellent synergistic therapeutic effects in in vivo and in vitro (Figure 3.10).

With the continuous advancement of research, the potential of upconversion materials in diagnosis and treatment has been explored. Recently, the unique role of UCNPs in the diagnostic treatment field has been attention and growing. In addition to the above several phototherapy methods,

UCNPs play an essential role in biological imaging, optogenetic modulation, multimodal therapy, and other fields, creating a new idea for biological diagnosis and

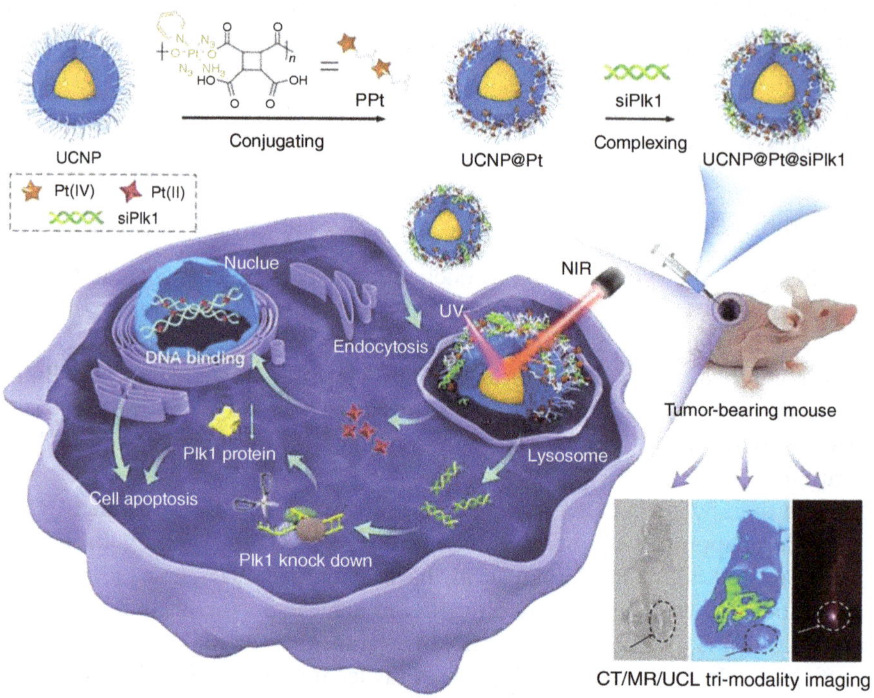

Figure 3.10 Schematic illustration of the NIR light-activatable multifunctional polyprodrug/siRNA-loaded UCNPs. Source: Reproduced with permission from Kuang et al. [143]/John Wiley & Sons.

treatment. The corresponding applications will be discussed in detail in the other chapters.

3.6 Conclusion and Perspective

This chapter reviewed the regulation of NIR-II luminescence and upconversion of rare-earth composites. It also discusses the synthesis of rare-earth upconversion light-emitting particles, luminescence enhancement techniques, and applications in biological therapy related to UCL. Specifically, the use of material informatics in UCL modulation was discussed, providing an alternate optimization method for UCL. Although upconversion and NIR-II luminescence modulation have made remarkable progress in recent years, there are still enormous challenges to be overcome for their luminescence optimization and further application. Great attention should be paid to these challenges. First of all, studies adjust the concentrations and species of ions. In actuality, luminescence modulation can have more parameters than this. For example, it can be modified according to the structure of micronanoparticles, crystal properties, and other aspects. The outcomes can then be utilized as material informatics input data to get the best possible results. Second, in addition to light intensity alone, material informatics can also be used to forecast additional qualities such as durability, color, and luminosity for better performance prediction. Lastly, a wider range of biological applications can be investigated. For example, biological detection can reach even better sensitivity and linear interval of detection by improving luminous performance.

References

1 Lv, R.C., Gai, S.L., Dai, Y.L. et al. (2013). Highly uniform hollow GdF_3 spheres: controllable synthesis, tuned luminescence, and drug-release properties. *ACS Applied Materials & Interfaces* 5: 10806–10818.

2 Lv, R.C., Gai, S.L., Dai, Y.L. et al. (2014). Surfactant-free synthesis, luminescent properties, and drug-release properties of LaF_3 and $LaCO_3F$ hollow microspheres. *Inorganic Chemistry* 53: 998–1008.

3 Lv, R.C., Yang, G.X., Dai, Y.L. et al. (2014). Self-produced bubble-template synthesis of La_2O_3:Yb/Er@Au hollow spheres with markedly enhanced luminescence and release properties. *CrystEngComm* 16: 9612–9621.

4 Lv, R.C., Yang, G.X., Gai, S.L. et al. (2014). Multifunctional $LaPO_4$:Ce/Tb@Au mesoporous microspheres: synthesis, luminescence and controllable light triggered drug release. *Rsc Advances* 4: 63425–63435.

5 Lv, R.C., Raab, M., Wang, Y.X. et al. (2022). Nanochemistry advancing photon conversion in rare-earth nanostructures for theranostics. *Coordination Chemistry Reviews* 460: 214486.

6 Prodi, L., Rampazzo, E., Rastrelli, F. et al. (2015). Imaging agents based on lanthanide doped nanoparticles. *Chemical Society Reviews* 44: 4922–4952.

7 Levy, E.S., Tajon, C.A., Bischof, T.S. et al. (2016). Energy-Looping nanoparticles: harnessing excited-state absorption for deep-tissue imaging. *ACS Nano* 10: 8423–8433.

8 Su, Q., Han, S., Xie, X. et al. (2012). The effect of surface coating on energy migration-mediated upconversion. *Journal of the American Chemical Society* 134: 20849–20857.

9 Dong, N.N., Pedroni, M., Piccinelli, F. et al. (2011). NIR-to-NIR two-photon excited CaF_2:Tm^{3+},Yb^{3+} nanoparticles: multifunctional nanoprobes for highly penetrating fluorescence bio-imaging. *ACS Nano* 5: 8665–8671.

10 Zhang, Y. and Liu, X. (2013). Shining a light on upconversion. *Nature Nanotechnology* 8: 702–703.

11 Xing, H., Bu, W., Ren, Q. et al. (2012). A $NaYbF_4$: Tm^{3+} nanoprobe for CT and NIR-to-NIR fluorescent bimodal imaging. *Biomaterials* 33: 5384–5393.

12 Xie, X., Gao, N., Deng, R. et al. (2013). Mechanistic investigation of photon upconversion in Nd^{3+}-sensitized core-shell nanoparticles. *Journal of The American Chemical Society* 135: 12608–12602.

13 Chan, E.M., Gargas, D.J., Schuck, P.J., and Milliron, D.J. (2012). Concentrating and recycling energy in lanthanide codopants for efficient and spectrally pure emission: the case of $NaYF_4$:Er^{3+}/Tm^{3+} upconverting nanocrystals. *The Journal of Physical Chemistry B* 116: 10561–10570.

14 Zhou, B., Yan, L., Huang, J. et al. (2020). NIR II–responsive photon upconversion through energy migration in an ytterbium sublattice. *Nature Photonics* 14: 760–766.

15 Lv, R., Wang, Y., Lin, B. et al. (2021). Targeted luminescent probes for precise upconversion/NIR II luminescence diagnosis of lung adenocarcinoma. *Analytical Chemistry* 93: 4984–4992.

16 Zhou, B., Shi, B., Jin, D., and Liu, X. (2015). Controlling upconversion nanocrystals for emerging applications. *Nature Nanotechnology* 10: 924–936.

17 Chen, G., Qiu, H., Prasad, P.N., and Chen, X. (2014). Upconversion nanoparticles: design, nanochemistry, and applications in theranostics. *Chemical Reviews* 114: 5161–5214.

18 Wang, J., Deng, R., MacDonald, M.A. et al. (2014). Enhancing multiphoton upconversion through energy clustering at sublattice level. *Nature Materials* 13: 157–162.

19 Liu, Y., Tu, D., Zhu, H. et al. (2010). A strategy to achieve efficient dual-mode luminescence of Eu^{3+} in lanthanides doped multifunctional $NaGdF_4$ nanocrystals. *Advanced Materials* 22: 3266–3271.

20 Wong, K.L., Law, G.L., Kwok, W.M. et al. (2005). Simultaneous observation of green multiphoton upconversion and red and blue nlo processes from polymeric terbium(III) complexes. *Angewandte Chemie International Edition* 44: 3436–3439.

21 Law, G.L., Kwok, W.M., Wong, W.T. et al. (2007). Terbium luminescence sensitized through three-photon excitation in a self-assembled unlinked antenna. *The Journal of Physical Chemistry B* 111: 10858–10861.

22 Zhang, Y.W., Sun, X., Si, R. et al. (2005). Single-crystalline and monodisperse LaF_3 triangular nanoplates from a single-source precursor. *Journal of the American Chemical Society* 127: 3260–3261.

23 Yin, A., Zhang, Y., Sun, L., and Yan, C. (2010). Colloidal synthesis and blue based multicolor upconversion emissions of size and composition controlled monodisperse hexagonal $NaYF_4$:Yb, Tm nanocrystals. *Nanoscale* 2: 953–959.

24 Ye, X., Collins, J.E., Kang, Y. et al. (2010). Morphologically controlled synthesis of colloidal upconversion nanophosphors and their shape-directed self-assembly. *Proceedings of the National Academy of Sciences of the United States of America* 107: 22430–22435.

25 Feng, M., Lv, R., Xiao, L. et al. (2018). Highly erbium-doped nanoplatform with enhanced red emission for dual-modal optical-imaging-guided photodynamic therapy. *Inorganic Chemistry* 57: 14594–14602.

26 Wang, X., Zhuang, J., Peng, Q., and Li, Y.D. (2005). A general strategy for nanocrystal synthesis. *Nature* 437: 121–124.

27 Li, N., Tan, H., Xie, S. et al. (2020). Hexadecylpyridinium chloride mediated hydrothermal synthesis of $NaYF_4$:Yb,Er nanocrystal and their luminescent properties. *Journal of Nanoscience and Nanotechnology* 20: 1866–1872.

28 Mousavand, T., Ohara, S., Naka, T. et al. (2010). Organic-ligand-assisted hydrothermal synthesis of ultrafine and hydrophobic ZnO nanoparticles. *Journal of Materials Research* 25: 219–223.

29 Xu, D., Liu, J., Wang, Y. et al. (2020). Black phosphorus nanosheet with high thermal conversion efficiency for photodynamic/photothermal/immunotherapy. *ACS Biomaterials Science and Engineering* 6: 4940–4948.

30 Rafique, R., Baek, S.H., Le, M.T.P. et al. (2019). A facile hydrothermal synthesis of highly luminescent $NaYF_4:Yb^{3+}/Er^{3+}$ upconversion nanoparticles and their biomonitoring capability. *Materials Science and Engineering C–Materials for Biological Applications* 99: 1067–1074.

31 Gao, Y., Li, R., Zheng, W. et al. (2019). Broadband NIR photostimulated luminescence nanoprobes based on $CaS:Eu^{2+}$, Sm^{3+} nanocrystals. *Chemical Science* 10: 5452–5460.

32 Amouroux, B., Roux, C., Marty, J.D. et al. (2019). Importance of the mixing and high-temperature heating steps in the controlled thermal coprecipitation synthesis of sub-5-nm $Na(Gd–Yb)F_4:Tm$. *Inorganic Chemistry* 58: 5082–5088.

33 Wang, F., Deng, R., and Liu, X. (2014). Preparation of core-shell $NaGdF_4$ nanoparticles doped with luminescent lanthanide ions to be used as upconversion-based probes. *Nature Protocols* 9: 1634–1644.

34 Zhou, Y., Chen, Y., He, H. et al. (2019). A homogeneous DNA assay by recovering inhibited emission of rare earth ions-doped upconversion nanoparticles. *Journal of Rare Earths* 37: 11–18.

35 Liu, X. and Lin, J. (2008). $LaGaO_3:A$ (A = Sm^{3+} and/or Tb^{3+}) as promising phosphors for field emission displays. *Journal of Materials Chemistry* 18: 221–228.

36 Li, G., Li, C., Zhang, C. et al. (2009). Tm^{3+} and/or Dy^{3+} doped LaOCl nanocrystalline phosphors for field emission displays. *Journal of Materials Chemistry* 19: 8936–8943.

37 Liu, X., Yan, L., and Lin, J. (2009). Synthesis and luminescent properties of $LaAlO_3:RE^{3+}$ (RE = Tm, Tb) nanocrystalline phosphors via a Sol-Gel process. *The Journal of Physical Chemistry C* 113: 8478–8483.

38 Liu, X., Li, C., Quan, Z. et al. (2007). Tunable luminescence properties of $CaIn_2O_4:Eu^{3+}$ phosphors. *The Journal of Physical Chemistry C* 111: 16601–16607.

39 Geng, D., Li, G., Shang, M. et al. (2012). Nanocrystalline $CaYAlO_4:Tb^{3+}/Eu^{3+}$ as promising phosphors for full-color field emission displays. *Dalton Transactions* 41: 3078–3086.

40 Mi, C., Tian, Z., Cao, C. et al. (2011). Novel microwave-assisted solvothermal synthesis of $NaYF_4:Yb,Er$ upconversion nanoparticles and their application in cancer cell imaging. *Langmuir* 27: 14632–14637.

41 Zhao, E., Lam, J.W.Y., Hong, Y. et al. (2013). How do substituents affect silole emission? *Journal of Materials Chemistry C* 1: 5661–5668.

42 Lu, W., Guo, N., Jia, Y. et al. (2013). Tunable color of $Ce^{3+}/Tb^{3+}/Mn^{2+}$–coactivated $CaScAlSiO_6$ via energy transfer: a single-component red/white-emitting phosphor. *Inorganic Chemistry* 52: 3007–3012.

43 Park, W.B., Singh, S.P., and Sohn, K.S. (2014). Discovery of a phosphor for light emitting diode applications and its structural determination,

$Ba(Si,Al)_5(O,N)_8:Eu^{2+}$. *Journal of the American Chemical Society* 136: 2363–2373.

44 Ren, F., Ding, L., Liu, H. et al. (2018). Ultra-small nanocluster mediated synthesis of Nd^{3+}-doped core-shell nanocrystals with emission in the second near-infrared window for multimodal imaging of tumor vasculature. *Biomaterials* 175: 30–43.

45 Shanmugam, V., Selvakumar, S., and Yeh, C.S. (2014). Near–infrared light-responsive nanomaterials in cancer therapeutics. *Chemical Society Reviews* 43: 6254–6287.

46 Connally, R., Jin, D., and Piper, J. (2006). High intensity solid-state UV source for time-gated luminescence microscopy. *Cytometry Part A* 69A: 1020–1027.

47 Heng, K.L., Jin, H.M., Li, Y., and Wu, P. (1999). Computer aided design of NiMH electrodes. *Journal of Materials Chemistry* 9: 837–843.

48 Yan, L., Zhou, B., Song, N. et al. (2018). Self-sensitization induced upconversion of Er^{3+} in core-shell nanoparticles. *Nanoscale* 10: 17949–17957.

49 Cao, X., Zhang, Y., Li, J., and Hui, C. (2020). Composition design of aluminum alloys optimizing stress corrosion cracking resistance using machine learning. *Materials Research Express* 7: 046506.

50 Nelson, J. and Sanvito, S. (2019). Predicting the Curie temperature of ferromagnets using machine learning. *Physical Review Materials* 3: 104405.

51 Halder, A., Rom, S., Ghosh, A., and Saha-Dasgupta, T. (2020). Prediction of the properties of the rare-earth magnets Ce_2Fe_{17}–xCo_xCN: a combined machine-learning and ab initio study. *Physical Review Applied* 14: 034024.

52 Sohn, K.S., Lee, J.M., and Shin, N.S. (2003). A search for new red phosphors using a computational evolutionary optimization process. *Advanced Materials* 15: 2081–2084.

53 Sohn, K.S., Park, D.H., Cho, S.H. et al. (2006). Computational evolutionary optimization of red phosphor for use in tricolor white LEDs. *Chemistry of Materials* 18: 1768–1772.

54 Chen, L., Chen, K.J., Lin, C.C. et al. (2010). Combinatorial approach to the development of a single mass $YVO_4:Bi^{3+}$, Eu^{3+} phosphor with red and green dual colors for high color rendering white light-emitting diodes. *Journal of Combinatorial Chemistry* 12: 587–594.

55 Park, W.B., Shin, N., Hong, K.P. et al. (2012). A new paradigm for materials discovery: heuristics-assisted combinatorial chemistry involving parameterization of material novelty. *Advanced Functional Materials* 22: 2258–2266.

56 Sohn, K.S., Kim, B.I., and Shin, N. (2004). Genetic algorithm-assisted combinatorial search for new red phosphors of high efficiency at soft ultraviolet excitation. *Journal of the Electrochemical Society* 151: H243–H248.

57 Lv, R., Xiao, L., Jiang, X. et al. (2018). Optimization of red luminescent intensity in eu^{3+}-doped lanthanide phosphors using genetic algorithm. *ACS Biomaterials Science and Engineering* 4: 4378–4384.

58 Lv, R., Xiao, L., Wang, Y. et al. (2019). Searching for the optimized luminescent lanthanide phosphor using heuristic algorithms. *Inorganic Chemistry* 58: 6458–6466.

59 Hamza, M.F., Yap, H.J., and Choudhury, I.A. (2017). Recent advances on the use of meta-heuristic optimization algorithms to optimize the type-2 fuzzy logic systems in intelligent control. *Neural Computing and Applications* 28: 979–999.

60 Hopper, E. and Turton, B.C.H. (2001). A review of the application of meta-heuristic algorithms to 2d strip packing problems. *Artificial Intelligence Review* 16: 257–300.

61 Ozdemir, G. and Karaboga, N. (2019). A review on the cosine modulated filter bank studies using meta-heuristic optimization algorithms. *Artificial Intelligence Review* 52: 1629–1653.

62 Katoch, S., Chauhan, S.S., and Kumar, V. (2021). A review on genetic algorithm: past, present, and future. *Multimedia Tools and Applications* 80: 8091–8026.

63 Sakurai, M., Wang, R., Liao, T. et al. (2020). Discovering rare-earth-free magnetic materials through the development of a database. *Physical Review Materials* 4: 114408.

64 Sun, J.S., Li, S.W., Shi, L.L. et al. (2015). Experimental optimal design of the Er^{3+}/Yb^{3+} codoped $BaGd_2ZnO_5$ phosphor and its upconversion luminescence properties. *Acta Physica Sinica* 64: 243301.

65 Zhao, Y., Yang, F., Sun, J.S. et al. (2019). Experimental optimal design of Er^{3+}/Yb^{3+} co-doped $Ba_5Gd_8Zn_4O_{21}$ phosphor and red upconversion luminescence properties. *Acta Physica Sinica* 68: 213301–213309.

66 Wang, Y., Lu, W.D., Lin, B. et al. (2020). An optimized lanthanide-chlorophyll nanocomposite for dual-modal imaging-guided surgery navigation and anti-cancer theranostics. *Biomaterials Science* 8: 1270–1278.

67 Thangaraj, R., Pant, M., Abraham, A., and Bouvry, P. (2011). Particle swarm optimization: hybridization perspectives and experimental illustrations. *Applied Mathematics and Computation* 217: 5208–5226.

68 Yang, F., Wang, Y., Jiang, X. et al. (2020). Optimized multimetal sensitized phosphor for enhanced red up-conversion luminescence by machine learning. *ACS Combinatorial Science* 22: 285–296.

69 Zhang, D., Dong, Y., Li, D. et al. (2021). Growth regularity and phase diagrams of $NaLu_{0.795-x}Y_xF_4$ upconversion nanocrystals synthesized by automatic nanomaterial synthesizer. *Nano Research* 14: 4760–4767.

70 Mitchell, M.J., Billingsley, M.M., Haley, R.M. et al. (2021). Engineering precision nanoparticles for drug delivery. *Nature Reviews Drug Discovery* 20: 101–124.

71 Vlerken, V.L.E., Vyas, T.K., and Amiji, M.M. (2007). Poly(ethylene glycol)-modified nanocarriers for tumor-targeted and intracellular delivery. *Pharmaceutical Research* 24: 1405–1414.

72 Chen, F., Wang, G., Griffin, J.I. et al. (2017). Complement proteins bind to nanoparticle protein corona and undergo dynamic exchange in vivo. *Nature Nanotechnology* 12: 387–393.

73 Verma, S.K., Jha, E., Panda, P.K. et al. (2017). Molecular aspects of core-shell intrinsic defect induced enhanced antibacterial activity of ZnO nanocrystals. *Nanomedicine* 13: 43–68.

74 Yang, Y. and Yu, C. (2016). Advances in silica based nanoparticles for targeted cancer therapy. *Nanomedicine-Nanotechnology Biology and Medicine* 12: 317–332.

75 Brannon-Peppas, L. and Blanchette, J.O. (2012). Nanoparticle and targeted systems for cancer therapy. *Advanced Drug Delivery Reviews* 64: 206–212.

76 Chen, W., Xie, Y., Wang, M., and Li, C. (2020). Recent advances on rare earth upconversion nanomaterials for combined tumor near-infrared photoimmunotherapy. *Frontiers in Chemistry* 8: 596658.

77 Zheng, K., Loh, K.Y., Wang, Y. et al. (2019). Recent advances in upconversion nanocrystals: expanding the kaleidoscopic toolbox for emerging applications. *Nano Today* 29: 100797.

78 Wang, Z. and Xing, B. (2020). Near-infrared multipurpose lanthanide-imaging nanoprobes. *Chemistry – An Asian Journal* 15: 2076–2091.

79 Liang, G., Wang, H., Shi, H. et al. (2020). Recent progress in the development of upconversion nanomaterials in bioimaging and disease treatment. *Journal of Nanobiotechnology* 18: 154.

80 Bagheri, A., Arandiyan, H., Boyer, C., and Lim, M. (2016). Lanthanide-doped upconversion nanoparticles: emerging intelligent light-activated drug delivery systems. *Advanced Science* 3: 1500437.

81 Gao, C.J., Zheng, P.R., Liu, Q.X. et al. (2021). Recent advances of upconversion nanomaterials in the biological field. *Nanomaterials* 2: 11515.

82 DaCosta, M.V., Doughan, S., Han, Y., and Krull, U.J. (2014). Lanthanide upconversion nanoparticles and applications in bioassays and bioimaging: a review. *Analytica Chimica Acta* 832: 1–33.

83 Zhu, X., Zhang, J., Liu, J., and Zhang, Y. (2019). Recent progress of rare-earth doped upconversion nanoparticles: synthesis, optimization, and applications. *Advanced Science* 6: 1901358.

84 Kachynski, A.V., Pliss, A., Kuzmin, A.N. et al. (2014). Photodynamic therapy by in situ nonlinear photon conversion. *Nature Photonics* 8: 455–461.

85 Song, J., Qu, J., Swihart, M.T., and Prasad, P.N. (2016). Near-IR responsive nanostructures for nanobiophotonics: emerging impacts on nanomedicine. *Nanomedicine-Nanotechnology Biology and Medicine* 12: 771–788.

86 Sun, X., Sun, J., Dong, B. et al. (2019). Noninvasive temperature monitoring for dual-modal tumor therapy based on lanthanide-doped up-conversion nanocomposites. *Biomaterials* 201: 42–52.

87 Kim, S., Ohulchanskyy, T.Y., Pudavar, H.E. et al. (2007). Organically modified silica nanoparticles co-encapsulating photosensitizing drug and aggregation-enhanced two-photon absorbing fluorescent dye aggregates for two-photon photodynamic therapy. *Journal of the American Chemical Society* 129: 2669–2675.

88 Chatterjee, D.K., Fong, L.S., and Zhang, Y. (2008). Nanoparticles in photodynamic therapy: an emerging paradigm. *Advanced Drug Delivery Reviews* 60: 1627–1637.

89 Wang, C., Cheng, L., and Liu, Z. (2013). Upconversion nanoparticles for photodynamic therapy and other cancer therapeutics. *Theranostics* 3: 317–330.

90 Tian, J., Huang, B., Nawaz, M.H., and Zhang, W. (2020). Recent advances of multi-dimensional porphyrin-based functional materials in photodynamic therapy. *Coordination Chemistry Reviews* 420: 213410.

91 Chen, J., Li, S., Liu, X. et al. (2021). Transforming growth factor-beta blockade modulates tumor mechanical microenvironments for enhanced antitumor efficacy of photodynamic therapy. *Nanoscale* 13: 9989.

92 Chen, Q., Ma, X., Xie, L. et al. (2021). Iron-based nanoparticles for MR imaging-guided ferroptosis in combination with photodynamic therapy to enhance cancer treatment. *Nanoscale* 13: 4855–4870.

93 Hu, C., He, X., Chen, Y. et al. (2021). Metformin mediated PD–L1 downregulation in combination with photodynamic-immunotherapy for treatment of breast cancer. *Advanced Functional Materials* 31: 2007149.

94 Lee, H.R., Kim, D.W., Jones, V.O. et al. (2021). Sonosensitizer-functionalized graphene nanoribbons for adhesion blocking and sonodynamic ablation of ovarian cancer spheroids. *Advanced Healthcare Materials* 10: 2001368.

95 Shen, Y., Wu, T., Wang, Y. et al. (2021). Nucleolin-targeted ratiometric fluorescent carbon dots with a remarkably large emission wavelength shift for precise imaging of cathepsin b in living cancer cells. *Analytical Chemistry* 93: 4042–4050.

96 Zhang, Z., Wang, R., Luo, R. et al. (2021). An activatable theranostic nanoprobe for dual-modal imaging-guided photodynamic therapy with self-reporting of sensitizer activation and therapeutic effect. *ACS Nano* 15: 5366–5383.

97 Zheng, K., Liu, X., Liu, H. et al. (2021). Novel pH-triggered doxorubicin-releasing nanoparticles self-assembled by functionalized beta-cyclodextrin and amphiphilic phthalocyanine for anticancer therapy. *ACS Applied Materials & Interfaces* 13: 10674–10688.

98 Xu, D., Yang, F., Qu, D. et al. (2020). Transferred Photothermal to photodynamic therapy based on the marriage of ultrathin titanium carbide and up-conversion nanoparticles. *Langmuir* 36: 13060–13069.

99 Feng, L., He, F., Liu, B. et al. (2016). g-C_3N_4 Coated upconversion nanoparticles for 808 nm near-infrared light triggered phototherapy and multiple imaging. *Chemistry of Materials* 28: 7935–7946.

100 Di, Z., Liu, B., Zhao, J. et al. (2020). An orthogonally regulatable DNA nanodevice for spatiotemporally controlled biorecognition and tumor treatment. *Science Advances* 6: 9381.

101 Feng, M., Wang, Y., Lin, B. et al. (2020). Degradable pH-responsive NIR–II imaging probes based on a polymer-lanthanide composite for chemotherapy. *Dalton Transactions* 49: 9444–9453.

102 Pinto, A., Marangon, I., Mereaux, J. et al. (2021). Immune reprogramming precision photodynamic therapy of peritoneal metastasis by scalable stem-cell-derived extracellular vesicles. *ACS Nano* 15: 3251–3263.

103 Yakavets, I., Francois, A., Lamy, L. et al. (2021). Effect of stroma on the behavior of temoporfin-loaded lipid nanovesicles inside the stroma-rich head and neck carcinoma spheroids. *Journal of Nanobiotechnology* 19: 3.

104 Yuan, Z., Fan, G., Wu, H. et al. (2021). Photodynamic therapy synergize with PD–L1 checkpoint blockade for immunotherapy of colorectal cancer by multifunctional nanoparticle. *Molecular Therapy* 29: 2931–2948.

105 Idris, N.M., Gnanasammandhan, M.K., Zhang, J. et al. (2012). In vivo photodynamic therapy using upconversion nanoparticles as remote-controlled nanotransducers. *Nature Medicine* 18: 1580–1585.

106 Roy, I., Ohulchanskyy, T.Y., Pudavar, H.E. et al. (2003). Ceramic-based nanoparticles entrapping water-insoluble photosensitizing anticancer drugs: a novel drug-carrier system for photodynamic therapy. *Journal of the American Chemical Society* 125: 7860–7865.

107 Lv, K., Yao, L., Fu, X. et al. (2023). Indocyanine green-equipped upconversion nanoparticles/CeO_2 trigger mutually reinforced dual photodynamic therapy. *Nano Today* 52: 101964.

108 Lv, R., Feng, M., Xiao, L. et al. (2018). Multilevel nanoarchitecture exhibiting biosensing for cancer diagnostics by dual-modal switching of optical and magnetic resonance signals. *ACS Applied Bio Materials* 1: 1505–1502.

109 Liu, K., Liu, X., Zeng, Q. et al. (2012). Covalently assembled NIR nanoplatform for simultaneous fluorescence imaging and photodynamic therapy of cancer cells. *ACS Nano* 6: 4054–4062.

110 Jiang, X., Wang, Y., Xu, D. et al. (2020). Lanthanide-based nanocomposites for photothermal therapy under near-infrared laser: relationship between light and heat, biostability, and reaction temperature. *Langmuir* 36: 4033–4043.

111 Gao, G., Jiang, Y.W., Guo, Y. et al. (2020). Enzyme-mediated tumor starvation and phototherapy enhance mild-temperature photothermal therapy. *Advanced Functional Materials* 30: 1909391.

112 Guo, T., Tang, Q., Guo, Y. et al. (2021). Boron quantum dots for photoacoustic imaging-guided photothermal therapy. *ACS Applied Materials & Interfaces* 13: 306–302.

113 Li, Z.H., Chen, Y., Sun, Y., and Zhang, X.Z. (2021). Platinum-doped Prussian blue nanozymes for multiwavelength bioimaging guided photothermal therapy of tumor and anti-inflammation. *ACS Nano* 15: 5189–5200.

114 Pang, H., Tian, C., He, G. et al. (2021). NIR-absorbing Prussian blue nanoparticles for transarterial infusion photothermal therapy of VX_2 tumors implanted in rabbits. *Nanoscale* 13: 8490–8497.

115 Zhao, Y., He, Z., Zhang, Q. et al. (2021). 880 nm NIR-triggered organic small molecular-based nanoparticles for photothermal therapy of tumor. *Nanomaterials* 2: 773.

116 Ai, X., Ho, C.J.H., Aw, J. et al. (2016). In vivo covalent cross-linking of photon-converted rare-earth nanostructures for tumour localization and theranostics. *Nature Communications* 7: 10432.

117 Ding, B., Shao, S., Xiao, H. et al. (2019). $MnFe_2O_4$-decorated large-pore mesoporous silica-coated upconversion nanoparticles for near-infrared light-induced and O_2 self-sufficient photodynamic therapy. *Nanoscale* 11: 14654–14667.

118 Wang, J., Yao, C., Shen, B. et al. (2019). Upconversion-magnetic carbon sphere for near infrared light-triggered bioimaging and photothermal therapy. *Theranostics* 9: 608–619.

119 Fakayode, O.J., Tsolekile, N., Songca, S.P., and Oluwafemi, O.S. (2018). Applications of functionalized nanomaterials in photodynamic therapy. *Biophysical reviews* 10: 49–67.

120 Ma, S., Wang, L., Liu, Z. et al. (2021). "One stone, two birds": engineering 2-D ultrathin heterostructure nanosheet BiNS@$NaLnF_4$ for dual-modal computed tomography/magnetic resonance imaging guided, photonic synergetic theranostics. *Nanoscale* 13: 185–194.

121 Bian, H., Ma, D., Zhang, X. et al. (2021). Tailored engineering of novel xanthonium polymethine dyes for synergetic PDT and PTT triggered by 1064 nm laser toward deep-seated tumors. *Small* 17: 2100398.

122 Mei, S., Zhou, J., Sun, H.T. et al. (2021). Networking state of ytterbium ions probing the origin of luminescence quenching and activation in nanocrystals. *Advanced Science* 8: 2003325.

123 Yang, C., Zhu, Y., Li, D. et al. (2021). Red phosphorus decorated TiO_2 nanorod mediated photodynamic and photothermal therapy for renal cell carcinoma. *Small* 17: 2101837.

124 Liu, H., Zhang, J., Jia, Y. et al. (2022). Theranostic nanomotors for tumor multimode imaging and photothermal/photodynamic synergistic therapy. *Chemical Engineering Journal* 442: 135994.

125 Tan, S., Li, D., and Zhu, X. (2020). Cancer immunotherapy: pros, cons and beyond. *Biomedicine & Pharmacotherapy* 124: 109821.

126 Borcoman, E., Kanjanapan, Y., Champiat, S. et al. (2019). Novel patterns of response under immunotherapy. *Annals of Oncology* 30: 385–396.

127 Mpekris, F., Voutouri, C., Baish, J.W. et al. (2020). Combining microenvironment normalization strategies to improve cancer immunotherapy. *Proceedings of the National Academy of Sciences of the United States of America* 117: 3728–3737.

128 Wallis, C.J.D., Butaney, M., Satkunasivam, R. et al. (2019). Association of patient sex with efficacy of immune checkpoint inhibitors and overall survival in advanced cancers a systematic review and meta-analysis. *JAMA Oncology* 5: 529–536.

129 Lv, R., Yang, P., He, F. et al. (2015). An imaging-guided platform for synergistic photodynamic/photothermal/chemo-therapy with pH/temperature-responsive drug release. *Biomaterials* 63: 115–127.

130 Lv, R., Yang, P., Hu, B. et al. (2017). In situ growth strategy to integrate up-conversion nanoparticles with ultrasmall CuS for photothermal theranostics. *ACS Nano* 11: 1064–1072.

131 Lin, B., Wu, J., Wang, Y. et al. (2021). Peptide functionalized upconversion/NIR II luminescent nanoparticles for targeted imaging and therapy of oral squamous cell carcinoma. *Biomaterials Science* 9: 1000–1007.

132 Wu, J., Liu, J., Lin, B. et al. (2021). Met-targeted dual-modal MRI/NIR II imaging for specific recognition of head and neck squamous cell carcinoma. *ACS Biomaterials Science & Engineering* 7: 1640–1650.

133 Chen, M. and Chen, Q. (2020). Biomaterial-assisted photoimmunotherapy for cancer. *Biomaterials Science* 8: 5846–5858.

134 Yan, S., Zeng, X., Tang, Y.A. et al. (2019). Activating antitumor immunity and antimetastatic effect through polydopamine-encapsulated core-shell upconversion nanoparticles. *Advanced Materials* 31: 1905825.

135 Lin, B., Liu, J., Wang, Y. et al. (2020). Enhanced upconversion luminescence-guided synergistic antitumor therapy based on photodynamic therapy and immune checkpoint blockade. *Chemistry of Materials* 32: 4627–4640.

136 Chu, H., Zhao, J., Mi, Y. et al. (2019). NIR-light-mediated spatially selective triggering of anti-tumor immunity via upconversion nanoparticle-based immunodevices. *Nature Communications* 10: 2839.

137 Qin, X., Liu, J., Xu, Y. et al. (2020). Mesoporous bi-containing radiosensitizer loading with DOX to repolarize tumor-associated macrophages and elicit immunogenic tumor cell death to inhibit tumor progression. *ACS Applied Materials & Interfaces* 12: 31225–31234.

138 Mintzer, M.A. and Simanek, E.E. (2009). Nonviral vectors for gene delivery. *Chemical Reviews* 109: 259–302.

139 Lai, W.F., Andrey, L.R., and Wong, W.T. (2017). Molecular design of upconversion nanoparticles for gene delivery. *Chemical Science* 8: 7339–7358.

140 Ozcan, G., Ozpolat, B., Coleman, R.L. et al. (2015). Preclinical and clinical development of siRNA-based therapeutics. *Advanced Drug Delivery Reviews* 87: 108–119.

141 Jayakumar, M.K.G., Bansal, A., Huang, K. et al. (2014). Near-infrared-light-based nano-platform boosts endosomal escape and controls gene knockdown in vivo. *ACS Nano* 8: 4848–4858.

142 Tan, S.J., Kiatwuthinon, P., Roh, Y.H. et al. (2011). Engineering nanocarriers for siRNA delivery. *Small* 7: 841–856.

143 Kuang, G., Lu, H., He, S. et al. (2021). Near-infrared light-triggered polyprodrug/siRNA loaded upconversion nanoparticles for multi-modality imaging and synergistic cancer therapy. *Advanced Healthcare Materials* 10: 2100938.

4

Composites Based on Lanthanide-Doped Upconversion Nanomaterials and Metal-Organic Frameworks: Fabrication and Bioapplications

Ze Yuan and Xiaoji Xie

Nanjing Tech University (NanjingTech), School of Flexible Electronics (Future Technologies) & Institute of Advanced Materials (IAM), 30 South Puzhu Road, Nanjing, 211816, China

4.1 Introduction

Lanthanide-doped upconversion nanomaterials (UCNMs) are a class of near-infrared (NIR) light-excitable luminescent materials with anti-Stokes emission ranging from deep-ultraviolet (UV) to NIR [1, 2]. UCNMs feature long-lived, sharp, tunable, and stable emissions, as well as the NIR excitation-enabled high-penetration depth, and thus are powerful tools for bioapplications, such as optical imaging [1–5]. UCNMs can also be used for magnetic resonance (MR), X-ray computed tomography (CT), and single-photon emission computed tomography (SPECT) imaging, if paramagnetic Gd^{3+}, lanthanide ions with high atomic numbers (e.g. Yb^{3+} and Lu^{3+}), and radioactive lanthanide ions (e.g. ^{153}Sm) are doped, respectively, offering a versatile platform for multimodal bioimaging [3–5]. By harnessing energy transfer (ET) between UCNMs and disease indicators, such as ions and biomolecules, UCNMs can remotely diagnose some early-stage diseases with high sensitivity and minimal injury [6–8]. Despite the advances already achieved in biochemistry, UCNMs are daunted by limited biological functions. For example, UCNMs have inappreciable therapeutic ability, thereby impeding their use as theranostic agents that can realize diagnosis and therapy simultaneously [6–8]. Additionally, most UCNMs are solid matters showing unideal loading capacity to carry probes or drugs [9–11]. To promote the development of UCNMs in biochemistry, developing UCNM-based composites by integrating UCNMs with other functional materials is an elegant solution [9–13].

To obtain the UCNM-based composite, a multiskilled material that can hybridize with UCNMs, load functional species, mediate interactions between different components, and add functions to the composite is highly desired [9–11]. Metal-organic frameworks (MOFs) represent one class of such versatile materials [11, 14–16]. MOFs are crystalline materials with reticular structures built by multidentate organic ligands and metal ions or clusters through self-assembly [17, 18]. Enabled by the diverse building blocks and their unique assembly motifs, MOFs embrace

Photofunctional Nanomaterials for Biomedical Applications, First Edition.
Edited by Chunxia Li and Jun Lin.

tailorable and controllable pores throughout the skeleton, possessing huge and tailorable capacity to accommodate diverse cargoes (e.g. drugs) [17, 18]. MOFs are also armed by abundant unsaturated sites on their surface, which provide potential room to interact with UCNMs and other materials through coordination or electrostatic interactions [17, 18]. Importantly, some MOFs can serve as functional species, such as therapeutic materials, if their building blocks are deliberately designed [14–16]. For instance, porphyrin-based MOFs can be used for photodynamic therapy due to the generation of reactive oxygen species (ROS) by porphyrin molecules under light activation, and simultaneously, metal ions in these MOFs can act like enzymes to enhance the chemotherapy [19, 20]. Moreover, the large number of MOFs allows the fabrication of diverse UCNM-MOF composites with desired properties for different purposes [17]. Collectively, the exceptional capacity and customized properties of MOFs make them natural multifunctional materials, distinguishing them from their counterparts (e.g. SiO_2, TiO_2, and polymers) [11, 14–17]. The combination of UCNMs with MOFs is thus gaining enormous attention [11, 14–16].

In fact, UCNM-MOF composites can compensate for the limitations of both parent materials and incorporate their merits [11, 14–16]. Furthermore, the composites can afford novel functions through synergetic effects or interactions between the two materials [11, 14–16]. For example, recent studies show that UCNM-MOF composites can remotely control therapy at intended locations with high accuracy and noninvasiveness, by virtue of the highly penetrable NIR light and ET between MOFs and UCNMs [14–16]. In addition, UCNM-MOF composites can facilely afford multimodal imaging and therapy when MOFs and UCNMs are rationally designed and integrated. It should be mentioned that UCNMs or MOFs alone can hardly achieve these functions [14–16]. On account of this, UCNM-MOF composites are promising candidates for the next generation of biomaterials [14–16]. In this chapter, we aim to provide an overview of recent advances in UCNM-MOF composites, with an emphasis on the fabrication and bioapplications of newly developed UCNM-MOF composites.

4.2 Fabrications of Composites

Benefiting from the advances in the controllable synthesis of heterogeneous nanomaterials, several strategies have been developed for the fabrication of varied UCNM-MOF composites thus far, providing a kaleidoscopic toolbox for different bioapplications [14–16]. According to the configuration of reported UCNM-MOF composites, these developed strategies can be generally classified into three types, which are in situ encapsulation, partial embedment, and interfacial attachment [14–16]. In this section, we briefly discuss the fabrication process, mechanisms, merits, and limitations of the three types of strategies.

4.2.1 In Situ Encapsulation

In situ encapsulation typically refers to the in situ growth (or assembly) of one material around a presynthesized material, forming architectures such as core–shell structure [21, 22]. For in situ encapsulation-based fabrication of UCNM-MOF composites, UCNMs are commonly presynthesized as the seeds, and MOFs then heterogeneously nucleate and grow on UCNMs. In addition, homogenous nucleation of MOFs and successive absorption of UCNMs during the growth of MOFs also affords a successful encapsulation [23]. Both isotropic and anisotropic encapsulations can occur in this in situ process, depending on the interactions between UCNMs and MOFs. Isotropic encapsulation can yield perfect core–shell structured composites due to the isotropic interactions between UCNMs and MOFs. Meanwhile, anisotropic encapsulation can produce eccentric core–shell structured composites when the interactions between UCNMs and MOFs are moderate or anisotropic [24].

However, pristine UCNMs and MOFs barely show mutual affinities with each other because of their mismatched crystal structures and incompatible interfaces [14–16]. In particular, the crystalline lattices and components of MOFs and UCNMs are substantially different, since MOFs are coordinative compounds framed by metal ions and organic linkers, and UCNMs are typically ionic crystals consisting of inorganic ions [1, 2, 17, 18]. Moreover, hydrophobic ligands capped on pristine UCNMs, prepared through widely used coprecipitation and hydrolysis methods, can block the wetting and absorption of UCNMs by MOF in the polar reaction mixture [25]. Consequently, the surface engineering of UCNMs for enhancing the interactions between UCNMs and MOFs generally becomes a prerequisite for in situ encapsulation [23, 24]. In this regard, functional ligands, such as polymers, silica (SiO_2), and polydentate chelating ligands, have been modified on UCNMs to assist the encapsulation of MOFs on UCNMs [15, 16]. By means of in situ encapsulations, some sophisticatedly structured UCNM-MOF composites have been fabricated [14–16].

Polymers are able to stretch their flexible long-chain body over the surface of nanomaterials and expose abundant coordinative groups outward, imparting good hydrophilicity to nanomaterials and providing enormous anchoring sites for nodes and ligands of MOFs [21–24]. Thus, polymers can buffer the mismatch between MOFs and UCNMs once decorated on UCNMs, allowing the nucleation and growth of MOF on UCNMs, which have motivated the fabrication of various core–shell structured UCNM-MOF composites [15, 23, 24].

Until now, polyvinyl pyrrolidone (PVP) has been the most frequently used polymer to assist in the synthesis of core–shell structured UCNM-MOF composites [15, 23, 24]. In 2012, for the first time, Lu and coworkers used PVP to encapsulate $NaYF_4$:Yb/Er upconversion nanoparticles in a zeolitic imidazolate framework (ZIF), ZIF-8 [23]. Specifically, they first decorated PVP molecules on the surface of $NaYF_4$:Yb/Er upconversion nanoparticles through a one-step ligand-exchange method and then transferred the PVP-modified upconversion nanoparticles to a methanol solution containing 2-methylimidazole and Zn^{2+}. After the mixture reacted for 24 hours, they obtained $NaYF_4$:Yb/Er encapsulated ZIF-8 core–shell nanocomposites. They also found that the distribution of nanoparticles in ZIF-8 could be modulated by changing reaction conditions. They described the encapsulation process as the successive adsorption of PVP-modified nanoparticles on the continuously growing ZIF-8. In this process, PVP enhanced the affinity between

UCNMs and ZIF-8 through weak coordination interactions between pyrrolidone rings (C=O) and Zn^{2+}, and perhaps also through hydrophobic interactions between apolar groups of PVP and 2-methylimidazole. Importantly, this method can be extended to synthesize other functional nanomaterial@ZIF-8 core–shell nanocomposites with controllable content and spatial distribution.

In addition to the adsorption-induced encapsulation manner, PVP-assisted growth of MOF on UCNMs can follow a heterogeneous nucleation route [24, 26]. For example, PVP-modified $NaYF_4$:Yb/Tm nanoplates can serve as seeds for the heterogeneous nucleation and layer-by-layer growth of MIL-53(Fe) (MIL represents Materials of Institute Lavoisier) MOF [26]. In this scenario (Figure 4.1), the PVP-modified nanoplates can enrich Fe^{3+} by coordinative pyrrolyl groups of PVP, and subsequently capture terephthalic acid organic linkers, initiating the nucleation of MIL-53(Fe) on the nanoplates and avoiding homogeneous nucleation. Once a thin MOF layer forms, it can be easily thickened through layer-by-layer growth. Through heterogeneous nucleation and successive adsorption methods, PVP-modified UCNMs have been successfully coated with diverse MOFs, including ZIF, MIL, and UiO (Universitetet i Oslo) series [23, 24, 26, 27]. It should be pointed out that the anisotropic encapsulation also exists in the PVP-assisted coating of MOF on UCNMs, which has been exemplified by the eccentric $NaYF_4$@Fe-MIL-101-NH_2 core–shell composites [24].

Notably, many other hydrophilic polymers, such as polyethyleneimine (PEI), polyacrylic acid (PAA), and polyethylene glycol (PEG), can also assist the isotropic coating of MOFs on UCNMs [28–31]. In particular, PAA-modified upconversion nanoparticles can work as substrates for the wrap of ZIF-8, UiO-66, and HKUST-1 (The Hong Kong University of Science and Technology) MOFs thanks to the plentiful carboxylic acid groups in PAA [29–31]. PEG-phosphate can displace the pristine oleate ligands on $NaYF_4$:Yb/Tm UCNMs and offer nucleation sites for the growth of UiO-66-NH_2 [32]. Interestingly, some polymers can afford additional functions

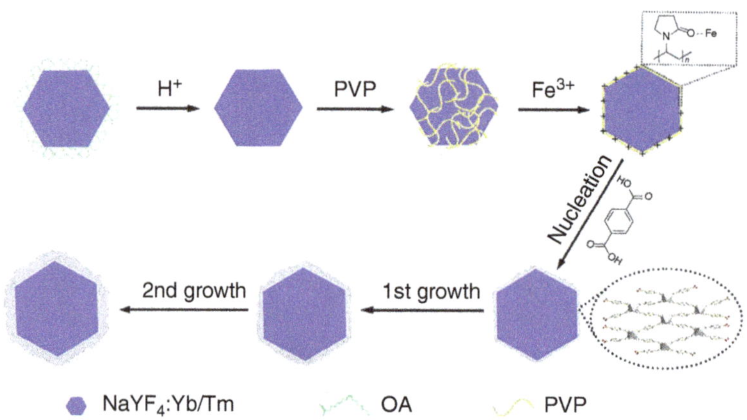

Figure 4.1 Schematic illustration of the formation process of the upconversion nanoparticle@MIL-53(Fe) composite. Source: Reproduced with permission from [26], 2017 / American Chemical Society.

in addition to the coating of the MOF shell. As exemplified by the PEI-involved synthesis of the $NaGdF_4$:Yb/Tm@$NaGdF_4$-g-C_3N_4@ZIF-8 multicomponent composite, PEI capped on $NaGdF_4$:Yb/Tm@$NaGdF_4$ upconversion nanoparticles not only work as a buffer layer for the nucleation of ZIF-8 but also can electrostatically attract charged g-C_3N_4 nanosheets [28].

Polymers, possessing flexible long-chain structures and ample coordinative groups, are excellent glues to chemically integrate UCNMs with MOFs. Sometimes, they can also bring UCNM-MOF composites additional functions [28, 33]. However, polymer-assisted fabrication of the UCNM-MOF composite still has limitations. One major problem is that the roles of polymers in fabricating the composite are not well-disclosed or comprehensive. Particularly, the formation mechanisms, proposed by reported studies, fail to explain the fact that most polymers can only afford the growth of limited kinds of MOFs on UCNMs [14–16]. It is also difficult to answer why the same polymer-modified UCNMs can yield different conformations of UCNM-MOF composites (e.g. core–shell, multiple cores-shell, and Janus structures) [15]. Another obvious challenge is that the energy and electron transfer between MOFs and UCNMs can be blocked by the densely coated insulating polymer layer [15, 16, 22]. This can lead to declined performances of the composites in applications [15, 16, 22]. Novel functional polymers need to be developed, and deeper investigations on the formation mechanism of the polymer-assisted synthesis of UCNM-MOF composites are necessary.

SiO_2 can serve as a hard template for the nucleation and growth of MOF [21, 22]. Generally, SiO_2 is biocompatible and almost transparent, and thus it is regarded as a bio-safe substrate with negligible influence on the upconversion luminescence of UCNMs [34]. Furthermore, SiO_2 can exhibit a porous structure with high tunability, enabling the loading of other functional materials [35]. Thus, the SiO_2-assisted fabrication of UCNM-MOF composites has gained burgeoning interest [15, 16]. UCNM-MOF composites obtained through SiO_2-assisted synthesis can feature hierarchical structures and superior properties, such as enhanced adsorption capacity and selectivity, compared with their counterparts without SiO_2 [15, 16].

In a typical SiO_2-assisted synthesis of UCNM-MOF composites, UCNMs are first decorated by a SiO_2 layer via a well-established sol-gel method [36]. The resulting UCNM@SiO_2 nanomaterials are then incubated in the solution, dissolving one building block (metal ions or organic linkers) of the target MOF, wherein the SiO_2 layer can act like a sponge to absorb the building block. The target MOF can grow on UCNM@SiO_2 nanomaterials upon the addition of the other building blocks through the in situ reaction of building blocks. Following this process, several MOFs, including MIL-100(Cu/Fe), ZIF-90, and ZIF-8, have been coated on UCNM@SiO_2 nanomaterials, yielding core–shell-shell structured architectures [37–40].

Nevertheless, the above-mentioned synthesis can incur self-nucleation of MOFs, possibly because of the unideal affinity between the SiO_2 layer and MOFs, as well as the leakage of absorbed building blocks from the SiO_2 layer before their reaction into MOFs [36]. To alleviate this problem, further functionalization of the SiO_2 layer with other functional groups, such as $-NH_2$ and –COOH group, after obtaining UCNM@SiO_2 nanomaterials is recently demonstrated

[41–43]. These groups can harvest and chemically anchor building blocks of MOFs (e.g. metal ions or organic linkers), significantly improving the affinity of MOFs and UCNM@SiO_2 nanomaterials. By virtue of the modified strategy, UCNM@SiO_2@HKUST-1 and UCNM@SiO_2@ZIF-8 composites have been facilely achieved without obvious self-growth of MOFs [41–43]. Notably, the functional group-modified SiO_2 layer also can stably load drugs through the well-documented EDC/NHS (EDC: 1-ethyl-3-(3-dimethylaminopropyl)carbodiimide hydrochloride, NHS: *N*-hydroxysuccinimide) coupling method, which provides additional functions to UCNM-MOF composites [15, 41].

Very recently, a novel UCNM-MOF composite, in which many upconversion nanoparticles are sandwiched by a SiO_2 core and a MOF shell, has been fabricated by Tang and coworkers (Figure 4.2) [41]. They first obtained the SiO_2@upconversion nanoparticle core-satellite superstructure by conjugating $-NH_2$ functionalized SiO_2 microspheres with PAA-modified $NaYF_4$@$NaYF_4$:Yb/Er@$NaYF_4$ core–shell-shell upconversion nanoparticles through the EDC/NHS coupling. Then, they modified the SiO_2@upconversion nanoparticle superstructure with dodecyltrimethylammonium bromide and synthesized SiO_2@upconversion nanoparticle@HKUST-1 by repeatedly reacting the modified SiO_2@upconversion nanoparticle superstructure with copper acetate and 1,3,5-benzenetricarboxylic acid. Compared with the UCNM@SiO_2@MOF composite, the newly developed structure can encapsulate plenty of UCNMs in a single composite, ensuring strong upconversion luminescence during applications.

SiO_2-assisted encapsulation is attracting growing attention due to the porous structure and biocompatibility of SiO_2, but it is challenged by the tedious fabrication

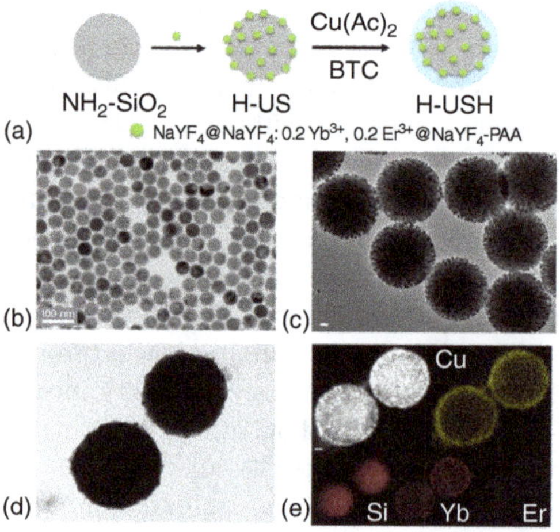

Figure 4.2 (a) Schematic diagram for synthesizing H-USH. (b–d) Transmission electron microscopy images of (b) upconversion nanoparticles, (c) H-US and (d) H-USH. (e) Element distributions of H-USH (scale bar = 100 nm). H-US denotes upconversion nanoparticle-modified SiO_2 microspheres, and H-USH denotes MOF-modified H-US. Source: Reproduced with permission from [41], 2023 / American Chemical Society.

procedure. Also, the SiO_2 layer spatially separates MOF and UCNMs, which may impede the energy flux between them.

Alternatively, polydentate small molecules, which have multiple chelating sites capable of binding to both rare earth ions on UCNMs and metal ions of MOFs, can assist the formation of UCNM-MOF composites [44–48]. Generally, they can trigger the coating of MOFs on UCNMs following a similar process and mechanisms to those adopted in polymer-assisted encapsulation [15, 16]. Compared with polymers, polydentate small molecules potentially have weaker interference to the transfer of energy or electrons between UCNMs and MOFs because of their much shorter backbones [44–48]. To date, polydentate small molecule-assisted coating of MOFs on UCNMs has been extensively investigated [15, 16, 44–48]. One prototype of UCNM-MOF composites obtained through this method is the $NaYF_4$:Yb/Er@porous coordination network (PCN) composite reported in 2019 (Figure 4.3) [44]. Specifically, 3,4-dihydroxyhydrocinnamic acid (DHCA) small molecule was modified on the surface of $NaYF_4$:Yb/Er upconversion nanoparticles through ligand exchange. Due to the double carboxyl groups in DHCA, the DHCA-modified UCNMs can chelate with Zr^{4+}, and further coordinate with 5,10,15,20-tetrakis(4-carboxyphenyl)porphyrin (TCPP) to form small PCN-222 clusters, allowing the growth of PCN-222 around UCNMs. Transmission electron microscopy analysis of the intermediates extracted at different reaction time reveals that the overgrowth of PCN-222 follows the Ostwald ripening-mediated grafting mechanism. Interestingly, the DHCA-assisted PCN-222 coating is also applicable to other nanomaterials (e.g. iron oxide, gold, silver, and UiO-66), and the thickness of PCN-222 can be facilely tuned.

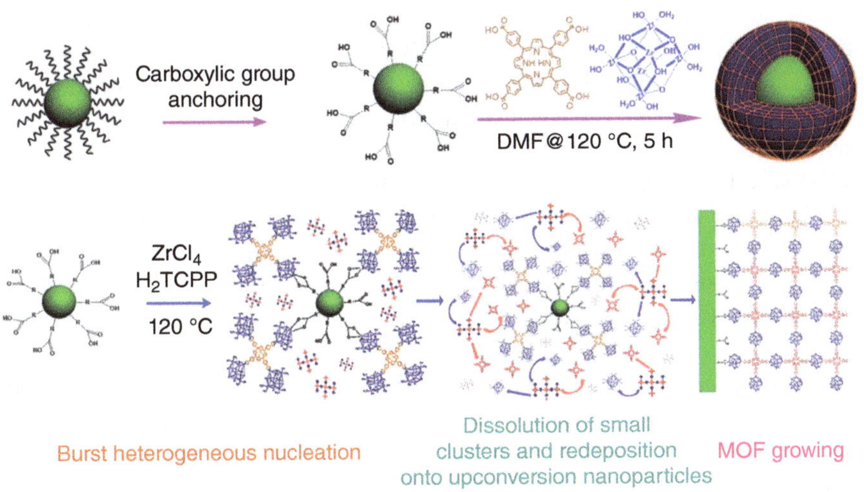

Figure 4.3 Schematic illustration of heterogeneous nucleation and Ostwald ripening-mediated grafting of PCN-222 on upconversion nanoparticles. Source: Reproduced with permission from [44], 2019 / American Chemical Society.

The use of DHCA provides a new avenue to construct the UCNM-MOF composite, while the precise grafting of MOFs on UCNMs is still daunted by the uncontrollable nucleation/growth rate of MOFs. To gain more accurate control of the coating of MOFs, a double solvent reaction system was integrated into the DHCA-assisted encapsulation [45]. In particular, DHCA-modified UCNMs and MOF precursors are reacted in a mixed solvent containing dimethylformamide (DMF) and ethanol with a specific ratio. In the mixed solvent, DMF mainly determines the formation of the MOF matrix, while ethanol can change the coordination environment of metal ions for preferential nucleation of MOFs. By careful control of the reaction parameters, the rate of MOF formation can be adjusted precisely, inducing a very efficient coating (a yield of 98%) of MOF on UCNMs. Moreover, the thickness of the MOF shell can be precisely adjusted from a few to tens of nanometers simply by changing the amount of UCNM seeds.

In addition to DHCA, other polydentate small molecule-based coating of MOF on UCNMs has also been demonstrated [46–48]. For example, citric acid holding triple carboxyl groups can anchor on the surface of $NaGdF_4$:Yb/Er@$NaGdF_4$ upconversion nanoparticles and concurrently provide nucleation sites for the subsequent growth of PCN-224 [46]. It should be pointed out that both citric acid and DHCA are foreign to MOFs and UCNMs, which potentially may hinder the interactions between MOFs and UCNMs. Recent works smartly addressed this problem by capping linkers of MOFs on UCNMs, and the modified UCNMs can be easily coated by corresponding MOFs in the presence of MOF precursors [47, 48]. Using this strategy, UCNMs were facilely wrapped by Eu-BDC (BDC: 1,3-benzenedicarboxylic acid) MOF and MIL-100(Fe) only using precursors of MOFs [47, 48].

Polydentate small molecule-assisted encapsulation has recently gained a lot of interest because of its simple and easily conducted procedure. However, its wide use is constrained by its limited applicability. Currently explored polydentate small molecules, such as citric acid and DHCA, can only enable the coating of specific MOFs on UCNMs [16]. Using linkers of MOFs as capping ligands is a potential solution that also affords the merit of clean fabrication, while successful demonstrations are still scarce. Moreover, the encapsulation mechanisms also need to be comprehensively investigated to guide the expansion of this strategy.

Collectively, in situ encapsulation is one of the most-used strategies to fabricate UCNM-MOF composites with the core–shell structure [15, 16]. This method commonly relies on the extensive wrapping of UCNMs by surface agents, including polymers, SiO_2, and polydentate small molecules. It has the advantage of producing small core–shell composites with uncompromised porosity of MOFs, which is vital for drug delivery and cell imaging [44–48]. In addition, the resulting core–shell structured composites potentially have very high ET efficiency between UCNMs and MOFs because of the full cover of MOFs, advantaging the photo-stimulated events, such as phototriggered drug release and therapy [44–48]. However, the indispensable capping agents may considerably decline the ET efficacy in practical applications. Besides, in situ encapsulation commonly requires tedious fabrication, which can encounter self-nucleation of MOFs, thus challenging a wider application of this method.

4.2.2 Partial Embedment

Partially embedding UCNMs into the MOF matrix is an alternative to constructing functional UCNM-MOF composites [46, 49, 50]. Different from in situ encapsulation, this method typically produces composites with other structures, such as Janus and core-satellite, majorly depending on the surface properties of UCNMs and MOFs [46, 49, 50]. Until now, partial embedment has mostly been used to fabricate UCNM-PCN composites [46, 49].

Similar to in situ encapsulation, partial embedment of UCNMs in MOFs can also be induced by capping ligands of UCNMs, and heterodimers with Janus structure are commonly obtained [46, 49]. PVP is one of the most frequently used capping ligands for the preparation of partial embedment Janus UCNM-MOF composites [46, 49]. Li and coworkers reported the UCNM-MOF heterodimer by selective anisotropic growth of PCN-224 onto PVP-capped upconversion nanoparticles (Figure 4.4) [46, 49]. In their fabrication, the capping layer of PVP can stabilize the upconversion nanoparticles in the reaction solution and harvest Zr^{4+} ions onto the surface of upconversion nanoparticles to facilitate the subsequent growth of PCN-224. Importantly, theoretical calculations by density functional theory revealed that PVP could preferentially bind to the (001) facet of upconversion nanoparticles, resulting in the growth of PCN-224 on the corresponding facet, which should account for the formation of Janus rather than core–shell structure. When the upconversion nanoparticles were not modified by PVP, no heterodimers were formed, and self-nucleation of MOFs significantly occurred. Moreover, their recent study found that when the citric acid small molecule was modified on upconversion nanoparticles, core–shell structured composites instead of heterodimers were obtained (Figure 4.4) [46]. These experiments confirm the indispensable role of PVP in this strategy. On the basis of this method, different upconversion heterodimers based on PCN-224 have been synthesized.

However, the multidentate sites of PVP and incomplete coating of MOF increase the chances for the formation of trimers or even tetramers. Such multiple growth of MOFs onto the same single UCNM can result in apparent aggregation of the prepared composites. Moreover, both the usable capping ligands and the resulting UCNM-MOF heterodimers are still very limited.

Capping ligand-mediated fabrication generally demands premodification of the UCNMs, which is complicated and can be deleterious to the interaction between the components [50]. To circumvent this problem, the fabrication of a core-satellite UCNM-MOF composite through electrostatic attraction-induced partial embedment without the assistance of any capping agents has been demonstrated (Figure 4.5) [50]. Briefly, to fabricate the composite, ligand-free upconversion nanoparticles directly reacted with MOF precursors. During the reaction, the homogenous nucleation of MOF generated small MOF crystals, accompanied by the attachment of positively charged upconversion nanoparticles to the surface of negatively charged MOF crystals. With the continuous growth of MOF crystals, upconversion nanoparticles were partially embedded in MOF crystals, just like streets paved with cobblestones. Intriguingly, this strategy can partially embed

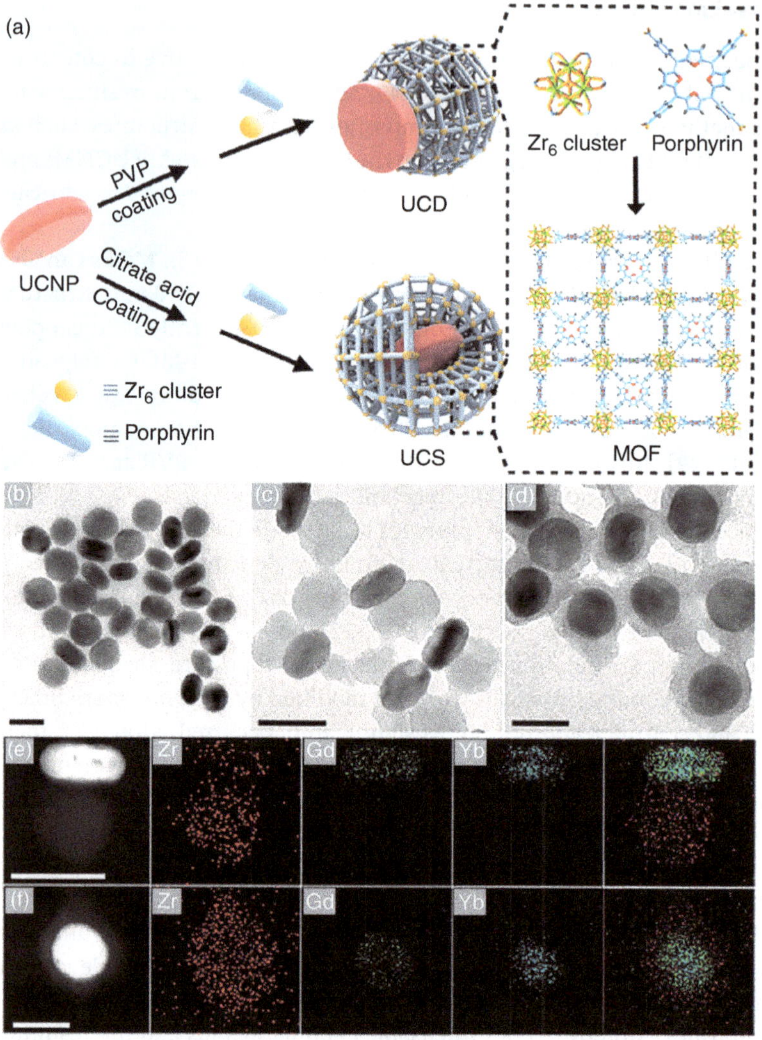

Figure 4.4 (a) Schematic illustration of the synthesis of upconversion nanoparticle-MOF heterodimers (UCDs) and core–shell upconversion nanoparticle@MOFs (UCSs) through the conditional surface engineering of upconversion nanoparticles (UCNPs). The structure of porphyrinic MOFs is illustrated in the right column. (b–d) Transmission electron microscopy images of (b) UCNPs, (c) UCDs, and (d) UCSs (scale bar = 50 nm). (e,f) Elemental mapping images of a single (e) UCD and (f) UCS (scale bar = 50 nm). Source: Reproduced with permission from [46], 2020 / American Chemical Society.

various UCNMs (e.g. $NaYF_4$, $LiYF_4$, and $NaGdF_4$) in MOFs with different components (e.g. UiO-66, UiO-66-NH_2, MOF-801, and PCN-223) and morphologies (e.g. octahedron, spindle, and sphere), proving its good generality. It also allows the further design of multifunctional nanocomposites with sophisticated structures, such as a sandwiched MOF@upconversion nanoparticles@MOF structure (Figure 4.5).

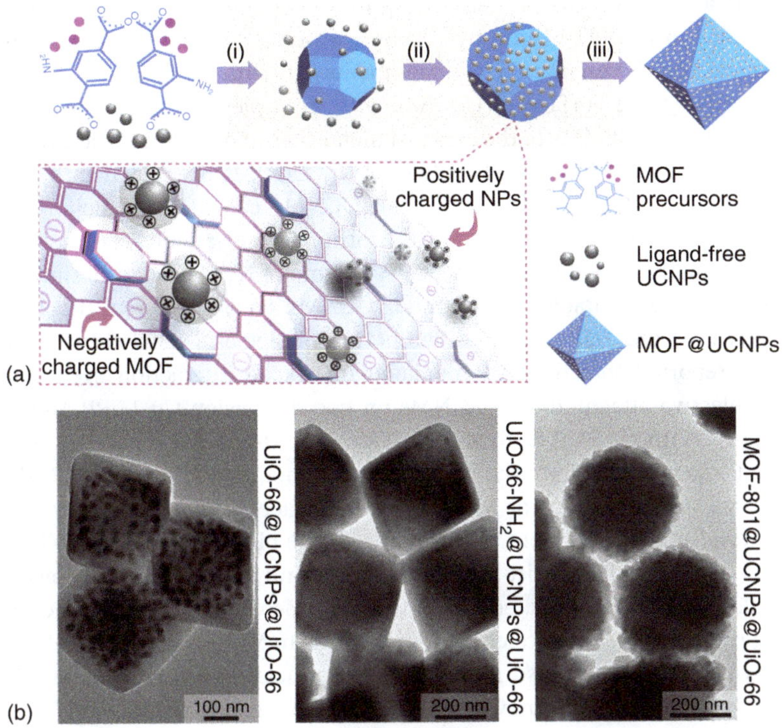

Figure 4.5 (a) Schematic illustration of the fabrication of upconversion nanoparticles (UCNPs) and MOF composites. (b) Transmission electron microscopy images of the obtained MOF@nanoparticles@MOF core–shell-shell sandwiched composites. Source: Reproduced with permission from [50], 2018 / American Chemical Society.

Electrostatic attraction-induced partial embedment provides a new way, free from capping ligands, to fabricate UCNM-MOF composites. It also can uniformly arrange a large number of UCNMs on MOFs, imparting strong upconversion luminescence and potential designability to the resulting composites. However, the size of the resulting core-satellite composite is relatively large for bioapplications.

Taken together, the partial embedment strategy elevates the controllability of the structure and morphology of UCNM-MOF composites. Nevertheless, capping ligands and the relatively low interfacial contact area may considerably compromise the ET efficiency between UCNMs and MOFs. Electrostatic attraction-induced partial embedment may, to a certain extent, alleviate the issue, but its control fabrication toward small-sized composites is still challenging.

4.2.3 Interfacial Attachment

Interfacial attachment briefly means that UCNMs and MOFs are only attached together, but not one totally or partially in the body of another [51–59]. This attachment typically relies on the interfacial interaction between UCNMs and MOFs, and such interactions can be established by the precise surface engineering of both UCNMs and MOFs [51–59]. According to the types of interactions, interfacial attachment can be divided into two classes, which are ligand-mediated and electrostatic attraction-driven interfacial attachments [51–59].

Capping ligands on the surface of UCNMs with multiple anchoring sites can direct the attachment of UCNMs to the exposed metal sites of presynthesized MOFs [51–58]. For example, when the PVP-coated $NaYF_4$ UCNMs are added into the solution containing MIL-53(Fe) octahedra, the UCNMs can facilely anchor on the concave surface of MIL-53(Fe) octahedra, forming core-satellite composites, due to the coordinative capability of the PVP polymer [51]. Through a similar PVP-mediated interfacial attachment method, UCNMs have also been attached to MIL-101-NH_2(Fe) and ZIF-8 with different morphologies [52]. Recently, Li and coworkers reported that the PVP-mediated interfacial interaction even can guide a facet-selective attachment of UCNMs on two-dimension Cu-TCPP MOFs (Figure 4.6) [54]. Without PVP, both the UCNMs and MOFs are separately dispersed without assembly. Differently, using the PVP-modified UCNMs, UCNMs can well disperse on the surface of MOFs with high facet selectivity. To further understand the selective attachment, they calculated the binding energy of PVP to different facets of UCNMs by density functional theory and found that PVP showed preferential adsorption onto the lateral facet (100) of upconversion nanoparticles, thus directing the facet-selective attachment of UCNMs to MOFs via (100) facets.

Besides polymers, DNA can also mediate the interfacial attachment of MOFs and UCNMs through the well-known Watson-Crick base pairing [56–58]. The first attempt was reported by He and coworkers [56]. Briefly, they conjugated two complementary single-stranded DNA to the surface of amine-modified MOFs and UCNMs through click reaction, yielding DNA-modified MOFs and UCNMs, respectively. Then, they incubated the DNA-modified MOFs with DNA-modified UCNMs, allowing them to interact with each other through the hybridization of complementary DNA strands and form a core-satellite structured UCNM-MOF

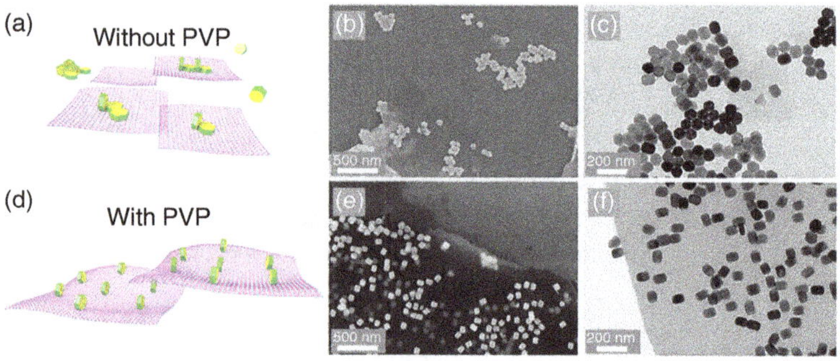

Figure 4.6 (a) Illustration of the composite synthesized with bare upconversion nanoparticles and the associated (b) scanning electron microscopy and (c) transmission electron microscopy images. (d) Illustration of the upconversion nanoparticle-MOF composite assembled with PVP-modified upconversion nanoparticles and the associated (e) scanning electron microscopy and (f) transmission electron microscopy images. Source: Reproduced with permission from [55], 2021 / John Wiley & Sons.

composite. Precise control of the number of UCNMs attached to MOFs was achieved by simply varying the molar ratio of UCNMs to MOFs.

Electrostatic attraction is another interfacial interaction that is commonly used to attach UCNMs to MOFs [59]. As demonstrated in 2020, UiO-68-NH_2 MOFs and $NaGdF_4$:Yb/Er@$NaGdF_4$:Nd/Yb upconversion nanoparticles can be facilely integrated through electrostatic attraction [59]. Specifically, MOFs and upconversion nanoparticles were first prepared separately. The pristine oleate ligands on the surface of upconversion nanoparticles were then removed through HCl treatment, releasing the positively charged surface of upconversion nanoparticles. Subsequently, negatively charged MOFs were incubated with positively charged upconversion nanoparticles under ultrasonication and stirring. In this process, the oppositely charged two materials would attach to each other because of the electrostatic attraction, yielding uniform UCNM-MOF core-satellite structured composites. A close check on the composite revealed that nearly 34 upconversion nanoparticles were evenly attached on the surface of single MOFs, conferring the composite with strong upconversion emissions. However, limited UCNM-MOF composites have been fabricated using this method. Moreover, the long-term stability of the composites obtained through this method needs to be monitored, as the electrostatic attraction may not be so stable [60].

4.3 Bioapplications

Since UCNM-MOF composites inherit the properties of both UCNMs and MOFs, they principally are competent to implement all the applications that UCNMs and MOFs are capable of achieving [11]. Thus, UCNM-MOF composites can be facilely applied for bioapplications, such as imaging and drug delivery [11, 16]. Notably, the synergetic effects between UCNMs and MOFs (or between UCNMs and cargoes in MOFs) can further broaden the applications of UCNM-MOF composites, which also makes UCNM-MOF composites superior to their parent materials in some fields, such as multimode therapy and imaging [15]. In the following section, we discuss the cutting-edge bioapplications of UCNM-MOF composites, including therapy, bioimaging, and biosensing [11, 15, 16].

4.3.1 Therapy

Disease control is a pivotal but challenging task for the public health system, and therapeutic agents can help to fulfill this task by strengthening immunity, slowing disease progression, prolonging survival, and even completely curing some diseases [61]. UCNM-MOF composites can serve as therapeutic agents [11]. In a typical use of UCNM-MOF composites as the therapeutic agent, MOFs majorly account for the therapy via releasing preloaded drugs or generating therapeutic species through chemical reactions due to their high porosity and activity [11, 21, 22]. UCNMs seldom possess therapeutic effects because of their inert nature but can provide photonic stimulus to therapeutic MOFs, rendering light-activatable therapy [1, 2, 11].

To some extent, UCNM-MOF composites are ideal for on-demand therapy with high accuracy, alleviated side effects, and customized schedules [11, 15, 16]. UCNM-MOF composite-based therapy can even be more precise and less harmful by harnessing the luminescence of UCNMs for targeting during therapy [11, 15, 16]. Furthermore, UCNM-MOF composites are able to fight against deep-seated diseases in the human body thanks to the highly penetrative NIR excitation light [11, 15, 16]. In fact, UCNM-MOF composites have been demonstrated to be promising in treating diverse cancers, and different therapeutic strategies, such as photodynamic therapy (PDT), photochemical therapy, photothermal therapy, immunotherapy, and chemotherapy, have been developed based on UCNM-MOF composites [11, 15, 16].

PDT is a promising clinical treatment that combats skin diseases and cancers by generating cytotoxic ROS [60]. Compared to conventional therapeutic modalities, it has advantages, such as small long-term side effects, minimal invasiveness, high selectivity, and easy operation [62]. PDT typically relies on photosensitizers and a light source [62]. Specifically, photosensitizers can generate ROS under light irradiation. The generated ROS can directly kill lesional cells and may also damage blood vessels around target tissues, which prevents the tissue from receiving nutrition via blood [62].

Traditional photosensitizers are organic molecules, such as porphyrin, porphine, and phthalocyanine derivates [62, 63]. They have advanced the development and clinicalization of PDT but remain suffering from poor water-solubility, low ROS generation capacity, and aggregation-induced quenching effect [64]. Fortunately, MOF-based photosensitizers, using photosensitive ligands as organic linkers, can give a decent solution [64]. In particular, the ordered arrangement of ligands in MOFs can prevent the aggregation of photosensitizers, guaranteeing a high ROS generation capacity [64]. MOF-based photosensitizers can also feature high long-term stability and good biocompatibility [64]. Nevertheless, most MOF-based photosensitizers can only be activated by UV/visible light, which shows shallow penetration depth and is potentially harmful to normal tissues under long-term exposure [11, 15, 64]. In this context, UCNM-coupled MOF photosensitizers can realize NIR-driven PDT [11].

UCNM-MOF composites involving porphyrinic MOFs are the most representative composites for PDT [11]. In 2017, He and co-workers carried out a study on NIR-driven PDT using a core-satellite UCNM-MOF composite (Figure 4.7a) [56]. The composite was fabricated through DNA-mediated interfacial attachment of multiple $NaYF_4$:Yb/Er upconversion nanoparticles on a single PCN-224 MOF nanocrystal. In the core-satellite composite, PCN-224 can absorb green and red upconversion emissions of upconversion nanoparticles under 980 nm excitation, producing $^1O_2$ for therapy. The $^1O_2$ production can be modulated by controlling the loading amount of upconversion nanoparticles on MOF, reaching a yield of 57.9% in a composite with an upconversion nanoparticle-to-MOF ratio of 12. For comparison, the mixture of the same amount of MOF and upconversion nanoparticles yielded significantly less $^1O_2$ (25.1%). The in vitro study of PDT showed a decrease of 63.7% in cell viability after 20 min 980 nm irradiation in the presence of the composites, which was much greater than that observed in the control experiment conducted

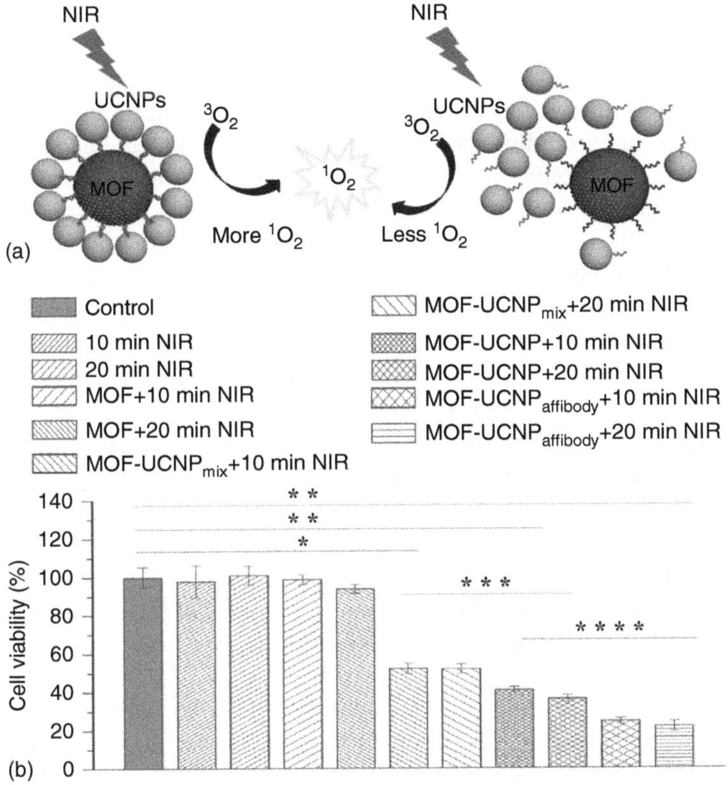

Figure 4.7 (a) Illustration of self-assembled MOF-upconversion nanoparticle (UCNP) core-satellite composites (left) and random mixtures of MOFs and UCNPs (right) for PDT. (b) Quantitative analysis of MDA-MB-468 cell viabilities with different treatments. Data are averages ± one SD ($n > 3$ experimental replicates). $^*p < 10^{-4}$, $^{**}p < 10^{-6}$ relative to control (black). $^{***}p < 10^{-5}$, $^{****}p < 10^{-3}$. Source: Reproduced with permission from [56], 2017 / John Wiley & Sons.

with a simple mixture of upconversion nanoparticles and MOFs (Figure 4.7b). These results clearly prove the PDT effect of PCN-224@$NaYF_4$:Yb/Er core-satellite composite. Notably, the PCN-224@$NaYF_4$:Yb/Er core-satellite composite showed negligible cytotoxicity to MDA-MB-468 cells, indicating good biocompatibility of the composite. Unfortunately, the performance of the composite for in vivo PDT is not evaluated, and the heating effect from 980 nm excitation light could cause thermal damage in practical therapy [65]. In addition, the attachment motif of UCNMs and MOFs is nonideal for the generation of $^1O_2$ as the ET from upconversion nanoparticles to MOFs is distance sensitive [66]. The developed PDT system also lacks spatial control over the photodynamic effect and, therefore, possesses limited therapeutic efficiency.

Recently, an 808 nm light-activable system based on a UCNM-MOF Janus composite was demonstrated for organelle-targeting PDT both in vivo and in vitro, with high efficiency and no obvious heating damage [67]. In this study, the PDT targeted mitochondria, the cell's powerhouse and weapon store for cellular suicide, to regulate the levels of $^{1}O_2$ in mitochondria and induce mitochondrial dysfunctions or trigger mitochondria-mediated apoptosis, thus enhancing therapeutic efficiency. Specifically, the presynthesized Janus composite was functionalized by triphenylphosphine molecules, a mitochondria-targeting ligand, onto the surface of the MOF domain through coordination interactions between exposed metal ions and –COOH of triphenylphosphine molecules. The triphenylphosphine-modified composites can precisely dwell among mitochondria, ensuring a 1.9-fold higher photodynamic cytotoxicity to 4T1 cells than that of pristine composites under 808 nm excitation. The in vivo study further confirmed the greater PDT effect of triphenylphosphine-modified composites than the unmodified counterpart. Further mechanism study uncovered that $^{1}O_2$ generated by Janus composites under NIR excitation can cause the depolarization of mitochondrial membrane potential, triggering the release of apoptotic factors, the activation of caspase-3, and eventually the boost of cell apoptosis. Notably, the use of Nd^{3+}-sensitized UCNMs enables the system to work under 808 nm light excitation, minimizing the heating effect of irradiation to water in cells and thus protecting the normal tissues. Generally, this attempt gives implications to highly specific NIR-driven PDT with improved efficiency and minimal thermal damage. However, it could be daunted by the tumor hypoxia ubiquitously existing in solid tumors because the yield of $^{1}O_2$, generated through the route used in this study, is highly related to the surrounding O_2 level [28, 68].

To gain a step forward in efficient PDT under a hypoxia environment, UCNM-MOF composites that can produce ROS independent of O_2 concentrations have been designed [28, 68]. A prototypical example is the core-multi-shelled composite consisting of TiO_2 nanoparticles, PCN-224 MOF, and $NaYF_4$:Yb/Tm upconversion nanoparticles (Figure 4.8) [68]. In this composite, TiO_2 nanoparticles serve as a photosensitizer that can produce ·OH and $O_2{\cdot}^-$ without strong dependence on surrounding O_2 level, and PCN-224 MOF works as another photosensitizer capable of converting $^{3}O_2$ into $^{1}O_2$.

Both photosensitizers can be activated by upconversion luminescence of upconversion nanoparticles under NIR excitation. Benefiting from the dual-photosensitizer-induced PDT, the ternary composite showed higher cytotoxicity (over 70% cell death) than the TiO_2-free binary composite and greater photoinduced ablation of the hypoxic tumors under NIR irradiation. It is worth mentioning that some other nanoparticle-based photosensitizers, such as graphitic-phase carbon nitride nanosheets and carbon dots, have also been introduced to UCNM-MOF composites for synergistic PDT against hypoxic tumor cells [28].

PDT that uses different photosensitizers to simultaneously initiate different routes for the production of ROS can, to some extent, overcome hypoxia during therapy. However, the exocytosis of PDT materials and consumption of ROS by intracellular organics such as glutathione may considerably decline the therapeutic effect [15]. Moreover, the loading of massive photosensitizers can significantly darken the brightness of the UCNMs, disabling the upconversion luminescence imaging based on the composites [28, 68].

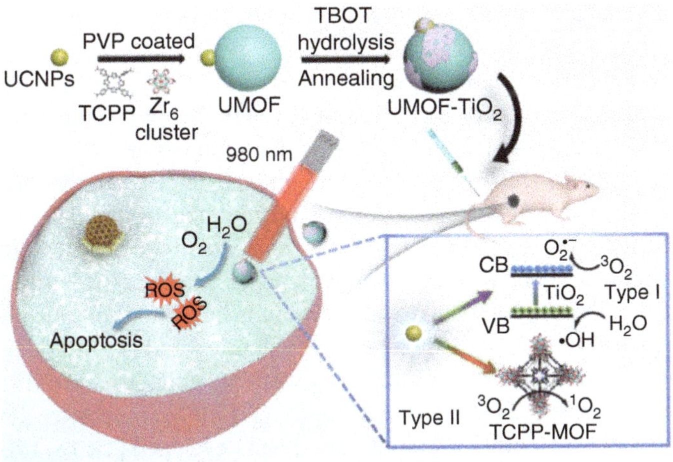

Figure 4.8 Illustration of the preparation and PDT process of the UCNM-MOF-TiO_2 composites. Source: Reproduced with permission from [68], 2020 / American Chemical Society.

Elevating the O_2 level in lesional cells is an effective alternative to address hypoxia in PDT [39, 45, 69–71]. As a result, UCNM-MOF composites that can decompose endogenous H_2O_2 that is excessively produced by tumor cells into O_2 have emerged [45, 69, 70]. For example, Cai and coworkers designed a catalase-modified UCNMs/methylene blue@ZIF-8 core–shell composite for NIR/H_2O_2-responsive PDT against hypoxic tumors [69]. In this design, they employed methylene blue as the photosensitizer to generate $^1O_2$, catalase as the H_2O_2 catalyst to convert H_2O_2 for O_2 supply, UCNMs as the NIR light transducer to trigger PDT, and ZIF-8 as the container to hold and protect methylene blue, catalase, as well as UCNMs. Under a condition with a low O_2 content, the designed UCNMs/methylene blue@ZIF-8@catalase composite showed higher toxicity as compared to the catalase-free composite because the continuous supply of O_2 through H_2O_2 decomposition alleviates hypoxia in cancer cells. Similarly, some other UCNM-MOF composites, such as Fe_3O_4@MIL-100(Fe)@$NaYF_4$:Yb/Tm@$NaGdF_4$ and UCNM@PCN-222(Fe)@Au, that can generate O_2 using endogenous H_2O_2 as fuel have also been designed for phototriggered treatments of hypoxic tumors [45, 70].

Another strategy to guarantee the O_2 supply during PDT is directly delivering molecular O_2 to tumors [15, 39]. For example, ZIF-90, an excellent O_2 absorber, was used to fabricate UCNM-MOF composites as O_2-loading agents for highly efficient PDT, even in a hypoxia environment [39]. The composite can rapidly lift the dissolved O_2 in a buffer solution with a pH of 5.5 but barely change the O_2 content in the solution with a pH of 7.4, indicating its high O_2 uptake capacity and pH-responsive O_2 release ability. Rose Bengal, a photosensitizer, was also loaded in the composite for the ROS generation by absorbing the green upconversion emission of UCNMs. Consequently, this ZIF-90-based UCNM-MOF composite can allow a quick release of preloaded O_2 for enhanced PDT under NIR excitation, once entering the acidic tumor.

Notably, PDT alone is not powerful enough to combat tumors [15, 16]. In the common case, PDT is only able to treat very local tumors as the diffusion distance and lifespan of generated ROS are short [15]. PDT is ineffective for deep-seated tumors inaccessible to light since PDT requires the trigger of light [11]. Moreover, the efficacy of PDT is limited by intrinsic biological responses and complex tumor microenvironments, such as the exocytosis of PDT agents and the consumption of ROS by endogenous reactive organics [15]. The long-existing hypoxia also remains hard to overcome during PDT [16]. Multimodal therapy can avoid the limitations of a single therapy due to the engagement of diverse therapeutic modalities, and the interplay of involved therapeutic modalities often results in appreciably enhanced curative effects [11, 15, 16]. On the basis of this consideration, UCNM-MOF composites for the combined therapy of PDT and other therapies have been developed [11, 15, 16], and other therapeutic modalities can be photochemical, immune, chemo, gaseous, and starvation therapies [15, 16].

UCNM-MOF composites for PDT and photochemical therapy are representative of UCNM-MOF composites for dual-modal therapy. In 2018, the Fe_3O_4@MIL-100(Fe)@upconversion nanoparticles heterojunction composite was developed for the combined therapy of photochemical therapy and PDT [70]. When this composite is excited by 980 nm light, upconversion nanoparticles, $NaYF_4$:Yb/Tm@$NaGdF_4$:Yb, can give intense UV/vis upconversion emission, pumping electrons of MIL-100(Fe) to Fe_3O_4 and leaving holes in the valence band of MIL-100(Fe). The photo-generated electrons and holes both can initiate PDT by reacting with endogenous H_2O_2 and H_2O, respectively. Parallelly, Fe^{3+} ions spreading throughout the MIL-100(Fe) framework can act as photochemical therapy agents that strengthen PDT by producing more ROS. It should be mentioned that Fenton reagents, such as Fe^{3+}, can decompose H_2O_2 for the production of $\cdot OH$ and $O_2\cdot^-$ via a Fenton reaction, strengthening the total ROS generation and enhancing the therapeutic effect [71]. Since Fenton reactions are barely affected by O_2 molecules, such photochemical therapy can work effectively in a hypoxia environment [70]. Thanks to the photochemical therapy-reinforced PDT and high content of H_2O_2 in tumors, the heterojunction composite can effectively fight against HeLa cells in vitro and U14 tumors in mice. Noticeably, the efficacy of Fenton reactions is closely related to the pH value of a system and may reduce both at high and low pH. Thus, an adequate control of pH is advisable for an effective photochemical therapy-coupled PDT [71].

UCNM-MOF composites that can couple PDT with immunotherapy (or immune therapy) have also been developed [72]. Immunotherapy generally works by awakening the immune system to recognize and attack cancer cells [73]. There are different types of agents for immunotherapy, including monoclonal antibodies, immunologic checkpoint inhibitors, tumor vaccines, recombinant cytokines, and CAR-T cells [73, 74]. Among them, immunologic checkpoint blockade programmed cell death protein-1 or its ligand represents a paradigm shift in immunotherapy, as it focuses on the disinhibition of native immune responses instead of the activation of the immune system [74]. In a recent example, UCNM-MOF composites, which integrated with the anti-programmed death-ligand 1 (α-PD-L1) for both PDT and immunotherapy, were established [72]. Under 980 nm excitation, the $NaLnF_4$ upconversion nanoparticles in the composite can convey the upconverted photon energy to the porphyrinic MOF through Förster resonance energy transfer (FRET), generating $^1O_2$ for PDT. Meanwhile, the employed α-PD-L1 can free the inhibited immune responses of T cells to tumors, thus increasing T cell-mediated tumor cytotoxicity and eventually killing tumor cells. This immunotherapy-PDT dual-modal therapy can simultaneously eradicate primary tumors and restrict distant tumors via the abscopal effect, affording an effective inhibition rate of 95% to the bilateral tumor-bearing mice. Interestingly, the $NaLnF_4$ upconversion nanoparticles are able to emit highly penetrative 1550 nm downshifted emission for tracking the location and distribution of the composites during in vivo therapy.

Chemotherapy is another treatment commonly integrated with PDT during the use of UCNM-MOF composites for synergetic dual-modal therapy [49, 75–79]. In fact, chemotherapy itself is one of the most often used strategies to treat cancers in clinical therapy [79]. It utilizes chemo drugs, such as doxorubicin and 5-fluorouracil, to kill cancers and prevent the spread of cancers by blocking the proliferation of target cells [79]. Since the highly porous MOFs are ideal drug containers, UCNM-MOF composites can be facilely armed by chemo drugs for simultaneous chemotherapy and PDT [49, 77–79]. For instance, under the irradiation of 980 nm light, the doxorubicin-loaded $NaYF_4$:Yb/Er-PCN-224 composite showed higher cytotoxicity against 4T1 cells and evoked a lower level of healthy cells (13.44%) in Annexin V-FITC/PI apoptosis detection assay as compared to its doxorubicin-free counterpart (Figure 4.9a,b) [49]. In vivo tests further revealed the high effectiveness of tumor suppression of the doxorubicin-loaded composite (Figure 4.9c). In another study, $NaYF_4$:Yb/Er@Al-MOF composite accommodating doxorubicin hydrochloride also showed excellent antitumor effects under NIR irradiation [76]. It is worth mentioning that other chemo drugs, such as curcumin and 5-fluorouracil, can also couple with UCNM-MOF composites to enhance the therapeutic effects of UCNM-MOF composites [77, 78]. Despite the exciting advances, we would like to note that chemotherapy is commonly a systemic treatment of the whole body, possibly poisoning healthy cells [79]. Although intended therapy based on light- and pH-controlled chemo drug release has been developed to alleviate the side effects, drug leakage may still occur during delivery and therapy, especially under the complex microenvironment of tumors [15, 16].

Considering that cancer cells are greedy for glucose to survive and proliferate, starvation therapy based on glucose deprivation is emerging as an effective and safe reinforcement for cancer treatments [80]. Starvation therapy typically operates on the basis of the consumption of glucose by glucose oxidase [80]. Importantly, the H_2O_2 byproduct of glucose decomposition can produce O_2 once encountering catalase, elevating the O_2 level [16]. Since a high O_2 level can help ROS production during PDT, the combination of starvation therapy and PDT often results in a synergetic antitumor effect stronger than the simple sum of the two therapies [16, 81]. Accordingly, upconversion nanoparticle-ZIF-8 composites modified with glucose

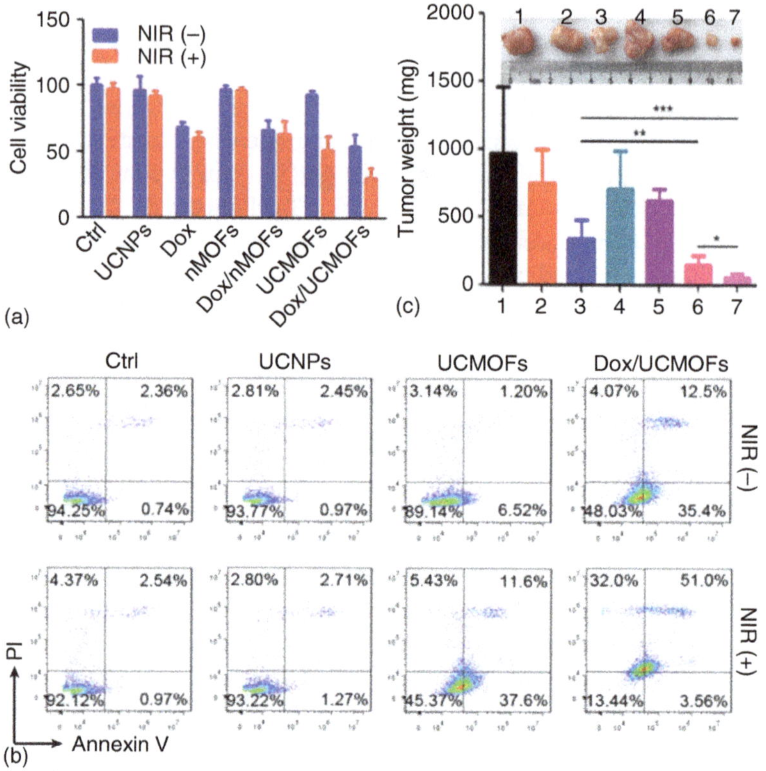

Figure 4.9 (a) Cell viability of 4T1 cells with different treatments (Data: means ± SD; $N = 4$). (b) Flow cytometry analysis of 4T1 cell apoptosis induced by different treatments. (c) Final weights of tumor tissues 14 days after treatment (Data: means ± SD; $N = 5$; $^{*}P < 0.05$, $^{**}P < 0.01$). Inset: representative images of the tumors at day 14 (1: saline, 2: NIR, 3: Dox, 4: UCMOFs, 5: Dox/UCMOFs, 6: UCMOFs + NIR, 7: Dox/UCMOFs + NIR). Ctrl, UCNPs, Dox, nMOFs, and UCMOFs denote control, upconversion nanoparticles, doxorubicin, nanoscale MOFs, and UCNM-MOF composite, respectively. (+) and (–) refer to with and without NIR irradiation, respectively. Source: Reproduced with permission from [49], 2017 / American Chemical Society.

oxidase and catalase were designed for synergetic PDT and starvation therapy [81]. The co-existence of glucose oxidase and catalase can trigger a self-accelerated catalytic cascade reaction that is capable of boosting glucose consumption and producing numerous O_2, which thus can amplify the therapeutic efficacy of both PDT and starvation therapy. The designed composite can kill 77% of HeLa cells in vitro. Interestingly, the composite was also a self-propelled nanomotor powered by H_2O_2 decomposition, which rendered it increased cellular uptake ability during therapy. Notably, in addition to glucose oxidase, glucose oxidase-like nanozyme can also be integrated with UCNM-MOF composites to enable starvation therapy [45]. For example, ultrasmall Au nanoparticles, a glucose oxidase-like nanozyme, have been modified on a UCNM-MOF composite composed of the $NaYF_4$@$NaYbF_4$:Er@$NaYF_4$ upconversion nanoparticle core and the Zr-TCPP(Fe)

MOF shell for synergetic PDT and starvation therapy [45]. The in vivo experiment indicated that the Au-modified composite induced nearly 100% tumor eradication on day 8 with no tumor recurrence.

In order to further enhance the therapeutic effect, UCNM-MOF composites for multimodal therapy that combine PDT and more than one other treatments have also been developed [82]. As a paradigm, Wang and coworkers fabricated a Janus UCNM-MOF composite for simultaneous PDT, photochemical therapy, and sonodynamic therapy (Figure 4.10a) [82]. Specifically, they integrated $NaYF_4$:Yb/Tm@$NaYF_4$:Yb@$NaNdF_4$ upconversion nanoparticles with porphyrin MOF PCN-224(Fe), forming the Janus composite. Under the excitation of 808 nm light, the upconversion nanoparticles can upconvert the photonic energy and then transfer it to the porphyrin MOF segment for PDT. The ultrasound with strong tissue-penetrating ability was also applied to activate the composite, boosting the generation of ROS and triggering sonodynamic therapy. Concurrently, Fe^{3+} ions in the MOF can work as a tri-functional reagent, which supplies O_2 by catalyzing H_2O_2, preserves ROS via glutathione depletion, and generates Fe^{2+} to trigger chemodynamic therapy by producing ·OH radical. The cooperation of three types of therapy, together with the alleviated hypoxia and glutathione depletion, confers this UCNM-MOF composite with excellent antitumor performance. An almost complete

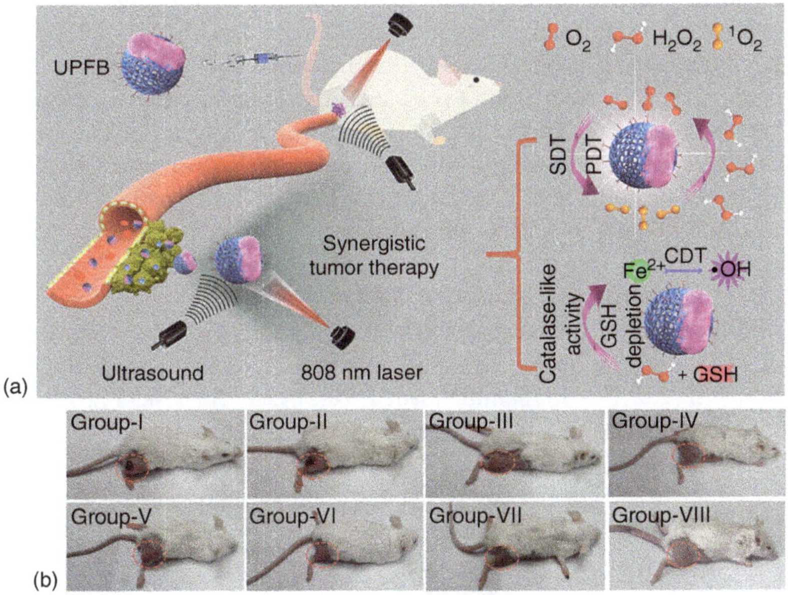

Figure 4.10 (a) Schematic Illustration of antitumor mechanism of the Janus UCNM-MOF composite. (b) The digital photos of tumors extracted from mice after different treatments for 14 days. The groups were (1) control, (2) 808 nm laser, (3) US, (4) UPFB + DMTU, (5) UPFB + DMTU + NIR, (6) UPFB + DMTU + US, (7) UPFB + DMTU + NIR + US, and (8) UPFB + NIR + US. US, UPFB, and DMTU denote ultrasound, Janus UCNM-MOF composite, and UPFB + *N,N′*-dimethylthiourea, respectively. Source: Reproduced with permission from [83], 2021 / American Chemical Society.

tumor regression was observed in U14 tumor-bearing Kunming mice after the mice were treated using this composite (Figure 4.10b). Using a similar all-in-one strategy, doxorubicin-loaded $NaGdF_4$:Yb/Tm@$NaGdF_4$:Yb/Nd@MIL-100(Fe) composites, capable of PDT, photothermal therapy, and chemotherapy, have also been prepared for synergistic tumor treatment [48].

In addition to PDT and PDT-coupled multimodal therapy, UCNM-MOF composites can also be applied to other types of therapies. For example, Yang and coworkers constructed a nitrosothiol-loaded $NaYF_4$:Yb/Tm@ZIF-8 composite for NO therapy of spinal cord injury [83]. They found that the upconverted UV light of $NaYF_4$:Yb/Tm can induce the photo-cleavage of the S-NO bond in nitrosothiol, releasing NO. The generated NO can then activate the differentiation pathway for a markedly pronounced outgrowth of neuronal processes in both PC12 cells and dorsal root ganglion neurons. In another interesting work, it was demonstrated that UCNM-MOF composite can release NO and then enhance wound healing of infectious diabetic ulcers [33]. In the fabricated UCNM-MOF composite, photo-reactive sodium nitroprusside serves as the NO donor, Er^{3+}-activated UCNM offers light stimulus, the degradable polydopamine works as the ROS-responsive chemical and reactive sodium nitroprusside protector, two organic dyes, Cy7 and IR786s, function as ROS indicators for diagnosis, and ZIF-8 is used as a substrate to host above materials. Under NIR irradiation, UCNM can emit green upconversion luminescence to activate the decomposition of sodium nitroprusside into NO. The elevated NO level can inhibit ubiquitination and interaction with the von Hippel Lindau tumor suppressor protein to stabilize hypoxia-inducible factor-1α level in cells, enhancing angiogenesis in wound sites and promoting vascular endothelial growth factor secretion and cell migration. Furthermore, surface-loaded dyes (Cy7 and IR786s) could detect wound infection through the reaction with ROS produced during the bacterial invasion. Notably, UCNM-MOF composites, possessing both NO-based therapy and PDT, also have been developed for antibacterial applications [84].

UCNM-MOF composites can be applied for ion-interference therapy as well [85]. It should be mentioned that ion-interference therapy regulates metal ions to fight against cancers, as metal ions play essential roles in many biological processes [86]. In 2021, a berbamine-loaded $NaYF_4$:Yb/Tm@$NaYF_4$@ZIF-82 composite that can regulate the intracellular Ca^{2+} level was developed for cancer treatment [85]. When the composite is exposed to NIR light, $NaYF_4$:Yb/Tm@$NaYF_4$ upconversion nanoparticles in the composite can generate upconverted photons, and subsequently, the ZIF-82 MOF can produce NO due to the photoreaction of 2-nitroimidazole ligands and release preloaded berbamine through skeleton degradation. The generated NO then can rapidly diffuse and invade the overexpressed ryanodine receptors in breast cancer cells for protein S-nitrosylation, allowing Ca^{2+} generation. Meanwhile, the released berbamine can bind to Ca^{2+} pumps in the cell membrane, inhibiting Ca^{2+} excretion. Collectively, Ca^{2+} can be overloaded in tumor cells, which upregulates the oxidative stress to induce intrinsic cell apoptosis. Both in vitro and in vivo studies confirmed the appreciable antitumor effect of this composite.

4.3.2 Bioimaging

Biological imaging is a fundamental but powerful technique to visualize the structure of the body, organs, tissues, cells, organelles, and even proteins [87]. It provides real-time information about various physiological processes if proper imaging agents are employed [87]. Interestingly, UCNM-MOF composites can work as imaging agents in diverse imaging modes [15, 16]. UCNMs in the UCNM-MOF composites are a typical material with anti-Stokes luminescence that is suitable for optical imaging with high stability, resolution, and deep penetration [11]. The tunable luminescence of UCNMs can even reach the NIR range, which lowers the background signal and deepens the penetration [1, 2]. UCNMs can also be used for other imaging modes, such as CT, MR, and SPECT imaging, if their component and structures are delicately designed [3–5]. Meanwhile, the MOF domain in UCNM-MOF composites can possess imaging ability [88]. Specifically, paramagnetic metal-containing MOFs can function as MR imaging agents, and MOFs involving emissive organic linkers can work for optical imaging [88]. The porous skeleton and high surface area also enable MOFs to host diverse imaging agents, further broadening the imaging applications [88]. The excellent imaging ability of both UCNMs and MOFs as well as the synergetic interactions between them, endow UCNM-MOF composites with great potential in bioimaging not only for understanding biological processes but also for guiding the treatment of diseases [15, 16].

UCNM-MOF composite-based luminescent imaging is one of the most common imaging modes based on UCNM-MOF composites [15, 16]. Typically, upconversion luminescence inherited from UCNMs can be observed in most UCNM-MOF composites, regardless of their structure and fabrication method [15, 16]. As a result, the location of UCNM-MOF composites in biosystems, such as cells, can be directly observed simply using commercialized microscopy equipped with NIR excitation sources [68]. In addition to upconversion luminescence-based optical imaging, luminescent imaging using UCNM-MOF composites as imaging agents can also be achieved using the luminescence from the MOF section [68]. For example, the UCNM-PCN-224 composite can emit characteristic luminescence of porphyrin organic linkers of the PCN-224 MOF and be used for imaging tumors of mice, confirming its potential for imaging-guided cancer treatments [68]. However, luminescent imaging still has some limitations. For example, light scattering widely exists in luminescent imaging and thus would compromise the imaging resolution [3–5]. Moreover, the spatial resolution in three dimensions is poor for traditional luminescent imaging, resulting in fragmentary information [89]. Although the point-to-point scanning mode of luminescent imaging can reconstruct the volumetric structure, it is time-consuming and unsuitable for moving targets [90]. Most importantly, the penetration depth of usable excitations and emissions remains shallow, disabling luminescent imaging of deep-seated targets [3–5].

To overcome the limitations of traditional luminescent imaging, UCNM-MOF composites that are capable of other imaging modes, such as MR imaging, have been constructed [24]. In 2015, a polyethylene glycol-folic acid-modified $NaYF_4$:Yb/Er@Fe-MIL-101-NH_2 core–shell composite was studied for targeted

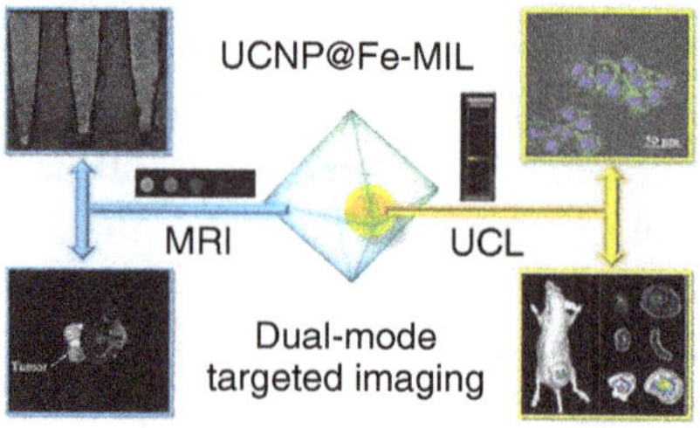

Figure 4.11 Dual-mode targeted imaging based on the $NaYF_4$:Yb/Er@Fe-MIL-101-NH_2 core–shell composite (UCNP@Fe-MIL). Source: Reproduced with permission from [24], 2015 / John Wiley & Sons.

luminescence/MR dual-mode imaging both in vitro and in vivo (Figure 4.11) [24]. An intense upconversion luminescence signal was observed in the tumor site after intravenous injection of this composite, revealing its luminescent imaging ability. Meanwhile, due to the Fe-based MOF, this composite can serve as an MR imaging contrast agent in vivo. The T_2-weighted MR imaging was conducted on a KB tumor-bearing mouse using a 3.0 T human MR imaging scanner, and the tumors with the composites showed obvious darkening effects. Taken together, the core–shell UCNM-MOF composite is a promising imaging agent for luminescence/MR dual-mode imaging. In another study, it was demonstrated that upconversion luminescence/MR dual-mode imaging could also be achieved using Gd^{3+}-doped UCNMs, wherein the UCNMs could offer both luminescent and MR imaging functions [77]. Besides, UCNM-MOF composites for CT and positron emission tomography imaging have also been developed via doping heavy metal ions and adding radioactive elements, respectively [45, 70]. Notably, MR imaging uses strong magnetic fields to image tissues by measuring the relaxation of water protons, which involves no X-ray or ionizing radiation and thus is less harmful than CT and positron emission tomography imaging [15].

Recently, the multimodal imaging strategy that combines more than two imaging modes in a single system is becoming an unparalleled imaging tool for cutting-edge theranostic studies [48]. UCNM-MOF composites can be a good candidate for multimodal imaging. For example, a smart UCNM-MOF composite, $NaGdF_4$:Yb/Tm@$NaGdF_4$:Yb/Nd@MIL-100(Fe), for upconversion luminescence/CT/MR trimodal imaging was developed by Yang and coworkers [48]. In this composite, the core–shell structured UCNM accounts for efficient upconversion luminescence imaging signal under the excitation of 808 nm light. Heavy Gd and Yb elements, showing strong X-ray attenuation, guarantee high imaging contrast for CT imaging. Gd^{3+} and Fe^{3+} provide T_1/T_2-weighted MR imaging signals. Thus, the fabricated composite is simultaneously applicable for luminescent, CT, and MR imaging. Compared to single and dual-mode imaging systems, this trimodal imaging platform potentially provides a more precise diagnostic tool for disease treatment.

4.3.3 Biosensing

Biosensors are essential to understand biological processes for health monitoring and disease treatment [6–8]. UCNM-MOF composites can be ideal platforms for biosensing. In particular, UCNMs in the composites can provide sharp, tunable, stable, and penetrative optical signals, which is favorable for both in vitro and in vivo detection [6–8]. MOFs can allow the accumulation of analytes because of their porous structure and abundant interaction sites [17, 18]. MOFs can also provide vast room to accommodate analyte identifiers and catchers, which can improve the sensitivity and selectivity of sensing [17, 18]. These merits of MOFs and UCNMs have considerably advanced the development of UCNM-MOF composites in biosensing [15, 16].

UCNM-MOF composite-based biosensors typically rely on the optical signal changes caused by targeted analytes for sensing. One of the common changes induced by targeted analytes is the change in the ET between UCNMs and MOFs [7]. For instance, a FRET-based detection of bisphenol A in high-salt foods was achieved using a DNA-assembled $NaYF_4$:Yb/Er-PCN-224 composite [58]. Specifically, when the bisphenol A aptamer-modified UCNMs were mixed with the complementary DNA-coated MOFs in solution, hybridization of DNA can yield the UCNM-MOF composites, quenching the luminescence of UCNMs through FRET. When bisphenol A was added, the aptamer-modified UCNMs competitively recognized the analyte, bisphenol A, impeding the assembly of UCNMs and MOFs. Consequently, FRET was blocked, and the luminescence intensity of UCNMs remained stable. By monitoring the luminescent signals, the UCNM-MOF sensors can quantify the bisphenol A level in a commercial dink. Similarly, using the FRET strategy, a UCNM-MOF composite, composed of $NaYF_4$:Yb/Tm and MIL-101(Cr), was used for sensitive detection of T-2 toxin with the assistance of the DNA-mediate sensor-analyte recognition [57].

In addition to DNA, molecular imprinting technology can also help UCNM-MOF-based sensors selectively detect targets through FRET. In 2016, a $NaYF_4$:Yb/Er/HKUST-1/molecularly imprinted polymer sensor was developed for Bovine hemoglobin detection [31]. In this sensor, the molecularly imprinted polymer layer is an important identification unit that can specifically capture Bovine hemoglobin for subsequent FRET-based sensing. This layer was prepared by employing Bovine hemoglobin as the template, *N*-isopropyl acrylamide as the thermo-sensitive functional monomer, and *N*,*N*-methylenebisacrylamide as the cross-linker. During sensing, Bovine hemoglobin can linearly quench the green upconversion emission of the composite sensor, enabling quantification detection with a detection limit of $0.03\,mg\,L^{-1}$ (S/N = 3). Noticeably, electrostatic attraction of analytes onto the UCNM-MOF composites can also trigger a sensing process, which has been illustrated by the $NaErF_4$:Tm@SiO_2@ZIF-8 composite sensor for gallic acid detection [43], but may be interfered by other charged species in the sensing system [6–8].

Importantly, UCNM-MOF composite can allow the fabrication of sensors with multisignals for sensing. Sensors with multisignals for detection can improve both sensitivity and selectivity [41]. Representatively, Fang and coworkers developed a luminescent/photothermal/electrochemical triple-signal sensor for H_2S sensing based on a UCNM-MOF composite (Figure 4.12) [41]. This composite features a multilayer structure, wherein $NaYF_4$@$NaYF_4$:Yb/Er@$NaYF_4$ UCNMs are sandwiched by a SiO_2 core and a HKUST-1 MOF shell. When the sensor encounters H_2S,

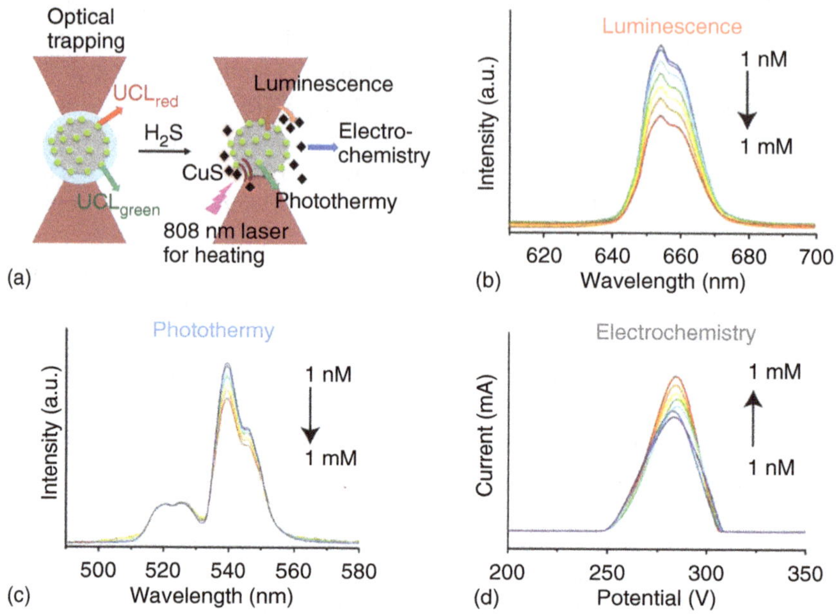

Figure 4.12 (a) Schematic diagram of H_2S detection by the triple-signal sensor. (b) Luminescent spectra, (c) thermal-sensitive upconversion luminescence spectra, and (d) square-wave voltammetry curves of the triple-signal sensor in response to different concentrations of H_2S. Source: Reproduced with permission from [41], 2023 / American Chemical Society.

the Cu^{2+} in the MOF layer can combine with S^{2-} to produce CuS. The generated CuS can quench the 654 nm red upconversion emission of UCNMs, change the intensity ratio of 524 and 545 nm emission through thermal effect, and vary the current difference in the test solution. Accordingly, the luminescent, photothermal, and electrochemical triple-signal detection strategy can be established (Figure 4.12). Through the linear relationship between three detection signals and the H_2S concentration, it was verified that the proposed detection platform can accurately and reliably detect H_2S in vitro.

4.4 Conclusion and Perspectives

In conclusion, we give a fundamental introduction to UCNM-MOF composites in this chapter, revolving around the fabrication and bioapplications of UCNM-MOF composites. Recent studies have proved that UCNM-MOF composites can possess diverse configurations, including the core–shell, core-multishell, Janus, core-satellites, and heterostructure, thanks to the flexible affinity between UCNMs and MOFs. The merits of both UCNMs and MOFs, together with the synergetic interactions between UCNMs and MOFs, make UCNM-MOF composites versatile in various bioapplications including therapy, bioimaging, and biosensing. Thus,

UCNM-MOF composites are gaining more and more research interest. Despite the progress already achieved, there still exist some challenges to be addressed. We here propose some directions for future studies as follows.

First of all, a comprehensive understanding of the forming mechanisms of differently structured UCNM-MOF composites is urgently needed. Current proposed mechanisms can only partially explain the formation of different UCNM-MOF composites. It also remains unclear why the same surface modification of UCNMs results in differently structured UCNM-MOF composites, when different MOFs are engaged. Moreover, we can only design the structure and morphology of very limited UCNM-MOF composites. To overcome the above obstacles, systematic studies on the interface of modified UCNMs, the growth mechanisms of MOFs, as well as the heterogeneous nucleation and self-assembly of MOFs and UCNMs are required.

The enhancement of the ET efficiency between UCNMs and MOFs (or cargos in MOFs) in UCNM-MOF composites is desired to improve the therapy performance, imaging resolution, and probing sensitivity of UCNM-MOF composites. To achieve this goal, amplifying the brightness of involved UCNMs is essential, and it relies on the optimization of the dopant and structure of corresponding UCNMs. An extensive study on how different structures of composites affect the ET between UCNMs and MOFs can guide the design of UCNM-MOF composites and the improvement of ET. Integrating multiple photosensitive species in the composites to absorb all the luminescence of UCNMs is an alternative to improve ET efficiency.

Last but not least, UCNM-MOF composites that can work under more penetrative excitation (e.g. X-ray light) and give more penetrative luminescence (e.g. emission in NIR-II window) need to be explored as this type of composites can meet the requirements of optical therapy, imaging, and sensing on deep-seated targets.

References

1 Zheng, B., Fan, J., Chen, B. et al. (2022). Rare-earth doping in nanostructured inorganic materials. *Chemical Reviews* 122: 5519–5603.

2 Cheng, X., Zhou, J., Yue, J. et al. (2022). Recent development in sensitizers for lanthanide-doped upconversion luminescence. *Chemical Reviews* 122: 15998–16050.

3 Mettenbrink, E.M., Yang, W., and Wilhelm, S. (2022). Bioimaging with upconversion nanoparticles. *Advanced Photonics Research* 3: 2200098.

4 Ansari, A.A., Parchur, A.K., Thorat, N.D., and Chen, G. (2021). New advances in pre-clinical diagnostic imaging perspectives of functionalized upconversion nanoparticle-based nanomedicine. *Coordination Chemistry Reviews* 440: 213971.

5 Sun, Y., Zhu, X., Peng, J., and Li, F. (2013). Core–shell lanthanide upconversion nanophosphors as four-modal probes for tumor angiogenesis imaging. *ACS Nano* 7: 11290–11300.

6 Zhang, M., Wang, N., and Li, Z. (2022). Recent advances in chromophore-assembled upconversion nanoprobes for chemo/biosensing. *TrAC Trends in Analytical Chemistry* 151: 116602.

7 Ansari, A.A., Thakur, V.K., and Chen, G. (2021). Functionalized upconversion nanoparticles: new strategy towards FRET-based luminescence bio-sensing. *Coordination Chemistry Reviews* 436: 213821.

8 Arai, M.S. and de Camargo, A.S.S. (2021). Exploring the use of upconversion nanoparticles in chemical and biological sensors: from surface modifications to point-of-care devices. *Nanoscale Advances* 3: 5135–5165.

9 Zhang, Y., Zhu, X., and Zhang, Y. (2021). Exploring heterostructured upconversion nanoparticles: from rational engineering to diverse applications. *ACS Nano* 15: 3709–3735.

10 Wen, S., Zhou, J., Schuck, P.J. et al. (2019). Future and challenges for hybrid upconversion nanosystems. *Nature Photonics* 13: 828–838.

11 Wei, M., Yuan, Z., Lu, M. et al. (2021). Recent advance in lanthanide doped upconversion nanoparticle–metal organic framework composites. *Chemical Journal of Chinese Universities* 42: 2313–2323.

12 Liu, B., Sun, J., Zhu, J. et al. (2020). Injectable and NIR-responsive DNA–inorganic hybrid hydrogels with outstanding photothermal therapy. *Advanced Materials* 32: 2004460.

13 Zhang, D., Peng, R., Liu, W. et al. (2021). Engineering DNA on the surface of upconversion nanoparticles for bioanalysis and therapeutics. *ACS Nano* 15: 17257–17274.

14 Liu, Q., Wu, B., Li, M. et al. (2022). Heterostructures made of upconversion nanoparticles and metal–organic frameworks for biomedical applications. *Advanced Science* 9: 2103911.

15 Du, J., Jia, T., Zhang, J., and Chen, G. (2023). Heterostructures combining upconversion nanoparticles and metal–organic framework: fundamental, classification, and theranostic applications. *Advanced Optical Materials* 11: 2202122.

16 Zhang, L., Liu, M., Fang, Z., and Ju, Q. (2022). Synthesis and biomedical application of nanocomposites integrating metal–organic frameworks with upconversion nanoparticles. *Coordination Chemistry Reviews* 468: 214641.

17 Freund, R., Zaremba, O., Arnauts, G. et al. (2021). The current status of MOF and COF applications. *Angewandte Chemie International Edition* 60: 23975–24001.

18 Tang, J. and Yamauchi, Y. (2016). MOF morphologies in control. *Nature Chemistry* 8: 638–639.

19 Yu, W., Zhen, W., Zhang, Q. et al. (2020). Porphyrin-based metal–organic framework compounds as promising nanomedicines in photodynamic therapy. *ChemMedChem* 15: 1766–1775.

20 Wang, Z., Sun, Q., Liu, B. et al. (2021). Recent advances in porphyrin-based MOFs for cancer therapy and diagnosis therapy. *Coordination Chemistry Reviews* 439: 213945.

21 Yu, J., Mu, C., Yan, B. et al. (2017). Nanoparticle/MOF composites: preparations and applications. *Materials Horizons* 4: 557–569.

22 Chen, L., Luque, R., and Li, Y. (2017). Controllable design of tunable nanostructures inside metal–organic frameworks. *Chemical Society Reviews* 46: 4614–4630.

23 Lu, G., Li, S., Guo, Z. et al. (2012). Imparting functionality to a metal–organic framework material by controlled nanoparticle encapsulation. *Nature Chemistry* 4: 310–316.

24 Li, Y., Tang, J., He, L. et al. (2015). Core–shell upconversion nanoparticle@metal–organic framework nanoprobes for luminescent/magnetic dual-mode targeted imaging. *Advanced Materials* 27: 4075–4080.

25 Luo, Z., Yi, Z., and Liu, X. (2023). Surface engineering of lanthanide nanoparticles for oncotherapy. *Accounts of Chemical Research* 56: 425–439.

26 Li, M., Zheng, Z., Zheng, Y. et al. (2017). Controlled growth of metal–organic framework on upconversion nanocrystals for NIR-enhanced photocatalysis. *ACS Applied Materials & Interfaces* 9: 2899–2905.

27 Zhang, Y., Liu, H., Ning, L. et al. (2021). A novel core–shell upconversion nanoparticles@zirconium-based metal organic framework fluorescent nanoprobe for efficient continuous detection of trace methylene blue and ferrous ions. *Talanta* 224: 121853.

28 Yang, D., Yang, G., Gai, S. et al. (2017). Multifunctional theranostics for dual-modal photodynamic synergistic therapy via stepwise water splitting. *ACS Applied Materials & Interfaces* 9: 6829–6838.

29 Ghosh, S., Gul, A.R., Xu, P. et al. (2022). Target delivery of photo-triggered nanocarrier for externally activated chemo-photodynamic therapy of prostate cancer. *Materials Today Chemistry* 23: 100688.

30 Zhao, X., Li, Y., Du, L. et al. (2021). Soft X-Ray stimulated lanthanide@MOF nanoprobe for amplifying deep tissue synergistic photodynamic and antitumor immunotherapy. *Advanced Healthcare Materials* 10: 2101174.

31 Guo, T., Deng, Q., Fang, G. et al. (2016). Upconversion fluorescence metal–organic frameworks thermo-sensitive imprinted polymer for enrichment and sensing protein. *Biosensors and Bioelectronics* 79: 341–346.

32 Zuo, C., Guo, Y., Li, J. et al. (2021). A nanoprobe for fluorescent monitoring of microRNA and targeted delivery of drugs. *RSC Advances* 11: 8871–8878.

33 Yang, Y., Huang, K., Wang, M. et al. (2021). Ubiquitination flow repressors: enhancing wound healing of infectious diabetic ulcers through stabilization of polyubiquitinated hypoxia-inducible factor-1α by theranostic nitric oxide nanogenerators. *Advanced Materials* 33: 2103593.

34 Li, X., Wang, L., Fan, Y. et al. (2012). Biocompatibility and toxicity of nanoparticles and nanotubes. *Journal of Nanomaterials* 2012: 548389.

35 Zhang, S., Chu, Z., Yin, C. et al. (2013). Controllable drug release and simultaneously carrier decomposition of SiO_2-drug composite nanoparticles. *Journal of the American Chemical Society* 135: 5709–5716.

36 Liu, J.N., Bu, W.B., and Shi, J.L. (2015). Silica coated upconversion nanoparticles: a versatile platform for the development of efficient theranostics. *Accounts of Chemical Research* 48: 1797–1805.

37 Chu, H., Li, Y., Wang, C. et al. (2022). MOF-coated upconversion nanoconstructs for synergetic photo-chemodynamic/oxygen-elevated photodynamic therapy. *Dalton Transactions* 51: 16336–16343.

38 Yang, L., Zhu, S., He, Z. et al. (2022). Aqueous-phase synthesis of upconversion metal–organic frameworks for ATP-responsive in situ imaging and targeted combinational cancer therapy. *Chinese Chemical Letters* 33: 314–319.

39 Xie, Z., Cai, X., Sun, C. et al. (2018). O_2-loaded pH-responsive multifunctional nanodrug carrier for overcoming hypoxia and highly efficient chemo-photodynamic cancer therapy. *Chemistry of Materials* 31: 483–490.

40 Fang, W.K., Liu, L., Zhang, L.L. et al. (2021). Detection of amyloid β oligomers by a fluorescence ratio strategy based on optically trapped highly doped upconversion nanoparticles-SiO_2@metal–organic framework microspheres. *Analytical Chemistry* 93: 12447–12455.

41 Fang, W.K., Xu, D.D., Liu, D. et al. (2023). Combining upconversion luminescence, photothermy, and electrochemistry for highly accurate triple-signal detection of hydrogen sulfide by optically trapping single microbeads. *Analytical Chemistry* 95: 5443–5453.

42 Meng, J., Bao, W., Liu, M. et al. (2024). MOFs-based nanoagents enable sequential damage to cancer-associated fibroblast and tumor cells for phototriggered tumor microenvironment regulation. *Small* 20: 2304491.

43 Zhang, Y., Ning, L., Gao, D. et al. (2021). A highly sensitive upconversion nanoparticles@zeolitic imidazolate frameworks fluorescent nanoprobe for gallic acid analysis. *Talanta* 233: 122588.

44 Liu, Y., Yang, Y., Sun, Y. et al. (2019). Ostwald ripening-mediated grafting of metal–organic frameworks on a single colloidal nanocrystal to form uniform and controllable MXF. *Journal of the American Chemical Society* 141: 7407–7413.

45 He, L., Ni, Q., Mu, J. et al. (2020). Solvent-assisted self-assembly of a metal–organic framework based biocatalyst for cascade reaction driven photodynamic therapy. *Journal of the American Chemical Society* 142: 6822–6832.

46 Shao, Y., Liu, B., Di, Z. et al. (2020). Engineering of upconverted metal–organic frameworks for near-infrared light-triggered combinational photodynamic/chemo-/immunotherapy against hypoxic tumors. *Journal of the American Chemical Society* 142: 3939–3946.

47 Wei, X., Zhao, G., Feng, P. et al. (2022). Core–shell lanthanide-doped nanoparticles@Eu-MOF nanocomposites for anticounterfeiting applications. *ACS Applied Nano Materials* 5: 1161–1168.

48 Yang, D., Xu, J., Yang, G. et al. (2018). Metal–organic frameworks join hands to create an anti-cancer nanoplatform based on 808 nm light driving up-conversion nanoparticles. *Chemical Engineering Journal* 344: 363–374.

49 Li, Y., Di, Z., Gao, J. et al. (2017). Heterodimers made of upconversion nanoparticles and metal–organic frameworks. *Journal of the American Chemical Society* 139: 13804–13810.

50 Yuan, Z., Zhang, L., Li, S. et al. (2018). Paving metal–organic frameworks with upconversion nanoparticles via self-assembly. *Journal of the American Chemical Society* 140: 15507–15515.

51 Li, M., Wang, J., Zheng, Y. et al. (2017). Anchoring $NaYF_4$:Yb,Tm upconversion nanocrystals on concave MIL-53(Fe) octahedra for NIR-light enhanced photocatalysis. *Inorganic Chemistry Frontiers* 4: 1757–1764.

52 Ren, H., Bai, R., Huang, F. et al. (2022). Decorating upconversion nanoparticles mediated by both active and inert shells on NH_2-MIL-101(Fe) for the photocatalytic degradation of antibiotics and organic dyes. *Crystal Growth & Design* 22: 4864–4873.

53 Jiang, J., Bai, R., Liu, C. et al. (2023). Construction of full solar-light driven $NaYF_4$:Yb,Tm,Er@$NaYF_4$:Eu/NH_2-MIL-101(Fe) composite photocatalyst toward degradation of antibiotics: introducing the energy trapping centers. *Journal of Alloys and Compounds* 935: 167963.

54 Feng, Y., Li, X., Lu, S. et al. (2023). Modulator-directed assembly of hybrid composites based on metal–organic frameworks and upconversion nanoparticles. *Nano Research* 16: 1482–1490.

55 Li, Z., Gao, H., Shen, R. et al. (2021). Facet selectivity guided assembly of nanoarchitectures onto two-dimensional metal–organic framework nanosheets. *Angewandte Chemie International Edition* 60: 17564–17569.

56 He, L., Brasino, M., Mao, C. et al. (2017). DNA-assembled core-satellite upconverting-metal–organic framework nanoparticle superstructures for efficient photodynamic therapy. *Small* 13: 1700504.

57 Zhao, X., Wang, Y., Li, J. et al. (2021). A fluorescence aptasensor for the sensitive detection of T-2 toxin based on FRET by adjusting the surface electric potentials of UCNPs and MIL-101. *Analytica Chimica Acta* 1160: 338450.

58 Xu, Z., Zhang, L.W., Long, L.L. et al. (2020). Metal organic frame-upconverting nanoparticle assemblies for the FRET based sensor detection of bisphenol A in high-salt foods. *Frontiers in Bioengineering and Biotechnology* 8: 626269.

59 Li, Z., Qiao, X., He, G. et al. (2020). Core-satellite metal–organic framework@upconversion nanoparticle superstructures via electrostatic self-assembly for efficient photodynamic theranostics. *Nano Research* 13: 3377–3386.

60 Segets, D., Marczak, R., Schäfer, S. et al. (2011). Experimental and theoretical studies of the colloidal stability of nanoparticles–a general interpretation based on stability maps. *ACS Nano* 5: 4658–4669.

61 Zhao, F., Ma, M.L., and Xu, B. (2009). Molecular hydrogels of therapeutic agents. *Chemical Society Reviews* 38: 883–891.

62 Dougherty, T.J. and Marcus, S.L. (1992). Photodynamic therapy. *European Journal of Cancer* 28: 1734–1742.

63 Ormond, A.B. and Freeman, H.S. (2013). Dye sensitizers for photodynamic therapy. *Materials* 6: 817–840.

64 Lismont, M., Dreesen, L., and Wuttke, S. (2017). Metal–organic framework nanoparticles in photodynamic therapy: current status and perspectives. *Advanced Functional Materials* 27: 1606314.

65 Xie, X., Gao, N., Deng, R. et al. (2013). Mechanistic investigation of photon upconversion in Nd^{3+}-sensitized core–shell nanoparticles. *Journal of the American Chemical Society* 135: 12608–12611.

66 Ray, P.C., Fan, Z., Crouch, R.A. et al. (2014). Nanoscopic optical rulers beyond the FRET distance limit: fundamentals and applications. *Chemical Society Reviews* 43: 6370–6404.

67 Liu, C., Liu, B., Zhao, J. et al. (2020). Nd^{3+}-sensitized upconversion metal–organic frameworks for mitochondria-targeted amplified photodynamic therapy. *Angewandte Chemie International Edition* 59: 2634–2638.

68 Shi, Z., Zhang, K., Zada, S. et al. (2020). Upconversion nanoparticle-induced multimode photodynamic therapy based on a metal–organic framework/titanium dioxide nanocomposite. *ACS Applied Materials & Interfaces* 12: 12600–12608.

69 Cai, H.J., Shen, T.T., Zhang, J. et al. (2017). A core–shell metal-organic-framework (MOF)-based smart nanocomposite for efficient NIR/H_2O_2-responsive photodynamic therapy against hypoxic tumor cells. *Journal of Materials Chemistry B* 5: 2390–2394.

70 Wang, X., Xu, J., Yang, D. et al. (2018). Fe_3O_4@MIL-100(Fe)-UCNPs heterojunction photosensitizer: rational design and application in near infrared light mediated hypoxic tumor therapy. *Chemical Engineering Journal* 354: 1141–1152.

71 Chamarro, E., Marco, A., and Esplugas, S. (2001). Use of Fenton reagent to improve organic chemical biodegradability. *Water Research* 35: 1047–1051.

72 Wang, Q.X., Yang, Y.F., Yang, X.F. et al. (2022). Upconverted/downshifted $NaLnF_4$ and metal–organic framework heterostructures boosting NIR-II imaging-guided photodynamic immunotherapy toward tumors. *Nano Today* 43: 101439.

73 Mellman, I., Coukos, G., and Dranoff, G. (2011). Cancer immunotherapy comes of age. *Nature* 480: 480–489.

74 Bonnet, S. (2018). Why develop photoactivated chemotherapy? *Dalton Transactions* 47: 10330–10343.

75 Alard, E., Butnariu, A.B., Grillo, M. et al. (2020). Advances in anti-cancer immunotherapy: Car-T cell, checkpoint inhibitors, dendritic cell vaccines, and oncolytic viruses, and emerging cellular and molecular targets. *Cancers* 12: 1826.

76 Lu, J., Zhu, X., Li, M. et al. (2021). Engineering near-infrared-excitable metal–organic framework for tumor microenvironment responsive therapy. *ACS Applied Bio Materials* 4: 6316–6325.

77 Ling, D., Li, H., Xi, W. et al. (2020). Heterodimers made of metal–organic frameworks and upconversion nanoparticles for bioimaging and pH-responsive dual-drug delivery. *Journal of Materials Chemistry B* 8: 1316–1325.

78 Wen, X., Liu, N., Ren, J. et al. (2022). In situ synthesis of a functional ZIF-8 nanocomposite for synergistic photodynamic–chemotherapy and pH and NIR-stimulated drug release. *New Journal of Chemistry* 46: 6966–6970.

79 DeVita, V.T. and Chu, E. (2008). A history of cancer chemotherapy. *Cancer Research* 68: 8643–8653.

80 Yu, S., Chen, Z., Zeng, X. et al. (2019). Advances in nanomedicine for cancer starvation therapy. *Theranostics* 9: 8026.

81 You, Y., Xu, D., Pan, X., and Ma, X. (2019). Self-propelled enzymatic nanomotors for enhancing synergetic photodynamic and starvation therapy by self-accelerated cascade reactions. *Applied Materials Today* 16: 508–517.

82 Wang, Z., Liu, B., Sun, Q. et al. (2021). Upconverted metal–organic framework Janus architecture for near-infrared and ultrasound co-enhanced high performance tumor therapy. *ACS Nano* 15: 12342–12357.
83 Jiang, Y., Fu, P., Liu, Y. et al. (2020). Near-infrared light-triggered NO release for spinal cord injury repair. *Science Advances* 6: eabc3513.
84 Sun, J., Fan, Y., Ye, W. et al. (2021). Near-infrared light triggered photodynamic and nitric oxide synergistic antibacterial nanocomposite membrane. *Chemical Engineering Journal* 417: 128049.
85 Chu, X., Jiang, X., Liu, Y. et al. (2021). Nitric oxide modulating calcium store for Ca^{2+}-initiated cancer therapy. *Advanced Functional Materials* 31: 2008507.
86 Carter, K.P., Young, A.M., and Palmer, A.E. (2014). Fluorescent sensors for measuring metal ions in living systems. *Chemical Reviews* 114: 4564–4601.
87 Berezin, M.Y. and Achilefu, S. (2010). Fluorescence lifetime measurements and biological imaging. *Chemical Reviews* 110: 2641–2684.
88 Wang, H.S. (2017). Metal–organic frameworks for biosensing and bioimaging applications. *Coordination Chemistry Reviews* 349: 139–155.
89 Schermelleh, L., Ferrand, A., Huser, T. et al. (2019). Super-resolution microscopy demystified. *Nature Cell Biology* 21: 72–84.
90 Lu, Y., Lu, J., Zhao, J. et al. (2014). On-the-fly decoding luminescence lifetimes in the microsecond region for lanthanide-encoded suspension arrays. *Nature Communications* 5: 3741.

5

Lanthanide-Doped Nanomaterials for Luminescence Biosensing and Biodetection

Zhijie Ju[1,2], Peng Zhao[1], and Renren Deng[1,2]

[1] *Zhejiang University, The First Affiliated Hospital, School of Medicine, Department of Medical Oncology, 79 Qingchun Road, Hangzhou 310003, P. R. China*
[2] *Zhejiang University, School of Materials Science and Engineering, State Key Laboratory of Silicon and Advanced Semiconductor Materials, Institute for Composites Science Innovation, 388 Yuhangtang Road, Hangzhou 310058, P. R. China*

5.1 Introduction

In recent decades, lanthanide-doped nanoparticles have garnered significant attention in the field of biosensing due to their unique properties, including a wide range of absorption and luminescence options from near-infrared (NIR) to ultraviolet (UV), large anti-Stokes shift, excellent photostability, and low toxicity. Particularly, considerable attention has been drawn on real-time biosensing and the development of optical bioprobes for emerging biological and biomedical applications. Lanthanide-based luminescence transducers typically rely on changes in absorption and luminescence signals to detect and measure biological events. In this chapter, we will begin by introducing some fundamental concepts related to optical biosensing and lanthanide luminescence before delving into the principles of biosensor design. Finally, we will discuss recent advances relevant to this exciting field of study.

5.2 Basics of Optical Bioprobe and Lanthanide-Doped Nanoparticles

5.2.1 Design Considerations for Bioprobe Development

Generally, an optical bioprobe should efficiently transduce changes in biological processes into optical signals with high sensitivity. These probes are designed to sense various parameters, such as temperature and pH, or detect specific molecules and ions. Temperature alterations influence phonon activity, resulting in distinct spectra in a multilevel luminophore. Sensing specific substances in the environment is more delicate. A common design strategy involves leveraging energy transfer

Photofunctional Nanomaterials for Biomedical Applications, First Edition.
Edited by Chunxia Li and Jun Lin.

between the luminophore and molecules whose optical characteristics are influenced by the surrounding environment. With proper designs, the incorporation of the target substance with the bioprobe allows for the detection of luminescence spectra or lifetime changes.

After transducing these changes, the subsequent step involves conveying the signal. However, most biological tissues are not transparent in visible-light spectral range. Hence, selecting a wavelength with deep penetration as the detection signal is preferable. The photon signal undergoes deduction due to absorption and scattering of biological tissues. The autofluorescence may also account for the increase in background noise when short-wavelength-light excitation is adopted. Short-wavelength photons are particularly susceptible to scattering, following the principles of Rayleigh scattering theory. Therefore, long-wavelength NIR photons are more suitable as carriers for bioprobe detection. For transparent samples, visible photons are acceptable for microscopic detection. Traditional bioprobes commonly employ ultraviolet or blue light for excitation, which suffers from autofluorescence. Upconversion nanoprobes employing longer-wavelength NIR excitation are capable of transforming infrared photons into visible photons, thus greatly improving the signal-to-noise ratio.

5.2.2 Characteristics of Lanthanide-Doped Nanoparticles

Trivalent lanthanide ions, except La^{3+} and Lu^{3+}, exhibit narrow-band absorption and emission characteristics originating from 4f–4f electron transitions. Due to the abundance of 4f energy levels and extended excited state lifetimes, lanthanide ions find widespread use in upconversion and downshifting materials, spanning the spectral regions of ultraviolet, visible, and NIR lights [1–5] (Figure 5.1).

To illustrate the principles of 4f–4f luminescence, let's consider the example of Er^{3+} ions (Figure 5.2). When excited with 800 nm light, the Er^{3+} ions occupy the $^4I_{9/2}$ excited state. From this state, the ions may nonradiatively transit to the $^4I_{11/2}$ state, subsequently emitting a 980 nm photon. Alternatively, the $^4I_{9/2}$ state may nonradiatively transit to the $^4I_{13/2}$ state, emitting a 1550 nm photon. These processes exemplify downshifting emission. In scenarios where the intensity of the excited light is sufficiently high, and another 800 nm photon is absorbed by the excited Er^{3+} before it decays to the ground state, the Er^{3+} ions may be pumped into the $^4S_{3/2}$ state, resulting in the emission of a 540 nm photon. This phenomenon achieves upconverted emission. The hierarchical energy levels, resembling a ladder, prove advantageous for upconversion and downshifting processes, observed in Er^{3+}, Tm^{3+}, Nd^{3+}, Dy^{3+}, Ho^{3+}, and Pr^{3+}. The ions responsible for emitting light are referred to as activators. However, achieving satisfactory efficiency in upconversion and downshifting often necessitates more than relying solely on activator ions.

On the other hand, the small oscillation strength of 4f–4f transitions means that high doping concentration is often necessary to augment the absorption of a single nanoparticle and enhance luminescence intensity. However, employing highly doped lanthanide ions can lead to concentration quenching [7]. Concentration

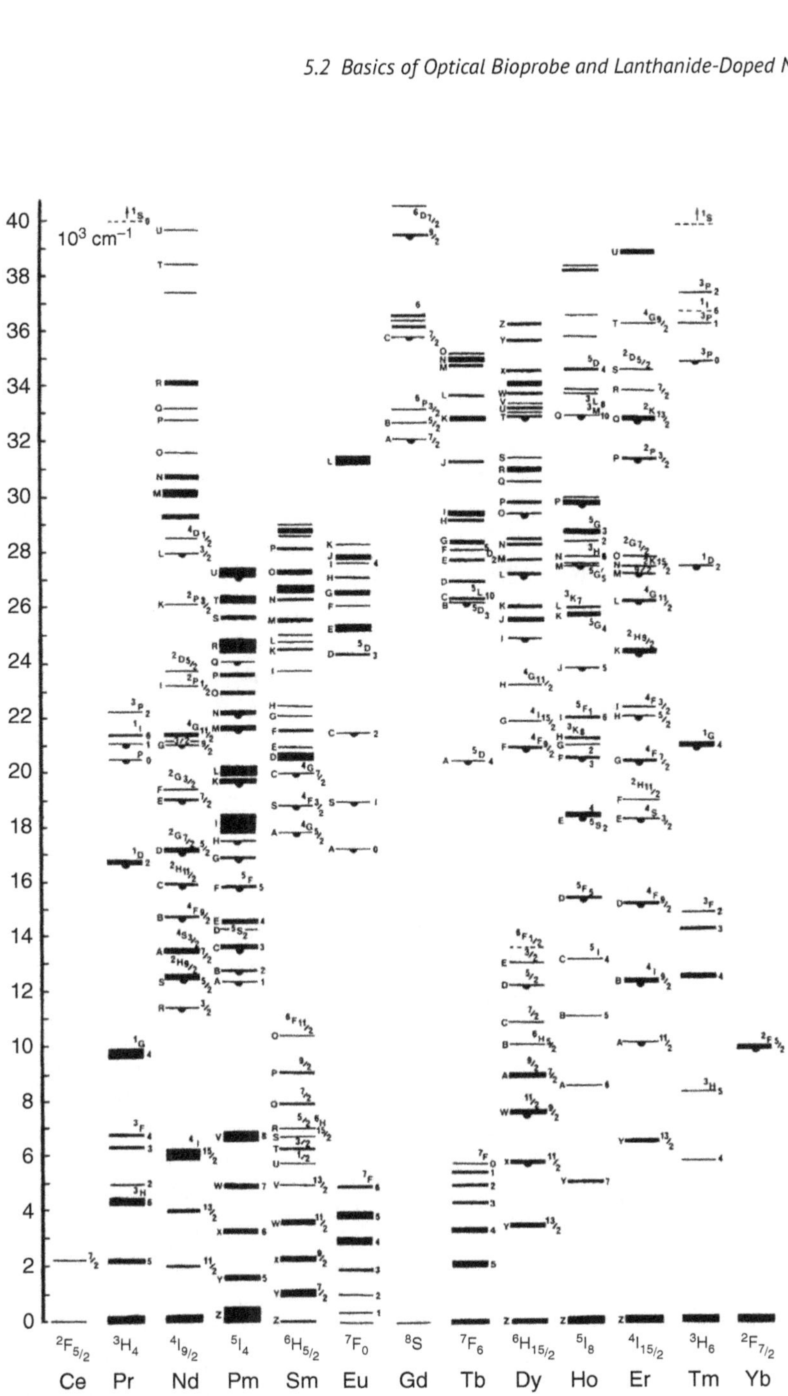

Figure 5.1 Observed energy levels of the rare earth ions. The thickness of each level represents the total crystal field splitting in $LaCl_3$. A pendant semicircle indicates that this level fluoresces in the $LaCl_3$ structure. Source: Reproduced with permission from [6], 1968 / Optical Society of America.

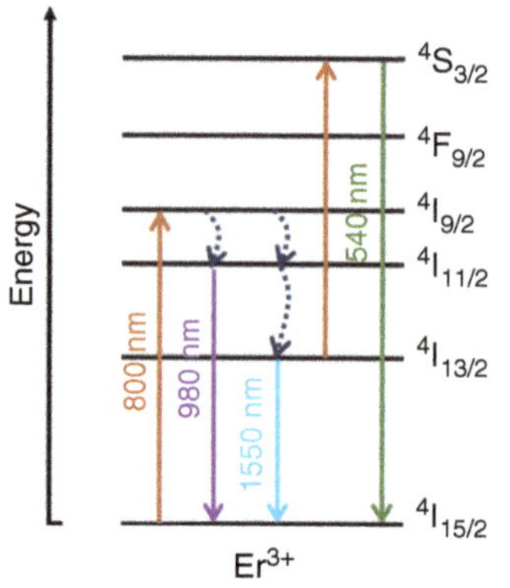

Figure 5.2 Energy diagram showing the processes for downshifting and upconversion emission of Er^{3+}.

quenching occurs when emitters with similar energy levels are in close proximity. In this situation, the excited states of one emitter can transfer to many other ground-state emitters, ultimately resulting in quenching by a single emitter plagued by a quencher (Figure 5.3a). The primary quenchers of nanoparticles are typically found on their surfaces, where crystal defects are abundant. Additionally, the high-energy vibrations of solvent molecules, such as C–H, O–H, and N–H, can also impair the excited state lifetime. This process is called multiphonon relaxation. The rate of multiphonon relaxation k_{MPR} in a host lattice is described as

$$k_{\mathrm{MPR}} \propto \exp\left(-\beta \frac{\Delta E}{\hbar\omega_{\mathrm{max}}}\right)$$

where β is a constant determined by the host, ΔE is the energy difference between the excited level and the next lower energy level of the excited emitter, and $\hbar\omega_{max}$ is the highest-energy vibrational mode of the host lattice. Ideally, a nanocrystal without any quenching sites can achieve a high level of brightness, even with high doping [10]. Thus, enhancing crystal quality emerges as a key solution to this challenge. A common approach involves growing an inert shell to diminish the occurrence of surface quenching. To address defects within the crystal structure, the use of anhydrous chemicals during synthesis effectively reduces the presence of O–H vibration quenchers [11, 12]. Additionally, a post-annealing process can eliminate vacancies in nanocrystals, significantly improving luminescence intensity by reconstructing the lattice structure [13].

Cross-relaxation is typically detrimental for upconversion but may be beneficial for downshifting emission. In a cross-relaxation process, a higher excited state decays to a lower intermediate state, releasing energy that excites another ground-state ion to a lower excited state (Figure 5.3b). Lower excited states are more susceptible to multiphonon relaxation, making it challenging for them to store the energy needed to be pumped to higher levels. Consequently, upconverted emission may be impaired, while downshifting emission could be enhanced.

To mitigate cross-relaxation, a sensitizer is introduced to the nanoparticles. A high concentration of sensitizer with a relatively simple energy-level structure is doped into the nanoparticle to enhance absorption. The excited sensitizers then

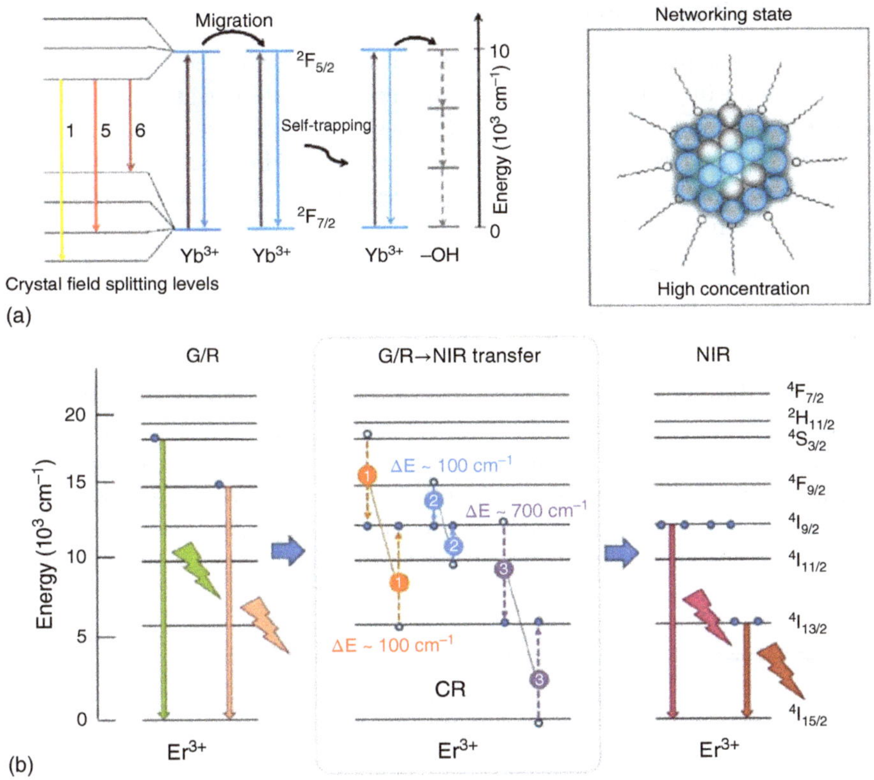

Figure 5.3 Schematic diagrams indicating concentration quenching and cross-relaxation. (a) Concentration quenching of Yb^{3+} network caused by surface multiphonon relaxation. Advanced Science published by Wiley-VCH GmbH. (b) Typical cross-relaxation processes involved in green/red(G/R) emission of Er^{3+}, which are turned into NIR emission after cross-relaxation (CR). The arrows of the same color represent pairs of cross-relaxation electron transitions, and ΔE denotes the energy difference of a pair of transitions. Source: [8, 9], 2022 / John Wiley & Sons / CC BY 4.0.

transfer energy to the activators. Consequently, only a small amount of activators is required to gather the energy from the excited sensitizers, resolving the issue of cross-relaxation in upconversion.

Yb^{3+}, Nd^{3+}, and Er^{3+} are commonly used sensitizers for lanthanide-doped nanoparticles, which can be excited by 980, 808, and 1530 nm photons, respectively. With rational designs, these nanoparticles can emit luminescence spanning a broad spectrum of wavelengths, ranging from near-infrared to ultraviolet, endowing them with great versatility as bioprobes. Downshifting NIR luminescence is suitable for deep penetration. On the other hand, upconversion visible luminescence realizes low noise microscopic detection. Lanthanide-doped nanoparticles are also biocompatible, photostable, and can be easily surface-functionalized, making them ideal for biosensing applications. Lanthanide-doped nanoparticles have been loaded with different molecules to achieve multifunctionality.

5.2.3 NIR Biological Windows

In the design of bioprobes for in vivo applications, considering penetration depth is crucial due to the impact of absorption attenuation and scattering disturbance on the loss of excitation and signal lights. NIR-light-based bioprobes are more immune to tissue absorption and scattering, making them widely used in optical biodetection in vivo. Within the NIR range, three wavelength regions with significant penetrability have been identified [14–16]:

(1) NIR-I region: 700–900 nm;
(2) NIR-II region: 1000–1700 nm;
(3) NIR-III region: 2080–2340 nm.

Figure 5.4 illustrates the NIR optical properties of water and various biological tissues [18]. The arrows point out the water absorption peaks at ~980, ~1200, ~1450, and ~1930 nm. In practical applications, the excitation wavelength should avoid strong absorption, while the signal wavelength may benefit from mild absorption. Mild absorption at the signal wavelength reduces the intensity of scattered light, thereby improving the signal-to-noise ratio [19, 20]. This is crucial as scattered light

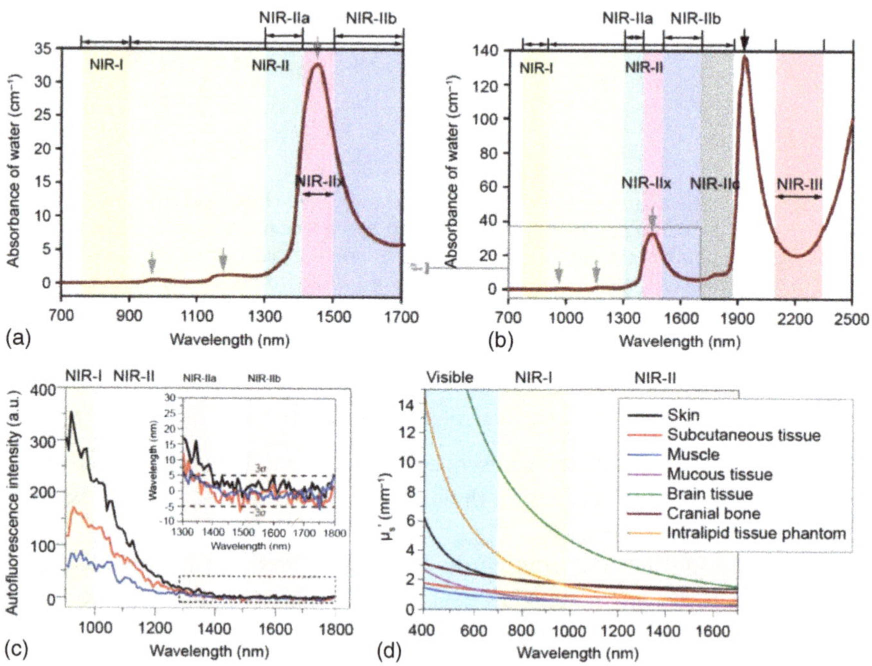

Figure 5.4 The light absorption spectra of water within (a) 700–1700 nm and (b) 700–2500 nm. The gray arrows pointed out the absorption peaks. (c) Autofluorescence spectra of ex vivo mouse liver (black), spleen (red), and heart tissue (blue) under 808 nm excitation light, showing the absence of autofluorescence in the >1500 nm NIR-II window. (d) Reduced scattering coefficients of different biological tissues and of intralipid scattering tissue phantom as a function of wavelength in the 400–1700 nm region. Source: [16, 17], 2017/ Springer Nature / CC by 4.0.

is prone to loss during a longer travel path. Emitted signal lights are typically not strong, mitigating concerns about the photothermal effect caused by signal light absorption. Both scattering and autofluorescence should be minimized at both excitation and signal wavelengths.

Abundant research has been carried out in NIR-I region. Though the water and tissue absorption in NIR-I region is low, this region suffers from relatively strong scattering and autofluorescence of tissue. The NIR-II region has attracted more attention recently. The absorption peak at 1400–1500 nm is moderate and will not hamper the signal collection. Suppressed scattering and autofluorescence in this region provide great signal quality. The NIR-III region, while promising with low scattering, still requires the development of cheaper detectors for practical implementation.

5.2.4 Energy Transfer: A Key Factor in Biodetection

In luminescence bioprobes, energy transfer (ET) is commonly categorized into two types: radiative and nonradiative. Radiative ET occurs when the light emitted by donors is re-absorbed by acceptors, where the photon serves as the transfer medium. Because photons travel through the entire solid angle, the efficiency of radiative ET is typically low, as ET only occurs at the angle where the acceptors are located. Therefore, the efficiency of radiative ET is usually very low. Mostly, nonradiative ET is the predominant mechanism. Nonradiative ET process utilizes electric dipoles (or multipoles) or electrons as the transfer medium. As energy must be conserved, nonradiative ET typically occurs between resonant states, with any energy differential being covered by phonons. In biodetection, the rate of the ET process can be adjusted by two factors, which are (i) the density of states capable of interaction, ρ and (ii) the coupling strength of the initial and final states, W_{DA}. The ρ is given by

$$\rho = \int f_D(E) f_A(E) \mathrm{d}E$$

where $f_D(E)$ and $f_A(E)$ are the normalized emission spectrum of the donor and the normalized absorption spectrum of the acceptor, respectively [21].

The nonradiative ET is commonly classified into Förster resonance energy transfer (FRET) and Dexter energy transfer (DET), based on different types of coupling. FRET relies on electrodynamic interaction, where the strength of interaction increases with the magnitude of the dipole–dipole coupling. If the dipoles of the donor and acceptor can be approximated as point dipoles, the coupling strength $W_{DA\text{-}FET}$ decreases with the inverse sixth power of the distance R_{DA} between the donor and the acceptor as

$$W_{DA-FET} \propto R_{DA}{}^{-6}$$

Higher-order interaction, such as electric dipole and electric quadrupole interaction, may also facilitate FRET. In these cases, the coupling strength decreases more quickly with the inverse eight (or larger) power of R_{DA}.

DET, on the other hand, is based on exchange interactions and occurs when the wave functions of the donor and acceptor pair overlap. For the coupling strength of DET, the following formula applies:

$$W_{\mathrm{DA-DET}} \propto \exp(-2R_{\mathrm{DA}}/L)$$

where L is the effective Bohr radius of the system. DET is not dependent on electric dipoles and can occur between nonemissive states, such as triplet states, whose electric dipoles are nearly zero. The drawback of DET is that it decreases rapidly with R_{DA} that the interacting distance is usually within 1 nm. For FRET, the interacting distance can be expended to 10 nm.

Considering both the density of states (ρ) and coupling strength (W_{DA}), the rate of FRET is given by [22]:

$$k_{\mathrm{FRET}}(R) = \frac{3c^4\hbar^4\sigma_A}{4\pi n^4\tau_{\mathrm{D}}R^6}\int \frac{f_{\mathrm{D}}(E)f_{\mathrm{A}}(E)}{E^4}\mathrm{d}E$$

where R is the distance between the donor and the acceptor, n is the refractive index of the host, τ_D is the radiative lifetime of the donor in the absence of the acceptor, σ_A is the absorption cross-section of the acceptor, and $f_{\mathrm{D}}(E)$ and $f_{\mathrm{A}}(E)$ are the normalized emission spectrum of the donor and the normalized absorption spectrum of the acceptor. The rate of DET is given by [22]:

$$k_{\mathrm{DET}}(R) = \frac{2\pi}{\hbar}K^2\exp\left(-\frac{2R}{L}\right)\int f_{\mathrm{D}}(E)f_{\mathrm{A}}(E)\mathrm{d}E$$

where K is a constant related to the overlap of the wave function, L is the effective Bohr radius of the system. More details about the fundamentals of FRET and DET can be found in references [21–25].

In practical applications, R is often the most critical factor, and other factors can be simplified as one constant. In a luminescent process involving FRET and DET, with sufficient acceptors, the concentration of the excited donor $[D^*]$ following excitation can be described by the following formula:

$$[D^*] = -k_{\mathrm{D}}[D^*] - \frac{k_{\mathrm{FRET0}}}{R^6}[D^*] - k_{\mathrm{DET0}}\exp\left(-\frac{2R}{L}\right)[D^*] - k_q[D^*]$$

where k_{FRET0} and k_{DET0} are constants related to FRET and DET, $k_{\mathrm{D}} = \frac{1}{\tau_{\mathrm{D}}}$, k_q is the rate constant of any other de-excitation pathways excluding ET.

While both radiative and nonradiative ET result in altered spectra of the donor-acceptor pair, only nonradiative ET leads to change in the lifetime of the donor luminescence. The variation in ET efficiency is triggered by the spectra overlap or distance between the donor-acceptor pair (Figure 5.5). These two factors form the basis of ET-based biodetection. For most ET-based bioprobes, the signal is collected in three primary ways:

(1) Detect the change in luminescence intensity of the donor;
(2) Detect the change in luminescence intensity of the acceptor;
(3) Detect the change in luminescence lifetime of the donor.

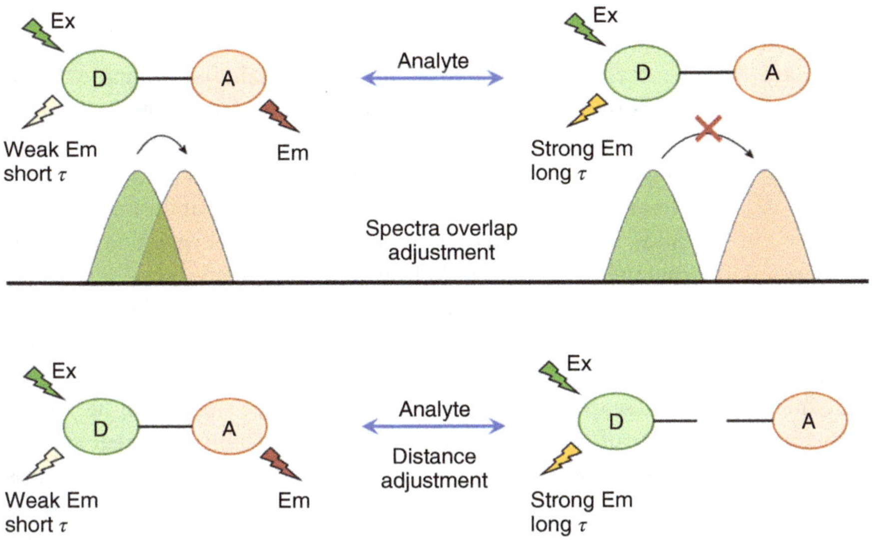

Figure 5.5 Schematic diagrams of the ET process in bioprobe design. The terms Ex, Em, and τ refer to excitation, emission, and luminescence lifetime, respectively. The analyte may alter the emission spectrum of the donor or the absorption spectrum of the acceptor. Additionally, the distance between the donor and acceptor molecules can be modified by the analyte, which can in turn affect the efficiency of the ET process.

5.3 Synthesis and Functionalization of Lanthanide-Dope Nanocrystals

5.3.1 Design and Synthesis of Core–Shell Structured Nanocrystals

Lanthanide ions are characterized by long luminescence lifetimes, typically on the order of microseconds to milliseconds. This prolonged luminescence lifetime makes their excited states highly susceptible to the surrounding environment, often resulting in nonradiative relaxation processes that quench the excited states. Consequently, the design of a suitable structure to protect the excited states of lanthanide ions, while maintaining high-ET efficiency requires meticulous consideration.

The ideal host for lanthanide ions should possess low phonon energy to minimize multiphonon relaxation. Additionally, the local environment surrounding lanthanide ions plays a crucial role in influencing emission intensity. To increase the 4f–4f transition probabilities, a less symmetric crystal with large crystal field strength is preferred. In addition, nanocrystals often exhibit a significant number of surface imperfections, necessitating a protective shell to mitigate surface quenching. The ability to synthesis intricate core–shell structure offers more possibility for multifunctional biosensor. Currently, hexagonal phase $NaREF_4$ (RE stands for Sc, Y, and lanthanides) is commonly employed as a host for lanthanide-doped nanomaterials. In certain cases, cubic phase $NaREF_4$ may show better performance for certain upconversion or downshifting emissions [26, 27]. In this chapter, we mainly focus on $NaREF_4$ nanocrystals (NCs), of which synthetic and surface modification approaches have been well developed.

There are various techniques available for synthesizing lanthanide-doped nanocrystals, such as thermal coprecipitation, thermal decomposition, hydrothermal/solvothermal method, liquid–solid–solution phase transfer, sol–gel, and so on. For a detailed introduction to synthetic approaches, we recommend a few review articles [3, 28, 29]. Among these techniques, thermal coprecipitation and thermal decomposition methods stand out as the most widely employed due to their convenience in producing intricate core–shell structured nanocrystals.

The dilemma faced by $NaREF_4$ NCs lies in the fact that approaches aimed to enhancing lanthanide luminescence may inadvertently compromise ET efficiency [30, 31]. Large size of NCs and thick inert shell coating are effective in preventing surface quenching, as research has shown that dark surface radius can be 1.7 nm for sub-10-nm NCs without inert shell [32]. However, only activators near surface respond to ET efficiently. A viable solution involves delicately refining the thickness of the protecting shell to achieve the ideal detection sensitivity. Besides, synthesizing sub-10-nm active core or adopting an inert core inside an active thin shell will improve the portion of activators that participated in ET.

For upconverted ET, energy migration of Tb^{3+} and Gd^{3+} can greatly improve the biodetection sensitivity based on improved ET [33, 34]. Energy migration has been observed in many lanthanide ions including $Yb^{3+}(^2F_{5/2}\rightarrow{}^2F_{7/2})$, $Gd^{3+}(^6P_{7/2}\rightarrow{}^8S_{7/2})$, $Tb^{3+}(^5D_4\rightarrow{}^7F)$, and $Er^{3+}(^4I_{13/2}\rightarrow{}^4I_{15/2})$. In a nanocrystal, the energy migration allows the excited state in the center of a nanocrystal to travel several nanometers toward the surface. The excited states of Tb^{3+} and Gd^{3+} are relatively insensitive to multiphonon relaxation due to their large energy gaps between emitting energy levels. Highly doped Tb^{3+} or Gd^{3+} can be used as the outermost shell to achieve effective ET. However, the migrated energy of Yb^{3+} and Er^{3+} (amount to the energy of 980 and 1550 nm photon, respectively) is close to that of some high-energy molecule vibrations. Therefore, surface passivation is necessary for nanocrystals with highly doped Yb^{3+} or Er^{3+} [8, 10].

Dye sensitization is an effective strategy for enhancing lanthanide emission [30, 35, 36]. The dye, with singlet or triplet energy levels matching the lanthanide 4f energy level, is attached to the surface of nanocrystals. The excited state energy is transferred to the lanthanide through FRET or DET. Leveraging the large absorption cross-section of dyes, dye sensitization significantly improves the emission intensity of both upconversion and downshifting. The sensitizing dye may also interfere with other functional molecules for biodetection, complicating the component design of the bioprobe.

5.3.1.1 Design of Upconversion Nanoparticles (UCNPs)

In most UCNPs, Yb^{3+} has large absorption at 980 nm and only two energy levels for f–f transition, making it a commonly used sensitizer. The widely adopted composition for upconversion is 20%Yb^{3+} and 2%Er^{3+} doped $NaYF_4$, which has green and red upconversion emissions. This composition is ideal for large NCs (larger than 20 nm diameter) and normal intensity of excitation (~100 W cm^{-2}).

However, for excitation intensities ranging from 10^5 to 10^6 W cm^{-2}, as utilized in single molecule detection and imaging, the optimal doping concentration of Er^{3+} can be extended to 20% [32]. Given the susceptibility of Er^{3+} upconversion to surface quenchers, a thin protective shell proves beneficial without significantly impeding the FRET process [31, 37]. To confine Er^{3+} within the energy transfer (ET) radius, a small Er-doped core is preferable. Another effective approach is to grow a thin Er-doped shell layer on the core without Er [38, 39], further enhancing detection sensitivity.

For violet upconversion needed for some high-energy molecule sensitization, Yb^{3+} and Tm^{3+} doped $NaYF_4$ or $NaGdF_4$ are used. Due to severe cross-relaxation, the doping concentration of Tm^{3+} is typically limited to below 1%. Though the upconversion efficiency of Tm^{3+} is much lower than Er^{3+}, Tm^{3+} is able to take advantage of energy migration based on Gd^{3+} or Tb^{3+}. The migration layer also serves as protective layer, which greatly improves ET efficiency. Tb^{3+} can also be excited by simultaneous sensitization of two excited Yb^{3+}, but the efficiency is poorer than lanthanides with ladder-like energy levels.

Ho^{3+} is also a green and red upconversion activator, though its efficiency is lower than Er^{3+}. Ho^{3+} is not so widely used in biodetection as Er^{3+}. One advantage of Ho^{3+} is that the red-green upconversion ratio can be modulated by adjusting the pulse width of infrared laser beams, which might facilitate the single wavelength excited double-channel detection [40, 41].

Yb^{3+} acts as the sensitizer with the highest upconversion efficiency. However, the 980 nm excitation wavelength overlaps with water absorption, leading to a significant photothermal effect. In recent years, increased attention has been directed toward Nd^{3+} sensitized 800 nm excitation and Er^{3+} sensitized 1550 nm excitation [42, 43]. In these nanocrystals, the Yb sublattice facilitates energy migration, delivering pump energy and preventing cross-relaxation between the sensitizer and activator (such as Er^{3+} and Tm^{3+}). The multi-shell structure can be generalized into two main components: (i) Nd^{3+} (or Er^{3+}) sensitized Yb^{3+} energy migration and (ii) Yb^{3+} sensitized traditional upconversion. The latter layer is typically the outer layer if activator-to-molecule ET is required.

5.3.1.2 Design of Downshifting Nanoparticles (DSNPs)

For DSNPs, the emission wavelength can be fine-tuned across a broad-spectrum region from visible to NIR. While Eu^{3+}, Tb^{3+}, Sm^{3+}, and Dy^{3+} serve as emitters in the visible region, NIR-emissive DSNPs are more competitive for biological applications and are often doped with Nd^{3+}, Er^{3+}, Ho^{3+}, Tm^{3+}, Pr^{3+}, and Yb^{3+}, covering a wavelength range from 800 to 2100 nm. Unlike UCNPs, cross-relaxation is not a major concern for DSNPs, allowing for design of highly doped DSNPs where emitter ions can also act as sensitizers themselves, such as Nd^{3+} and Er^{3+}. While the downshifting emissions of Ho^{3+}, Tm^{3+}, and Pr^{3+} are usually sensitized by Yb^{3+}, multiphonon relaxation and energy migration to surface quenchers, remain significant challenges for DSNPs.

Inert-shell protection becomes more important for near-infrared DSNPs due to the high-energy-vibration quenchers in solvent molecules. Johnson et al.

demonstrated that, in $NaY_{1-x}Er_xF_4$@$NaLuF_4$ core–shell nanocrystal, the downshifting emission at 1550 nm can be enhanced under both 808 and 980 nm excitation by increasing the concentration of Er^{3+} up to $x = 1$, when the nanocrystals are protected by the $NaLuF_4$ shell [10]. Many downshifting activators also have upconversion emissions. To suppress the energy loss of upconversion, Ce^{3+} doping is a traditional strategy, especially for Er^{3+} downshifting [44–46]. The energy spacing between $^2F_{5/2}$ and $^2F_{7/2}$ levels of Ce^{3+} is about 2300 cm^{-1} (c. 0.29 eV), which facilitates the nonradiative downshifting of Er^{3+} from $^4I_{11/2}$ to $^4I_{13/2}$ (energy difference about 3700 cm^{-1}). The concentration of Ce^{3+} needs to be carefully controlled; otherwise, it will strongly quench both upconversion and downshifting emissions.

5.3.2 Functionalization of Lanthanide-Doped Nanoparticles (LnNPs)

The majority of LnNPs, prepared through thermal coprecipitation and thermal decomposition methods, are inherently hydrophobic and lack functional moieties, rendering them unsuitable for direct use in biological applications. Consequently, surface modification plays a pivotal role in facilitating the chemical attachment of biomolecules and ensuring dispersion stability in buffer solutions. Here, we provide brief introductions to several frequently employed surface modification strategies:

5.3.2.1 Amphiphilic Polymer Absorption

The hydrophobic ends of amphiphilic polymers are absorbed on the hydrophobic surface of LnNPs. The hydrophilic groups help the dispersion of LnNPs in aqueous solution and provide functional groups (such as $-COOH$, $-NH_2$, and $-SH$) for the attachment of biomolecules. The frequently used amphiphilic polymers are polyethyleneglycol (PEG), polyethylenimine (PEI), polyvinylpyrrolidone (PVP), etc. Many of them have derivatives adapting to different applications.

5.3.2.2 Ligand Removal

A straightforward strategy involves removing the surface ligands initially and subsequently absorbing other molecules through electrostatic interaction or covalent binding. Two widely adopted methods for ligand removal are strong acid treatment [47, 48] and $NOBF_4$ treatment [49]. Strong acid treatment involves the protonation of oleate, resulting in the coordination of Ln^{3+} ions on the positively charged surface with water. The ligand-free LnNPs can then be dispersed in water, methanol, and other solvents with high dielectric constants. Subsequently, other ligands such as poly(acrylic acid) (PAA), PEI, PVP, and biotin can be attached to the ligand-free LnNPs for enhanced biocompatibility. In $NOBF_4$ treatment, the original surface ligands are replaced by BF_4^-, stabilizing the ligand-free LnNPs in dimethylformamide (DMF). The obtained LnNPs can then be functionalized using various ligand molecules, similar to the acid treatment process.

5.3.2.3 Ligand Exchange

Oleic acid (OA) commonly serves as the original surface ligand for most LnNPs. OA can be effectively replaced by hydrophilic ligands exhibiting stronger coordination with Ln^{3+}. Duong et al. systematically investigated the competitive adsorption of phosphate, carboxylic acid, and sulfonic acid onto the surface of UCNPs [50]. They synthesized three di-block copolymers comprising poly(ethylene glycol) methyl ether acrylate and a polymer block with phosphate, carboxylic, or sulfonic acid anchoring groups. Phosphate ligands demonstrated superior ligand replacement

5.3.2.4 Surface Silanization

Silica shells can be grown on both hydrophilic and hydrophobic LnNPs to improve their dispersion in complex aqueous media [51, 52]. The silica shell coating provides significant chemical stability and prevents nanoparticle aggregation. Silanes with functional groups can improve biocompatibility and facilitate the conjugation of other biomolecules.

5.4 Applications of Luminescence Biosensing and Biodetection

LnNPs have found diverse applications in sensing and detection due to their unique luminescence properties. Benefiting from deep penetration depth of NIR light and facile surface functionalization, LnNPs have demonstrated remarkable sensitivity in various in vivo sensing applications. In this section, we will explore the design principles of LnNP-based bioprobes for temperature sensing, pH sensing, and the detection of protein, nucleic acid, small molecules, and ions.

5.4.1 Temperature Sensing

In vivo temperature sensing within the physiological range (20–50 °C) is important for monitoring microenvironments and controlled photothermal therapy. LnNPs, with their long lifetime and abundant energy levels in the 4f–4f transition, exhibit a sensitive response to temperature. Six parameters reflecting luminescence changes include intensity, band-shape, spectral position, polarization, lifetime, and bandwidth. Among these, changes in intensity and lifetime are commonly used parameters in lanthanide-doped nanothermometers. Unlike pH sensing and the detection of target molecules or ions, temperature sensing relies on lanthanides themselves, and molecular coupling is unnecessary.

The temperature-sensitive luminescence of lanthanides is influenced by several mechanisms, including population redistribution due to Boltzmann statistics, activation of the multiphonon decay probability, ET between emitting ions or quenching centers, phonon-assisted Auger conversion processes, and thermal enhancement of ET processes between rare earth ions and the host energy levels or charge transfer states [53]. Due to the entangled interaction of these processes, it is sometimes difficult to elucidate the mechanism of a lanthanide-doped nanothermometer. Nevertheless, for practical applications, the relationship between luminescence and temperature can be determined experimentally, even in the absence of a clear theoretical model. Among these mechanisms, the simplest involves population redistribution due to Boltzmann statistics emanating from the thermally coupled energy levels E_1 and E_2. The absolute temperature can be deduced according to the Boltzmann distribution equation:

$$\frac{I_2}{I_1} = Ae^{-\frac{E_2-E_1}{k_BT}}$$

where I_1 and I_2 are the intensity of luminescence from energy levels E_1 and E_2, respectively, k_B is the Boltzmann constant, T is the absolute temperature, and $A = W_0/W_R$ is the ratio between the nonradiative rates (W_0) and radiative rates (W_R). Reported thermally coupled energy-level pairs include Er^{3+} ($^2H_{11/2}$ and $^4S_{3/2}$) [54], Nd^{3+} ($^4F_{5/2}$ and $^4F_{3/2}$) [55], and Eu^{3+} ($^5D_1$ and $^5D_0$) [56], with only Nd^{3+} ($^4F_{5/2}$ and $^4F_{3/2}$) emission falling within the NIR biological window.

Several indicators can be used to compare the performance of different thermometers. For instance, absolute sensitivity (S_a) is defined as

$$S_a = \left|\frac{\partial Q}{\partial T}\right|$$

and relative sensitivity (S_r), defined as

$$S_r = \left|\frac{1}{Q}\frac{\partial Q}{\partial T}\right|$$

can be used to quantify the change in luminescence (Q; intensity, lifetime, spectral position, etc.) at different temperature T.

While nanothermometers emitting visible light are well-established, recent attention has shifted to emitting probes at NIR biological window for subcutaneous thermal sensing and imaging. To this regard, Nd^{3+} is a potential emitter at both the NIR-I and NIR-II biological windows. In additional to Boltzmann statistics emitting of Nd^{3+} at the NIR-I biological window, Nd^{3+} collaborating with Yb^{3+} also has thermal-sensitive emission based on ET in between. Ximendes et al. reported that the net $Nd^{3+} \rightarrow Yb^{3+}$ ET efficiency depends on the balance between direct ET and phonon-assisted $Nd^{3+} \leftarrow Yb^{3+}$ back ET (Figure 5.6a) [57]. Nd^{3+} serves as the sole sensitizer, and the excitation wavelength is at 790 nm. The intensity ratio ($Q = I_{Nd}/I_{Yb}$) of Nd^{3+} at around 1300 nm ($^4F_{3/2} \rightarrow {}^4I_{13/2}$) and Yb^{3+} at around 1000 nm ($^2F_{5/2} \rightarrow {}^2F_{7/2}$) shows a linear calibration relation with temperature, achieving a thermal sensitivity of $0.41 \pm 0.01\%$ °C^{-1} in the LaF_3:Nd^{3+}@ LaF_3:Yb^{3+} nanothermometers. When Nd^{3+} and Yb^{3+} are co-doped in the core, the sensitivity is not ideal due to the dominant $Nd^{3+} \leftarrow Yb^{3+}$ back ET process.

Ximendes et al. also demonstrated a more sensitive nanothermometer based on ET of $Tm^{3+} \rightarrow Yb^{3+} \rightarrow Er^{3+}$ (Figure 5.6b) [58]. They designed core–shell LaF_3:Er^{3+}, Yb^{3+}@ LaF_3:Yb^{3+}, Tm^{3+} DSNPs. Under 690 nm excitation of Tm^{3+}, the intensity ratio ($Q = I_{Yb}/I_{Er}$) of Yb^{3+} at around 1000 nm ($^2F_{5/2} \rightarrow {}^2F_{7/2}$) and Er^{3+} at around 1550 nm ($^4I_{13/2} \rightarrow {}^4I_{15/2}$) exhibits the best sensitivity as high as 5% °C^{-1}. Despite the greater penetration of NIR lights through biological tissues, variation in absorption and scattering at different wavelengths can cause bias or inaccuracies in sensing. To address this issue, Qiu et al. designed a PbS quantum dots (QDs) and UCNPs co-assembled thermometer that both parts show emission around 810 nm

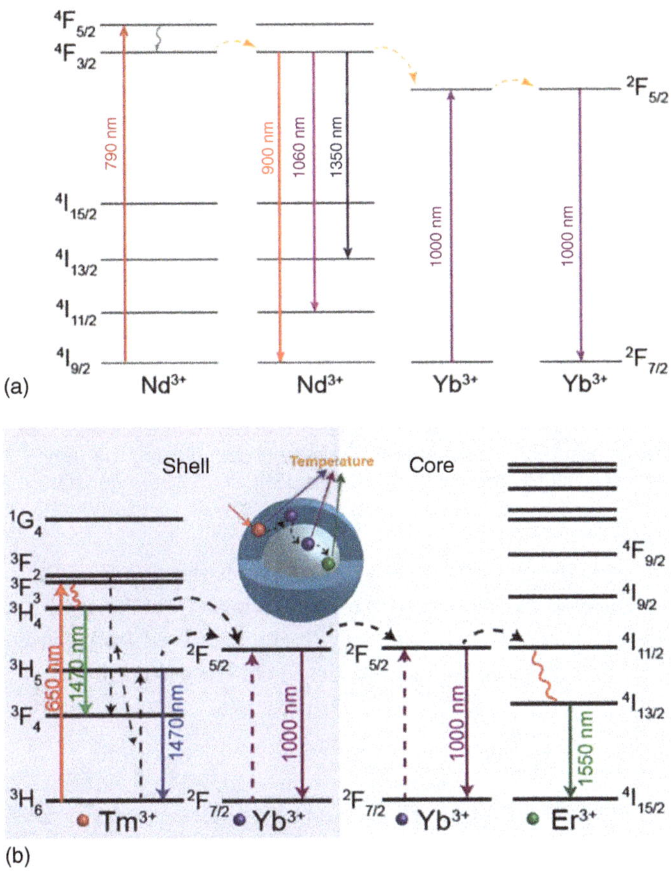

Figure 5.6 (a) Simplified energy scheme of the Nd^{3+} and Yb^{3+} emitting centers representing excitation and radiative decays (full lines), nonradiative decays (curved line), and possible ion–ion energy transfer paths (dashed lines). (b) Simplified energy scheme of the Tm^{3+}, Yb^{3+}, and Er^{3+} emitting centers representing excitation and radiative decays (full lines), nonradiative decays (curved line), and possible cross-relaxations and ion–ion energy transfer paths (dashed lines). Source: Reproduced with permission from [57, 58], 2016 / American Chemical Society. 2007 / John Wiley & Sons.

under 865 nm excitation [59]. The two components exhibit different luminescence lifetimes, enabling their differentiation through time-resolved technology. The structure of UCNPs comprises $NaYbF_4$:0.5%Tm^{3+}@$NaYF_4$:10%Yb^{3+}@$NaYF_4$:50%Nd^{3+}, with Nd^{3+} serving as the sensitizer. In the in vitro study, the luminescence intensity of both QDs and UCNPs showed temperature dependence. The total intensity (I_{sum}) of UCNPs and QDs was detected by a real-time spectrum. Then, a 20 μs delayed time-resolved spectrum was acquired to decode the luminescence intensity of Tm^{3+}(I_{Tm}) (Figure 5.7c). By evaluating the ratio between $I_{sum} - I_{Tm}$ and I_{Tm}, a thermal sensitivity of 5.6% K^{-1} and a thermal resolution of ~0.5 K were obtained at around 45 °C. In simulated biological tissue environments using pork tissue slice of varying thickness, the ratio $(I_{sum} - I_{Tm})/I_{Tm}$ remained nearly

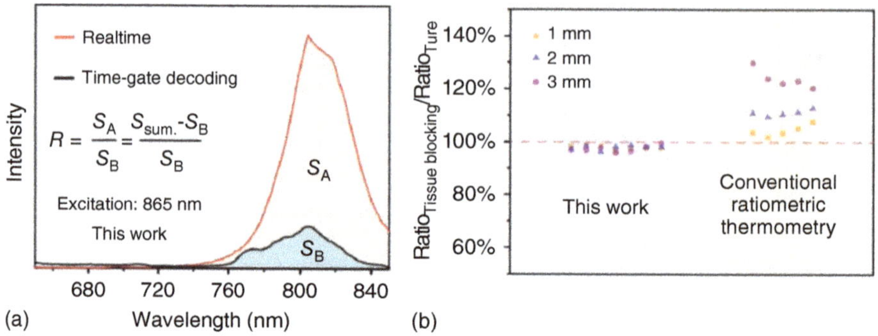

Figure 5.7 (a) Upconversion emission signals decoding of UCNPs using time-resolved spectrometer. The integral area of the spectrum (S) is proportional to the intensity (I). (b) Ratio values deviation due to blocking with pork tissue slice of different thickness compared with ratio signals measured without pork tissue slice blocking using the same wavelength or conventional ratiometric thermometry. Source: [59], 2023 / Springer Nature / CC BY 4.0.

unaffected by the tissue covering under different temperatures ranging from 20 to 50 °C. In contrast, conventional ratiometric thermometry based on the 1280 nm emission of PbS (I_{PbS}) and the 980 nm emission of Yb (I_{Yb}) suffered from biased sensing, with a deviation of up to 130% under thick tissue covering (Figure 5.7b).

5.4.2 pH Sensing

pH serves as a crucial indicator for monitoring metabolic processes within cells, and abnormal pH variations have been associated with various diseases, including cancer, neurological disorders, myopathies, and cardiovascular diseases. The development of intracellular fluorescent pH sensors suitable for microscopy holds significant practical applications.

The fundamental mechanism of pH sensing involves the detection of proton or hydroxide ions. Numerous pH-sensitive organic dyes exhibit changes in absorption or luminescence spectra based on the surrounding pH. However, traditional organic probes excited by short-wavelength light often encounter issues such as photobleaching or autofluorescence background. In this context, NIR-excited UCNPs present themselves as promising candidates for intracellular pH sensing.

The upconverted emission intensity changes in response to alterations in the absorption properties of pH-sensitive organic dyes on the surface of UCNPs. This change in intensity results from either reabsorption (i.e. inner filter effect) or FRET [60]. The multiple emission peaks of UCNPs facilitate the utilization of long-wavelength pH-insensitive luminescence as a reference signal for ratiometric calculations. This becomes particularly advantageous in intracellular sensing scenarios where luminescence intensity may be influenced by the transparency of different samples, enhancing accuracy. Additionally, UCNPs are often coated with PEI or other cationic polymers to facilitate cellular uptake. The high positive zeta potential (>30 meV) of these coatings aids in stabilizing colloids in aqueous media.

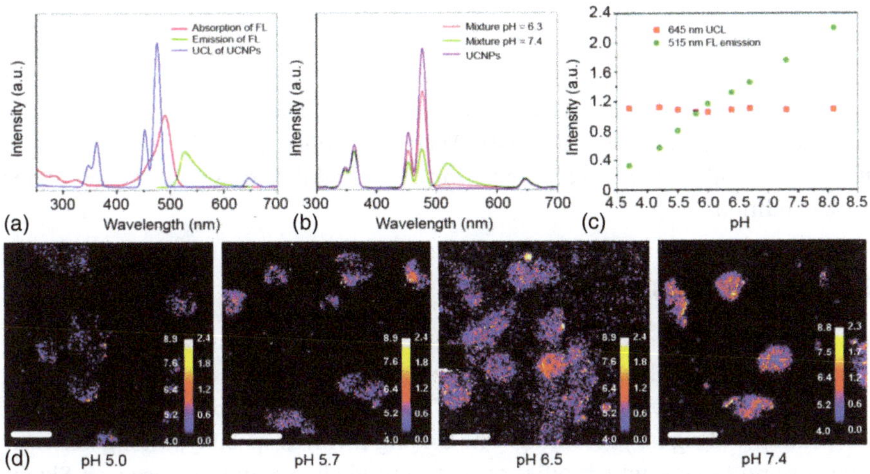

Figure 5.8 (a) Overlaid spectra of the upconversion luminescence of the UCNPs (808 nm excitation, 90 W cm^{-2}) (blue line) and the absorption (red line) and emission (excitation 475 nm Xe light) (green line) of FL (aqueous solution). (b) Photoluminescence spectra of the mixture of UCNPs and FL in water under 808 nm NIR irradiation (90 W cm^{-2}) in different pH environments. (c) Evolution of the signal intensity of UCNP-sensitized FL fluorescence at 515 nm (the area was taken from 500 to 610 nm) and UCNPs UCL at 645 nm in the system under different pH environments. (d) Representative confocal laser scanning microscopy ratiometric images obtained for samples incubated with UCNPs-FL probes and then washed and incubated with phosphate-buffered saline (PBS) at different pH values. From left to right, pH 5.0, 5.7, 6.5, and 7.4. Scale bar, 20 μm in all cases. Calibration bar: right, intensity scale; left, pH scale Source: Reproduced with permission from [61], 2017 / The Royal Society of Chemistry.

As an illustrative example, we consider the pH imaging probe reported by Du et al. [61]. This probe combines $NaGdF_4$:Yb^{3+}, Tm^{3+}@$NaGdF_4$:Nd^{3+}, Yb^{3+}@$NaYF_4$ UCNPs, and fluorescein (FL) through a PEI biopolymer. Under 980 or 808 nm excitation, the UCNPs emit blue light around 475 nm, which is absorbed by FL under basic conditions (Figure 5.8a). FL exists in its highly fluorescent green-colored open form in basic solutions and changes into its spirocyclic form (from colorless to nonfluorescent) upon the addition of H^+. FL is covalently attached to the surface of UCNPs to avoid fluorescence intensity fluctuations due to distance fluctuations between UCNPs and FL. The luminescence spectra at different pH levels are depicted in Figure 5.8b. The 645 nm UC emission is insensitive to pH and serves as a reference signal (Figure 5.8c). The change in the relative intensity of green and red emission enables the generation of ratiometric pH-weighted images (Figure 5.8d), although improvements in emission intensity are needed for enhanced resolution.

Most LnNP pH probes operate based on similar mechanisms. In addition, green upconversion emission of Er^{3+} has also been integrated with sensitive dyes. Tsai et al. utilized two anthraquinone dyes to quench the green emission band for pH detection, while using the insensitive red emission band of Er^{3+} as an internal reference [60]. Their work revealed that most quenching was attributed to reabsorption, influenced by the short Förster distance and the thick PEI interlayer between Er^{3+}

in the LnNPs and the dyes. To enhance sensitivity, they suggested increasing the proportion of near-surface Er^{3+} and directly attaching dyes to the surface of LnNPs.

Another critical consideration is the potential change in the zeta potential of LnNPs in aqueous solutions with varying pH, which potentially leads to the aggregation of LnNP probes [62–64]. Delicate surface potential adjustment is required to stabilize the LnNPs, ensuring the reliability of LnNP-probes under diverse conditions.

5.4.3 Detection of Biomolecules

Early diagnosis of malignant diseases such as cancer is crucial for increasing survival chances, but detecting specific biomarkers of these diseases in body fluid requires highly sensitive biodetectors [65]. The biodetection process consists of two steps: finding and signal processing. The finding involves specific binding between biosensors and targets, which relies on a structural change in the biosensor. Signal processing involves transforming the structural change into luminescence signal change.

Nucleic acids, which carry genetic information vital for biological processes, are essential biomarkers. Detecting aberrant nucleic acid is one of the important tools in disease diagnosis. Based on complementary pairing, the nucleic acid targets are mainly RNA and single-stranded DNA. The common design is the assembling (or disassembling) of the donor–acceptor ET structure, the components of which are connected by nucleic acid strands that response to the target. Li et al. designed DNA-driven gold nanoparticle-UCNP (Au-UCNP) pyramids for real-time intracellular microRNA (miRNA) monitoring [66]. The pyramid structure comprises two UCNPs and two gold nanoparticles (Au-NPs) connected by DNA strands (Figure 5.9b). Some parts of the DNA frame are noncomplemented parts. The miRNA recognition sequences are embedded in each side of the DNA frame. The pyramids display strong plasmonic circular dichroism (CD) at 521 nm because of the connection of two AuNPs and significant UC luminescence quenching

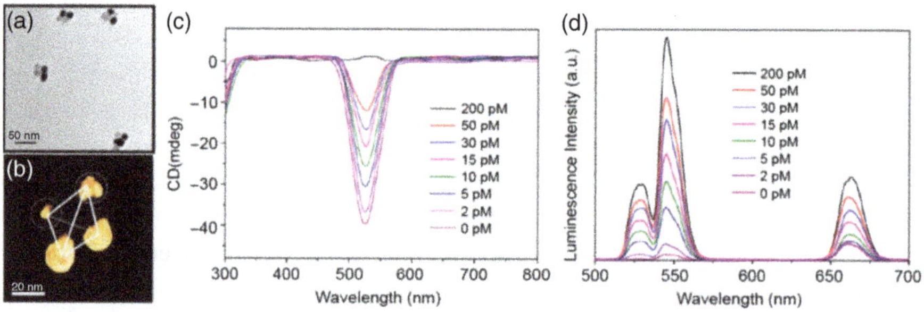

Figure 5.9 (a) TEM images of pyramids assembled with Au NPs and UCNPs in 10 mM Tris buffer. Scale bar 50 nm. (b) 3D reconstruction cryo-TEM tomography image of pyramids. Scale bar 20 nm. (c) CD and (d) luminescence spectra of Au-UCNP pyramids responded to different concentrations of miR-21 (0, 2, 5, 10, 15, 30, 50, or 200 pM) in vitro upon irradiation with 980 nm NIR laser (0.8W). Source: Reproduced with permission from [66], 2015 / American Chemical Society.

due to FRET between the Au NPs and UCNPs. In the presence of the target miRNA (MiR-21), the DNA frame of the pyramid will be complementary to the recognition sequence on each side, leading to complete dissociation of the pyramid structure. The Au NPs and UCNPs in a pyramid are then separated from one another. The chiroplasmonic activities of pyramid disappear, and the luminescence of the UCNPs is restored. Experimental results showed outstanding CD intensity linearity ranging from 0.073 to 43.65 fmol/10 μg RNA with a limit of detection (LOD) of 0.03 fmol/10 μg RNA. The luminescence intensity change also showed good detection sensitivity ranging from 0.16 to 43.65 fmol/10 μgRNA with a LOD of 0.12 fmol/10 μg RNA.

Detection sensitivity is important for intracellular miRNAs detection. Zhang et al. demonstrated a target cycling amplification strategy using one target miRNA to repeatedly open multiple donor-acceptor pairs, achieving a high detection signal gain [39]. A lock-like DNA (LLD) was designed by a hairpin DNA (H1) hybridizing with a bolt DNA (bDNA) sequence (Figure 5.10a). The LLD has target

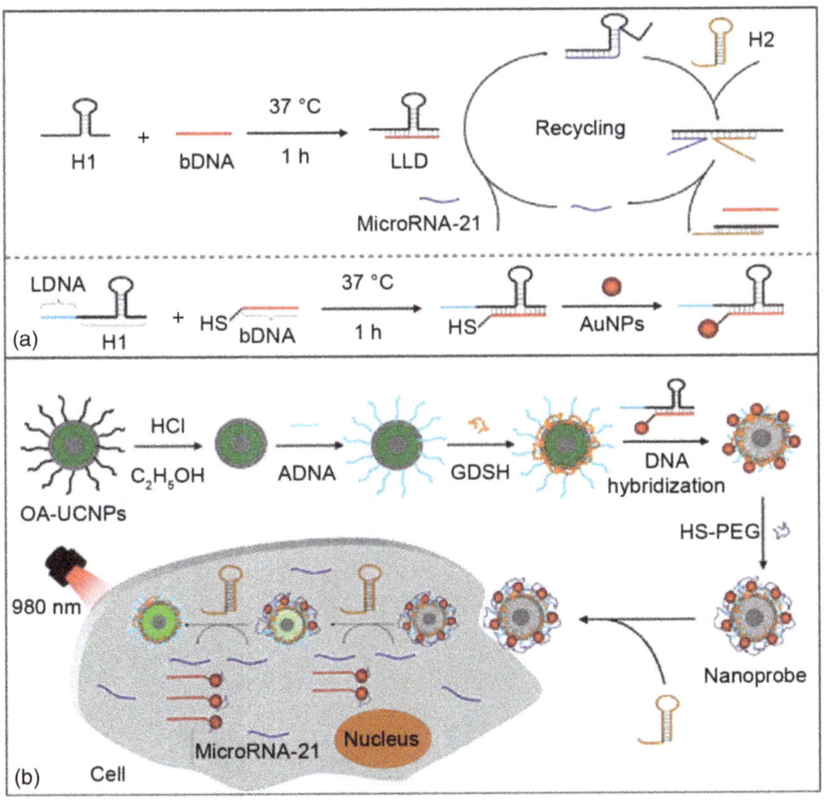

Figure 5.10 Schematic illustration of the preparation of (a) LLD and (b) upconversion nanoprobe, and application in monitoring microRNA-21 in living cells. H1, hairpin DNA; bDNA, bolt DNA; LLD, lock-like DNA; H2, fuel hairpin DNA; LDNA, linker DNA; AuNPs, gold nanoparticles; ADNA, anchor DNA; GDSH, Glycerol phosphate disodium salt hydrate; HS-PEG, thiol-functionalized polyethylene glycol. Source: Reproduced with permission from [39], 2018 / Wiley-VCH Verlag GmbH & Co. KGaA.

cycling amplification capacity with the aid of fuel hairpin DNA strands (H2). The nanoprobe, a UCNP with gold nanoparticles tethered on the surface through LLD, featured UC luminescence quenching by AuNPs through FRET. In their design, the UC activators of the UCNPs ($NaYF_4$@$NaYF_4$:Yb, Er@$NaYF_4$) were incorporated into the internal shell layer to enhance quenching efficiency. A single copy of the target miRNA can repeatedly unlock the LLD, triggering multiple AuNPs to move away from the UCNP with the assistance of H2, leading to high signal gain. The detection limit was about 1000 times lower than that of previously reported upconversion nanoprobes without signal amplification.

UCNPs can also be employed as triggers for miRNA detection with high spatiotemporal control. Many bioprobes rely on passive interactions with miRNAs, responding to encountered targets in transit. Achieving spatiotemporally controlled process enhances the effective utilization of miRNA probes for in vivo applications. Zhao et al. developed a NIR-light-activatable nanodevice for miRNA sensing using UNCPs [67]. They first designed a UV light-responsive DNA probe by incorporating a photocleavable (PC) bond into the hairpin loop of a molecular beacon, which rendered it temporarily inert to recognize the target miRNA (Figure 5.11). This DNA probe is equipped with UCNPs, which transduce NIR light into UV light.

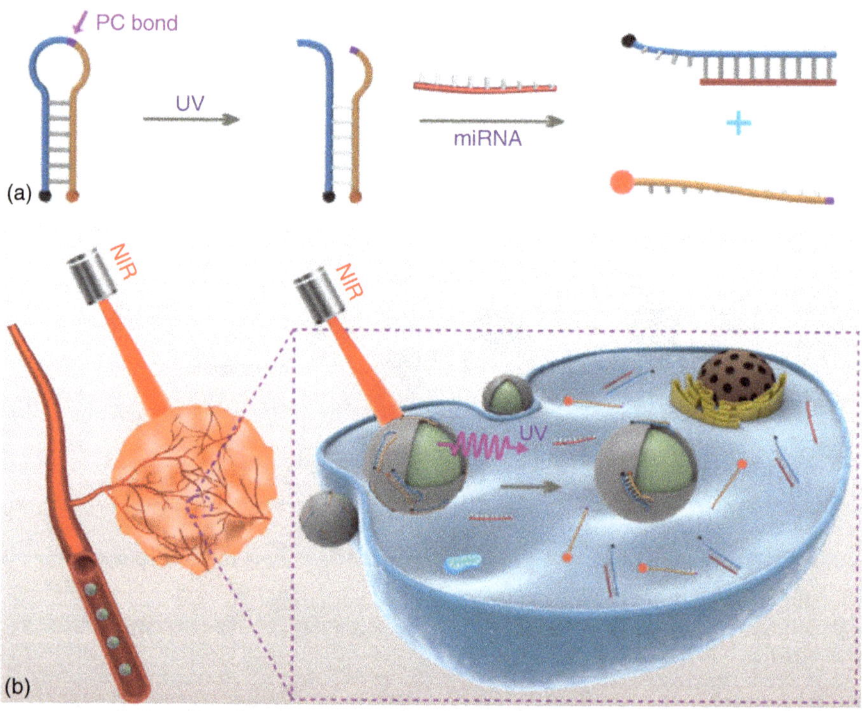

Figure 5.11 Working principle of the activatable DNA nanodevice for NIR light-controlled miRNA sensing and imaging. (a) Schematic illustration of the design features of the DNA molecular beacon. (b) Schematic of DNA nanodevice for NIR light-activated miRNA sensing in vivo. *Source:* Reproduced with permission from [67], 2019 / American Chemical Society.

The probe was labeled with fluorophore Cy5 and black hole quencher (BHQ2) at opposite ends, creating a low-fluorescence background through FRET. Upon NIR light irradiation, the photolysis of the PC bond, triggered by upconverted UV light, shifts the initial equilibrium. This enables the dose-dependent displacement of the quencher-labeled strand with target miRNA, leading to the separation of the Cy5-labeled strand from the BHQ2 and the restoration of Cy5 luminescence. The researchers demonstrated that the nanodevice could be easily delivered into cancer cells both in vitro and in vivo. This approach allows for precise control and activation of miRNA detection and can be combined with chain reaction for spatiotemporally resolved signal amplification [68].

Protein is another type of important biomarkers in clinical diagnosis, with specific binding commonly conducted through monoclonal antibodies and nucleic acid aptamers. Aptamers, a class of functional oligonucleotides, exhibit strong affinity and specificity comparable to monoclonal antibodies [69]. Qu et al. designed Au–Au–UCNP trimers for the detection of alpha-fetoprotein (AFP) and mucoprotein 1 (mucin-1), which are both significant disease biomarkers [70]. The aptamers of these two molecules have been used to functionalize the Au nanoparticles (AuNPs) and UCNPs, as shown in Figure 5.12. AFP-aptamer-functionalized UCNPs, mucin-1-aptamer-functionalized Au NPs, and AFP and mucin-1 complementary-modified AuNPs were assembled in the absence of the AFP and mucin-1 targets, causing the formation of Au–Au–UCNP trimers. The two Au NPs form intense electromagnetic hotspots, generating strong Raman signal for 4-aminothiophenol (4-ATP). The luminescence of UCNP in the trimer is quenched by AuNPs in the assembly. As the concentration of AFP or mucin-1 increased, aptamers tended to bind with their respective targets, leading to the decrease in the trimer concentration. AFP-attached UCNPs then freely disperse in solution, restoring the average distance between AuNPs and UCNPs beyond effective FRET

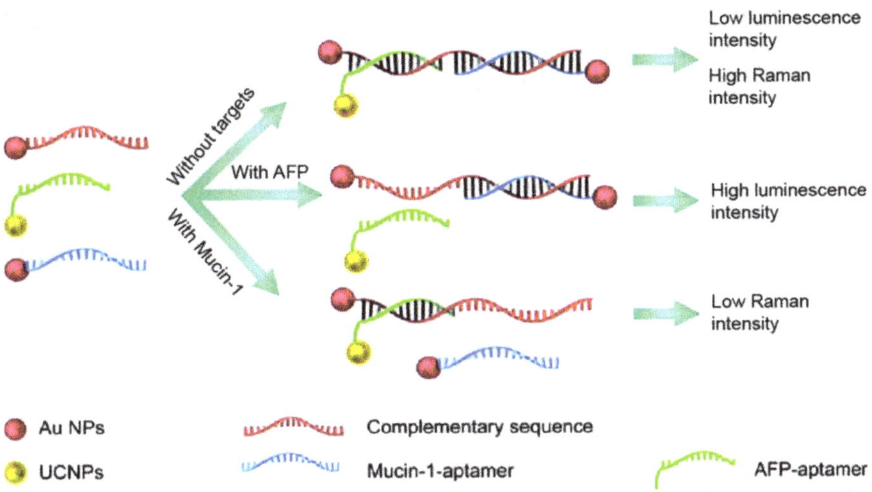

Figure 5.12 Scheme for detection of AFP and mucin-1 based on Au-Au-UCNP-trimers. Source: Reproduced with permission from [70], 2017 / The Royal Society of Chemistry.

and thereby restoring upconversion luminescence. For mucin-1, the disconnection of AuNP dimers destroyed the hotspots, reducing Raman signal intensity. Both luminescence and Raman intensity exhibited an excellent linear relationship with AFP and mucin-1 concentration, respectively. The LOD was 4.1 aM for mucin-1 and 0.059 aM for AFP. Similarly, Hao et al. utilized AuNP-UCNP pyramid based on aptamers to detect thrombin and prostate-specific antigen (PSA) [71]. The LOD for thrombin and PSA reached 57 and 0.032 aM, respectively.

Monoclonal antibodies are also frequently used in specifically protein target binding, particularly in dissociation-enhanced lanthanide fluoroimmunoassays (DELFIA). DELFIA utilizes a nonluminescent lanthanide (Ln^{3+}) chelate as a molecular probe to label the analyte. Upon reacting with a weakly acidic enhancer solution, the chelate transforms into a highly luminescent Ln^{3+} micelles. Conventional Ln^{3+} chelate suffers from low labeling ratios (typically 10–30 Ln^{3+} labels per biomolecule). Zhou et al. demonstrated that using LnNPs provides much more Ln^{3+} for dissolution, leading to enhanced luminescence and improved detection sensitivity [72]. The study aimed for carcinoembryonic antigen (CEA), a glycoprotein associated with colorectal cancer. Biotinylated LnNPs were bound to biotinylated anti-CEA antibodies through biotin–avidin–biotin interaction. The capture anti-CEA antibody was immobilized on the microplate well (Figure 5.13). The CEA targets caused LnNPs to be bound to the microplate for enhancement. The enhanced luminescence is based on the "antenna effect." The original pure LnNPs (for example, $NaEuF_4$) suffer from concentration quenching and small absorption cross-section. The ligand molecules have a strong affinity for Ln^{3+} ions and are able to dissolve the LnNPs to form many chelates dispersed in solution. In each chelate, the ligand efficiently absorbs light and sensitizes the luminescence of Ln^{3+} through nonradiative ET. Because of the high Ln^{3+} labeling ratio, the LOD was reduced to 0.5 fM, which is a significant improvement compared to commercial DELFIA. Besides, this method can be applied to detect circulating tumor cells (CTCs) in cancer patients. EpCAM is highly expressed on the cancer-cell surface, which could be recognized by anti-EpCAM-antibody-modified $NaEuF_4$ LnNPs. Blood CTCs ($\geq$10 cells ml^{-1}) can be detected with a detection rate of 93.9 % (14/15 patients) [73].

Fan et al. demonstrated in vivo detection and imaging of tumors using time-domain multiplexing [74]. They employed Nd^{3+} sensitizers and Er^{3+} activators in nanoparticles that absorb 808 nm excitation and emit luminescence centered at 1525 nm. By controlling the thickness of the inert shell layer, they achieved a series of lifetime populations for multiplexed in vivo imaging (Figure 5.14). Three-lifetime populations of Er^{3+} doped nanoparticles were conjugated with primary antibodies against three routinely examined biomarkers for breast cancers: estrogen receptor, progesterone receptor, and human epidermal growth factor receptor-2. With one injection of the mixed LnNPs, three biomarker detections were resolved simultaneously using a pattern recognition algorithm.

5.4.4 Detection of Small Molecules and Ions

LnNPs have emerged as a versatile toolbox for sensing various ions and small molecules. One common approach involves using UCNPs to sensitize a chemically

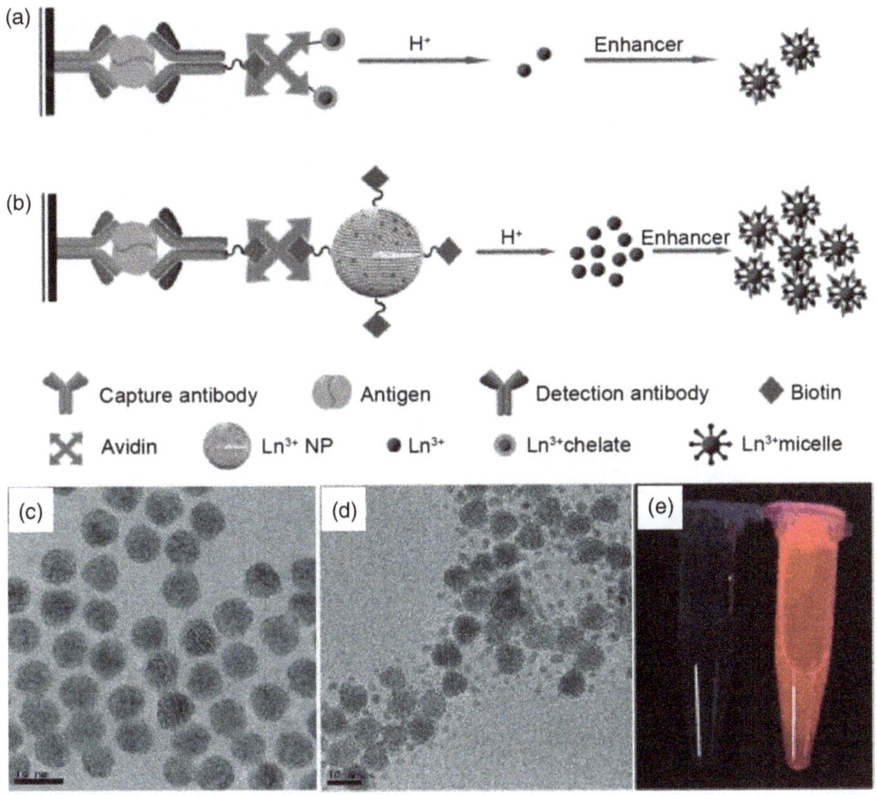

Figure 5.13 Schematic representation of (a) conventional DELFIA immunoassay based on Ln^{3+} chelates and (b) the proposed dissolution-enhanced luminescent bioassay based on inorganic LnNPs. TEM images of (c) ligand-free $NaEuF_4$ LnNPs and (d) ligand-free $NaEuF_4$ LnNPs one minute after their addition to the enhancer solution (scale bars = 10 nm). (e) Images showing the photoluminescence of ligand-free $NaEuF_4$ LnNPs in PBS (left) and the enhancer solution (right) under UV-lamp illumination at $\lambda = 300$ nm. Source: Reproduced with permission from [72], 2014 / Wiley-VCH Verlag GmbH & Co. KGaA.

sensitive dye, excited by blue or violet light in traditional fluorescent analysis. In recent years, there has been a significant focus on in vivo detection using downshifting nanoparticles working in NIR range. In this section, we introduce several examples to showcase different detection methods.

As a typical example, Li et al. demonstrated the detection of ClO^- based on downshifting sensitized NIR dye emission [75]. They used multi-shell structure $NaYbF_4{:}Er^{3+}@NaYF_4{:}Yb^{3+}@NaYF_4{:}Nd^{3+}$ nanoparticles to sensitize the 1525 nm emission from Er^{3+} under 808 nm excitation of Nd^{3+}. The NIR cyanine dye Cy925 was designed to detect ClO^-. Cy925, with an absorption peak at 890 nm, could be sensitized by 808 nm excited Nd^{3+}. The emission peak of Cy925 is at 925 nm, and upon encountering ClO^-, the conjugated structure of Cy925 is irreversibly destroyed, leading to a decrease in emission at 925 nm (Figure 5.15). Using 925 nm as the detection signal and 1525 nm as the reference signal, they achieved the in vivo imaging of ClO^- concentration in mice limbs with arthritis under 808 nm excitation.

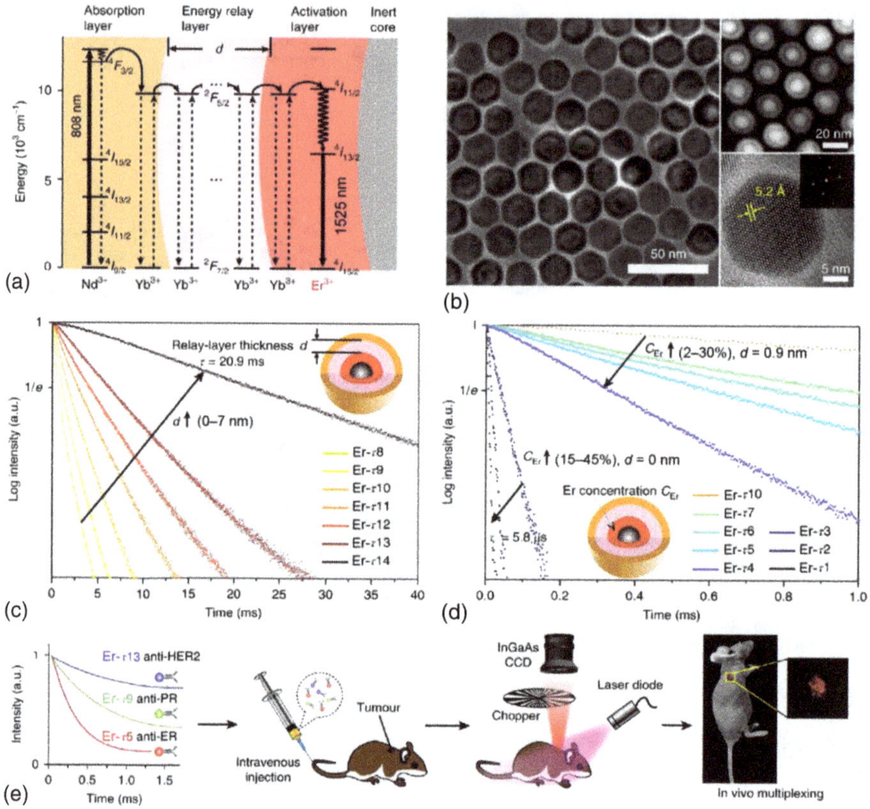

Figure 5.14 (a) Energy-level diagram illustrating the luminescence process of the core–multi-shell nanoparticles. (b) Transmission electron microscopy (TEM) image (left), high-angle annular dark-field scanning transmission electron microscopy image (top right), high-resolution TEM image, and the corresponding fast Fourier transform (bottom right) of the $NaGdF_4$@$NaGdF_4$:Yb^{3+},Er^{3+}@$NaYF_4$:Yb^{3+}@$NaNdF_4$:Yb^{3+} nanoparticles. (c) Luminescence decay curves measured at 1525 nm from the as-prepared Er^{3+} doped nanoparticles with energy relay shells of increasing thickness d from 0 to 7 nm (identical composition). (d) Luminescence decay curves of the nanoparticles with incremental Er^{3+} doping concentration C_{Er} from 2% to 30% for $d = 0.9$ nm and from 15% to 45% for $d = 0$ nm. (e) Schematics illustrating animal experiment procedures. Three batches of Er^{3+} doped nanoparticles exhibiting distinct lifetimes are conjugated to three antibodies (anti-ER, anti-PR and anti-HER2), respectively, and intravenously injected into the mouse via tail vein. Lifetime-resolved imaging is then performed to quantify the biomarker expressions on the tumor. Source: Reproduced with permission from [74], 2023 / Springer Nature.

The development of stable, responsive, and highly emissive NIR-II molecules is challenging, which has led to the adoption of absorption competition-induced emission (ACIE) as a practical sensing strategy. ACIE involves the use of a chemically responsive molecule attached to the surface of DSNPs in high concentration to absorb the exciting light, thereby preventing the lanthanide from emitting. When the target molecule interacts with the sensitive molecule, the absorption changes and the lanthanide emission is recovered. Wang et al. designed a

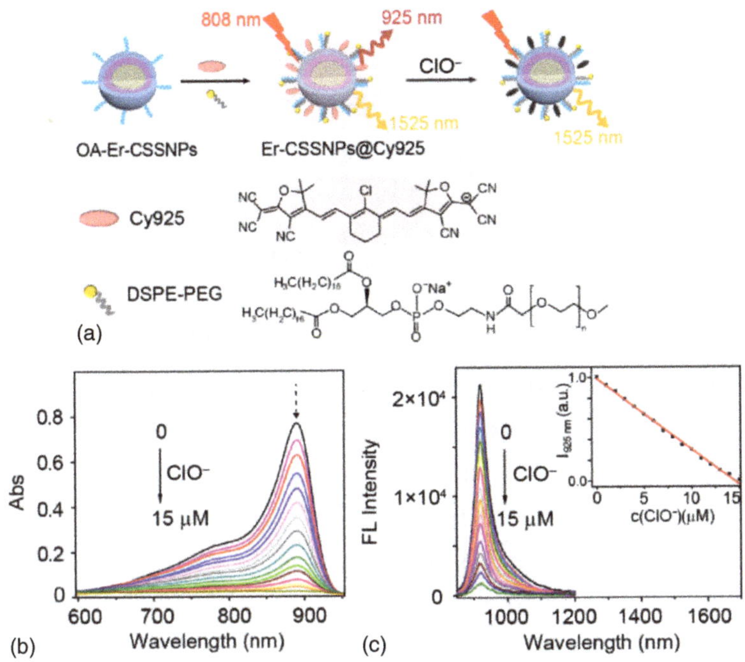

Figure 5.15 (a) Scheme of the designed composite probe Er-CSSNPs@Cy925. The oleic acid ligands on the surface of DSNPs were combined with the Cy925 and the DSPE-PEG through the hydrophilic–hydrophobic interaction to obtain stabilized water-soluble probes. (b) Absorption spectra of Cy925 (5 μM) upon addition of 15 μM ClO^- in a mixed solution of EtOH and water (v/v, 1 : 1). (c) Fluorescence spectrum of Er-CSSNPs@Cy925 upon addition of ClO^- (from 0 to 24 μM) and the ratio of the NIR emission ($I_{925\,nm}/I_{1525\,nm}$) as a function of the ClO^- concentration with the 808 nm excitation (75 mW cm^{-2}). Source: Reproduced with permission from [75], 2019 / American Chemical Society.

downshifting nanoprobe responding to hypochlorous acid (HOCl) based on ACIE [76]. This nanoprobe demonstrated a detection limit down to 500 nM and utility for high-resolution (micrometer scale) ratiometric sensing of lymphatic inflammation in vivo. The downshifting $NaY_{0.5}E_{r0.5}F_4$@$NaYF_4$ nanoparticles, excited by 808 and 980 nm, emit at 1550 nm. Cy7.5 fluorophores encapsulated into nanomicelles were conjugated with DSNP@N3 on the DSNP's surface. HOCl bleaches Cy7.5, allowing the 1550 nm emission under 808 nm excitation to recover. The 980 nm-excited 1550 nm emission serves as a reference signal (Figure 5.16). Liu et al. utilized the ACIE of an H_2S-sensitive dye attached to $NaGdF_4$:2%Nd^{3+}@$NaGdF_4$ nanoparticles to monitor endogenous H_2S in lipopolysaccharide-induced liver inflammation [77]. The competitive absorption of dye and Nd^{3+} results in highly sensitive luminescence intensity at 1060 nm in the presence of H_2S.

Besides ACIE, another viable approach involves dye-sensitized emission. As a typical example, Wang et al. employed the dye 4-nitrophenol-Cy7 (NPh) attached to $NaYF_4$:20%Yb^{3+}, 2%Er^{3+}@$NaYF_4$:30%Nd^{3+} upconversion nanoparticles (UCNPs) for in vivo ratiometric imaging of glutathione [78]. NPh strongly absorbs 808 nm light but exhibits no fluorescence due to intramolecular photoinduced electron

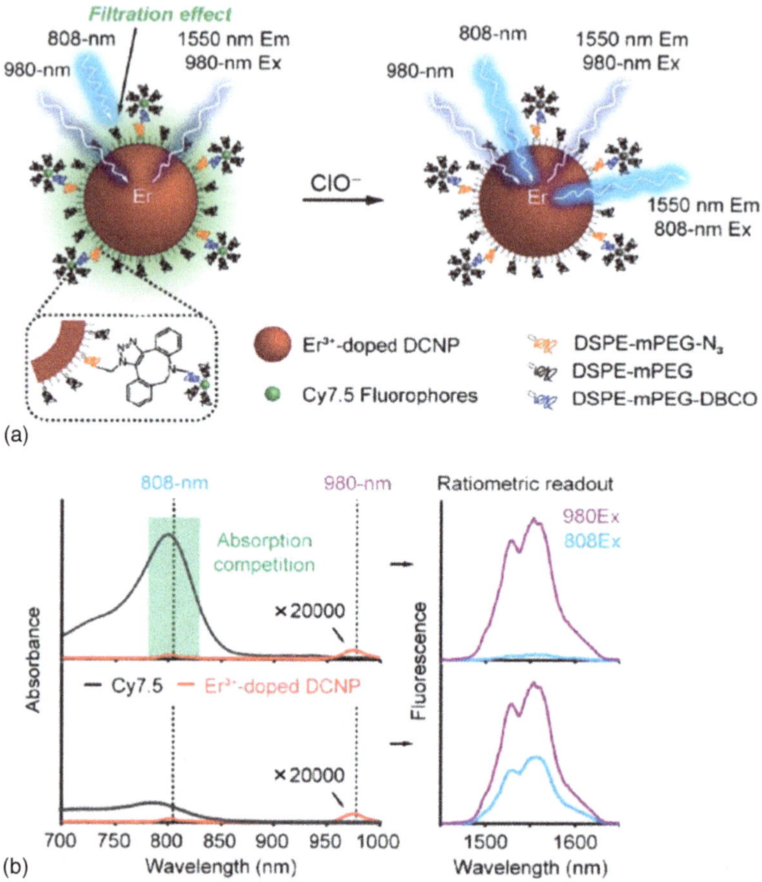

Figure 5.16 Schematic illustration showing the ratiometric response of DSNP@Cy7.5 to HOCl based on an ACIE mechanism. When excitation is performed in the absorption overlapping region (808 nm), fluorescence of DSNPs are quenched due to the energy filtration of excitation light by strong absorbing Cy7.5 (the upper diagram of (b)). The quenching process is reversed after the degradation of Cy7.5 by HOCl oxidation (the lower diagram of (b)). Source: Reproduced with permission from [76], 2019 / American Chemical Society.

transfer. Therefore, without glutathione, the 1550 nm emission of Er^{3+} is markedly weakened under 808 nm excitation. In the presence of glutathione, the NPh molecule reacts with glutathione to form Cy7-SG, which has a similar absorption profile at 808 nm but is luminescent at 820 nm. Therefore, Cy7-SG sensitizes Nd^{3+} and enhances the 1550 nm emission. As both NPh and Cy7-SG exhibit no absorption at 980 nm, the 980 nm-excited 1550 nm emission remains insensitive to glutathione, serving as an internal reference. Thus, the ratiometric nanoprobe was successfully implemented for the detection of glutathione with a low detection limit, allowing high-resolution imaging of glutathione in vivo.

In addition, the strategy of NIR-excited upconversion and subsequent dye quenching has gained widespread application in detection. For instance, Liu et al.

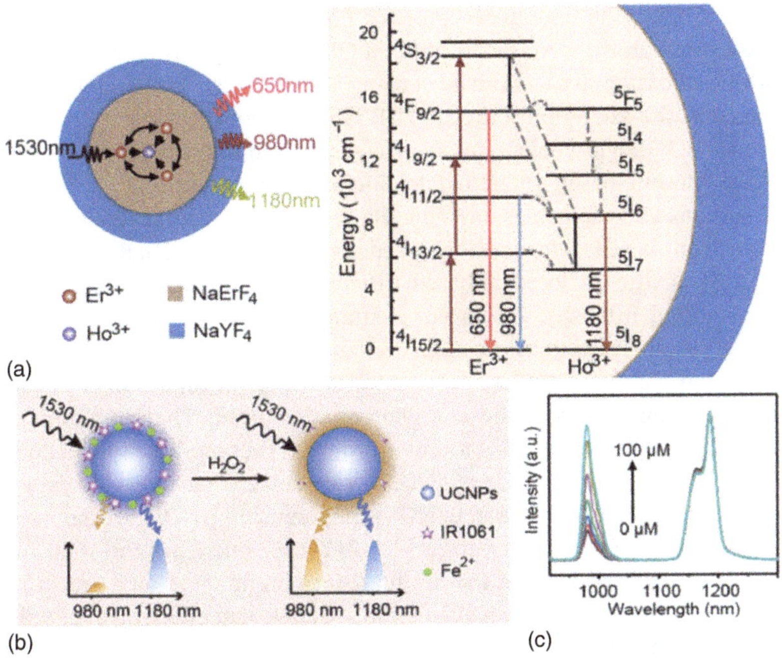

Figure 5.17 (a) Energy transfer mechanism in the $NaErF_4$:2 %Ho@$NaYF_4$ core–shell UCNPs. (b) Illustration of ratiometric fluorescent sensor for H_2O_2. (c) Upconversion emission spectra of UCNPs in presence of different concentrations of H_2O_2. Source: Reproduced with permission from [79], 2018 / John Wiley & Sons.

reported Er^{3+} sensitized upconversion from 1530 to 1180 nm for H_2O_2 sensing [79]. When excited at 1530 nm, $NaErF_4$:2%Ho@$NaYF_4$ UCNPs showed upconversion emission at 980 nm (from Er^{3+}) and 1180 nm (from Ho^{3+}). A H_2O_2 sensitive dye IR1061, with strong absorption in the range of 800–1100 nm, is bound to the surface of UCNPs. IR1061 quenched the 980 nm emission while leaving the 1180 nm emission unaffected. Upon exposure to •OH generated by H_2O_2, IR1061 can be destructed, leading to the recovery of 980 nm emission (Figure 5.17). Based on the ratiometric fluorescent intensity of 980 and 1180 nm of the UCNPs, a microneedle patch was prepared and subjected to in vivo dynamic inflammation sensing at very high resolution ($200 \times 200\ mm^2$). In another example, Yuan et al. demonstrated the use of MnO_2 nanosheets as quenchers of upconversion emission on $NaYF_4$:Yb, Tm@$NaYF_4$ UCNP's surface for the detection of H_2O_2 and glucose in blood [80]. They presumed that the recovery of fluorescence was attributed to the H_2O_2-mediated reduction of MnO_2 to Mn^{2+}, leading to decomposition of the MnO_2 nanosheets. This nanoprobe is also able to detect glucose in blood based on the enzymatic conversion of glucose by glucose oxidase to form H_2O_2.

The combination of sensitive luminescence quenching of dyes or QDs with UCNPs also represents a viable approach for detection. In these schemes, UCNPs function as energy donors for sensitive luminescence. For example, Xu et al. developed a FRET biosensor, employing $NaYF_4$:Yb^{3+}/Tm^{3+} UCNPs as the energy donor and CdTe QDs as the energy acceptor for Pb^{2+} detection. The presence of Pb^{2+} induces a change in the surface structure of QDs, leading to the luminescent quenching of QDs emission. An advantage of using UCNPs as the energy donor lies in the ability to shift the excitation source to NIR range, effectively suppressing the self-luminescence of biological substrates such as serum and tissues [81].

Luminescence lifetime-based in situ detection has found application in the analysis of small ions and molecules, owing to its independence from tissue transparency. In spectrum-based detection, the need for a reference signal arises for quantitative analysis due to location-based differences in the tissue, resulting in varying transmitted intensity. However, accurate analysis in spectrum-based nanoprobe design is challenging as the detection and reference signals must have similar transmission characteristics across a broad wavelength range. In contrast, lifetime-based detection requires only a single emission peak. The alteration in lifetime induced by nonradiative ET is a prerequisite for lifetime-based detection, distinguishing it from techniques such as ACIE. For instance, Zhao et al. developed in situ peroxynitrite ($ONOO^-$) detection in NIR-II region using $ONOO^-$-responsive NIR-II dye, MY-1057, as the FRET acceptor [82]. MY-1057 quenches the 1060 nm emission of Nd^{3+} through FRET, and the luminescence lifetime is shortened from 305 ± 3 to 203 ± 2 μs after MY-1057 encapsulation. $ONOO^-$ can efficiently degrade MY-1057; thus, the luminescence intensity and lifetime are recovered (Figure 5.18). The $ONOO^-$ concentration at tumor lesions can be quantitatively measured based on the in vitro standard curve. Another benefit of lifetime-based bioprobes is that the pulsed excitation of lifetime detection and imaging mitigates thermal accumulation and reduces the risk of tissue damage compared to conventional intensity imaging.

5.5 Integrated Devices for Point-of-Care Testing

Lanthanide-doped nanomaterials provide a powerful toolbox for detecting a variety of disease biomarkers. However, many of the applications mentioned above require bulky instruments and professional operation. To enable real-time monitoring of biochemical indexes, there has been a trend toward decentralized diagnostic analysis, known as point-of-care testing. This type of testing can be conducted at a patient's home using a highly integrated small device that includes the excitation light source, optic system, detector, and data analysis software. The complex testing process is simplified into a few instructions that can be easily followed by patients.

You et al. developed a highly sensitive household prognosis platform for heart failure diagnosis, integrating a smartphone-based reader with multiplexed upconversion fluorescent lateral flow strip (LFS) [83]. Two target antigens available in the serum, brain natriuretic peptide (BNP) and suppression of tumorigenicity 2 (ST2), are used for diagnosis. UCNPs with green and blue fluorescence were used for multiplexed detection of ST2 and BNP antigens in test line1 and test line2, respectively. The excess UCNPs probes without binding analytes would react with

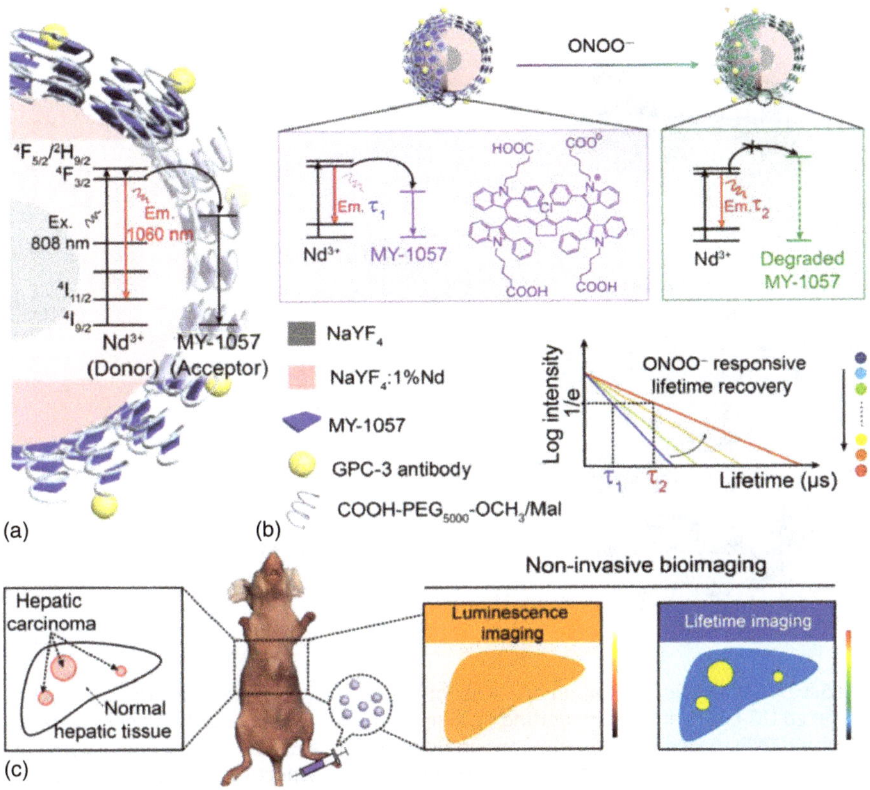

Figure 5.18 (a) Scheme of the $ONOO^-$-responsive nanosensor DSNP@MY-1057-GPC-3. (b) In the presence of $ONOO^-$, the structure of the energy acceptor MY-1057 degrades sensitively, leading to the lifetime recovery in NIR-II region. (c) Illustration of noninvasive NIR-II intensity- and lifetime-based imaging for hepatocellular carcinoma (HCC) mice after administration of DSNP@MY-1057-GPC-3 nanosensor. As ONOO- exists in the microenvironment of tumors. HCC lesions could be distinguished from normal hepatic tissue from lifetime imaging, while the intensity-based luminescence imaging fails. Source: Reproduced with permission from [82], 2020 / John Wiley & Sons.

the goat antimouse immunoglobulin G (IgG) and be trapped in the control line, as shown in Figure 5.19. The power supply, optical lenses, and laser are all integrated into the small box. The upconverted light can be detected by the camera of a smart phone, and the results are obtained from the smartphone app designed for this system. They detected 38 clinical serum samples using the platform and successfully detected the minimal concentration of 29.92 ng ml^{-1} for ST2 and 17.46 pg ml^{-1} for BNP in these clinical samples.

Integrated devices for in situ detection have also been demonstrated. Hou et al. designed a swallowable X-ray dosimeter for the simultaneous real-time monitoring of absolute absorbed radiation dose and of changes in pH and temperature in the gastrointestinal tract (Figure 5.20) [84]. The capsule comprises a flexible optical fiber encapsulated with $NaLuF_4$:Tb^{3+}@$NaYF_4$ persistent nanoscintillators, a pH-sensitive polyaniline film, a temperature sensor, and a wireless

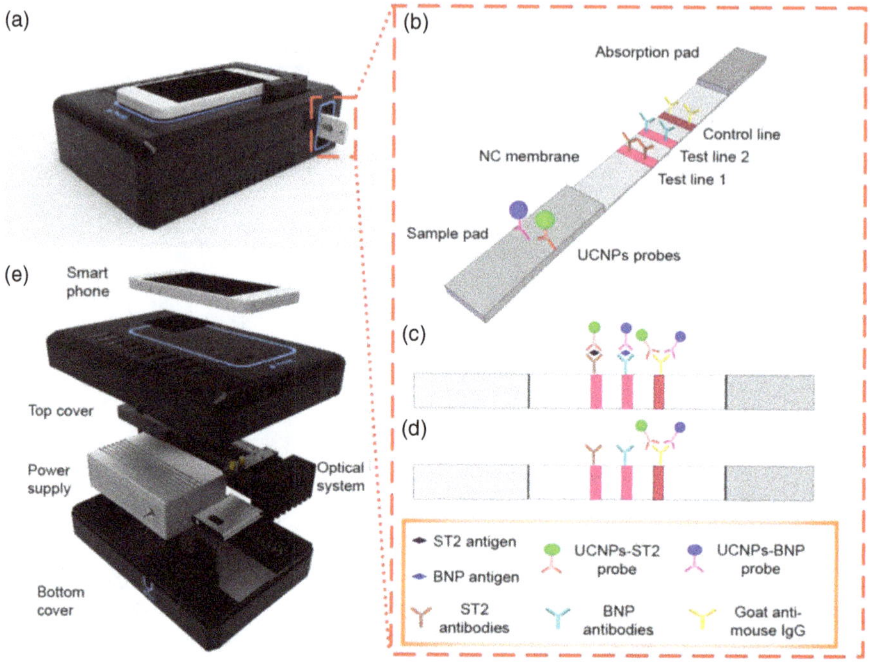

Figure 5.19 Schematic illustration of household fluorescent LFS platform. The multiplexed UC-LFS platform integrating (a) a smartphone-based portable reader and (b) a dual-color UCNPs-based LFS. (c) When two analytes of BNP and ST2 are present in a sample, the dual-color UCNPs would be captured in two test lines, respectively. (d) Otherwise, no UCNPs would be trapped in test lines. Dual-color UCNPs would be captured in control lines in two cases. (e) The details of smartphone-based reader are displayed in the enlarged view. Source: Reproduced with permission from [83], 2017 / American Chemical Society.

miniaturized luminescence readout system. To ensure the solution inside the capsule is constantly refreshed, a fluidic module with multiple inlets is employed. The radioluminescence and afterglow intensities are used for X-ray dosimetry. The afterglow of Tb^{3+} is utilized to track pH changes after irradiation due to the pH-sensitive absorption of the polyaniline films. As a result, the transmitted light exhibits a different green/blue color ratio as it passes through the pH-sensing film immersed in various pH solutions and finally reaches the pH sensor. The detected photoelectric signals can be wirelessly transmitted to a mobile phone for easy viewing.

The integration of on-chip lasers, complementary metal-oxide-semiconductor (CMOS) detectors, waveguide systems, and microcontrollers is critical for the miniaturization of integrated LnNP devices, which is still a relatively new area. Modern CMOS technology not only enables spectral detection [85, 86] (when combined with filters, gratings, or cavities) but also facilitates time-resolved measurements at a microsecond time frame [86, 87], which is suitable for the long-lifetime luminescence of lanthanide ions. To meet the trend of point-of-care testing, further development of on-chip bioprobes requires multidisciplinary collaboration.

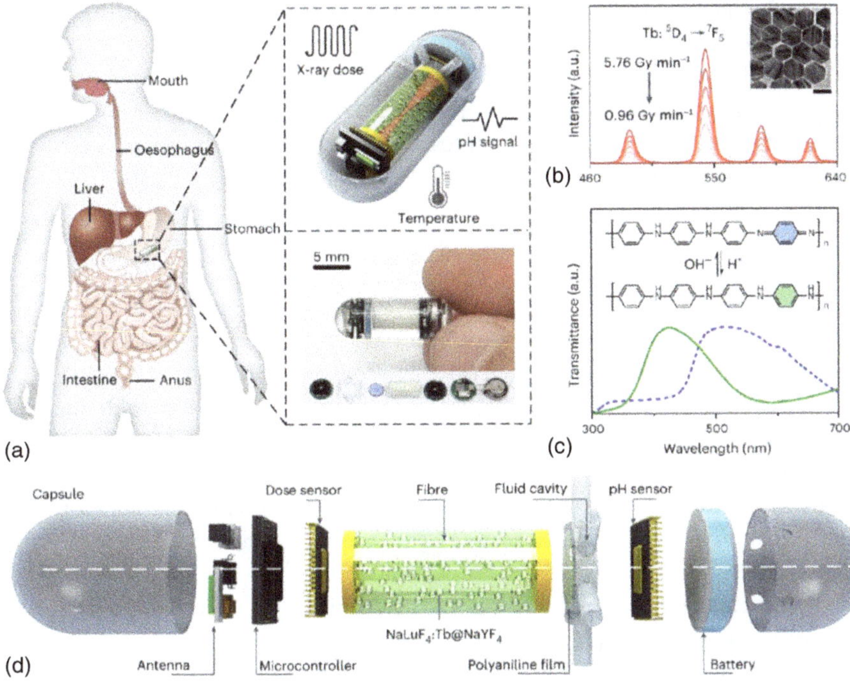

Figure 5.20 Design of a swallowable multimodal X-ray dosimeter. (a) The dosimeter in schematic and real view, with the capacity for simultaneous in situ detection of absorbed radiation dose, pH, and temperature in the gastrointestinal tract. (b) Radioluminescence spectra of $NaLuF_4$:Tb^{3+}@$NaYF_4$ nanoscintillators irradiated with different dose rates (0.96–5.76 Gy min^{-1}) at room temperature (inset: TEM image of as-synthesized nanoscintillators). Scale bar, 100 nm. (c) Transmission spectra of the polyaniline film recorded in acidic and basic solutions. (d) The all-in-one capsule dosimeter in exploded three-dimensional view. From left to right, the capsule consists of an antenna, an onboard microcontroller, a color contrast sensor for dose detection, an optical fiber embedding persistent $NaLuF_4$:Tb^{3+}@$NaYF_4$ nanoscintillators, a pH-responsive polyaniline film, a multichannel fluid cavity, a color contrast sensor for pH detection and a silver oxide battery. The radioluminescence propagates to both ends of the fiber in the form of total internal reflection. The dose is determined by the total intensity of the dose sensor, which measures light directly from the optical fiber, and the pH is measured by the color contrast ratio of the pH sensor, which measures light after it passes through the polyaniline film. Photoelectric signals are processed by an integrated detection circuit and transmitted wirelessly to a mobile device for convenient reading. Source: Reproduced with permission from [84], 2023 / Springer Nature.

5.6 Summary

Benefiting from the chemical and photostability, excitation and emission in the biological window, and facile functionalization strategies, lanthanide-doped nanomaterials have emerged as highly promising bioprobes. A variety of sensing and detecting systems have already been demonstrated to have high sensitivity. The main drawback of LnNPs is their weak absorption and low efficiency compared to QDs and organic molecule counterparts. As a result, it is still worthwhile to investigate the design of appropriate core–shell compositions and the refinement of crystal quality to enhance the emission intensity of lanthanide-doped nanoparticles. The construction of superior bioprobes necessitates collaborative efforts among researchers from diverse disciplines, including materials science, chemistry, biology, electronics, and computer science.

References

1 Renata, R. (2015). Optical properties of lanthanides in condensed phase, theory and applications. *AIMS Materials Science* 2: 37–60.

2 Dong, H., Sun, L.D., and Yan, C.H. (2015). Energy transfer in lanthanide upconversion studies for extended optical applications. *Chemical Society Reviews* 44: 1608–1634.

3 Zhao, X., He, S., and Tan, M.C. (2016). Design of infrared-emitting rare earth doped nanoparticles and nanostructured composites. *Journal of Materials Chemistry C* 4: 8349–8372.

4 Wei, H.L., Zheng, W., Zhang, X. et al. (2023). Tuning near-infrared-to-ultraviolet upconversion in lanthanide-doped nanoparticles for biomedical applications. *Advanced Optical Materials* 11: e2201716.

5 Dong, H., Du, S.R., Zheng, X.Y. et al. (2015). Lanthanide nanoparticles: from design toward bioimaging and therapy. *Chemical Reviews* 115: 10725–10815.

6 Dieke, G.H. and Robert, A. (1969). Satten spectra and energy levels of rare earth ions in crystals. *Physics Bulletin* 20: 525.

7 Wen, S., Wen, S.H., Zhou, J. et al. (2018). Advances in highly doped upconversion nanoparticles. *Nature Communications* 9: 2415.

8 Mei, S., Zhou, J., Sun, H.T. et al. (2021). Networking state of ytterbium ions probing the origin of luminescence quenching and activation in nanocrystals. *Advanced Science* 8: 2003325.

9 Tu, L., Wu, K., Luo, Y. et al. (2023). Significant enhancement of the upconversion emission in highly Er^{3+}-doped nanoparticles at cryogenic temperatures. *Angewandte Chemie International Edition* 62: e202217100.

10 Johnson, N.J.J., He, S., Diao, S. et al. (2017). Direct evidence for coupled surface and concentration quenching dynamics in lanthanide-doped nanocrystals. *Journal of the American Chemical Society* 139: 3275–3282.

11 Feng, Y., Li, Z., Li, Q. et al. (2021). Internal OH^- induced cascade quenching of upconversion luminescence in $NaYF_4$:Yb/Er nanocrystals. *Light: Science & Applications* 10: 105.

12 Frencken, A.L., Blackburn, A.M., and van Veggel, F.C.J.M. (2022). The internal structure of lanthanide-doped nanoparticles and the effect of high-temperature annealing on their luminescent properties. *The Journal of Physical Chemistry C* 126: 16341–16348.

13 Chen, J., Liang, L., Tan, S. et al. (2023). Volumetric nanocrystal lattice reconstruction through dynamic metal complex docking. *Nano Letters* 23: 7221–7227.

14 Smith, A.M., Mancini, M.C., and Nie, S. (2009). Second window for in vivo imaging. *Nature Nanotechnology* 4: 710–711.

15 Fan, Y. and Zhang, F. (2019). A new generation of NIR-II probes: lanthanide-based nanocrystals for bioimaging and biosensing. *Advanced Optical Materials* 7: 1801417.

16 Feng, Z., Tang, T., Wu, T. et al. (2021). Perfecting and extending the near-infrared imaging window. *Light: Science & Applications* 10: 197.

17 Hong, G., Antaris, A.L., and Dai, H. (2017). Near-infrared fluorophores for biomedical imaging. *Nature Biomedical Engineering* 1: 0010.

18 Bashkatov, A.N., Genina, E.A., Kochubey, V.I., and Tuchin, V.V. (2005). Optical properties of human skin, subcutaneous and mucous tissues in the wavelength range from 400 to 2000 nm. *Journal of Physics D: Applied Physics* 38: 2543.

19 Mehbuba, T., Nathaniel, J.H., Ali, S. et al. (2016). Absorption-induced image resolution enhancement in scattering media. *ACS Photonics* 3: 1787–1793.

20 Carr, J.A., Marianne, A., Franke, D., and Bawendi, M.G. (2018). Absorption by water increases fluorescence image contrast of biological tissue in the shortwave infrared. *Proceedings of the National Academy of Sciences* 115: 9080–9085.

21 Ronda, C.R. (2007). Emission and excitation mechanisms of phosphors. In: *Luminescence: From Theory to Applications. © 2008* (ed. C. Ronda), 14–19. Wiley-VCH Verlag GmbH & Co. KgaA.

22 Dexter, D.L. (1953). A theory of sensitized luminescence in solids. *The Journal of Chemical Physics* 21: 836–850.

23 Olaya-Castro, A. and Scholes, G.D. (2011). Energy transfer from Förster–Dexter theory to quantum coherent light-harvesting. *International Reviews in Physical Chemistry* 30: 49–77.

24 Liu, X. and Qiu, J. (2015). Recent advances in energy transfer in bulk and nanoscale luminescent materials: from spectroscopy to applications. *Chemical Society Reviews* 44: 8714–8746.

25 Förster, T. (1948). Zwischenmolekulare energiewanderung und fluoreszenz. *Annalen der Physik* 437: 55–75.

26 Yang, Y., Chen, Y., Pei, P. et al. (2023). Fluorescence-amplified nanocrystals in the second near-infrared window for in vivo real-time dynamic multiplexed imaging. *Nature Nanotechnology.* 18: 1195–1204.

27 Cai, Z., Qin, Y., Guo, X. et al. (2021). High-efficiency and water-quenching-resistant Tb^{3+}-based nanoparticles for single-particle imaging. *Nanophotonics* 10: 1845–1852.

28 Wang, F. and Liu, X. (2009). Recent advances in the chemistry of lanthanide-doped upconversion nanocrystals. *Chemical Society Reviews* 38: 976–989.

29 Yan, C., Zhao, H., Perepichka, D.F., and Rosei, F. (2016). Lanthanide ion doped upconverting nanoparticles: synthesis, structure and properties. *Small* 12: 3888–3907.

30 Garfield, D.J., Borys, N.J., Hamed, S.M. et al. (2018). Enrichment of molecular antenna triplets amplifies upconverting nanoparticle emission. *Nature Photonics* 12: 402–407.

31 Pini, F., Soriano, L.F., Andrigo, V. et al. (2023). Optimizing upconversion nanoparticles for FRET biosensing. *ACS Nano* 17: 4971–4984.

32 Gargas, D.J., Chan, E.M., Ostrowski, A.D. et al. (2014). Engineering bright sub-10-nm upconverting nanocrystals for single-molecule imaging. *Nature Nanotechnology* 9: 300–305.

33 Deng, R., Wang, J., Chen, R. et al. (2016). Enabling förster resonance energy transfer from large nanocrystals through energy migration. *Journal of the American Chemical Society* 138: 15972–15979.

34 Zhou, J., Li, C., Li, D. et al. (2020). Single-molecule photoreaction quantitation through intraparticle-surface energy transfer (i-SET) spectroscopy. *Nature Communications* 11: 4297.

35 Wei, W., Chen, G., Baev, A. et al. (2016). Alleviating luminescence concentration quenching in upconversion nanoparticles through organic dye sensitization. *Journal of the American Chemical Society* 138: 15130–15133.

36 Cheignon, C., Kassir, A.A., Soro, L.K., and Charbonnière, L.J. (2022). Dye-sensitized lanthanide containing nanoparticles for luminescence based applications. *Nanoscale* 14: 13915–13949.

37 Rabouw, F.T., Prins, P.T., Pedro, V.-D. et al. (2018). Quenching pathways in $NaYF_4$:Er^{3+},Yb^{3+} upconversion nanocrystals. *ACS Nano* 12: 4812–4823.

38 Wrobel, A.P., Kotulska, A.M., Lahtinen, S. et al. (2022). Engineering the compositional architecture of core-shell upconverting lanthanide-doped nanoparticles for optimal luminescent donor in resonance energy transfer: the effects of energy migration and storage. *Small* 18: 2200464.

39 Zhang, K., Song, S., Huang, S. et al. (2018). Lighting up microRNA in living cells by the disassembly of lock-like DNA-programmed UCNPs-AuNPs through the target cycling amplification strategy. *Small* 14: 1802292.

40 Deng, R., Qin, F., Chen, R. et al. (2015). Temporal full-colour tuning through non-steady-state upconversion. *Nature Nanotechnology* 10: 237–242.

41 Huang, K., Qiu, H., Zhang, X. et al. (2023). Orthogonal trichromatic upconversion with high color purity in core-shell nanoparticles for a full-color display. *Angewandte Chemie International Edition* 62: e202218491.

42 Zhou, B., Yan, L., Huang, J. et al. (2020). NIR II-responsive photon upconversion through energy migration in an ytterbium sublattice. *Nature Photonics* 14: 760–766.

43 Sun, T., Chen, B., Guo, Y. et al. (2022). Ultralarge anti-Stokes lasing through tandem upconversion. *Nature Communications* 13: 1032.

44 Zhong, Y., Ma, Z., Zhu, S. et al. (2017). Boosting the down-shifting luminescence of rare-earth nanocrystals for biological imaging beyond 1500 nm. *Nature Communications* 8: 737.

45 Wang, W., Feng, Z., Li, B. et al. (2021). Er^{3+} self-sensitized nanoprobes with enhanced 1525 nm downshifting emission for NIR-IIb in vivo bio-imaging. *Journal of Materials Chemistry B* 9: 2899–2908.

46 Ren, F., Liu, H., Zhang, H. et al. (2020). Engineering NIR-IIb fluorescence of Er-based lanthanide nanoparticles for through-skull targeted imaging and imaging-guided surgery of orthotopic glioma. *Nano Today* 34: 100905.

47 Kong, W., Sun, T., Chen, B. et al. (2017). A general strategy for ligand exchange on upconversion nanoparticles. *Inorganic Chemistry* 56: 872–877.

48 Bogdan, N., Vetrone, F., Ozin, G.A., and Capobianco, J.A. (2011). Synthesis of ligand-free colloidally stable water dispersible brightly luminescent lanthanide-doped upconverting nanoparticles. *Nano Letters* 11: 835–840.

49 Dong, A., Ye, X., Chen, J. et al. (2011). A generalized ligand-exchange strategy enabling sequential surface functionalization of colloidal nanocrystals. *Journal of the American Chemical Society* 133: 998–1006.

50 Duong, H.T.T., Chen, Y., Tawfik, S.A. et al. (2018). Systematic investigation of functional ligands for colloidal stable upconversion nanoparticles. *RSC Advances* 8: 4842–4849.

51 Horno, E.U., Kyriazi, M.E., and Kanaras, A.G. (2021). A method for the growth of uniform silica shells on different size and morphology upconversion nanoparticles. *Nanoscale Advances* 3: 3522–3529.

52 Piao, Y., Burns, A., Kim, J. et al. (2008). Designed fabrication of silica-based nanostructured particle systems for nanomedicine applications. *Advanced Functional Materials* 18: 3745–3758.

53 Jaque, D. and Vetrone, F. (2012). Luminescence nanothermometry. *Nanoscale* 4: 4301–4326.

54 Vetrone, F., Naccache, R., Zamarrón, A. et al. (2010). Temperature sensing using fluorescent nanothermometers. *ACS Nano* 4: 3254–3258.

55 Balabhadra, S., Debasu, M.L., Brites, C.D.S. et al. (2015). Boosting the sensitivity of Nd^{3+}-based luminescent nanothermometers. *Nanoscale* 7: 17261–17267.

56 Senapati, S. and Nanda, K.K. (2017). Red emitting Eu:ZnO nanorods for highly sensitive fluorescence intensity ratio based optical thermometry. *Journal of Materials Chemistry C* 5: 1074–1082.

57 Ximendes, E.C., Santos, W.Q., Rocha, U. et al. (2016). Unveiling in vivo subcutaneous thermal dynamics by infrared luminescent nanothermometers. *Nano Letters* 16: 1695–1703.

58 Ximendes, E.C., Rocha, U., Sales, T.O. et al. (2017). In vivo subcutaneous thermal video recording by supersensitive infrared nanothermometers. *Advanced Functional Materials* 27: 1702249.

59 Qiu, X., Zhou, Q., Zhu, X. et al. (2020). Ratiometric upconversion nanothermometry with dual emission at the same wavelength decoded via a time-resolved technique. *Nature Communications* 11: 4.

60 Tsai, E.S., Himmelstoß, S.F., Wiesholler, L.M. et al. (2019). Upconversion nanoparticles for sensing pH. *Analyst* 144: 5547–5557.

61 Du, S., Gil, J.H., Dong, H. et al. (2017). Design and validation of a new ratiometric intracellular pH imaging probe using lanthanide-doped upconverting nanoparticles. *Dalton Transactions* 46: 13957–13965.

62 Näreoja, T., Deguchi, T., Christ, S., and Peltomaa, R. (2017). Ratiometric sensing and imaging of intracellular pH using polyethylenimine-coated photon upconversion nanoprobes. *Analytical Chemistry* 89: 1501–1508.

63 Arpp, R., Näreoja, T., Nylund, S. et al. (2014). Photon upconversion sensitized nanoprobes for sensing and imaging of pH. *Nanoscale* 6: 6837–6843.

64 Tsai, E.S., Joud, F., Wiesholler, L.M. et al. (2020). Upconversion nanoparticles as intracellular pH messengers. *Analytical and Bioanalytical Chemistry* 412: 6567–6581.

65 Ludwig, J.A. and Weinstein, J.N. (2005). Biomarkers in cancer staging, prognosis and treatment selection. *Nature Reviews Cancer* 5: 845–856.

66 Li, S., Xu, L., Ma, W. et al. (2016). Dual-Mode ultrasensitive quantification of microRNA in living cells by chiroplasmonic nanopyramids self-assembled from gold and upconversion nanoparticles. *Journal of the American Chemical Society* 138: 306–312.

67 Zhao, J., Chu, H., Zhao, Y. et al. (2019). A NIR light gated DNA nanodevice for spatiotemporally controlled imaging of microrna in cells and animals. *Journal of the American Chemical Society* 141: 7056–7062.

68 Chu, H., Zhao, J., Mi, Y. et al. (2019). Near-Infrared light-initiated hybridization chain reaction for spatially and temporally resolved signal amplification. *Angewandte Chemie International Edition* 58: 14877–14881.

69 Ni, S., Zhuo, Z., Pan, Y. et al. (2021). Recent progress in aptamer discoveries and modifications for therapeutic applications. *ACS Applied Materials & Interfaces* 13: 9500–9519.

70 Qu, A., Wu, X., Xu, L. et al. (2017). SERS- and luminescence-active Au–Au–UCNP trimers for attomolar detection of two cancer biomarkers. *Nanoscale* 9: 3865–3872.

71 Hao, T., Wu, X., Xu, L. et al. (2017). Ultrasensitive detection of prostate-specific antigen and thrombin based on gold-upconversion nanoparticle assembled pyramids. *Small* 13: 1603944.

72 Zhou, S., Zheng, W., Chen, Z. et al. (2014). Dissolution-enhanced luminescent bioassay based on inorganic lanthanide nanoparticles. *Angewandte Chemie International Edition* 53: 12498–12502.

73 Guo, H., Song, X., Lei, W. et al. (2019). Direct detection of circulating tumor cells in whole blood using time-resolved luminescent lanthanide nanoprobes. *Angewandte Chemie International Edition* 58: 12195–12199.

74 Fan, Y., Wang, P., Lu, Y. et al. (2018). Lifetime-engineered NIR-II nanoparticles unlock multiplexed in vivo imaging. *Nature Nanotechnology* 13: 941–946.

75 Cao, C., Zhou, X., Xue, M. et al. (2019). Dual near-infrared-emissive luminescent nanoprobes for ratiometric luminescent monitoring of ClO^- in living organisms. *ACS Applied Materials & Interfaces* 11: 15298–15305.

76 Wang, S.L., Lu, F., El-Toni, A.M. et al. (2019). In vivo high-resolution ratiometric fluorescence imaging of inflammation using NIR-II nanoprobes with 1550 nm emission. *Nano Letters* 19: 2418–2427.

77 Liu, Q., Zhong, Y., Su, Y. et al. (2021). Real-time imaging of hepatic inflammation using hydrogen sulfide-activatable second near-infrared luminescent nanoprobes. *Nano Letters* 21: 4606–4614.

78 Wang, C., Lin, H., Ge, X. et al. (2021). Dye-sensitized downconversion nanoprobes with emission beyond 1500 nm for ratiometric visualization of cancer redox state. *Advanced Functional Materials* 31: 2009942.

79 Liu, L., Wang, S., Zhao, B. et al. (2018). Er^{3+} sensitized 1530 nm to 1180 nm second near-infrared window upconversion nanocrystals for in vivo biosensing. *Angewandte Chemie International Edition* 57: 7518–7522.

80 Yuan, J., Cen, Y., Kong, X.J. et al. (2015). MnO^2-nanosheet-modified upconversion nanosystem for sensitive turn-on fluorescence detection of H_2O_2 and glucose in blood. *ACS Applied Materials & Interfaces* 7: 10548–10555.

81 Xu, S., Zhu, Y., Xu, W. et al. (2014). A novel upconversion, fluorescence resonance energy transfer biosensor (FRET) for sensitive detection of lead ions in human serum. *Nanoscale* 6: 12573–12579.

82 Zhao, M., Li, B., Wu, Y. et al. (2020). A tumor-microenvironment-responsive lanthanide–cyanine FRET sensor for NIR-II luminescence-lifetime in situ imaging of hepatocellular carcinoma. *Advanced Materials* 32: 2001172.

83 You, M., Lin, M., Gong, Y. et al. (2017). Household fluorescent lateral flow strip platform for sensitive and quantitative prognosis of heart failure using dual-color upconversion nanoparticles. *ACS Nano* 11: 6261–6270.

84 Hou, B., Yi, L., Hu, D. et al. (2023). A swallowable X-ray dosimeter for the real-time monitoring of radiotherapy. *Nature Biomedical Engineering* 7 (10): 1242–1251.

85 Ho, D., Noor, M.O., Krull, U.J. et al. (2013). CMOS spectrally-multiplexed FRET-on-a-chip for DNA analysis. *IEEE Transactions on Biomedical Circuits and Systems* 7: 643–654.

86 Wei, L., Doughan, S., Han, Y. et al. (2014). The Intersection of CMOS microsystems and upconversion nanoparticles for luminescence bioimaging and bioassays. *Sensors* 14: 16829–16855.

87 Tyndall, D., Rae, B.R., Li, D.D. et al. (2012). A high-throughput time-resolved mini-silicon photomultiplier with embedded fluorescence lifetime estimation in 0.13 μm CMOS. *IEEE Transactions on Biomedical Circuits and Systems* 6: 562–570.

6

Rare Earth Luminescent Nanomaterials for Gene Delivery

Jiajun Li and Tao Zhang

South China Normal University, College of Biophotonics, Guangdong Provincial Key Laboratory of Laser Life Science, MOE Key Laboratory of Laser Life Science and Institute of Laser Life Science, School of Optoelectronic Science and Engineering, No. 55, West of Yat-sen Avenue, Guangzhou 510631, P.R. China.
Email: zt@scnu.edu.cn

6.1 Introduction

In 1990, scientists at the University of Pennsylvania started the first successful clinical trial of gene therapy on a four-year-old girl who was diagnosed with severe combined immunodeficiency, in which case a retroviral vector was used to transfer a normal copy of adenosine deaminase into her T cells. Since then, numerous efforts have been made for the approval of gene therapy products in recent years, ushering humanity into a new era of gene therapy. Summarizing the trends of gene therapy, particularly the gene delivery vectors based on rare earth materials, provides valuable information for future research, improved study designs, and realistic policy-making for the therapeutic modality.

Gene therapy generally involves delivering a functional gene into specific host cells to modify or manipulate the gene expression to compensate for the missing or mutated endogenous counterpart and produce missing functional proteins [1, 2]. The therapeutic strategies normally refer to using coding DNA (such as pDNA) to express the gene of interest and synthesize the target proteins, using functional RNA (such as siRNA) to mediate gene silencing, or using CRISPR-Cas gene editing technology to knock out, inhibit or activate specific genes [3]. As a radical treatment technology to correct abnormal genes, gene therapy could break through the difficulties faced by traditional drugs, exhibiting great potential in the diagnosis and treatment of major diseases such as cancers. To achieve therapeutic effects, gene therapy relies on vectors to deliver the required genes to target cells [4]. However, the systematic delivery of genes efficiently and accurately remains a significant challenge. The collision between electronegative nucleic acids and the equally electronegative cell membrane of host cells results in very low efficiency of direct gene delivery. The key to solving this problem is the design and development of precise gene delivery vectors. The emergence of viral vectors has made gene therapy a reality [5]. Researchers and clinicians have developed various viral vector-based genetic products, which

Photofunctional Nanomaterials for Biomedical Applications, First Edition.
Edited by Chunxia Li and Jun Lin.

have saved many patients. Virus is an effective form of gene delivery because the virus structure prevents the degradation of DNA via the liposome. The delivery protocols contain an ex-vivo manner that allows the genetic materials to be implanted into a person's cells, cultivated, manipulated, and then returned, and an in vivo manner that allows the working gene to be delivered by the vectors to the cells in the human body directly [6]. The former method is considered relatively safe but need reliable techniques, and only a small percentage of implanted cells remain viable at present. The directly delivering technique is still the major and advanced strategy at present. However, these genetic materials based on viral vectors often suffer from host immune responses, as the mammalian immune system has evolved to recognize infectious agents, which largely limits the current basic research and in-depth development of clinical applications of gene therapy [7].

Given the immunogenicity and cytotoxicity of viral vectors, nonviral vectors have exhibited great potential due to its reduced pathogenicity, low cost, and ease of production [8, 9]. The major advantage of using nonviral vectors is its biosafety, which has invested the nonviral vectors significant attention in recent years. Efforts in efficiency, specificity, gene expression duration, and safety have led to an increased number of nonviral vector products. As an emerging kind of nonviral vector, nanocarriers play an irreplaceable role in gene delivery due to their high degree of modification and biosafety [10]. Nanovectors can increase the gene loading capacity through the large specific surface area of mesoporous/hollow nanostructures. Moreover, they can also accumulate in the tumor lesions through leaky tumor vasculature, the so-called hyperpermeability, and enhanced permeability and retention (EPR) effect of solid tumors. By virtue of these advantages, researchers have designed numerous nanovectors such as liposome nanoparticles, polymer nanoparticles, carbon nanotubes, gold nanoparticles, magnetic nanoparticles, and rare-earth-based nanomaterials [11].

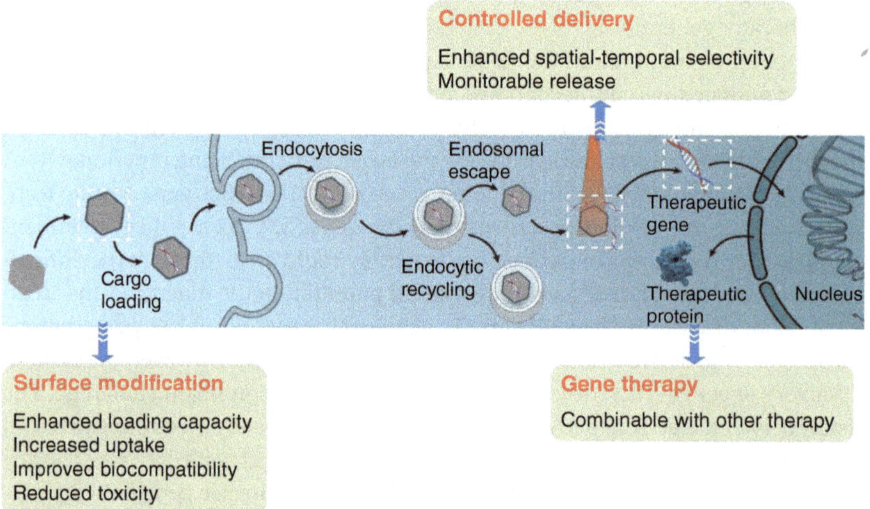

Figure 6.1 Rare earth-based nanovectors for gene delivery.

Among rare earth luminescent nanomaterials, especially the lanthanide-doped upconversion nanoparticles (UNCPs), gene delivery carriers exhibit unique advantages as they can not only monitor the process of gene therapy but can also control the processes of transport, targeting, and release, which can meet the precise and controllable requirements for clinical application of gene therapy [12, 13]. Thereby, the main content of this chapter is the design and application of rare earth luminescent nanomaterials for gene delivery (Figure 6.1).

6.2 UCNPs Nanovectors

UCNPs, as emerging multifunctional nanoparticles, exhibit unique anti-Stokes luminescent properties due to the long-lived emission generated by the doped lanthanide ions through forbidden 4f-4f transition between the intermediate energy states [14, 15]. Near-infrared (NIR) excitation enables deep tissue penetration, minimizing interference from endogenous autofluorescence and providing a higher signal-to-noise ratio for imaging, diagnosis, and therapy, which is an advantage not possessed by other carriers. In general, monitoring the gene delivery/therapeutic effects and spatiotemporal control of gene release are two basic functions of UCNPs-based gene delivery systems.

Since gene delivery is a multi-stage process, the first step is to prevent nucleic acid molecules from being degraded by enzymes in complex physiological environment before reaching the target site [16]. After the vector reaches the cell, it needs to undergo multiple processes including endocytosis, internalization, and endosomal escape to deliver the aimed genes. Only when the delivered genes are successfully released into the cytoplasm can gene therapy be truly effective [17]. In addition, another factor determining the efficacy of gene therapy is the loading amount of nucleic acids. According to the gene cellular delivery process, the following will summarize and discuss several aspects including surface modification, spatiotemporal control of gene delivery, and gene therapy.

6.3 Surface Modification

6.3.1 Silica

Silica-based nanocarriers could offer several advantages in terms of high loading capacity, easy functionalization and well biocompatibility. Silica nanoparticles, particularly porous nanoparticles, have attracted interest as delivery platforms due to their large surface area and ordered porous channels that can be used to encapsulate molecules [18]. Encapsulating cargoes, including the nucleic acids, in a silica shell could protect them from the surrounding aqueous medium and enhance the solubility and stability of the cargoes without affecting the physical and chemical properties of the core nanoparticles themselves [19]. The mesoporous silica nanoparticles (MSNs) have exhibited their potential in addressing key therapeutic issues, including drug solubility and prolonged drug release rates. In addition, the hydroxyl groups on the surface of silica have enabled MSNs to be easily conjugated with functional agents [20–22]. Moreover, the thickness of the silica shell can be varied by adjusting the concentration of the silane, which is important for drug delivery and bioassays, since the thickness of the shell is directly related to energy transfer efficiency between the luminescent nanocore and the loaded agents via luminescence resonance energy transfer (LRET) mechanism, which has strict requirements on the distance [23, 24].

Due to the well-established construction of silica-coated nanoparticles and the suitability of silica shells for both hydrophobic and hydrophilic surfaces, the MSNs have grown rapidly for surface modifications [25–27]. The surface silanization of UCNPs ($NaYF_4$:Yb, Tm) has been pioneeringly demonstrated for the delivery of genes by Zhang et al (Figure 6.2). They demonstrated that the as-prepared nanovector could protect the nucleic acid from degradation and control gene expression under NIR excitation. In-depth photoactivation through tissue phantoms and in vivo activation of caged nucleic acids for green fluorescent protein

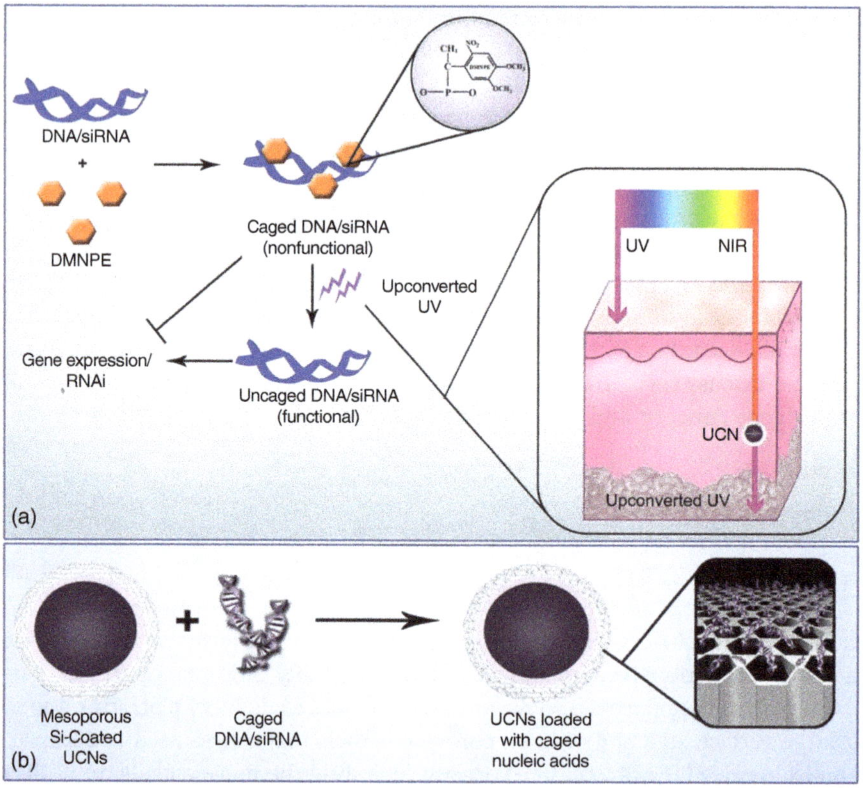

Figure 6.2 Schematic illustration of silica-coated UNCPs nanovector for delivery of plasmid DNA and siRNA. (a) Penetration depth of UV and NIR light in the skin. (b) Loading of caged plasmid DNA/siRNA into the mesopores of UCNs. Source: Reproduced with permission from Jayakumar et al. [28], 2012/National Academy of Science.

(GFP) expression was accomplished [28]. This work has actually started a new direction for UCNPs-based materials in their biological applications.

In addition, by further modifying the silica surface, photosensitive azophenyl caps were conjugated to UNCPs to achieve the light-controlled release of nucleic acids [29]. In short, surface silanization may hold promise for engineering the surface properties of UCNP-based gene nanovectors in future therapeutic and diagnostic applications.

6.3.2 Cationic Polymers

Cationic polymer is a kind of common polymer material, including polyethyleneimine (PEI), polyurethane (PBAE), chitosan (chitosan), polypropyleneamine (PAH), diethylaminoethyl dextran (DEAE-dextran), and polyamine dendrimers (PAMAM) [30]. Among, PEI is the most extensively studied nonviral gene vector and is often used as a positive control in the development of gene vectors. The basic principle of PEI-based vector for gene delivery is that there are many amino groups on the PEI molecules, and under physiological conditions, these protonated amino groups can interact with the negatively charged nucleic acids and protect them from nuclease degradation. The electrostatic interaction between the cationic polymer and the negatively charged phosphate groups forms a cationic polymer-nucleic acid complex, which can adhere to the cell surface and enter the cells through endocytosis. Upon entry into the endosome, the protonation of the amine results in an influx of counterions and a decrease in cell osmotic pressure, which expands the vesicle. At the same time, the repulsion between protonated amines causes conformational changes in PEI, allowing the nucleic acids released from the vector [31].

To increase the loading efficiency of UCNPs nanovectors, researchers have constructed a series of PEI-based delivery strategies. For example, Wang et al. prepared a multifunctional gene vector based on UCNPs through a ligand exchange protocol. The PEI-coated gene vector ($NaGdF_4$:Yb, Er) could transfer the enhanced green fluorescent protein (EGFP) pDNA and bcl-2 siRNA into HeLa cells to achieve gene introduction and silencing [32]. However, based on this strategy, PEI can also induce high toxicity by breaking up the membranous structures of cells (such as the cell membrane and mitochondrial membrane, etc.), limiting its applications [33].

Thereby, further modifications have been applied to PEI-based carriers to reduce their toxicity [34]. He et al. reported a Gd^{3+} containing UCNP carrier UCNP-PEG@×PEI modified with a dual-type polymer polyethylene glycol (PEG) and two layers of PEI. Compared to UCNP-PEG@1×PEI with only one layer of PEI, the two-layer PEI-modified nanovector exhibited higher gene transfection efficiency and lower cell toxicity. Unlike free PEI, which can only function in serum-free culture medium, UCNP-PEG@2×PEI cannot only transfect cells in serum-free culture medium, but the addition of serum could enhance the gene transfection efficiency, which possibly was due to the protein corona of the nanoparticles triggering receptor-mediated endocytosis of the carrier [35]. Based on the previous reports, Wang et al. developed a new UCNP nanovector

UCNP@2×CA@2×PEI through layer-by-layer self-assembly of citric acid (CA) and two layers of PEI. The introduction of CA could enhance the aqueous solubility of the nanoparticles, reducing obviously the cytotoxicity of PEI. In short, the carrier shows excellent photostability, low toxicity, and good stability under physiological or serum conditions, successfully transfected plasmid DNA encoding luciferase and GFP into HeLa cells [36]. The layer-by-layer assembly strategy have also followed by a UCNP/PAA/PEI/siRNA nanovector [37]. The hydrophobic oleic acid (OA) was exchanged with hydrophilic PAA, and the resulting negatively charged PAA layer attracted the positively charged PEI through electrostatic interaction, which then formed stable complexes with negatively charged nucleic acids. The carrier successfully delivered MDR1-siRNA into paclitaxel-resistant ovarian cancer cells and tracked the intracellular fate of MDR1-siRNA.

In addition to PEI, other cationic polymers have also been reported to modify UCNPs to enhance gene loading efficiency. Bai et al constructed a pDNA-UCNPs@MFAP nanocapsule by the self-assembly of a positively charged multifunctional amphiphilic polymer (MFAP) based on the polymerization of aminolyzing polysuccinimide and *N*-(3-aminopropyl) imidazole with the polyethylene glycol-polylactic-glycolic acid (PEG-PLGA) polymer. The negatively charged plasmid DNA (pDNA) can be absorbed on the surface of the nanocapsules through electrostatic interaction. The UCL emission of UCNPs could serve as an indicator for gene delivery and gene transfection. The nanovector exhibits high gene loading capacity and biocompatibility [38].

6.4 Increasing Endosomal Escape

In the process of gene delivery via nanovector, endosomal escape of the vector is a key step. Therefore, a major obstacle faced by gene vectors is their low endosomal escape efficiency, which may cause the degradation of genes by lysosomal acids or extracellular excretion by endosomal circulation [39], which is also an important reason for low transfection efficiency. To improve intracellular gene delivery efficiency, photochemical internalization (PCI) has been developed as an effective technique, which utilizes photosensitizers present in endosomes to destroy endosomal membranes by locally generating reactive oxygen species (ROS) to achieve endosomal escape [40–43].

By virtue of the PCI stagey, Jayakumar et al. developed a gene delivery system based on UCNPs ($NaYF_4$:Yb, Tm@$NaYF_4$:Yb, Er), where the UCNPs surface was coated with mesoporous silica and loaded with two different photoactivatable molecules, photomorpholine, and meso-tetraphenylporphyrin. The UV/VIS emission of UCNPs could simultaneously activate both photosensitizers. And the endosome-localized meso-tetraphenylporphyrin performs PCI to destroy the endosomal membrane and enhance cellular uptake [44]. Moreover, the photomorpholine could knock down the expression of STAT3 in tumors, achieving excellent therapeutic effects in both in vitro and in vivo mouse melanoma models that are difficult to transfect with standard transfection reagents [45].

DNAzyme, a deoxyribozyme, can catalyze the cleavage of specific phosphodiester bonds, and the DNAzyme targeting disease-related mRNA is a tool for mediating gene silencing [46]. Thereby, a NIR light-controlled DNAzyme delivery platform UP-Dz was engineered, which consists of a DNAzyme targeting survivin, the photosensitizer rose bengal (RB), and core-shell UCNPs ($NaGdF_4$:Yb, Er@$NaGdF_4$). Then, the DNAzyme was loaded through the PEI-coated UCNPs via electrostatic interactions. When UP-Dz is trapped in endosomes under 980 nm light irradiation, the green light emission of UCNPs could activate RB to generate ROS, destroying the endosomal membrane and enhancing endosomal escape [47]. In the HeLa tumor-bearing mouse model, UP-Dz showed efficient gene delivery and significant antitumor effects.

6.5 Controlling Delivery Strategy

Controlling release is the essence of precision genetic engineering. Stimulus-responsive vectors can effectively deliver nucleic acids to target sites by responding to specific stimuli, thereby maximizing the effectiveness of gene therapy. After years of efforts, researchers have developed nanovectors that can be activated by numerous endogenous stimuli (pH, ROS, GSH, enzymes, etc.) and exogenous stimuli (light, heat, ultrasound, magnetic field, etc.) [48–50]. In contrast, one advantage of the UCNPs-based gene delivery system is that it can achieve light-controlled release of genes while avoiding the biological toxicity of UV. This is due to the unique upconversion UV/VIS emission of UCNPs (this localized UV does not reduce cell viability), while most traditional photolabile groups only respond to UV light. Utilizing the upconversion process to activate or silence the expression of target genes in a temporally and spatially specific manner provides a way to develop more precise therapies and a tool for fundamental research on signal transduction. Not only that, recent studies have also taken advantage of the orthogonal emission phenomenon of UCNPs to achieve programmable and orthogonal photoactivation of multiple processes, which are beyond the capabilities of traditional gene delivery methods.

Photolabile groups are the basis for photocontrolled release. Photocleavage of small molecules or polymeric backbones is the general method to trigger nucleic acid drug release and gene expression, which can also be achieved through photoisomerization of specific agents. Photocleavage methods include: (i) directly or indirectly (such as through ROS generated by photosensitizers) cleaving the bond between the molecule and the carrier; (ii) changing the surface charge of the carrier to cause electrostatic repulsion between the molecule and the carrier; (iii) destroying the carrier complex itself. In addition, the activity of nucleic acids is shielded by photocaging compounds, and then specific light is used to destroy the photocage groups to restore the activity of nucleic acids. This also belongs to the category of light control of genes.

6.5.1 Photodegradable Polymers

Remote activation of the release of substances from nanocarriers through light-initiated disruption of polymeric micelles or vesicles is a common strategy for light-controlled delivery [51, 52]. The upconversion luminescence properties of UCNPs can convert NIR light into UV light required for photodegradation of photolabile polymers, which can reduce damage to biological cells. Given this, Zhang et al. demonstrated a phototearable tape-wrapped nanocapsule for siRNA-controlled delivery [53]. The surface of UCNPs ($NaYF_4$:Yb, Tm@$NaYbF_4$) is first coated with a mesoporous silica shell, and PEG was then covalently connected to the UCNPs through a photocleavable linker to form a photocleavable polymer "ribbon." siRNA can be loaded on the silica surface through electrostatic adsorption, and controlled release of siRNA can be achieved through UV light from UCNPs. At the same time, in order to improve the ability of endosomal escape, the photosensitizer hibiscus A (HA) is loaded in the pores of mesoporous silica and activated by blue light of UCNP to generate ROS, which can not only destroy the endosomal membrane to enhance release but also destroy cellular organelles to assist in apoptosis.

6.5.2 Changes in Carrier Surface Charge

Unlike traditional chemical drugs, genetic drugs must be delivered intracellularly to be effective. Since electronegative nucleic acids repel the equally electronegative cell membrane, effectively shielding nucleic acids is an aspect that needs to be considered in carrier design. The strong electrostatic interaction between cationic polyelectrolytes and nucleic acids can increase the loading efficiency of nucleic acids and prevent the dissociation of the complex. Although a variety of light-induced charge-reversal methods have been successfully used for the controlled release of genes, most photoresponsive groups are excited by UV or visible light and have limited tissue penetration [54].

By virtue of the upconverted UV light, the UCNPs ($NaYF_4$:Yb, Tm) are coated with a silicon layer to form Si-UCNPs to increase biocompatibility and active sites for chemical modification [55]. The negatively charged siRNA is then covalently bonded to the Si-UCNPs via a photoactive *o*-nitrobenzyl thiol linker. Under 980 nm laser irradiation, the UV emission of UCNPs cuts the photosensitive linker, causing the positively charged amino groups to detach from the nanoparticle surface, resulting in the effective release of siRNA for subsequent effects. Through in vitro agarose gel electrophoresis (AGE) and intracellular imaging, the above-mentioned carrier system can effectively eliminate membrane barriers, achieve targeted gene delivery simply and effectively, and reduce photodamage of living cells.

In a similar manner, Fan et al. reported a light-induced charge-reversal cationic conjugated polyelectrolyte (CPE) brush encapsulating upconversion nanoparticles UCNP@CCPEB, which integrated the siRNA carrier and a photosensitizer into one molecule to achieve controlled release of siRNA and coordinated photodynamic therapy [56]. In this story, CPE has a conjugated main chain and charged flexible side chain, which is used as a photosensitizer; the photodegradable group o-nitrobenzyl is introduced into CPE to obtain a quaternized CPE brush (CPEB).

Since CPEB has a large number of positive charges for efficient siRNA loading, the UV emission of UCNP converts CCPEB into zwitterionic conjugated polyelectrolyte brush (ZCPEB), triggering the release of siRNA. At the same time, ZCPEB avoids the aggregation of the conjugated skeleton and maintains the PDT effect. Under 980 nm light irradiation, UCNP@CCPEBPDT demonstrated efficient singlet oxygen generation and achieved 80% siRNA release.

Cellular gene editing has been developed as a very powerful technique to study gene function, disease modeling, and disease therapy in molecular biology and biomedicine. Clustered regularly interspaced short palindromic repeat (CRISPR)/CRISPR associated protein 9 (Cas9), which consists of an endonuclease and a single-guide RNA (sgRNA), has emerged as a robust and versatile genome-editing platform [57, 58]. However, the large size of such plasmids (~10,000 bp) results in relatively low transfection efficiency [59, 60]. Given this, Pan et al and Wu et al developed UCNP-based NIR light-responsive nanovectors for the delivery of Cas9 ribonucleoprotein (RNP) and CRISPR/Cas9 plasmid, respectively. In the Pan story, UCNPs were coated with photocleavable 4-(hydroxymethyl)-3-nitrobenzoic acid (ONA) and PEI to afford the nanovector UCNPs-Cas9@PEI [61]. The upconverted UV light under 980 nm irradiation could induce ONA cleavage to release Cas9 RNP on demand to perform gene editing. Intratumoral injection of UCNPs-Cas9@PEI targeting Polo-like kinase 1 (PLK-1) sgRNA induced controllable knockout of the Plk1 gene and successfully inhibited the proliferation of tumor cells in vitro and in vivo (Figure 6.3). In the

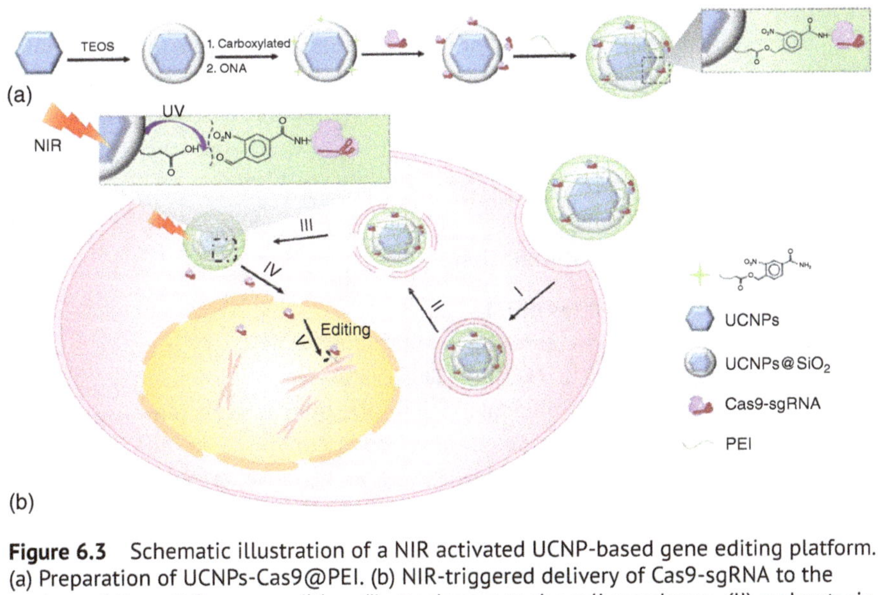

Figure 6.3 Schematic illustration of a NIR activated UCNP-based gene editing platform. (a) Preparation of UCNPs-Cas9@PEI. (b) NIR-triggered delivery of Cas9-sgRNA to the nucleus of the cell for gene editing: (I) attachment to the cell membrane, (II) endocytosis, (III) endosome escape, (IV) release from particles and step into the nucleus, and (V) search for the target DNA locus and initiate the DNA double-strand break for genome editing. Source: Reproduced with permission from Pan et al. [61], 2019/American Association for the Advancement of Science.

Wu side, a charge-reversal polymer was employed to the UCNPs for the delivery of Cas9/sgRNA plasmid directly [62].

Under 980 nm NIR irradiation, the UV emission destroys the linkers in the polymer to expose carboxyl groups, converting the original cationic polymer into an anionic polymer and releasing the plasmid through electrostatic repulsion. In xenograft tumor models, the agent effectively inhibited the expression of Plk-1, thereby effectively inhibiting tumor growth.

6.5.3 Photoisomerization

Photoisomerization is a phenomenon in which a compound changes its spatial conformation and transforms into isomers under certain conditions (heating or illumination). Reversible photoisomerization is the structural basis of photoswitching. Since azobenzene, spiropyran, and dithienethylene structures can reversibly change conformation under UV or visible light irradiation, they are often used in the design of drug/gene controllable release [63, 64].

Trans-azobenzene has a strong host-guest interaction with cyclodextrin (CD), and under UV light irradiation, the azobenzene could change from the thermodynamically stable trans-isomer to the metastable cis form, thus inhibiting the host-guest interactions with CD due to the changes in its polarity and steric hindrance [65, 66]. Based on this principle, Chen et al. reported a UCNPs-based siRNA delivery nanovector by functionalizing UCNPs with β-cyclodextrin (β-CD) and iRNA-azobenzene via host-guest interactions [67]. Under NIR irradiation, the UV light emitted by UCNPs caused Azo-siRNA to photoisomerize into the cis configuration, destroying the stability of the host-guest interaction and ultimately leading to the release of siRNA. The experiments selected triple-negative breast cancer (TNBC) cells with overexpression of EGFR as the research object. After NIR irradiation, siRNA was used in TNBC cells with GFP expressing, and specific GFP gene silencing was observed.

6.5.4 Microenvironments Stimulation

6.5.4.1 Reactive Oxygen Species (ROS)

Endogenous ROS in cells include hydroxyl radicals (OH), superoxide anions (O_2^-), and hydrogen peroxide (H_2O_2). These ROS can, for example, oxidize borate esters/acids, sulfides, thioketals, selenium, tellurium, proline oligomers, and etc., which could trigger intracellular decomposition of modifications and drug release from the nanocarriers [68]. Song et al. developed a UCNP-based optogenetic nanosystem UCNPs-TPP/TK-siRNA [69]. First, the photosensitizing gene KillerOrange-Mito (mtKO) with mitochondrial targeting properties was transfected into cancer cells in advance. Then, the mitochondrial targeting group triphenylphosphine (TPP) was linked on the surface of UCNPs. Under 980 nm light irradiation, the blue light emission of UCNPs could activate the photosensitizer to generate ROS through the LRET process from UCNPs to mtKO to perform PDT. Since the Bcl-2 specific siRNA is modified onto the UCNPs surface through the

ROS-sensitive group thioketal (TK), ROS will subsequently stimulate the release of siRNA to achieve cascade gene therapy.

To spatially and temporally control the release of siRNA to achieve efficient gene therapy, a new kind of nanovector with UCNPs stacked with three shells, namely upconverting nanoonions (UCNOs), was designed and constructed by He et al. [70]. The outer shells consist of a photosensitizer RB-conjugated PEI (PEI-RB), cell-penetrating peptide R8-modified diselenium bond-linked PEI (PEI-SeSe), and hyaluronic acid (HA). The negatively charged outer layer of HA can prevent UCNOs from being degraded during delivery and can also specifically target tumor cells through the cell membrane receptor CD44. After entering the tumor microenvironment, the outer layer of HA is degraded by hyaluronidase (HAase) to expose positively charged R8, which promotes cellular internalization and lysosomal escape. The therapeutic siRNA is electrostatically adsorbed on the middle layer of PEI-SeSe. Under 980 nm light irradiation, the blue light emission core of UCNPs activates the RB in the inner PEI-RB to generate ROS, resulting in the decomposition of the middle layer PEI-SeSe and, ultimately, the release of siRNA.

6.5.4.2 Matrixmetallo Proteinases (MMPs)

Other endogenous stimulus, such as matrixmetallo proteinase-2 (MMP-2), was also employed for the trigger of gene release. Han et al. designed a UCNP-siRNA-PAA-AS1411 nanocage system for siRNA delivery [71]. This system used surface-modified polyacrylic acid (PAA) to connect single-stranded DNA, while carboxyl-modified double-stranded DNA connects UCNPs and the nanocage surface. PAA is divided into two groups (PAA-1 and PAA-2), and PPA-1 is linked to DNA-1 (AS1411) and DNA-3. Another group, PAA-2, is linked to DNA-1 and DNA-4. Finally, since DNA-2 is the complementary antisense strand of DNA-1, the hybridization of DNA-1 and DNA-2 could connect PAA-1 and PAA-2 on the surface of UCNPs. The UCNPs nanocage is finally sealed through the complementary base pairing of DNA-3 and DNA-4 with DNA-5. The tumor-responsive release is achieved by modifying DNA-5 with an MMP-2 cleavable peptide. VEGF siRNA is then loaded into the nanocage. When the nanocage reaches the tumor tissue, the overexpression of the MMP-2 enzyme results in the cleavage of the MMP-2 peptide. The decomposition of the peptide triggers the release of siRNA, and at the same time, AS1411 on the surface of the UCNP nanocage improves the tumor-targeting ability.

Parkinson's disease (PD) is one of the major neurodegenerative diseases. Dopaminergic neurons in PD patients are selectively apoptotic, resulting in motor deficits [72]. Some studies suggest that MMP3 can be used as a biomarker for early PD detection and a target for clinical intervention [73]. To this end, Li et al. designed a UCNP/AIEgen-based nanoprobe UCNP-peptide-AIEgen, which can modulate and monitor the inflammatory response of dopaminergic neurons. UCNPs ($NaYF_4$:Yb, Tm) coated with silica (UCNP@SiO_2) were linked to CKKRGD peptide via PEG connectomes (MAL-PEG4-NHS) to enhance cellular uptake and provide a positive charge to increase siRNA loading efficiency (Figure 6.4) [74].

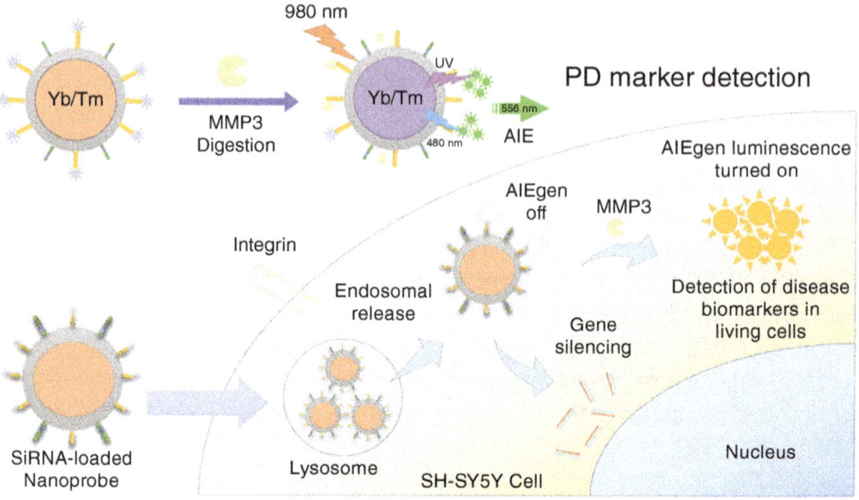

Figure 6.4 Schematic illustration of UCNP-peptide-AIEgen nanoprobe for the detection and inhibition of biomarkers in PD models. Source: Reproduced with permission from Li et al. [74], 2021/American Chemical Society.

The peptide-AIEgen reporting unit is also connected to the UCNP@SiO_2 surface. After the nanoprobe is taken up by dopaminergic neurons, MMP3 upregulated due to inflammatory response would cleave the MMP3 substrate junction on the probe, triggering the release and aggregation of AIEgen in the aqueous environment. UV/blue light emission from UCNPs transferred energy to AIEgen in the aggregated form under 980 nm light, resulting in 556 nm emission, which monitored the upregulation of enzymes under pathological conditions in real time, thus evaluating the inflammatory response of dopaminergic neurons. At the same time, the nanoprobe delivered MMP3-targeting siRNA to inflammatory catecholaminergic SH-SY5Y cells, delayed and inhibited the MMP3 activity, and successfully suppressed MMP3-induced inflammatory neural response.

6.5.5 Light Cage

In the light-caging strategy, the bioactive molecule is covalently attached to a photoprotective protecting group (PPG), rendering it inactive. Upon irradiation with specific wavelengths, the molecules are activated in an irreversible manner and are free to perform their biological functions locally. The UCNPs-assisted photoinitiated payload release via photocleavage is designed based on the integration of PPG within the drug carrier structure. The PPG used in this system is cleavable under UV/visible light irradiation, and this special photosensitive property provides spatially and temporally controlled "on-demand" drug delivery [75, 76].

4,5-Dimethoxy-2-nitroacetophenone (DMNPE) can attach to the DNA phosphate backbone, thereby shielding plasmid DNA or siRNA, and this binding can be undone by UV light. Zhang et al. were the first to report a remotely light-controlled gene expression with the UCNPs-based PPG systems [28]. Plasmid DNA encoding green fluorescent protein (GFP) and siRNA targeting GFP formed complexes with DMNPE to block their respective functions. UCNP ($NaYF_4$:Yb, Tm) serves as the luminescent core, which is covered with a layer of mesoporous silica to load the DMNPE-nucleic acid complex. Under 980 nm light irradiation, the UV emission of UCNPs unlocks the photocage of the DMNPE-nucleic acid complex and releases it. Cells were transfected with nanovectors loaded with a photocaged GFP plasmid, and the corresponding pattern was subsequently formed after irradiation of a template with an NUS pattern, demonstrating that the technology can activate genes with very high spatial and temporal control.

With a similar light-caging strategy, Li et al. also developed a nanocomplex (UCNP-Peptide-AIE-siRNA) to control and monitor stem cell differentiation [77]. In this report, UCNP ($NaYF_4$:Yb, Tm@$NaYF_4$) is used as the luminescent core, and mesoporous silica is modified on its surface. The amino groups on the silica surface are used to further modify the RGD peptide and peptide-AIEgen on the UCNPs surface to form UCNP-Peptide-AIE. The mesoporous silica structure of UCNP-Peptide-AIE can physically adsorb the DMNPE/siRNA complex, ultimately forming UCNP-Peptide-AIE-siRNA. The RGD peptide modified on the surface of UCNP-Peptide-AIE-siRNA is a sensitive peptide sequence of the osteoblast marker matrix metalloproteinase 13 (MMP 13), which is high-expressed in differentiated stem cells. Under 980 nm excitation, the UV emission of UCNP unlocks the photocage of the DMNPE/siRNA complex, releases siRNA for PPARγ targeting, and induces the differentiation of human mesenchymal stem cells (hMSC). Moreover, when hMSCs are successfully differentiated, the peptide–AIE linker is hydrolyzed by MMP13 to release AIEgen. AIEgen aggregates in an aqueous environment, resulting in enhanced luminescence at 560 nm, enabling monitoring of the differentiation process.

6.5.6 Orthogonal Control

Spatiotemporal controllability is the main advantage of rare-earth-based nanovectors. However, traditional UCNPs are difficult to apply to the time-specific photoactivation required for various biological processes due to their symbiotic multi-wavelength emission. Orthogonally emitting UCNPs (OUCNPs) can control the independent emission of different lanthanide activators by changing external excitations, and can modularly assemble multiple functions to achieve programmable and orthogonal photoactivation of multiple molecules [78, 79]. To some extent, thereby, OUCNPs is an ideal platform for gene delivery and the subsequent gene-based multiple theranostics.

UCNPs have been used in NIR-based PDT with relative success, but its effectiveness as a standalone treatment modality is relatively low. The main reason might be due to the excretion of nanoparticles by cells and the cell's inherent resistance to free radicals. The two issues can be addressed by using photochemical internalization for endosomal escape and preventing them from being excreted by the cell. And the inherent resistance to free radicals can be addressed by knocking out the superoxide dismutase 1(SOD1) gene responsible for destroying them. However, it is very

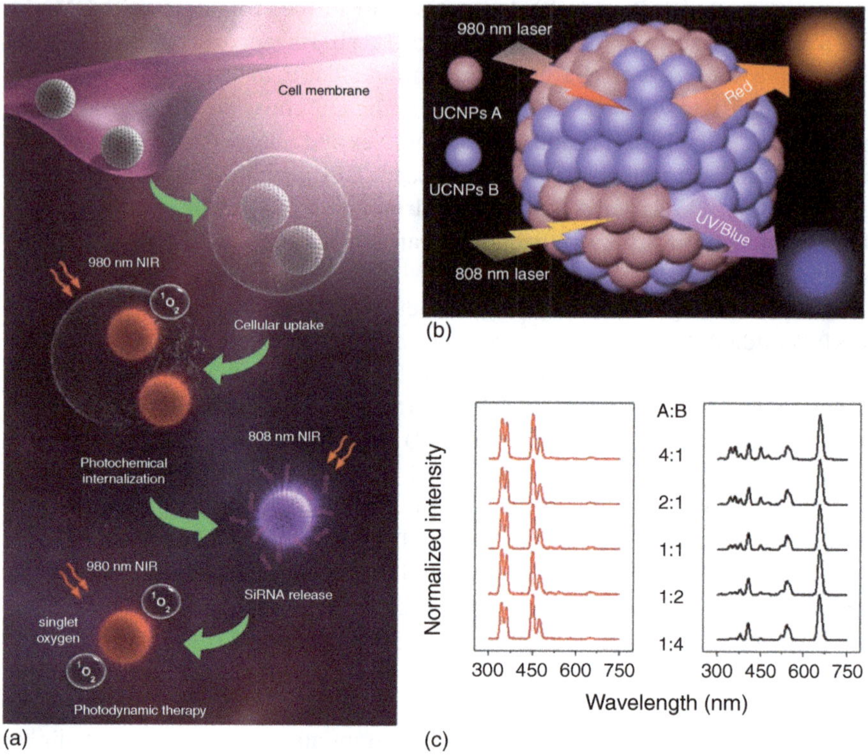

Figure 6.5 Schematic illustration of upconversion superballs for programmable photoactivation of therapeutics. (a) Schematic illustration of orthogonal excitation of photosensitizers and azobenzene-based caps for endosomal escape, siRNA release and photodynamic therapy. (b) Schematic illustration of OP-SBs with luminescent photos taken by smart phone camera. (c) Normalized 808 nm laser (left) and 980 nm laser (right) excited upconversion emission spectra of superball with different mixing ratio between A and B. Source: Zhang et al. [29], 2019/Springer Nature/CC BY 4.0.

important that these processes are activated in a sequential manner, which requires programmable and orthogonal light-activation capabilities that are not available in traditional UCNPs. Given this, Zhang et al. modularized the assembly of UCNPs that can be individually stimulated by 980/808 nm light, enhancing endosomal escape with photochemical internalization (PCI) and knocking down superoxide dismutase-1 (SOD1) to increase sensitivity to reactive oxygen species (ROS), respectively, significantly improving the efficacy of photodynamic therapy (Figure 6.5) [29]. They prepared the nanovector (OP-SB) via loading photosensitizers (zinc phthalocyanine, ZnPc) for PCI/PDT and SOD1 siRNA for gene knockdown. The release of siRNA in OP-SB is controlled by an azobenzene photoactivated at 808 nm irradiation. ZnPc was activated by OP-SB at 980 nm irradiation. Short exposure to 980 nm will result in PCI, and longer exposure to 980 nm will result in PDT. The authors actually demonstrated the high controllability of UCNPs as promising nanovector.

Multidrug resistance (MDR) is a major obstacle to cancer chemotherapy [80]. The multidrug resistance-related protein 1 (MRP1) is closely related to adverse clinical outcomes in cancer, and MRP1-mediated drug resistance can be reversed by down-regulating the MRP1 gene through gene silencing methods mediated by CRISPR-Cas13d [81, 82]. However, Cas13d-mediated gene knockdown often takes a relatively long time to achieve effective downregulation of the corresponding target protein. If chemotherapy prodrugs are activated at the same time, the efficacy will inevitably be reduced. Therefore, precise activation of Cad13d and prodrugs in a highly spatiotemporal manner is necessary to achieve synergistic therapeutic effects. To this end, Liu et al. proposed the concept of an orthogonal NIR-activated CRISPR-Cas13d nanoprodrug by introducing green light-activated Pt(IV) prodrugs and UV-activated Cas13d gene editing tools into orthogonal emitting core/multiple-shell UCNPs to maximize Pt(IV) prodrug activation and ROS production with the assistance of Cas13d-mediated attenuation of drug resistance [83]. Among them, the activation of the prodrug is achieved by the cleavage of the Pt—O bond under irradiation of 980 nm light; on the other hand, the activation of the CRISPR/Cas13d gene editing system is achieved by the cleavage of the photocleavable linker under irradiation of 808 nm light. This orthogonally activated nanoprodrug promotes the activation of Pt(IV) prodrug and ROS production and massively accumulates drug molecules with the help of Cas13d-mediated attenuation of drug resistance, effectively eradicating the primary tumor and preventing further liver metastasis.

6.5.7 Release Monitoring

Monitoring the gene delivery process could provide more support for precise spatial and temporal control of the gene release. UCNPs-based materials are well known for its intriguing optical properties, endowing the UCNPs-based nanovectors with robust potentials for monitoring gene delivery through upconversion luminescence imaging. Jiang et al. modified SiO_2-coated UCNPs ($NaYF_4$:Yb, Tm) with the antifolate antibodies and anti-Her2 monoclonal antibodies for the target of HT-29 cells with folate receptors and SK-BR-3 cells with Her2 receptors. Then, GL3 siRNA was loaded in the nanovector. Under 980 nm light irradiation, the nanovector was able to track the uptake of siRNA in real time and achieved a 45% downregulation of luciferase gene expression. However, the system did not track the release of genes [84]. To this end, the team modified the vector further in a follow-up study by embedding dye BOBO-3 in the siRNA as an acceptor of UCNP phosphor. The gene release can be visualized by fluorescence changes and quantified by LRET efficiency calculations [85].

For a further step, Wang et al. developed a UCNPs-HA-based microneedle (MN) for transforming transdermal siRNA delivery of growth faction-β-type I receptor (TGF-βRI), which could also visualize the effect of siRNA therapy and monitored the depth of MN insertion [86]. The platform consists of mesoporous silica ($mSiO_2$)-coated UCNPs (UCNP@$mSiO_2$) and a peripheral HA matrix. Among them, the $mSiO_2$ shell is used to load and protect the siRNA. When the MN passes through the cuticle, the HA dissolves, affording the UCNPs release. Subsequently, some UCNP@mSiO_2 was absorbed and internalized by fibroblasts in the dermis, and siRNA was released from the mesopore layer to inhibit the expression of TGF-βRI mRNA and repair abnormal scar tissue. The depth of MN insertion was measured by the ratio of UCL red to blue emission under 980 nm light irradiation (I654/I540). For MN with shorter tips, it exhibits higher values of I654/I540. SiRNA therapy was evaluated by overexpression of connective tissue growth factor (CTGF) in abnormal scar tissue by loading molecular beacons (MB) of GAPDH and CTGF mRNA into UCNP@mSiO_2. The signal of CTGF was standardized with the signal of GAPDH, so as to avoid the error caused by the degradation of MB by intracellular nuclease. The therapeutic effect of siRNA could be detected by the expression of CTGF mRNA.

6.6 Gene Therapy and Syndication

Gene delivery systems developed using lanthanide-doped UCNPs have achieved many meaningful results; however, there are still some problems. For example, genetic payload is important for the expression of genes and the effectiveness of treatments. The payload of current nanocarriers is generally low, and merely increasing the concentration of nanocarriers can induce high toxicity in normal tissues. Therefore, in order to further improve the therapeutic effect, combining gene therapy based on UCNP nanovectors with other therapeutics is an effective solution. In addition to the above-mentioned, UCNP-based nanovectors have also integrated multiple treatments as follows.

6.6.1 Phototherapy

Photodynamic therapy (PDT), a noninvasive light treatment method, has become a clinically approved palliative care technology, especially showing unique advantages in personalized cancer medicine. PDT uses laser light to activate photosensitizers and surrounding molecular oxygen to produce ROS for the destruction of pathological tissues, revealing that the efficacy of the modality is relatively low as it highly depends on the available oxygen in tumors. Wang et al. reported an imaging-guided cancer combination therapy system UCNPs-CE6-siRNA [87]. UCNP-PEG@2PEI ($NaGdF_4$:Yb, Er) was synthesized using a multilayer polymer-coated UCNPs, and then loaded with the photosensitizer Ce6 and PLK1-targeting siRNA through a layer-by-layer strategy. Under excitation of 980 nm light, the resonance energy is transferred from UCNPs to Ce6, producing cytotoxic singlet oxygen, while the remaining upconversion luminescence is used for imaging. On the other hand, co-delivered siRNA induced Plk1 silencing and subsequently significant cancer cell apoptosis. The combination of PDT with gene therapy significantly enhanced the cancer cell killing effect.

Photothermal therapy (PTT), which kills tumors by thermal ablation, is another promising phototreatment approach for orthotopic tumors [89, 90]. The efficacy of PTT, however, is highly limited by the heat shock protein 70 (HSP70), which is present in almost all organisms and protects cells from heat damage. Therefore, the

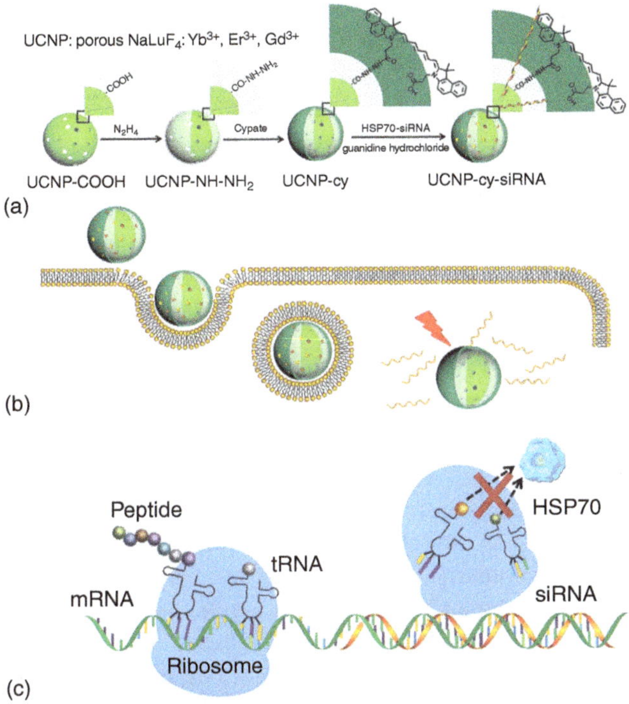

Figure 6.6 Schematic illustration of UCNP-cy-siRNA for programmed delivery of HSP70 siRNA for gene silencing and photothermal ablation. (a) Schematic drawing of UCNP-COOH loading with cypate and HSP70-siRNA (UCNP-cy-siRNA). (b) Schematic drawing of endocytosis process of UCNP-cy-siRNA and the release of HSP70-siRNA upon irradiation. (c) Mechanism of the HSP70-siRNA inhibiting the synthesis of HSP70. Source: Reproduced with permission from Wang et al. [88], 2016 / John Wiley & Sons.

siRNA that targets HSP70 (HSP70-siRNA) can inhibit the activity of HSP70 and increase the effectiveness of photothermal therapy [91]. Wang et al. developed a conjugate porous upconversion nanocomposite UCNP-CY-SiRNA ($NaLuF_4$:Gd,Yb, Er) for synergistic gene therapy and photothermal therapy (Figure 6.6) [88]. Porous UCNP is synthesized using the microemulsion method, and its large surface area and porous structure can adsorb more siRNA. Photothermic agent Cypate is connected to via hydrazine bond. Under 980 nm light irradiation, the emission at 652 nm of UCNPs transfers energy to Cypate and thus converts it into heat, performing PTT. At the same time, the intracellular release of HSP70-siRNA can reduce the temperature required for PTT to kill cancer cells and increase the effect of PTT. Gd^{3+}-doped UCNPs also impart the ability of UCNP-Cy-siRNA to guide the therapeutics via UCL/MRI dual-modal imaging.

6.6.2 Chemotherapy

Cisplatin, as a first-line anticancer drug, has been widely accepted and achieved great success in clinical practice. However, the resistance of platinum drugs can lead to serious consequences of tumor recurrence and even death of patients [92, 93]. Apart from the above-mentioned Cas13d-mediated attenuation of drug resistance with UCNPs, the delivery of siRNA targeting eIF 4E is also one way to overcome chemotherapy resistance since eukaryotic translation initiation factor(eIF) 4E is an important marker for the prognosis of many human cancers. Given this, Teng et al. reported a UCNP-based carrier system loaded with both eIF4E siRNA and Pt(IV) (UCNP@PEI-Pt-PEG@siRNA) [94]. UCNPs nanocrystal ($NaGdF_4$:Yb, Er) acts as a luminescent core with a carrier template, coated with PEI on its surface, and connected Pt(IV). Negatively charged eIF4E siRNA is loaded by binding with the positively charged PEI. After the nanocarriers enter the cells, Pt(IV) is reduced to cisplatin by a large amount of reducing agents such as ascorbic acid and glutathione contained in tumor cells, damaging DNA. Moreover, siRNA can also be released in cancer cells through the protonated PEI, resulting in silencing the cisplatin-resistant eIF4E gene and increasing the sensitivity of cancer cells to platinum-based chemotherapy.

Polyprodrug strategies can deliver high doses of the parent drug at the lesion site with covalently linked duplicate prodrug units. In antitumor drugs based on polyprodrug nanoplatforms, photosensitive platinum (IV) complexes (Pt(IV) prodrugs) are virtually nontoxic in the dark environment, while they can be activated as highly toxic Pt(II) compounds under UV or green light. So, a controllable nanovector (PUCNP@Pt@siPlk1) was developed for the co-delivery of Pt(IV) and Polo-like kinase 1 (siPlk1) [95]. The PEI-coated UCNP (PUCNP, $NaGdF_4$:Yb, Tm) serves as the luminescence core and drug carrier, and the photoactivated Pt(IV)-backbone polymer (PPt) was connected to the surface of PUCNP to obtain PUCNP@Pt. Then siRNA targeting Polo-like kinase 1(Plk1) was bound with PEI on the carrier surface to obtain PUCNP@Pt@siPlk1. On the one hand, due to the introduction of the Gd^{3+} ion, PUCNP@Pt@siPlk1 can be used for T1-weighted three-modal imaging of MRI, CT, and UCL. On the other hand, PPt on the surface of PUCNP@Pt@siPlk1 can be activated by UV emission of UCNP and release highly toxic Pt(II) complex, and the fracture of PPt can also accelerate the release of siPlk1, thus achieving excellent synergistic therapeutic effect.

6.7 Other Lanthanide-Based Nanovectors

In addition to UCNPs, other rare-earth-based materials, such as rare earth ion-coordinated cationic polymers, rare earth-doped phosphate nanoparticles, and rare earth-functionalized carbon-based nanomaterials, are also developed as gene delivery carriers.

Introducing rare earth ions into traditional cationic polymers can endow the carrier additional imaging capabilities. Reineke et al. took the lead in designing a polymeric gene carrier with terbium (Tb^{3+}) and gadolinium (Gd^{3+}) implanted. In this story, the electropositive oligoethyleneimine units on the polymer backbone are combined with pDNA and compressed into nanoparticles to protect nucleic acids from nuclease damage. And the polyamide unit provides fluorescence and

MRI capabilities through coordination with Tb^{3+} and Gd^{3+}, thereby effectively monitoring intracellular delivery of pDNA [96]. On this basis, the binding and dissociation of the polyplex were achieved through lanthanide resonance energy transfer (LRET) between the tetramethylrhodamine (TMR) labeled on pDNA and Tb^{3+} chelated with polyamide units. Meanwhile, in order to improve the gene transfection profile, the number of imine groups of the oligoethyleneimine unit was optimized to six [97]. The research group also found that introducing trehalose into the polymer backbone could increase the siRNA delivery and gene knockdown efficiency to U87 cells with luciferase-expressing stably [98].

Cationic polymers with reducible groups in the main chain have been found to exhibit better transfection efficiency and less cytotoxicity, which is due to the presence of intracellular glutathione that can decompose these groups, resulting in the degradation of polymer within cells, thereby reducing its toxicity [99, 100]. To this end, Wang et al. designed a gene delivery carrier GdCPUAs based on reducible cationic poly(urethane amide) (CPUA). The tertiary amine group in its structure is responsible for binding to pDNA, and the ethylene triamine pentaacetic acid (DTPA) moiety coordinated with Gd^{3+} ions endows the carrier to serve as a T1-MRI contrast agent. Moreover, the disulfide bonds within the molecule provide the carrier to decompose in the intracellular reducing environment to release genes. Under the optimal formulation, the transfection efficiency of GdCPUAs was higher than that produced by using 25 kDa branched PEI as a positive control, and the cytotoxicity was lower [101]. Lin et al. further used Gdplex, a cationic polymer carrier with reducing groups, to deliver pDNA systemically through intravenous injection, which significantly inhibited the growth of SKOV-3 tumor xenografts in the BALB/c nude mouse model, indicating that the feasibility of this strategy [102]. The high gene transfectivity of lanthanide coordination polymers is also related to the special biological properties of rare earth ions themselves. For example, Nd^{3+} can enhance the energy metabolism of cells to produce more ATP, and enhance cellular uptake by increasing the expression of caveolin, and regulate cellular pathways by regulating CaM [103]. With the extremely high charge-to-volume ratio, direct treatment of cells with La^{3+} can improve the transfection efficiency of adenoviral vectors [104].

Furthermore, in order to enable the lanthanide coordination polymer vector to be used for intracerebral gene delivery, Hart et al. designed a Gd^{3+}-labeled liposome complex LPD for the penetration of the blood-brain barrier by introducing the receptor-targeting DNA-binding peptide P [105]. The optimal formula is a carrier with 15% Gd^{3+} chelating lipids, which has the best peptide-mediated targeted transfection efficiency. The vector could simultaneously deliver the contrast agents and therapeutic genes, achieving real-time monitoring of brain gene therapy.

Hollow nanomaterials can protect active cargo from the external environment and provide abundant surface sites for further functionalization, which have been widely used as drug delivery carriers [106]. The lanthanide-doped macroporous hollow nanoparticles ($GdPO_4$:Eu HMNs) were then developed, and the large pores and cavity structures could protect pDNA from degradation during delivery and improve transfection efficiency. The lanthanide ions Eu^{3+} and Gd^{3+} are used to provide luminescence and MRI properties for tracking nanovectors [107].

Carbon-based nanomaterials have also been widely used as gene carrier materials due to their high biocompatibility, water dispersibility, chemical inertness, and easy functionalization [108]. Carbon nanodots (CD) are a type of carbon-based nanomaterials that possess special optical properties due to quantum confinement. For example, reducing the size of carbon nanodots can increase the band gap energy, resulting in blue-shifted photoluminescence [109]. Based on this, Yu et al. prepared a new type of carbon dot L-CD/C-CD using Gd^{3+} complex, cationic polymer, and CA as starting materials, integrating gene delivery and multimodal (MR/FL) imaging [110]. They found that the in vitro transfection efficiency of L-CD under 10% serum conditions was 74 times that of PEI 25 kDa, while the longitudinal relaxation rate of L-CD/C-CD was 3–15 times higher than the clinically used Gd reagent Gd-DTPA. In addition, the CD produces higher accumulation at the tumor site through the EPR effect due to its smaller particle size, allowing clearer in vivo tumor-specific MRI.

Nanographene oxide (NGO) has been developed alternatively by researchers as a carrier for drug loading and delivery due to its ultra-high surface area and unique planar π–π domains [111]. The lanthanide-NGO was also reported as a delivery system Gd-NGO/Let-7g/EPI, which is composed of PAMAM dendrimers and Gd-functionalized nanographene oxide (NGO), simultaneously loaded with Let-7g miRNA and anticancer drug epirubicin (EPI) [112]. The positively charged surface of Gd-NGO can adsorb EPI and interact with Let-7gmiRNA to increase loading efficiency and efficiently deliver them into human glioblastoma U87 cells, damaging DNA and knocking down the expression ofRas oncogene family. Moreover, Gd-NGO/Let-7g/EPI can be used as an MRI contrast agent to determine the location and opening extent of blood-brain barrier, and can also quantify the drug delivery in tumor tissues.

6.8 Perspective

The currently available nonviral vectors are still far away from fulfilling ideal vector properties for clinical applications. This has led to a research focus on the development of new suitable vector delivery system. The main technical limitations or critical steps to achieve successful gene therapy are divided into efficiency of vector transport and offloading into target cells, persistence, activity, immune response, regulatory and ethical issues, and commercialization. These different stages pose significant challenges for gene therapy in effectively treating the diseases.

Among gene delivery is an emerging application field of rare earth luminescent nanomaterials. Compared with traditional gene delivery carriers, the unique advantage of rare-earth-based materials is that they can provide imaging monitoring and spatiotemporal control for the gene delivery process with their rich and tunable spectral properties. In recent years, researchers have devoted themselves to vector design, loading methods, release strategies, etc., and developed many excellent rare earth luminescent nanovectors for gene delivery. Among the various rare-earth-based gene carriers, the UCNP nanovectors are undoubtedly dominant due to their unique chemical, photophysical, and biological properties. Even though there is still significant room for improvement in terms of rare earth-based nanovectors, including the material design and modification, delivery accuracy and efficacy, and biosecurity, the following aspects can be taken into consideration: (i) orthogonalized multifunctional design; (ii) NIR-II excited nanovectors for deep control; and (iii) ultra-small nanovectors for better metabolism. In short, the development of more ideal nanovectors requires the cross-fusion of multi-disciplines and different research directions. In the future, as progress is gradually made, gene carriers based on rare earth luminescent materials will go beyond the material preparation and truly become a new paradigm in the field of gene delivery.

References

1. Naldini, L. (2015). Gene therapy returns to centre stage. *Nature* 526: 351–360.
2. Dunbar, C.E., High, K.A., Joung, J.K. et al. (2018). Gene therapy comes of age. *Science* 359: eaan4672.
3. Cox, D.B.T., Platt, R.J., and Zhang, F. (2015). Therapeutic genome editing: prospects and challenges. *Nature Medicine* 21: 121–131.
4. Wang, D., Tai, P.W., and Gao, G. (2019). Adeno-associated virus vector as a platform for gene therapy delivery. *Nature Reviews Drug Discovery* 18: 358–378.
5. Li, C. and Samulski, R.J. (2020). Engineering adeno-associated virus vectors for gene therapy. *Nature Reviews Genetics* 21: 255–272.
6. Bulcha, J.T., Wang, Y., Ma, H. et al. (2021). Viral vector platforms within the gene therapy landscape. *Signal Transduction and Targeted Therapy* 6: 53.
7. Shirley, J.L., de Jong, Y.P., Terhorst, C., and Herzog, R.W. (2020). Immune responses to viral gene therapy vectors. *Molecular Therapy* 28: 709–722.
8. Mintzer, M.A. and Simanek, E.E. (2009). Nonviral vectors for gene delivery. *Chemical Reviews* 109: 259–302.
9. Yin, H., Kanasty, R.L., Eltoukhy, A.A. et al. (2014). Non-viral vectors for gene-based therapy. *Nature Reviews Genetics* 15: 541–555.
10. Lostalé-Seijo, I. and Montenegro, J. (2018). Synthetic materials at the forefront of gene delivery. *Nature Reviews Chemistry* 2: 258–277.
11. Wang, H.X., Li, M., Lee, C.M. et al. (2017). CRISPR/Cas9-based genome editing for disease modeling and therapy: challenges and opportunities for nonviral delivery. *Chemical Reviews* 117: 9874–9906.
12. Lai, W.F., Rogach, A.L., and Wong, W.T. (2017). Molecular design of upconversion nanoparticles for gene delivery. *Chemical Science* 8: 7339–7358.
13. Lee, G. and Park, Y.I. (2018). Lanthanide-doped upconversion nanocarriers for drug and gene delivery. *Nanomaterials* 8: 511.
14. Zhou, B., Shi, B., Jin, D., and Liu, X. (2015). Controlling upconversion nanocrystals for emerging applications. *Nature Nanotechnology* 10: 924–936.

15 Chen, G., Qiu, H., Prasad, P.N., and Chen, X. (2014). Upconversion nanoparticles: design, nanochemistry and applications in theranostics. *Chemical Reviews* 114 (10): 5161–5214.

16 Haasteren, J., Li, J., Scheideler, O.J. et al. (2020). The delivery challenge: fulfilling the promise of therapeutic genome editing. *Nature Biotechnology* 38: 845–855.

17 Stalder, L., Heusermann, W., Sokol, L. et al. (2013). The rough endoplasmatic reticulum is a central nucleation site of siRNA-mediated RNA silencing. *The EMBO Journal* 32: 1115–1127.

18 Piao, Y., Burns, A., Kim, J. et al. (2008). Designed fabrication of silica-based nanostructured particle systems for nanomedicine applications. *Advanced Functional Materials* 18: 3745–3758.

19 Wang, F. and Liu, X. (2009). Recent advances in the chemistry of lanthanide-doped upconversion nanocrystals. *Chemical Society Reviews* 38: 976–989.

20 Biju, V. (2014). Chemical modifications and bioconjugate reactions of nanomaterials for sensing, imaging, drug delivery and therapy. *Chemical Society Reviews* 43: 744–764.

21 Deng, G., Markowitz, M.A., Kust, P.R., and Gaber, B.P. (2000). Control of surface expression of functional groups on silica particles. *Materials Science and Engineering: C* 11: 165–172.

22 An, Y., Chen, M., Xue, Q., and Liu, W. (2007). Preparation and self-assembly of carboxylic acid-functionalized silica. *Journal of Colloid and Interface Science* 311: 507–513.

23 Liu, J., Bu, W., Zhang, S. et al. (2012). Controlled synthesis of uniform and monodisperse upconversion core/mesoporous silica shell nanocomposites for bimodal imaging. *Chemistry--A European Journal* 18: 2335–2341.

24 Bhuckory, S., Lahtinen, S., Hoysniemi, N. et al. (2023). Understanding FRET in upconversion nanoparticle nucleic acid biosensors. *Nano Letters* 23: 2253–2261.

25 Li, Z. and Zhang, Y. (2006). Monodisperse silica-coated polyvinylpyrrolidone/ $NaYF_4$ nanocrystals with multicolor upconversion fluorescence emission. *Angewandte Chemie International Edition* 118: 7896–7899.

26 Li, C., Yang, D., Ma, P.A. et al. (2013). Multifunctional upconversion mesoporous silica nanostructures for dual modal imaging and in vivo drug delivery. *Small* 9: 4150–4159.

27 Sedlmeier, A. and Gorris, H.H. (2015). Surface modification and characterization of photon-upconverting nanoparticles for bioanalytical applications. *Chemical Society Reviews* 44 (6): 1526–1560.

28 Jayakumar, M.K.G., Idris, N.M., and Zhang, Y. (2012). Remote activation of biomolecules in deep tissues using near-infrared-to-UV upconversion nanotransducers. *Proceedings of the National Academy of Sciences* 109: 8483–8488.

29 Zhang, Z., Jayakumar, M.K.G., Zheng, X. et al. (2019). Upconversion superballs for programmable photoactivation of therapeutics. *Nature Communications* 10: 4586.

30 Morille, M., Passirani, C., Vonarbourg, A. et al. (2008). Progress in developing cationic vectors for non-viral systemic gene therapy against cancer. *Biomaterials* 29: 3477–3496.

31 Boussif, O., Lezoualc'h, F., Zanta, M.A. et al. (1995). A versatile vector for gene and oligonucleotide transfer into cells in culture and in vivo: polyethylenimine. *Proceedings of the National Academy of Sciences* 92: 7297–7301.

32 Wang, L., Liu, J., Dai, Y. et al. (2014). Efficient gene delivery and multimodal imaging by lanthanide-based upconversion nanoparticles. *Langmuir* 30 (43): 13042–13051.

33 Penacho, N., Simoes, S., and de Lima, M.C.P. (2009). Polyethylenimine of various molecular weights as adjuvant for transfection mediated by cationic liposomes. *Molecular Membrane Biology* 26: 249–263.

34 Varga, C.M., Tedford, N.C., Thomas, M. et al. (2005). Quantitative comparison of polyethylenimine formulations and adenoviral vectors in terms of intracellular gene delivery processes. *Gene Therapy* 12: 1023–1032.

35 He, L., Feng, L., Cheng, L. et al. (2013). Multilayer dual-polymer-coated upconversion nanoparticles for multimodal imaging and serum-enhanced gene delivery. *ACS Applied Materials and Interfaces* 5: 10381–10388.

36 Wang, Y., Cao, P., Li, S. et al. (2018). Layer-by-layer assembled PEI-based vector with the upconversion luminescence marker for gene delivery. *Biochemical and Biophysical Research Communications* 503: 2504–2509.

37 Lin, M., Gao, Y., Diefenbach, T.J. et al. (2017). Facial layer-by-layer engineering of upconversion nanoparticles for gene delivery: near-infrared-initiated fluorescence resonance energy transfer tracking and overcoming drug resistance in ovarian cancer. *ACS Applied Materials and Interfaces* 9: 7941–7949.

38 Bai, X., Xu, S., Liu, J., and Wang, L. (2016). Upconversion luminescence tracking of gene delivery via multifunctional nanocapsules. *Talanta* 150: 118–124.

39 Huotari, J. and Helenius, A. (2011). Endosome maturation. *The EMBO Journal* 30: 3481–3500.

40 Berg, K., Kristian Selbo, P., Prasmickaite, L. et al. (1999). Photochemical internalization: a novel technology for delivery of macromolecules into cytosol. *Cancer Research* 59: 1180–1183.

41 Adigbli, D.K. and MacRobert, A.J. (2012). Photochemical internalisation: the journey from basic scientific concept to the threshold of clinical application. *Current Opinion in Pharmacology* 12: 434–438.

42 Chen, H., Xiao, L., Anraku, Y. et al. (2014). Polyion complex vesicles for photoinduced intracellular delivery of amphiphilic photosensitizer. *Journal of the American Chemical Society* 136: 157–163.

43 Han, K., Lei, Q., Jia, H.Z. et al. (2015). A tumor targeted chimeric peptide for synergistic endosomal escape and therapy by dual-stage light manipulation. *Advanced Functional Materials* 25: 1248–1257.

44 Tallafuss, A., Gibson, D., Morcos, P. et al. (2012). Turning gene function ON and OFF using sense and antisense photo-morpholinos in zebrafish. *Development* 139: 1691–1699.

45 Jayakumar, M.K.G., Bansal, A., Huang, K. et al. (2014). Near-infrared-light-based nano-platform boosts endosomal escape and controls gene knockdown in vivo. *ACS Nano* 8: 4848–4858.
46 Wang, Y., Nguyen, K., Spitale, R.C., and Chaput, J.C. (2021). A biologically stable DNAzyme that efficiently silences gene expression in cells. *Nature Chemistry* 13: 319–326.
47 Chen, Y., Zhao, R., Li, L., and Zhao, Y. (2022). Upconversion luminescence-boosted escape of DNAzyme from endosomes for enhanced gene-silencing efficacy. *Angewandte Chemie International Edition* 61: e202206485.
48 Chen, G., Abdeen, A.A., Wang, Y. et al. (2019). A biodegradable nanocapsule delivers a Cas9 ribonucleoprotein complex for in vivo genome editing. *Nature Nanotechnology* 14: 974–980.
49 Chen, X., Chen, Y., Xin, H. et al. (2020). Near-infrared optogenetic engineering of photothermal nanoCRISPR for programmable genome editing. *Proceedings of the National Academy of Sciences* 117: 2395–2405.
50 Wang, J., He, X., Shen, S. et al. (2018). ROS-sensitive cross-linked polyethylenimine for red-light-activated siRNA therapy. *ACS Applied Materials and Interfaces* 11: 1855–1863.
51 Yan, B., Boyer, J.C., Branda, N.R., and Zhao, Y. (2011). Near-infrared light-triggered dissociation of block copolymer micelles using upconverting nanoparticles. *Journal of the American Chemical Society* 133: 19714–19717.
52 Lee, H.I., Wu, W., Oh, J.K. et al. (2007). Light-induced reversible formation of polymeric micelles. *Angewandte Chemie International Edition* 46: 2453–2457.
53 Zhang, Y., Ren, K., Zhang, X. et al. (2018). Photo-tearable tape close-wrapped upconversion nanocapsules for near-infrared modulated efficient siRNA delivery and therapy. *Biomaterials* 163: 55–66.
54 Kostiainen, M.A., Smith, D.K., and Ikkala, O. (2007). Optically triggered release of DNA from multivalent dendrons by degrading and charge-switching multivalency. *Angewandte Chemie International Edition* 46: 7600–7604.
55 Yang, Y., Liu, F., Liu, X., and Xing, B. (2013). NIR light controlled photorelease of siRNA and its targeted intracellular delivery based on upconversion nanoparticles. *Nanoscale* 5: 231–238.
56 Zhao, H., Hu, W., Ma, H. et al. (2017). Photo-induced charge-variable conjugated polyelectrolyte brushes encapsulating upconversion nanoparticles for promoted siRNA release and collaborative photodynamic therapy under NIR light irradiation. *Advanced Functional Materials* 27: 1702592.
57 Hsu, P.D., Lander, E.S., and Zhang, F. (2014). Development and applications of CRISPR-Cas9 for genome engineering. *Cell* 157: 1262–1278.
58 Xue, W., Chen, S., Yin, H. et al. (2014). CRISPR-mediated direct mutation of cancer genes in the mouse liver. *Nature* 514: 380–384.
59 Chu, V.T., Weber, T., Wefers, B. et al. (2015). Increasing the efficiency of homology-directed repair for CRISPR-Cas9-induced precise gene editing in mammalian cells. *Nature Biotechnology* 33: 543–548.

60 Wang, P., Zhang, L., Zheng, W. et al. (2018). Thermo-triggered release of CRISPR-Cas9 system by lipid-encapsulated gold nanoparticles for tumor therapy. *Angewandte Chemie International Edition* 57: 1491–1496.

61 Pan, Y., Yang, J., Luan, X. et al. (2019). Near-infrared upconversion–activated CRISPR-Cas9 system: a remote-controlled gene editing platform. *Science. Advances* 5: eaav7199.

62 Wu, Y., Zheng, J., Zeng, Q. et al. (2020). Light-responsive charge-reversal nanovector for high-efficiency in vivo CRISPR/Cas9 gene editing with controllable location and time. *Nano Research* 13: 2399–2406.

63 Jiang, J., Tong, X., Morris, D., and Zhao, Y. (2006). Toward photocontrolled release using light-dissociable block copolymer micelles. *Macromolecules* 39: 4633–4640.

64 Aujard, I., Benbrahim, C., Gouget, M. et al. (2006). o-Nitrobenzyl photolabile protecting groups with red-shifted absorption: syntheses and uncaging cross-sections for one-and two-photon excitation. *Chemistry--A European Journal* 12: 6865–6879.

65 Bandara, H.D. and Burdette, S.C. (2012). Photoisomerization in different classes of azobenzene. *Chemical Society Reviews* 41 (5): 1809–1825.

66 Wang, Y., Ma, N., Wang, Z., and Zhang, X. (2007). Photo-controlled reversible supramolecular assembly of an azobenzene-containing surfactant with α-cyclodextrin. *Angewandte Chemie International Edition* 46: 2823–2826.

67 Chen, G., Ma, B., Xie, R. et al. (2018). NIR-induced spatiotemporally controlled gene silencing by upconversion nanoparticle-based siRNA nanocarrier. *Journal of Controlled Release* 282: 148–155.

68 Liang, J. and Liu, B. (2016). ROS-responsive drug delivery systems. *Bioengineering & Translational Medicine* 1: 239–251.

69 Song, X., Li, F., Tian, F. et al. (2023). Upconversion nanoparticle-based optogenetic nanosystem for photodynamic therapy and cascade gene therapy. *Acta Biomaterialia* 157: 538–550.

70 He, Y., Guo, S., Wu, L. et al. (2019). Near-infrared boosted ROS responsive siRNA delivery and cancer therapy with sequentially peeled upconversion nano-onions. *Biomaterials* 225: 119501.

71 Han, Y., Yang, Y., Sun, Q. et al. (2022). Dual-targeted lung cancer therapy via inhalation delivery of UCNP-siRNA-AS1411 nanocages. *Cancer Biology & Medicine* 19: 1047.

72 Dauer, W. and Przedborski, S. (2003). Parkinson's disease: mechanisms and models. *Neuron* 39: 889–909.

73 Choi, D.H., Kim, J.H., Seo, J.H. et al. (2014). Matrix metalloproteinase-3 causes dopaminergic neuronal death through Nox1-regenerated oxidative stress. *PLoS One* 9: e115954.

74 Li, R., Li, Y., Mu, M. et al. (2021). Multifunctional nanoprobe for the delivery of therapeutic siRNA and real-time molecular imaging of Parkinson's disease biomarkers. *ACS Applied Materials and Interfaces* 13: 11609–11620.

75 Brieke, C., Rohrbach, F., Gottschalk, A. et al. (2012). Light-controlled tools. *Angewandte Chemie International Edition* 51: 8446–8476.

76 Ellis-Davies, G.C. (2007). Caged compounds: photorelease technology for control of cellular chemistry and physiology. *Nature Methods* 4: 619–628.

77 Li, J., Leung, C.W.T., Wong, D.S.H. et al. (2017). Photocontrolled SiRNA delivery and biomarker-triggered luminogens of aggregation-induced emission by up-conversion $NaYF_4$: Yb^{3+} Tm^{3+}@ SiO_2 nanoparticles for inducing and monitoring stem-cell differentiation. *ACS Applied Materials and Interfaces* 11: 22074–22084.

78 Zhang, Z. and Zhang, Y. (2021). Orthogonal emissive upconversion nanoparticles: material design and applications. *Small* 17: 2004552.

79 Yang, Y., Huang, J., Wei, W. et al. (2022). Switching the NIR upconversion of nanoparticles for the orthogonal activation of photoacoustic imaging and phototherapy. *Nature Communications* 13: 3149.

80 Ward, R.A., Fawell, S., Floch, N. et al. (2020). Challenges and opportunities in cancer drug resistance. *Chemical Reviews* 121 (6): 3297–3351.

81 Hanssen, K.M., Haber, M., and Fletcher, J.I. (2021). Targeting multidrug resistance-associated protein 1 (MRP1)-expressing cancers: beyond pharmacological inhibition. *Drug Resistance Updates* 59: 100795.

82 Vaghari-Tabari, M., Hassanpour, P., Sadeghsoltani, F. et al. (2022). CRISPR/Cas9 gene editing: a new approach for overcoming drug resistance in cancer. *Cellular and Molecular Biology Letters* 27: 49.

83 Liu, Z., Feng, Z., Chen, M. et al. (2023). An orthogonally activatable CRISPR-Cas13d nanoprodrug to reverse chemoresistance for enhanced chemo-photodynamic therapy. *Chemical Science* 14: 4102–4113.

84 Jiang, S., Zhang, Y., Lim, K.M. et al. (2009). NIR-to-visible upconversion nanoparticles for fluorescent labeling and targeted delivery of siRNA. *Nanotechnology* 20: 155101.

85 Jiang, S. and Zhang, Y. (2010). Upconversion nanoparticle-based FRET system for study of siRNA in live cells. *Langmuir* 26: 6689–6694.

86 Wang, M., Han, Y., Yu, X. et al. (2020). Upconversion nanoparticle powered microneedle patches for transdermal delivery of siRNA. *Advanced Healthcare Materials* 9: 1900635.

87 Wang, X., Liu, K., Yang, G. et al. (2014). Near-infrared light triggered photodynamic therapy in combination with gene therapy using upconversion nanoparticles for effective cancer cell killing. *Nanoscale* 6: 9198–9205.

88 Wang, L., Gao, C., Liu, K. et al. (2016). Cypate-conjugated porous upconversion nanocomposites for programmed delivery of heat shock protein 70 small interfering RNA for gene silencing and photothermal ablation. *Advanced Functional Materials* 26: 3480–3489.

89 Hirsch, L.R., Stafford, R.J., Bankson, J.A. et al. (2003). Nanoshell-mediated near-infrared thermal therapy of tumors under magnetic resonance guidance. *Proceedings of the National Academy of Sciences* 100: 13549–13554.

90 Nam, J., Son, S., Ochyl, L.J. et al. (2018). Chemo-photothermal therapy combination elicits anti-tumor immunity against advanced metastatic cancer. *Nature Communications* 9 (1): 1074.

91 Huang, Q., Hu, J.K., Lohr, F. et al. (2000). Heat-induced gene expression as a novel targeted cancer gene therapy strategy. *Cancer Research* 60: 3435–3439.

92 Dasari, S. and Tchounwou, P.B. (2014). Cisplatin in cancer therapy: molecular mechanisms of action. *European Journal of Pharmacology* 740: 364–378.

93 Pan, C., Kang, J., Hwang, J.S. et al. (2021). Cisplatin-mediated activation of glucocorticoid receptor induces platinum resistance via MAST1. *Nature Communications* 12: 4960.

94 Teng, B., Yu, C., Zhang, X. et al. (2017). Upconversion nanoparticles loaded with eIF4E siRNA and platinum (IV) prodrug to sensitize platinum based chemotherapy for laryngeal cancer and bioimaging. *Journal of Materials Chemistry B* 5: 307–317.

95 Kuang, G., Lu, H., He, S. et al. (2021). Near-infrared light-triggered polyprodrug/siRNA loaded upconversion nanoparticles for multi-modality imaging and synergistic cancer therapy. *Advanced Healthcare Materials* 10: 2100938.

96 Bryson, J.M., Fichter, K.M., Chu, W.J. et al. (2009). Polymer beacons for luminescence and magnetic resonance imaging of DNA delivery. *Proceedings of the National Academy of Sciences* 106: 16913–16918.

97 Kelkar, S.S., Xue, L., Turner, S.R., and Reineke, T.M. (2014). Lanthanide-containing polycations for monitoring polyplex dynamics via lanthanide resonance energy transfer. *Biomacromolecules* 15: 1612–1624.

98 Xue, L., Kelkar, S.S., Wang, X.L. et al. (2015). A theranostic polycation containing trehalose and lanthanide chelate domains for siRNA delivery and monitoring. *RSC Advances* 5: 74102–74106.

99 Son, S., Namgung, R., Kim, J. et al. (2012). Bioreducible polymers for gene silencing and delivery. *Accounts of Chemical Research* 45: 1100–1112.

100 Lin, C. and Engbersen, J.F. (2009). The role of the disulfide group in disulfide-based polymeric gene carriers. *Expert Opinion on Drug Delivery* 6: 421–439.

101 Gao, X., Wang, G., Shi, T. et al. (2016). Biodegradable gadolinium-chelated cationic poly (urethane amide) copolymers for gene transfection and magnetic resonance imaging. *Materials Science and Engineering: C* 65: 181–187.

102 Li, D., Jiang, W., Xu, X. et al. (2020). Hierarchically assembled theranostic nanopolyplexes enable robust tumor gene therapy and magnetic resonance imaging in vivo. *Journal of Biomedical Nanotechnology* 16: 125–135.

103 Wang, Q., Jin, W., Wu, G. et al. (2014). Rare-earth-incorporated polymeric vector for enhanced gene delivery. *Biomaterials* 35: 479–488.

104 Palmer, G.D., Stoddart, M.J., Gouze, E. et al. (2008). A simple, lanthanide-based method to enhance the transduction efficiency of adenovirus vectors. *Gene Therapy* 15: 357–363.

105 Writer, M.J., Kyrtatos, P.G., Bienemann, A.S. et al. (2012). Lipid peptide nanocomplexes for gene delivery and magnetic resonance imaging in the brain. *Journal of Controlled Release* 162: 340–348.

106 Lin, W.I., Lin, C.Y., Lin, Y.S. et al. (2013). High payload Gd (III) encapsulated in hollow silica nanospheres for high resolution magnetic resonance imaging. *Journal of Materials Chemistry B* 1: 639–645.

107 Li, Z., Tao, Y., Huang, S. et al. (2013). Lanthanide-based hollow mesoporous nanoparticles: a novel multifunctional platform for simultaneous gene delivery and cell imaging. *Chemical Communications* 49: 7129–7131.

108 Baker, S.N. and Baker, G.A. (2010). Luminescent carbon nanodots: emergent nanolights. *Angewandte Chemie International Edition* 49: 6726–6744.

109 Zhu, S., Song, Y., Zhao, X. et al. (2015). The photoluminescence mechanism in carbon dots (graphene quantum dots, carbon nanodots, and polymer dots): current state and future perspective. *Nano Research* 8: 355–381.

110 He, X., Luo, Q., Zhang, J. et al. (2019). Gadolinium-doped carbon dots as nano-theranostic agents for MR/FL diagnosis and gene delivery. *Nanoscale* 11: 12973–12982.

111 Jain, V.P., Chaudhary, S., Sharma, D. et al. (2021). Advanced functionalized nanographene oxide as a biomedical agent for drug delivery and anti-cancerous therapy: a review. *European Polymer Journal* 142: 110124.

112 Yang, H.W., Huang, C.Y., Lin, C.W. et al. (2014). Gadolinium-functionalized nanographene oxide for combined drug and microRNA delivery and magnetic resonance imaging. *Biomaterials* 35: 6534–6542.

7

Biosafety of Rare-Earth-Doped Nanomaterials

Yang Li and Guanying Chen

Harbin Institute of Technology, School of Chemistry and Chemical Engineering, 92 Xidazhi Street, Harbin 150001, China

Lanthanide-doped nanocrystals represent a class of rare-earth upconversion nanomaterials (UCNPs) with exceptional functionality and capable of transforming low-energy photons into higher-energy counterparts. Numerous characteristics distinguish upconversion luminescent materials from conventional down-conversion luminescent materials, such as organic dyes, fluorescent proteins, and quantum dots. These include long luminous life, adjustable emission spectrum, narrow emission bandwidth, and high emission stability. In particular, these materials are less affected by background fluorescence during imaging, making them highly useful for biomedical applications. The biosafety of these nanoparticles is therefore of great importance and has garnered significant attention. This chapter discusses their internalization, biodistribution, excretion, and cytotoxicity, as well as their toxic effects on living organisms.

7.1 Internalization of UCNPs into Cells

In cells and animals, nanomaterial localization and distribution determine the exposed organelles and therefore influence the toxic effects [1]. Biosafety assessments should thus evaluate the internalization and distribution sites of UCNPs within cells. Cells use the plasma membrane to define their boundaries and maintain their vital intracellular environment. As opposed to small molecules, nanomaterials normally undergo endocytosis after they have been internalized through the plasma membrane. Endocytosed nanomaterials cannot reach the cytosol after they have been endocytosed [2].

Researchers have previously shown that pinocytosis is a common method of capturing nanomaterials in cells [2, 3]. Endocytosis manifests through four primary mechanisms: macropinocytosis, clathrin-mediated endocytosis, caveolae-mediated endocytosis, and clathrin- and caveolae-dependent endocytosis. The internalization of UCNPs into cells, including both cancerous and normal cells, has been widely documented. However, not all UCNPs enter cells. The cellular uptake of UCNPs

Photofunctional Nanomaterials for Biomedical Applications, First Edition.
Edited by Chunxia Li and Jun Lin.

depends on various factors, such as their size, surface ligand properties, and incubation conditions.

The primary mode of endocytosis between charged UCNPs and the negatively charged plasma membranes of cells typically involves electrostatic interactions. This interaction affects the cytotoxicity and uptake efficiency of UCNPs. Evidence suggests that positively charged $NaYF_4$: Yb, Er nanoparticles are taken up by cells much more efficiently than their negatively charged counterparts. Positively charged $NaYF_4$: Yb, Er nanoparticles can be observed inside cells. Negatively charged polyacrylic acid (PAA)-modified $NaYF_4$: Yb, Er nanoparticles did not produce strong fluorescence signals inside cells under the same conditions as neutral polyvinyl pyrrolidone (PVP)-modified $NaYF_4$: Yb, Er nanoparticles. Inductively coupled plasma mass spectrometry (ICP-MS) analysis of further quantitative data revealed that polyethylene glycol (PEI)-UCNPs endocytosed by cells more efficiently than PVP-UCNPs after 24 hours of incubation [4].

Hyeon and coworkers reported that the majority of PEG-phospholipid-coated $NaYF_4$: Yb, Er nanoparticles or their intracellular aggregates exhibit slight spatial fluctuations [5]. The trajectory of UCNPs within the cell comprises various dynamic stages and exhibits varying transmission speeds. UCNPs are actively transported by dynamic protein on the microtube or muscle protein in cells. Microtario-dependent dynamic proteins, such as dynamic protein and driving protein, are the main causes of active transport. In further research, the internal distribution transfer of the $NaYF_4$ of the polyethyleneglycol (PEG)–phospholipid coating is mainly promoted by microtubule-dependent dynamic protein. These proteins play a stronger role in the early stages of granular transmission than dynamicin [6].

There is a scarcity of comprehensive investigations on the mechanisms underlying the internalization of UCNPs. Findings from experiments involving the inhibition of endocytosis have revealed that PEG-UCNPs measuring 35 nm in diameter exhibit a notable capacity to interfere with endocytosis processes upon exposure to cytochalasin D, a compound known to disrupt actin filaments. PEG-UCNPs remain near the plasma membrane and are not internalized [6]. Wong and colleagues developed plasmids encoding red fluorescent protein-labeled RFP clathrin and RFP vesicles, which were subsequently expressed in HeLa cells. Following the incubation of UCNPs with HeLa cells, it was observed that PEI-UCNPs consistently co-localized with RFP clathrin across all time intervals, while no co-localization was observed with RFP vesicles. The study observed the transportation of PEI-UCNPs with clathrin-coated vesicles from the cell periphery toward the periphery of the nuclear membrane. These findings suggest that clathrin-mediated endocytosis is likely the primary mode of entry for PEI-UCNPs into HeLa cells [4].

The internalization rate of UCNPs into SK-BR-3 cells shows a gradual increase. When $NaGdF_4$: Yb, Er UCNPs were incubated with SK-BR-3 cells for two hours, a subdued fluorescence emission was detected; however, this fluorescence intensity significantly intensified after an extended incubation period of four hours (Figure 7.1). In a separate study, it was demonstrated that a total of 1.3×10^4 UCNPs were internalized and subsequently quantified through ICP [7].

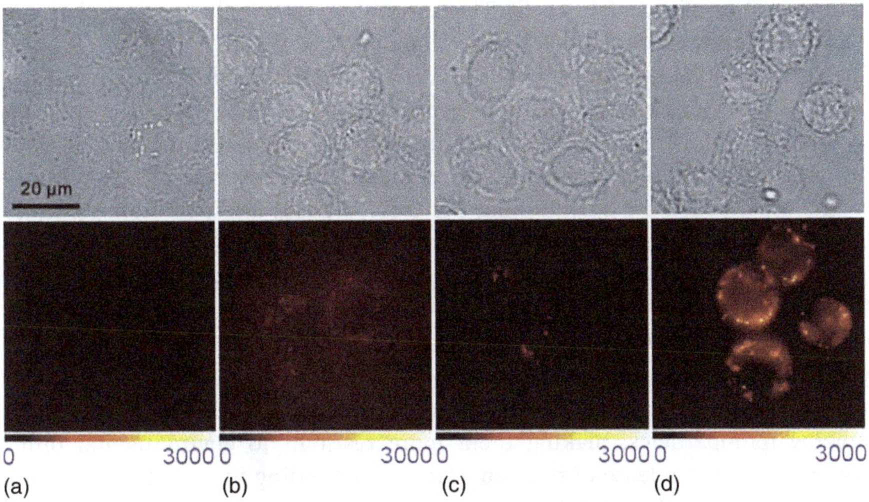

Figure 7.1 The SK-BR-3 cells absorb 20 nm UCNP particles. The upper row depicts the bright field images, while the lower row shows the luminescence images generated by a 980 nm laser excitation. (a) Cells were incubated without UCNPs, (b) incubation time was 30 minutes, (c) 2 hours, (d) 4 hours. Source: Reprinted with permission from Park et al. [7]/John Wiley & Sons.

7.2 Distribution of UCNPs

Caenorhabditis elegans is a multicellular eukaryotic organism with fully developed organs like intestines, muscles, subcutaneous tissues, sex glands, and nervous systems. Zebrafish, which are tropical freshwater fish, serve as significant vertebrate model organisms in scientific studies. *C. elegans* and zebrafish, as model organisms, have transparent bodies that allow for efficient light penetration. This makes them excellent models for studying the biodistribution and metabolism of nanomaterials.

Yan and coworkers found that when they fed *C. elegans* a combination of PEI-UCNPs for 2, 6, and 12 hours, they discovered that most of the PEI-UCNPs were situated in the intestinal tract of the worms [8]. PEI-UCNPs were discovered exclusively in the intestinal tract, rather than inside the cells. This suggests that PEI-UCNPs may lack the capability to traverse the intestinal cell membrane and gain entry into the systemic circulation. There were no notable variations in the absorption of PEI-UCNPs between males and females. Similar findings have been reported by other research groups [9, 10]. Similarly, Cui and coworkers vertically injected LaF_3: Yb, Er@SiO_2 solution vertically into the pectoral fin of zebrafish. After 24 hours, they observed a significant accumulation of UCNPs@SiO_2 in the intestinal tract, while no substantial amount of UCNPs@SiO_2 was detected in zebrafish cells [11].

After injecting UCNP intravenously, the distribution of UCNP in vivo was recorded in mice. The liver and spleen were primarily identified as the final deposition sites for UCNPs, along with ultrasmall nanoparticles. This is attributed to their uptake by the mononuclear phagocyte system, a component of the immune system consisting of phagocytic cells. The distribution ratio in these organs depends on particles' size, surface characters, injection time, and so on.

The surface functional group of the UCNPs is important for ensuring that the UCNPs' surface can be made hydrophilic or modified to allow for easy dispersion in chemically significant biological settings [12–22]. Surface functional groups allow their utilization in imaging and medicinal applications [23–26]. Moreover, the organic coating surrounding the UCNPs could assist in their ability to enter cell membranes and travel through the bloodstream [27–30]. The size and texture of UCNPs are important factors that affect how nanoparticles interact with biomolecules in a specific way [31]. The enhanced reactivity and catalytic efficiency of nanoparticles can be attributed to their effective surface area [32]. The free radicals play a crucial role in the biotoxicity of UCNPs [33]. The shell can effectively reduce or eliminate the adverse effects, while also potentially stabilizing nanoparticles, making them more resistant to environmental influences, reducing the release of toxic substances, and aiding in their tissue-specific targeting [34–36]. Different polymer nanoparticles were altered by adding a layer of lectins. The modified nanoparticles were able to specifically attach to tumor cells that had sialic acid molecules on their surface. This characteristic makes the modified nanoparticles a perfect tool for detecting cancer cells [37, 38]. The UCNPs can be coated with substances containing hydrophilic groups, including PEG [21, 29, 39–42], PEI [22], chitosan [43], silica, phospholipids, and protein compounds [12, 16, 17, 21, 22, 29, 44–48]. Moreover, nanoparticles can have their surface modified with these types of modifiers, which can alter their characteristics or aid in the gathering and movement of particular attributes [49, 50].

The typical surface polymeric coating found in UCNPs is utilized to enhance their dispersion in the aqueous system [51–54] and the ability to reduce biotoxicity due to toxic metals [42, 48, 55–70]. To enhance stability and prevent desalination and degradation from oxidation and photolysis, the cores of UCNPs are coated with a polymer or silica layer [42, 50, 64, 67, 69–72]. Thus, the UCNP core minimizes the release of metal ions, thereby reducing their potential toxicity [73, 74].

The process of altering the surface of UCNPs on silica involves changing the hydrophobic groups to hydrophilic ones. By adjusting the quantity of tetraethyl orthosilicate (TEOS) under alkaline conditions and utilizing Igepal CO520 as a surfactant, the sol-gel technique can produce a compact silica layer with a thickness ranging from 20 to 100 nm. Hydrophobic UCNPs mixed with cetyltrimethylamine bromide (CTAB) can be enclosed in mesoporous silica shells. Moreover, the process of chemical etching can create a distinctive composite of hollow UCNPs coated with silica in an egg-yolk structure, providing a larger specific surface area and a cavity suitable for a variety of applications. Silica's chemically active surface can be easily customized with functional groups through silanization to suit different needs in nanomedicine. The safety and toxicology of UCNPs@SiO_2 are crucial due to their active properties of silica. Studies have demonstrated the uniform distribution of silica-coated $NaYF_4$: Yb, Er in a rat model [75]. The nanomaterials were exposed to mesenchymal stem cells derived from bone marrow (BMSCs) and skeletal myoblasts from rats. The morphology of BMSCs started to proliferate and form small colonies after 4 days, and the myoblast cells reached confluency at 30 days. Even at the highest concentrations (100 $\mu g\,ml^{-1}$), these cells managed to survive quite well. After intravenous injection, the determination of yttrium through ICP showed that the highest concentration was found in the lung at 29.2 and 18.6 $mg\,lg^{-1}$; the heart had the second highest concentration at 18.0 and 10.9 $mg\,lg^{-1}$ at the time points of 10 and 30 minutes post-injection, respectively. By seven days following the injection, the majority of the nanocrystals had been removed. The results indicate the UCNPs' excellent biocompatibility and endorse the potential applications of UCNP@SiO_2 in cellular and living body-level imaging technologies.

Among the nanoparticles approved by the US Food and Drug Administration (FDA), it has been observed that PEG is the surface ligand with the greatest influence on their pharmacokinetics. The HOOC–PEG–COOH compound possesses the capability to modify the hydrophobic surface of the UCNPs [76]. Moreover, it was discovered that PEGylation improved the uptake of UCNPs; their 35 nm size facilitated successful entry into the cytoplasm in a study on inhibition. The addition of cytochalasin D disrupts the actin filaments involved in endocytosis, thereby affecting the process of the nanoparticles' endocytosis [6]. After 5 hours incubation, the 50 nm $NaYF_4$: Yb, Er@PEG nanoparticles exhibit the most intense luminescence within cells. After 6 hours, the UCNPs can be effectively excreted by exocytosis. The inclusion of PEG can also prolong the time that nanoparticles remain in the bloodstream. The Vivo experiment [77] shows that the blood radioactive value decreased from $10.34 \pm 0.70\%$ ID g^{-1} to $6.84 \pm 1.0\%$ ID g^{-1} in 30 minutes. The outstanding biocompatibility of PEG-UCNPs is further supported and validated by the prolonged half-life of 4.3 ± 0.6 hours observed in the second phase.

The toxicity and uptake efficiency of UCNPs are known to be influenced by the electrostatic interactions between the UCNPs and the cell membrane, which result from their opposite charges. The negatively charged PAA-$NaYF_4$: Yb, Er only show a limited number of bright spots when illuminated in cells. The positively charged PEI-UCNPs (51.1 mV) show the greatest absorption by cells [4]. In the past, the FDA has mandated that drugs used for diagnosis and treatment, which are administered through injection, should be broken down and eliminated from the body. The liver and kidneys are crucial in this elimination process. The majority of nanomaterials are excreted through the biliary system and eliminated by the kidneys. These nanomaterials do not enter cells and are not involved in cellular breakdown. Imaging technology can effectively track the bile excretion of PAA-UCNPs [25]. The UCL signal of PAA-UCNPs (15 $mg\,kg^{-1}$) in the intestines at 800 nm confirmed their presence at seven days, with the signal remaining detectable up to 69 days. This indicates that nanomaterials were eliminated through hepatobiliary transport. After 115 days, the analysis results of major organs revealed no notable variances in cardiac muscle, liver hepatocytes, and lung samples compared to the control group. Similarly, the PEI labeled (α-$NaYbF_4$: Tm)/CaF_2 was used to label rat mesenchymal stem cells (rMSC) to investigate cytotoxicity and track osteogenic and adipogenic differentiation [78]. While the PEI-UCNPs did not cause significant harm to rMSCs initially, with 90% viability after hours, their viability dropped to 60% after co-incubation, which could limit their potential biological applications.

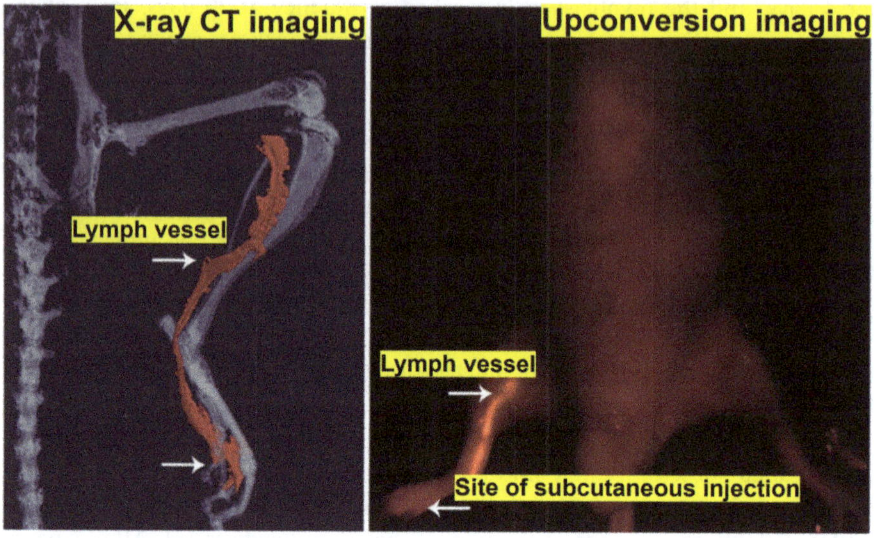

Figure 7.2 The behavior of subcutaneously injected cit-UCNPs for 30 minutes. Source: Reprinted with permission from Sun et al. [80]/Elsevier.

Additionally, rMSCs labeled with PEI-UCNPs seemed to be less effective. These findings highlight the necessity for more in-depth studies on PEI-coated UCNPs.

An alternative approach for drug delivery involves utilizing intra-arterial injection. The PEG-$NaYF_4$: Yb, Tm@SiO_2 predominantly accumulates in the liver and spleen. The accumulation of PEG-UCNP@SiO_2 within the MCF-7 tumor model exhibited nearly a threefold increase following intra-arterial administration in comparison to intravenous delivery [79]. While the utilization of intra-arterial injection in clinical trials is not a novel approach, its application for the delivery of UCNPs offers promising prospects for enhancing the efficacy of tumor therapy. It should be noted that UCNPs injected subcutaneously can enter the lymphatic system (Figure 7.2) [80]. The large size cit-UCNPs have a smaller diffusion coefficient and slower flow velocity in the lymphatic vessels compared to small molecule drugs. In addition, cit-UCNPs injected intradermally or subcutaneously do not distribute to the main organs of mice within a few hours after injection [81].

7.3 Excretion Behavior of UCNPs

Over the course of history, the FDA of the United States stipulates that drugs, especially diagnostic substances, must be completely cleared from the body within a reasonable time. It is widely recognized that the majority of medicines are predominantly excreted through either the hepatobiliary or renal pathways. In terms of hepatobiliary excretion, the main pathway for eliminating them from the body is through bile excretion. Bile is a liquid produced by the liver. It carries waste and excess cholesterol, fat, and other substances, transporting them from the liver to the intestinal tract for excretion. Nanomaterials can be taken up by the liver and excreted with bile through passive diffusion or receptor-mediated mechanisms. This excretion mechanism helps eliminate nanomaterials from the body and reduces their retention time in other parts of the body. Renal excretion occurs primarily through glomerular filtration and tubular secretion to remove substances from the body. Nanomaterials are primarily absorbed by renal epithelial cells and eliminated through urine via passive diffusion or receptor-mediated mechanisms. The elimination of UCNPs through biliary excretion was observed through UCL imaging. By seven days post-injection of PAA-UCNPs, the intestinal fluorescence suggested that the clearance method of UCNPs was via hepatobiliary transport. Subsequently, the fluorescence signal in the intestine remained consistent for 90 days. The majority of UCNPs were eliminated by day 115 [25]. Another study indicates that the biliary-excreted UCNPs exhibited no alterations in comparison to the pre-injection groups. The PEG-UCNPs found in the fecal matter displayed consistent size characteristics [82]. The PEG-UCNPs: ^{153}Sm were detected in the bladder between 30 minutes and 6 hours post-injection. Additionally, this nanomaterial was identified in the intestines ($0.21 \pm 0.04\%$ ID g^{-1}) one hour after injection [77]. These results suggest that the UCNP sizes could affect excretion behavior. However, the ultrasmall Gd_2O_3@SiO_2 particles are still deposited in the liver and lungs. The nanoparticles were observed to remain in the bladder one hour post intravenous injection, suggesting prompt excretion via the renal system (Figure 7.3) [83].

The excretion time of nanoparticles varies depending on their size and the route of excretion. Small nanoparticles that are excreted through the renal pathway are quickly removed from the body within a short period ranging from hours to days. Nevertheless, in the case of sizable nanoparticles eliminated via the biliary pathway, the duration of excretion is extended, spanning from weeks to months, and, in some instances, even years. Therefore, the excretion time of nanoparticles is a crucial factor to consider when assessing their potential toxicity and environmental impact. The PAA-$NaYF_4$: Yb, Er with a size of 11.5 nm have a complete excretion time of 115 days [25]. As shown in Figure 7.4, during 0.5~12 hours, strong fluorescence is mainly located in the liver and kidney. UCNP was excreted after 24 hours, resulting in weak fluorescence in the intestines. The 5.1 nm $NaGdF_4$ exhibits a brief half-life of 1.4 days, whereas the 18.5 nm $NaGdF_4$: Yb, Er possesses a longer half-life of seven days. Additionally, certain studies suggest that the elimination of UCNPs from the lungs typically occurs at a faster rate compared to clearance from the liver and spleen. SPECT imaging suggests that the elimination kinetics of $NaLuF_4$: Yb, Tm, revealing a more rapid clearance from the lungs (82.2%) in comparison to the liver (69.9%) and the spleen (40%) [84]. To the best of our knowledge, PEI-$NaYF_4$: Yb, Er@SiO_2 was cleared within seven days which has the most rapid speed of clearance [75, 85].

7.4 The Toxic Effect of Cell Incubated with UCNPs

Cytotoxicity is a process in which cells or chemicals cause cell death, regardless of the mechanisms involved, such as apoptosis or necrosis. It can cause changes in

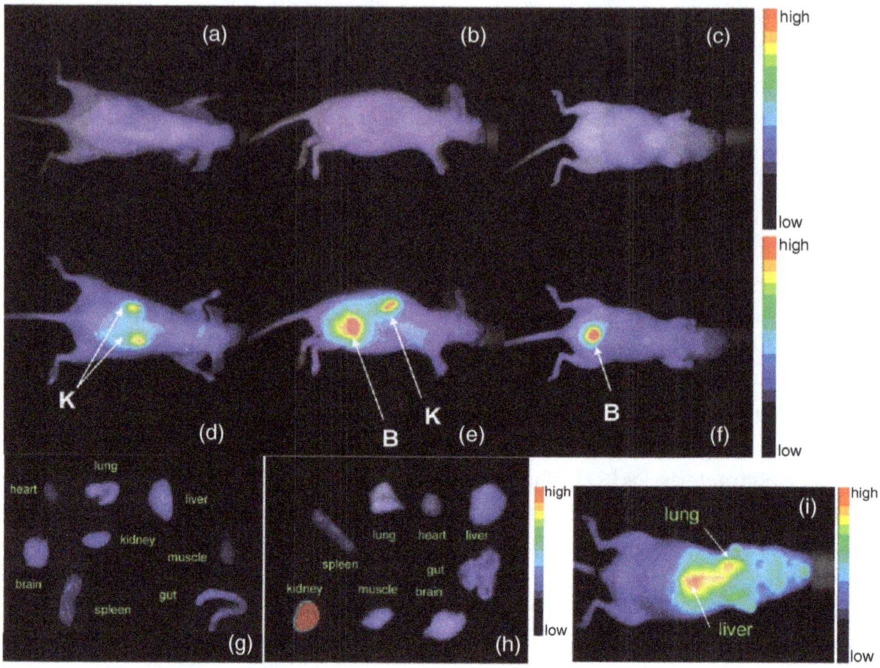

Figure 7.3 Fluorescence imaging (a,b,c) was performed before and (d,e,f) three hours after the injection of GadoSiPEG$_2$C. The images show the kidneys (K) and bladder (B). Fluorescence reflectance imaging of major organs was conducted after dissection (g) of a control mouse (no particle injection) and (h) of the nude mouse visualized in pictures (a–f). (i) The image shows mice after the injection of GadoSi$_2$C (particles without PEG). Source: Reproduced with permission from Bridot et al. [83]/American Chemical Society.

cellular metabolism. In many biochemical research fields, such as drug screening, biomaterial development, and food safety research, cytotoxicity testing is not only a fundamental detection method but also an essential indicator. The common methods include MTT (3-(4,5-dimethylthiazol-2-yl)-2,5-diphenyltetrazolium bromide), MTS (3-(4,5-dimethylthiazol-2-yl)-5-(3-carboxymethoxyphenyl)-2-(4-sulfophenyl)-2*H*-tetrazolium, inner salt), and CCK-8 assays. In the majority of instances, more than 75% of cells exhibited viability, suggesting that the cytotoxic impact of UCNPs on cell viability was minimal under the specified experimental conditions [74, 86–92]. In rare-earth nanocrystals, the luminescence has a long duration and narrow emission peaks due to the long fluorescence lifetime (typically in the millisecond range) of the rare-earth ion's f–f transition energy level. Utilizing various rare-earth ions provides opportunities to achieve multicolor luminescence and tailor the spectral properties. Rare-earth nanocrystals possess characteristics that make them useful for live cell imaging, allowing for the tracking of dynamic changes and cell states within cells. Real-time tracking of subtle changes in the distribution of transplanted cells was achieved [93]. When exposed to a culture medium in vitro, mesenchymal stem cells (MSCs) that have been marked with PEI and chemically linked to α-$NaYbF_4$: Tm@CaF_2 can differentiate into osteogenic

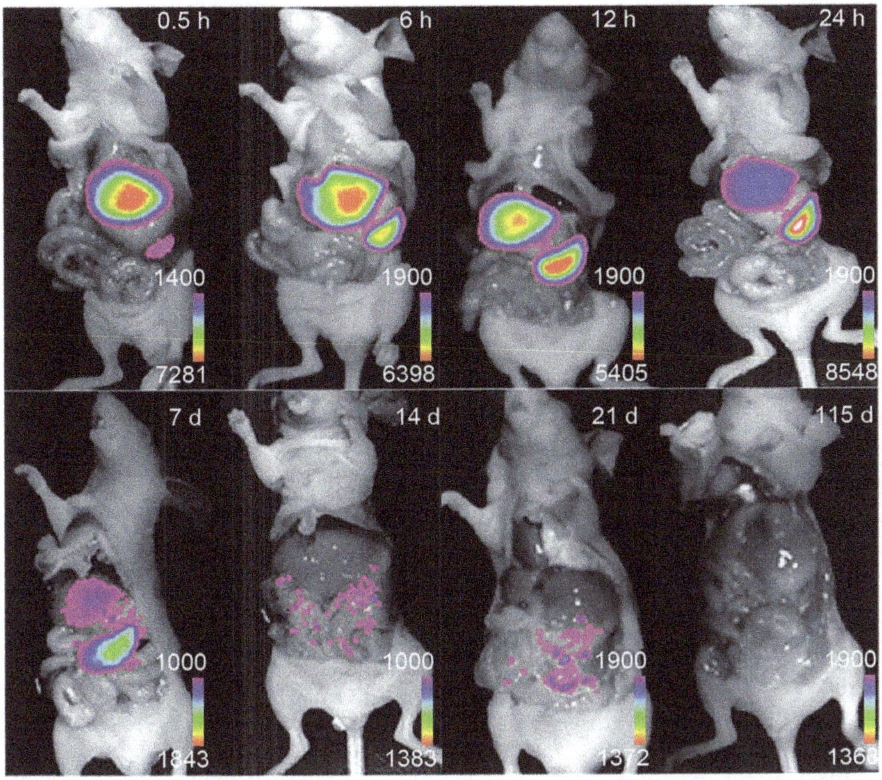

Figure 7.4 The UCL imaging of mice was conducted following the intravenous injection of PAA-$NaYF_4$: Yb, Er nanoparticles from 0.5 hours to 115 days. Source: Reproduced with permission from Xiong et al. [25]/Elsevier.

and adipogenic lineages [78]. However, it appears that the osteogenic effect of these labeled rat MSCs is weaker compared to unlabeled MSCs. In addition, the cit-$NaLuF_4$: Yb, Tm at 400 mg ml^{-1} slightly affected the viability of JEG-3 cells and significantly inhibited their metastatic ability. Various signaling pathways can regulate the growth and spread of cancer cells [94], and this suggests that suppressing the capacity for metastasis does not always necessitate the suppression of cellular proliferation. Hence, additional research is required to explore the intricate mechanisms that underlie these findings and to pinpoint potential therapeutic approaches that can efficiently address both tumor growth and proliferation.

Autophagy involves the engulfment of self-proteins or organelles in the cytoplasm. These proteins or organelles are subsequently encapsulated into vesicles, which then fuse with lysosomes to form autophagosomes. The autophagosomes then degrade the encapsulated contents. This process enables the metabolic needs of the cell itself and the renewal of certain organelles [95, 96]. HeLa cells that were genetically modified to express GFP-LC3 (a hybrid protein consisting of green fluorescent protein (GFP) and microtubule-associated light chain 3 (LC3) protein) exhibited noticeable formation of GFP-LC3 dots after exposure to $NaYF_4$: Yb, Er after co-cultured for

24 hours. This observation suggests the activation of autophagy within the treated cells. Moreover, the autophagy phenomenon also occurred when cells were treated with $NaYF_4$: Yb, Tm [97], NdO_2 [98], and CeO_2 [99]. Nevertheless, the findings indicated that the autophagy triggered was contingent upon the concentration and surface-coating utilized [100].

7.5 Toxic Effect of UCNPs In Vivo

Assessing the biosafety of UCNPs is crucial for medical applications. Although lanthanide elements cannot naturally become part of any biological molecule, they can be introduced into living organisms under certain circumstances. To evaluate the toxicity of lanthanide elements, numerous studies have been conducted. These studies primarily focus on the effects of lanthanide elements on different organisms, including mice, *C. elegans* worms, and zebrafish embryos. In these studies, the growth, development, and survival of the organisms are typically observed to assess the toxicity of UCNPs. The mice treated with $NaYF_4$: Yb, Er@SiO_2 showed no notable variations in body weight compared to groups without treatment [75]. The results of the Yttrium analysis on biodistribution in mice organs suggest that most silica-coated UCNPs were cleared from the body within seven days.

Changes in blood indicators can indicate anemia, inflammation, infection, tumor, parasitic infection, tuberculosis, and other diseases. Therefore, utilizing blood indicators to detect animal models after treatment with upconversion nanocrystals can effectively assess the animal's physical condition and predict the biological toxicity of UCNPs. Liu et al found that intravenous administration of PAA-UCNPs (35 nm) and PEG-UCNP (30 nm) did not result in significant hepatic toxicity. This conclusion was drawn based on the assessment of various hepatic markers including alanine aminotransferase, alkaline phosphatase, aspartate aminotransferase, globulin, and total protein levels. The blood urea levels in the treated mice were the same as those in the control group [52]. While the findings presented suggest that UCNPs exhibit biocompatibility, further substantiation is required to confirm their safety. $NaLuF_4$: Yb, Tm showed significant toxicity in the low-dose group, with no observable effects. The toxic effects were eliminated 90 days after injecting $NaLuF_4$: Yb, Tm, and the organ functions gradually returned to normal [101].

Caenorhabditis elegans is small in size, short in lifespan, and can develop rapidly. It has a small genome and highly conserved genes. By applying different compounds to it, researchers can study the effect of drugs, which has significant reference value for drug screening [102–105]. After feeding $NaYF_4$: Yb, Tm, the protein expression, lifespan, egg production, and viability were the same as in the control group [8].

The zebrafish genome is highly conserved with the human genome, with approximately 87% of human homologous genes having corresponding genes in the zebrafish genome. Furthermore, some disease-related genes in zebrafish are conserved as much as 99% with human genes [106]. Moreover, adult zebrafish can survive and reproduce without a thymus and lymphocyte production, making them an ideal model for studying the immune system. Zebrafish embryos develop rapidly,

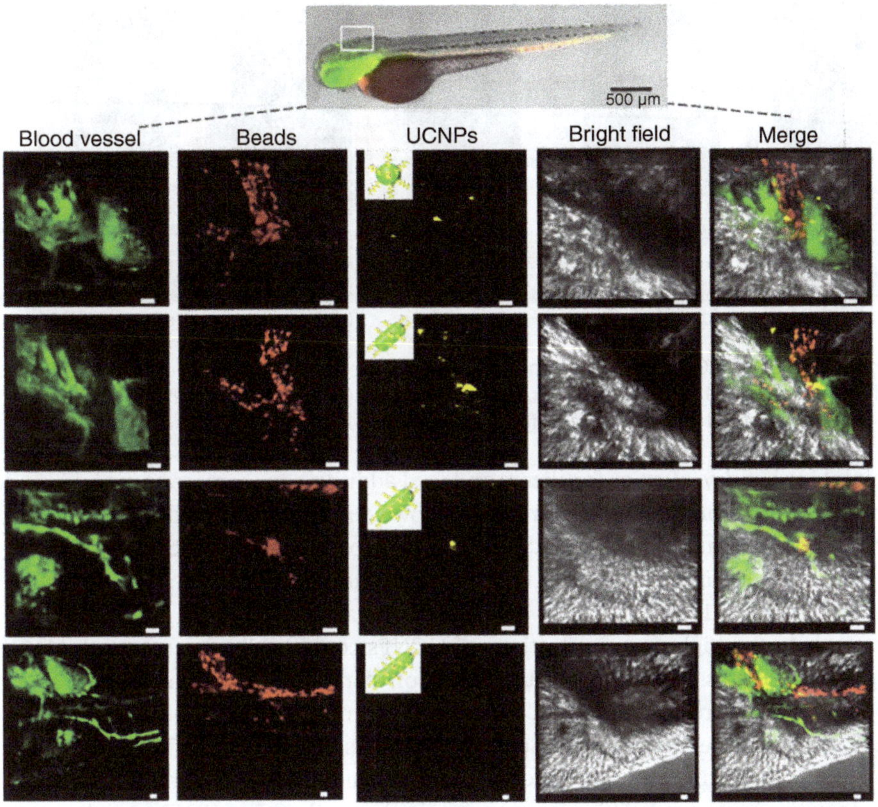

Figure 7.5 Confocal microscope images of zebrafish injected with PEG-UCNPs. The blood vessels in the zebrafish brain and fluorescent beads within the blood vessels were shown in the first column and the second column, respectively. The signal of UCNPs was displayed in the third column. The fourth column and the fifth column images were the bright field images and merged images (scale bar: 20 μm). Source: Reproduced with permission from Fu et al. [110]/Elsevier.

taking only 24 hours from fertilized eggs to become complete embryos. Three to 5 days after fertilization, each fish can swim and forage freely, and the internal organs, which are mainly similar to those of humans, have been established. Moreover, zebrafish are transparent during the first 7 days of development, which enables direct observation of internal organs. Therefore, zebrafish are also excellent candidates for biological models [107–109]. Figure 7.5 describes the transparent zebrafish larvae PEG-UCNPs at 5 days post-fertilization. The UCNPs were directly injected into the blood vessel to evaluate their ability to cross the blood–brain barrier and accumulation in the brain. The study expresses the explicitly green fluorescent protein (GFP) in the blood vessel (Tg(-3mnx1:TagGFP)) to visualize these blood vessels and track the accumulation of UCNPs in real time [110].

Materials containing lanthanide elements are widely distributed in the environment, largely because of their extensive utilization in diverse industrial and commercial applications. These materials are also widely used in agriculture and

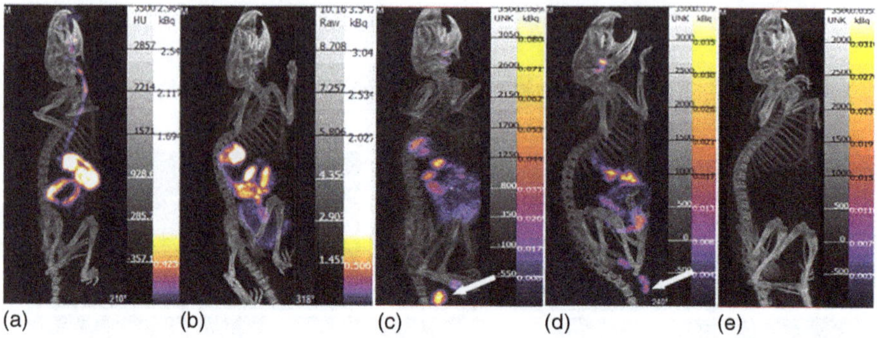

Figure 7.6 SPECT/CT imaging of Kunming mice after feeding 153 Sm-labeled cit-UCNP at different times, (a) 1 hours, (b) 2.5 hours, (c) 13.5 hours, (d) 19.5 hours, and (e) 43 hours. Source: Reproduced with permission from Peng et al. [111]/Springer Nature.

animal husbandry, often as fertilizers and feed additives, to increase crop yields and improve animal production performance. However, these lanthanide elements will eventually accumulate in the bodies of animals or humans. Hence, investigating the toxic effects of UCNP in plants is crucial. When incubated at a concentration of UCNP below 100 $\mu g\, ml^{-1}$, the development state of the beans was promoted. In contrast, the higher concentrations of UCNPs could inhibit bean growth [111]. Figure 7.6 shows the single-photon emission computed tomography and X-ray computed tomography (SPECT/CT) imaging of mice fed with cit-UCNP-treated bean sprouts. The ^{153}Sm-labeled cit-UCNP shows a clear signal in the stomach and intestine during the first several hours. From 13.5 to 43 hours, the signal in the intestine weakened. This phenomenon indicates that the UCNP were excreted through the feces.

Although the potential mechanism of accelerated growth of bean sprouts under the influence of UCNP concentration is not clear, this research provides preliminary verification and reference for the biological safety of UCNP in biological environments.

7.6 Conclusions and Prospects

The UCNPs have surfaced as a promising advancement in the field of biology owing to their distinctive optical characteristics, including extended fluorescence duration, minimal spectral width, and resilience against photobleaching. The issue of biosecurity is a primary concern as it hinders the progress of utilizing rare-earth nanocrystals for medical applications. Herein, we focus on the representative research concerning the biosafety of rare-earth doped nanomaterials. Over the last 10 years, notable advancements have been achieved in elucidating the mechanisms of cellular internalization, cytotoxic effects, biodistribution, and metabolic pathways of rare-earth nanocrystals within the body. However, there are still some issues that require further research, which will undoubtedly help enhance the applicability of rare-earth nanocrystals in clinical medicine.

The majority of nanomaterials are taken up through processes such as endocytosis and pinocytosis, and subsequently sequestered within endolysosomes. The UCNPs' cellular uptake efficiency is severely affected by the surface modification. The positive charge promotes the endocytosis of the UCNPs because the membrane structure carries a negative charge. Moreover, the cellular endocytosis efficiency of UCNPs is notably influenced by their size. In forthcoming times, research is expected to focus on studying the shapes and types of different UCNPs and conducting correlation studies on the effects of various subcellular organelles after cellular endocytosis.

The extended retention time of UCNPs in the body can last for several weeks to months before complete excretion, there is a consequent prolongation of the coexistence period between UCNPs and vital immune regulatory organs like the liver, lungs, and kidneys. To ensure high clearance rates and avoid associated toxicity, the particle size of UCNPs should be maintained within a relatively narrow range. However, ultrasmall particle sizes (<5 nm) of UCNPs exhibit weak fluorescence intensity due to lattice defects. Therefore, it is inappropriate to ignore their effect as fluorescent probes in the body, especially for rapid elimination. When employing UCNPs as agents for multimodal imaging, it is crucial to minimize their dimensions to mitigate potential toxicity risks. For large-sized UCNPs that cannot be excreted through the kidneys, they can be slowly cleared through the biliary tract. Therefore, future research can concentrate on enhancing surface properties to facilitate the faster elimination of UCNPs through the biliary pathway, thereby reducing their biological toxicity even more.

Research has indicated that the toxicity of UCNPs varies based on concentration levels, underscoring the significance of identifying the precise concentration range employed in in vivo studies. On this basis, further optimization of the quantum yield is necessary [112, 113]. The primary determinants influencing the upconversion luminescence observed in rare-earth nanocrystals are nonradiative decay processes [114] and emission cross-section [115] of lanthanide ions. Hence, the enhancement of upconversion efficiency can be achieved through adjusting particle size [116–118], laser annealing [119], and so on. This is undoubtedly important for the clinical translation of UCNPs.

References

1 Lewinski, N., Colvin, V., and Drezek, R. (2008). Cytotoxicity of nanoparticles. *Small* 4: 26–49.

2 Conner, S.D. and Schmid, S.L. (2003). Regulated portals of entry into the cell. *Nature* 422: 37–44.

3 Verma, A. and Stellacci, F. (2010). Effect of surface properties on nanoparticle–cell interactions. *Small* 6: 12–21.

4 Jin, J., Gu, Y.J., Man, C.W.Y. et al. (2011). Polymer-coated $NaYF_4$:Yb^{3+}, Er^{3+} upconversion nanoparticles for charge-dependent cellular imaging. *ACS Nano* 5: 7838–7847.

5 Nam, S.H., Bae, Y.M., Park, Y.I. et al. (2011). Long-term real-time tracking of lanthanide ion doped upconverting nanoparticles in living cells. *Angewandte Chemie International Edition* 50: 6093–6097.

6 Bae, Y.M., Park, Y.I., Nam, S.H. et al. (2012). Endocytosis, intracellular transport, and exocytosis of lanthanide-doped upconverting nanoparticles in single living cells. *Biomaterials* 33: 9080–9086.

7 Park, Y.I., Kim, J.H., Lee, K.T. et al. (2009). Nonblinking and nonbleaching upconverting nanoparticles as an optical imaging nanoprobe and T1 magnetic resonance imaging contrast agent. *Advanced Materials* 21: 4467–4471.

8 Zhou, J.C., Yang, Z.L., Dong, W. et al. (2011). Bioimaging and toxicity assessments of near-infrared upconversion luminescent $NaYF_4$: Yb, Tm nanocrystals. *Biomaterials* 32: 9059–9067.

9 Jang, G.H., Hwang, M.P., Kim, S.Y. et al. (2014). A systematic in-vivo toxicity evaluation of nanophosphor particles via zebrafish models. *Biomaterials* 35: 440–449.

10 Chen, J., Guo, C., Wang, M. et al. (2011). Controllable synthesis of $NaYF_4$: Yb, Er upconversion nanophosphors and their application to in vivo imaging of *Caenorhabditis elegans*. *Journal of Materials Chemistry* 21: 2632–2638.

11 Wang, K., Ma, J., He, M. et al. (2013). Toxicity assessments of near-infrared upconversion luminescent LaF_3: Yb, Er in early development of zebrafish embryos. *Theranostics* 3: 258.

12 Das, G.K., Stark, D.T., and Kennedy, I.M. (2014). Potential toxicity of up-converting nanoparticles encapsulated with a bilayer formed by ligand attraction. *Langmuir* 30: 8167–8176.

13 Sikora, B., Kowalik, P., Mikulski, J. et al. (2017). Mammalian cell defence mechanisms against the cytotoxicity of $NaYF_4$:(Er,Yb,Gd) nanoparticles. *Nanoscale* 9: 14259–14271.

14 Yang, P.P., Gai, S.L., and Lin, J. (2012). Functionalized mesoporous silica materials for controlled drug delivery. *Chemical Society Reviews* 41: 3679–3698.

15 Li, C.X., Hou, Z.Y., Dai, Y.L. et al. (2013). A facile fabrication of upconversion luminescent and mesoporous core- shell structured beta-$NaYF_4$:Yb^{3+}, Er^{3+}@$mSiO_2$ nanocomposite spheres for anti-cancer drug delivery and cell imaging. *Biomaterials Science* 1: 213–223.

16 Lucky, S.S., Idris, N.M., Li, Z.Q. et al. (2015). Titania coated upconversion nanoparticles for near-infrared light triggered photodynamic therapy. *ACS Nano* 9: 191–205.

17 Tian, G., Zheng, X.P., Zhang, X. et al. (2015). TPGS-stabilized $NaYbF_4$:Er upconversion nanoparticles for dual-modal fluorescent/CT imaging and anticancer drug delivery to overcome multi-drug resistance. *Biomaterials* 40: 107–116.

18 Lucky, S.S., Idris, N.M., Huang, K. et al. (2016). In vivo biocompatibility, biodistribution and therapeutic efficiency of titania coated upconversion nanoparticles for photodynamic therapy of solid oral cancers. *Theranostics* 6: 1844–1865.

19 Generalova, A.N., Rocheva, V.V., Nechaev, A.V. et al. (2016). PEG–modified upconversion nanoparticles for in vivo optical imaging of tumors. *RSC Advances* 6: 30089–30097.

20 Portioli, C., Pedroni, M., Benati, D. et al. (2016). Citrate–stabilized lanthanide–doped nanoparticles: brain penetration and interaction with immune cells and neurons. *Nanomedicine* 11: 3039–3051.

21 Gao, Y.L., Liu, L., Shen, B. et al. (2017). Amphiphilic PEGylated lanthanide–doped upconversion nanoparticles for significantly passive accumulation in the peritoneal metastatic carcinomatosis models following intraperitoneal administration. *ACS Biomaterials Science & Engineering* 3: 2176–2184.

22 Yu, J., Yin, W.Y., Peng, T. et al. (2017). Biodistribution, excretion, and toxicity of polyethyleneimine modified $NaYFW_4$: Yb, Er upconversion nanoparticles in mice via different administration routes. *Nanoscale* 9: 4497–4507.

23 Wang, B., Song, J., Yuan, H. et al. (2014). Multicellular assembly and light–regulation of cell–cell communication by conjugated polymer materials. *Advanced Materials* 26: 2371–2375.

24 Pudovkin, M.S., Zelenikhin, P.V, Shtyreva, V.V., Shtyreva, V., Koryakovtseva, D., Evtugyn, V., Salnikov, V., Nizamutdinov, A. and Semashko, V. (2019) Cellular uptake and cytotoxicity of unmodified Pr^{3+}:LaF_3 nanoparticles. *Journal of Nanoparticle Research*, 21, 184.

25 Xiong, L.Q., Yang, T.S., Yang, Y. et al. (2010). Long–term in vivo biodistribution imaging and toxicity of polyacrylic acid–coated upconversion nanophosphors. *Biomaterials* 31: 7078–7085.

26 Zhao, J., Zhang, X., Fang, L. et al. (2020). Iridium(III) complex–derived polymeric micelles with low dark toxicity and strong NIR excitation for phototherapy and chemotherapy. *Small* 16: 2000363.

27 Wysokinska, E., Cichos, J., Ziolo, E. et al. (2016). Cytotoxic interactions of bare and coated $NaGdF_4$:Yb^{3+}:Er^{3+} nanoparticles with macrophage and fibroblast cells. *Toxicology in Vitro* 32: 16–25.

28 Khabir, Z., Guller, A.E., Rozova, V.S. et al. (2019). Tracing upconversion nanoparticle penetration in human skin. *Colloids and Surfaces B: Biointerfaces* 184: 110480.

29 Li, Q., Wang, Z., Chen, Y., and Zhang, G.X. (2017). Elemental bio–imaging of PEGylated $NaYF_4$:Yb/Tm/Gd upconversion nanoparticles in mice by laser ablation inductively coupled plasma mass spectrometry to study toxic side effects on the spleen, liver and kidneys. *Metallomics* 9: 1150–1156.

30 Feng, L.L., Gai, S.L., He, F. et al. (2017). Multifunctional mesoporous ZrO_2 encapsulated upconversion nanoparticles for mild NIR light activated synergistic cancer therapy. *Biomaterials* 147: 39–52.

31 Petri-Fink, A., Steitz, B., Finka, A. et al. (2008). Effect of cell media on polymer coated superparamagnetic iron oxide nanoparticles (SPIONs): colloidal stability, cytotoxicity, and cellular uptake studies. *European Journal of Pharmaceutics and Biopharmaceutics* 68: 129–137.

32 Liu, G., Liu, N., Zhou, L.Z. et al. (2015). NIR-responsive polypeptide copolymer upconversion composite nanoparticles for triggered drug release and enhanced cytotoxicity. *Polymer Chemistry* 6: 4030–4039.

33 Patel, B., Gupta, N., and Ahsan, F. (2015). Particle engineering to enhance or lessen particle uptake by alveolar macrophages and to influence the therapeutic outcome. *European Journal of Pharmaceutics and Biopharmaceutics* 89: 163–174.

34 Elbakry, A., Wurster, E.C., Zaky, A. et al. (2012). Layer-by-layer coated gold nanoparticles: size-dependent delivery of DNA into cells. *Small* 8: 3847–3856.

35 Guller, A.E., Nadort, A., Generalova, A.N. et al. (2018). Rational surface design of upconversion nanoparticles with polyethylenimine coating for biomedical applications: better safe than brighter? *ACS Biomaterials Science & Engineering* 4: 3143–3153.

36 Asadi, M., Ghahari, M., Hassanzadeh-Tabrizi, S.A. et al. (2020). Studying the toxicity effects of coated and uncoated $NaLuF_4$: Yb^{3+}, Tm^{3+} upconversion nanoparticles on blood factors and histopathology for Balb/C mice's tissue. *Materials Research Express* 6: 125421.

37 Cho, J., Kushiro, K., Teramura, Y., and Takai, M. (2014). Lectin-tagged fluorescent polymeric nanoparticles for targeting of sialic acid on living cells. *Biomacromolecules* 15: 2012–2018.

38 Chavez-Garcia, D., Juarez-Moreno, K., Campos, C.H. et al. (2018). Upconversion rare earth nanoparticles functionalized with folic acid for bioimaging of MCF-7 breast cancer cells. *Journal of Materials Research* 33: 191–200.

39 Feng, Y., Chen, H.D., Ma, L.N. et al. (2017). Surfactant-free aqueous synthesis of novel Ba_2GdF_7:Yb^{3+}, Er^{3+}@PEG upconversion nanoparticles for in vivo trimodality imaging. *ACS Applied Materials & Interfaces* 9: 15096–15102.

40 Hao, W., Liu, D., Wang, Y. et al. (2018). Dual-stimuli responsive nanoparticles (UCNP-CD@APP) assembled by host-guest interaction for drug delivery. *Colloids and Surfaces A: Physicochemical and Engineering Aspects* 537: 446–451.

41 Zhang, X.Y., Guo, Z., Zhang, X. et al. (2019). Mass production of poly(ethylene glycol) monooleate-modified core-shell structured upconversion nanoparticles for bio-imaging and photodynamic therapy. *Scientific Reports* 9: 5212.

42 Mohan, M. and Poddar, R. (2021). Polymerically engineered upconversion nanoparticles (UCNPs) as contrast agent for functionally modified optical coherence tomography (OCT). *Materials Science and Engineering: C* 121: 111841.

43 Duong, H.V., Chau, T.T.L., Dang, N.T.T. et al. (2018). Biocompatible chitosan-functionalized upconverting nanocomposites. *ACS Omega* 3: 86–95.

44 Xia, T., Kovochich, M., Liong, M. et al. (2009). Polyethyleneimine coating enhances the cellular uptake of mesoporous silica nanoparticles and allows safe delivery of siRNA and DNA constructs. *ACS Nano* 3: 3273–3286.

45 Kembuan, C., Oliveira, H., and Graf, C. (2021). Effect of different silica coatings on the toxicity of upconversion nanoparticles on RAW 264.7 macrophage cells. *Beilstein Journal of Nanotechnology* 12: 35–48.

46 Dhanya, C.R., Jeyaraman, J., Janeesh, P.A. et al. (2016). Bio-distribution and in vivo/in vitro toxicity profile of PEGylated polymer capsules encapsulating $LaVO_4$:Tb^{3+} nanoparticles for bioimaging applications. *RSC Advances* 6: 55125–55134.

47 Kostiv, U., Rajsiglova, L., Luptakova, D. et al. (2017). Biodistribution of upconversion/magnetic silica-coated $NaGdF_4$:Yb^{3+}/Er^{3+} nanoparticles in mouse models. *RSC Advances* 7: 45997–46006.

48 Ansari, A.A., Parchur, A.K., Thorat, N.D., and Chen, G.Y. (2021). New advances in pre-clinical diagnostic imaging perspectives of functionalized upconversion nanoparticle-based nanomedicine. *Coordination Chemistry Reviews* 440: 213971.

49 Hühn, D., Kantner, K., Geidel, C. et al. (2013). Polymer-coated nanoparticles interacting with proteins and cells: focusing on the sign of the net charge. *ACS Nano* 7: 3253–3263.

50 Zhou, M.Z., Ge, X.Q., Ke, D.M. et al. (2019). The bioavailability, biodistribution, and toxic effects of silica-coated upconversion nanoparticles in vivo. *Frontiers in Chemistry* 7: 218.

51 Shan, J.N., Chen, J.B., Meng, J. et al. (2008). Biofunctionalization, cytotoxicity, and cell uptake of lanthanide doped hydrophobically ligated $NaYF_4$ upconversion nanophosphors. *Journal of Applied Physics* 104: 094308.

52 Cheng, L., Yang, K., Shao, M.W. et al. (2011). In vivo pharmacokinetics, long-term biodistribution and toxicology study of functionalized upconversion nanoparticles in mice. *Nanomedicine* 6: 1327–1340.

53 Wu, S.J., Duan, N., Wang, Z.P., and Wang, H.X. (2011). Aptamer-functionalized magnetic nanoparticle-based bioassay for the detection of ochratoxin a using upconversion nanoparticles as labels. *Analyst* 136: 2306–2314.

54 Liu, C.H., Wang, Z., Wang, X.K., and Li, Z.P. (2011). Surface modification of hydrophobic $NaYF_4$:Yb,Er upconversion nanophosphors and their applications for immunoassay. *Science China-Chemistry* 54: 1292–1297.

55 Liu, F.Y., He, X.X., Liu, L. et al. (2013). Conjugation of $NaGdF_4$ upconverting nanoparticles on silica nanospheres as contrast agents for multi-modality imaging. *Biomaterials* 34: 5218–5225.

56 Lee, J., Lee, T.S., Ryu, J. et al. (2013). RGD peptide-conjugated multimodal $NaGdF_4$:Yb^{3+}/Er^{3+} nanophosphors for upconversion luminescence, MR, and PET imaging of tumor angiogenesis. *Journal of Nuclear Medicine* 54: 96–103.

57 Tawfik, S.M., Sharipov, M., Huy, B.T. et al. (2018). Naturally modified nonionic alginate functionalized upconversion nanoparticles for the highly efficient targeted pH-responsive drug delivery and enhancement of NIR-imaging. *Journal of Industrial and Engineering Chemistry* 57: 424–435.

58 Liu, L., Zhang, H., Wang, Z.X., and Song, D.Q. (2019). Peptide-functionalized upconversion nanoparticles-based FRET sensing platform for Caspase-9 activity detection in vitro and in vivo. *Biosensors & Bioelectronics* 141: 111403.

59 He, S.Q., Song, J., Liu, J.F. et al. (2019). Enhancing photoacoustic intensity of upconversion nanoparticles by photoswitchable azobenzene-containing polymers for dual NIR-II and photoacoustic imaging in vivo. *Advanced Optical Materials* 7: 1900045.

60 Yang, L.M., Sun, H.F., Wang, X. et al. (2019). An aptamer based aggregation assay for the neonicotinoid insecticide acetamiprid using fluorescent upconversion nanoparticles and DNA functionalized gold nanoparticles. *Microchimica Acta* 186: 1–11.

61 Liu, M.L., Wei, J.Y., Wang, Y. et al. (2019). Dopamine-functionalized upconversion nanoparticles as fluorescent sensors for organophosphorus pesticide analysis. *Talanta* 195: 706–712.

62 Wang, M., Zhang, T., Hu, Y.J. et al. (2019). In situ synthesis of dicarboxylic acid functionalized upconversion nanoparticles for bioimaging applications. *Chemphotochem* 3: 145–150.

63 Sun, Q.Q., Chen, W.L., Wang, M. et al. (2023). A "chase and block" strategy for enhanced cancer therapy with hypoxia-promoted photodynamic therapy and autophagy inhibition based on upconversion nanocomposites. *Advanced Healthecare Materials* 12: 2301087.

64 Reddy, K.L., Sharma, P.K., Singh, A. et al. (2019). Amine-functionalized, porous silica-coated $NaYF_4$:Yb/Er upconversion nanophosphors for efficient delivery of doxorubicin and curcumin. *Materials Science and Engineering: C* 96: 86–95.

65 Li, X., Liu, L., Fu, Y. et al. (2020). Peptide-enhanced tumor accumulation of upconversion nanoparticles for sensitive upconversion luminescence/magnetic resonance dual-mode bioimaging of colorectal tumors. *Acta Biomaterialia* 104: 167–175.

66 Chen, Y., D'amario, C., Gee, A. et al. (2020). Dispersion stability and biocompatibility of four ligand-exchanged $NaYF_4$: Yb, Er upconversion nanoparticles. *Acta Biomaterialia* 102: 384–393.

67 Ansari, A.A., Thakur, V.K., and Chen, G. (2021). *Coordination Chemistry Reviews* 436: 213821.

68 Lin, B., Wu, J., Wang, Y. et al. (2021). Peptide functionalized upconversion/NIR II luminescent nanoparticles for targeted imaging and therapy of oral squamous cell carcinoma. *Biomaterials Science* 9: 1000–1007.

69 Liu, L., Li, X., Zhang, H. et al. (2021). Six-in-one peptide functionalized upconversion@polydopamine nanoparticle-based ratiometric fluorescence sensing platform for real-time evaluating anticancer efficacy through monitoring caspase-3 activity. *Sensors and Actuators B: Chemical* 333: 129554.

70 Zhang, L., Jin, D., and Stenzel, M.H. (2021). Polymer-functionalized upconversion nanoparticles for light/imaging-guided drug delivery. *Biomacromolecules* 22: 3168–3201.

71 Chavez-Garcia, D., Juarez-Moreno, K., Campos, C.H. et al. (2018). Cytotoxicity, genotoxicity and uptake detection of folic acid-functionalized green upconversion nanoparticles Y_2O_3/Er^{3+}, Yb^{3+} as biolabels for cancer cells. *Journal of Materials Science* 53: 6665–6680.

72 Zhang, J., Liu, F., Li, T. et al. (2015). Surface charge effect on the cellular interaction and cytotoxicity of $NaYF_4:Yb^{3+}$, $Er^{3+}@SiO_2$ nanoparticles. *RSC Advances* 5: 7773–7780.

73 Guller, A.E., Generalova, A.N., Petersen, E.V. et al. (2015). Cytotoxicity and non-specific cellular uptake of bare and surface-modified upconversion nanoparticles in human skin cells. *Nano Research* 8: 1546–1562.

74 Himmelstoss, S.F. and Hirsch, T. (2019). Long-term colloidal and chemical stability in aqueous media of $NaYF_4$-type upconversion nanoparticles modified by ligand-exchange. *Particle & Particle Systems Characterization* 36: 1900235.

75 Jalil, R.A. and Zhang, Y. (2008). Biocompatibility of silica coated $NaYF_4$ upconversion fluorescent nanocrystals. *Biomaterials* 29: 4122–4128.

76 Yi, G.S. and Chow, G.M. (2006). Synthesis of hexagonal-phase $NaYF_4$:Yb, Er and NaYF4: Yb, Tm nanocrystals with efficient up-conversion fluorescence. *Advanced Functional Materials* 16: 2324–2329.

77 Cao, T., Yang, Y., Sun, Y. et al. (2013). Biodistribution of sub–10 nm PEG–modified radioactive/upconversion nanoparticles. *Biomaterials* 34: 7127–7134.

78 Zhao, L., Kutikov, A., Shen, J. et al. (2013). Stem cell labeling using polyethylenimine conjugated (α–$NaYbF_4$:Tm^{3+})/CaF_2 upconversion nanoparticles. *Theranostics* 3: 249–257.

79 Zhu, X., Da Silva, B., Zou, X. et al. (2014). Intra–arterial infusion of PEGylated upconversion nanophosphors to improve the initial uptake by tumors in vivo. *RSC Advances* 4: 23580–23584.

80 Sun, Y., Peng, J., Feng, W., and Li, F.Y. (2013). Upconversion nanophosphors NaLuF4: Yb, Tm for lymphatic imaging in vivo by real–time upconversion luminescence imaging under ambient light and high–resolution X–ray CT. *Theranostics* 3: 346.

81 Cao, T., Yang, Y., Gao, Y. et al. (2011). High–quality water–soluble and surface–functionalized upconversion nanocrystals as luminescent probes for bioimaging. *Biomaterials* 32: 2959–2968.

82 Liu, C., Gao, Z., Zeng, J. et al. (2013). Magnetic/upconversion fluorescent $NaGdF_4$: Yb, Er nanoparticle–based dual–modal molecular probes for imaging tiny tumors in vivo. *ACS Nano* 7: 7227–7240.

83 Bridot, J.-L., Faure, A.-C., Laurent, S. et al. (2007). Hybrid gadolinium oxide nanoparticles: multimodal contrast agents for in vivo imaging. *Journal of the American Chemical Society* 129: 5076–5084.

84 Sun, Y., Liu, Q., Peng, J. et al. (2013). Radioisotope post–labeling upconversion nanophosphors for in vivo quantitative tracking. *Biomaterials* 34: 2289–2295.

85 Chatterjee, D.K., Rufaihah, A.J., and Zhang, Y. (2008). Upconversion fluorescence imaging of cells and small animals using lanthanide doped nanocrystals. *Biomaterials* 29: 937–943.

86 Gao, G., Zhang, C., Zhou, Z. et al. (2013). One–pot hydrothermal synthesis of lanthanide ions doped one–dimensional upconversion submicrocrystals and their potential application in vivo CT imaging. *Nanoscale* 5: 351–362.

87 Chen, Q., Wang, C., Cheng, L. et al. (2014). Protein modified upconversion nanoparticles for imaging-guided combined photothermal and photodynamic therapy. *Biomaterials* 35: 2915–2923.

88 Zeng, S., Tsang, M-K., Chan, C-F., Wong, K.L., Fei, B. and Hao, J.H. (2012) Dual-modal fluorescent/magnetic bioprobes based on small sized upconversion nanoparticles of amine–functionalized $BaGdF_5$: Yb/Er. *Nanoscale*, 4, 5118–24.

89 Chávez–García, D. and Juarez–Moreno, K. (2023). Nanotoxicological assessments of upconversion nanoparticles. *IntechOpen* https://doi.org/10.5772/intechopen.111883.

90 Liu, C., Sun, S., and Mao, J. (2023). Water–soluble Yb^{3+}, Er^{3+} codoped $NaYF_4$ nanoparticles induced SGC–7901 cell apoptosis through mitochondrial dysfunction and ROS–mediated ER stress. *Human & Experimental Toxicology* 42: 1–10.

91 Jayamole, A.A., Ganeshan, J.E., Sundaram, T. et al. (2023). Rapid and one–step mechanochemical ligand exchange of Yb^{3+}/Er^{3+} co–doped $NaGdF_4$ upconversion nanoparticles for efficient MR and CT imaging. *Zeitschrift für Physikalische Chemie* 237: 617–644.

92 Ajithkumar, G., Yoo, B., Goral, D.E. et al. (2013). Multimodal bioimaging using a rare earth doped Gd_2O_2 S: Yb/Er phosphor with upconversion luminescence and magnetic resonance properties. *Journal of Materials Chemistry B* 1: 1561–1572.

93 Idris, N.M., Li, Z., Ye, L. et al. (2009). Tracking transplanted cells in live animal using upconversion fluorescent nanoparticles. *Biomaterials* 30: 5104–5113.

94 Asati, A., Santra, S., Kaittanis, C., and Perez, J.M. (2010). Surface–charge–dependent cell localization and cytotoxicity of cerium oxide nanoparticles. *ACS Nano* 4: 5321–5331.

95 Mizushima, N., Levine, B., Cuervo, A.M., and Klionsky, D.J. (2008). Autophagy fights disease through cellular self-digestion. *Nature* 451: 1069–1075.

96 Levine, B. (2007). Autophagy and cancer. *Nature* 446: 745–747.

97 Pan, H., Wang, H., Yu, J. et al. (2019). Near–infrared light remotely up–regulate autophagy with spatiotemporal precision via upconversion optogenetic nanosystem. *Biomaterials* 199: 22–31.

98 Chen, Y., Yang, L., Feng, C., and Wen, L.P. (2005). Nano neodymium oxide induces massive vacuolization and autophagic cell death in non–small cell lung cancer NCI–H460 cells. *Biochemical and Biophysical Research Communications* 337: 52–60.

99 Song, W., Soo, L.S., Savini, M. et al. (2014). Ceria nanoparticles stabilized by organic surface coatings activate the lysosome–autophagy system and enhance autophagic clearance. *ACS Nano* 8: 10328–10342.

100 Zhang, Y., Zheng, F., Yang, T. et al. (2012). Tuning the autophagy–inducing activity of lanthanide–based nanocrystals through specific surface–coating peptides. *Nature Materials* 11: 817–826.

101 Song, J. and Sun, G. (2015). Analysis of application of antibiotics in the outpatient and emergency pharmacy of the Fifth People's Hospital of Shanghai, Fudan University during 2012–2014. *Evaluation and Analysis of Drug–Use in Hospitals of China* 15: 1459–1462.

102 Giunti, S., Andersen, N., Rayes, D., and De Rosa, M.J. (2021). Drug discovery: insights from the invertebrate *Caenorhabditis elegans*. *Pharmacology Research & Perspectives* 9: e00721.

103 Bulterijs, S. and Braeckman, B.P. (2020). Phenotypic screening in *C. elegans* as a tool for the discovery of new geroprotective drugs. *Pharmaceuticals* 13: 164.

104 Ahamefule, C.S., Qin, Q., Odiba, A.S. et al. (2020). Caenorhabditis elegans-based *Aspergillus fumigatus* infection model for evaluating pathogenicity and drug efficacy. *Frontiers in Cellular and Infection Microbiology* 10: 320.

105 Zhang, A., Hsiung, K.C., Kern, C.C. et al. (2023). Unraveling effects of anti-aging drugs on *C. elegans* using liposomes. *GeroScience* 45: 1583–1603.

106 Macrae, C.A. and Peterson, R.T. (2015). Zebrafish as tools for drug discovery. *Nature Reviews Drug Discovery* 14: 721–731.

107 Chakraborty, C., Hsu, C.H., Wen, Z.H. et al. (2009). Zebrafish: a complete animal model for in vivo drug discovery and development. *Current Drug Metabolism* 10: 116–124.

108 Rosa, J.G.S., Lima, C., and Lopes-Ferreira, M. (2022). Zebrafish larvae behavior models as a tool for drug screenings and pre-clinical trials: a review. *International Journal of Molecular Sciences* 23: 6647.

109 Ling, D., Chen, H., Chan, G., and Lee, S.M.Y. (2022). Quantitative measurements of zebrafish heartrate and heart rate variability: a survey between 1990–2020. *Computers in Biology and Medicine* 142: 105045.

110 Fu, L., Shi, B., Wen, S. et al. (2022). Aspect ratio of PEGylated upconversion nanocrystals affects the cellular uptake in vitro and in vivo. *Acta Biomaterialia* 147: 403–413.

111 Peng, J., Sun, Y., Liu, Q. et al. (2012). Upconversion nanoparticles dramatically promote plant growth without toxicity. *Nano Research* 5: 770–782.

112 Li, H., Tan, M., Wang, X. et al. (2020). Temporal multiplexed in vivo upconversion imaging. *Journal of the American Chemical Society* 142: 2023–2030.

113 Jones, C.M., Biner, D., Misopoulos, S. et al. (2021). Optimized photoluminescence quantum yield in upconversion composites considering the scattering, inner-filter effects, thickness, self-absorption, and temperature. *Scientific Reports* 11: 13910.

114 Zhai, X., Li, Y., Zhao, W. et al. (2023). One-pot synthesis of hexagonal $NaLuF_4$:Yb,Er microcrystals with enhanced upconversion emission and high production yield. *Journal of Rare Earths* 41: 498–506.

115 Fu, L., Wu, Y., and Fu, T. (2022). Determination of absorption cross-section of RE^{3+} in upconversion powder materials: application to β-$NaYF_4$: Er^{3+}. *Journal of Luminescence* 245: 118758.

116 Wang, F., Wang, J., and Liu, X. (2010). Direct evidence of a surface quenching effect on size-dependent luminescence of upconversion nanoparticles. *Angewandte Chemie International Edition* 49: 7456–7460.

117 Huang, F., Labrador-Páez, L., Ågren, H. et al. (2023). Transient energy trapping as a size-conserving surface passivation strategy for producing bright ultrasmall upconversion nanoprobes. *Nano Energy* 105: 108015.

118 Zhang, Y., Yu, Z., Li, J. et al. (2017). Ultrasmall–superbright neodymium–upconversion nanoparticles via energy migration manipulation and lattice modification: 808 nm–activated drug release. *ACS Nano* 11: 2846–2857.

119 Fu, H., Hu, C., Liu, J. et al. (2022). An overview of boosting lanthanide upconversion luminescence through chemical methods and physical strategies. *CrystEngComm* 24: 7698–7717.

8

Design and Construction of Photosensitizers for Photodynamic Therapy of Tumor

Ruohao Zhang[1,2], *Jing Feng*[1,2], *Yifei Zhou*[1,2], *Jitong Gong*[1,2], *and Hongjie Zhang*[1,2,3]

[1] *Chinese Academy of Sciences, Changchun Institute of Applied Chemistry, State Key Laboratory of Rare Earth Resource Utilization, 5625 Renmin Street, Changchun, 130022, P. R. China*
[2] *University of Science and Technology of China, School of Applied Chemistry and Engineering, 96 Jinzhai Road, Hefei, 230026, P. R. China*
[3] *Tsinghua University, Department of Chemistry, 30 Shuangqing Road, Beijing, 100084, P. R. China*

8.1 Introduction

Photodynamic therapy (PDT) emerged gradually in the 1980s and has been widely studied and applied in the treatment of tumors due to its excellent selectivity and safety, making it a strong candidate for traditional treatment options [1–3]. As a noninvasive phototherapy, PDT uses the light of a specific wavelength to excite the photosensitizer (PS) to generate cytotoxic reactive oxygen species (ROS) in target tissues, thereby destroying the life activities of tumor cells and promoting tumor cell death. There are several types of ROS, including hydroxyl radical (•OH), superoxide radical anion ($O_2^{\bullet-}$), singlet oxygen ($^1O_2$), and hydrogen peroxide (H_2O_2). ROS is highly reactive and could react with a variety of endogenous substances in tumor cells to disrupt life activities [4]. For example, •OH can undergo nonspecific oxidation reactions with almost any surrounding biologically active substances, thereby destroying the vital activities of tumor cells [5, 6]. Although H_2O_2 is less reactive overall, it can selectively oxidize methionine (Met) and cysteine (Cys) residues in certain proteins [7, 8]. Thanks to the high reactivity of ROS, the diffusion distance of ROS in cells and tissues is very short (<50 nm) [2, 9]. Therefore, it is difficult for ROS to cause damage to the extracellular matrix, nerves, and surrounding normal tissue cells. Compared with traditional surgical treatment, PDT could treat tumors noninvasively, greatly reducing the secondary harm to patients caused by surgical trauma. In addition, compared with radiotherapy or chemotherapy, ROS-based PDT could effectively reduce the multidrug resistance of tumor cells and reduce the side effects and discomfort caused by traditional treatment methods, which could improve the adaptability of patients [10, 11]. Importantly, PDT could implement safe and dynamic treatment of target tissues.

Photofunctional Nanomaterials for Biomedical Applications, First Edition.
Edited by Chunxia Li and Jun Lin.

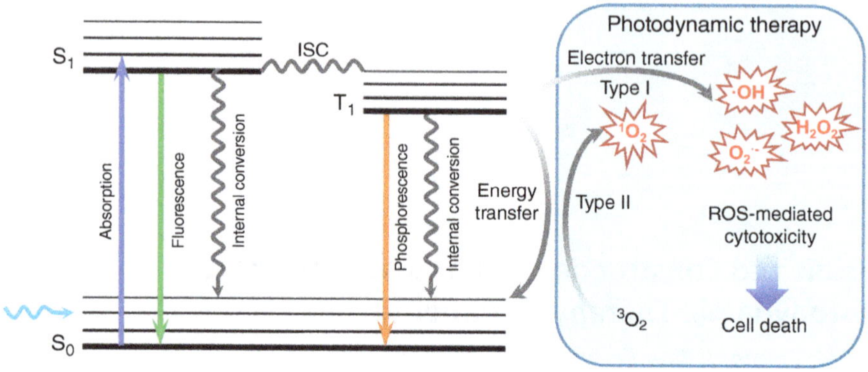

Figure 8.1 Simplified Jablonski diagram illustrating the photophysical and photochemical basis of PDT. The ground state, excited singlet state, and excited triplet state are abbreviated as S_0, S_1, and T_1, respectively. Intersystem crossing is abbreviated as ISC.

PDT is a treatment method that disrupts important activities of tumor cells by exciting PSs to produce ROS [5]. For organic PSs (Table 8.1), as shown in Figure 8.1, the generation of ROS first requires active molecules to be excited to the singlet excited state (S_1) by the light with a specific wavelength. However, S_1 is unstable and has a relatively short energy level lifetime (~ns). Therefore, S_1 will undergo intersystem crossing (ISC) to form a more stable excited triplet state (T_1) with a long lifetime (~μs). According to the different ROS production mechanisms, PDT could be divided into two types of reactions. When T_1 may react with intracellular substrates (such as nucleic acids, proteins, and lipids) through the electron transfer mechanism to form free radicals, which then react with H_2O or O_2 to generate ROS (such as •OH, $O_2^{\bullet-}$, and H_2O_2). This process is defined as the type I PDT process. On the other hand, T_1 may undergo a series of photochemical reactions (such as phosphorescence emission) to transfer energy to O_2 molecules to generate $^1O_2$, which is a type II PDT process. For semiconductor nanomaterials, it has been reported that semiconductors could generate electron–hole pairs under light irradiation, thereby directly reacting with H_2O and O_2 to produce ROS (•OH and $O_2^{\bullet-}$) [10, 12]. It is worth noting that most organic PSs undergo type II reactions to generate $^1O_2$.

The effect of PDT depends on the oxygen level in tumor tissue. As is known, the tumor microenvironment (TME) is often hypoxic [13, 14]; thus, it is difficult for most PSs to exert their optimal properties. Some researchers seek to combine enzymes that improve hypoxic conditions with PSs to continuously generate oxygen in situ during PDT process and improve the hypoxic state of the TME. It has also been reported that Ru(II)–porphyrin complex can produce phototoxicity without oxygen dependence and achieve oxygen-free PDT [15, 16]. In addition, the therapeutic effect of PDT depends on the choice of excitation light. Generally, the light absorption of common organic PS molecules is in the ultraviolet (UV) or visible light region [10]. This part of the light has shallow tissue penetration, making it difficult to perform PDT on deep tumors [17, 18]. Another strategy proposes that a micro-light source device could be implanted deep into the tissue for responsive light using a minimally invasive

Table 8.1 Summary of organic photosensitizers.

Photosensitizer	Description	Tumor model[a)]	Treatment parameters[b)]	References
Porphyrin	Porphyrin lipid nanoemulsions with paclitaxel	KB s.c. tumors; i.v. injection	PDT + chemotherapy; 671 nm, 100 mW cm^{-2}, 500 s	[34]
	Metal-free porphyrin linked covalently to Mn-porphyrin	HT-29 s.c. tumors; i.v. injection	PDT; 650 nm, 200 mW, 30 min	[38]
	Charge-switchable, four-armed polymeric TAPP	CT-26 s.c. tumors; i.v. injection	PDT; 670 nm, 0.5 W cm^{-2}, 5 min	[35]
	Alternating copolymer photosensitizer	A549 s.c. tumors; i.v. injection	PDT; 650 nm, 200 mW cm^{-2}, 5 min	[36]
Phthalocyanine	Thioketal-linked diamine-condensed polyphthalocyanine	H22 s.c. tumors; i.v. injection	PDT + SDT; SDT: ultrasound, 1.0 MHz, 3 W cm^{-2}, 5 min PDT: 690 nm, 280 mW cm^{-2}, 15 min	[41]
	$ZnPcN_4$-albumin dimers	HT-29 tumors; i.v. injection	PDT; 655 nm, 20 mW, 30 min	[43]
	ZnPc-Ganetesplib	4T1 s.c. tumors; i.v. injection	PDT; 671 nm, 500 mW cm^{-2}, 10 min	[54]
BODIPY	Aza-BODIPY	Hela s.c. tumors; i.v. injection	PDT + PTT; 730 nm, 500 mW cm^{-2}, 5 min	[55]
	2-Pyridone and BODIPY	B16-F10 s.c. tumors; –	PDT; 660 nm, 0.5 W cm^{-2}, 5 min	[56]
	DPP-BODIPY	Hela s.c. tumors; i.v. injection	PDT; 540 nm	[60]
Indocyanine dye	Fe_3O_4@PB/PEI/ICG	Hela s.c. tumors; i.v. injection	PDT + PTT; 808 nm, 2 W cm^{-2}, 10 min	[68]
	ICG-SF_6 nanobubbles with free O_2	Cal27 s.c. tumors; intratumoral injection	PDT; 808 nm, 0.5 W cm^{-2}, 10 min	[68]
AIEgen	AIEgen-exosome-dexamethasone nanovesicle	4T1 s.c. tumors; i.v. injection	PDT; 532 nm, 0.5 W cm^{-2}, 20 min	[76]
	Reticulated AIEgen polymer	4T1 s.c. tumors; intratumoral injection	PDT; 520 nm, 0.1 W cm^{-2}, 30 min	[77]

(Continued)

Table 8.1 (Continued)

Photosensitizer	Description	Tumor model[a)]	Treatment parameters[b)]	References
Ru(II) complex	Dinuclear ruthenium complexes	A375 s.c. tumors; intratumoral injection	PDT; 808 nm, 0.1 W cm^{-2}, 5 min	[82]
	Ru acceptors and fluorescent ligands	A549 s.c. tumors; intratumoral injection	PDT; 808 nm, 1 W cm^{-2}, 5 min	[81]
	Ru(II)-arene complexes	MDA-MB-231 s.c. tumors; i.v. injection	PDT + PTT; 808 nm, 0.5 W cm^{-2}, 10 min	[83]
Ir(III) complex	Engineered bacterial with IrCl photosensitizer	4T1 s.c. tumors; i.v. injection	PDT; 520 nm, 1 W cm^{-2}, 10 min	[100]
	BODIPY-Ir micelle	4T1 orthotopic tumors; i.v. injection	PDT; 660 nm, 0.6 W cm^{-2}, 5 min	[101]
MOF	Hf-porphyrin nanoscale MOF	SQ20B s.c. tumors; intratumoral injection	PDT; 640 nm, 100 mW cm^{-2}, 30 min	[103]
	Albumin/sulfonamide stabilized Fe^{3+}-porphyrin MOF	4T1 s.c. tumors; i.v. injection	PDT + PTT; PDT: 660 nm, 50 mW cm^{-2}, 10 min; PTT: 660 nm, 1 W cm^{-2}, 5 min	[109]
COF	Porphyrin-based COF nanodots	H22 s.c. tumors; i.v. injection	PDT; 638 nm, 90 mW, 15 min	[111]
HOF	HOF integrating pyrene and doxorubicin	Hela cells (in vitro)	PDT + chemotherapy; 400 nm, 30 min	[112]

a) s.c., subcutaneous; i.v., intravenous; –, not mentioned.
b) PDT, photodynamic therapy; SDT, sonodynamic therapy; PTT, photothermal therapy.

incision [19, 20]. Compared with traditional surgery, this strategy with small incision is safer and more controllable. Moreover, researchers often use near-infrared (NIR) lasers for phototherapy because they have greater tissue penetration than UV and visible light [21–23]. Using NIR light in the transparent window of biological tissue (e.g. 808 nm laser) to excite PSs has shown great advantages in PDT of tumors. Upconversion nanomaterials could emit upconversion luminescence (UCL) under the irradiation of NIR laser, which could excite the photosensitizer to generate ROS for PDT [24, 25]. Alternatively, X-ray can be used to excite photosensitizers to produce ROS [26, 27]. Three strategies require a specific light source (such as laser with a specific wavelength), which requires a certain economic and equipment cost in the early stage of PDT. Moreover, due to the high reactivity of ROS, it is difficult to circulate and metabolize in the body as ordinary therapeutic drugs. PDT can only produce therapeutic effect in the light-irradiated area, sufficient ROS needs to accumulate in the tumor area to effectively kill tumor cells and curb tumor recurrence. Moreover, common organic photosensitizers (such as Ce6) are often hydrophobic and require molecular modification or modification of functional molecules to improve their water solubility [28, 29]. Therefore, photosensitizers need to be effectively accumulated in the tumor tissue in the early stage of treatment to achieve a significant therapeutic effect. Therefore, researchers prepare small-sized nanocomposite photosensitizers, which can circulate stably in the body for a long time. Based on the enhanced permeability and retention effect of tumor tissues or the modification of active targeting molecules or proteins, the photosensitizers could be efficiently accumulated into tumor tissue for PDT [30–33]. The photosensitizers could target the tumor passively or actively to achieve effective therapeutic concentration.

Here, we outline the basic principles of PDT and the cellular mechanisms activated during treatment. Additionally, we present the advantages of PDT in comparison with existing treatment options. Despite these advantages, several limitations still exist in the development of PDT, prompting researchers to propose innovative solutions. In this chapter, we present photosensitizers categorized by their unique characteristics, such as organic molecules, metal complexes, organic frameworks, carbon nanomaterials, silicon-based materials, simple substances, metal oxides, and lanthanide upconversion nanoparticles-based photosensitizers. Furthermore, we select notable recent studies to illustrate the design concept and logic, as well as key mechanisms for therapeutic applications. Guidelines and examples are presented to aid future research in the development of photosensitizers, as well as to discuss combination therapies and potential applications. The purpose of this section is to present recent developments that aid researchers in addressing specific limitations of current PDT, offer guidance and support, and compile and summarize prior research.

8.2 Small Molecule Photosensitizers

8.2.1 Porphyrins

Porphyrin photosensitizers are a family of macrocyclic compounds with 18 π-electron aromaticity, including chlorin e6 (Ce6), protoporphyrin IX disodium

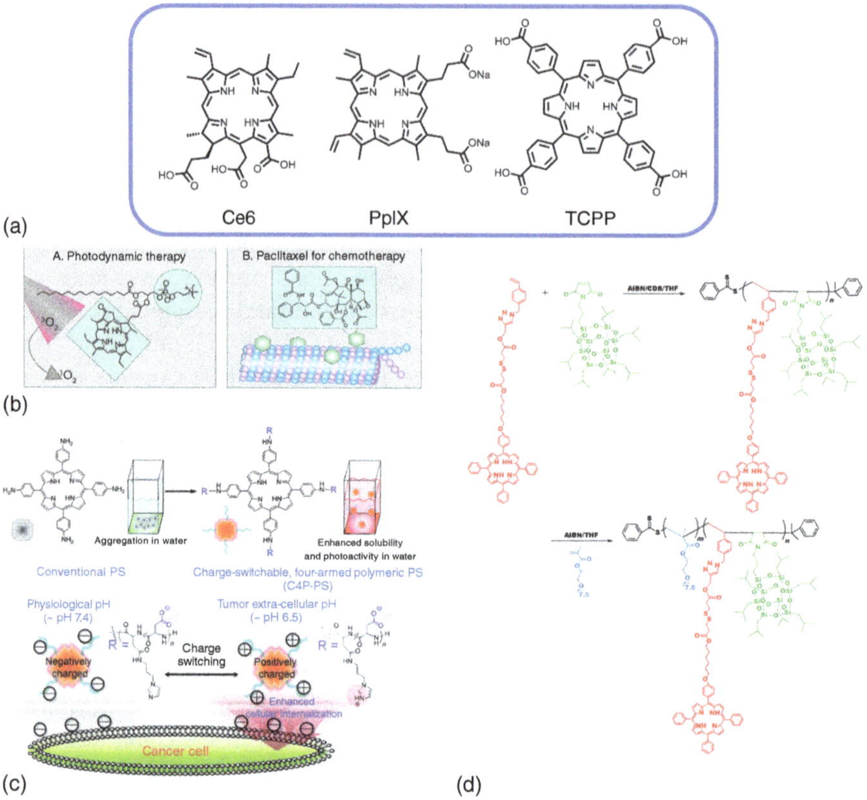

Figure 8.2 Porphyrin photosensitizers. (a) Chemical structures of conventional porphyrin photosensitizers. (b) Schematic diagram of PLNE-PTX. Source: Adapted from Chang et al. [34]. (c) Schematic diagram of charge-transferable C4P-PS. Source: Reproduced with permission from Lee et al. [35]/Royal Society of Chemistry. (d) Synthetic steps of P(MIPOSS-*alt*-VBTPP)-*b*-POEGMA. Source: Reproduced with permission from Jin et al. [36]/John Wiley & Sons.

salt (PpIX), and tetrakis (4-carboxyphenyl) porphyrin (TCPP) (Figure 8.2a). Porphyrin derivatives with strong $^1O_2$ generation efficiency and excellent fluorescence properties have unique advantages in PDT. Photofrin is a widely used commercial porphyrin derivative photosensitizer in the 1990s and is the first generation of photosensitizer for PDT [37]. Unfortunately, the excitation wavelengths of porphyrin derivative photosensitizers are in the UV (π-π^*, ~400 nm) or visible (four Q bands) regions, which are so shallow in tissue penetration depth.

Since porphyrins are lipid-soluble molecules, Zheng and Chen and coworkers used porphyrins and lipids to stabilize the oil/water interface and prepared porphyrin lipid nanoemulsions loaded with paclitaxel (PLNE-PTX) to achieve synergistic chemotherapy and PDT of tumors (Figure 8.2b) [34]. The modification of lipids could increase the water solubility of porphyrin, and paclitaxel (a chemotherapeutic drug) could be effectively loaded; thereby, the tumor can be effectively suppressed through synergistic treatment. The Ligand compound of Mn-porphyrin is a nanoplatform that enables integrated magnetic resonance imaging and PDT of tumors [38]. Metal-free porphyrin is covalently linked to Mn-porphyrin to avoid premature release during systemic cycling. The organic combination of the two components can increase the effective accumulation within the tumor. In order to achieve the water solubility and stability required for circulation in organisms, porphyrin molecules could be modified directly in addition to combining existing porphyrin molecules with other water-soluble molecules. In addition to responding only to light, the development of TME-responsive photosensitizers is critical. Na and coworkers performed ring-opening polymerization of α-amino acid *N*-carboxylic anhydride using *meso*-tetra(4-aminophenyl) porphyrin as an initiator (Figure 8.2c). And then charge-transferable photosensitizer (C4P-PS) was prepared by introducing a pH-responsive ionizable group (imidazole group) through the reaction [35, 39]. Compared with conventional photosensitizers, C4P-PS could change surface charge with pH due to the modification of the imidazole group, resulting in highly soluble and photoactivatable C4P-PS in water. C4P-PS is negatively charged under physiological conditions (pH 7.4). It turns positively charged in tumor tissues (pH 6.5) through protonation of the imidazole group. Zhang and coworkers designed an alternating copolymer photosensitizer (P(MIPOSS-*alt*-VBTPP)-*b*-POEGMA) via alternating reversible addition-fragmentation chain transfer polymerization of 4-vinylbenzyl-terminated tetraphenylporphyrin (VBTPP) with maleimide isobutyl POSS (Figure 8.2d) [36]. The alternating structure of P(MIPOSS-*alt*-VBTPP)-*b*-POEGMA based on three-dimensional POSS units reduces the aggregation between porphyrin units, which greatly increases the yield of ROS. Moreover, the nanoparticles are formed by amphiphilic block copolymers, which makes the porphyrin-based photosensitizer highly biocompatible and improves cellular uptake.

8.2.2 Phthalocyanines

Phthalocyanines (Pcs) are second-generation photosensitizers. Compared with porphyrins, phthalocyanines have defined molecular structures and excellent photochemical properties. Phthalocyanines have high extinction coefficients in the red-light wavelength region. Moreover, when phthalocyanine is coordinated with metal ions (Zn, Al, Si, etc.), the lifetime of T_1 energy level could increase, and sufficient $^1O_2$ could be generated [40]. However, phthalocyanines easily accumulate in water, which is a major obstacle to tumor treatment in vivo.

Zheng and coworkers prepared a water-soluble polyphthalocyanine (pPC-TK) nanosensitizer system through condensation with a thioketal-linked diamine (Figure 8.3a) [41]. Due to electrostatic and hydrophobic interactions, pPC-TK could self-assemble into nanoparticles in water. The thioketal linkers undergo cleavage when exposed to endogenous ROS in tumor tissues [44–47]. pPC-TK could efficiently generate ROS under the irradiation of 690 nm light or ultrasonic. Huang and coworkers reported a Zn(II)-phthalocyanine (ZnPc)-doxorubicin (DOX) prodrug with a fibroblast activation protein (FAP)-responsive short peptide (ZnPc)-TSGP-DOX nanoconjugate for chemo-PDT, as shown in Figure 8.3b [42].

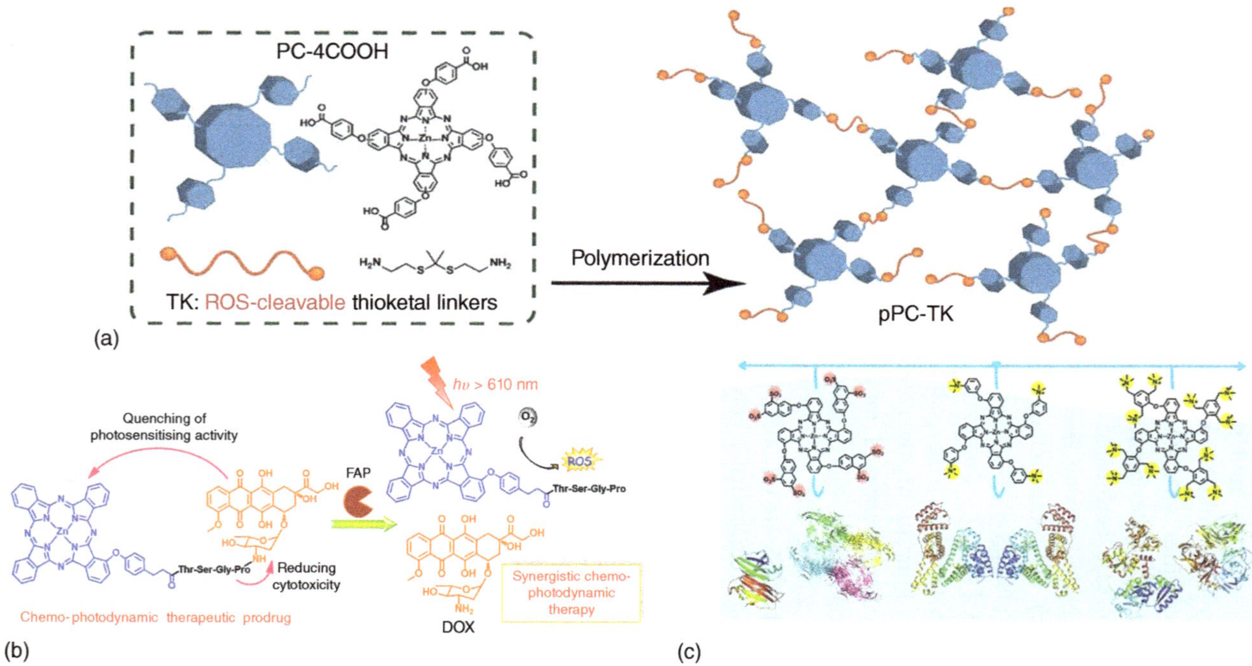

Figure 8.3 Phthalocyanine photosensitizers. (a) Synthetic schematic diagram of pPC-TK. Source: Reproduced with permission from Liu et al. [41]/John Wiley & Sons. (b) Synthetic schematic diagram of ZnPc-TSGP-DOX. Source: Reproduced with permission from Ke et al. [42]/Elsevier. (c) Synthetic schematic diagram of phthalocyanine-albumin derivatives with different charges. Source: Li et al. [43]/Ivyspring/CC-BY 4.0.

FAP is a transmembrane serine protease that is highly expressed on the cancer-associated fibroblasts of most human epithelial tumors, while it is rarely expressed in normal tissues [48–50]. Therefore, FAP has been recognized as a potential target for the diagnosis and treatment of a variety of tumors. Using a FAP-sensitive peptide covalently conjugated to DOX, ZnPc-TSGP-DOX nanoconjugate could effectively deliver chemotherapy and PDT that is responsive to active tumors. The modification of the targeting peptide plays a key role in targeting tumors, and the use of nonspecific albumin transport is also a good option. Yoon and coworkers designed and synthesized a series of phthalocyanine derivatives with various anionic or cationic groups (Figure 8.3c) [43]. Two positively charged phthalocyanines selectively bind to albumin dimers, whereas negatively charged phthalocyanines could bind to both albumin monomers and dimers. Through albumin transport, these phthalocyanine compounds, especially $ZnPcN_4$-albumin, were able to efficiently accumulate in tumor tissues and exhibit excellent PDT effect. Zhao et al. conjugated ZnPc to ganetesplib (a heat shock protein 90 inhibitor, Hsp90 inhibitor) [51–53], which could be actively targeted into tumor cells [54]. A large amount of ROS could be generated under 671 nm irradiation, and the Hsp90 inhibitor could hinder cell proliferation and induce apoptosis of tumor cells, resulting in highly selective PDT of tumors.

8.2.3 BODIPYs

Boron dipyrromethene (BODIPY) is another common type of PSs. The rationally designed BODIPY dyes have strong NIR absorption and narrow emission bands, which can realize imaging-guided PDT. However, BODIPYs are also strongly hydrophobic photosensitizers; therefore, BODIPYs need to be modified for further use in PDT. The primary approach is to utilize surfactants or proteins to facilitate the delivery of the photosensitizer complex to the tumor tissue.

Zhao and Liu et al. synthesized a series of aza-BODIPY photosensitizers substituted with halogens, followed by encapsulation and assembly using the hydrophilic polymer DSPE-mPEG5000 to obtain nanoparticles (IABNs) with excellent biocompatibility [55]. IABNs not only generate ROS under 730 nm irradiation but also have high photothermal conversion efficiency (34.8%), enabling synergistic photothermal-PDT for Hela tumors. Besides, since the efficiency of type II PDT is dependent on oxygen concentration but TME is hypoxic, She and Yang et al. synthesized the novel PS (PEG-*b*-PAsp-BODIPY) by coupling 2-pyridone and BODIPY and surface modification with polyethyleneglycol-*b*-poly(aspartic acid) [56]. The 2-pyridone group enables continuous chemodynamic therapy by reacting with $^1O_2$ to form endoperoxides under light irradiation and then continuously and reversibly releasing $^1O_2$ in the dark. Therefore, the 2-pyridone group plays an important role in PS modification [57–59]. In order to enhance the ROS-generating ability of BODIPY, Dong and coworkers coupled two PSs (diketopyrrolopyrrole (DPP) and BODIPY) by π-bridge of benzene [60]. This donor–acceptor–donor (D–A–D) structured organic photosensitizer (DPPBDPI) significantly improves fluorescence quantum yield (5.0%) and $^1O_2$ yield (80%). Due to its excellent fluorescence imaging capabilities, DPPBDPI enables fluorescence imaging diagnostic-guided PDT for Hela tumors.

8.2.4 Indocyanine Dyes

Indocyanine dye family contains some popular molecules with strong absorption in the NIR region, such as indocyanine green (ICG), IR-780, and IR-825. ICG, is a widely used medical contrast agent for NIR fluorescence imaging with good biosafety [61], which has been approved by U.S. Food and Drug Administration (FDA) [61–63]. In addition, ICG can generate ROS and heat under the irradiation of NIR laser, enabling synergistic treatment of tumors with PDT and photothermal therapy (PTT) [64]. Nevertheless, ICG is often prone to agglomeration and blood clearance and needs to be compounded with other carriers to enhance the PDT effect. ICG is negatively charged and, therefore, can be easily modified through electrostatic interaction to fabricate micelles or loaded onto other inorganic nanomaterials [65–67].

Xue et al. encapsulated and loaded ICG with positively charged polyethyleneimine (PEI) using Prussian blue (PB) as a carrier on the surface of Fe_3O_4 (Fe_3O_4@PB/PEI/ICG) [68]. Because of the modified Fe_3O_4, the nanocomposites can be magnetically targeted to tumors. ICG serves as PS loading on the nanocomposite by electrostatic interaction, while PB acts as a photothermal reagent in the NIR region, realizing photothermal–photodynamic synergistic treatment of tumor. Fe_3O_4@PB/PEI/ICG can effectively ablate tumors by raising the temperature at the tumor issue by ~14 °C and generating a large amount of ROS under 808 nm laser irradiation. In addition to the modification with polymer, to improve the efficacy of ICG, Gu and Huang and coworkers used SF_6 nanobubbles (NBs) to stabilize ICG molecules in an aqueous solution [69]. Compared with free ICG, the yield of $^1O_2$ is significantly increased to 1.6% due to the combination of self-assembled ICG-NBs-O_2 with dissolved oxygen. Under NIR laser irradiation, the yield of $^1O_2$ yield of ICG-NBs-O_2 was increased by eight times compared with free ICG. ICG-NBs-O_2 has a significant ablation therapeutic effect on Cal27 tumors.

8.2.5 AIEgens

Most organic PSs show aggregation-caused quenching (ACQ) in aqueous solutions, resulting in a reduction in their ROS generation effect. Aggregation-induced emission (AIE) technology has been recognized by the International Union of Pure and Applied Chemistry as one of the Top Ten Emerging Technologies in Chemistry for the year 2020 [70]. Because fluorescent conjugated aromatic groups tend to aggregate in water, π-π stacking effect leads to fluorescence quenching, called ACQ, which prevents fluorescence detection probes from being widely used in biological applications based on aqueous environments. However, in 2001, AIE was first discovered and defined by Tang's group, that is, fluorescent groups aggregate at high concentrations, leading to the enhanced fluorescence signal [71]. There are two mechanisms of AIE luminogens (AIEgens) identified by Tang's group [72, 73]. (i) Restriction on intramolecular rotation (RIR). Tetraphenylethene (TPE) does not emit light in dilute solution at high) concentrations. Because of the phenyl RIR, the energy of incident light cannot be transferred to the environment through backbone movement, and the energy can only be released through fluorescence. (ii) Restriction on intramolecular vibration. Since the nonradiative energy consumption pathway of molecular vibrations is limited, 10,10′,11,11′-tetrahydro-5,5′-bidibenzo[*a*,*d*][7]annulenylidene (THBA) has the property of AIE at high concentration.

Since the TME is often hypoxic, Tang et al. used dexamethasone (DEX) to ameliorate vasculopathy and extracellular matrix in tumor tissues to improve PDT effect [74–76]. The co-preparation of AIEgen PS ((*E*)-4-(2-(7-(diphenylamino)-9-ethyl-9*H*-carbazol-2-yl)vinyl)-1-methylpyridin-1-ium hexafluorophosphate, DCPy) with tumor exosomes (EXO) and DEX into nanovesicles enables enhanced penetration and efficient PDT in tumor tissue [76]. This biomimetic nanovesicle could generate ROS under 532 nm laser excitation in a constant stream and could better inhibit 4T1 tumor growth with the help of DEX. In addition to single AIEgen molecular PS, Song and coworkers synthesized a reticulated AIEgen polymer PS via a ring-opening polymerization reaction [77]. The PS exhibits excellent fluorescence in the >650 nm region and $^1O_2$ generation ability due to the severe restriction of intramolecular vibrations upon polymerization. Besides, the PS was further modified with quaternary ammonium salt groups to improve the water solubility and biocompatibility. This cationic AIEgen polymer photosensitizer can effectively ablate 4T1 tumors under 520 nm laser excitation. To enhance the ROS yield of AIEgen photosensitizers, He and Cheng and coworkers designed the donor–acceptor (D–A) strength of AIEgen photosensitizer with a molecular engineering perspective and successfully used thiophene and quinoline to modulate the strength of the intramolecular charge transfer (ICT) effect, reducing the energy gap and promoting the intersystem crossing from S_1 to T_1 [78]. Molecular engineering provides an approach to design molecules from the nature of electronic behavior through the luminescence mechanism of organic molecules. Then, they introduced a positively charged pyridinium derivative to target mitochondria for PDT. The PS could generate large amounts of $^1O_2$ under white light excitation and kill MCF-7 and MDA-MB-231 cells.

8.3 Metal Complexes

8.3.1 Ru(II) Complexes

Most of the existing PSs require the excitation of tissue-penetrating shallow visible light; therefore, there is an urgent need to improve this phenomenon through molecular engineering. Due to the coordination of heavy atoms, the PDT effect could be enhanced by ISC process in the system [79, 80]. ISC could be enhanced by the introduction of Ru(II), and the ring structure has a large molecular volume, effectively reducing the $S_1 - T_1$ energy gap. It enables the excitation light wavelength to be shifted toward the NIR region; thus, it can penetrate deeper into the tissue.

Sun and Kim et al. constructed a series of supramolecular PSs of ruthenium complexes (RuA–RuD, Figure 8.4a) through molecular engineering [81]. Among them, RuD has absorption around 850 nm and can emit fluorescence at 1115 nm, which has the potential for PDT to tumors deep in the tissue. The cytotoxicity of traditional Ru complexes hinders the antitumor applications of PSs in vivo [84, 85]. However, in the absence of laser irradiation, the dark toxicity of RuD is negligible, while RuD can effectively ablate A549 tumors under 808 nm laser irradiation. Reinforced PDT can be achieved by selecting novel ligands. Gao and coworkers utilized a newly developed bridging ligand (pppp) to connect a binuclear Ru(II) complex featuring polypyridyl groups (Ru1–8), as shown in Figure 8.4b [82]. This strategy effectively lowered the excitation energy level of PS, owing to the strong electron-donating properties of pppp. Additionally, with the tert-butyl substitution, Ru3 can specifically target mitochondria and does not have any adverse effect on the genetic material in the nucleus, thus decreasing the dark cytotoxicity of the complex. Using an 808 nm laser at low power density ($0.1\ W/cm^2$), Ru3 could efficiently inhibit A375 tumors based on the PDT effect. Moreover, the aggregation state of Ru complexes in aqueous solution also affects PDT at the same time. Gou and coworkers self-assembled a novel supramolecular PS using arene-Ru(II) complex with modification of Pluronic F127 (RuDA, as shown in Figure 8.4c) [83]. Due to the octahedral-like spatial site resistance of Ru(II)–arene complexes, H-aggregation in water can be effectively prevented. This ultimately results in J-aggregates with longer absorption wavelength. Hence, it is crucial to encapsulate Ru complex PS with a surfactant [86]. Besides, triphenylamine (TPA) coupled with [1,2,5]thiadiazolo-[3,4-*i*]dipyrido[*a*,*c*]phenazine (TDP) in the D–A structure is also a widely used PTT agent [87–90]. RuDA enables photothermal and photodynamic synergistic treatment of MDA-MB-231 tumors under 808 nm laser irradiation.

8.3.2 Ir(III) Complexes

Iridium(III) complexes, another major branch of transition metal complex PSs, have also been investigated [91–93]. Due to the hypoxic tropism and motility of some anaerobic bacteria, they are able to target and penetrate microenvironmentally hypoxic tumors [94, 95]. And with chemical modifications, various types of bacteria can be used in tumor therapy [96–99]. Xia and coworkers synthesized an Ir(III) complex linked to chloroalkane to form a PS IrCl, which was then modified onto the surface of *Escherichia coli* Nissle 1917 to prepare an engineering bacterial PS (Ir-HEcN) [100]. Ir-HEcN enables both type I and type II PDT. It could achieve PDT for 4T1 tumors with multiple ROS ($^1O_2$, $O_2^{\bullet-}$, and •OH) under 520 nm irradiation and stimulate antitumor immune responses, effectively preventing tumor growth and recurrence. Thanks to the electron-donating ability and π-conjugation nature of the methoxy-styryl units of BODIPY, Sun and coworkers coordinated Ir(III) with BODIPY to form a neutral micellar photosensitizer (BODIPY-Ir) [101]. Due to the electron-donating ability of BODIPY, BODIPY-Ir has strong absorption in the near- and far-infrared region. BODIPY-Ir has a relatively long lifetime in the T_1 energy level and could undergo nonradiative decay; therefore, it could generate ROS and

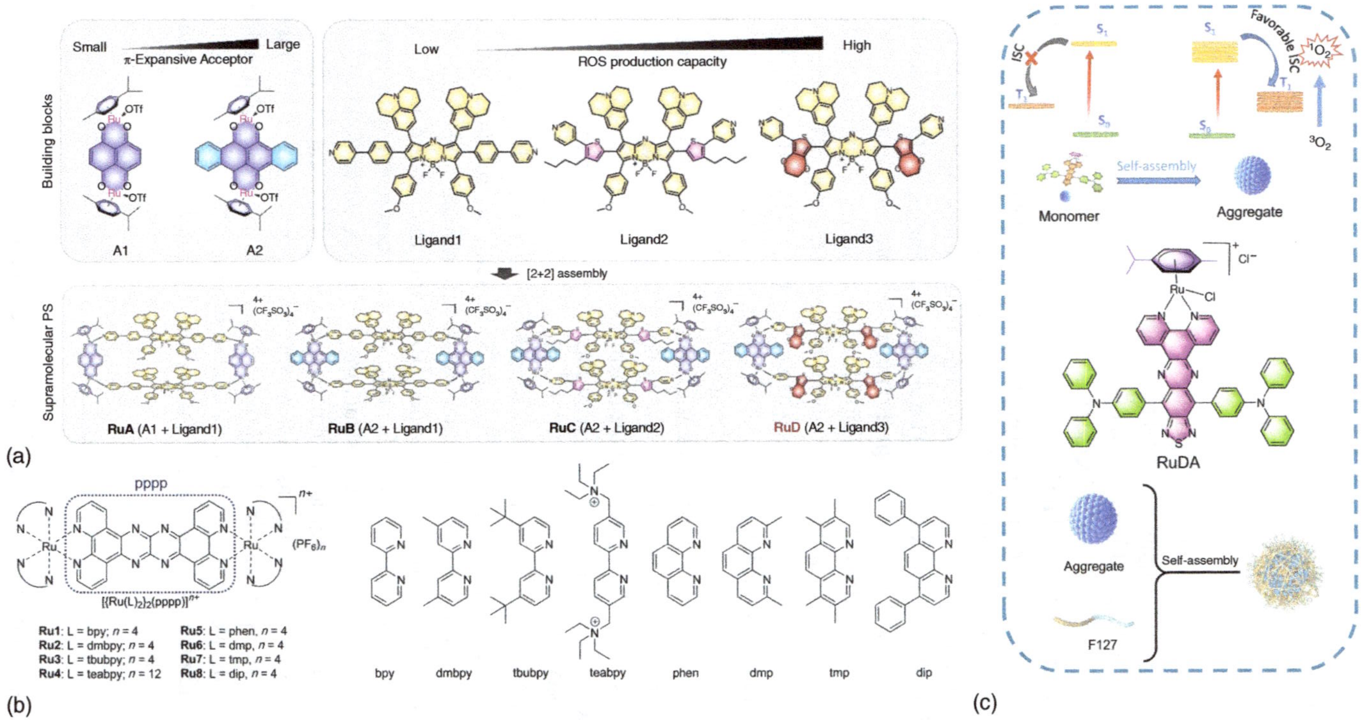

Figure 8.4 (a) Illustration of metallacycle-based supramolecular PSs with two Ru acceptors A1/A2 and fluorescent ligands 1/2/3. Source: Reproduced from Tu et al. [81]/John Wiley & Sons. (b) Molecular structure and designs of dinuclear Ru complexes with bridging ligands. Source: Reproduced from Wang et al. [82]/John Wiley & Sons. (c) Chemical structure, synthesis, and schematic illustration of RuDA in monomer and aggregate forms for PDT. Source: Xu et al. [83]/Springer Nature/CC-BY 4.0.

photothermal effects under 660 nm irradiation. BODIPY-Ir was encapsulated into micelles (Micelle-Ir) by using functional poly(caprolactone)-modified polyethylene glycol (PEG-PCL). Both BODIPY-Ir and Micelle-Ir have good anti-4T1 tumor effects in vitro and in vivo.

8.3.3 MOFs

Metal-organic frameworks (MOFs) are primarily created by joining metal nodes (or metal clusters) and organic connectors (usually carboxyl-containing or nitrogen-containing) via ligand bonding to create one-, two-, or three-dimensional networks. A vast array of MOF with diverse structures and attributes can be created by combining different metals and linkers. To create intrinsic photodynamic MOFs for the application of PDT, photo-responsive groups are used as the organic linkers [102]. Under laser irradiation, these MOFs can directly actualize PDT.

Porphyrins and their derivatives such as carboxyl derivatives are the most used organic ligands for photodynamic MOFs. The channels and holes in MOFs facilitate easy diffusion of $^3O_2$ and $^1O_2$. The appropriate metal ion coordination such as Hf^{4+} could enhance $^1O_2$ production capacity due to the heavy atom effect. Lin's group published the first research on porphyrin-based MOFs for PDT. They built DBP-UiO with 77 wt% loading of 5,15-di(*p*-benzoato) porphyrin (H_2DBP) ligand and Hf^{4+} nodes [103]. High Z-ionic Hf^{4+} coordination encourages the PDT system to interlineally cross, including the red shifting of the absorption peaks of DBP-UiO and the production of $^1O_2$ at least twice that of H_2DBP. Thanks to the unique porous structure of MOFs, the generated $^1O_2$ could diffuse with ease and thus kill cancer cells. Tetrakis(4-carboxyphenyl)-porphyrin (TCPP) ligands and Zr clusters, primarily octahedral Zr_6 clusters, construct the porous coordination network (PCN) (Figure 8.5a) [104]. This type of MOF is quickly rising to the top of the photodynamic MOF market because of its superior biodegradability, stability in aqueous solution, and the ability to generate ROS. In addition, the ROS creation of photosensitizer ligands can be quenched by the coordination of TCPP with other metal ions, such as iron [106], copper [107], and manganese ions [108]. ROS generation of these MOFs is significantly reduced compared with that of free photosensitizer ligands. The MOF structure will break down following the interaction, releasing active metal ions, photosensitizer ligands, and medications to achieve chemodynamic therapy as well as imaging effects since these metal ions can react to various stimuli in the TME. Fe^{3+}-porphyrin MOFs stabilized by albumin/sulfonamide were produced by Xu and colleagues. These MOFs can be served as multifunctional photosensitizer for magnetic resonance imaging-guided synergistic PDT and PTT [109].

8.3.4 COFs

Some distinctive and noticeable characteristics of covalent organic frameworks (COFs) are their high surface area, superior crystallinity, uniform pores with adjustable sizes and shapes, and variety of building components [110]. Using

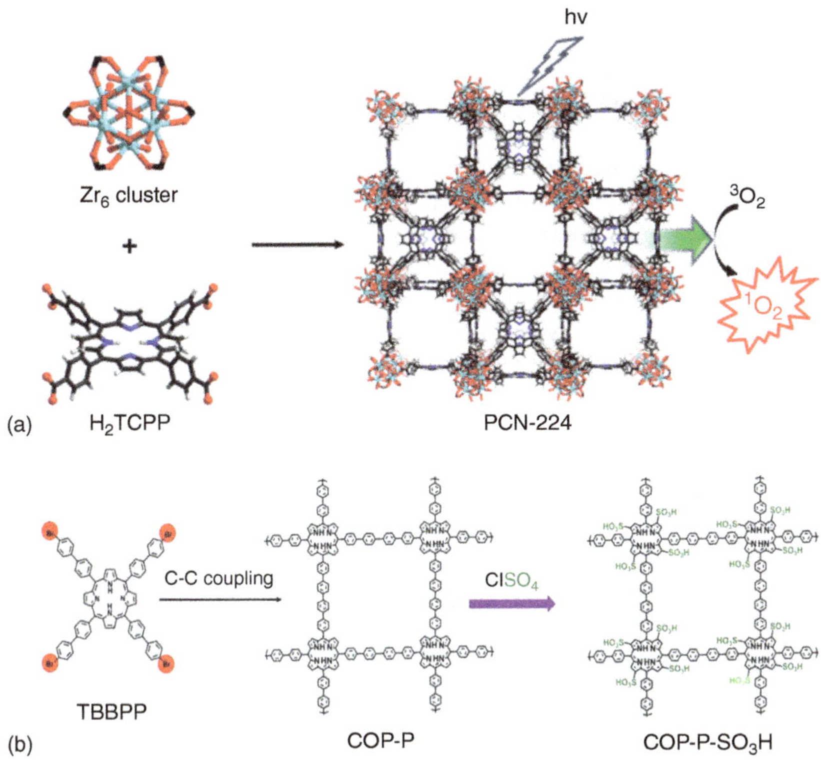

Figure 8.5 (a) Illustration of PCN-224 structure. 6-connected Zr6 cluster ($Zr_6O_4(OH)_4(H_2O)_6(OH)_6(COO)_6$), tetratopic linker (tetrakis (4-carboxyphenyl) porphyrin (H_2TCPP)), and 3D nanoporous framework of PCN-224. Source: Reproduced with permission from Park et al. [104]/American Chemical Society. (b) Synthetic schematic diagram of sulfonated porphyrin-based covalent organic polymers. Source: Reproduced from Xiang et al. [105]/ American Chemical Society.

porphyrin derivatives as building blocks to construct porphyrin-based COFs for PDT, it could avoid the aggregation of the porphyrin macrocycles in an aqueous solution and increase the therapeutic efficiency. Chen and coworkers synthesized the fully conjugated 2D COPs (COP-P-SO_3H) by covalently bonding porphyrin macrocycles together, as shown in Figure 8.5b [105]. Through protonic doping and ICT, the sulfonic acid groups in the porphyrin not only increase the water dispersibility of COP-P-SO_3H but also dramatically lower its bandgap. The optical adsorption of COP-P-SO_3H enhances and redshifts because of its lowering bandgap. When exposed to NIR light, COP-P-SO_3H has a greater yield of $^1O_2$ than clinically approved photosensitizer (PpIX). Qu and colleagues also reported similar work, wherein ultrasmall porphyrin-based COF nanodots were synthesized using a straightforward liquid exfoliation technique [111]. PEG-decorated COF nanodots, also known as COF nanodots-PEG, exhibit excellent PDT efficiency both in vitro and in vivo. The well-isolated porphyrin molecules enhanced the formation of light-triggered ROS.

8.3.5 HOFs

Compared with MOF and COF, the use of hydron-bonded organic framework (HOF) itself for PDT is rarely reported due to the incredibly weak H-bonds that are required to build and maintain the HOF structure. Because of their numerous H-bonds and metal-free nature, HOFs have unique advantages such as gentle synthetic conditions, solution processability, facile regeneration, and excellent biocompatibility. Liu and colleagues developed an ultra-robust HOF (PFC-1) by using 1,3,6,8-tetrakis (*p*-benzoic acid) pyrene as the structural unit. The chemical and nanostructural stabilities of PFC-1 were improved by the numerous H-bonds and strong π–π interactions between the two pyrene molecules [112]. Additionally, PFC-1 has outstanding drug carrier activity, with a high DOX loading efficiency of 26.5 wt%. Meanwhile, PFC-1 is an intriguing PS for PDT due to the periodic incorporation of pyrene. The DOX-loaded PFC-1 demonstrated synergistic computed tomography (CT) and PDT with significant biocompatibility and excellent therapeutic efficacy in vitro.

8.4 Inorganic Photosensitizers

8.4.1 Carbon-Based Photosensitizers

Carbon-based PSs include carbon dots, fullerene, graphene, and carbon nitrides. These carbon-based PSs possess favorable attributes such as chemical inertness, easy functionalization, high photostability, and good biocompatibility. Based on these merits, researchers have developed various carbon-based PSs for tumor treatment.

Zhang's group developed a C_3N_4 nano-photosensitizer (PCCN) to solve the problem of insufficient ROS production in PDT resulting from hypoxic TME in PDT (Figure 8.6a) [113]. PCCN enables the generation of O_2 and reacts to generate large amounts of ROS in response to 630 nm laser irradiation. It is commonly acknowledged that TME has a partial pressure of oxygen below 2.5 mmHg, significantly constraining the efficacy of PDT [115]. C_3N_4 is known for its excellent biosafety due to the absence of heavy metal elements. Besides, C_3N_4 can catalyze water cracking to produce O_2 under light [116, 117]. Moreover, carbon dots are modified to increase the absorption of C_3N_4 in the red-light region. Due to the modification of organic photosensitizer (PpIX) and tumor-targeting peptide, PCCN enables antitumor PDT under 630 nm laser irradiation. In order to significantly enhance the effect of PDT, the location of the photosensitizer is critical [118]. Therefore, Du's group developed Se- and N-doped carbon dots (Se/N-CDs) capable of localizing the nuclei of tumor cells [119]. Positively charged N-doped carbon dots could effectively localize with the negatively charged nuclear membrane of the nucleus via electrostatic interaction. Se-doped carbon dots with rigid structures could also effectively enhance the ISC process, increasing the PDT effect. Then, Se/N-CDs could bind to RNA within tumor cells and transport it near the nucleus, generating large amounts of ROS under 550 nm laser irradiation, thereby killing tumor cells via

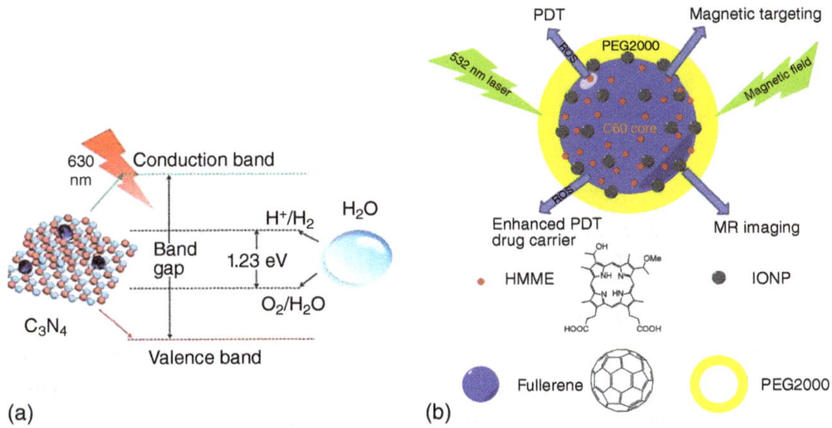

Figure 8.6 (a) Schematic illustration of the C_3N_4-mediated water splitting process. Source: Reproduced with permission from Zheng et al. [113]/American Chemical Society. (b) Scheme of C60-IONP-PEG/HMME and its biofunctions. Source: Reproduced with permission from Shi et al. [114]/Elsevier.

PDT. Other carbon-based nanoparticles, in addition to carbon nitrides and carbon dots, are commonly utilized as biosafe carriers. Zhang's group modified iron oxide nanoparticles (IONPs) onto the fullerene (C_{60}) surface and further used PEG to improve the water solubility and biocompatibility (Figure 8.6b) [114]. Therefore, a drug delivery system (C60-IONP-PEG) with magnetic targeting and MRI properties was successfully prepared. Then, the drug delivery system was conjugated with the photodynamic anticancer drug (hematoporphyrin monomethyl ether, abbreviated as HMME). C60-IONP-PEG/HMME could achieve effective targeting and PDT for B16-F10 tumors under 532 nm laser irradiation.

8.4.2 Silicon-Based Photosensitizers

Because of the undesirable water solubility and biocompatibility of most organic photosensitizers, tumor cells are prone to fail to take up photosensitizers directly, and PDT cannot be performed [120, 121]. Porous silica materials, especially mesoporous silica, are widely used as extremely biocompatible drug delivery systems due to the adjustable pore volume, surface modifiability, and ability to be loaded with various drugs or photosensitizers.

Inspired by fluorescein isothiocyanate (FITC)-labeled mesoporous silica nanoparticles (MSNs) [122, 123], Chen and co-workers prepared Ce6-loaded mesoporous silica nanorods (CMSRs) by coupling the carboxyl group of Ce6 with 3-aminopropyltriethoxysilane (APTES) [124]. With the loading strategy, the problem of leakage of photosensitizers or other drugs during in vivo circulation caused by ordinary loading onto MSNs could be effectively prevented [125]. And then, the chemotherapeutic drug DOX was loaded onto CMSRs. With the modification of lactobionic acid, CMSR-DOX could be actively targeted to HepG2 tumor for chemo-photodynamic synergistic therapy under 660 nm laser irradiation.

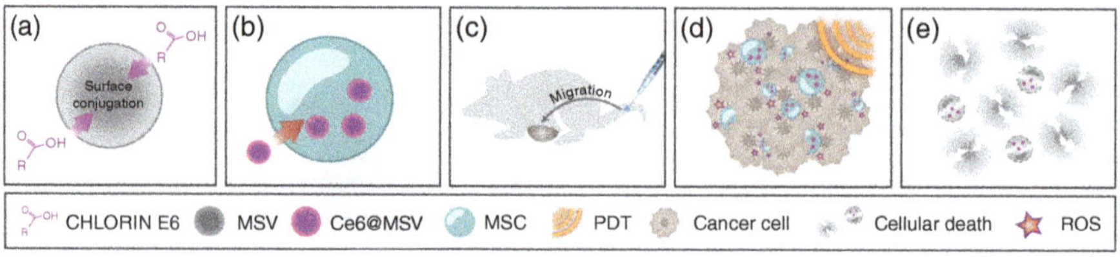

Figure 8.7 Schematic illustration of PDT with a Trojan-horse-like approach. Source: Reproduced with permission from Nakki et al. [126]/American Chemical Society.

In addition to modifying lactobionic acid to achieve active targeting, Martinez and co-workers chose another strategy in which they conjugated Ce6 to the multistage nanovectors (MSV) of mesoporous silica (Figure 8.7) [126]. Since mesenchymal stem cells (MSC) can actively migrate toward the inflamed tissue at the tumor site [127], MSC were used to deliver Ce6@MSV in a Trojan horse-like strategy. In vivo tropism and targeting kinetics of MSC+Ce6@MSV were confirmed in orthotopic 4T1 tumor-bearing mice. MSC + Ce6@MSV could perform effective PDT of tumor under 405-nm laser irradiation.

8.4.3 Simple Substance Photosensitizers

Inorganic simple substance photosensitizers (including black phosphorus and some noble metal nanoparticles conjugated with photosensitizers) have excellent optical properties and can be excited by light to generate $^1O_2$. In particular, their surfaces could be conjugated and modified with other organic functional molecules, and they have attracted much interest in PDT. In addition, many simple substance photosensitizers have photothermal properties, enabling synergistic photothermal and photodynamic antitumor therapy.

It is well known that type II PDT process is restrained in the hypoxic TME because its efficiency depends on oxygen concentration. Cyanobacteria are single-celled prokaryotes with a long history of producing oxygen through photosynthesis, similar to green plants. Certain species of cyanobacteria, for example, *Synechococcus elongatus* PCC7942, exhibit high biocompatibility and are utilized in various biomedical applications [128, 129]. Huo and Shi et al. modified photosynthesizing cyanobacteria onto two-dimensional black phosphorus nanosheets (BPNSs) to improve the therapeutic efficacy of PDT (Figure 8.8a) [130]. BPNS could independently generate $^1O_2$ under 660 nm laser excitation. BPNSs hybridized with cyanobacteria (Cyan@BPNSs) could address the hypoxia in TME and produce more ROS for PDT, thus inhibiting 4T1 tumor growth. Besides, black phosphorus (BP) also has a photothermal conversion effect. Liu's group modified BP with PEGylated hyaluronic acid (HA) to enable photothermal and photodynamic synergistic therapy (Figure 8.8b) [131]. HA modification could not only enhance the cellular uptake of HA-BP but also repolarize tumor-associated macrophages from M2 to M1 phenotype, which significantly improves tumor immunotherapy [132, 133]. They observed the downregulation of CD206 (a marker for M2 macrophages) and the upregulation of CD86 (a marker for M1 macrophages) upon treatment with HA-BP.

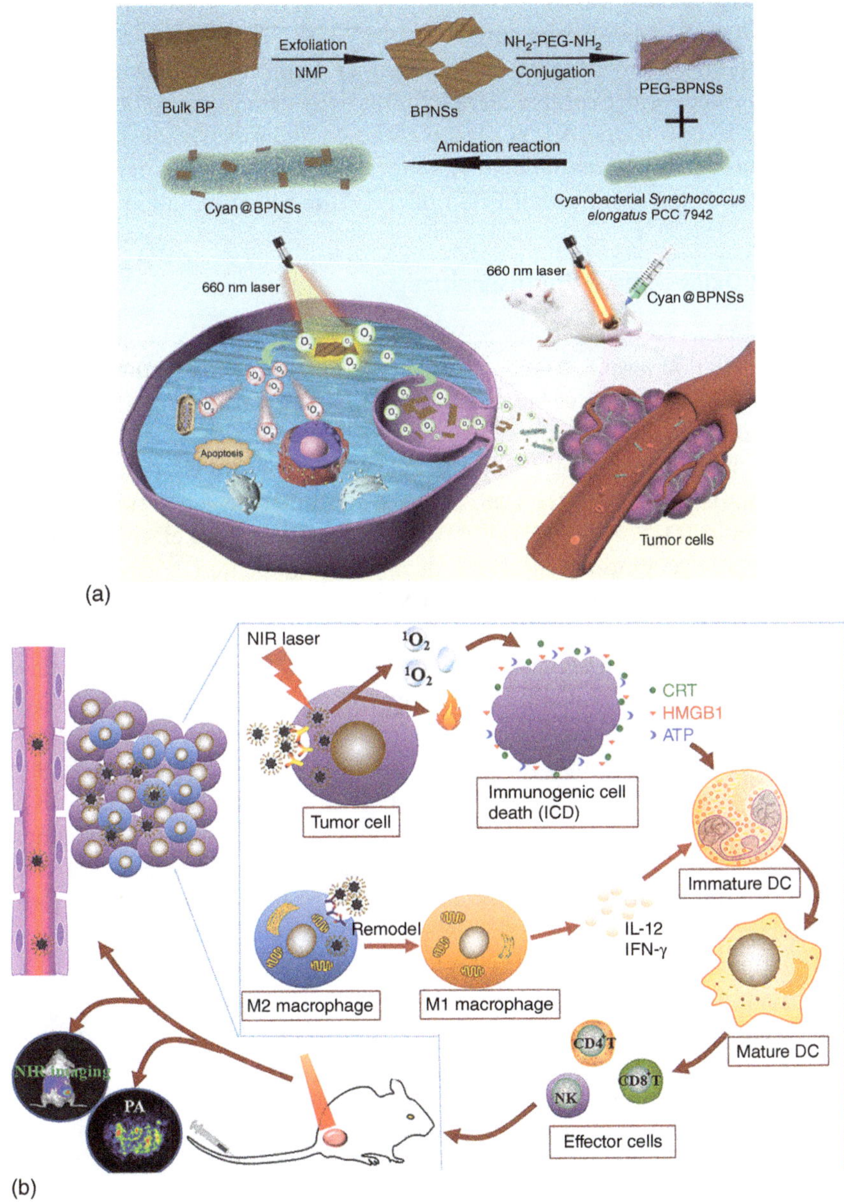

Figure 8.8 (a) Fabrication and photosynthesis-promoted PDT schematics of Cyan@BPNSs. Source: Reproduced with permission from Qi et al. [130]/John Wiley & Sons. (b) The synthetic scheme of HA-BP nanoparticles and the function of HA-BP nanoparticles in vivo. Source: Zhang et al. [131]/Elsevier/CC-BY 4.0.

It implies that HA-BP could target tumor tissues and facilitate the initiation of ICD phototherapy, resulting in a robust immune response that hinders tumor growth and avoids tumor recurrence. In addition to BP, several noble metal nanoparticles conjugated with photosensitizers could also facilitate PDT. Wang and coworkers conjugated pyropheophorbide photosensitizer with mitochondria-targeting aptamers to produce water-soluble targeted photosensitizer (ApPCs) [134]. DNA- and RNA-gold nanoclusters (Au NCs) assemblies could be biosynthesized in situ in the TME using ApPCs as templates [135]. Au NCs-ApPCs can target tumor mitochondria and enable antitumor PDT under 660 nm laser irradiation.

8.4.4 Metal Oxides-Based Photosensitizers

Metal oxides such as TiO_2, Bi_2O_3, and $SnWO_4$, with excellent ROS generation ability under light irradiation have been reported. Triggered by the energy of the high photons, the electrons in the valence band (VB) of the metal oxide are facilitated to cross the bandgap and reach the conduction band (CB). In this process, electron–hole pairs can be generated to promote the formation of charge-separated states and facilitate electron transfer reactions with substrates or molecular oxygen, resulting in the production of ROS ($O_2^{\bullet-}$ and •OH) [136, 137]. Depending on the electronic structure, the ROS photogeneration capacity of metal oxide nanomaterial depends on the electron–hole separation ability, exhibiting type I ROS-producing behavior, which is promising for tumor treatment.

As one of the most classic semiconductor metal oxides, TiO_2 nanomaterials have the advantages of high photostability, good photocatalytic activity, low cost, and relatively low toxicity [138]. The bandgap width of TiO_2 is between 3.00 and 3.20 eV, which limits its light absorption in the UV region. Photons with energies higher than the bandgap width could excite the electronic transition of TiO_2 from VB to CB, resulting in charge separation and participating in the surface redox reaction with the substrate to form various ROS. Due to the high oxidation potential of TiO_2 (3.00 V), H_2O and OH^- adsorbed on the surface of TiO_2 are easily oxidized to form •OH. At the same time, electron acceptors, such as O_2 and H^+, can directly receive an electron from the photosensitized TiO_2 to produce $O_2^{\bullet-}$ and H_2O_2 [139].

Yang's group demonstrated that an easy-to-prepare TiO_2@red phosphorus (RP) nanorods with core-shell structure could effectively achieve a synergistic effect on PDT and PTT of clear cell renal cell carcinoma (Figure 8.9A) [140], in which the photogenerated electrons generated by the excited RP shell can be rapidly transferred to TiO_2 and generate ROS for achieving PDT effect. After five minutes of 808 nm laser irradiation, most of the renal tubular epithelial cells attached to nanorods were killed by local heat and ROS produced by TiO_2@RP nanorods, while normal renal cells still maintained high activity. As shown in Figure 8.9B, Qiu et al. demonstrated that Bi/BiO_X-based lateral nano-heterostructures prepared by hydrothermal regioselective oxidation methods could produce robust ROS upon irradiation in an O_2-independent environment [141]. Bi/BiO_X can have both types of PDT at the same time. Bi in the core is excited by light and transfers energy to O_2 in the ground state to generate $^1O_2$, thus, realizing type II PDT. Thanks to a suitable

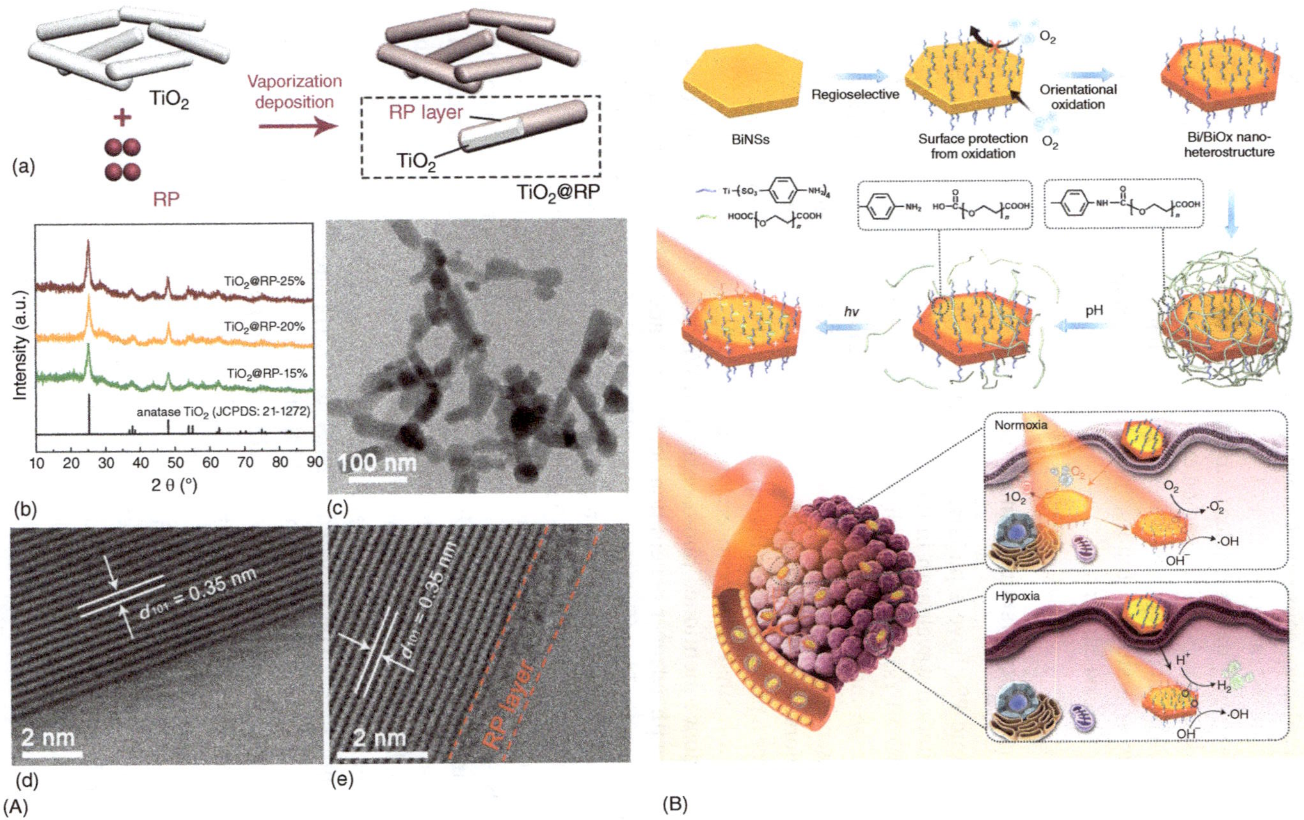

Figure 8.9 (A) Schematic illustration, XRD patterns, and TEM images of the synthesis of TiO_2@RP heterostructure. Source: Reproduced with permission from Yang et al. [140]/John Wiley & Sons. (B) Lateral nano-heterostructure synthesis and PDT treatment schematic. Source: Reproduced with permission from Qiu et al. [141]/John Wiley & Sons.

band arrangement in Bi/BiO_X lateral nano-heterojunction, type I PDT could also be activated through the electron transfer process under light irradiation. Whether under normoxic or hypoxic conditions, the results show that at the anode VB of BiO_X, the OH^- in the aqueous solution are oxidized to •OH, while at the cathode Fermi level, O_2 and H^+ are reduced to $O_2^{\bullet-}$ and H_2, respectively. Carmen Seidl et al. prepared β-$SnWO_4$ nanoparticles as an inorganic PS for PDT [142]. Since the bandgap energy is 2.7 eV, β-$SnWO_4$ can be excited by blue light at a wavelength less than 550 nm. β-$SnWO_4$ could be an interesting addition to PDT and a weapon in the treatment of proximal surface tumors, including skin, esophageal, gastric, colon, and some types of breast cancer.

8.4.5 Lanthanide Upconversion Nanoparticles-Based PSs

In general, organic PSs have a photosensitive active conjugated structure, which makes their excitation wavelength in the UV-vis region (e.g., porphyrins, phthalocyanines, and BODIPYs). It is difficult to adjust the excitation wavelength of the organic PSs through molecular engineering without altering their active structure. Moreover, the tissue penetration depth of light in the UV-vis region is shallow. For deep-seated tumors, conventional organic photosensitizers cannot be sufficiently excited to produce a photodynamic therapeutic effect. In contrast, a more ingenious alternative is that lanthanide upconversion nanoparticles (UCNPs) could be combined with the above-mentioned photosensitizers. When irradiated by a NIR laser (808 or 980 nm) with a deeper penetration depth, Nd^{3+} and Yb^{3+} could transfer the energy to luminescent ions (Er^{3+} or Tm^{3+}) in UCNPs to emit stable upconversion luminescence (UCL) in the UV-visible region [143]. Finally, the energy of UCL could be transferred to the photosensitizers to generate ROS for antitumor PDT.

As biosafe and biocompatible nanoparticles, UCNPs can be complexed with various types of photosensitizers to form novel nano-photosensitive material. First, organic molecular photosensitizers can be coupled with UCNPs. As shown in Figure 8.10a, Zhang's group developed a composite photosensitizer based on UCNPs modified with Ce6 and AIEgen (UCNP-Ce6/AIEgen) [144]. Since Nd^{3+} can directly transfer electrons to the T_1 level of the photosensitizer (Ce6) after being excited at 808 nm, Ce6 can be activated to generate $^1O_2$ via direct lanthanide-triplet energy transfer (DTET) [147]. AIEgen could then adjust the emission wavelength of UCL to 650 nm, thereby effectively activating Ce6 to generate ROS via Förster resonance energy transfer (FRET). Therefore, UCNP-Ce6/AIEgen could effectively treat 4T1 tumor by generating ROS through dual-pathway energy transfer. Liu and co-workers developed a multifunctional DNA/UCNPs complex, enabling controlled delivery of CRISPR-Cas9, hemin, and protoporphyrin (PP) for synergistic PDT (Figure 8.10b) [145]. The G-quadruplex/hemin DNAzyme could catalyze the endogenous H_2O_2 to O_2, which could relieve hypoxia in tumors. Besides, Cas9 ribonucleoprotein could improve the sensitivity of tumor cells to ROS and enhance the synergistic PDT effect. Then, under the irradiation of NIR laser, UCNPs could emit UCL and excite PP to generate $^1O_2$. The integration of gene editing and PDT can effectively inhibit MCF-7 xenograft tumor growth. UCNPs can also be

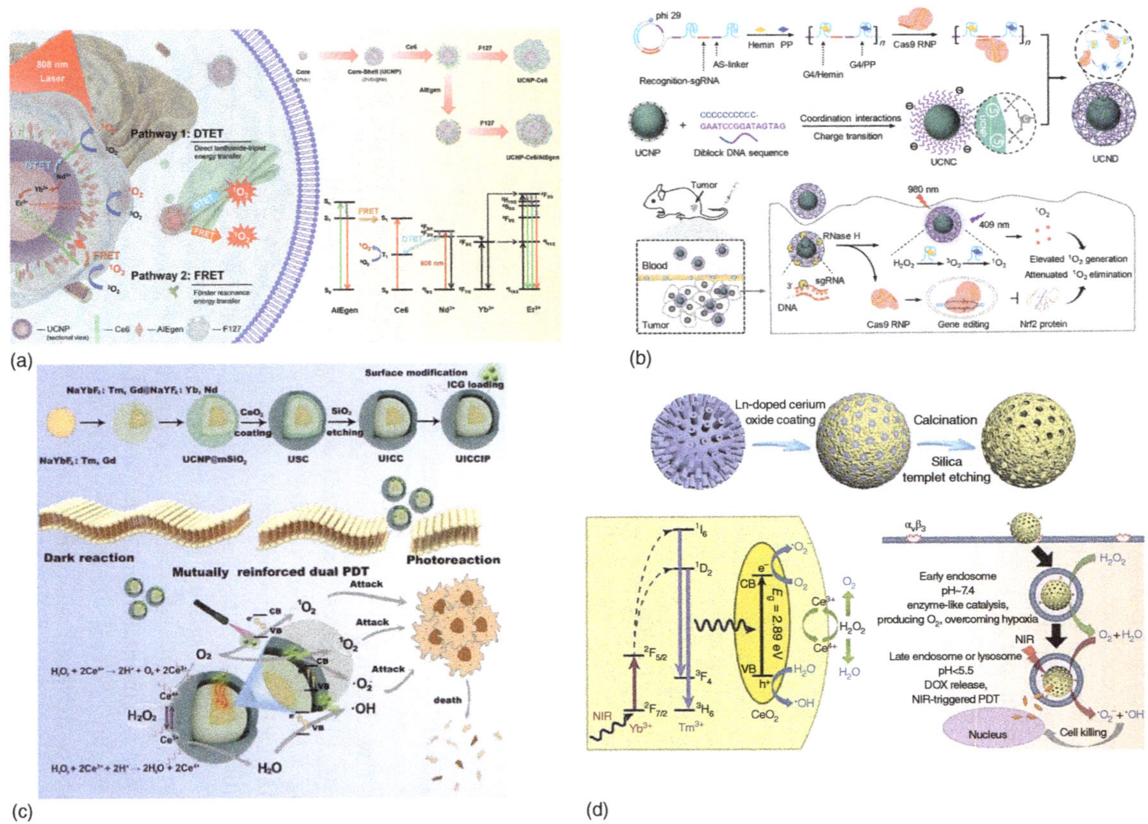

Figure 8.10 (a) Synthesis illustration and schematic depiction of UCNP-Ce6/AIEgen for dual-pathway reinforced PDT of tumor. Source: Reproduced with permission from Zhang et al. [144]/Springer Nature. (b) An RCA-based multifunctional DNA/UCNP complex for the controlled co-delivery of Cas9 RNP, hemin, and PP for synergistic PDT. Source: Reproduced with permission from Song et al. [145]/John Wiley & Sons. (c) A novel nanoplatform of indocyanine green-equipped UCNP/CeO_2 was fabricated for mutually reinforced dual PDT of tumor. Source: Reproduced with permission from Gao et al. [24]/Elsevier. (d) Schematic illustration of the synthesis process, catalysis mechanism, and hypothetical subcellular pathway actions for the Yb^{3+}/Tm^{3+}-codoped mesoporous CeO_2 hollow nanoparticles (Ce-UCNPs). Source: Reproduced with permission from Yao et al. [146]/John Wiley & Sons.

combined with MOF to achieve PDT [148, 149]. However, UCNPs modified with MOF will be introduced in a subsequent chapter and will not be repeated here.

For inorganic photosensitizers, UCNPs could also be composited with BP [150], MnO_2 [151], and others. Modification with MnO_2 can enhance the energy transfer efficiency to PDT within tumor cells, while BP can assist in photodynamic-photothermal synergistic therapy. CeO_2 is widely applied in PDT as an assistant with high reactivity in the rare earth family. A novel photosensitizer with UCNPs as the core and CeO_2 as the shell layer equipped with ICG (UICCIP) was developed for PDT (Figure 8.10c) [24]. Different valence states of CeO_2 have different enzyme-like properties [152]. Ce^{3+} shows oxidase-like activity to produce peroxides, whereas Ce^{4+} shows catalase-like activity and could catalyze endogenous H_2O_2 to produce O_2. Therefore, the change in the Ce^{3+}/Ce^{4+} ratio is critical for PDT [153]. UICCIP could catalyze endogenous H_2O_2 to produce O_2, which could solve the problem of oxygen depletion during PDT. Under 808 nm laser irradiation, the UCL of UCNPs could activate CeO_2 to generate ROS [154]. Moreover, the equipped ICG could generate ROS under 808 nm laser irradiation. Ultimately, this nanoplatform could achieve mutually reinforced dual PDT of tumor. In addition to single PDT, Zhang's group developed mesoporous cerium oxide capped with UCNPs (Ce-UCNP) as an inorganic PS in Figure 8.10d [146]. With the help of UCNPs, UV UCL could be generated under 980 nm laser irradiation, which could separate out the electron–hole pairs of CeO_2 to generate $O_2^{\bullet-}$ and •OH. Moreover, CeO_2 with mesoporous structure could load with DOX. After modification of Arg-Gly-Asp (RGD)-peptide, Ce-UCNP/DOX could achieve in situ response to the weakly acidic TME for synergistic chemo-photodynamic antitumor therapy.

8.5 Conclusions and Perspectives

Over the past few decades, PDT has emerged as an innovative approach to tumor treatment. PDT achieves tumor cell death by generating ROS from PSs in the targeting tumor tissue. This chapter provides an overview of various PSs, including organic molecules, metal complexes, inorganic semiconductors, and organic-inorganic nanocomposites. Regardless of the type of PS used, the generation of sufficient ROS within the tumor tissue is crucial for efficient PDT. However, conventional PSs often have various shortcomings, including a low ROS quantum yield, insufficient tumor accumulation, shallow penetration depth with the excitation light, and the inability of single PDT to effectively inhibit tumor growth. Therefore, novel PSs need to be designed and constructed from the following aspects.

(i) Improving quantum yield of ROS. For organic molecule PSs, researchers could focus on molecular design engineering to increase ISC and regulate the lifetime of the energy level, and enhance the energy transfer from the T_1 energy level to the ground state O_2 to generate more $^1O_2$. For inorganic semiconductors, the ROS yield could be enhanced by optimizing the particle size and morphology, doping heteroatom to change the bandgap, and so on.

(ii) Enhancing the tumor-targeting ability of photosensitizer. Due to the conjugated ring structure, many organic molecular photosensitizers exhibit low water solubility. This results in their clearance by the reticuloendothelial system, which hinders their circulation in vivo, efficient delivery to tumor tissue, and PDT. Modification with biocompatible carriers, such as surfactants, polymers, and silica, could extend the circulation time in the bloodstream. Functional modification of photosensitizers with targeting peptides could significantly increase the photosensitizer concentration within tumor tissue, leading to improved therapeutic effect of PDT.

(iii) Selecting the NIR excitation light with deep tissue penetration. Since the wavelength of excitation light is related to the inherent properties of photosensitizer, the modification of the photosensitizer itself remains the fundamental solution. The S_1 energy level of organic molecular photosensitizers could be lowered through molecular engineering. For inorganic semiconductor photosensitizers, doping heteroatoms or constructing heterostructures to narrow the bandgap width could achieve red-shifting of the excitation wavelength. A convenient alternative is to use NIR indocyanine dyes or UCNPs excited by NIR light to emit upconverted light in the UV-visible region to activate the photosensitizer, solving the problem of shallow penetration depth.

(iv) Improving the treatment efficiency through combinational/synergistic therapy. The effectiveness of PDT is limited by the fact that the TME is hypoxic and contains large amounts of reducing substances. The insufficient ROS generation and limited diffusion distance hinder the effective apoptosis of tumor cells. In addition to using metal complex photosensitizers that can achieve oxygen-free PDT and the choice of oxygen-rich carriers, the efficacy of PDT can be significantly improved by modification with catalytical active substances (e.g. enzymes or inorganic enzyme-like nanoparticles) that react specifically to the endogenous substances in tumors (e.g. high concentrations of H^+ and H_2O_2) to release reversibly oxygen. Besides, multiple synergistic therapies could effectively inhibit tumor growth and suppress tumor recurrence. Chemotherapy, photothermal, sonodynamic, and PDT have been widely studied as methods of efficiently destroying tumor tissue, preventing the development of drug resistance, and reducing the risk of recurrence. Another strategy for treating tumors is to elicit an immune response by modulating immune factors and activating immune cells to recognize and eliminate tumor cells.

References

1 Dougherty, T.J. (1984). Photodynamic therapy (PDT) of malignant tumors. *Critical Reviews in Oncology/Hematology* 2: 83–116.

2 Overchuk, M., Weersink, R.A., Wilson, B.C., and Zheng, G. (2023). Photodynamic and photothermal therapies: synergy opportunities for nanomedicine. *ACS Nano* 17: 7979–8003.

3 Li, S., Yang, F., Wang, Y. et al. (2023). Emerging nanotherapeutics for facilitating photodynamic therapy. *Chemical Engineering Journal* 451: 138621.

4 Sies, H., Belousov, V.V., Chandel, N.S. et al. (2022). Defining roles of specific reactive oxygen species (ROS) in cell biology and physiology. *Nature Reviews Molecular Cell Biology* 23: 499–515.

5 Murphy, M.P., Bayir, H., Belousov, V. et al. (2022). Guidelines for measuring reactive oxygen species and oxidative damage in cells and in vivo. *Nature Metabolism* 4: 651–662.

6 Halliwell, B., Adhikary, A., Dingfelder, M., and Dizdaroglu, M. (2021). Hydroxyl radical is a significant player in oxidative DNA damage in vivo. *Chemical Society Reviews* 50: 8355–8360.

7 Winterbourn, C.C. (2020). Hydrogen peroxide reactivity and specificity in thiol-based cell signalling. *Biochemical Society Transactions* 48: 745–754.

8 Lim, J.M., Kim, G., and Levine, R.L. (2019). Methionine in proteins: it's not just for protein initiation anymore. *Neurochemical Research* 44: 247–257.

9 Sharman, W.M., Allen, C.M., and van Lier, J.E. (2000). Role of activated oxygen species in photodynamic therapy. *Methods in Enzymology* 319: 376–400.

10 Lan, M., Zhao, S., Liu, W. et al. (2019). Photosensitizers for photodynamic therapy. *Advanced Healthcare Materials* 8: 1900132.

11 Li, X., Lee, S., and Yoon, J. (2018). Supramolecular photosensitizers rejuvenate photodynamic therapy. *Chemical Society Reviews* 47: 1174–1188.

12 Fan, W., Huang, P., and Chen, X. (2016). Overcoming the Achilles' heel of photodynamic therapy. *Chemical Society Reviews* 45: 6488–6519.

13 Gulzar, A., Xu, J., Wang, C. et al. (2019). Tumour microenvironment responsive nanoconstructs for cancer theranostic. *Nano Today* 26: 16–56.

14 He, H., Liu, L., Liang, R. et al. (2020). Tumor-targeted nanoplatform for in situ oxygenation-boosted immunogenic phototherapy of colorectal cancer. *Acta Biomaterialia* 104: 188–197.

15 Monro, S., Colon, K.L., Yin, H. et al. (2019). Transition metal complexes and photodynamic therapy from a tumor-centered approach: challenges, opportunities, and highlights from the development of TLD1433. *Chemical Reviews* 119: 797–828.

16 Mo, J., Mai Le, N.P., and Priefer, R. (2021). Evaluating the mechanisms of action and subcellular localization of ruthenium(II)-based photosensitizers. *European Journal of Medical Chemistry* 225: 113770.

17 Li, C., Chen, G., Zhang, Y. et al. (2020). Advanced fluorescence imaging technology in the near-Infrared-II window for biomedical applications. *Journal of the American Chemical Society* 142: 14789–14804.

18 Chen, C., Wang, F., Wen, S. et al. (2018). Multi-photon near-infrared emission saturation nanoscopy using upconversion nanoparticles. *Nature Communications* 9: 3290.

19 Kim, A., Zhou, J., Samaddar, S. et al. (2019). An implantable ultrasonically-powered micro-light-source (microLight) for photodynamic therapy. *Scientific Reports* 9: 1395.

20 Shi, Z., Gao, X., Cai, X. et al. (2023). Fully implantable and retrievable upconversion waveguides for photodynamic therapy in deep tissue. *Advanced Optical Materials* 11: 2300689.

21 Tang, Y., Pei, F., Lu, X. et al. (2019). Recent advances on activatable NIR-II fluorescence probes for biomedical imaging. *Advanced Optical Materials* 7: 1900917.

22 Yu, Z., Eich, C., and Cruz, L.J. (2020). Recent advances in rare-earth-doped nanoparticles for NIR-II imaging and cancer theranostics. *Frontiers in Chemistry* 8: 496.

23 Smith, A.M., Mancini, M.C., and Nie, S. (2009). Bioimaging: second window for in vivo imaging. *Nature Nanotechnology* 4: 710–711.

24 Lv, K., Yao, L., Fu, X. et al. (2023). Indocyanine green-equipped upconversion nanoparticles/CeO_2 trigger mutually reinforced dual photodynamic therapy. *Nano Today* 52: 101964.

25 Gao, X., Feng, J., Lv, K. et al. (2023). Engineering CeO_2/CuO heterostructure anchored on upconversion nanoparticles with boosting ROS generation-primed apoptosis-ferroptosis for cancer dynamic therapy. *Nano Research* 16: 5322–5334.

26 He, L., Yu, X., and Li, W. (2022). Recent progress and trends in X-ray-induced photodynamic therapy with low radiation doses. *ACS Nano* 16: 19691–19721.

27 Song, L., Li, P.P., Yang, W. et al. (2018). Low-dose X-ray activation of W(VI)-doped persistent luminescence nanoparticles for deep-tissue photodynamic therapy. *Advanced Functional Materials* 28: 1707496.

28 Jin, C.S., Cui, L., Wang, F. et al. (2014). Targeting-triggered porphysome nanostructure disruption for activatable photodynamic therapy. *Advanced Healthcare Materials* 3: 1240–1249.

29 Kim, J., Tung, C.H., and Choi, Y. (2014). Smart dual-functional warhead for folate receptor-specific activatable imaging and photodynamic therapy. *Chemical Communications* 50: 10600–10603.

30 Li, X., Yu, S., Lee, D. et al. (2018). Facile supramolecular approach to nucleic-acid-driven activatable nanotheranostics that overcome drawbacks of photodynamic therapy. *ACS Nano* 12: 681–688.

31 Sun, W., Li, S., Haupler, B. et al. (2017). An amphiphilic ruthenium polymetallodrug for combined photodynamic therapy and photochemotherapy in vivo. *Advanced Materials* 29: 1603702.

32 Cheng, K., Yang, X.Q., Zhang, X.S. et al. (2018). High-security nanocluster for switching photodynamic combining photothermal and acid-induced drug compliance therapy guided by multimodal active-targeting imaging. *Advanced Functional Materials* 28: 1803118.

33 Li, X., Xu, W., Yang, Z. et al. (2023). A lipid droplet-targeted multifunctional AIE-active fluorescent probe for hydrogen peroxide detection and imaging-guided photodynamic therapy. *Sensors and Actuators B* 375: 132892.

34 Chang, E., Bu, J., Ding, L. et al. (2021). Porphyrin-lipid stabilized paclitaxel nanoemulsion for combined photodynamic therapy and chemotherapy. *Journal of Nanobiotechnology* 19: 154.

35 Lee, C.-S., Park, W., Jo, Y.U., and Na, K. (2014). A charge-switchable, four-armed polymeric photosensitizer for photodynamic cancer therapy. *Chemical Communications* 50: 4354–4357.

36 Jin, J., Zhu, Y., Zhang, Z., and Zhang, W. (2018). Enhancing the efficacy of photodynamic therapy through a porphyrin/POSS alternating copolymer. *Angewandte Chemie International Edition* 57: 16354–16358.

37 Dougherty, T.J., Gomer, C.J., Henderson, B.W. et al. (1998). Photodynamic therapy. *Journal of the National Cancer Institute* 90: 889–905.

38 Liang, X., Li, X., Jing, L. et al. (2014). Theranostic porphyrin dyad nanoparticles for magnetic resonance imaging guided photodynamic therapy. *Biomaterials* 35: 6379–6388.

39 Nakanishi, M., Park, J.-S., Jang, W.-D. et al. (2007). Study of the quantitative aminolysis reaction of poly(β-benzyl l-aspartate) (PBLA) as a platform polymer for functionality materials. *Reactive and Functional Polymers* 67: 1361–1372.

40 Jiang, Z., Shao, J., Yang, T. et al. (2014). Pharmaceutical development, composition and quantitative analysis of phthalocyanine as the photosensitizer for cancer photodynamic therapy. *Journal of Pharmaceutical and Biomedical Analysis* 87: 98–104.

41 Liu, S., Ma, J., Xue, E.Y. et al. (2023). Polymeric phthalocyanine-based nanosensitizers for enhanced photodynamic and sonodynamic therapies. *Advanced Healthcare Materials* 12: 2300481.

42 Ke, M.R., Chen, S.F., Peng, X.H. et al. (2017). A tumor-targeted activatable phthalocyanine-tetrapeptide-doxorubicin conjugate for synergistic chemo-photodynamic therapy. *European Journal of Medical Chemistry* 127: 200–209.

43 Li, X., Jeong, K., Lee, Y. et al. (2019). Water-soluble phthalocyanines selectively bind to albumin dimers: a green approach toward enhancing tumor-targeted photodynamic therapy. *Theranostics* 9: 6412–6423.

44 Wilson, D.S., Dalmasso, G., Wang, L. et al. (2010). Orally delivered thioketal nanoparticles loaded with TNF-alpha-siRNA target inflammation and inhibit gene expression in the intestines. *Nature Materials* 9: 923–928.

45 Ding, F., Li, F., Tang, D. et al. (2022). Restoration of the immunogenicity of tumor cells for enhanced cancer therapy via nanoparticle-mediated copper chaperone inhibition. *Angewandte Chemie International Edition* 61: e202203546.

46 Xie, J., Yao, Y., Wang, S. et al. (2022). Alleviating oxidative injury of myocardial infarction by a fibrous polyurethane patch with condensed ROS-scavenging backbone units. *Advanced Healthcare Materials* 11: 2101855.

47 Wang, Y., Hu, L.F., Cui, P.F. et al. (2021). Pathologically responsive mitochondrial gene therapy in an allotopic expression-independent manner cures Leber's hereditary optic neuropathy. *Advanced Materials* 33: 2103307.

48 Garin-Chesa, P., Old, L.J., and Rettig, W.J. (1990). Cell surface glycoprotein of reactive stromal fibroblasts as a potential antibody target in human epithelial cancers. *Proceedings of the National Academy of Sciences of the United States of America* 87: 7235–7239.

49 Scanlan, M.J., Raj, B.K., Calvo, B. et al. (1994). Molecular cloning of fibroblast activation protein alpha, a member of the serine protease family selectively expressed in stromal fibroblasts of epithelial cancers. *Proceedings of the National Academy of Sciences of the United States of America* 91: 5657–5661.

50 Liu, R., Li, H., Liu, L. et al. (2012). Fibroblast activation protein: a potential therapeutic target in cancer. *Cancer Biology & Therapy* 13: 123–129.

51 Kamal, A., Thao, L., Sensintaffar, J. et al. (2003). A high-affinity conformation of Hsp90 confers tumour selectivity on Hsp90 inhibitors. *Nature* 425: 407–410.

52 Whitesell, L. and Lindquist, S.L. (2005). HSP90 and the chaperoning of cancer. *Nature Reviews Cancer* 5: 761–772.

53 Shimamura, T., Perera, S.A., Foley, K.P. et al. (2012). Ganetespib (STA-9090), a nongeldanamycin HSP90 inhibitor, has potent antitumor activity in in vitro and in vivo models of non-small cell lung cancer. *Clinical Cancer Research* 18: 4973–4985.

54 Huang, L., Wei, G., Sun, X. et al. (2018). A tumor-targeted Ganetespib-zinc phthalocyanine conjugate for synergistic chemo-photodynamic therapy. *European Journal of Medical Chemistry* 151: 294–303.

55 Zhao, M., Xu, Y., Xie, M. et al. (2018). Halogenated aza-BODIPY for imaging-guided synergistic photodynamic and photothermal tumor therapy. *Advanced Healthcare Materials* 7: 1800606.

56 Zhang, Y.-C., She, L., Xu, Z.-Y. et al. (2022). A BODIPY-modified polymeric micelle for sustaining enhanced photodynamic therapy. *Chinese Chemical Letters* 33: 3277–3280.

57 Turan, I.S., Yildiz, D., Turksoy, A. et al. (2016). A bifunctional photosensitizer for enhanced fractional photodynamic therapy: singlet oxygen generation in the presence and absence of light. *Angewandte Chemie International Edition* 55: 2875–2878.

58 Zou, J., Zhu, J., Yang, Z. et al. (2020). A phototheranostic strategy to continuously deliver singlet oxygen in the dark and hypoxic tumor microenvironment. *Angewandte Chemie International Edition* 59: 8833–8838.

59 Xiao, W., Wang, P., Ou, C. et al. (2018). 2-Pyridone-functionalized aza-BODIPY photosensitizer for imaging-guided sustainable phototherapy. *Biomaterials* 183: 1–9.

60 Zou, J., Yin, Z., Wang, P. et al. (2018). Photosensitizer synergistic effects: D-A-D structured organic molecule with enhanced fluorescence and singlet oxygen quantum yield for photodynamic therapy. *Chemical Science* 9: 2188–2194.

61 Sheng, Z., Hu, D., Xue, M. et al. (2013). Indocyanine green nanoparticles for theranostic applications. *Nano-Micro Letters* 5: 145–150.

62 Zhang, R.R., Schroeder, A.B., Grudzinski, J.J. et al. (2017). Beyond the margins: real-time detection of cancer using targeted fluorophores. *Nature Reviews Clinical Oncology* 14: 347–364.

63 Vahrmeijer, A.L., Hutteman, M., van der Vorst, J.R. et al. (2013). Image-guided cancer surgery using near-infrared fluorescence. *Nature Reviews Clinical Oncology* 10: 507–518.

64 Liu, Y., Zhi, X., Yang, M. et al. (2017). Tumor-triggered drug release from calcium carbonate-encapsulated gold nanostars for near-infrared photodynamic/photothermal combination antitumor therapy. *Theranostics* 7: 1650–1662.
65 Altinoglu, E.I., Russin, T.J., Kaiser, J.M. et al. (2008). Near-infrared emitting fluorophore-doped calcium phosphate nanoparticles for in vivo imaging of human breast cancer. *ACS Nano* 2: 2075–2084.
66 Lee, C.H., Cheng, S.H., Wang, Y.J. et al. (2009). Near-infrared mesoporous silica nanoparticles for optical imaging: characterization and in vivo biodistribution. *Advanced Functional Materials* 19: 215–222.
67 Ito, T., Iida-Tanaka, N., and Koyama, Y. (2008). Efficient in vivo gene transfection by stable DNA/PEI complexes coated by hyaluronic acid. *Journal of Drug Targeting* 16: 276–281.
68 Xue, P., Yang, R., Sun, L. et al. (2018). Indocyanine green-conjugated magnetic Prussian blue nanoparticles for synchronous photothermal/photodynamic tumor therapy. *Nano-Micro Letters* 10: 74.
69 Yang, L., Huang, B., Hu, S. et al. (2022). Indocyanine green assembled free oxygen-nanobubbles towards enhanced near-infrared induced photodynamic therapy. *Nano Research* 15: 4285–4293.
70 Gomollón-Bel, F. (2020). Ten chemical innovations that will change our world. *Chemistry International* 42: 3–9.
71 Luo, J., Xie, Z., Lam, J.W. et al. (2001). Aggregation-induced emission of 1-methyl-1,2,3,4,5-pentaphenylsilole. *Chemical Communications* 1740–1741.
72 Mei, J., Hong, Y., Lam, J.W. et al. (2014). Aggregation-induced emission: the whole is more brilliant than the parts. *Advanced Materials* 26: 5429–5479.
73 Gao, M., Wu, Y., Chen, B. et al. (2015). Di(naphthalen-2-yl)-1,2-diphenylethene-based conjugated polymers: aggregation-enhanced emission and explosive detection. *Polymer Chemistry* 6: 7641–7645.
74 Gould, S.J., Booth, A.M., and Hildreth, J.E. (2003). The Trojan exosome hypothesis. *Proceedings of the National Academy of Sciences of the United States of America* 100: 10592–10597.
75 Martin, J.D., Panagi, M., Wang, C. et al. (2019). Dexamethasone increases cisplatin-loaded nanocarrier delivery and efficacy in metastatic breast cancer by normalizing the tumor microenvironment. *ACS Nano* 13: 6396–6408.
76 Zhu, D., Duo, Y., Suo, M. et al. (2020). Tumor-exocytosed exosome/aggregation-induced emission luminogen hybrid nanovesicles facilitate efficient tumor penetration and photodynamic therapy. *Angewandte Chemie International Edition* 59: 13836–13843.
77 Cong, Z., Xie, S., Jiang, Z. et al. (2022). In vivo photodynamic therapy based on near-infrared AIE cationic polymers. *Chemical Engineering Journal* 431: 133748.
78 Yuan, G., Lv, C., Liang, J. et al. (2021). Molecular engineering of efficient singlet oxygen generators with near-infrared AIE features for mitochondrial targeted photodynamic therapy. *Advanced Functional Materials* 31: 2104026.
79 Jia, P.-P., Xu, L., Hu, Y.-X. et al. (2020). Orthogonal self-assembly of a two-step fluorescence-resonance energy transfer system with improved photosensitization

efficiency and photooxidation activity. *Journal of the American Chemical Society* 143: 399–408.

80 Zhu, H., Li, Q., Shi, B. et al. (2020). Dual-emissive platinum(II) metallacage with a sensitive oxygen response for imaging of hypoxia and imaging-guided chemotherapy. *Angewandte Chemie International Edition* 59: 20208–20214.

81 Tu, L., Li, C., Xiong, X. et al. (2023). Engineered metallacycle-based supramolecular photosensitizers for effective photodynamic therapy. *Angewandte Chemie International Edition* 62: e202301560.

82 Wang, M.F., Yang, R., Tang, S.J. et al. (2022). In vivo realization of dual photodynamic and photothermal therapy for melanoma by mitochondria targeting dinuclear ruthenium complexes under civil infrared low-power laser. *Angewandte Chemie International Edition* 61: e202208721.

83 Xu, G., Li, C., Chi, C. et al. (2022). A supramolecular photosensitizer derived from an arene-Ru(II) complex self-assembly for NIR activated photodynamic and photothermal therapy. *Nature Communications* 13: 3064.

84 Casini, A., Woods, B., and Wenzel, M. (2017). The promise of self-assembled 3D supramolecular coordination complexes for biomedical applications. *Inorganic Chemistry* 56: 14715–14729.

85 Reichardt, C., Schneider, K.R.A., Sainuddin, T. et al. (2017). Excited state dynamics of a photobiologically active Ru(II) dyad are altered in biologically relevant environments. *Journal of Physical Chemistry A* 121: 5635–5644.

86 Villemin, E., Ong, Y.C., Thomas, C.M., and Gasser, G. (2019). Polymer encapsulation of ruthenium complexes for biological and medicinal applications. *Nature Reviews Chemistry* 3: 261–282.

87 Antaris, A.L., Chen, H., Cheng, K. et al. (2015). A small-molecule dye for NIR-II imaging. *Nature Materials* 15: 235–242.

88 Qi, J., Li, J., Liu, R. et al. (2019). Boosting fluorescence-photoacoustic-raman properties in one fluorophore for precise cancer surgery. *Chem* 5: 2657–2677.

89 Gao, S., Wei, G., Zhang, S. et al. (2019). Albumin tailoring fluorescence and photothermal conversion effect of near-infrared-II fluorophore with aggregation-induced emission characteristics. *Nature Communications* 10: 2206.

90 Liu, S., Zhou, X., Zhang, H. et al. (2019). Molecular motion in aggregates: manipulating TICT for boosting photothermal theranostics. *Journal of the American Chemical Society* 141: 5359–5368.

91 Kuang, S., Wei, F., Karges, J. et al. (2022). Photodecaging of a mitochondria-localized iridium(III) endoperoxide complex for two-photon photoactivated therapy under hypoxia. *Journal of the American Chemical Society* 144: 4091–4101.

92 Kuang, S., Sun, L., Zhang, X. et al. (2020). A mitochondrion-localized two-photon photosensitizer generating carbon radicals against hypoxic tumors. *Angewandte Chemie International Edition* 59: 20697–20703.

93 Wang, W.J., Ling, Y.Y., Zhong, Y.M. et al. (2022). Ferroptosis-enhanced cancer immunity by a ferrocene-appended iridium(III) diphosphine complex. *Angewandte Chemie International Edition* 61: e202115247.

94 Duong, M.T.-Q., Qin, Y., You, S.-H., and Min, J.-J. (2019). Bacteria-cancer interactions: bacteria-based cancer therapy. *Experimental & Molecular Medicine* 51: 1–15.

95 Zhou, S., Gravekamp, C., Bermudes, D., and Liu, K. (2018). Tumour-targeting bacteria engineered to fight cancer. *Nature Reviews Cancer* 18: 727–743.

96 Shi, L., Liu, X., Li, Y. et al. (2022). Living bacteria-based immune-photodynamic therapy: metabolic labeling of *Clostridium butyricum* for eradicating malignant melanoma. *Advanced Science* 9: 2105807.

97 Sun, R., Liu, M., Lu, J. et al. (2022). Bacteria loaded with glucose polymer and photosensitive ICG silicon-nanoparticles for glioblastoma photothermal immunotherapy. *Nature Communications* 13: 5127.

98 Chu, B., Yang, Y., Tang, J. et al. (2022). Trojan nanobacteria system for photothermal programmable destruction of deep tumor tissues. *Angewandte Chemie International Edition* 61: e202208422.

99 Li, J., Xia, Q., Guo, H. et al. (2022). Decorating bacteria with triple immune nanoactivators generates tumor-resident living immunotherapeutics. *Angewandte Chemie International Edition* 61: e202202409.

100 Lin, W., Liu, Y., Wang, J. et al. (2023). Engineered bacteria labeled with iridium(III) photosensitizers for enhanced photodynamic immunotherapy of solid tumors. *Angewandte Chemie International Edition* 62: e202310158.

101 Liu, B., Jiao, J., Xu, W. et al. (2021). Highly efficient far-red/NIR-absorbing neutral Ir(III) complex micelles for potent photodynamic/photothermal therapy. *Advanced Materials* 33: 2100795.

102 Zheng, Q., Liu, X., Zheng, Y. et al. (2021). The recent progress on metal-organic frameworks for phototherapy. *Chemical Society Reviews* 50: 5086–5125.

103 Lu, K., He, C., and Lin, W. (2014). Nanoscale metal-organic framework for highly effective photodynamic therapy of resistant head and neck cancer. *Journal of the American Chemical Society* 136: 16712–16715.

104 Park, J., Jiang, Q., Feng, D. et al. (2016). Size-controlled synthesis of porphyrinic metal-organic framework and functionalization for targeted photodynamic therapy. *Journal of the American Chemical Society* 138: 3518–3525.

105 Xiang, Z., Zhu, L., Qi, L. et al. (2016). Two-dimensional fully conjugated polymeric photosensitizers for advanced photodynamic therapy. *Chemistry of Materials* 28: 8651–8658.

106 He, L., Ni, Q., Mu, J. et al. (2020). Solvent-assisted self-assembly of a metal-organic framework based biocatalyst for cascade reaction driven photodynamic therapy. *Journal of the American Chemical Society* 142: 6822–6832.

107 Han, D., Han, Y., Li, J. et al. (2020). Enhanced photocatalytic activity and photothermal effects of Cu-doped metal-organic frameworks for rapid treatment of bacteria-infected wounds. *Applied Catalysis B* 261: 118248.

108 He, M., Chen, Y., Tao, C. et al. (2019). Mn-porphyrin-based metal-organic framework with high longitudinal relaxivity for magnetic resonance imaging guidance and oxygen self-supplementing photodynamic therapy. *ACS Applied Materials & Interfaces* 11: 41946–41956.

109 Zhu, W., Liu, Y., Yang, Z. et al. (2018). Albumin/sulfonamide stabilized iron porphyrin metal organic framework nanocomposites: targeting tumor hypoxia by carbonic anhydrase IX inhibition and T_1-T_2 dual mode MRI guided photodynamic/photothermal therapy. *Journal of Materials Chemistry B* 6: 265–276.

110 Feng, J., Ren, W.-X., Kong, F., and Dong, Y.-B. (2021). Recent insight into functional crystalline porous frameworks for cancer photodynamic therapy. *Inorganic Chemistry Frontiers* 8: 848–879.

111 Zhang, Y., Zhang, L., Wang, Z. et al. (2019). Renal-clearable ultrasmall covalent organic framework nanodots as photodynamic agents for effective cancer therapy. *Biomaterials* 223: 119462.

112 Yin, Q., Zhao, P., Sa, R.J. et al. (2018). An ultra-robust and crystalline redeemable hydrogen-bonded organic framework for synergistic chemo-photodynamic therapy. *Angewandte Chemie International Edition* 57: 7691–7696.

113 Zheng, D.W., Li, B., Li, C.-X. et al. (2016). Carbon-dot-decorated carbon nitride nanoparticles for enhanced photodynamic therapy against hypoxic tumor via water splitting. *ACS Nano* 10: 8715–8722.

114 Shi, J., Yu, X., Wang, L. et al. (2013). PEGylated fullerene/iron oxide nanocomposites for photodynamic therapy, targeted drug delivery and MR imaging. *Biomaterials* 34: 9666–9677.

115 Brown, S.B., Brown, E.A., and Walker, I. (2004). The present and future role of photodynamic therapy in cancer treatment. *Lancet Oncology* 5: 497–508.

116 Palao, E., Slanina, T., Muchová, L. et al. (2015). Transition-metal-free co-releasing BODIPY derivatives activatable by visible to NIR light as promising bioactive molecules. *Journal of the American Chemical Society* 138: 126–133.

117 Wang, Y., Song, S., Liu, J. et al. (2014). ZnO-functionalized upconverting nanotheranostic agent: multi-modality imaging-guided chemotherapy with on-demand drug release triggered by pH. *Angewandte Chemie International Edition* 54: 536–540.

118 Tejedor-Estrada, R., Nonell, S., Teixido, J. et al. (2012). An artificial neural network model for predicting the subcellular localization of photosensitisers for photodynamic therapy of solid tumours. *Current Medicinal Chemistry* 19: 2472–2482.

119 Xu, N., Du, J., Yao, Q. et al. (2020). Precise photodynamic therapy: penetrating the nuclear envelope with photosensitive carbon dots. *Carbon* 159: 74–82.

120 Mishra, P.P. and Datta, A. (2006). Difference in the effects of surfactants and albumin on the extent of deaggregation of purpurin 18, a model of hydrophobic photosensitizer. *Biophysical Chemistry* 121: 224–233.

121 Vrouenraets, M.B., Visser, G.W.M., Loup, C. et al. (2000). Targeting of a hydrophilic photosensitizer by use of internalizing monoclonal antibodies: a new possibility for use in photodynamic therapy. *International Journal of Cancer* 88: 108–114.

122 Huang, X., Li, L., Liu, T. et al. (2011). The shape effect of mesoporous silica nanoparticles on biodistribution, clearance, and biocompatibility in vivo. *ACS Nano* 5: 5390–5399.

123 Li, W.-Q., Sun, L.-P., Xia, Y. et al. (2018). Preoccupation of empty carriers decreases endo-/lysosome escape and reduces the protein delivery efficiency of mesoporous silica nanoparticles. *ACS Applied Materials & Interfaces* 10: 5340–5347.

124 Ma, T., Zhang, Q., Xuan, Q. et al. (2021). pH/near-infrared light dual activated Ce6-doped silicon nanoparticles with tumor chemo-photodynamic synergistic therapy for improving efficiency of monotherapy. *Chemical Engineering Journal* 424: 130536.

125 Zhang, Y., Huang, F., Ren, C. et al. (2017). Targeted chemo-photodynamic combination platform based on the DOX prodrug nanoparticles for enhanced cancer therapy. *ACS Applied Materials & Interfaces* 9: 13016–13028.

126 Näkki, S., Martinez, J.O., Evangelopoulos, M. et al. (2017). Chlorin e6 functionalized theranostic multistage nanovectors transported by stem cells for effective photodynamic therapy. *ACS Applied Materials & Interfaces* 9: 23441–23449.

127 Kim, N. and Cho, S.-G. (2013). Clinical applications of mesenchymal stem cells. *Korean Journal of Internal Medicine* 28: 387–402.

128 Bižić, M., Klintzsch, T., Ionescu, D. et al. (2020). Aquatic and terrestrial cyanobacteria produce methane. *Science Advances* 6: eaax5343.

129 Zhang, Y., Liu, H., Dai, X. et al. (2020). Cyanobacteria-based near-infrared light-excited self-supplying oxygen system for enhanced photodynamic therapy of hypoxic tumors. *Nano Research* 14: 667–673.

130 Qi, F., Ji, P., Chen, Z. et al. (2021). Photosynthetic cyanobacteria-hybridized black phosphorus nanosheets for enhanced tumor photodynamic therapy. *Small* 17: 2102113.

131 Zhang, X., Tang, J., Li, C. et al. (2021). A targeting black phosphorus nanoparticle based immune cells nano-regulator for photodynamic/photothermal and photo-immunotherapy. *Bioactive Materials* 6: 472–489.

132 Song, M., Liu, T., Shi, C. et al. (2016). Correction to bioconjugated manganese dioxide nanoparticles enhance chemotherapy response by priming tumor-associated macrophages toward M1-like phenotype and attenuating tumor hypoxia. *ACS Nano* 10: 3872.

133 Rayahin, J.E., Buhrman, J.S., Zhang, Y. et al. (2015). High and low molecular weight hyaluronic acid differentially influence macrophage activation. *ACS Biomaterials Science & Engineering* 1: 481–493.

134 Xiong, H., Ye, J., Wang, M. et al. (2022). In-situ bio-assembled specific Au NCs-Aptamer-Pyro conjugates nanoprobe for tumor imaging and mitochondria-targeted photodynamic therapy. *Biosensors & Bioelectronics* 218: 114763.

135 Wang, M., Chen, Y., Cai, W. et al. (2020). In situ self-assembling Au-DNA complexes for targeted cancer bioimaging and inhibition. *Proceedings of the National Academy of Sciences of the United States of America* 117: 308–316.

136 Medhi, R., Marquez, M.D., and Lee, T.R. (2020). Visible-light-active doped metal oxide nanoparticles: review of their synthesis, properties, and applications. *ACS Applied Nano Materials* 3: 6156–6185.

137 Zhang, C., Zhao, K., Bu, W. et al. (2014). Marriage of scintillator and semiconductor for synchronous radiotherapy and deep photodynamic therapy with diminished oxygen dependence. *Angewandte Chemie International Edition* 54: 1770–1774.

138 Wei, W. and Zhu, Y. (2019). TiO_2@perylene diimide full-spectrum photocatalysts via semi-core-shell structure. *Small* 15: 1903933.

139 Chen, D., Xu, Q., Wang, W. et al. (2021). Type I photosensitizers revitalizing photodynamic oncotherapy. *Small* 17: 2006742.

140 Yang, C., Zhu, Y., Li, D. et al. (2021). Red phosphorus decorated TiO_2 nanorod mediated photodynamic and photothermal therapy for renal cell carcinoma. *Small* 17: 2101837.

141 Qiu, M., Wang, D., Huang, H. et al. (2021). A regioselectively oxidized 2D Bi/BiO_x lateral nano-heterostructure for hypoxic photodynamic therapy. *Advanced Materials* 33: 2102562.

142 Seidl, C., Ungelenk, J., Zittel, E. et al. (2016). Tin tungstate nanoparticles: a photosensitizer for photodynamic tumor therapy. *ACS Nano* 10: 3149–3157.

143 Liu, B., Li, C., Yang, P. et al. (2017). 808-nm-light-excited lanthanide-doped nanoparticles: rational design, luminescence control and theranostic applications. *Advanced Materials* 29: 1605434.

144 Zhang, R., Lu, Y., Zhou, Y. et al. (2024). Utilizing dual-pathway energy transfer in upconversion nanoconjugates for reinforced photodynamic therapy. *Nano Research* 17: 2941–2948.

145 Song, N., Fan, X., Guo, X. et al. (2024). A DNA/upconversion nanoparticle complex enables controlled co-delivery of CRISPR-Cas9 and photodynamic agents for synergistic cancer therapy. *Advanced Materials* 2309534.

146 Yao, C., Wang, W., Wang, P. et al. (2018). Near-infrared upconversion mesoporous cerium oxide hollow biophotocatalyst for concurrent pH-/H_2O_2-responsive O_2-evolving synergetic cancer therapy. *Advanced Materials* 30: 1704833.

147 Zheng, B., Zhong, D., Xie, T. et al. (2021). Near-infrared photosensitization via direct triplet energy transfer from lanthanide nanoparticles. *Chem* 7: 1615–1625.

148 Zhang, X., Cui, J., Liu, J. et al. (2023). Dual ligand-assisted assembly of metal-organic frameworks on upconversion nanoparticles for NIR photodynamic therapy against hypoxic tumors. *Journal of Materials Chemistry B* 11: 9516–9524.

149 Chen, Y., Yang, Y., Du, S. et al. (2023). Mitochondria-targeting upconversion nanoparticles@MOF for multiple-enhanced photodynamic therapy in hypoxic tumor. *ACS Applied Materials & Interfaces* 15: 35884–35894.

150 Zhang, Q., Wang, W., Zhang, M. et al. (2020). A theranostic nanocomposite with integrated black phosphorus nanosheet, Fe_3O_4@MnO_2-doped upconversion nanoparticles and chlorin for simultaneous multimodal imaging, highly efficient photodynamic and photothermal therapy. *Chemical Engineering Journal* 391: 123525.

151 Zhao, M., Zhuang, H., Li, B. et al. (2023). In situ transformable nanoplatforms with supramolecular cross-linking triggered complementary function for enhanced cancer photodynamic therapy. *Advanced Materials* 35: 2209944.

152 Celardo, I., Pedersen, J.Z., Traversa, E., and Ghibelli, L. (2011). Pharmacological potential of cerium oxide nanoparticles. *Nanoscale* 3: 1411–1420.

153 Tian, Z., Liu, H., Guo, Z. et al. (2020). A pH-responsive polymer-CeO_2 hybrid to catalytically generate oxidative stress for tumor therapy. *Small* 16: 2004654.

154 Tanaka, A., Hashimoto, K., and Kominami, H. (2012). Preparation of Au/CeO_2 exhibiting strong surface plasmon resonance effective for selective or chemoselective oxidation of alcohols to aldehydes or ketones in aqueous suspensions under irradiation by green light. *Journal of the American Chemical Society* 134: 14526–14533.

9

Persistent Luminescent Materials for Optical Information Storage Applications

Cunjian Lin[1,2], Yixi Zhuang[1], and Rong-Jun Xie[1]

[1] *Xiamen University, College of Materials, 361005, Simingnan-Road 422, Xiamen, China*
[2] *Japan Advanced Institute of Science and Technology (JAIST), Graduate School of Advanced Science and Technology, 923-1292, 1-1 Asahidai, Nomi, Ishikawa, Japan*

9.1 Introduction

Persistent luminescent (PersL) materials are able to store a portion of the photonic energy within traps under the excitation of high-energy electromagnetic rays (X-ray or β-ray), ultraviolet light, or visible light, and then slowly emit photons with thermal or optical stimulations [1]. In recent decades, researchers have been continuously exploring new inorganic systems, leading to the development of numerous new high-performance PersL materials. Meanwhile, the mechanism of PersL in inorganics has been investigated in-depth, and some consensus has been reached. Due to the unique delayed emission feature, PersL materials have found extensive applications in the fields of night vision security [2, 3], bioluminescence imaging [4–13], alternating current light-emitting diodes (AC-LEDs) [14–19], optical information storage [20], and optical anti-counterfeiting [21, 22]. It is noteworthy that the application of PersL materials is closely related to the trap depth. Applications such as night vision security and bioluminescence imaging require PersL materials that emit photons moderately at room temperature or within the biological temperature range, resulting in these materials with typical trap depth ranging from 0.6 to 1.0 ev. In the case of AC-LEDs applications, PersL materials are primarily used to fill in nonemissive gaps with a period of approximately 10 ms, and thus, the trap depths are generally less than 0.6 eV. For optical information storage applications, to prevent the charge carriers from escaping from the traps and achieve a high information retention rate at room temperature, PersL materials are generally required to show deep-trap depths larger than 1.0 eV.

The demand for information storage capacity is growing rapidly both in daily data storage and advanced scientific research. With the widespread use of multimedia technology and the continuous improvement of network transmission speeds, the global generation of data has reached an astounding level. Current estimates

Photofunctional Nanomaterials for Biomedical Applications, First Edition.
Edited by Chunxia Li and Jun Lin.

suggest that annual data production reaches the zettabyte (10^{21} bytes) scale, still doubling every two years. In scientific research, mega-scientific projects such as cosmic ray detection, genomics, and materials genome engineering heavily depend on the acquisition of long-term, high-density data, as well as the efficient processing and analysis of big data. Optical information storage is currently one of the three main information storage technologies, with applications such as CDs and Blu-ray DVDs already widespread. Compared to magnetic information storage and solid-state information storage, optical information storage offers advantages such as excellent rewritability, low background noise, large storage capacity, low storage energy consumption, and high design ability, presenting the development of multidimensional optical information storage technology [23, 24]. However, suffering from intense optical diffraction, it is limited in the improvement in storage capacity for traditional planar optical data storage technology, and makes optical information storage lag behind the other two mainstream information storage technologies. Recently, researchers have been dedicated to developing new multidimensional optical information storage technologies based on the two-dimensional foundation, including the addition of spatial dimension (i.e. three-dimensional volume storage or holographic storage) [25–28], intensity dimension (i.e. multilevel gray value) [29, 30], wavelength dimension (i.e. wavelength-division multiplexing) [31–35], polarization dimension [31, 36, 37], angular momentum dimension [38, 39], and temporal domain dimension [40, 41], to achieve a substantial increase in the capacity of optical information storage.

PersL materials, also referred to as electron-trapping optical storage materials, have been applied to information storage and high-energy radiation dosimetry since the early 1980s [42, 43]. Importantly, photons released under thermal stimulation or photo-stimulation can carry operable optical information, such as emission intensity and wavelength, enabling PersL materials to possess multidimensional optical information storage capabilities. The advancement of electron-trapping optical storage materials and technology, capable of storing large amounts of information with high retention rates, represents a crucial direction in current research on PersL materials [20, 44, 45].

In recent years, researchers have made significant progress in the field of PersL materials, and several papers have provided systematic summaries and overviews of relevant material systems, luminescence mechanisms, and application advancements [46–53]. This chapter primarily highlights the key developmental milestones and the latest research achievements in deep-trap PersL materials for optical information storage applications. This chapter begins with the fundamental mechanisms of information storage and readout based on charge carrier capture and release in deep-trap PersL materials. Subsequently, it provides a comprehensive classification and an introduction of deep-trap PersL materials, along with an overview of their research progress in optical information storage applications. Finally, it offers a perspective on the future development of deep-trap PersL materials.

9.2 Luminescent Mechanism of Persistent Luminescent Materials with Deep Traps

Great progress has been made in understanding the mechanism behind the PersL phenomenon recently. It is widely accepted that the PersL in most inorganics originates from thermodynamically metastable states formed when excited charge carriers (electrons or holes) are trapped by point defects in the materials. These metastable states need to overcome a certain energy barrier to release the charge carriers, and the activation energy is defined as trap depth. This metastable state can release the charge carriers through thermal activation, which is known as thermally stimulated luminescence (thermoluminescence, TL), or through excitation by light, known as photo-stimulated luminescence (PSL). The excitation sources could be either high-energy electromagnetic rays (X-ray or β-ray), ultraviolet light, or visible light. The excitation process can originate from band-to-band excitation in the host, where the electrons are directly excited from the valence band (VB) to the conduction band (CB), or start from luminescent centers, where electrons are excited from the ground state of the luminescent centers to an excited state and then transferred to the CB.

After the electrons in the CB are captured by shallow traps (generally considered to be less than a subjectively defined value e.g. 1.0 eV), they could return to the CB through thermal activation (such as at room temperature), and finally recombine with the luminescent centers and emit PersL. This process represents the typical electron transition path. In contrast, if the traps in the materials are deeper than 1.0 eV, the electrons stored in the deep traps would not be energetic enough to overcome the energy barrier and escape from the traps at room temperature. In this case, only under high-energy-oriented excitation, such as intense photon-stimulation (photon energy greater than 1.0 eV) or higher temperature thermal activation (typically above 200 °C), can endow the electrons with enough energy to release from the traps. Thus, deep-trap PersL materials exhibit a metastable state when exposed to high-energy light, which is preserved at room temperature and returns to the initial state during the process of photo-stimulation or high-temperature activation. These characteristics match well with the requirements of optical information storage (Figure 9.1a).

As shown in Figure 9.1b, the write-in information in the deep-trap PersL materials represents receiving high-energy light store energy and remaining in a metastable state. During the information readout process, only the written points can generate emitted photons. Due to the rich optical information, such as emission wavelength and intensity, three-dimensional (3D) or higher-dimensional information retrieval could be realized by scanning with optical or thermal activation. Moreover, the written points can release all the charge carriers and completely return to the initial unwritten state after being processed at sufficiently high temperatures, making the electron-capture-based optical information storage technology highly reproducible. Therefore, using deep-trap PersL materials as the storage materials enables the

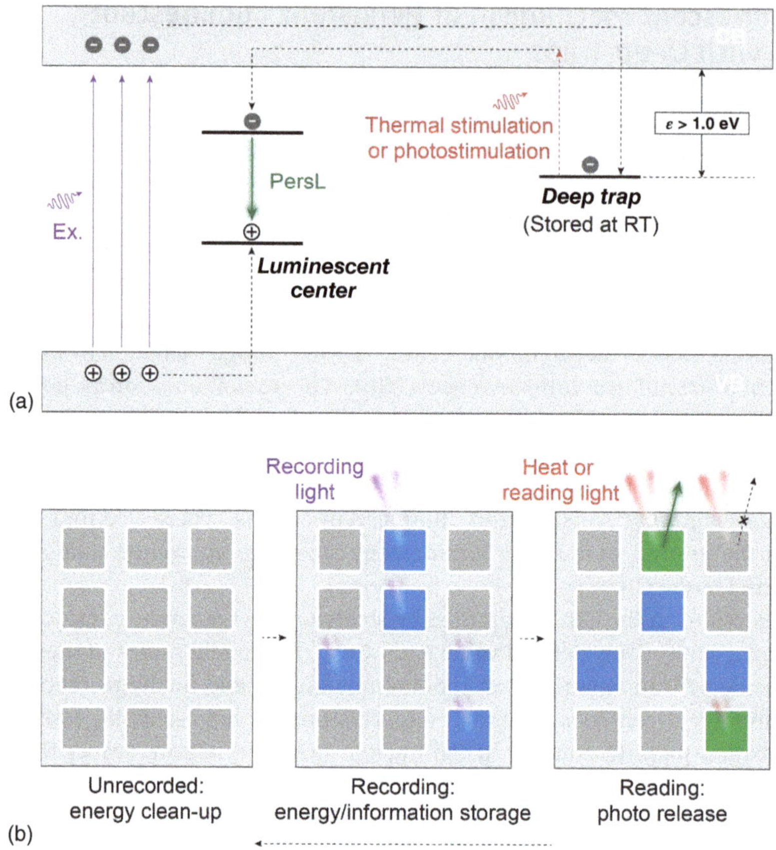

Figure 9.1 (a) Semantics of energy-level model of deep trap persistent luminescence materials. (b) Schematics of optical information storage by using deep trap persistent luminescent materials.

construction of a next-generation multidimensional optical information storage system with a large capacity.

9.3 Persistent Luminescent Materials with Deep Traps

PersL materials generally consist of a matrix (or host), luminescent centers, and traps, all of which have a significant impact on the PersL characteristics. The matrix composition can significantly influence the band structure of materials, including the bandgap width and the relative relationship between the matrix band structure and the luminescent center energy levels, as well as the coordination environment of the luminescent centers, including coordination number, symmetry, and crystal field strength. These factors can alter the excitation wavelength, emission wavelength, and trap depth of the materials. In this chapter, deep-trap PersL materials are categorized into four types based on the major compositions of the matrix, including (i) halide or oxyhalide compounds, (ii) sulfides, (iii) oxides, and (iv) nitride or oxynitride compounds.

The chemical composition, usable excitation light sources, PersL wavelength (emission peak), and trap depth (or TL peak temperature) of these materials are summarized in Table 9.1.

In Table 9.1, the trap depth (measured in eV) primarily reflects the storage stability of trapped charge carriers at room temperature. If the trap depth of a material is not provided in the literature, the peak temperature of the TL peak temperature (measured in K) is used to indicate the storage stability. Notably, the peak temperature of the thermally stimulated luminescence curve may vary when the same materials are irradiated with light of different wavelengths or when different heating rates are applied after irradiation. Additionally, different estimation methods of the trap depth, including the commonly used methods such as the deconvolution method [131], initial rise method [132], Randall-Wilkins model [133], full width at half maximum method [134], and trap depth distribution analysis method [135], may yield different estimates of the trap depth for the same TL glow curve.

9.3.1 Halides or Oxyhalides

Halides (metal halides) generally have a large bandgap width, and they are typically excited by using high-energy rays as light source for energy storage. The stringent excitation conditions make halide-based deep-trap PersL materials primarily applicable in high-end medical and military fields, such as radiation dosimeters and infrared detectors.

MFX is a typical X-ray excited deep-trap PersL material, among which *M* represents one or more alkali-earth metals such as Ba, Sr, or Ca, and *X* represents one or more halogens such as Cl, Br, or F. *M* can be replaced by divalent or trivalent rare earth elements such as Eu^{2+}, Sm^{2+}, or Sm^{3+} as luminescent centers. As early as the 1980s, researchers from Fuji Photo Film Co. Ltd. in Japan embarked on the exploration and investigation of $BaFX:Eu^{2+}$ single crystals, focusing on their scanning laser-stimulated luminescence (SLSL) studies [42]. For example, they used X-ray radiation on $BaFCl:Eu^{2+}$ single crystals to induce two absorption bands within the range of 430–630 nm. Notably, due to the significant anisotropy of $BaFCl:Eu^{2+}$ single crystals, the main absorption peak is located about 550 nm when the X-rays are perpendicular to the *c*-axis, while it appears around 440 nm when the X-rays are parallel to the *c*-axis [43]. The researchers used a light beam within the range of 430–630 nm as the excitation source to irradiate the $BaFCl:Eu^{2+}$ single crystals, resulting in ultraviolet emission at 385–405 nm and a similar excitation spectrum to the absorption spectrum. In addition, the researchers proposed an energy band model to explain the PSL in the X-ray-irradiated $BaFX:Eu^{2+}$ phosphor. According to their model, the X-ray-irradiated $BaFX:Eu^{2+}$ crystals created a monovalent F^+ color center ($V_F^{\bullet}$), which acted as a metastable trap state for capturing the electron from Eu^{2+} luminescent center. When exposed to excitation light, the metastable state, represented as $[V_F^{\bullet} + e']$, released the trapped electrons, which subsequently returned to the luminescent center to produce PersL [136]. The *M*FX crystals, as the

Table 9.1 Representative deep-trap PersL materials and their major properties.

Group	Composition	Excitation source used in literatures	Emission peak (nm)	Trap depth (eV) or TL peak (K)	Ref. No.
Halides	$BaFX:Eu^{2+}$ (X = Cl, Br, F)	X-ray	385–405	2.0–2.5 eV	[42, 43]
	$BaFCl:Sm^{3+}/Sm^{2+}$	UV	688	3.1 eV	[54–56]
	$KBr:In^{+}$	X-ray	428, 517	2.1 eV	[57]
	$RbBr:Ga^{+}$	X-ray	560	1.85 eV	[58]
	$Cs_2NaYF_6:Ce^{3+}$	X-ray	360	2.25 eV	[59, 60]
	$NaLuF_4:Ln^{3+}$	X-ray	350–800	0.5–0.9 eV	[61]
	$NaYF_4:Ln^{3+}$	X-ray	480–1060	0.73–1.05 eV	[62]
	$NaMgF_3:Tb^{3+}$	X-ray	488, 548, 584, 621	1.08 eV	[63]
	$CsCdCl_3:Mn^{2+}, R^{4+}$ (R = Ti, Zr, Hf, Sn)	X-ray	570–605	1.41 eV	[64]
	$KZnF_3:Mn^{2+}$	X-ray	587	0.52–0.96 eV	[65]
	$NaLuF_4:Mn^{2+}$	X-ray	525	0.75 eV	[66]
	$Ba_2B_5O_9Br:Eu^{2+}$	X-ray	420	1.19 eV	[67]
	$Ca_{2-x}Sr_xBO_3Cl:Eu^{2+}, Dy^{3+}$	UV	420, 585	0.6–1.0 eV	[68]
Sulfides	$CaS:Eu^{2+}, Sm^{3+}$	UV to visible light	643	1.1 eV	[69]
	$SrS:Eu^{2+}, Sm^{3+}$	UV to visible light	600	1.1 eV	[70]
	$Y_2O_2S:Eu^{3+}, Mg^{2+}, Ti^{4+}$	UV	625	0.88 eV	[71, 72]
	CaZnOS (activator/trap = Pr^{3+}, Tb^{3+}, Ho^{3+}, Dy^{3+}, Tm^{3+}, Yb^{3+}, Nd^{3+}, or Er^{3+}) (trap filler= Cu^{+}, Y^{3+}, Pb^{2+})	UV	400–850	~0.68 eV	[73]

Oxides	$Lu_2O_3:Tb^{3+}$	UV	550	0.7–1.0 eV	[74]
	$Lu_2O_3:Pr^{3+}, Hf^{4+}$	X-ray or UV	621	1.14, 1.69, 2.11 eV	[75]
	$Al_2O_3:C$	β-ray or γ-ray	420	460 K	[76, 77]
	$MgO:Tb^{3+}$	β-ray	550	573 K	[78]
	$Ba_2SiO_4:Eu^{2+}, Ln^{3+}$ (Ln = Ho, Dy, Tm)	UV	504	0.7–1.0 eV	[79, 80]
	$BaSi_2O_5:Eu^{2+}, Nd^{3+}$	UV	515	1.29 eV	[81]
	$Sr_3SiO_5:Eu^{2+}, Dy^{3+}$	UV	570	1.05 eV	[82]
	β-$Sr_2SiO_4:Eu^{2+}, Tm^{3+}$	UV	470, 540	1.35 eV	[83]
	$LiYSiO_4:Ce^{3+}$	X-ray or UV	405	520 K	[84]
	$LiLuSiO_4:Ce^{3+}, Tm^{3+}$	β-ray or UV	410	0.7 eV	[85]
	$Y_2GeO_5:Pr^{3+}$	UV	492, 621	0.9, 1.31 eV	[86]
	$Zn_2GeO_4:Mn^{2+}$	UV	536	345 K	[87, 88]
	$MgGeO_3:Mn^{2+}, Yb^{3+}$	UV	677	0.99 eV	[89]
	$MgGeO_3:Mn^{2+}, Eu^{3+}$	UV	677	1.49 eV	[89]
	$Mg_3Y_2Ge_3O_{12}:Tb^{3+}$	UV	382, 414, 435, 484, 541, 583, 617	0.91–1.24 eV	[90]
	$BaZrGe_3O_9:Pr^{3+}$	UV	615	0.85–0.98 eV	[91]
	$NaLuGeO_4:Bi^{3+}, Cr^{3+}$	UV	400	439 K	[92]
	$LiScGeO_4:Bi^{3+}$	UV	365	408, 521 K	[93]
	Mg_2SnO_4	UV	498	383, 441, 486 K	[94]
	$Mg_2SnO_4:Mn^{2+}$	UV	497	0.71, 0.91 eV	[95]
	$Mg_2SnO_4:Cr^{3+}$	UV to blue light	800	0.5–1.3 eV	[96]
	$CaSnO_3:Bi^{2+}$	UV to near-infrared (NIR)	810	0.8, 1.4 eV	[97]

(Continued)

Table 9.1 (Continued)

Group	Composition	Excitation source used in literatures	Emission peak (nm)	Trap depth (eV) or TL peak (K)	Ref. No.
	$CaSnSiO_5:Dy^{3+}$	UV	426, 488, 580, 666	420, 520 K	[98, 99]
	$Sr_2SnO_4:Eu^{3+}, Nd^{3+}$	UV	595	0.65–0.99 eV	[100]
	$Zn_2SnO_4:Cr^{3+}, Eu^{3+}$	UV	800	317, 364, 401 K	[101, 102]
	$12CaO{\cdot}7Al_2O_3:Eu^{2+}, Mn^{2+}$	UV	444	0.64, 0.86 eV	[103]
	$12CaO{\cdot}7Al_2O_3:Tb^{3+}$	X-ray or UV	541	0.72, 0.95 eV	[104]
	$LiGa_5O_8:Cr^{3+}$	UV	716	420, 490 K	[44]
	$LiGa_5O_8:Mn^{2+}$	UV	510, 625	0.85–1.27 eV	[45]
	$Zn_{1-2x}Li_xGa_{2+x}O_4:Cr^{3+}$	UV or X-ray	709	0.70–0.72 eV	[105]
	$BaGa_2O_4:Pr^{3+}$	UV	485, 628,651	0.65, 0.75 eV	[106]
	$BaGa_2O_4:Pr^{3+}, Zn^{2+}$	UV	485, 628,651	0.66, 0.70, 0.76 eV	[106]
	$MgGa_2O_4:Fe^{3+}$	UV	703	0.67, 0.95 eV	[107]
	$Zn(Ga_{1-x}Al_x)_2O_4:Cr^{3+}, Bi^{3+}$	UV	695	333–573 K	[108]
	$Y_3Al_{5-x}Ga_xO_{12}:Ce^{3+}, Cr^{3+}$	UV to blue light	525–555	0.41–1.2 eV	[109, 110]
	$Y_3Al_{5-x}Ga_xO_{12}:Ce^{3+}, V^{3+}$	UV to blue light	525–555	1.13–1.62 eV	[111]
	$Y_3Al_{5-x}Ga_xO_{12}:Cr^{3+}$	UV	690	297–545 K	[112]
	$Y_3Al_2Ga_3O_{12}:Pr^{3+}$	UV to visible light	487, 607	1.15, 1.54	[113]
	$Gd_3Al_{5-x}Ga_xO_{12}:Cr^{3+}, Eu^{3+}$	UV	695, 713	355–498 K	[114]
	$Ca_4Ti_3O_{10}:Pr^{3+}, Y^{3+}$	UV to visible light	612	0.9–0.98 eV	[115]
	$CaZrO_3$	UV	410	1.13, 1.55 eV	[116]
	Ca-Al-Si-O:$Ce^{3+}/Tb^{3+}/Pr^{3+}$ glass	NIR laser	B/G/R	—	[117]
	Zn-Si-B-O:Mn^{2+} glass	UV	590	>360 K	[118]
	Zn-Si-B-O:Mn^{2+}, Yb^{3+} glass	UV	605, 980	0.80, 0.98 eV	[119]

Nitride	$CaAlSiN_3:Eu^{2+}, Tm^{3+}$	UV	635	330–430 K	[120]
	$(Ca_{1-x}Sr_x)_2Si_5N_8:Eu^{2+}, Tm^{3+}$	UV	604–630	0.64, 0.72 eV	[121, 122]
	$SrLiAl_3N_4:Eu^{2+}$	UV	650	0.47, 0.81 eV	[123]
	$CaSi_{10\text{-}n}Al_{2+n}O_nN_{16\text{-}n}: xEu^{2+}$ (Ca-αSialon:Eu^{2+}, $n = 0\sim1$, $x = 0.1\%\sim8\%$)	UV	595	0.65, 0.85 eV	[124]
	$Ba[Mg_2Al_2N_4]:Eu^{2+}, Tm^{3+}$	visible light	710, 800	100–400 K	[125]
	$SrSi_2O_2N_2:Eu^{2+}, Ln^{3+}$	UV or blue light	540	0.90–1.18 eV	[20, 126]
	(Ln = Dy, Ho, Er)				
	$SrSi_2O_2N_2:Yb^{2+}, Ln^{3+}$	UV or blue light	620	0.90–1.17 eV	[20, 126]
	(Ln = Dy, Ho, Er)				
	$BaSi_2O_2N_2:Eu^{2+}$	UV	494	0.67–0.82 eV	[127]
	$YSiO_2N:Ce^{3+}, Sm^{3+}$	UV	400, 427	1.10 eV	[128]
	$YSiO_2N:Ce^{3+}, Tm^{3+}$	UV	400, 427	0.80 eV	[128]
	$Si_{6\text{-}z}Al_zO_zN_{8\text{-}z}:Eu^{2+}$ (β-Sialon:Eu^{2+}, $0 < z \leq 4.2$)	UV	537	0.28–1.23 eV	[129]
	$(Ba_{1-x}Sr_x)AlSi_5O_2N_7:Yb^{2+}, Yb^{3+}$	UV or blue light	664, 985	0.84, 1.32 eV	[130]

main constituent materials of the first-generation X-ray imaging plates, have been widely used in early medical X-ray fluoroscopy, X-ray dosimetry, and crystal structure analysis. They exhibit advantages such as high sensitivity, reusability, and ease of digital integration but also have drawbacks such as low spatial resolution, rapid signal attenuation, and limited reuse times [136, 137].

Alkaline-earth metal halide BaFCl:Sm^{3+}/Sm^{2+} nanocrystals were systematically investigated by Riesen for optical data storage application [54–56]. The valence state of the nanocrystals was switched from Sm^{3+} to Sm^{2+} with exposure to a UV-C lamp (185 nm). Subsequently, a characteristic emission signal of Sm^{2+} appeared at 688 nm under 405 nm readout light irradiation. Bleaching of these nanocrystals with $Sm^{2+} \rightarrow Sm^{3+}$ process occurred when increasing the readout power at 405 nm from 50 to 220 μW, resulting in the erasure of the initially written information data [55]. The process of $Sm^{3+} \rightarrow Sm^{2+}$ conversion in BaFCl is shown to be tunable over a large dynamic range by adjusting the power of short-wavelength ultraviolet light and, therefore, could serve as a platform for reversible ultra-high-density multilevel encoding [56].

Alkali metal halides also exhibit X-ray-charged electron storage capacity and photo-stimulation efficiency comparable to the above-mentioned BaFBr:Eu^{2+} or BaFCl:Eu^{2+} compounds. Thanks to the cubic-phase crystals, alkali metal halides can mitigate photon scattering in polycrystalline states. After being irradiated with X-rays, the formed F^+ color center ($V_F^{\bullet}$) and In^+ ions ($In_K^{\times}$) in KBr:In^+ may serve as electron and hole traps, respectively. Under 580 nm (2.1 eV, corresponding to the F band) excitation, dual-emission peaks at 428 nm (2.9 eV) and 517 nm (2.4 eV) were observed [57]. Before X-ray irradiation, no luminescence was observed when exciting the RbBr:Ga^+ samples with light in the 600–800 nm range. However, after X-ray irradiation, a new absorption band at 670 nm (1.85 eV) emerged, indicating the presence of electron trap centers known as F^+ color centers ($V_F^{\bullet}$). Additionally, the PersL peaked at 560 nm could be attributed to the hole trap centers identified as Ga^+ centers ($Ga_{Rb}^{\times}$) [58]. Furthermore, Cs_2NaYF_6:Ce^{3+} showed an absorption band at 550 nm (2.25 eV) and photoluminescence at 360 nm (3.44 eV) upon X-ray irradiation. The absorption band was likely attributed to the F^+ color center ($V_F^{\bullet}$) as well [59, 60].

In general, microscale halide polycrystalline materials exhibit a significant scattering effect for light transmission. To suppress the light scattering, Chen and coworkers reported a class of cubic-phase $NaLuF_4$:Ln^{3+} nanoparticles with high spatial resolution of X-ray imaging [61]. $NaLuF_4$:Tb^{3+}@$NaYF_4$ nanoscintillators, consisting of uniform spherical particles with an average size of 27 nm, exhibited enhanced luminescence due to core–shell encapsulation. These nanoparticles demonstrated a detectable strong PersL intensity even after exposure to low-dose X-ray irradiation and being stored at room temperature for 30 days. Moreover, by doping different lanthanide ions into $NaLuF_4$:Ln^{3+}@$NaYF_4$ nanoparticles, it is possible to achieve multicolor radioluminescence modulation ranging from the ultraviolet-visible to the near-infrared. The trap states with relatively shallow trap depths ranging from 0.5 to 0.9 eV were formed through elastic collisions of large-momentum X-ray photons with small fluoride ions. These collisions resulted in the creation of fluoride vacancies as electron traps and interstitials as hole traps. Importantly, $NaLuF_4:Ln^{3+}@NaYF_4$ nanoparticles offer several advantages, including extremely low X-ray detection thresholds, high luminescence efficiency, and uniform dispersion in organic matrices such as polydimethylsiloxane (PDMS). These nanoparticles hold great potential for applications in next-generation flexible, highly sensitive, and high spatial resolution X-ray delay imaging, as well as wearable X-ray detection devices. At nearly the same time, Zhuang and Xie reported core-shell encapsulated PersL-enhanced $NaYF_4:Ln^{3+}@NaYF_4$ nanoparticles [62]. By introducing different Ln^{3+} ions, the PersL emissions of $NaYF_4:Ln^{3+}@NaYF_4$ nanoparticles can be achieved in green (Tb^{3+}), white (Dy^{3+}), red (Ho^{3+}), orange (Eu^{3+}), or near-infrared (Nd^{3+}). Upon X-ray irradiation, numerous ionized electrons are generated through cascading collisions of highly energetic electrons. The Ln^{3+} (i.e. Tb^{3+}) was recognized as electron traps and activators consequently. Furthermore, different types of optical inks were prepared, each containing well-dispersed PersL nanoparticles with specific Ln^{3+} ions such as Tb^{3+}, Dy^{3+}, and Ho^{3+} (Figure 9.2a,b). These inks were then utilized to print three images on a sapphire substrate using high-precision inkjet printing technology. The encoded information was readout using wavelength decoding, enabling a multidimensional optical information storage (Figure 9.2c–f). Zhuang and Xie also reported another core-shell $NaMgF_3:Tb^{3+}@NaMgF_3$ nanoparticles with nanocubic morphology, excellent aqueous dispersibility, and light-storage ability [63]. According to the proposed mechanism, Tb^{3+} ions can serve as both emitting and

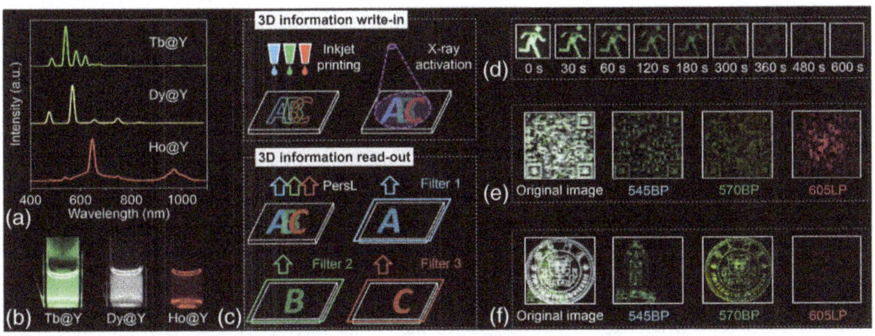

Figure 9.2 (a) Persistent luminescent spectra of the $NaYF_4:Tb^{3+}@NaYF_4$ (Tb@Y), $NaYF_4:Dy^{3+}@NaYF_4$ (Dy@Y) and $NaYF_4:Ho^{3+}@NaYF_4$ (Ho@Y) nanoparticles after X-ray irradiation. (b) Persistent luminescent images of the Tb@Y, Dy@Y, and Ho@Y nanoparticles dispersed in cyclohexane. (c) Schematic illustration of the application in 3D optical information storage. The upper and lower figures illustrate the information write-in and readout processes, respectively. Three independent overlapping patterns were printed on a glass sheet. The luminescent patterns were activated after X-ray irradiation. Mixed information was read directly by an image sensor, and the patterns could be correctly decomposed by using specific optical filters. (d) Persistent luminescent images of a printed pattern (emergency exit sign) with a delay time from 0 to 600 seconds. The pattern was printed using the $NaYF_4:Tb^{3+}@NaYF_4$ ink and irradiated by the X-ray source for five minutes. (e–f) The original and monochromatic images after passing through 545 band pass (545BP), 570BP, and 605 long pass (605LP) filters. Source: [63] / Springer Nature / CC BY 4.0.

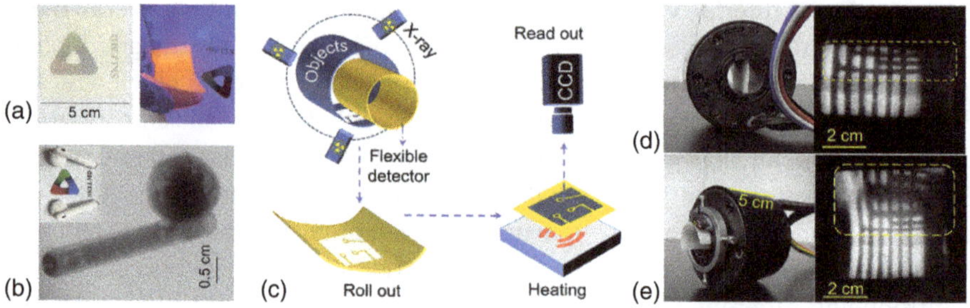

Figure 9.3 (a) Photographs of a PMMA-polymer film containing 40 wt% $CsCdCl_3$:5%Mn^{2+}, 0.1%Zr^{4+} crystals under ambient and UV light conditions. (b) X-ray images of an earphone, with an inset in the top panel displaying the earphone under ambient light. (c) A schematic diagram illustrating nondestructive inspection of 3D curved objects facilitated by the PDMS-polymer film containing 10 wt% $CsCdCl_3$:5%Mn^{2+}, 0.1%Zr^{4+} crystals. The film is initially inserted into the target curved object, and the X-ray source is turned to excite the film, creating a latent image. After turning off the X-rays, the detector is placed on a hot substrate for thermal stimulation and reading the latent images. (d) Imaging of an annulus electric conduction link using the flat-panel PDMS-polymer film (5 cm × 2.5 cm). (e) Curved planar reformation of the same object using a highly flexible PDMS-polymer film (5 cm × 5 cm). The image readout temperature is 100 °C. The X-ray operation parameters in these measurements were set to a voltage of 50 kV with a tube current of 0.2 mA, resulting in an X-ray dose rate of 20.12 mGy s^{-1}. Source: Reproduced with permission from [64], 2023 / John Wiley & Sons.

trap centers in the $NaMgF_3$ host during X-ray irradiation, with the average trap depth reaching 1.08 eV. These remarkable features make the developed nanoparticles suitable for use in optical information storage and information encryption applications.

Xia and coworkers have rationally designed and synthesized doped halide perovskite $CsCdCl_3$:Mn^{2+}, R^{4+} (R = Ti, Zr, Hf, and Sn) for use as X-ray storage phosphors, and the capability has been significantly enhanced by trap management via Mn^{2+} site occupation manipulation and heterovalent substitution [64]. Moreover, X-ray imaging with a resolution of 12.5 lp mm^{-1} has been demonstrated in the PMMA-based detector, allowing for convenient 3D X-ray imaging of curved objects in a time-lapse fashion (Figure 9.3).

Recently, optical information storage using deep-trap PersL fluorides has been extended to a wider range of applications. For example, time–temperature indicators (TTIs) have the capability to convert time–temperature history into readily accessible information, enabling the monitoring of the quality of perishable foods, pharmaceuticals, or specialty chemicals throughout the entire transport and storage process. Current TTI technologies, such as diffusion-based TTIs and polymerization-activated TTIs, suffer from limitations in recyclability, sensitivity, and environmental tolerance, which restrict their broad adoption. Zhuang and Xie et al. developed light-storage-based TTI technology that exhibits exceptional recyclability, reliability, and environmental stability [65]. As shown in Figure 9.4, PersL materials serve as the basis for TTI design, relying on the principle that the release rate of light-induced trapped carriers in these materials correlates with

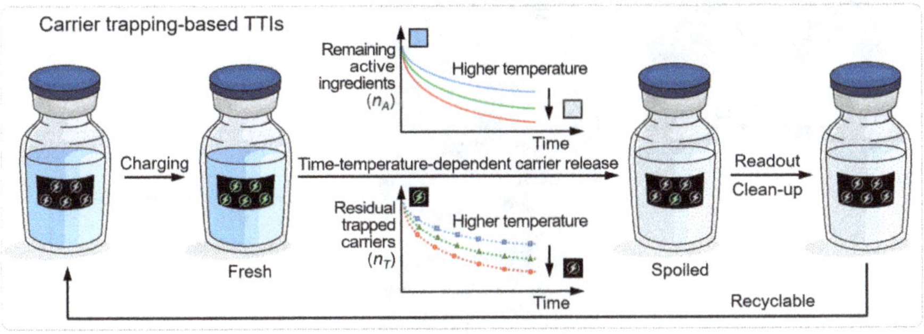

Figure 9.4 Schematics of light-storage-based TTIs. The release of trapped carriers is closely dependent on the storage temperature and time. The number of residual trapped carriers (n_T) can be readout by using a TL method. The light-storage-based TTIs are recyclable. The top inset data of the remaining active ingredient (n_A) were cited for comparison [38, 138]. Source: Reproduced with permission from [65], 2023 / John Wiley & Sons.

storage time and temperature. This correlation allows us to assess product freshness based on the number of residual carriers. The $KZnF_3:Mn^{2+}$ fluoride submicron particles with a broad trap distribution from 0.52 to 0.96 eV were suitable for potential TTI applications.

In another interesting example, the optical information storage by PersL fluorides is applied to latent fingerprint (LFP) recognition. In general, an ideal latent LFP recognition technology should encompass comprehensive details, spanning from Level 1 (overall fingerprint shape, core point, delta, etc.) to Level 2 (ending, bifurcation, island, short ridge, etc.) and finally to Level 3 (including details such as ridge width, edge shape, and pore size). Zhuang and Xie et al. developed $NaLuF_4:Mn^{2+}$ nanoparticles with intense green PersL and functionalized surfaces for time-gated imaging of LFPs [66]. The nanoparticles exhibited tight chemical bonding (–COOH) with amino acids of fingerprints, giving intense grayscale values either at ridges or at furrows (Figure 9.5a,b). The high signal-to-noise ratio allowed for detailed analysis of LFPs, covering details from Level 1 to Level 3 (Figure 9.5c,d).

Halogen oxides and halides have similar physicochemical properties and have been widely used in X-ray detection and light energy storage. Meijerink and Blasse reported $Ba_2B_5O_9Br:Eu^{2+}$ exhibited an absorption band at around 620 nm and PSL at 420 nm after X-ray excitation [67]. The TL peaks were observed around 330 and 415 K, confirming the presence of traps. The X-ray storage and photoluminescence mechanism is like those of $BaFX:Eu^{2+}$ materials. The main electron and hole traps were attributed to bromine vacancies ($V_{Br}^{\bullet}$) and Eu^{2+} ions ($Eu_{Ba}^{\times}$), respectively. Wang and coworkers have reported a series of $Ca_{2-x}Sr_xBO_3Cl:Eu^{2+}$, Dy^{3+} samples with white light PersL (420 nm from Sr^{2+} solid solution and 585 nm from doped Eu^{2+} ions) [68]. As the Sr^{2+} solid solution concentration increases, there is a significant improvement in the trap depth and width of the samples, which results in PersL duration of up to 46 hours and improved mechanoluminescence (ML) properties.

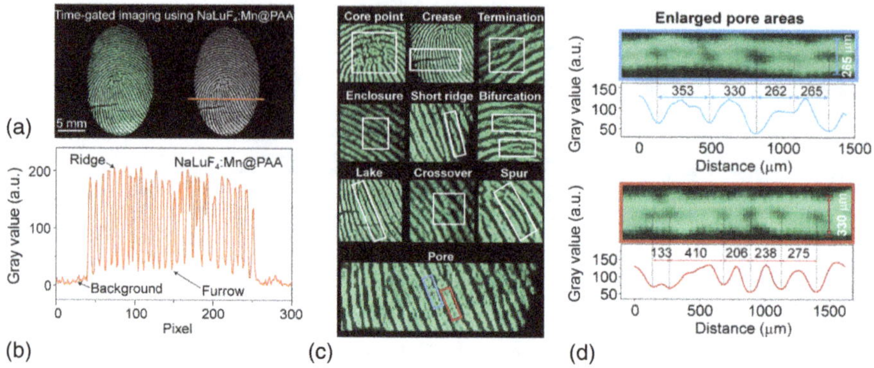

Figure 9.5 (a) Photographs and grayscale images of LFPs deposited on glass by using $NaLuF_4$:Mn@PAA. (b) The variations in grayscale values along the highlighted straight line in (a). (c) The enlarged details of LFPs, including Level 1, Level 2, and Level 3, were developed using the $NaLuF_4$:Mn@PAA nanoparticles. (d) The enlarged pore areas and their corresponding grayscale value distribution. Source: Reproduced with permission from [66], 2022 / American Chemical Society.

9.3.2 Sulfides

Sulfide PersL materials were among the earliest to be discovered and studied. By doping with lanthanide ions or transition metal ions, sulfides with different emission wavelengths and trap depths can be obtained. The sulfides with deep traps are suitable for optical information storage, including CaS:Eu^{2+}, Sm^{3+} and SrS:Eu^{2+}, Sm^{3+}. In 1993, Tamura et al. from Japan NTT Opto-Electronics Laboratories reported the photoluminescence and infrared-stimulated luminescence properties of CaS:Eu^{2+}, Sm^{3+} [69]. The sulfide stored light energy with the excitation wavelength range from 220 to 650 nm, which coincided with the $4f^7$ to $4f^65d^1$ and $4f^66s^1$ transitions of Eu^{2+} ion. Under near-infrared (950–1300 nm) photo-stimulation, the phosphor emitted deep-red light with a main peak at 643 nm, which coincided with the photoluminescence of Eu^{2+} ion. The UV-visible light excitation spectra (250–550 nm) and near-infrared stimulated luminescent spectra (800–1700 nm) of the SrS:Eu^{2+}, Sm^{3+} were similar to those of CaS:Eu^{2+}, Sm^{3+}. Due to the relatively weaker crystal field strength of SrS, the peak wavelength of SrS:Eu^{2+}, Sm^{3+} was blue-shifted to around 600 nm [70]. Among them, Sm^{3+} ions stored the migrated electrons and acted as electron traps ($Sm_{Ca}^{\cdot}$or $Sm_{Sr}^{\cdot}$), and Eu^{2+} ion acted as luminescent centers [139]. Notably, higher doping concentrations of Sm^{3+} ions might result in Sm^{3+} ion-contributed emissions (with peak wavelengths at 560 or 600 nm) [70, 140].

In the late twentieth century, optical storage and photoluminescent properties of halide and sulfide materials met the requirements for data storage density, data transfer speed, and data erasure speed of computer processors at that time and were used in computer logic and computation design. In 1990s, Jutamulia et al. designed a spatial light modulator for parallel Boolean logic operations using $(Ca_xSr_{1-x})S$:Eu^{2+}, Ce^{3+}, Sm^{3+} as the information recording material [141].

They verified 16 kinds of parallel Boolean logic in a 2×2 plane matrix using ultraviolet light as the write-in beam and near-infrared light as the readout beam and achieved a storage density of 40 000 bit cm^{-2}, which was a relatively high value at that time. However, some main drawbacks limit the further development of these sulfide PersL materials. Firstly, the luminescent performance (especially the PersL brightness) lags behind later-developed silicate or aluminate PersL materials such as $SrAl_2O_4$:Eu^{2+}, Dy^{3+}. Additionally, sulfides lack chemical stability and are prone to oxidation and moisture absorption, making them unsuitable for long-term storage. Moreover, the production process releases harmful gases, posing a significant environmental threat. These issues have led to the gradual replacement of sulfide PersL materials by other categories of materials.

Liu and coworkers reported Y_2O_2S:Eu^{3+}, Mg^{2+}, Ti^{4+} nanoparticles with uniformity and well-dispersed distribution ranging from 30 to 50 nm in size [71]. Emission peaks in the spectrum are attributed to Eu^{3+} ions transition and the electron traps formed by the co-doped Mg^{2+} and Ti^{4+} ions [72]. As shown in Figure 9.6, Wang and coworkers developed systematic tuning of PersL in a wurtzite CaZnOS crystal through synergistic defect engineering [73]. The authors introduce a strategy for deliberate PersL control in a single material system, simultaneously managing activator/trap states and trap-filling processes through combinatorial doping. This design approach yields a diverse array of emitters with distinct electronic

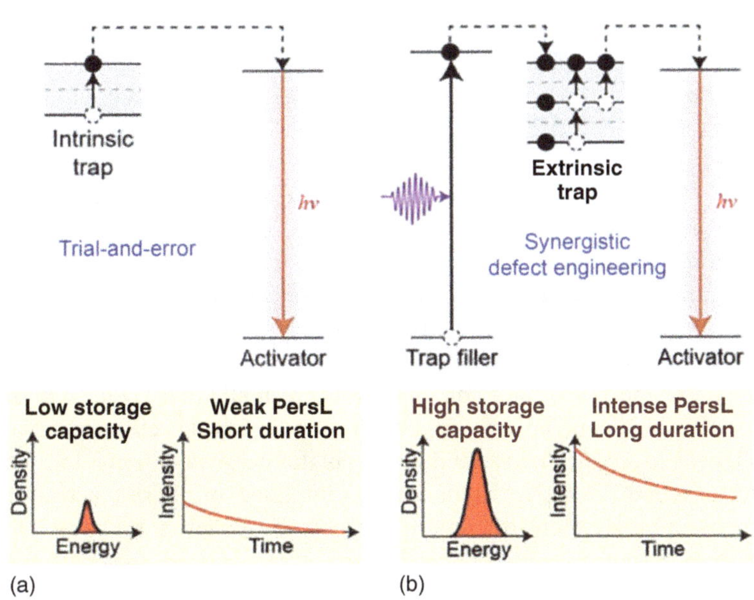

Figure 9.6 (a) Trial-and-error method for discovering Persistent luminescent phosphors, relying on inefficient cooperations between intrinsic host lattice traps and activators, resulting in low storage capacity and poor Persistent luminescent. (b) Synergistic defect engineering method for designing Persistent luminescent phosphors, which exploits simultaneous regulation of the activator/trap state and trap-filling process, leading to high storage capacity and enhanced Persistent luminescent. Source: Reproduced with permission from [73], 2023 / John Wiley & Sons.

transitions, including ns^2-nsnp, 3d–3d, 4f–4f, and donor–acceptor recombination, and creates PersL properties with unprecedented stimulus-responsive features.

9.3.3 Oxides

Compared to fluorides and sulfides, oxide PersL materials typically have a moderately narrow bandgap width. Most oxides can be charged and store energy through UV radiation, even visible light. Moreover, oxide phosphors also demonstrate good chemical stability, ease of synthesis, preparation, and storage and are suitable for making ceramic materials. In recent years, the research on deep-trap PersL materials has primarily focused on oxides.

9.3.3.1 Monobasic Cation Oxide

Zych et al. reported Lu_2O_3:Tb^{3+} with photo-stimulation (980 and 647 nm) and thermal stimulation (300 °C) luminescence upon UV irradiation [74]. During vacuum sintering, oxygen vacancies were created, which could serve as traps to store electrons that escaped from Tb^{3+} ions, leading to the formation of F^+ and F metastable states. The holes left in Tb^{3+} ions could either form a Tb^{4+} center or escape the terbium and get trapped as V_k centers. When NIR stimulation was applied to the materials, they could release partial electrons from the F centers, which subsequently recombined with Tb^{4+} ion to give Tb^{3+} emission. More energetic red light could liberate the other electrons from the F^+ center, resulting in additional intense PersL emission at ~550 nm. Holes trapped in V_k luminescent centers could recombine with electrons from the CB, especially at higher temperatures. Wiatrowska and Zych further investigated Lu_2O_3:Pr^{3+}, Hf^{4+} phosphor with PersL emission at 621 nm, which could effectively store energy under X-ray or UV light irradiation [75]. The main TL peak was located at ~520 K, and the relative trap depth was estimated to be approximately 1.69 eV. Meanwhile, a 30% fading of TL intensity was recorded after nine months at room temperature upon 254 nm irradiation. The hole traps may originate from the luminescent center Pr^{3+} ($Pr_{Lu}^{\times}$) and interstitial oxygen (O_i''). Because Hf^{4+} is smaller and Pr^{3+} is larger than Lu^{3+}, aggregated $Hf_{Lu}^{\cdot}$ defect led to intense electron trap-induced PersL intensity.

Al_2O_3:C crystal exhibits a low radiation detection threshold (10^{-6} Gy for γ-rays) and has been extensively studied as a high-energy radiation detection material [76, 77]. The TL peak of Al_2O_3:C crystal induced by high-energy radiation is located at 460 K, and the center wavelength of photo-stimulated or thermo-activated PersL emission is approximately 420 nm. It is generally believed that the primary electron trap responsible for optical storage in Al_2O_3:C is the F^+ color center ($V_O^{\cdot}$), and the TL process involves the electron escaping from the trap and recombining with another F^+ color center to form an excited-state F color center ($[V_O^{\times}]^*$) that returns to the ground state, resulting in emission. Bos et al. reported another high-performance, high-energy radiation detection material MgO:Tb^{3+} phosphor [78]. The main TL peak of this material excited by β-ray is located at 573 K, and the blue and green emission originate from Tb^{3+} electron transitions. MgO:Tb^{3+} exhibits a high TL intensity and superior sensitivity to β-radiation compared to Al_2O_3:C under the same conditions (approximately 1.7 times), with a wide linear range (10^{-3}–10^4 Gy). Only 7% fading of TL intensity was obtained after being exposed to high-energy radiation and kept at room temperature for 36 hours.

9.3.3.2 Silicate/Germanate/Stannate

Deep-trap PersL materials in silicates are usually achieved by using Eu^{2+} or Ce^{3+} ions as luminescent centers and doping another trivalent rare earth ion as trap centers. Jin and coworkers reported a series of green PersL materials in Ba_2SiO_4:Eu^{2+}, Ln^{3+} [79, 80]. Ba_2SiO_4:Eu^{2+}, Ho^{3+} exhibited a peak temperature of approximately 350 K in TL glow curves, and the estimated average trap depth was about 0.72 eV. Ba_2SiO_4:Eu^{2+}, Ho^{3+} maintained above 2.97 mcd m^{-2} even 24 hours after the removal of the UV irradiation sources. Due to the tailoring of trap distribution via Ho^{3+} co-doping, Ba_2SiO_4:Eu^{2+}, Ho^{3+} could store charge carriers under UV excitation and maintained it at room temperature for 15 days, and showed observable green luminescence by 980 nm photo-stimulation [79]. Various traps, i.e. Dy^{3+} and Tm^{3+} ions, were intentionally introduced and presented discrete energy levels of 0.84 and 1.02 eV, respectively. Moreover, the TL intensity showed an obvious dependence on the co-doping concentrations of different Ln^{3+} ions at room temperature [80]. Lin et al. prepared a novel $BaSi_2O_5$:Eu^{2+}, Nd^{3+} PersL phosphor and further introduced into phosphor-in-glass (PiG) as light-storage materials [81]. $BaSi_2O_5$:Eu^{2+}, Nd^{3+} phosphor contained deep traps (1.29 eV) in a narrow distribution (0.16 eV), resulting in a remarkably low signal attenuation (74% signal remained after delay for 48 hours). Zhang and coworkers systemically compared the PersL and 808 nm PSL characteristics of Sr_3SiO_5:Eu^{2+} and Sr_3SiO_5:Eu^{2+}, Dy^{3+} [82]. The introduction of Dy^{3+} ions into Sr_3SiO_5:Eu^{2+} can generate intense deep trap (near 1.05 eV) and greatly prolong the PersL lifetime up to six hours after UV irradiation. Other reported silicate deep-trap PersL materials include Sr_2SiO_4:Eu^{2+}, Tm^{3+}, $LiYSiO_4$:Ce^{3+}, $LiLuSiO_4$:Ce^{3+}, and Tm^{3+} [83–85].

In the group of germanate materials, Y_2GeO_5:Pr^{3+} consists of two isolated traps, with the shallow trap located at approximately 0.9 eV and the deep traps narrowly distributed at approximately 1.31 eV. The isolated deeper traps, which would not be influenced by the fading of the shallow traps, were rather stable under thermal stimulation at room temperature and then emitted PersL by NIR photo-stimulation [86]. PersL nanoparticles, capable of emitting light even after excitation, have gained importance in biomedicine for their ability to eliminate tissue autofluorescence. Yuan and coworkers developed a direct synthesis of Zn_2GeO_4:Mn PersL nanorods [87]. By adjusting the pH of the hydrothermal reaction system, precise tunability of the emission wavelength of Zn_2GeO_4:Mn nanorods was obtained. These nanorods achieved strong PersL even within 30 minutes of hydrothermal treatment. Importantly, aptamer-guided Zn_2GeO_4:Mn bioprobes were developed for serum lysozyme analysis, showing promise in cancer diagnostics. Ueda and coworkers studied the effect of different trivalent rare earth ions on the trap distribution of $MgGeO_3$:Mn^{2+}, Ln^{3+} [89]. After co-doping with Yb^{3+} and Eu^{3+}, the peak temperatures of the TL spectra were located at 335 and 502 K, respectively, corresponding to trap depths of 0.99 and 1.49 eV, which could be a good candidate for the photo-stimulated functions in optical information storage applications. Krieke et al. reported a novel garnet $Mg_3Y_2Ge_3O_{12}$:Tb^{3+} with bright blue to green tunable emission [90]. The relative intensity ratio of the $^5D_3 \rightarrow {}^7F_J$ transitions was observed to decrease as the Tb^{3+} content increased, leading to the change in the emission color. The introduction of Tb^{3+} into the garnet structure reduces the presence of paramagnetic hole-type centers in the host material, suggesting that Tb^{3+} ions function as hole traps. The tunable multicolor emission of $Mg_3Y_2Ge_3O_{12}$:Tb^{3+} highlights the potential of these materials for advanced anti-counterfeiting applications. Other reported germanate deep-trap PersL materials include $BaZrGe_3O_9$:Pr^{3+}, $NaLuGeO_4$:Bi^{3+}, Cr^{3+}, and $LiScGeO_4$:Bi^{3+} [91–93].

For stannate materials, Wang et al. reported a kind of nondoped Mg_2SnO_4 phosphor with green PersL at 498 nm, which was attributed to the recombination of F centers with holes. The decay curves revealed that the PersL was governed by a tunneling mechanism, and it proved the presence of different introduced trap cluster defects in Mg_2SnO_4 (such as [$Sn_{Mg}^{\cdot\cdot}$-O_i''], [$Sn_{Mg}^{\cdot\cdot}$-2e$'$], and [$Sn_{Mg}^{\cdot\cdot}$-2e$''$]) [94]. Park et al. reported the self-reduction behavior of Mn in Mg_2SnO_4 host lattices prepared using the solid-state reaction technique in an air atmosphere. The self-reduction process of Mn can be suppressed by the co-doping of Li^+, resulting in the coexistence of Mn^{2+} and Mn^{4+} in Mg_2SnO_4 host lattices with multicolored emissions [95]. The authors broaden the potential applications of the prepared phosphors in anti-counterfeiting, fingerprint identification, and optical information storage. Lin and coworkers developed Mg_2SnO_4:Cr^{3+} as a broadband NIR PersL phosphor with peak wavelength at around 800 nm [96]. Doped with Cr ions in a tervalence state, the disordered (Mg, Sn) sites with octahedral coordination give rise to multiple Cr^{3+} centers, and the broadband luminescence originates from the $^4T_2(^4F) \rightarrow {}^4A_2$ transition at sites with intermediate crystal field strength. The distribution of trap depths is associated with the random distribution of Mg^{2+} and Sn^{4+} ions in the second coordination sphere of the Cr^{3+} centers, rather than being attributed to oxygen-related defects. When screening new Cr^{3+} activated materials for PersL and PSL properties, consider three key factors: suitable octahedral sites for Cr^{3+} doping (e.g. Zn^{2+}, Mg^{2+}, Al^{3+}, Ga^{3+}, Ge^{4+}, Ti^{4+}, and Sn^{4+}), a somewhat disordered second coordination sphere for distributing trapped levels, and appropriate band gaps. A larger band gap enhances deep trap distribution for long-term PersL and PSL, while a smaller band gap facilitates charging with low-energy light. Qiu and coworkers reported a novel paradigm of trap energy up-conversion-like NIR-to-NIR light rejuvenate PersL has been introduced in Bi^{2+}-doped $CaSnO_3$ phosphors and nanoparticles [97]. These materials demonstrate a unique phenomenon wherein carriers transition from deep-level energy traps to shallow-level traps when excited by low-energy NIR photons. This enables repetitive charging of the materials using deep-tissue penetrable NIR photons. The energy trapped in the deep-level traps can be stored for extended periods and then released as afterglow signals in the NIR range. This innovative approach paves the way for the development of NIR-absorbing and NIR-emitting PersL materials and nanoparticles that can be rejuvenated using nonharmful, deep-tissue penetrable NIR light.

Xu et al. reported $CaSnSiO_5:Dy^{3+}$ with white PersL, which was assigned to the characteristic transitions of $^4F_{9/2}{\rightarrow}^6H_J$ in Dy^{3+} ions. The trap distribution in $CaSnSiO_5:Dy^{3+}$ is closely related to the doping concentration of Dy^{3+} ions. When the Dy^{3+} ion concentration is high, two high-temperature peaks at 420 and 520 K in TL glow curves are significantly intensified, especially the latter one. The presence of deep traps could be able to immobilize the carriers permanently at room temperature and be released under 980 nm NIR stimulation for optical information storage applications [98, 99]. Zhang and coworkers reported 5 nm $ZnSn_2O_4:Cr^{3+}$, Eu^{2+} nanoparticles with NIR PersL at 800 nm for deep-tissue penetration (3 cm) and high signal-to-noise ratio (25.5) in vivo imaging [102]. $ZnSn_2O_4:Cr^{3+}$, Eu^{2+} nanoprobes, conjugated with folic acid (ZSO-FA), exhibited excellent in vitro and in vivo tumor targeting capability, facilitating accurate tumor diagnosis. Importantly, nanoprobes can be re-excited by using NIR light (808 nm) to achieve renewable NIR PersL in vivo, extending imaging time and enhancing signal-to-noise ratio for in vitro and in vivo applications.

9.3.3.3 Aluminate/Gallate

$12CaO{\cdot}7Al_2O_3:Eu^{2+}$, Mn^{2+} with unique nanocage structure was found to process outstanding PersL and PSL properties by Liu and coworkers [103]. Compared to $12CaO{\cdot}7Al_2O_3:Eu^{2+}$, the introduction of Mn^{2+} ions increased the state density of shallow traps (0.64 eV) and formed new deep-trap levels (0.86 eV) in the material, resulting in the increased PersL duration and enlarged emission intensity under 980 nm photo-stimulation. They proposed that the extra-framework oxygen vacancies ($V_O^{\cdot\cdot}$) or F^+-like centers associated with the host played a crucial role as dominant electron-trapping centers in $12CaO{\cdot}7Al_2O_3:Eu^{2+}$, Mn^{2+}. Furthermore, they studied a green-emitting nanocage aluminate material $12CaO{\cdot}7Al_2O_3:Tb^{3+}$ with two typical trap depths at around 0.72 and 0.95 eV [104]. The contents of encaged anions in lanthanide-doped phosphors, including O^{2-}, H^-, OH^-, and e^-, could be modified by different treatments (such as reduction, ultraviolet irradiation, or thermal oxidation). Thereby, the fewer encaged anions cause shallower trap depths, shorter the X-ray storage time and lower the quality of X-ray imaging. Specifically, after heating in air at 400 °C for two hours (thermal oxidation treatment), the quantity of OH^- in the $12CaO{\cdot}7Al_2O_3:Tb^{3+}$ nanocage significantly decreases, achieving a spatial resolution of around 15 lp mm^{-1}, which is a potential candidate material for computed radiography system and erasable/rewritable optical memory materials.

Pan and coworkers reported a NIR PersL gallate material, $LiGa_5O_8:Cr^{3+}$, with peak wavelength at 716 nm [44]. The TL glow curve consists of two broad bands with maxima at around 420 and 490 K, which correspond to shallow and deep traps, respectively. A new optical information readout form was proposed to make the phosphor repeatedly obtain an intense NIR PersL when an ultraviolet light (250–360 nm) preirradiated $LiGa_5O_8:Cr^{3+}$ powders in a period of more than 1000 h. The TL results indicated that the recharging excitation energy caused electrons in deeper traps to transfer to shallower traps within the CB, leading to the reappearance of NIR PersL at room temperature. This readout method has promising applications in high signal-to-noise ratio optical information storage and biomedical imaging applications. Besides, transparent glass-ceramic embedded with photo-stimulated $LiGa_5O_8$:Mn^{2+} nanocrystals as a new optical storage material was developed by Lin et al. through a heat-treatment-induced in situ precipitation method [45]. The different thermal treatment processes could alter the PersL and PSL spectra, allowing for tuning the emission color from red (625 nm) to green (510 nm). The depths of the newly generated traps in the transparent glass ceramic were evaluated to be 0.85–1.27 eV for the $LiGa_5O_8$:Mn^{2+} nanocrystals and 0.89–1.64 eV for glass, which originated from the intrinsic defects in $LiGa_5O_8$:Mn^{2+} nanocrystals (e.g. oxygen vacancies, anti-site defects, and/or defect clusters) and defects at the glass-crystalline interface (e.g. dangling bonds in the glass network), respectively. It is worth mentioning that the excellent transparency of the prepared bulk material makes 3D volumetric optical data storage possible using laser direct writing technology, providing an important material for enhancing storage capacity and improving information security. Lin and coworkers reported a series of NIR-emitting solid solutions $Zn_{1-2x}Li_xGa_{2+x}O_4$:Cr^{3+} with tunable emission in both the deep-red and NIR regions, which was attributed to typical Cr^{3+} emission [105]. The emission was achieved by crystal field engineering by replacing [Zn^{2+}–Zn^{2+}] units with [Li^+–Ga^{3+}]. Wang and coworkers reported a color-tunable PersL phosphor, $BaGa_2O_4$:Pr^{3+}, Zn^{2+}, which originates from energy transfer from host luminescence to Pr^{3+} and its cross-relaxation [106]. Pr^{3+} and Zn^{2+} occupying Ba^{2+} and Ga^{3+} sites, respectively, result in *c*-axis cell expansion, potentially leading to the generation of defects like oxygen vacancies $V_O^{\bullet\bullet}$, $Pr_{Ba}^{\bullet}$, and $Zn_{Ga}^{\bullet}$.

The spinel configuration is often considered a favorable host candidate when designing PersL materials because it readily produces vacancies and anti-site defects. Dong and coworkers introduced a novel strategy for defect enrichment, utilizing the crystal structure of a near-inverse spinel to ensure a higher density of defects and activate NIR PersL from Fe^{3+} ions doped in $MgGa_2O_4$ [107]. Moreover, TL experiments were conducted to investigate trapping and de-trapping processes, indicating that the near-inverse spinel structure could reduce accelerates carrier transfer and forms trap states within the bandgap. The high stability of the PersL in different environments shows a great potential application in biological imaging.

$ZnGa_2O_4$:Cr^{3+} phosphor possesses excellent NIR PersL characteristics at room temperature with a trap depth of approximately 0.46 eV. Also, $ZnGa_2O_4$:Cr^{3+} phosphor can be easily prepared as nanoparticles and used for surface modification and functional design, which is considered as promising optical materials for *in vivo* imaging applications [5–13]. Zhuang et al. proposed a band engineering approach to control the trap depth by shifting the energy-level location of the CB in $Zn(Ga_{1-x}Al_x)_2O_4$:Cr^{3+}, Bi^{3+} solid solution [108]. As the molar mass ratio of Al increased, the CB bottom gradually upward shifted, resulting in an increasing energy difference with the energy level of electron traps (positive valence anti-site defects $Ga_{Zn}^{\bullet}$). At the same time, the peak temperature of thermal release luminescence in the material increases from 333 ($x = 0\%$) to 573 K ($x = 33\%$).

Band engineering has subsequently been applied to other aluminate-based solid solution PersL materials. Ueda et al. developed yellow persistent phosphor of $Y_3Al_{5-x}Ga_xO_{12}$:Ce^{3+}, Cr^{3+} that could be charged by blue light (460 nm) and

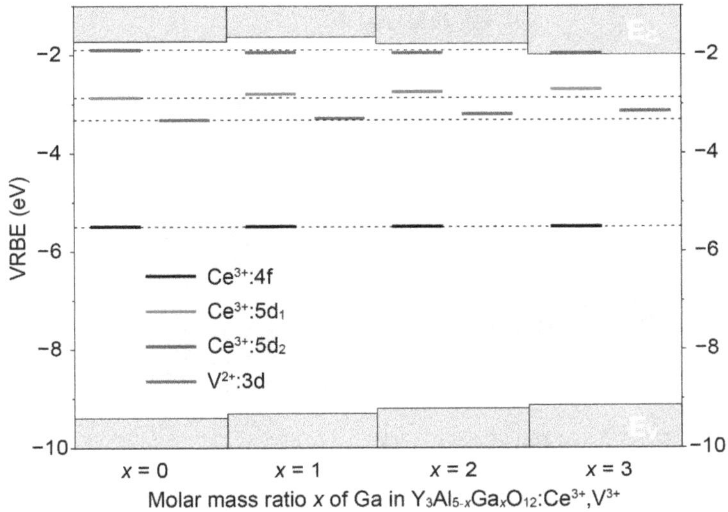

Figure 9.7 Stacked vacuum referred binding energy (VRBE) diagrams of $Y_3Al_{5-x}Ga_xO_{12}$:Ce^{3+}, V^{3+} ($x = 0, 1, 2$, and 3). The bottom of the conduction band (E_C) rose from 0 to 1 and dropped down from 1 to 3. The top of valence band (E_V) monotonously moved upward when x increased. Source: Reproduced with permission from [111], 2018 / American Chemical Society.

UV excitations [109, 110]. By varying the molar mass ratio of Ga, the TL glow peak temperature could be controlled from 150 to 400 K, and the corresponding trap depths increased from 0.41 to 1.2 eV. A larger trap depth was reported by Zhuang and coworkers in $Y_3Al_{5-x}Ga_xO_{12}$:Ce^{3+}, V^{3+} fluorescent particles-based PiG films [111]. As the molar mass ratio of Ga decreased from 3 to 0, the TL glow peak temperature increased from 410 to 545 K and then slightly decreased to 540 K. Thanks to the bandgap engineering effect, as shown in Figure 9.7, the trap depth in the phosphors and PiG films could be tailored from 1.13 to 1.62 eV by changing the energy gap between the bottom of the CB and the electron trap, which is also confirmed by theoretical calculations of the electronic band structure of $Y_3Al_{5-x}Ga_xO_{12}$ based on first principles [142].

Du et al. developed Pr^{3+}doped $Y_3Al_2Ga_3O_{12}$ transparent ceramic with in-line transmittance over 80% at 1800 nm, exhibiting robust red PersL after several minutes UV or blue light up-conversion charging [113]. Two trap centers involving intrinsic oxygen vacancies with varying trap depths were thoroughly investigated. Remarkably, the phosphor also exhibits photochromic characteristics with excellent resistance to fatigue. The excellent fatigue resistance is demonstrated by the reversibility of the greenish-reddish color change through alternating UV/blue light irradiation and thermal stimulation; additionally, the reddish color can persist for over three weeks (Figure 9.8a). This work paves the way for future applications in optical information encryption media and optical sensors (Figure 9.8b,c). Other garnet-type PersL materials that achieve trap depth modulation through band engineering can be found in $Y_3Al_{5-x}Ga_xO_{12}$:Cr^{3+} and $Gd_3Al_{5-x}Ga_xO_{12}$:Cr^{3+}, Eu^{3+} [112, 114].

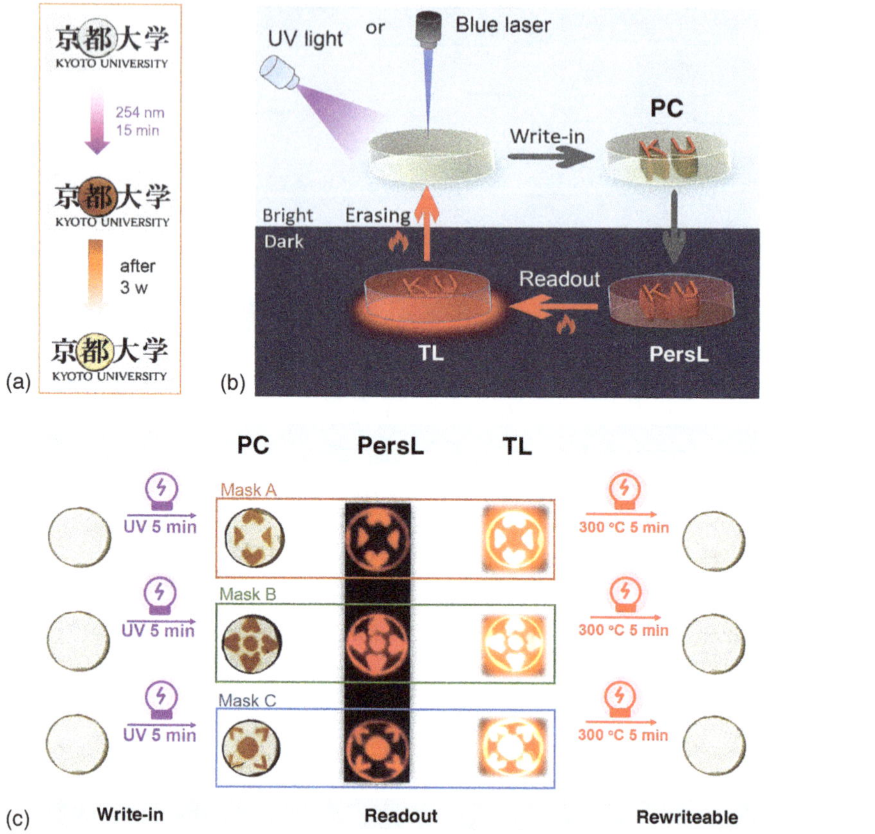

Figure 9.8 (a) Photographs of the $Y_3Al_2Ga_3O_{12}$:Pr^{3+} transparent ceramic sample before (top), after 15 minutes of UV (λ_{ex} = 254 nm) light charging (middle), and three weeks after UV light cessation (bottom), displaying discoloration. (b) Schematic representation of optical information write-in and readout processes. (c) Optical information display demonstrating write-in, readout, and bleaching through photochromism and Persistent luminescent in $Y_3Al_2Ga_3O_{12}$:Pr^{3+} transparent ceramic, achieved by alternating between 254 nm irradiation and a 300 °C thermal stimulation. Source: Reproduced with permission from [113], 2023 / John Wiley & Sons.

9.3.3.4 Titanate/Zirconate

Wang et al. reported red PersL excited by UV/visible light and PSL under the excitation of NIR laser (980 nm) in a titanate compound, $Ca_4Ti_3O_{10}$:Pr^{3+}, Y^{3+} [115]. The TL glow curves exhibited that the deep traps in the $Ca_4Ti_3O_{10}$:Pr^{3+}, Y^{3+} had a narrow distribution from 0.90 to 0.98 eV, which was qualified for maintaining the recorded data at room temperature for a long duration time.

Zhang and coworkers reported violet-blue PersL with a peak wavelength of approximately 410 nm in undoped zirconate compound $CaZrO_3$, which may originate from [CaO_7 $V_O^{\bullet\bullet}$]–[CaO_7 $V_O^{\bullet}$] complex clusters [116]. The number of shallow traps in the prepared $CaZrO_3$ was small, resulting in short-term PersL for 94 seconds at room temperature. The TL glow curves revealed that at least two types of deep traps with peak temperature at 445 and 494 K existed in $CaZrO_3$, and the trap depths were calculated to be 1.13 and 1.55 eV, respectively. Therefore, when stimulated by light at 980 nm, a significant Stokes shift emission close to 500 nm was obtained.

9.3.3.5 Oxide Glass

Compared to phosphors in microscale powders, oxide glass exhibits excellent bulk processability, which especially holds promise for large-scale 3D optical information storage. Qiu et al. observed bright PersL in rare earth-doped calcium aluminosilicate glasses (Ca–Al–Si–O glasses) under femtosecond laser irradiation at 800 nm [117]. By selecting appropriate glass compositions and species of rare earth ions, the emission color appeared blue (Ce^{3+}), green (Tb^{3+}), or red (Pr^{3+}). The optional 3D image patterns emitting PersL in various colors could be formed by moving the focal point of the laser. The origin of PersL in the Ca–Al–Si–O glass was attributed to the high transient power density of the femtosecond laser, which created defect centers and separated charge carriers to these defects. This research represents one of the pioneering works in the fabrication of rewritable 3D optical memory devices with both an ultrahigh recording speed and an ultrahigh storage density.

Su and coworkers discovered red PersL in Mn^{2+}-doped zinc borosilicate glasses (Zn–Si–B–O:Mn^{2+} glass) upon 254 nm UV irradiation [118]. When the bright red PersL induced by the removal of the UV lamp, the glass could re-emit PersL under the stimulation of lower energy light ranging from 300 to 800 nm after several weeks of storage. The researchers explained the shallow traps mainly contribute to PersL at room temperature, while deep traps store charge carriers and generate luminescence under optical stimulation. Subsequently, the influence of Yb^{3+} co-doping on the PersL and optical stimulated luminescence of Zn–Si–B–O:Mn^{2+} glass was investigated [119]. Compared with the glasses solely activated by Mn^{2+}, the red luminescent was largely improved in both brightness and duration time, but PSL is greatly depressed. When introducing Yb^{3+} ions, the alteration of the energy transfer channel reduced the trap depth. Additionally, energy transfer from Mn^{2+} to Yb^{3+} enabled strong NIR emission in Zn–Si–B–O:Mn^{2+}, Yb^{3+} glass.

Apart from the transparent glasses, other researchers have prepared a series of microcrystalline glass materials suitable for optically stimulated luminescence and optical information storage through thermal treatment of glasses [45, 143–145]. It should be noted that in most reported microcrystalline glass materials, efficient luminescent centers and major contributing defects originate from the crystalline phase, while the glass phase mainly serves as a binder to form the bulk structure.

9.3.4 Nitride or Oxynitrides

Due to the strong covalency and electron cloud rearrangement effect, nitride or oxynitride optical materials doped with Eu^{2+} or Ce^{3+} exhibit longer emission wavelength compared to the oxide phosphors with similar structure. Many important red and near-infrared luminescent materials have been discovered in nitride or oxynitride compounds.

$CaAlSiN_3$:Eu^{2+} is an important red nitride PerL material that has found widespread applications in phosphor-converted white LEDs due to its high luminescence efficiency and relatively short luminescence lifetime. Lei et al. were the first to observe bright red PersL and photo-stimulation characteristics of $CaAlSiN_3$:Eu^{2+}, which was prepared under a nitrogen atmosphere using a radiofrequency induction furnace. The PersL emission wavelength of $CaAlSiN_3$:Eu^{2+} was centered around 635 nm, and the TL peaks covered a wide temperature range (approximately 320–450 K), indicating the presence of multiple trap levels (shallow and deep traps). The electron traps were attributed to positive trap levels ($O_N^{\bullet}$) formed by the partial substitution of O^{2-} in N^{3-} sites at low Eu^{2+} concentration. The hole traps were assigned to $Eu_{Ca}^{\times}$ or V_{Ca}'' [120]. They also reported another red rare earth-doped nitridosilicates $(Ca_{1-x}Sr_x)_2Si_5N_8$:Eu^{2+}, Tm^{3+} (x = 0–1), which exhibited similar PersL and PSL behaviors with a maximum intensity at about x = 0.2 [121]. The emission bands of this material red-shifted from 604 to 630 nm with the increasing doping ratio of Ca^{2+} on account of the cation substitution effect. The traps were mainly located at two positions, 0.64 and 0.72 eV. With an increase in the molar mass ratio of Sr (x value), the trap density at 0.72 eV increases, leading to improve the properties of optically stimulated luminescence and optical storage. The researchers suggest that the electron traps in this material may mainly originate from Tm^{3+} substitution for Ca^{2+}/Sr^{2+} ions ($Tm_{Ca/Sr}^{\bullet}$) [122].

$SrLiAl_3N_4$:Eu^{2+} is an efficient red nitride phosphor material with excellent thermal stability and efficient narrow-band red emission (650 nm). Liu et al. reported that after 254 nm UV irradiation, $SrLiAl_3N_4$:Eu^{2+} exhibits the shallow trap (0.47 eV) and the deep trap (0.81 eV), which were responsible for PersL and PSL, respectively. The light energy could be released with 980 nm photo-stimulation even being stored for 15 days at room temperature. They verified that the trap states were due to N vacancies (V_N) with the aid of the density functional theory (DFT) calculations [123].

Ca-α-Sialon:Eu^{2+} (typical compounds with chemical formula $CaSi_{10-n}Al_{2+n}O_nN_{16-n}$:$Eu^{2+}$, n = 0–1) exhibits broad emission characteristics with emission wavelengths around in the yellow to red region, which is significant for high color render index white LED lighting. Liu et al. reported that PersL properties of Ca-α-Sialon:Eu^{2+} were strongly dependent on the Eu^{2+} and oxygen concentrations. The trap depth distribution of the optimal sample $CaSi_{10}Al_2N_{16}$:0.5% Eu^{2+} (n = 0) exhibited broad trap depth distribution, covering 0.6–1.4 eV, with trap densities mainly located at 0.65 and 0.85 eV. The continuous distribution of traps was verified to originate from oxygen vacancies ($V_O^{\bullet\bullet}$) and nitrogen vacancies ($V_N^{\bullet\bullet\bullet}$) by DFT calculations. Thanks to the deep traps as the majority in distribution of traps, $CaSi_{10}Al_2N_{16}$:0.5% Eu^{2+} phosphor was demonstrated to be a good candidate for optical information storage materials [124]. Ueda et al. reported a red light charged NIR PersL $Ba[Mg_2Al_2N_4]$:Eu^{2+}, Tm^{3+} phosphor [125]. Tm^{3+} serves as NIR emission center, expanding the PersL band from 600 to 830 nm through energy transfer from Eu^{2+} to Tm^{3+}.

In addition to the mentioned band gap engineering strategy, trap depth engineering by appropriately choosing trivalent lanthanide co-dopants as trap

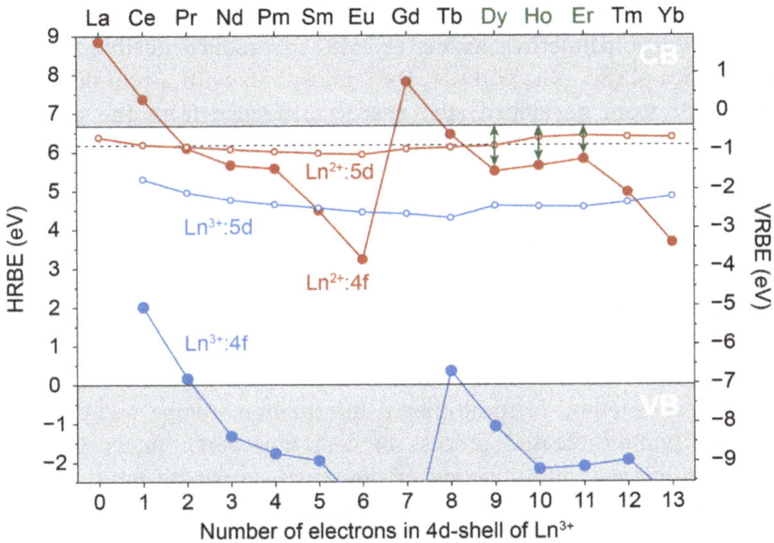

Figure 9.9 HRBE (left ordinate) and VRBE (right ordinate) schemes of Ln^{2+}/Ln^{3+}-doped $SrSi_2O_2N_2$. The 4f ground states (Ln^{2+}:4f) and the lowest 5d excited states (Ln^{2+}:5d) of Ln^{2+} are labeled by red dots and hollow circles. The 4f ground states (Ln^{3+}:4f) and the lowest 5d excited states (Ln^{3+}:5d) of Ln^{3+} are labeled by blue dots and hollow circles. The estimated trap depths are labeled by green double arrow line. Source: Adapted from [148].

centers in a certain host has emerged to be another important strategy to obtain required trap depths. However, there are still relatively few systematic studies on trap depth control by adopting trap depth engineering method, including $YPO_4:Ce^{3+}$, Ln^{3+} [146], $SrAl_2SiO_2N_3:Eu^{2+}$, Ln^{3+} [147], $SrSi_2O_2N_2:Eu^{2+}$, Ln^{3+} [148], and $SrSi_2O_2N_2:Yb^{2+}$, Ln^{3+} [148] (Ln^{3+} represents different trivalent rare earth ions).

Zhuang et al. reported systematic control of the trap depth in Ln^{2+}/Ln^{3+}-doped $SrSi_2O_2N_2$ ($Ln^{2+} = Yb^{2+}$, Eu^{2+}, and $Ln^{3+} = Dy^{3+}$, Ho^{3+}, Er^{3+}). The host-referred binding energy (HRBE) scheme in the $SrSi_2O_2N_2$ host was first established, in which Dy^{3+}, Ho^{3+}, Er^{3+}, Tm^{3+}, and other trivalent lanthanide ions were predicted to be the electron traps with a depth more than 0.6 eV [148]. As shown in Figure 9.9, the initial state of the electron trap is trivalent lanthanide ions ($Ln_{Sr}^{\cdot}$), and then divalent lanthanide ions ($Ln_{Sr}^{\times}$) were formed after capturing an electron. The established HRBE schemes provide important information on the energy-level locations for the whole family of rare earth-doped $SrSi_2O_2N_2$ phosphors. The trap depth is determined by the energy difference between the ground state of the $Ln_{Sr}^{\times}$ and the bottom of the CB. The trap depths were calculated as 1.17, 1.04, 0.90, and 1.55 eV by co-doping with Dy^{3+}, Ho^{3+}, Er^{3+}, and Tm^{3+}, respectively in $SrSi_2O_2N_2$ host according to TL measurements, which matched well with the predictions of the HRBE model [126]. These results demonstrated that the HRBE model could provide valuable guidance to modulate the trap depth in certain material systems and accelerate the development of new deep-trap PersL materials.

Subsequently, the group of Zhuang and Xie incorporated deep trap in nitride oxide PersL phosphors into polydimethylsiloxane (PDMS) to produce flexible optical storage films. A series of $(Sr_{1-x}Ba_x)Si_2O_2N_2:Dy^{3+}$ phosphors with a trap depth of approximately 1.2 eV were developed (the energy difference from the ground state of Dy^{2+} to the CB bottom). By modifying the types of luminescent centers (Yb^{2+}/Eu^{2+}) and the surrounding crystal field (high sensitivity of Eu^{2+}:5d→4f transition), the tunability of PersL color was also achieved. Optical data can be conveniently encoded onto the flexible film using laser direct writing technique (using a commercial 405 nm laser) and decoded by either heating or scanning with a 980 nm laser at room temperature. Moreover, they developed a bit-by-bit optical data storage and readout strategy to accurately retrieve the initially recorded information. Since the readout optical information possesses adjustable spectral characteristics, a novel method for multiplexed information storage was devised by controlling the photon-released process of deep-trap PersL materials and combining emission-intensity-multiplexing or emission-wavelength-multiplexing techniques [20]. Zhang and Li et al. developed RGB tricolors dynamic display of optical information encryption strategies by combing with red $Sr_2Si_5N_8:Eu^{2+}$, Dy^{3+}, green $SrSi_2O_2N_2:Eu^{2+}$, Dy^{3+} and blue $BaSi_2O_2N_2:Eu^{2+}$ nitride phosphors [127]. Kitagawa et al. reported that $YSiO_2N:Ce^{3+}$, Ln^{3+} (Ln = Sm^{3+} or Tm^{3+}) samples enhance the intensity of Ce^{3+}:5d→4f blue persistent luminescence [128]. In the thermoluminescence glow curves, the sample exhibits two glow peaks related to intrinsic defects and the Ln^{3+} traps.

Furthermore, Zhuang et al. reported a force-induced charge carrier storage (FICS) effect in deep-trap ML materials, i.e. $(Sr,Ba)Si_2O_2N_2:Eu^{2+}/Yb^{2+},Dy^{3+}$, which enabled storage of the applied mechanical energy (such as grinding or squeezing) in deep traps except for the above-mentioned optical excitation. The force loading may generate an instantaneous electric field within the crystal through triboelectricity or piezoelectricity, which excites the luminescent centers and stores some of the charge carriers in deep traps. The proposed FICS effect allows for distributed stress sensing in a nonreal-time mode, therefore expanding the application of deep-trap PersL materials to next-generation battery-free stress recording devices and mechanical energy storage and conversion systems [133].

Eu^{2+}-doped aluminum silicate oxy-nitride (β-Sialon:Eu^{2+}) has found extensive use in liquid-crystal display backlights owing to its exceptional green emission characteristics. Liu and coworkers reported a new PersL β-Sialon phosphor with the general formula $Eu_{0.015}Si_{5.5}Al_{0.485}O_{0.515}N_{7.485}$ [129]. The trap levels responsible for PersL are impurity levels induced by Si—O bonds located below the bottom of the CB on random substitution of Al–O for Si–N in β-Si_3N_4. Lv, Zhuang and Xie et al. reported Yb-doped $(Ba_{1-x}Sr_x)AlSi_5O_2N_7$ phosphors showing intense PersL in the visible and NIR regions simultaneously (a broad emission band centered at 664 nm and a sharp emission at 985 nm) [130]. Impressively, the radiance of PersL in the $Ba_{0.8}Sr_{0.2}AlSi_5O_2N_7:Yb^{2+}$ phosphor was approximately 4.82×10^{-3} mW Sr^{-1} m^{-2} after ceasing the UV irradiation for 60 min, which is roughly one-third of that observed in the well-known $ZnGa_2O_4:Cr^{3+}$ PersL phosphor. The average trap depths of the phosphor were estimated to be 0.84 and 1.32 eV using the Randall-Wilkins model.

9.4 Outlooks

Overall, this chapter focuses on deep-trap PersL materials for optical information storage applications. Based on their primary chemical compositions, these materials include halide or oxyhalide compounds, sulfides, oxides, and nitride or oxynitride compounds. This discussion primarily introduces the TL or PSL characteristics, spectral properties, trap characteristics, optical storage mechanisms, and exploration of applications in optical information storage of these four types of deep-traps PersL materials. It can be observed that deep-trap PersL materials have experienced rapid development in recent decades and have achieved many exciting breakthroughs. However, despite these advancements, there are still many pressing challenges to be addressed and overcome in material development, performance optimization, mechanism research, and storage applications.

(i) Deep-trap PersL materials with stable storage states need to be further developed. In recent years, a series of PersL materials with high-energy storage and luminescence efficiency under UV excitation have emerged in oxide and nitride (or oxynitride) systems. However, most of the reported materials have relatively shallow trap depths (or contain a portion of shallow traps), and the stored states of the materials are prone to spontaneous release of charge carriers and return to the initial state at room temperature, resulting in relatively low optical information retention rate. Effective control over trap depth and its distribution to obtain sufficiently large trap depths and narrower distribution characteristics is a key issue in the design of PersL materials and their application in information storage.

(ii) High-performance nanoscale PersL materials are urgently needed. For practical applications, sufficiently small particle size and uniform particle distribution are important prerequisites for achieving high-capacity storage and accurate signal recognition. Moreover, in bioimaging, clinical diagnosis, and radiation dosimetry. However, the majority of PersL materials are inorganic crystals with micrometer-sized dimensions, and only a few types of high-performance nanoscale PersL materials have been reported in gallates and fluorides. These types of nanomaterials have attracted significant attention and ongoing research upon their discovery.

(iii) In-depth analysis of the nature of trap structures is still insufficient. For many classical long-lasting phosphorescent material systems, researchers have conducted numerous experiments to demonstrate that doping specific elements or employing specific preparation techniques can significantly enhance the trap density of the materials. These enables researchers to establish correlations between the long-lasting phosphorescent characteristics and a certain 'possible trap structure' and provide valuable references for the design and fabrication of more unknown PersL materials. However, there are limitations in precise characterization techniques for defect structures. In-depth analysis of trap structures remains an important foundational scientific question.

(iv) Quantitative characterization of optical storage capacity is highly expected. As an optical information storage material, the optical storage capacity of deep-trap PersL materials is closely related to trap density. However, current research on trap characteristics primarily focuses on qualitative characterization of trap depth, making it difficult to compare trap densities between different materials. PersL (including TL and PSL) belongs to weak luminescence, and the luminescence intensity may undergo continuous dynamic changes. The quantitative characterization of PersL intensity and trap density poses significant technical challenges, and there is currently no widely recognized simple testing method. Future research will require the development of highly sensitive and standardized testing devices.

(v) The PersL wavelength range and spectral detection range should be further extended. With the advancement of research on PersL materials, particularly with the increasing demands in multidimensional optical information storage, advanced optical anti-counterfeiting, biomedical imaging, and other application fields, the emission wavelength range of PersL materials needs to extend to shorter and longer wavelengths. Although the reported luminescence of PersL materials already covers the spectral range of 300–1500 nm, many materials primarily emit in the visible region. There is a lack of materials that efficiently emit in the ultraviolet and infrared regions while possessing high trap densities and controllable trap depths, which are necessary for the application demands. Additionally, new emission wavelengths also impose higher requirements on photodetection devices, necessitating upgrades and modifications to traditional signal detection devices based on silicon-based photodetectors to expand their spectral detection range.

In summary, deep-trap PersL materials for optical information storage applications have become one of the research hotspots in the field of luminescent materials. Significant research progress has been made in deep-trap PersL materials in the past few years, while also giving rise to new questions and challenges. The coming years will further witness the rapid development of deep-trap PersL materials and their applications in optical information storage.

References

1 Pan, Z., Lu, Y.Y., and Liu, F. (2011). Sunlight-activated long-persistent luminescence in the near-infrared from Cr^{3+}-doped zinc gallogermanates. *Nature Materials* 11: 58–63.

2 Matsuzawa, T., Aoki, Y., Takeuchi, N., and Murayama, Y. (2019). A new long phosphorescent phosphor with high brightness $SrAl_2O_4$:Eu^{2+}, Dy^{3+}. *Journal of the Electrochemical Society* 143: 2670–2673.

3 Ueda, J., Miyano, S., and Tanabe, S. (2018). Formation of deep electron traps by Yb^{3+} codoping leads to super-long persistent luminescence in Ce^{3+}-doped yttrium aluminum gallium garnet phosphors. *ACS Applied Materials and Interfaces* 10: 20652–20660.

4 le Masne de Chermont, Q., Chaneac, C., Seguin, J. et al. (2007). Nanoprobes with near-infrared persistent luminescence for *in vivo* imaging. *Proceedings*

of the National Academy of Sciences of the United States of America 104: 9266–9271.

5 Maldiney, T., Bessiere, A., Seguin, J. et al. (2014). The *in vivo* activation of persistent nanophosphors for optical imaging of vascularization, tumours and grafted cells. *Nature Materials* 13: 418–426.

6 Li, Z., Zhang, Y., Wu, X. et al. (2015). Direct aqueous-phase synthesis of sub-10 nm "luminous pearls" with enhanced *in vivo* renewable near-infrared persistent luminescence. *Journal of the American Chemical Society* 137: 5304–5307.

7 Shi, J., Sun, X., Zhu, J. et al. (2016). One-step synthesis of amino-functionalized ultrasmall near infrared-emitting persistent luminescent nanoparticles for *in vitro* and *in vivo* bioimaging. *Nanoscale* 8: 9798–9804.

8 Wang, J., Ma, Q., Hu, X.X. et al. (2017). Autofluorescence-free targeted tumor imaging based on luminous nanoparticles with composition-dependent size and persistent luminescence. *ACS Nano* 11: 8010–8017.

9 Lin, X.H., Song, L., Chen, S. et al. (2017). Kiwifruit-like persistent luminescent nanoparticles with high-performance and in situ activable near-infrared persistent luminescence for long-term *in vivo* bioimaging. *ACS Applied Materials & Interfaces* 9: 41181–41187.

10 Zhou, Z., Zheng, W., Kong, J. et al. (2017). Rechargeable and LED-activated $ZnGa_2O_4$: Cr^{3+} near-infrared persistent luminescence nanoprobes for background-free biodetection. *Nanoscale* 9: 6846–6853.

11 Wang, J., Li, J., Yu, J. et al. (2018). Large hollow cavity luminous nanoparticles with near-infrared persistent luminescence and tunable sizes for tumor afterglow imaging and chemo-photodynamic therapies. *ACS Nano* 12: 4246–4258.

12 Liu, J.M., Zhang, D.D., Fang, G.Z., and Wang, S. (2018). Erythrocyte membrane bioinspired near-infrared persistent luminescence nanocarriers for *in vivo* long-circulating bioimaging and drug delivery. *Biomaterials* 165: 39–47.

13 Lv, Y., Ding, D., Zhuang, Y. et al. (2019). Chromium-doped zinc gallogermanate@zeolitic imidazolate framework-8: a multifunctional nanoplatform for rechargeable *in vivo* persistent luminescence imaging and pH-responsive drug release. *ACS Applied Materials & Interfaces* 11: 1907–1916.

14 Onushkin, G.A., Lee, Y.J., Yang, J.J. et al. (2009). Efficient alternating current operated white light-emitting diode chip. *IEEE Photonics Technology Letters* 21: 33–35.

15 Yeh, C.W., Li, Y., Wang, J., and Liu, R.-S. (2012). Appropriate green phosphor of $SrSi_2O_2N_2:Eu^{2+}$, Mn^{2+} for AC LEDs. *Optics Express* 20: 18031–18043.

16 Lin, H., Wang, B., Xu, J. et al. (2014). Phosphor-in-glass for high-powered remote-type white AC-LED. *ACS Applied Materials – Interfaces* 6: 21264–21269.

17 Lin, H., Xu, J., Huang, Q. et al. (2015). Bandgap tailoring *via* si doping in inverse-garnet $Mg_3Y_2Ge_3O_{12}:Ce^{3+}$ persistent phosphor potentially applicable in AC-LED. *ACS Applied Materials & Interfaces* 7: 21835–21843.

18 Liu, Y.F., Liu, P., Wang, L. et al. (2017). A two-step solid-state reaction to synthesize the yellow persistent $Gd_3Al_2Ga_3O_{12}:Ce^{3+}$ phosphor with an enhanced optical performance for AC-LEDs. *Chemical Communications* 53: 10636–10639.

19 Asami, K., Ueda, J., and Tanabe, S. (2018). Flicker suppression of ac driven white LED by yellow persistent phosphor of Ce^{3+}-Cr^{3+} co-doped garnet. *Journal of Science and Technology in Lighting* 41: 89–92.

20 Zhuang, Y., Wang, L., Lv, Y. et al. (2018). Optical data storage and multicolor emission readout on flexible films using deep-trap persistent luminescence materials. *Advanced Functional Materials* 28: 1705769.

21 Liu, Z., Zhao, L., Chen, W. et al. (2018). Multiple anti-counterfeiting realized in $NaBaScSi_2O_7$ with a single activator of Eu^{2+}. *Journal of Materials Chemistry C* 6: 11137–11143.

22 Long, Z., Wen, Y., Zhou, J. et al. (2019). No-interference reading for optical information storage and ultra-multiple anti-counterfeiting applications by designing targeted recombination in charge carrier trapping phosphors. *Advanced Optical Materials* 7: 1900006.

23 Gu, M., Li, X., and Cao, Y. (2014). Optical storage arrays: a perspective for future big data storage. *Light: Science & Applications* 3: 177.

24 Gu, M., Zhang, Q., and Lamon, S. (2016). Nanomaterials for optical data storage. *Nature Reviews Materials* 1: 16070.

25 Heanue, J.F., Bashaw, M.C., and Hesselink, L. (1994). Volume holographic storage and retrieval of digital data. *Science* 265: 749–752.

26 Cumpston, B.H., Ananthavel, S.P., Barlow, S. et al. (1999). Two-photon polymerization initiators for three-dimensional optical data storage and microfabrication. *Nature* 398: 51–54.

27 Kawata, S. and Kawata, Y. (2000). Three-dimensional optical data storage using photochromic materials. *Chemical Reviews* 100: 1777–1788.

28 Huang, L., Chen, X., Mühlenbernd, H. et al. (2013). Three-dimensional optical holography using a plasmonic metasurface. *Nature Communications* 4: 2808.

29 Zhang, C., Zhou, H.P., Liao, L.Y. et al. (2010). Luminescence modulation of ordered upconversion nanopatterns by a photochromic diarylethene: rewritable optical storage with nondestructive readout. *Advanced Materials* 22: 633–637.

30 Kallepalli, D.L., Alshehri, A.M., Marquez, D.T. et al. (2016). Ultra-high density optical data storage in common transparent plastics. *Scientific Reports* 6: 26163.

31 Wilson, O., Wilson, G.J., and Mulvaney, P. (2002). Laser writing in polarized silver nanorod films. *Advanced Materials* 14: 1000–1004.

32 Pham, H.H., Gourevich, I., Oh, J.K. et al. (2004). A multidye nanostructured material for optical data storage and security data encryption. *Advanced Materials* 16: 516–520.

33 Hirata, S., Lee, K.-S., and Watanabe, T. (2008). Reversible fluorescent on-off recording in a highly transparent polymeric material utilizing fluorescent resonance energy transfer (FRET) induced by heat treatment. *Advanced Functional Materials* 18: 2869–2879.

34 Genovese, D., Aliprandi, A., Prasetyanto, E.A. et al. (2016). Mechano- and photochromism from bulk to nanoscale: data storage on individual self-assembled ribbons. *Advanced Functional Materials* 26: 5271–5278.

35 Heydari, E., Sperling, J.R., Neale, S.L., and Clark, A.W. (2017). Plasmonic color filters as dual-state nanopixels for high-density microimage encoding. *Advanced Functional Materials* 27: 1701866.
36 Li, X., Chon, J.W., Wu, S. et al. (2007). Rewritable polarization-encoded multilayer data storage in 2,5-dimethyl-4-(p-nitrophenylazo)anisole doped polymer. *Optics Letters* 32: 277–279.
37 Li, X., Cao, Y., and Gu, M. (2011). Superresolution-focal-volume induced 3.0 tbytes/disk capacity by focusing a radially polarized beam. *Optics Letters* 36: 2510–2512.
38 Bozinovic, N., Yue, Y., Ren, Y. et al. (2013). Terabit-scale orbital angular momentum mode division multiplexing in fibers. *Science* 340: 1545–1548.
39 Ren, H., Li, X., Zhang, Q., and Gu, M. (2016). On-chip noninterference angular momentum multiplexing of broadband light. *Science* 352: 805–809.
40 Lu, Y., Zhao, J., Zhang, R. et al. (2013). Tunable lifetime multiplexing using luminescent nanocrystals. *Nature Photonics* 8: 32–36.
41 Sun, H., Liu, S., Lin, W. et al. (2014). Smart responsive phosphorescent materials for data recording and security protection. *Nature Communications* 5: 3601.
42 Sonoda, M., Takano, M., Miyahara, J., and Kato, H. (1983). Computed radiography utilizing scanning laser stimulated luminescence. *Radiology* 148: 833–838.
43 Takahashi, K., Miyahara, J., and Shibahara, Y. (2019). Photostimulated luminescence (PSL) and color centers in $BaFX : Eu^{2+} (X = Cl, Br, I)$ phosphors. *Journal of the Electrochemical Society* 132: 1492–1494.
44 Liu, F., Yan, W., Chuang, Y.J. et al. (2013). Photostimulated near-infrared persistent luminescence as a new optical read-out from Cr^{3+}-doped $LiGa_5O_8$. *Scientific Reports* 3: 1554.
45 Lin, S., Lin, H., Ma, C. et al. (2020). High-security-level multi-dimensional optical storage medium: nanostructured glass embedded with $LiGa_5O_8$: Mn^{2+} with photostimulated luminescence. *Light: Science and Applications* 9: 22.
46 Van den Eeckhout, K., Smet, P.F., and Poelman, D. (2010). Persistent luminescence in Eu^{2+}-doped compounds: a review. *Materials* 3: 2536–2566.
47 Smet, P.F., Botterman, J., Van den Eeckhout, K. et al. (2014). Persistent luminescence in nitride and oxynitride phosphors: a review. *Optical Materials* 36: 1913–1919.
48 Li, Y., Gecevicius, M., and Qiu, J. (2016). Long persistent phosphors from fundamentals to applications. *Chemical Society Reviews* 45: 2090–2136.
49 Sun, S.K., Wang, H.F., and Yan, X.P. (2018). Engineering persistent luminescence nanoparticles for biological applications: from biosensing/bioimaging to theranostics. *Accounts of Chemical Research* 51: 1131–1143.
50 Fan, X., Liu, Z., Yang, X. et al. (2019). Recent developments and progress of inorganic photo-stimulated phosphors. *Journal of Rare Earths* 37: 679–690.
51 Liang, L., Chen, N., Jia, Y. et al. (2019). Recent progress in engineering near-infrared persistent luminescence nanoprobes for time-resolved biosensing/bioimaging. *Nano Research* 12: 1279–1292.
52 Poelman, D., Van der Heggen, D., Du, J. et al. (2020). Persistent phosphors for the future: fit for the right application. *Journal of Applied Physics* 128: 240903.

53 Yuan, L., Jin, Y., Su, Y. et al. (2020). Optically stimulated luminescence phosphors: principles, applications, and prospects. *Laser and Photonics Reviews* 14: 2000123.

54 Riesen, H. and Kaczmarek, W.A. (2007). Efficient X-ray generation of Sm^{2+} in nanocrystalline $BaFCl/Sm^{3+}$: a photoluminescent X-ray storage phosphor. *Inorganic Chemistry* 46: 7235–7237.

55 Riesen, N., Pan, X., Badek, K. et al. (2018). Towards rewritable multilevel optical data storage in single nanocrystals. *Optics Express* 26: 25336–25346.

56 Riesen, H., Badek, K., Monro, T.M., and Riesen, N. (2016). Highly efficient valence state switching of samarium in BaFCl:Sm nanocrystals in the deep UV for multilevel optical data storage. *Optical Materials Express* 6: 3097–3108.

57 Rogulis, U., Tale, I., Hangleiter, T., and Spaeth, J.M. (1995). The photostimulation process in the X-ray storage phosphor KBr:In. *Journal of Physics: Condensed Matter* 7: 3129–3137.

58 Rogulis, U., Schweizer, S., Assmann, S., and Spaeth, J.M. (1998). Ga^{2+} hole centers and photostimulated luminescence in the x-ray storage phosphor $RbBr:Ga^{+}$. *Journal of Applied Physics* 84: 4537–4542.

59 Pawlik, T. and Spaeth, J.M. (1997). Investigation of the x-ray storage phosphors $Cs_2NaYF_6:Pr^{3+}$ or Ce^{3+}. *Journal of Applied Physics* 82: 4236–4240.

60 Pawlik, T., Spaeth, J.M., Otte, M., and Overhof, H. (2006). Endor-investigations of rare earth and transition metal ions in the cubic elpasolite crystal Cs_2NaYF_6. *Radiation Effects and Defects in Solids* 135: 49–54.

61 Ou, X., Qin, X., Huang, B. et al. (2021). High-resolution X-ray luminescence extension imaging. *Nature* 590: 410–415.

62 Zhuang, Y., Chen, D., Chen, W. et al. (2021). X-ray-charged bright persistent luminescence in $NaYF_4:Ln^{3+}$@$NaYF_4$ nanoparticles for multidimensional optical information storage. *Light: Science and Applications* 10: 132.

63 Wang, Y., Chen, D., Zhuang, Y. et al. (2021). $NaMgF_3:Tb^{3+}$@$NaMgF_3$ nanoparticles containing deep traps for optical information storage. *Advanced Optical Materials* 9: 2100624.

64 Zhou, X., Han, K., Wang, Y. et al. (2023). Energy-trapping management in x-ray storage phosphors for flexible 3d imaging. *Advanced Materials* 35: 2212022.

65 Song, Y., Du, J., Yang, R. et al. (2023). Recyclable time-temperature indicator enabled by light storage in particles. *Advanced Optical Materials* 11: 2202654.

66 Chen, W., Song, Y., Zhang, W. et al. (2022). Time-gated imaging of latent fingerprints with level 3 details achieved by persistent luminescent fluoride nanoparticles. *ACS Applied Materials and Interfaces* 14: 28230–28238.

67 Meijerink, A. and Blasse, G. (1991). Photostimulated luminescence and thermally stimulated luminescence of some new X-ray storage phosphors. *Journal of Physics D: Applied Physics* 24: 626–632.

68 Cao, J., Ding, S., Zhou, Y. et al. (2022). Enhancing the luminescence performance of $Ca_{2-x}Sr_xBO_3Cl:Eu^{2+}$, Dy^{3+} by substitution of Sr^{2+} for Ca^{2+}. *Journal of Materials Chemistry C* 10: 12266–12275.

69 Yasuaki Tamura, Y.T. and Atsushi Shibukawa, A.S. (1993). Optical studies of CaS:Eu, Sm infrared stimulable phosphors. *Japanese Journal of Applied Physics* 32: 3187.

70 Nanto, H., Sato, T., Miyazaki, M. et al. (1999). Advanced optical storage phosphor materials for erasable and rewritable optical memory utilizing photo-stimulated luminescence. *Proceeding of SPIE* 3802.

71 Li, W., Liu, Y., and Ai, P. (2010). Synthesis and luminescence properties of red long-lasting phosphor Y_2O_2S:Eu^{3+}, Mg^{2+}, Ti^{4+} nanoparticles. *Materials Chemistry and Physics* 119: 52–56.

72 Cui, C.E., Jiang, G., Huang, P. et al. (2014). Effect of Eu^{3+} concentration on the luminescence properties of Y_2O_2S:Eu^{3+}, Mg^{2+}, Ti^{4+} nanotubes. *Ceramics International* 40: 4725–4730.

73 Zhang, X., Suo, H., Wang, Y. et al. (2023). Systematic tuning of persistent luminescence in a quaternary wurtzite crystal through synergistic defect engineering. *Laser & Photonics Reviews* 17: 2300132.

74 Zych, E., Trojan-Piegza, J., Hreniak, D., and Strek, W. (2003). Properties of Tb-doped vacuum-sintered Lu_2O_3 storage phosphor. *Journal of Applied Physics* 94: 1318–1324.

75 Wiatrowska, A. and Zych, E. (2013). Traps formation and characterization in long-term energy storing Lu_2O_3:Pr, Hf luminescent ceramics. *The Journal of Physical Chemistry C* 117: 11449–11458.

76 Akselrod, M.S., Kravetsky, D.J., and Gotlib, V.I. (1990). Highly sensitive thermoluminescent anion-defective alpha-Al:C single crystal detectors. *Radiation Protection Dosimetry* 32: 15–20.

77 Yukihara, E.G., Whitley, V.H., Polf, J.C. et al. (2003). The effects of deep trap population on the thermoluminescence of Al_2O_3:C. *Radiation Measurements* 37: 627–638.

78 Bos, A.J., Prokic, M., and Brouwer, J.C. (2006). Optically and thermally stimulated luminescence characteristics of MgO:Tb^{3+}. *Radiation Protection Dosimetry* 119: 130–133.

79 Wang, C., Jin, Y., Lv, Y. et al. (2018). Trap distribution tailoring guided design of super-long-persistent phosphor Ba_2SiO_4:Eu^{2+}, Ho^{3+} and photostimulable luminescence for optical information storage. *Journal of Materials Chemistry C* 6: 6058–6067.

80 Liu, D., Yuan, L., Jin, Y. et al. (2019). Tailoring multidimensional traps for rewritable multilevel optical data storage. *ACS Applied Materials and Interfaces* 11: 35023–35029.

81 Lin, S., Lin, H., Huang, Q. et al. (2019). A photostimulated $BaSi_2O_5$:Eu^{2+}, Nd^{3+} phosphor-in-glass for erasable-rewritable optical storage medium. *Laser and Photonics Reviews* 13: 1900006.

82 Sun, X., Zhang, J., Zhang, X. et al. (2008). Long lasting yellow phosphorescence and photostimulated luminescence in Sr_3SiO_5: Eu^{2+} and Sr_3SiO_5: Eu^{2+}, Dy^{3+} phosphors. *Journal of Physics D: Applied Physics* 41: 195414.

83 Zhang, B., Yu, X., Wang, T. et al. (2015). Photostimulated and long persistent luminescence properties from different crystallographic sites of β-Sr_2SiO_4: Eu^{2+}, R^{3+} (R = Tm, Gd). *Journal of the American Ceramic Society* 98: 171–177.

84 Knitel, M.J., Dorenbos, P., and van Eijk, C.W.E. (1997). Photostimulated luminescence and thermoluminescence properties of $LiYSiO_4$:Ce. *Journal of Luminescence* 72-74: 765–766.

85 Dobrowolska, A., Bos, A.J.J., and Dorenbos, P. (2019). Synthesis optimization and charge carrier transfer mechanism in $LiLuSiO_4$:Ce, Tm storage phosphor. *Radiation Measurements* 127: 106147.

86 Wang, W., Yang, J., Zou, Z. et al. (2018). An isolated deep-trap phosphor for optical data storage. *Ceramics International* 44: 10010–10014.

87 Wang, J., Ma, Q., Zheng, W. et al. (2017). One-dimensional luminous nanorods featuring tunable persistent luminescence for autofluorescence-free biosensing. *ACS Nano* 11: 8185–8191.

88 Cong, Y., He, Y., Dong, B. et al. (2015). Long afterglow properties of Zn_2GeO_4:Mn^{2+}, Cr^{3+} phosphor. *Optical Materials* 42: 506–510.

89 Katayama, Y., Kayumi, T., Ueda, J. et al. (2017). The role of Ln^{3+} (Ln=Eu, Yb) in persistent red luminescence in $MgGeO_3$:Mn^{2+}. *Journal of Materials Chemistry C* 5: 8893–8900.

90 Krieke, G., Doke, G., Antuzevics, A. et al. (2022). Tunable persistent luminescence of novel $Mg_3Y_2Ge_3O_{12}$ garnet. *Journal of Alloys and Compounds* 922: 166312.

91 Tian, S., Liu, B., Zhao, L. et al. (2019). Red photo-stimulated luminescence from deep traps of $BaZrGe_3O_9$: Pr^{3+} for optical imaging application. *Journal of Alloys and Compounds* 800: 224–230.

92 Zou, Z., Tang, X., Wu, C. et al. (2018). How to tune trap properties of persistent phosphor: photostimulated persistent luminescence of $NaLuGeO_4$:Bi^{3+}, Cr^{3+} tailored by trap engineering. *Materials Research Bulletin* 97: 251–259.

93 Zhang, Y., Chen, D., Wang, W. et al. (2020). Long-lasting ultraviolet-a persistent luminescence and photostimulated persistent luminescence in Bi^{3+}-doped $LiScGeO_4$ phosphor. *Inorganic Chemistry Frontiers* 7: 3063–3071.

94 Zhang, J., Yu, M., Qin, Q. et al. (2010). The persistent luminescence and up conversion photostimulated luminescence properties of nondoped Mg_2SnO_4 material. *Journal of Applied Physics* 108: 123518.

95 Xue, J., Hu, T., Li, F. et al. (2023). Suppressed self-reduction of manganese in Mg_2SnO_4 via Li^+ incorporation with polychromatic luminescence for versatile applications. *Laser and Photonics Reviews* 17: 2200832.

96 Xie, W., Jiang, W., Zhou, R. et al. (2021). Disorder-induced broadband near-infrared persistent and photostimulated luminescence in Mg_2SnO_4:Cr^{3+}. *Inorganic Chemistry* 60: 2219–2227.

97 Chen, X., Li, Y., Huang, K. et al. (2021). Trap energy upconversion-like near-infrared to near-infrared light rejuvenateable persistent luminescence. *Advanced Materials* 33: 2008722.

98 Xu, X., He, Q., and Yan, L. (2013). White-light long persistent and photo-stimulated luminescence in $CaSnSiO_5$:Dy^{3+}. *Journal of Alloys and Compounds* 574: 22–26.

99 Xu, X., Yan, L., Yu, X. et al. (2013). Concentration-dependent effects of optical storage properties in CSSO:Dy. *Materials Letters* 99: 158–160.

100 Wang, C., Zheng, Z., Zhang, Y. et al. (2020). Modulating trap properties by Nd^{3+}-Eu^{3+} co-doping in Sr_2SnO_4 host for optical information storage. *Optics Express* 28: 4249–4257.

101 Zhang, Y., Huang, R., Lin, Z. et al. (2016). Co-dopant influence on near-infrared luminescence properties of Zn_2SnO_4:Cr^{3+}, Eu^{3+} ceramic discs. *Journal of Alloys and Compounds* 686: 407–412.

102 Li, J.L., Shi, J.P., Wang, C.C. et al. (2017). Five-nanometer $ZnSn_2O_4$:Cr,Eu ultra-small nanoparticles as new near infrared-emitting persistent luminescent nanoprobes for cellular and deep tissue imaging at 800 nm. *Nanoscale* 9: 8631–8638.

103 Yan, X.L., Liu, Y.X., Yan, D.T. et al. (2011). The effects of Mn^{2+} doping on the luminescence properties of $12CaO_7Al_2O$:Eu^{2+} nanocrystal phosphor. *Journal of Nanoscience Nanotechnology* 11: 9964–9969.

104 Li, S., Liu, Y., Liu, C. et al. (2017). Improvement of X-ray storage properties of $C_{12}A_7$:Tb^{3+} photo-stimulable phosphors through controlling encaged anions. *Journal of Alloys and Compounds* 696: 828–835.

105 Lyu, S., Zhou, P., Du, J. et al. (2022). Manipulating trap distribution and density by chemical unit cosubstitution for near-infrared persistent luminescent $Zn_{1-2x}Li_xGa_{2+x}O_4$:Cr^{3+} solid solutions. *Journal of Materials Chemistry C* 10: 18404–18414.

106 Ma, X., Feng, P., Wang, Y. et al. (2022). Design of efficient color-tunable long persistent luminescence phosphor $BaGa_2O_4$:Pr^{3+} and its performance enhancement via a trap-induced strategy. *Journal of Materials Chemistry C* 10: 1105–1117.

107 Zhou, Z., Zhang, S., Le, Y. et al. (2021). Defect enrichment in near inverse spinel configuration to enhance the persistent luminescence of Fe^{3+}. *Advanced Optical Materials* 10: 2101669.

108 Zhuang, Y., Ueda, J., Tanabe, S., and Dorenbos, P. (2014). Band-gap variation and a self-redox effect induced by compositional deviation in $Zn_xGa_2O_{3+x}$:Cr^{3+} persistent phosphors. *Journal of Materials Chemistry C* 2: 5502–5509.

109 Ueda, J., Kuroishi, K., and Tanabe, S. (2014). Bright persistent ceramic phosphors of Ce^{3+}-Cr^{3+}-codoped garnet able to store by blue light. *Applied Physics Letters* 104: 101904.

110 Ueda, J., Dorenbos, P., Bos, A.J.J. et al. (2015). Control of electron transfer between Ce^{3+} and Cr^{3+} in the $Y_3Al_{5-x}Ga_xO_{12}$ host via conduction band engineering. *Journal of Materials Chemistry C* 3: 5642–5651.

111 Li, W., Zhuang, Y., Zheng, P. et al. (2018). Tailoring trap depth and emission wavelength in $Y_3Al_{5-x}Ga_xO_{12}$:Ce^{3+}, V^{3+} phosphor-in-glass films for optical information storage. *ACS Applied Materials and Interfaces* 10: 27150–27159.

112 Xu, J., Ueda, J., Zhuang, Y. et al. (2015). $Y_3Al_{5-x}Ga_xO_{12}$:Cr^{3+}: a novel red persistent phosphor with high brightness. *Applied Physics Express* 8: 042602.

113 Du, Q., Ueda, J., Zheng, R., and Tanabe, S. (2023). Photochromism and long persistent luminescence in Pr^{3+}-doped garnet transparent ceramic via

UV or blue light up-conversion charging. *Advanced Optical Materials* 11: 2202612.

114 Xu, J., Ueda, J., and Tanabe, S. (2015). Design of deep-red persistent phosphors of $Gd_3Al_{5-x}Ga_xO_{12}:Cr^{3+}$ transparent ceramics sensitized by Eu^{3+} as an electron trap using conduction band engineering. *Optical Materials Express* 5: 963.

115 Wang, B., Li, X., Chen, Y. et al. (2018). Long persistent and photo-stimulated luminescence in Pr^{3+}-doped layered perovskite phosphor for optical data storage. *Journal of the American Ceramic Society* 101: 4598–4607.

116 Liu, X., Zhang, J., Ma, X. et al. (2013). Violet-blue up conversion photostimulated luminescence properties and first principles calculations of a novel un-doped $CaZrO_3$ phosphor for application in optical storage. *Journal of Alloys and Compounds* 550: 451–458.

117 Qiu, J., Miura, K., Inouye, H. et al. (1998). Femtosecond laser-induced three-dimensional bright and long-lasting phosphorescence inside calcium aluminosilicate glasses doped with rare earth ions. *Applied Physics Letters* 73: 1763–1765.

118 Li, C., Yu, Y., Wang, S., and Su, Q. (2003). Photo-stimulated long-lasting phosphorescence in Mn^{2+}-doped zinc borosilicate glasses. *Journal of Non-Crystalline Solids* 321: 191–196.

119 Li, C., Wang, J., Liang, H., and Su, Q. (2007). Near infrared long lasting emission of Yb^{3+} and its influence on the optical storage ability of Mn^{2+}-activated zinc borosilicate glasses. *Journal of Applied Physics* 101: 113304.

120 Wang, J., Zhang, H., Lei, B. et al. (2015). Enhanced photoluminescence and phosphorescence properties of red $CaAlSiN_3:Eu^{2+}$ phosphor via simultaneous UV-NIR stimulation. *Journal of Materials Chemistry C* 3: 4445–4451.

121 Wang, J., Zhang, H., Lei, B. et al. (2014). Red persistent and photo-stimulated luminescence properties of $SrCaSi_5N_8$: Eu^{2+}, Tm^{3+} solid solution. *Optical Materials* 36: 1855–1858.

122 Wang, J., Zhang, H., Lei, B. et al. (2015). Optical energy storage properties of $(Ca_{1-x}Sr_x)_2Si_5N_8$: Eu^{2+}, Tm^{3+} solid solutions. *Journal of the American Ceramic Society* 98: 1823–1828.

123 Zhang, S., Song, Z., Wang, S. et al. (2020). Red persistent and photostimulable phosphor $SrLiAl_3N_4:Eu^{2+}$. *Journal of Materials Chemistry C* 8: 4956–4964.

124 Wang, F., Guo, J., Wang, S. et al. (2020). Yellow persistent luminescence and electronic structure of Ca-α-Sialon: Eu^{2+}. *Journal of Alloys and Compounds* 821: 153482.

125 Ueda, J., Leaño, J.L., Richard, C. et al. (2019). Broadband near-infrared persistent luminescence of $Ba[Mg_2Al_2N_4]$ with Eu^{2+} and Tm^{3+} after red light charging. *Journal of Materials Chemistry C* 7: 1705–1712.

126 Zhuang, Y., Lv, Y., Wang, L. et al. (2018). Trap depth engineering of $SrSi_2O_2N_2:Ln^{2+}$, Ln^{3+} (Ln^{2+} = Yb, Eu; Ln^{3+} = Dy, Ho, Er) persistent luminescence materials for information storage applications. *ACS Applied Material & Interfaces* 10: 1854–1864.

127 Song, H., Zhang, R., Zhao, Z. et al. (2022). RGB tricolor and multimodal dynamic optical information encryption and decoding for anti-counterfeiting applications. *ACS Applied Materials & Interfaces* 14: 45562–45572.

128 Kitagawa, Y., Ueda, J., and Tanabe, S. (2022). Blue persistent phosphor of $YSiO_2N:Ce^{3+}$ developed by codoping Sm^{3+} or Tm^{3+} ions and thermoluminescence analysis of their trap distributions. *Physical Status Solidi A* 219: 2100670.

129 Wang, S., Liu, X., Qu, B. et al. (2019). Green persistent luminescence and the electronic structure of β-Sialon:Eu^{2+}. *Journal of Materials Chemistry C* 7: 12544–12551.

130 Lv, Y., Wang, L., Zhuang, Y. et al. (2017). Discovery of the Yb^{2+}-Yb^{3+} couple as red-to-NIR persistent luminescence emitters in Yb-activated $(Ba_{1-x}Sr_x)AlSi_5O_2N_7$ phosphors. *Journal of Materials Chemistry C* 5: 7095–7101.

131 Katayama, Y., Ueda, J., and Tanabe, S. (2014). Photo-electronic properties and persistent luminescence in Pr^{3+} doped $(Ca,Sr)TiO_3$ ceramics. *Journal of Luminescence* 148: 290–295.

132 Zhuang, Y., Ueda, J., and Tanabe, S. (2013). Tunable trap depth in $Zn(Ga_{1-x}Al_x)_2O_4$:Cr,Bi red persistent phosphors: considerations of high-temperature persistent luminescence and photostimulated persistent luminescence. *Journal of Materials Chemistry C* 1: 7849–7855.

133 Zhuang, Y., Tu, D., Chen, C. et al. (2020). Force-induced charge carrier storage: a new route for stress recording. *Light: Science and Applications* 9: 182.

134 Wang, Y. and Wang, L. (2007). Defect states in Nd^{3+}-doped $CaAl_2O_4:Eu^{2+}$. *Journal of Applied Physics* 101: 053108.

135 Van den Eeckhout, K., Bos, A.J.J., Poelman, D., and Smet, P.F. (2013). Revealing trap depth distributions in persistent phosphors. *Physical Review B* 87: 045126.

136 Takahashi, K. (2002). Progress in science and technology on photostimulable BaFX:Eu^{2+} (X = Cl, Br, I) and imaging plates. *Journal of Luminescence* 100: 307–315.

137 Schweizer, S. (2001). Physics and current understanding of X-ray storage phosphors. *Physica Status Solidi A* 187: 335–393.

138 Singh, R.K., Lund, D.B., and Buelow, F.H. (2006). Storage stability of intermediate moisture apples: kinetics of quality change. *Journal of Food Science* 48: 939–944.

139 Robins, L.H. and Tuchman, J.A. (1998). Photoluminescence studies of Sm^{2+} in the stimulable phosphor SrS:Eu, Sm. *Physical Review B* 57: 12094–12103.

140 Wu, J., Newman, D., and Viney, I.V.F. (1998). Study on relationship of luminescence in CaS:Eu, Sm and dopants concentration. *Journal of Luminescence* 99: 237–245.

141 Jutamulia, S., Storti, G.M., Lindmayer, J., and Seiderman, W. (1990). Use of electron trapping materials in optical signal processing: parallel Boolean logic. *Applied Optics* 29: 4806–4811.

142 Vrubel, I.I., Polozkov, R.G., Shelykh, I.A. et al. (2017). Bandgap engineering in yttrium-aluminum garnet with Ga doping. *Crystal Growth & Design* 17: 1863–1869.

143 Xu, J., Chen, Z., Gai, M. et al. (2020). Fabrication and OSL properties of Eu^{3+}-doped $NaCaPO_4$ glass-ceramics. *Materials Letters* 261: 126973.

144 Xu, J., Chen, Z., Gai, M. et al. (2020). Optically stimulated luminescence of Dy^{3+}-doped $NaCaPO_4$ glass-ceramics. *Journal of Rare Earths* 38: 927–932.

145 Okada, G., Shinozaki, K., Komatsu, T. et al. (2018). Tb^{3+}-doped BaF_2-Al_2O_3-B_2O_3 glass and glass-ceramic for radiation measurements. *Journal of Non-Crystalline Solids* 501: 111–115.

146 Bos, A.J.J., Dorenbos, P., Bessière, A. et al. (2011). Study of TL glow curves of YPO_4 double doped with lanthanide ions. *Radiation Measurements* 46: 1410–1416.

147 Zhuang, Y., Lv, Y., Li, Y. et al. (2016). Study on trap levels in $SrSi_2AlO_2N_3$:Eu^{2+}, Ln^{3+} persistent phosphors based on host-referred binding energy scheme and thermoluminescence analysis. *Inorganic Chemistry* 55: 11890–11897.

148 Dorenbos, P. (2013). Lanthanide 4f-electron binding energies and the nephelauxetic effect in wide band gap compounds. *Journal of Luminescence* 136: 122–129.

10

The Application of Ternary Quantum Dots in Tumor-Related Marker Detection, Imaging, and Therapy

Ling Yang[2], Xiaojiao Kang[3], Jun Lin[2], and Ziyong Cheng[1]

[1]Harbin Engineering University, College of Materials Science and Chemical Engineering, Key Laboratory of Superlight Materials and Surface Technology, 145 Nantong Street, Harbin 150001, China
[2]Chinese Academy of Sciences, Changchun Institute of Applied Chemistry, State Key Laboratory of Rare Earth Resource Utilization, 5625 Renmin street, Changchun 130022, China
[3]Dongguan University of Technology, School of Electrical Engineering and Intelligentization, No. 1, Daxue Rd, Songshan Lake, Dongguan 523808, China

10.1 Introduction

Quantum dots (QDs) are semiconductors with quasi-zero-dimensional characteristics, confining charge carriers in all three dimensions. This confinement restricts the ability of charge carriers to move freely when a radius is equal to or smaller than the exciton Bohr radius [1]. QDs typically exhibit spherical structures with particle sizes ranging from 1 to 10 nm. To date, various types of QDs have been synthesized through mature synthesis techniques, including II–VI, IV–VI, III–V, II–V, I–III–VI, and I–VI types [2–6]. Typical binary QDs include a series of II–VI semiconductor materials represented by cadmium sulfide (CdS), with unique photoelectric chemical properties such as broad excitation spectra, narrow emission spectra, large Stokes shift, and tunable color [7–13]. QDs have been widely used in the fields of photoelectrochemistry, batteries, and energy storage devices, but their application in the biological field is limited due to the high toxicity of heavy metals contained in them, so it is of significant importance to develop biocompatible and environment-friendly QD materials [14–19]. A new category of ternary QDs composed of I–III–VI group elements has gradually attracted the attention of researchers, including representative copper indium sulfide ($CuInS_2$), silver indium sulfide ($AgInS_2$), and silver indium selenide ($AgInSe_2$). These QDs address the issue of heavy metal elements contained in II–VI type QDs with high toxicity to the environment and biological systems, creating favorable foundation for the application of semiconductor QDs in the field of biomedicine [20–25].

For example, silver-based chalcogenide quantum dots can be classified into two types: binary QDs represented by silver sulfide and ternary QDs represented by silver indium sulfide. Binary QDs, represented by silver sulfide, are a type of inorganic

Photofunctional Nanomaterials for Biomedical Applications, First Edition.
Edited by Chunxia Li and Jun Lin.

nanocrystal material that exhibits UV-NIR-II optical absorption, high optical and chemical stability, and nontoxicity. Ag_2S, Ag_2Se, and Ag_2Te are narrow-bandgap semiconductor materials that exhibit broad absorption spectra. Compared to conventional QDs, they have lower solubility product constants, indicating that silver chalcogenide binary QDs are relatively stable [26–28]. Synthesized Ag_2S and Ag_2Se QDs have tunable emission spectra in the range of 510–1300 nm, making them suitable for in vivo noninvasive fluorescence imaging with better penetration depth and signal-to-noise ratio [28–30]. By changing particle size and designing core–shell structures, near-infrared QDs with tunable luminescence and high quantum efficiency can be designed and synthesized, further expanding their biomedical applications. However, in comparison with ternary QDs, the properties of binary QDs have many limitations, such as poor fluorescence intensity and low fluorescence quantum efficiency. Ternary QDs, represented by $AgInS_2$, exhibit stable optical properties, and their fluorescence quantum yield can be enhanced by performance tuning. The emission wavelength can also be controlled by adjusting the size of QDs to cover the visible to near-infrared light region [31–33]. Benefiting from their unique band transitions, they have good luminescence efficiency, while surface and internal defects can seriously affect their luminescence efficiency. Typically, surface defects can be reduced by changing the precursor ratio or other reaction conditions, or enhancing the quantum efficiency by designing core–shell structures. Due to the advancements of $AgInS_2$ QDs synthesis technology and ongoing enhancements of diverse properties, their potential applications in the field of biomedicine, including roles as biological probes and in bioimaging, have attracted widespread attention.

10.1.1 Fundamental Properties of QDs

QDs are semiconductor materials that belong to the nanoscale range and exhibit quantum confinement and quantum size effects [34]. As the size of the QDs changes, the electronic energy level or bandgap changes, and the absorption spectra of the QDs exhibit blue or red shifts with the decrease or increase of size, respectively, while the fluorescence emission spectra also change accordingly [35, 36]. Size regulation of QDs is one of the methods to change the bandgap of QDs, so semiconductor QDs with different fluorescence emissions can be obtained by adjusting the size, as shown in Figure 10.1 [37]. In addition, quantum dot materials have remarkable surface effects, and the ratio of surface atoms to total atoms of semiconductor nanomaterials increases abruptly with the decrease in size [38–40]. However, this phenomenon leads to the generation of numerous defects on the surface of QDs, which in turn enhances nonradiative transitions, weakens radiative transitions, and significantly reduces fluorescence intensity. Furthermore, QDs possess unique optical properties due to their special band structure, mainly manifested in the following aspects: (i) broad absorption and excitation wavelength ranges [41–43]. The QDs possess discrete energy levels, enabling them to absorb light across a range of wavelengths, spanning from ultraviolet to visible and even into the infrared spectrum. Correspondingly, the excitation wavelength

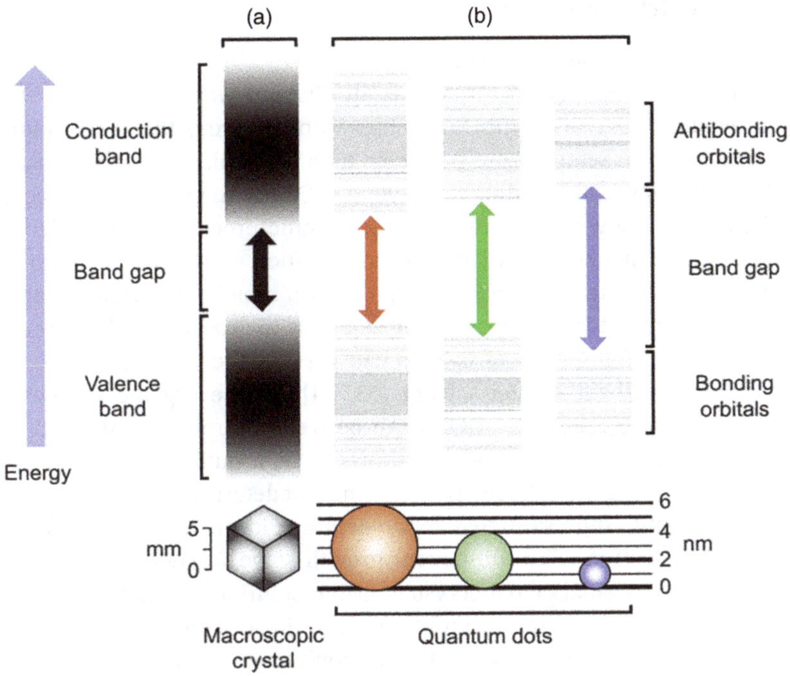

Figure 10.1 The bandgap of QDs becomes narrower with the increase in size, and vice versa, and the bandgap becomes wider. Source: Reproduced with permission from Ref. [37]; © 2010, Informa UK Limited.

demonstrates a broad absorption. The single excitation light source can activate QDs with different sizes, prompting the emissions of diverse fluorescence based on their unique properties, as long as the initial energy is higher than the energy of the bandgap; (ii) continuous tunability of fluorescence spectra [44–46]. QDs have discrete energy levels, and when electrons transition between these different energy levels, they emit light at various wavelengths. The size of QDs also plays a crucial role in determining the spacing of the energy level, which is closely related to the emitted fluorescence wavelength. Therefore, changing the size of QDs allows for continuous regulation of their fluorescence wavelength; (iii) good photostability [6, 47]. Organic small molecule fluorescence materials exhibit a photobleaching effect over time, limiting their applications. The photostability of quantum dot materials is dozens or even hundreds of times greater than that of organic small molecule fluorescence materials, which lays the foundation for their long-term tracking in organisms; (iv) large Stokes shift [6, 47–49]. The Stokes shift is the distance between the fluorescence emission spectrum and absorption spectrum. If the Stokes shift is small, the distance between the excitation spectrum and emission spectrum is very close, resulting in serious spectral overlap, which will cause fluorescence quenching and lead to low photoluminescence efficiency and strong fluorescence interference in fluorescence detection. Fortunately, QDs exhibit large Stokes shifts and do not display the aforementioned phenomenon.

10.1.2 Synthesis Methods of QDs

10.1.2.1 Metal-Organic Synthesis Method

The metal-organic synthesis method serves as an efficient way to prepare QDs, characterized by their ease of control and observation. During the preparation process, the high reaction temperature promotes high crystallization and reduces lattice defects in nanomaterials, leading to the yield of QDs with exceptional optical properties [50, 51]. As shown in Figure 10.2, the specific process of synthesizing QDs in organic system involves the thermal decomposition of precursor solutions in high-boiling-point organic solvents. The precursor rapidly decomposes and nucleates at high temperatures, with slow crystal growth leading to the formation of nanocrystals. The presence of ligands can prevent the continued growth of nuclei and stabilize them in the solvent [53, 54]. In the organic synthesis process, the nucleation and growth of QDs can be well controlled by the action of ligands and temperature regulation, resulting in high-quality quantum dot materials. Upon injecting the precursor into the predetermined temperature organic reaction solvent, supersaturated monomers are formed, which leads to an instantaneous nucleation. The subsequent growth stage follows the Ostwald ripening theory, whereby smaller nanocrystals in the solution gradually dissolve while larger nanocrystals continue to grow, resulting in a decrease in the total number of nanocrystals and a decrease in the solution's free energy [55]. However, the manipulation of this synthetic method is complex and requires inert gas protection during the entire experimental process. At the same time, this strategy resulted in the oil-based quantum dot materials due to the hydrophobic surface. To meet the requirements of biological experiments, it is necessary to remove the hydrophobic groups on the surface or connect two amphiphilic groups to make them water-soluble quantum dot materials. However, the luminescent and stable properties of QDs may be greatly reduced after water-soluble modification. Organic synthesis methods typically include one-pot synthesis, single-source precursor synthesis, and thermal injection synthesis [54, 56]. Bawendi and coworkers first investigated the thermal injection method for preparing QDs and proposed a simple

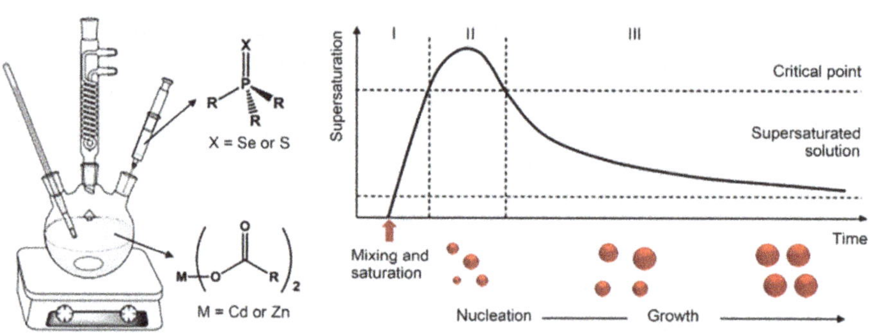

Figure 10.2 Schematic diagram of thermal injection methods and nucleation and growth processes of QDs. Source: Reproduced with permission from Ref. [52]; © 2010, Optica Publishing Group.

way to prepare high-quality cadmium-based semiconductor nanocrystals [50]. In the preparation process, dimethyl cadmium was used as the cadmium source, selenium powder was dissolved in a tri-*n*-octylphosphine solution as the selenium source, and tri-*n*-octylphosphine oxide and tri-*n*-octylphosphine were used as solvents. The reaction was carried out at 300 °C under inert gas protection, and the selenium source was then rapidly injected into the reaction system. Under high-temperature conditions, the monomer concentration was instantaneously saturated, resulting in the successful preparation of semiconductor nanocrystals with a narrow size distribution, uniform crystal structure, high monodispersity, and excellent luminescent properties.

In 2008, Lu and coworkers synthesized ternary $CuInS_2$ QDs with the wuerzite structure for the first time by organic-phase hot-injection method, but no fluorescence phenomenon due to the large size of nanoparticles [57]. Later, Peng and coworkers successfully synthesized monodisperse $CuInS_2$ QDs with adjustable size with improved methods by balancing the reactivity of cations in the precursor [58]. The fluorescence emission peak was highly adjustable, spanning from 500 to 950 nm, with a luminous efficiency of 30% after being modified with a ZnS shell. With the optimization and maturity of the experimental conditions, the fluorescence intensity of the synthesized nanoparticles has been rapidly enhanced, and the fluorescence quantum yield has also been greatly improved.

10.1.2.2 Hydrophilic Synthesis Method

Compared with organic synthesis methods, aqueous phase synthesis methods have the advantages of simple process conditions, low cost, safe experimental process, and direct applicability in the field of biology. Currently reported aqueous phase synthesis methods include direct heating method, hot-injection method, hydrothermal method, microwave radiation-assisted method, etc. [59–61]. The commonly used stabilizers for aqueous phase synthesis are small molecules with thiol groups, such as glutathione, cysteamine, and mercaptopropionic acid. Thiols can produce strong bonding forces with the metal elements on the surface of quantum dot materials, thereby stabilizing them in aqueous solvents [62–65]. The quantum efficiency of QDs prepared by direct heating method can reach 60%, and those prepared by the microwave method can reach 66%. QDs synthesized by aqueous phase synthesis methods have high biocompatibility and ideal quantum efficiency, which can be directly used in biological experiments, thus avoiding many experimental difficulties for researchers.

As shown in Figure 10.3, Xiong et al. presented a facile microwave-assisted synthesis route for the preparation of water-soluble and high-quality $CuInS_2/ZnS$ QDs with glutathione as the stabilizer. The as-prepared QDs exhibited excellent properties with low toxicity, tunable color, and water solubility [66]. Pan and coworkers fabricated the highly luminescent water-soluble green and red $AgInS_2/ZnS$ core/shell QDs in a large scale by using an electric pressure cooker as a reaction vessel [67]. The PLQY of these core/shell QDs can reach 18.4–39.1%. It is interesting that the authors realized the multigram scale preparation of water-soluble $AgInS_2/ZnS$ core/shell QDs by employing a household electric

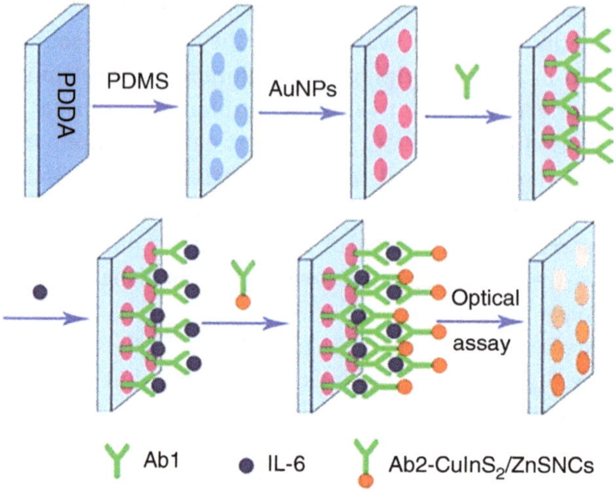

Figure 10.3 Schematic illustration of fluoroimmunoassay for IL-6 with aqueous synthesized $CuInS_2$/ZnS QDs. Source: Reproduced with permission from Ref. [66]; © 2013, American Chemical Society.

pressure cooker. The facile synthetic route has the conspicuous advantage of being greener for the production of nontoxic QDs on a large scale by one step.

Owing to the hydrophilic property, as well as the active functional groups on the surface of the prepared QDs, they are suitable for biological applications such as bioimaging or conjugating drugs.

10.1.2.3 Biosynthesis Method

Compared to in vitro synthesis, biogenic synthesis not only provides a sustainable preparation route, but also endows QDs with unique properties, which is advantageous for their application in biotic and abiotic systems. To date, biogenic QDs have been successfully synthesized in many living organisms ranging from prokaryotes to multicellular organisms. However, the uncontrollability of biogenic synthesis has limited its practical applications [68]. Overcoming these challenges requires a deeper understanding of the molecular mechanisms underlying biogenic synthesis and the development of effective regulatory strategies. As shown in Figure 10.4, Yu and coworkers successfully synthesized high-quality selenium sulfur cadmium semiconductor QDs in *Escherichia coli* by adding selenium source solvent and cadmium source solution into the culture medium during bacterial cultivation, stimulating the production of proteins containing reducing thiols involved in the metabolic process of bacteria, thereby promoting the successful assembly of biogenic QDs [69]. The synthesized biogenic QDs exhibit good optical properties and biocompatibility and have been successfully used for biological imaging of mouse cells and tumor tissues.

As biomacromolecules can regulate the formation process of inorganic minerals, many biomolecules, such as nucleic acids, peptides, proteins, and polysaccharides, have been applied in the synthesis of QDs [70, 71]. Fluorescent labeling of

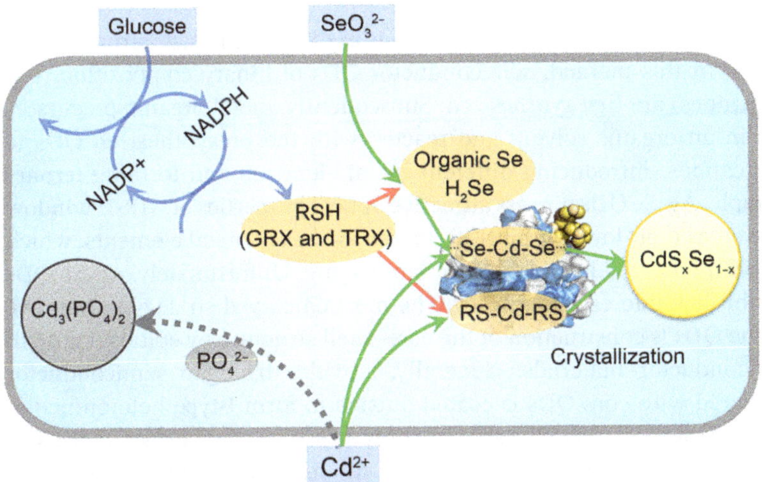

Figure 10.4 Schematic diagram of metabolic regulation of selenium and cadmium biotransformation pathways in *E. coli*. Source: Reproduced with permission from Ref. [69]; © 2019, American Chemical Society.

proteins can provide an effective means for exploring the functions of proteins in basic biological processes and achieving various biomedical applications, such as optical imaging and biosensing. Researchers have already synthesized protein-functionalized QDs that emit highly luminescent light in the red to near-infrared range at room temperature. Various proteins with a wide range of molecular weights and isoelectric points have been successfully used to construct protein-functionalized QDs [72]. The resulting protein-coupled QDs exhibit high quantum yield, high photostability, colloidal stability, and high functionalization efficiency. The attached proteins maintain their original biological activity and can be used to specifically label and image cancer cells.

10.1.3 Synthesis Methods of Ternary QDs

10.1.3.1 Hot-Injection Method

The hot-injection method is a widely used approach to synthesize ternary semiconductor QDs. In this method, metal-organic precursors (e.g. metal acetates or metal phosphine complexes) are injected into a high-boiling-point organic solvent at elevated temperatures. At the same time, a sulfur or selenium source is quickly injected, and a surfactant is added to control the growth and stabilization of the QDs. The high temperature ensures fast reaction kinetics, enabling precise control over the size and morphology of the QDs. Hu et al. successfully synthesized $AgInS_2$ QDs using the hot-injection method in their study [73]. They utilized metal acetates and a sulfur source to synthesize $AgInS_2$ QDs with varied sizes in octadecene. By adjusting the reaction time and temperature, they controlled the size of the QDs and characterized their properties.

10.1.3.2 Ion Exchange Method

The ion exchange method is another technique used to synthesize ternary semiconductor QDs. In this method, semiconductor QDs of binary composition (e.g. sulfides or selenides) are first synthesized. Subsequently, metal organic precursors are dissolved in an organic solvent and reacted with the presynthesized QDs to exchange the cations, introducing different metal elements and forming ternary QDs. For example, Ag_2Se QDs possess attractive optical properties of NIR-II window emission, as well as their low solubility and nontoxic heavy metal elements, which endow them with potential applications in bioimaging. Unfortunately, Ag_2Se QDs generally exhibit low fluorescent efficacy. The most employed strategy to improve the PLQY of the QDs is construction of the core–shell structure by epitaxial growth of shell semiconductor materials. Generally, a wider bandgap semiconductor material compared with core QDs is coated outside to form I-type heterojunction core–shell QDs. However, this strategy can only passivate the surface defects of Ag_2Se QDs the traps within the crystal lattice are still remained, leading to a limited improvement of luminescent intensity. Additionally, the lattice mismatching problem often results in unsuccessful epitaxial growth of shell.

In order to resolve this problem, Wang and coworkers constructed AgAuSe QDs via heteroatomic doping through ion exchange reaction [28]. The authors first synthesized the Ag_2Se QDs by hot-injection method in organic solvent. Then, the as-prepared QDs were immersed in the chloroform solution that contained $HAuCl_4{\cdot}3H_2O$/oleylamine and reacted for 12 hours. As a result, the photoluminescence of the obtained AgAuSe ternary QDs presented a significant blue shift from 1600 to 978 nm. Meanwhile, the PL intensity of AgAuSe QDs was remarkably enhanced by 400 times the original Ag_2Se QDs. The absolute PLQY of the AgAuSe QDs at 978 nm was achieved at 65.3%. This study demonstrated a successful instance of construction of ternary semiconductor QDs with high PLQYs (Figure 10.5).

10.1.3.3 Hydrothermal Method

The hydrothermal method involves the synthesis of nanomaterials under high-temperature and high-pressure conditions. It can also be utilized for the synthesis

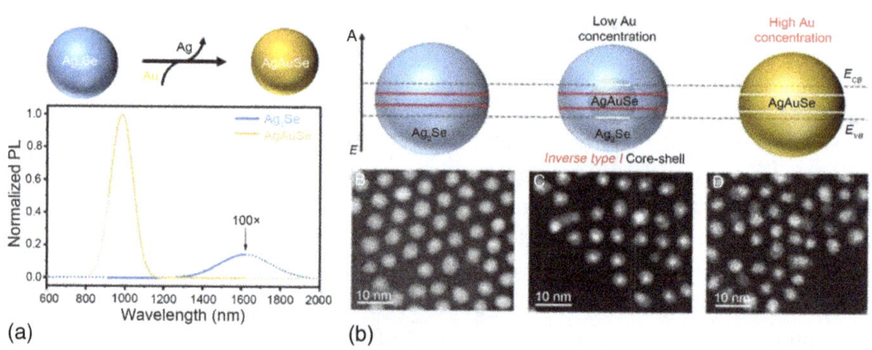

Figure 10.5 Comparison of fluorescence intensity before and after alloying Ag_2Se QDs and the effect of different Au doping ratio on the bandgap of QDs. Source: Reproduced with permission from Ref. [28]; © 2021, American Chemical Society.

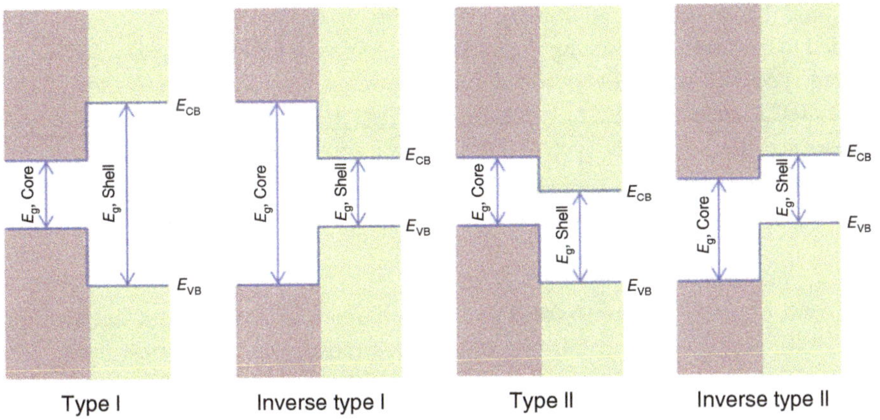

Figure 10.6 Types of core–shell QDs. Source: Reproduced with permission from Ref. [75]; © 2019, American Chemical Society.

of ternary semiconductor QDs. In this method, metal organic precursors and sulfur or selenium sources are dissolved in an aqueous solution. The reaction takes place under elevated temperature and pressure, allowing for the controlled growth of ternary QD with various sizes and morphologies. Wang et al. successfully synthesized $CuInS_2$ QDs using the hydrothermal method [74]. They dissolved metal copper complexes and a sulfur source in an aqueous solution and then conducted the reaction under high-temperature and high-pressure conditions to obtain $CuInS_2$ QDs. The synthesized QDs were characterized by their optical and electrical properties (Figure 10.6).

10.1.4 Performance Control of QDs

10.1.4.1 Core–Shell Structure

The surface effects of QDs have been found to significantly affect their fluorescence properties due to the presence of a large number of surface states. A strategy of using core–shell structures was proposed to coat the surface of core QDs with shell materials, which can effectively improve their surface properties, passivate surface defects, enhance quantum efficiency, and increase their biocompatibility [76–79]. Tan and coworkers synthesized NIR-emitting $CuInS_2$/ZnS core/shell QDs with a maximum photoluminescence quantum yield of 65% through a stepwise, consecutive solvothermal strategy [80]. The highly fluorescent core-shell $CuInS_2$/ZnS QDs had a remarkable enhancement of quantum yield (QY) in comparison to that of the 8.8% from $CuInS_2$ core only. This should be attributed to the passive surface defects of $CuInS_2$ core and the adjustment of the bandgap engineering for the core–shell heterojunction. The core–shell QDs can be classified into three types based on the relative position of their bandgap energy levels: type I, type II, inverted type I, and inverted type II (Figure 10.5) [75]. Type I QDs have a core bandgap smaller than that of the shell material [81–84]. For example, CdSe/ZnS is a typical type I core–shell QD material in which the core electrons cannot jump to the outer shell due to the limited domain effect. The charge carriers of the nucleus are confined to the interior, allowing the surface defects to be effectively passivated [85]. Inverted type I QDs possess a core bandgap energy level larger than that of the shell material, resulting in a redshift in the emission wavelength and an increased fluorescence lifetime [86]. Type II QDs have a core–shell conduction band and valence band energy levels that cross each other, causing a redshift in the emission position and effectively reducing the self-absorption of material.

10.1.4.2 Alloying

When two or more kinds of semiconductor nanocrystals fuse with each other to produce new alloy-type nanocrystals, the bandgap width of new materials will vary in response to the change of composition [87–89]. Alloying materials with different bandgaps have been widely used in the bandgap engineering of semiconductors.

10.1.4.3 Ioning

Ion doping refers to the introduction of transition metal or rare earth ions, including Cu^{2+}, Mn^{2+}, Yb^{3+}, etc., into the nanocrystal lattice. Doping metal ions into nanomaterials can adjust the crystal structure, electronic structure, and other properties. The doped ions form new energy levels within the original nanocrystal energy levels, resulting in new photoelectric properties of the nanocrystal materials [90, 91]. The advantage of doped quantum dot nanocrystals primarily stems from the fact that they not only preserve the merits of traditional QDs but also generate larger Stokes shift emissions than the intrinsic emission. This allows them to reach the fluorescence region that the intrinsic emission cannot attain. Moreover, this type of semiconductor nanocrystal exhibits exceptional thermal stability and minimally reduces the influence of temperature [92, 93].

10.1.5 Modification of QDs

10.1.5.1 Surfacing Ligand Molecular Exchange

The surface ligand exchange of QDs is a commonly used method for water-soluble modification, which involves replacing the hydrophobic groups on their surface with long-chain molecules that contain hydrophilic functional groups at both ends [30, 94, 95]. The hydrophilic long-chain molecule reagents typically have a thiol group at one end, which interacts strongly with the metal atoms on the surface of the QDs. The bond cooperation is often stronger than the action of hydrophobic groups, ensuring the effective modification of hydrophilic molecules [96, 97]. The other end of the molecule typically contains amino or carboxyl groups, depending on the desired application, to ensure the hydrophilicity of the modified materials. The specific process typically involves dispersing the hydrophobic QDs in an organic solvent, such as chloroform, and dispersing the hydrophilic reagents in a surfactant solvent, such as ethanol, followed by mixing and stirring. The strong bonding between the thiol group and the metal atoms on the surface of QDs replaces the original hydrophobic groups, causing a phase transition of the quantum dots transferring into the surfactant solvent. The resulting product is then subjected to solvent evaporation and washed with ethanol and pure water to obtain water-soluble QDs. These QDs can be well-dispersed in water and further functionalized according to the desired application.

10.1.5.2 Amphiphilic Organic Macromolecular Coating

Amphiphilic polymers refer to a class of molecules that can dissolve in both aqueous and organic solvents. Typically, one end of these polymers is hydrophobic, while the other end is hydrophilic [98]. The encapsulation principle is based on the similar solubility principle. In the same solvent, the hydrophobic end of the amphiphilic polymer can be embedded into the hydrophobic groups on the surface of QDs, while the hydrophilic end can be exposed to the aqueous phase, forming water-in-oil structure. This modification method can retain the original structure of QDs. During the modification process, ultrasonic treatment is often required to ensure the successful coupling between the hydrophobic groups.

10.1.6 Characterization of QDs

The synthesis of quantum dot materials through various preparation methods typically requires a series of tests, including determining their crystal structure, assessing the quality of their optical properties, and verifying the uniformity of their particle size. Common techniques for materials characterization include: (i) X-ray powder diffraction, which analyzes the distribution and intensity of X-ray diffraction from the regular arrangement of atoms in a crystal lattice to determine the phase of the sample and its quantitative composition; (ii) transmission electron microscopy, which uses an accelerated and focused electron beam projected onto a thin copper grid containing the sample to produce images that reveal the internal fine structure, chemical composition, crystal orientation, and electronic structure of materials; (iii) Ultraviolet-Visible-Near-Infrared absorption spectroscopy (UV-Vis-NIR) is a powerful technique to study the interaction of matter with electromagnetic radiation from UV to near-infrared wavelength range to detect the transition of valence electrons between energy levels and provides the valuable information about electronic structure; and (iv) fluorescence spectroscopy (FL), which measures the wavelength of light emitted by materials after absorbing photons and transitioning from an excited state to a stable state. Fluorescence spectroscopy yields two important parameters: fluorescence lifetime, which represents the average time that a particle remains in the excited state, and fluorescence quantum yield, which indicates the efficiency of fluorescence emission.

10.1.7 Biomedical Applications of QDs

Before the 1990s, research on QDs was primarily focused on optoelectronics.

However, in 1998, Alivisatos and Nie research groups nearly simultaneously published studies on the use of QDs as biological probes, thus opening the door to the application of QDs in the field of biomedical research [99, 100].

10.1.7.1 Biological Detection

In the field of biological detection, there are two important indicators for fluorescent detection: Stokes shift and fluorescence lifetime [94, 101]. A larger Stokes shift can effectively reduce the interference of biological autofluorescence and has advantages such as minimal damage to biological samples, strong sample penetration, and high detection sensitivity. Traditional fluorescent dyes generally have maximum absorption and emission wavelengths in the visible light region, with a smaller Stokes shift, which can easily lead to fluorescence quenching phenomena in a short period of time. However, QDs can regulate fluorescence through means such as particle size control, core–shell structure design, and alloying, and generally exhibit a larger Stokes shift. In fluorescence detection, markers with longer fluorescence lifetimes are often used, and time-resolved spectroscopy can separate them from biological autofluorescence. The autofluorescence of biological organisms is typically around 10 ns, while the fluorescence lifetime of inorganic nanomaterials can be around 100 ns. Shi et al. reported on the use of water-soluble $AgInS_2$ QDs for the specific detection of dopamine, synthesizing QDs fluorescence probe with high sensitivity and selectivity to dopamine using mercaptopropionic acid and polyethyleneimine as ligands [102]. Dopamine induces fluorescence quenching with the specific QDs fluorescence probe, allowing for the quantitative detection of dopamine. The extent of fluorescence quenching is directly related to the concentration of added dopamine. The experimental results show that the $AgInS_2$ QDs display high selectivity and sensitivity in detecting dopamine (Figure 10.7).

In our group, a highly sensitive fluorescence detection platform for lung cancer marker CYFRA21-1 DNA was designed based on the hybridization and accumulation technology of DNA functionalized QDs [103]. Ternary $AgInS_2$/ZnS core–shell QDs nanocrystals and magnetic nanoparticles of ferroferric oxide (Fe_3O_4) were synthesized, and the target DNA sequence was detected by using circulating tumor DNA (ctDNA) to trigger a hybridization chain reaction. In the presence of the target sequence, the three DNA hairpin structures are activated and opened successively according to the principle of base complementary pairing, resulting in the accumulation of QDs attached to the hairpin structures. With magnetic adsorption and a

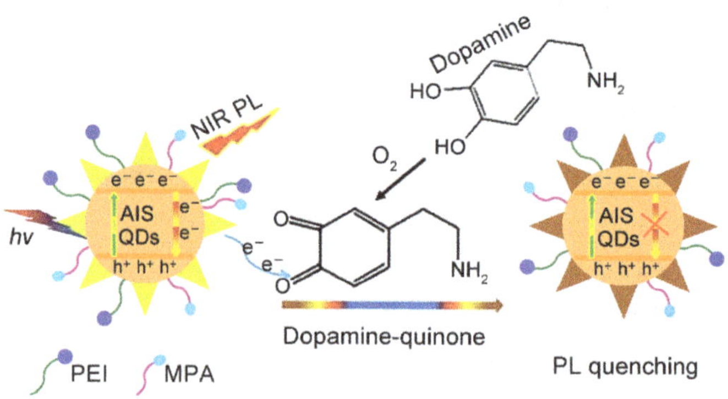

Figure 10.7 Illustration of $AgInS_2$ QDs-based fluorescence probe for dopamine detection. Source: Reproduced with permission from Ref. [102]; © 2001, American Institute of Physics.

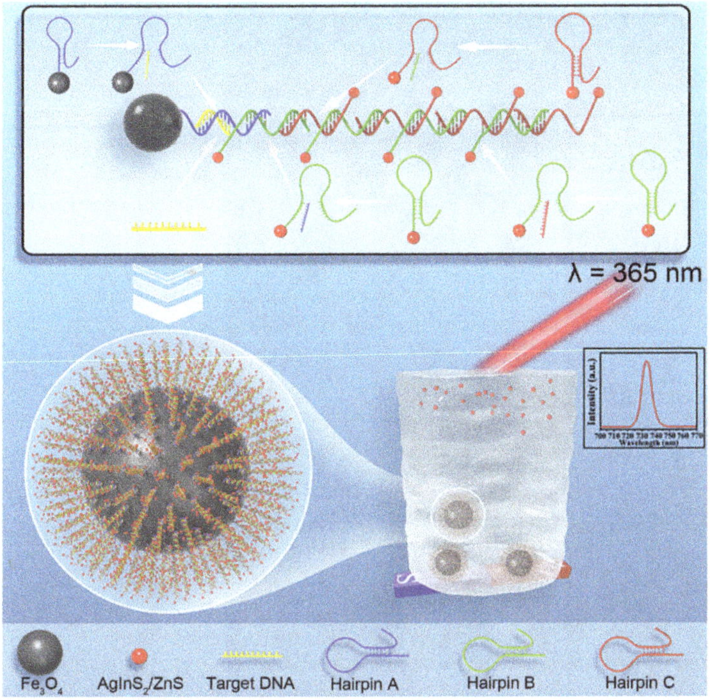

Figure 10.8 Schematic diagram of CYFRA21-1 DNA detection based on hybrid chain reaction. Source: Reproduced with permission from Ref. [103]; © 2022, The Royal Society of Chemistry.

large specific surface area, Fe_3O_4 can be used to fix and aggregate QDs attached to the single-chain DNA so that the fluorescence signal in the detection solution can be significantly changed. The concentration of ctDNA is closely related to the luminescence of the remaining quantum dots in the supernatant after magnetic adsorption, which can be used for quantitative analysis (Figure 10.8).

10.1.7.2 Cell Imaging

Quantum dot materials have the characteristics of high fluorescence intensity and strong photobleaching resistance and have excellent cell imaging performance [104–106]. Lin et al. synthesized a new type of green and environmental-friendly water-soluble $AgInS_2$-TG QDs using thioacetamide (TG) as ligands, silver nitrate, and indium nitrate tetrahydrate as metal precursors through a hydrothermal method. The results of cytotoxicity and cell imaging experiments showed that $AgInS_2$-TG QDs could penetrate the cell membrane and show strong fluorescence intensity in the cells, indicating that they can be used for cell staining and labeling. Moreover, they have good photobleaching resistance, making them an ideal fluorescent probe for cell imaging. Wu et al. used QDs conjugated with immunoglobulin and streptavidin to label surface markers of cancer cells, as well as stained cytoplasmic proteins such as actin, as shown in Figure 10.9 [107]. The signal strength of the labeling was higher than that of organic dyes, and the photo stability was higher.

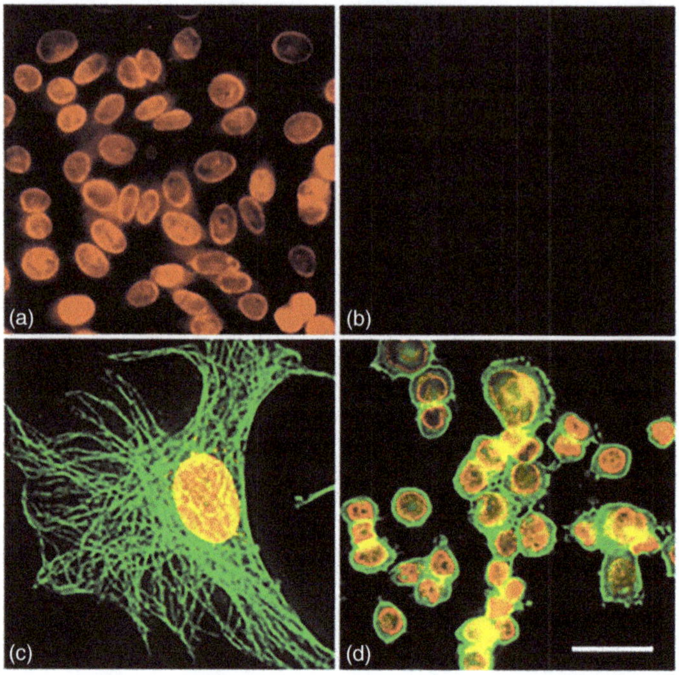

Figure 10.9 Nuclear antigen detection and double labeling. (a) Nuclear antigens in the nuclei of human epithelial cells were labeled. (b) Normal cells. (c) The stained 3T3 cell. (d) The stained SK-BR-3 cells. Scale bar, 50 μm for (a), (b), and (d), and 10 μm for (c). Source: Reproduced with permission from Ref. [107]; © 2003, Springer Nature Limited.

10.1.7.3 Live Imaging

Live imaging is essential for the early diagnosis and treatment of cancer. It provides a more intuitive and accurate means of tracking the biological processes within the subject being studied. Therefore, it plays an indispensable role in tumor diagnosis and surgery therapy [108]. The imaging methods for tumor diagnosis mainly include positron emission computed tomography (PET-CT), electron computed tomography, and magnetic resonance imaging. These techniques predominately utilize deep-penetrating radiation to visualize internal structures and effectively improve the detection rate of malignant tumor lesions [109–111]. However, these imaging techniques still have some drawbacks: (i) very few imaging methods can achieve millisecond-level time resolution and micron-level spatial resolution, resulting in poor monitoring capability of dynamic lesions; (ii) there is a high risk of radiation exposure during the imaging process, which is harmful to human health; and (iii) expensive equipment and high detection costs limit their practicality and popularity. Hence, live fluorescence imaging has emerged as a new imaging technique. Nevertheless, the fluorescence imaging technique faces a significant challenge due to the severe interference of tissue absorption, scattering, and spontaneous fluorescence. However, it is worth noting that biological tissue absorbs and scatters near-infrared light much less than visible light, especially

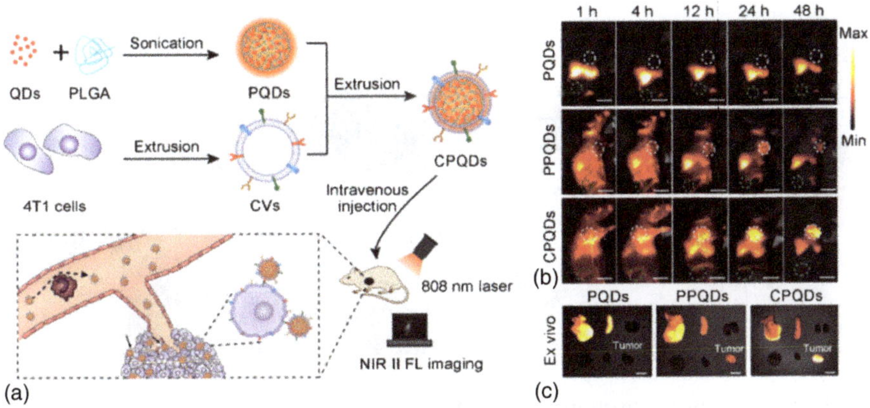

Figure 10.10 Schematic illustration of the nanobioprobe preparation and application. Source: Reproduced with permission from Ref. [116]; © 2019, Wiley-VCH Verlag GmbH & Co. KGaA.

in the near-infrared second window (NIR-II). The intrinsic fluorescence of the tissue is extremely low within this spectral region, and it is even possible to achieve true zero intrinsic fluorescence when the wavelength extends beyond 1500 nm [112–115].

Therefore, NIR-II imaging is more attractive in deep biological tissue imaging. QDs emitting in the NIR-II region demonstrate prominent optical characteristics such as high brightness and good photostability in vivo. Moreover, QDs fluorescence imaging exhibits excellent spatiotemporal resolution and greater penetration depth, becoming a promising fluorescence biomarker in this field. As shown in Figure 10.10, Pang and coworkers encapsulated Ag_2Te QDs with tumor cell membranes and poly(lactic-*co*-glycolic acid) (PLGA) to construct a biomimetic nano-biosensor [116]. This sensor retained the immune evasion and homologous targeting ability of tumor cell membranes, meanwhile enhancing the fluorescence intensity of QDs. With the active homologous tumor targeting ability of the source tumor cell membrane and the enhanced permeability and retention (EPR) effect, this nano-biosensor exhibited excellent performance in tumor imaging in vivo.

Lian et al. synthesized $CuInSe_2$/ZnS core–shell QDs with NIR emission fluorescence to detect the circulating tumor cells [76]. The photoluminescence quantum yield of NIR QDs can reach 21.8% after the coating of the ZnS shell, becoming the highest among Pb/Cd-free QDs reported so far. Furthermore, the synthesized nanoprobes also exhibit great potential in tumor bioimaging with a satisfactory signal-to-noise ratio, and the imaging of the hind limbs and organ blood vessels has a high resolution that can distinguish vessels with a diameter of 0.36 mm. It revealed that the NIR luminescent $CuInSe_2$ nanoprobes possess application prospects in the field of cancer diagnosis and imaging-guided surgery (Figure 10.11).

10.1.7.4 Tumor Therapy

So far, the application of ternary QDs in tumor therapy is still relatively limited. It has been reported that particle QDs have ideal photothermal (PTT) and

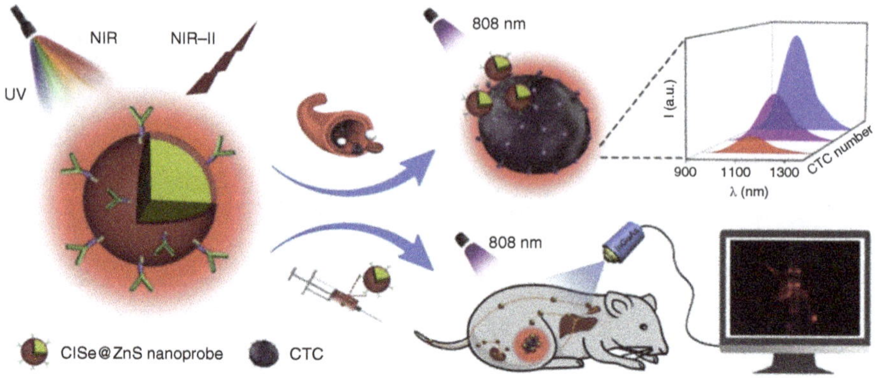

Figure 10.11 Schematic illustration of $CuInSe_2/ZnS$ nanoprobes for circulating tumor cells (CTCs) bioassay and tumor-targeted bioimaging. Source: Reproduced with permission from Ref. [76]; © 2020, Elsevier Ltd. All rights reserved.

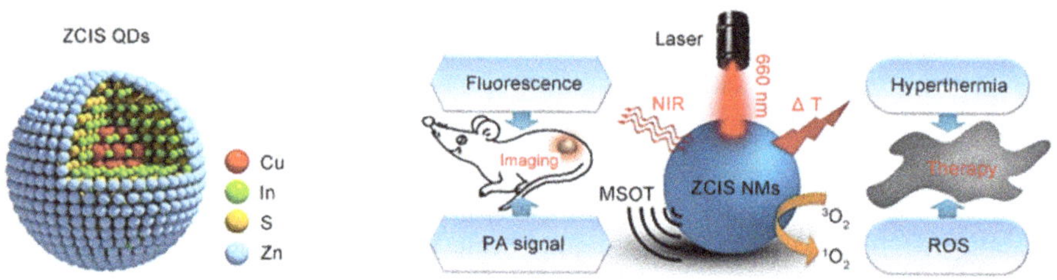

Figure 10.12 Scheme of "all-in-one" CuInS/ZnS QDs. Source: Reproduced with permission from Ref. [119]; © 2016, American Chemical Society.

photodynamic therapeutic (PDT) effects under laser irradiation [117, 118]. Lv et al. synthesized versatile "all-in-one" theranostic ternary CuInS/ZnS nanomedicines, exhibiting NIR imaging ability and synergistic PTT/PDT effects upon a single 660 nm laser irradiation [119]. The results have shown that the ternary QDs possess longer tumor retention times, higher tumor uptake, and deeper tumor penetration, exhibiting great potential as precision theranostic nanomedicines (Figure 10.12). However, the research on the use of ternary QDs as acoustic sensitizers for sonodynamic therapy (SDT) is still relatively scarce. SDT is a new type of treatment that differs from traditional tumor therapies such as chemotherapy and radiotherapy. It involves using ultrasound in combination with chemical sensitizing agents to generate reactive oxygen species (ROS) from oxygen or water in the surrounding environment [12, 120–122]. These ROS are oxygen-containing reactive chemical substances that can oxidize surrounding substrates and serve as highly reactive reagents to induce a series of irreversible damage in target tissues. High concentrations of ROS can induce cell apoptosis by directly oxidizing and damaging DNA molecules, cell membranes, and organelle membranes, leading to changes in the intracellular environment and attacking certain proteins [123]. Li et al. developed a red blood cell membrane-enveloped O_2 self-supplementing biomimetic nanoplatform based on Ag_2S QDs for enhanced tumor SDT [124].

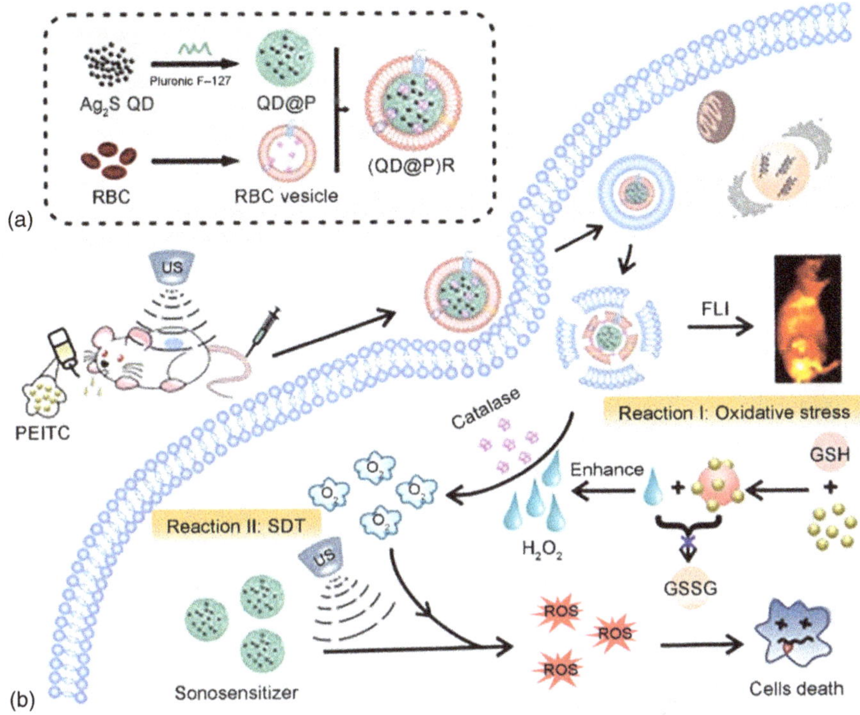

Figure 10.13 Schematic illustration of synthesis route of nanomedicine, tumor imaging, and therapy. Source: Reproduced with permission from Ref. [124]; © 2020, Ivyspring International Publisher.

For the first time, Ag_2S QDs were used as sonosensitizer to generate ROS under ultrasonic stimulation. Encapsulation of QDs in membranes not only prolonged the circulation time of the probe but also catalyzed endogenous H_2O_2 to ameliorate tumor hypoxia (Figure 10.13).

Photodynamic therapy, as a stimulus-responsive therapy, has been widely used in clinical settings and has ideal therapeutic effects on superficial lesions. However, its further clinical application and development have been severely hindered by the limitation of light penetration depth and accompanying phototoxicity [118, 125]. SDT is a new type of minimally invasive, safe, effective, and cost-effective specific therapy that has been developed based on the foundations of photodynamic therapy and ultrasound. In recent years, ternary silver-based QDs have exhibited a broad application prospect in the biomedical field. In particular, NIR-II QDs are well-suited for use in biological fluorescence imaging, as they can avoid interference from tissue scattering and autofluorescence. However, the lack of mechanism research on QDs used in the biomedical field has confined the application primarily to imaging, restricting their widespread use in diagnosis and tumor therapy.

In our group, we developed a research platform for neutrophil-mediated inactivated bacterial-loaded QDs for tumor imaging and SDT [126]. The NIR-emitting ternary luminescent AgAuSe QDs were synthesized. Staphylococcus aureus,

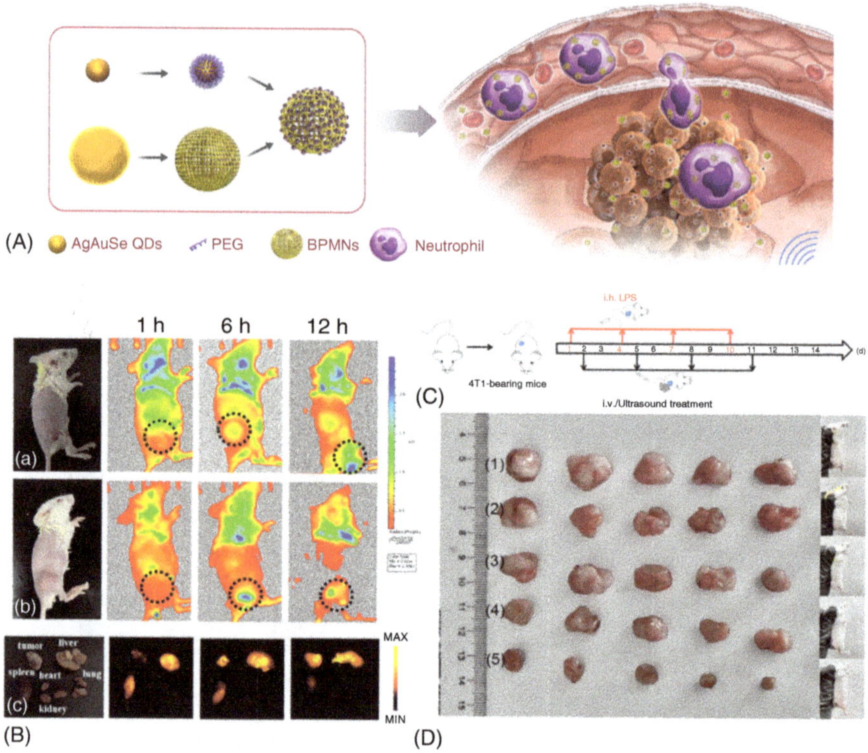

Figure 10.14 Schematic depiction of the assembling process and targeting tumor of BPMNs-AgAuSe, in vivo tumor imaging, and the ablation of tumor. Source: Reproduced with permission from Ref. [126]; © 2023, Wiley-VCH Verlag GmbH & Co. KGaA.

in its logarithmic growth phase, was treated with acid to produce inactivated bacterial-like particles that retained only peptidoglycan, and the QDs were then integrated onto the surface of the bacterial shell. Bacterial shells, as an immune adjuvant, can activate neutrophils in the body to phagocytosis. Due to the inflammatory responsiveness of neutrophils, neutrophils can act as carriers carrying bacterial-like particles containing QDs to actively migrate to the tumor site and release them at the tumor site, thus achieving accurate imaging of the tumor site. At the same time, we also found that AgAuSe QDs can achieve more electron-hole separation and generate ROS under the action of ultrasound to realize tumor SDT, as shown in Figure 10.14.

Furthermore, we developed an enhanced silver-based ternary QDs nanoplatform, which can realize the image-guided synergistic treatment mode of acoustic and chemodynamic therapy. First, water-soluble $AgInS_2$ trivalent QDs with good biosecurity were formed by the water phase synthesis method. Then, they were coupled and assembled with bovine serum albumin through the amino carboxyl condensation reaction between the surface ligands to form fluorescent-enhanced red light-emitting nanoclusters. Finally, the fluorescent-enhanced QDs were embedded in the cobalt organic framework (ZIF-67). In response to the weakly acidic tumor

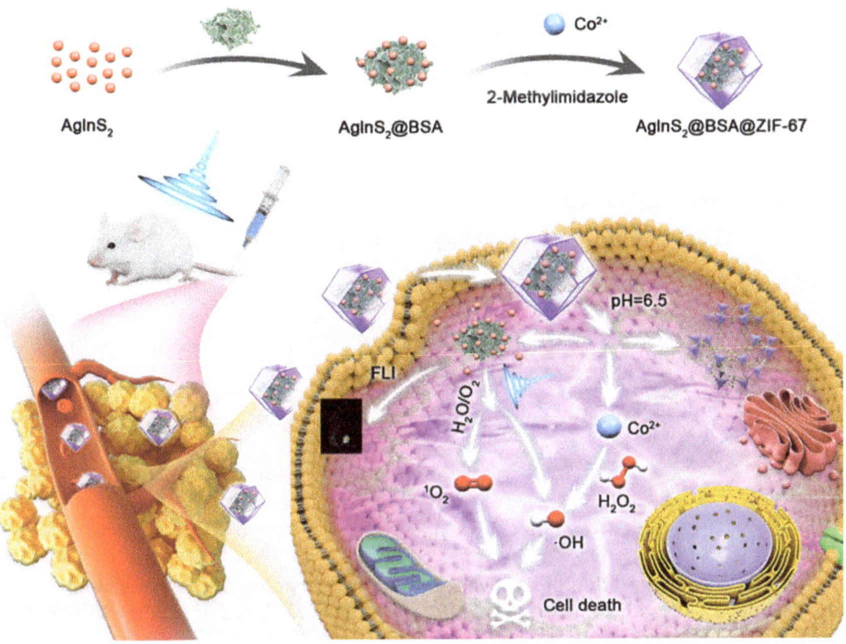

Figure 10.15 The preparation of $AgInS_2$@BSA@ZIF-67 nanocomposites and schematic diagram of treatment at tumor site.

microenvironment, the ZIF-67 framework can rapidly decompose and release Co^{2+}, which can catalyze the decomposition of overexpressed hydrogen peroxide (H_2O_2) in the tumor microenvironment to generate reactive oxygen hydroxyl radicals ($^{\cdot}OH$) through the Fenton-like reaction, thus inducing cell apoptosis. The QD nanoplatform can also serve as acoustic sensitizer to study the ability of QDs to produce ROS under ultrasonic radiation. This work expands the possibilities of ternary QDs as sonosensitive agents for SDT, and the biomedical applications of QDs are not limited to image-guided biofluorescent probes (Figure 10.15).

Compared with traditional treatment, immunotherapy attacks tumor cells by cultivating and treating immune cells in vivo, rather than attacking cancer cells by external forces. It has raised considerable concern due to alleviating the tumor load by removing a large number of tumor cells permanently. Immunotherapy has shown great application prospects in multiple treatment modalities, which can improve the therapeutic effect, prolong the survival time, and improve the living state of tumor patients, providing a powerful strategy for tumor therapy. Sun et al. synthesized a nanoparticle vaccine based on QDs to monitor vaccine dynamics and amplify immune response [127]. The prepared QDs exhibited excellent fluorescence efficiency, allowing for imaging of the trafficking dynamics. The vaccine formed after the modification and packaging of QDs enables the organism to generate a potent immune response and antitumor effect, which provides a satisfactory strategy to explore the vaccine efficacy and immune response (Figure 10.16).

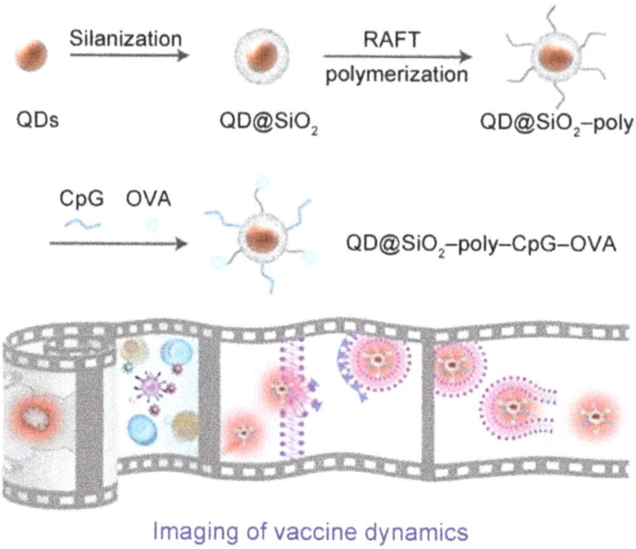

Figure 10.16 The preparation of nanovaccines and the imaging of the trafficking dynamics. Source: Reproduced with permission from Ref. [127]; © 2021, Wiley-VCH Verlag GmbH & Co. KGaA.

10.2 Conclusion

Quantum dots, as newly developed semiconductor nanomaterials, especially ternary QDs, are more suitable for biomedical research, benefiting from their excellent optical performance, tunable optical properties, and the ability to produce ROS with surrounding oxygen molecules under external stimulation for tumor therapy. Among them, near-infrared luminescent ternary QDs can eliminate the spontaneous fluorescence of biological tissue for deep imaging, which exhibits special practical value in biomedical imaging. Nevertheless, further application of ternary QDs still faces the following problems: organic-synthesized QDs cannot be directly applied to biological research; the fluorescence quantum yield of organic-phase QDs drops significantly after transferring to aqueous phase; the quantum efficiency of water-synthesized QDs is low, and their luminescence performance is weak; they cannot be directly applied to biological research due to size limitations, and rational design of nanocarriers is required for drug delivery; the research on the application of QDs for tumor treatment is scarce, and the mechanism of photodynamic/SDT is still unclear. Therefore, it is necessary to develop and design new environment-friendly pathways to prepare QDs with high quantum efficiency that can directly serve biological research. In order to construct an integrated imaging treatment platform, further exploration involving the optical absorption and fluorescence emission mechanism of ternary QDs is needed to strengthen the dynamics and mechanism research of ternary QDs in tumor treatment. In addition, the surface defects and suspended bonds of QDs can be used to design excellent photosensitizers/sonosensitizers for tumor treatment supported by other biocompatible carriers.

References

1 Yoffe, A.D. (1993). Low-dimensional systems: quantum size effects and electronic properties of semiconductor microcrystallites (zero-dimensional systems) and some quasi-two-dimensional systems. *Advances in Physics* 42: 173–262.

2 Chen, H.Q., Liu, L.L., Ma, A.Q. et al. (2021). Noninvasively immunogenic sonodynamic therapy with manganese protoporphyrin liposomes against triple-negative breast cancer. *Biomaterials* 269: 120639.

3 Peng, X.G. (2010). An essay on synthetic chemistry of colloidal nanocrystals. *Nano Research* 2: 425–447.

4 Bai, Z.L., Ji, W.Y., Han, D.B. et al. (2016). Hydroxyl-terminated $CuInS_2$ based quantum dots: toward efficient and bright light emitting diodes. *Chemistry of Materials* 28: 1085–1091.

5 Kroupa, D.M., Hughes, B.K., Miller, E.M. et al. (2017). Synthesis and spectroscopy of silver-doped PbSe quantum dots. *Journal of the American Chemical Society* 139: 10382–10394.

6 Santos, H.A.D., Zabala, G., Shen, Y.L. et al. (2020). Ultrafast photochemistry produces superbright short-wave infrared dots for low-dose *in vivo* imaging. *Nature Communications* 11: 2933.

7 Pan, D., Jiang, S., An, L., and Jiang, B. (2004). Controllable synthesis of highly luminescent and monodisperse CdS nanocrystals by a two-phase approach under mild conditions. *Advanced Materials* 16: 982–985.

8 Samadi, M. and A, Sadeghi-Maleki M.R. (2016). In-situ synthesis of high stable CdS quantum dots and their application for photocatalytic degradation of dyes. *Spectrochimica Acta. Part A: Molecular and Biomolecular Spectroscopy* 152: 156–164.

9 Hassanzadeh, J., Khataee, A.L., and R. (2017). Sensitive fluorescence and chemiluminescence procedures for methamphetamine detection based on CdS quantum dots. *Microchemical Journal* 132: 371–377.

10 McEwan, C., Owen, J., Stride, E. et al. (2015). Oxygen carrying microbubbles for enhanced sonodynamic therapy of hypoxic tumours. *Journal of Controlled Release* 203: 51–56.

11 Xing, X.J., Zhao, S.J., Xu, T. et al. (2021). Advances and perspectives in organic sonosensitizers for sonodynamic therapy. *Coordination Chemistry Reviews* 445: 214087.

12 Son, S.B., Kim, J.H., Wang, X.W. et al. (2020). Multifunctional sonosensitizers in sonodynamic cancer therapy. *Chemical Society Reviews* 49: 3244–3261.

13 Li, C.F., Zhang, K., Wang, P. et al. (2014). Sonodynamic antitumor effect of a novel sonosensitizer on S180 solid tumor. *Biopharmaceutics and Drug Disposition* 35: 50–59.

14 Albaladejo-Siguan, M., Becker-Koch, D., Taylor, A.D. et al. (2020). Efficient and stable PbS quantum dot solar cells by triple-cation perovskite passivation. *ACS Nano* 14: 384–393.

15 Xu, K.M., Zhou, W.J., and Ning, Z.J. (2020). Integrated structure and device engineering for high performance and scalable quantum dot infrared photodetectors. *Small* 16: e2003397.

16 Shi, G.Z., Wang, Y.J., Liu, Z.K. et al. (2017). Stable and highly efficient PbS quantum dot tandem solar cells employing a rationally designed recombination layer. *Advanced Energy Materials* 7: 1602667.

17 Wang, Y.W., Qi, H., Liu, Y. et al. (2021). The double-edged roles of ROS in cancer prevention and therapy. *Theranostics* 11: 4839–4857.

18 Jin, L., AlOtaibi, B., Benetti, D. et al. (2016). Near-Infrared colloidal quantum dots for efficient and durable photoelectrochemical solar-driven hydrogen production. *Advanced Science* 3: 1500345.

19 Zhuang, S.T. and Wang, J.L. (2021). Magnetic COFs as catalyst for fenton-like degradation of sulfamethazine. *Chemosphere* 264: 128561.

20 Tsolekile, N., Nahle, S., Zikalala, N. et al. (2020). Cytotoxicity, fluorescence tagging and gene-expression study of CuInS/ZnS QDS-meso (hydroxyphenyl) porphyrin conjugate against human monocytic leukemia cells. *Scientific Reports* 10: 4936.

21 Martynenko, I.V., Baimuratov, A.S., Weigert, F. et al. (2019). Photoluminescence of Ag-In-S/ZnS quantum dots: excitation energy dependence and low-energy electronic structure. *Nano Research* 12: 1595–1603.

22 Moodelly, D., Kowalik, P., Bujak, P. et al. (2019). Synthesis, photophysical properties and surface chemistry of chalcopyrite-type semiconductor nanocrystals. *Journal of Materials Chemistry C* 7: 11665–11709.

23 Aldakov, D., Lefrançois, A., and Reiss, P. (2013). Ternary and quaternary metal chalcogenide nanocrystals: synthesis, properties and applications. *Journal of Materials Chemistry C* 1: 3756–3776.

24 Zhong, H.Z., Bai, Z.L., and Zou, B.S. (2012). Tuning the luminescence properties of colloidal I-III-VI semiconductor nanocrystals for optoelectronics and biotechnology applications. *The Journal of Physical Chemistry Letters* 3: 3167–3175.

25 Liu, Y., Xie, X.Y., Cheng, C. et al. (2019). Strategies to fabricate metal–organic framework (MOF)-based luminescent sensing platforms. *Journal of Materials Chemistry C* 7: 10743–10763.

26 Zhang, J., Hao, G.Y., Yao, C.F. et al. (2016). Albumin-mediated biomineralization of paramagnetic NIR Ag_2S QDs for tiny tumor bimodal targeted imaging *in vivo*. *ACS Applied Materials & Interfaces* 8: 16612–16621.

27 Dong, B.H., Li, C.Y., Chen, G.C. et al. (2013). Facile synthesis of highly photoluminescent Ag_2Se quantum dots as a new fluorescent probe in the second near-infrared window for *in vivo* imaging. *Chemistry of Materials* 25: 2503–2509.

28 Yang, H.C., Li, R.F., Zhang, Y.J. et al. (2021). Colloidal alloyed quantum dots with enhanced photoluminescence quantum yield in the NIR-II Window. *Journal of the American Chemical Society* 143: 2601–2607.

29 Liu, Z.Y., Liu, A.A., Fu, H.H. et al. (2021). Breaking through the size control dilemma of silver chalcogenide quantum dots via trialkylphosphine-induced ripening: leading to Ag_2Te emitting from 950 to 2100 nm. *Journal of the American Chemical Society* 143: 12867–12877.

30 Zhu, C.N., Jiang, P., Zhang, Z.L. et al. (2013). Ag_2Se quantum dots with tunable emission in the second near-infrared window. *ACS Applied Materials & Interfaces* 5: 1186–1189.

31 Mrad, M., Chaabane, T.B., Rinnert, H., and Lavinia, B. (2020). Aqueous synthesis for highly emissive 3-mercaptopropionic acid-capped AIZS quantum dots. *Inorganic Chemistry* 59: 6220–6231.

32 Delices, A., Moodelly, D., Hurot, C. et al. (2020). Aqueous synthesis of DNA-functionalized near-infrared $AgInS_2$/ZnS core/shell quantum dots. *ACS Applied Materials & Interfaces* 12: 44026–44038.

33 Ren, S.Z., Wang, B., Zhu, X.H. et al. (2020). Oxygen self-sufficient core-shell metal-organic framework-based smart nanoplatform for enhanced synergistic chemotherapy and photodynamic therapy. *ACS Applied Materials & Interfaces* 12: 24662–24674.

34 Baskoutas, S. and Terzis, A.F. (2006). Size-dependent band gap of colloidal quantum dots. *Journal of Applied Physics* 99: 013708.

35 Park, Y.J., Oh, J.H., Han, N.S. et al. (2014). Photoluminescence of band gap states in $AgInS_2$ nanoparticles. *The Journal of Physical Chemistry C* 118: 25677–25683.

36 Assaf, B., Danie, S., and Gil, M. (2011). Size dependence of chiroptical activity in colloidal quantum dots. *ACS Nano* 5: 9034–9043.

37 Ioannou, D. and Griffin, D.K. (2010). Nanotechnology and molecular cytogenetics: the future has not yet arrived. *Nanotechnology Reviews* 1: 5117.

38 Owen, J. and Brus, L. (2017). Chemical synthesis and luminescence applications of colloidal semiconductor quantum dots. *Journal of the American Chemical Society* 139: 10939–10943.

39 Zhou, Y.F., Zhao, H.G., Ma, D.L., and Rosei, F. (2018). Harnessing the properties of colloidal quantum dots in luminescent solar concentrators. *Chemical Society Reviews* 47: 5866–5890.

40 Lu, H.P., Carroll, G.M., Neale, N.R., and Beard, M.C. (2019). Infrared quantum dots: progress, challenges, and opportunities. *ACS Nano* 13: 939–953.

41 Kim, J.Y., Voznyy, O., Zhitomirsky, D., and Sargent, E.H. (2013). 25th anniversary article: colloidal quantum dot materials and devices: a quarter-century of advances. *Advanced Materials* 25: 4986–5010.

42 Zhang, Y.J., Yang, H.C., An, X.Y. et al. (2020). Controlled synthesis of $Ag_2Te@Ag_2S$ core-shell quantum dots with enhanced and tunable fluorescence in the second near-infrared window. *Small* 16: e2001003.

43 Du, Y.P., Xu, B., Fu, T. et al. (2010). Near-Infrared photoluminescent Ag_2S quantum dots from a single source precursor. *Journal of the American Chemical Society* 132: 1470–1471.

44 Nagaoka, Y., Tan, R., Li, R.P. et al. (2018). Superstructures generated from truncated tetrahedral quantum dots. *Nature* 561: 378–382.

45 Kovalenko, M.V., Manna, L., Cabot, A. et al. (2015). Prospects of nanoscience with nanocrystals. *ACS Nano* 9: 1012–1057.

46 Nagaoka, Y., Zhu, H., Eggert, D., and Chen, O. (2018). Single-component quasicrystalline nanocrystal superlattices through flexible polygon tiling rule. *Science* 362: 1396–1400.

47 Zhang, Y. (2021). Controllable preparation and properties of silver-based NIR emitting quantum dots based on energy band regulation. Dissertation, University of Science and Technology of China.

48 Sun, D.Z. and Gang, O. (2013). DNA-functionalized quantum dots: fabrication, structural, and physicochemical properties. *Langmuir* 29: 7038–7046.

49 Son, J., Choi, D.S., Park, M. et al. (2020). Transformation of colloidal quantum dot: from intraband transition to localized surface plasmon resonance. *Nano Letters* 20: 4985–4992.

50 Murray, C.B., Norris, D.J., and Bawendi, M.G. (1993). Synthesis and characterization of nearly monodisperse CdE (E = S, Se, Te) semiconductor nanocrystallites. *Journal of the American Chemical Society* 115: 8706–8715.

51 Peng, Z.A. and Peng, X.G. (2001). Formation of high-quality CdTe, CdSe, and CdS nanocrystals using CdO as precursor. *Journal of the American Chemical Society* 123: 183–184.

52 Lim, J., Bae, W.K., Kwak, J. et al. (2012). Perspective on synthesis, device structures, and printing processes for quantum dot displays. *Optical Materials Express* 2: 594–628.

53 Yu, W.W., Falkner, J.C., Shih, B.S., and Colvin, V.L. (2004). Preparation and characterization of monodisperse PbSe semiconductor nanocrystals in a noncoordinating solvent. *Chemistry of Materials* 16: 3318–3322.

54 He, H., Lin, Y., Tian, Z.Q. et al. (2018). Ultrasmall Pb: Ag_2S quantum dots with uniform particle size and bright tunable fluorescence in the NIR-II window. *Small* 14: e1703296.

55 Reiss, P., Carriere, M., Lincheneau, C. et al. (2016). Synthesis of semiconductor nanocrystals, focusing on nontoxic and earth-abundant materials. *Chemical Reviews* 116: 10731–10819.

56 Gautam, U.K., Seshadri, R., Rajamathi, M. et al. (2001). A solvothermal route to capped CdSe nanoparticles. *Chemical Communications* 7: 629–630.

57 Pan, D.C., An, L.J., Sun, Z.M. et al. (2008). Synthesis of Cu–In–S ternary nanocrystals with tunable structure and composition. *Journal of the American Chemical Society* 130: 5620–5621.

58 Xie, R.G., Michael, R., and Peng, X.G. (2009). Formation of high-quality I–III–VI semiconductor nanocrystals by tuning relative reactivity of cationic precursors. *Journal of the American Chemical Society* 131: 5691–5697.

59 Qian, H.F., Dong, C.Q., Weng, J.F., and Ren, J.C. (2006). Facile one-pot synthesis of luminescent, water-soluble, and biocompatible glutathione-coated CdTe nanocrystals. *Small* 2: 747–751.

60 Shu, Y., Yan, J., Lu, Q. et al. (2020). Pb ions enhanced fluorescence of Ag_2S QDs with tunable emission in the NIR-II window: facile one pot synthesis and

their application in NIR-II fluorescent bio-sensing. *Sensors and Actuators B: Chemical* 307: 127593.

61 Li, L., Qian, H.F., and Ren, J.C. (2005). Rapid synthesis of highly luminescent CdTe nanocrystals in the aqueous phase by microwave irradiation with controllable temperature. *Chemical Communications* 4: 528–530.

62 Meladom, S.K., Arackal, S., Sreedharan, A. et al. (2018). Microwave assisted robust aqueous synthesis of Mn^{2+}-doped CdSe QDs with enhanced electronic properties. *RSC Advances* 8: 26771–26781.

63 Cao, J., Xue, B., Li, H. et al. (2010). Facile synthesis of high-quality water-soluble N-acetyl-L-cysteine-capped $Zn_{1-x}Cd_xSe$/ZnS core/shell quantum dots emitting in the violet-green spectral range. *Journal of Colloid and Interface Science* 348: 369–376.

64 Jiang, P., Zhu, C.N., Zhang, Z.L. et al. (2012). Water-soluble Ag_2S quantum dots for near-infrared fluorescence imaging *in vivo*. *Biomaterials* 33: 5130–5135.

65 Tan, L.J., Wan, A.J., Zhao, T.T. et al. (2014). Retraction of "Aqueous Synthesis of Multidentate-Polymer-Capping Ag_2Se Quantum Dots with Bright Photoluminescence Tunable in a Second Near-Infrared Biological Window". *ACS Applied Materials & Interfaces* 6: 6217–6222.

66 Xiong, W.W., Yang, G.H., Wu, X.C., and Zhu, J.J. (2013). Aqueous synthesis of color-tunable $CuInS_2$/ZnS nanocrystals for the detection of human interleukin 6. *ACS Applied Materials & Interfaces* 5: 8210–8216.

67 Wang, L., Kang, X.J., and Pan, D.C. (2016). High color rendering index warm white light emitting diodes fabricated from $AgInS_2$/ZnS quantum dot/PVA flexible hybrid films. *Physical Chemistry Chemical Physics* 18: 31634–31639.

68 Cui, R., Liu, H.H., Xie, H.Y. et al. (2009). Living yeast cells as a controllable biosynthesizer for fluorescent quantum dots. *Advanced Functional Materials* 19: 2359–2364.

69 Tian, L.J., Min, Y., Li, W.W. et al. (2019). Substrate metabolism-driven assembly of high-quality CdS_xSe_{1-x} quantum dots in *Escherichia coli*: molecular mechanisms and bioimaging application. *ACS Nano* 13: 5841–5851.

70 Deng, Z.T., Samanta, A., Nangreave, J. et al. (2012). Robust DNA-functionalized core/shell quantum dots with fluorescent emission spanning from UV-vis to near-IR and compatible with DNA-directed self-assembly. *Journal of the American Chemical Society* 134: 17424–17427.

71 Samanta, A., Deng, Z.T., Liu, Y., and Yan, H. (2013). A perspective on functionalizing colloidal quantum dots with DNA. *Nano Research* 6: 853–870.

72 He, X.W., Gao, L., and Ma, N. (2013). One-step instant synthesis of protein-conjugated quantum dots at room temperature. *Scientific Reports* 3: 2825.

73 Hu, Z.H., Chen, T., Xie, Z.X. et al. (2022). Emission tunable $AgInS_2$ quantum dots synthesized via microwave method for white light-emitting diodes application. *Optical Materials* 124: 111975.

74 Wang, Y., Yang, J., Gao, W.L. et al. (2014). Organic-free hydrothermal synthesis of chalcopyrite $CuInS_2$ and its photocatalytic activity for nitrate ions reduction. *Materials Letters* 137: 99–101.

75 Vasudevan, D., Gaddam, R.R., Trinchi, A., and Cole, I. (2015). Core–shell quantum dots: properties and applications. *Journal of Alloys and Compounds* 636: 395–404.

76 Lian, W., Tu, D.T., Hu, P. et al. (2020). Broadband excitable NIR-II luminescent nano-bioprobes based on $CuInSe_2$ quantum dots for the detection of circulating tumor cells. *Nano Today* 35: 100943.

77 Zhang, H., Hu, N., Zeng, Z.P. et al. (2019). High-efficiency green InP quantum dot-based electroluminescent device comprising thick-shell quantum dots. *Advanced Optical Materials* 7: 1801602.

78 Wang, Q.B., Liu, Y., Ke, Y.G., and Yan, H. (2008). Quantum dot bioconjugation during core-shell synthesis. *Angewandte Chemie International Edition* 47: 316–319.

79 Zeng, Y.M., Pan, L.J., Wang, J. et al. (2020). Interfacial synthesis of Ag_2S/ZnS core/shell quantum dots in a droplet microreactor. *ChemistrySelect* 5: 5889–5894.

80 Lim, L., Zhao, X., and Tan, Z. (2023). Non-toxic $CuInS_2$/ZnS colloidal quantum dots for near-infrared light-emitting diodes. *Advanced Materials* 35: 2301887.

81 Geyter, B.D., Justo, Y., Moreels, I. et al. (2010). The different nature of band edge absorption and emission in colloidal PbSe/CdSe core/shell quantum dots. *ACS Nano* 5: 58–66.

82 Nirmal, M., Dabbousl, B.O., Bawendi, M.G. et al. (1996). Fluorescence intermittency in single cadmium selenide nanocrystals. *Letters to Nature* 383: 31.

83 Girma, W.M., Fahmi, M.Z., Permadi, A. et al. (2017). Synthetic strategies and biomedical applications of I-III-VI ternary quantum dots. *Journal of Materials Chemistry B* 5: 6193–6216.

84 Reiss, P., Protiere, M., and Li, L. (2009). Core/Shell semiconductor nanocrystals. *Small* 5: 154–168.

85 Mews, A., Eychmuller, A., Giersig, M. et al. (1994). Preparation, characterization, and photophysics of the quantum dot quantum well system CdS/HgS/CdS. *The Journal of Physical Chemistry* 98: 934–941.

86 Little, R.B., El-Sayed, M.A., Bryant, G.W., and Burke, S. (2001). Formation of quantum-dot quantum-well heteronanostructures with large lattice mismatch: ZnS/CdS/ZnS. *The Journal of Chemical Physics* 114: 1813–1822.

87 Chen, Y.F., Xi, J.Y., Dumcenco, D.O. et al. (2013). Tunable band gap photoluminescence from atomically thin transition-metal dichalcogenide alloys. *ACS Nano* 7: 4610–4616.

88 Karmakar, G., Tyagi, A., Wadawale, A. et al. (2018). Synthesis, characterization and photo response behaviour of InSe and $CuInSe_2$ nanostructures using tris(5-methyl-2-pyridylselenolato) indium (III) as molecular precursor. *ChemistrySelect* 3: 10394–10401.

89 Makino, T., Segawa, Y., Kawasaki, M. et al. (2001). Band gap engineering based on $Mg_xZn_{1-x}O$ and $Cd_yZn_{1-y}O$ ternary alloy films. *Applied Physics Letters* 78: 1237–1239.

90 Tian, R., Shen, Z.Y., Zhou, Z.J. et al. (2019). Ultrasmall quantum dots with broad-spectrum metal doping ability for trimodal molecular imaging. *Advanced Functional Materials* 29: 1901671.

91 Rajesh, C., Phadnis, C., Sonawane, K.G., and Mahamuni, S. (2014). Generation of white light from co-doped (Cu and Mn) ZnSe QDs. *Journal of Experimental Nanoscience* 10: 1082–1092.

92 Arivarasan, A., Bharathi, S., Ezhilarasi, S. et al. (2019). Photovoltaic performances of Yb doped CdTe QDs sensitized TiO_2 photoanodes for solar cell applications. *Journal of Inorganic and Organometallic Polymers and Materials* 29: 859–868.

93 Hofman, E., Robinson, R.J., Li, Z.J. et al. (2017). Controlled dopant migration in CdS/ZnS core/shell quantum dots. *Journal of the American Chemical Society* 139: 8878–8885.

94 Zrazhevskiy, P., Sena, M., and Gao, X.H. (2010). Designing multifunctional quantum dots for bioimaging, detection, and drug delivery. *Chemical Society Reviews* 39: 4326–4354.

95 Medintz, L.L., Uyeda, H.T., Goldman, E.R., and Mattoussi, H. (2005). Quantum dot bioconjugates for imaging, labelling and sensing. *Nature Materials* 4: 435–446.

96 Willard, D.M., Carillo, L.L., Jung, J., and Orden, A.V. (2001). CdSe–ZnS quantum dots as resonance energy transfer donors in a model protein-protein binding assay. *Nano letters* 1: 469–474.

97 Uyeda, H.T., Medintz, L.L., Jaiswal, J.K. et al. (2005). Synthesis of compact multidentate ligands to prepare stable hydrophilic quantum dot fluorophores. *Journal of the American Chemical Society* 127: 3870–3878.

98 Xie, M., Luo, K., Huang, B.H. et al. (2010). PEG-interspersed nitrilotriacetic acid-functionalized quantum dots for site-specific labeling of prion proteins expressed on cell surfaces. *Biomaterials* 31: 8362–8370.

99 Bruchez, M. Jr., Moronne, M., Gin, P. et al. (1998). Semiconductor nanocrystals as fluorescent biological labels. *Science* 281: 2013–2016.

100 Chan, W. and Nie, S.M. (1998). Quantum dot bioconjugates for ultrasensitive nonisotopic detection. *Science* 281: 2016–2018.

101 Liu, C.X., Zhao, J., Zhang, R.R. et al. (2020). Development and application of fluorescence sensor and test strip based on molecularly imprinted quantum dots for the selective and sensitive detection of propanil in fish and seawater samples. *Journal of Hazardous Materials* 389: 121884.

102 Shi, H.X., Jia, L.C., Wang, C.J. et al. (2020). A high sensitive and selective fluorescent probe for dopamine detection based on water soluble $AgInS_2$ quantum dots. *Optical Materials* 99: 109549.

103 Yang, L., Ma, P.A., Chen, X.R. et al. (2022). High-sensitivity fluorescence detection for lung cancer CYFRA21-1 DNA based on accumulative hybridization of quantum dots. *Journal of Materials Chemistry B* 10: 1386–1392.

104 Maji, S.K. and Kim, D.H. (2018). $AgInS_2$-coated upconversion nanoparticle as a photocatalyst for near-infrared light-activated photodynamic therapy of cancer cells. *ACS Applied Bio Materials* 1: 1628–1638.

105 Bharali, D.J., Lucey, D.W., Jayakumar, H. et al. (2005). Folate-receptor-mediated delivery of InP quantum dots for bioimaging using confocal and two-photon microscopy. *Journal of the American Chemical Society* 127: 11364–11371.

106 Moulick, A., Heger, Z., Milosavljevic, V. et al. (2018). Real-time visualization of cell membrane damage using gadolinium-schiff base complex-doped quantum dots. *ACS Applied Materials and Interfaces* 10: 35859–35868.

107 Wu, X.Y., Liu, H.J., Liu, J.Q. et al. (2003). Immunofluorescent labeling of cancer marker Her2 and other cellular targets with semiconductor quantum dots. *Nature Biotechnology* 21: 41–46.

108 Zrazhevskiy, P. and Gao, X.H. (2009). Multifunctional quantum dots for personalized medicine. *Nano Today* 4: 414–428.

109 Iking, J., Staniszewska, M., Kessle, L. et al. (2021). Imaging inflammation with positron emission tomography. *Biomedicines* 9: 212.

110 Yuan, L., Song, Q., Liu, H.C. et al. (2023). Super-resolution computed tomography imaging spectrometry. *Photonics Research* 11: 212–224.

111 Richter, S., Winzeck, S., Kornaropoulos, E.N. et al. (2021). Neuroanatomical substrates and symptoms associated with magnetic resonance imaging of patients with mild traumatic brain injury. *JAMA Netw Open* 4: e210994.

112 Sun, C.X., Li, B.H., Zhao, M.Y. et al. (2019). J-Aggregates of cyanine dye for NIR-II *in vivo* dynamic vascular imaging beyond 1500 nm. *Journal of the American Chemical Society* 141: 19221–19225.

113 Yuan, Y., Feng, Z., Li, S.L. et al. (2022). Molecular programming of NIR-II b-emissive semiconducting small molecules for *in vivo* high-contrast bioimaging beyond 1500 nm. *Advanced Materials* 34: e2201263.

114 Hong, G.S., Robinson, J.T., Zhang, Y.J. et al. (2012). *In vivo* fluorescence imaging with Ag_2S quantum dots in the second near-infrared region. *Angewandte Chemie International Edition* 124: 9956–9959.

115 Welsher, K., Sherlock, S.P., and Dai, H.J. (2011). Deep-tissue anatomical imaging of mice using carbon nanotube fluorophores in the second near-infrared window. *Proceedings of the National Academy of Sciences of the United States of America* 108: 8943–8948.

116 Zhang, J.J., Lin, Y., Zhou, H. et al. (2019). Cell membrane-camouflaged NIR II fluorescent Ag_2Te quantum dots-based nanobioprobes for enhanced *in vivo* homotypic tumor imaging. *Advanced Healthcare Materials* 8: 1900341.

117 Hsu, C.Y., Chen, C.W., Yu, H.P. et al. (2013). Bioluminescence resonance energy transfer using luciferase-immobilized quantum dots for self-illuminated photodynamic therapy. *Biomaterials* 34: 1204–1212.

118 Shen, Y.Z., Sun, Y.D., Yan, R.Q. et al. (2017). Rational engineering of semiconductor QDs enabling remarkable $^1O_2$ production for tumor-targeted photodynamic therapy. *Biomaterials* 148: 31–40.

119 Lv, G.X., Guo, W.S., Zhang, W. et al. (2016). Near-Infrared emission CuInS/ZnS quantum dots: all-in-one theranostic nanomedicines with intrinsic fluorescence/photoacoustic imaging for tumor phototherapy. *ACS Nano* 10: 9637–9645.

120 Xu, H.Y., Zhang, X., Han, R.B. et al. (2016). Nanoparticles in sonodynamic therapy: state of the art review. *RSC Advances* 6: 50697–50705.

121 Zeng, W.W., Xu, Y., Yang, W.T. et al. (2020). An ultrasound-excitable aggregation-induced emission dye for enhanced sonodynamic therapy of tumors. *Advanced Healthcare Materials* 9: 2000560.

122 Wang, D., Cheng, D.B., Ji, L. et al. (2021). Precise magnetic resonance imaging-guided sonodynamic therapy for drug-resistant bacterial deep infection. *Biomaterials* 264: 120386.

123 Fan, W.P., Huang, P., and Chen, X.Y. (2016). Overcoming the Achilles' heel of photodynamic therapy. *Chemical Society Reviews* 45: 6488–6519.

124 Li, C., Yang, X.Q., An, J. et al. (2020). Red blood cell membrane-enveloped O_2 self-supplementing biomimetic nanoparticles for tumor imaging-guided enhanced sonodynamic therapy. *Theranostics* 10: 867–879.

125 Qian, Y., Wang, J.L., Bu, W.B. et al. (2023). Targeted implementation strategies of precise photodynamic therapy based on clinical and technical demands. *Biomaterials Science* 11: 704–718.

126 Yang, L., Yuan, M., Ma, P.A. et al. (2023). Assembling AgAuSe quantum dots with peptidoglycan and neutrophils to realize enhanced tumor targeting, NIR (II) imaging and sonodynamic therapy. *Small Methods* 7: 2201706.

127 Sun, J.L., Liu, F., Yu, W.Q. et al. (2021). Visualization of vaccine dynamics with quantum dots for immunotherapy. *Angewandte Chemie International Edition* 60: 24275–24283.

11

Nanomaterials-Induced Pyroptosis and Immunotherapy

Hao Chen[1,2], Binbin Ding[1], Jun Lin[1,2], and Ping'an Ma[1,2]

[1] *Chinese Academy of Sciences, Changchun Institute of Applied Chemistry, State Key Laboratory of Rare Earth Resource Utilization, 5625 Renmin Street, Changchun 130022, China*
[2] *University of Science and Technology of China, School of Applied Chemistry and Engineering, 96 Jinzhai Road, Hefei 230026, China*

11.1 Discovery and Definition of Pyroptosis

In 1992, Sansonetti and colleagues observed that *Shigella flexneri* infection led to macrophage death characterized by chromatin condensation, cytoplasmic blebbing, cytoplasmic vacuolization, endoplasmic reticulum expansion, and DNA fragmentation [1]. In 2000, Cookson's team noted that this cell death in infected macrophages differed from apoptosis [2]. Subsequently, in 2001, Cookson and Brennan officially termed this form of cell death as "pyroptosis" and described it as a pro-inflammatory programmed cell death (PCD) that autonomously eliminates potentially dangerous cells [3]. According to the definition of the Nomenclature Committee on Cell Death, pyroptosis is a type of PCD that involves the formation of plasma membrane pores by gasdermin proteins, often resulting from inflammatory caspase activation [4].

11.2 Mechanisms of Pyroptosis

Pyroptosis is a specific type of PCD that plays a role in maintaining intracellular homeostasis, and it is characterized by distinctive morphological changes. These changes include chromatin aggregation, cell swelling, and the formation of pores in the plasma membrane. As a consequence of these events, cytosolic contents and inflammatory factors such as interleukin 1β (IL-1β), interleukin 18 (IL-18), high mobility group protein 1 (HMGB1), adenosine triphosphate (ATP), and lactic dehydrogenase (LDH), among others, are released, leading to a local inflammatory response. These features distinguish pyroptosis from other forms of PCD and highlight its role in triggering inflammation within the cellular environment [5].

Photofunctional Nanomaterials for Biomedical Applications, First Edition.
Edited by Chunxia Li and Jun Lin.

11.2.1 Inflammasome and Pyroptosis

When cells are infected or externally stimulated and subsequently damaged, the inflammasome is triggered to be involved in the formation of proinflammatory cytokines, such as IL-1β, which is activated through caspase-1, and Martinon proposed in 2002 that the inflammasome operates upstream of caspase-1 [6, 7]. Nod-like receptor protein 1 (NLRP1) and various other proteins form a complex known as the inflammasome, which activates caspase-1 to form p20/p10. These cytokines further activate IL-1β and IL-18, a key pathway of pyroptosis. The mechanism underlying the activation of caspases to ultimately cell death during pyroptosis process has been addressed in subsequent studies.

11.2.2 Caspases, Gasdermins, and Pyroptosis

Caspase-1-mediated cytokine activation and pyroptosis have been extensively studied, and a large number of subsequent studies have progressively shown that, except caspase-1, multiple caspases are also involved in pyroptosis. In 2011, Dixit VM et al. noted that caspase-11 activates inflammasome via the nonclassical pathway. The caspase-1-knockout Strain 129 mice previously used for caspase-1 modeling studies is actually caspase-1/11 double knockouts. Exactly, whereas caspase-1 activation requires nod-like receptor protein 3 (NLRP3) and apoptosis-associated speck-like protein (ASC) to comprise inflammasome, the activation of caspase-11 does not need them. Caspase-11 releases IL-1β in response to ATP and monosodium urate, suggesting that caspase-11 is driven by noncanonical inflammasome [8]. In 2013, they further proposed that intracellular lipopolysaccharide (LPS) activates caspase-11 independently of the extracellular receptor toll-like receptor 4 (TLR4) [9]. The extensive research on caspases has facilitated the study of pyroptosis downstream, leading to a rapid clarification of the pyroptosis mechanisms.

In 2015, Shao et al. found that the cleavage of gasdermin D (GSDMD) by caspase-4/11 resulted in pyroptosis. At the same time, Han and coworkers respectively reported that the GSDMD functions as downstream protein of caspase-4/11 [10–12]. There was the initial observation of the downstream proteins of caspase-4/11. In detail, Shao et al. selected caspase-1/11-mediated pyroptosis cells and found that these cells were resistant to pyroptosis when GSDMD was defective. In contrast, when the pyroptosis precondition was activated, GSDMD would be cleaved to the N-terminal (GSDMD-N) and the C-terminal (GSDMD-C) structural domains. As a result, the binding of the GSDMD-N structural domain to membrane phospholipids leads to cellular perforation, resulting in bubbling and swelling of the cell, which was the biggest difference from other PCDs [13]. In 2017, Shao and coworkers further found that not only inflammatory caspases (caspase-1/11) induce pyroptosis, but apoptotic caspase-3 also leads to pyroptosis [14]. They found that gasdermin E (GSDME), also known as DFNA5, could convert apoptosis induced by chemotherapy drugs to pyroptosis. In cells with high GSDME expression, GSDME is cleaved by caspase-3 to GSDME-N, and GSDME-N perforates membranes, thus inducing pyroptosis. Additionally, the

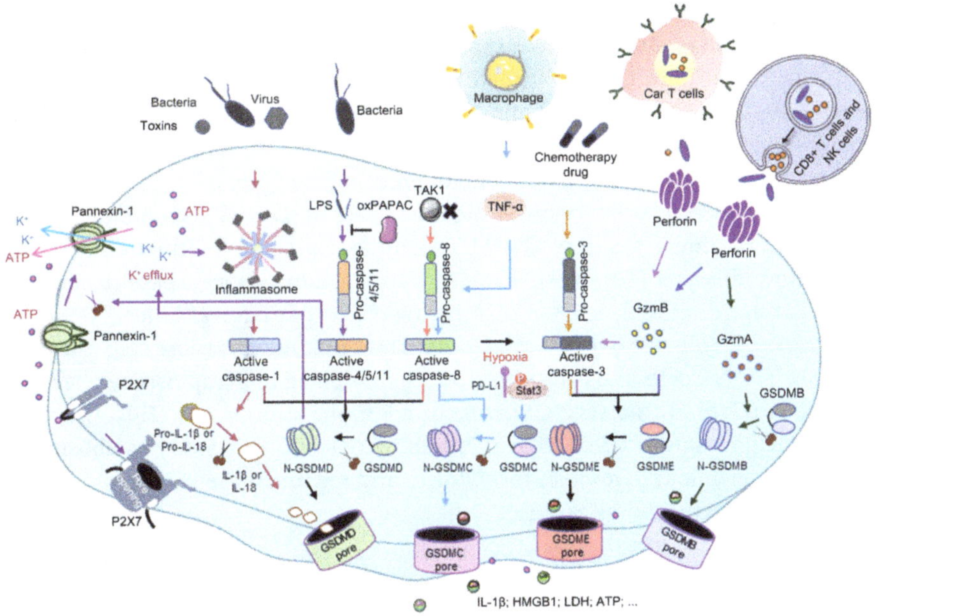

Figure 11.1 Molecular mechanism of pyroptosis. In the canonical pathway, PAMPs and damage-associated molecular patterns (DAMPs) receive intracellular signaling molecule stimulation and assemble with pro-caspase-1 and ASC to form inflammasomes and active caspase-1. Cleaved-caspase-1 cleaves GSDMD and pro-IL-1β/18. GSDMD- perforates the cell membrane by forming nonselective pores, further causing water influx, lysis, and death. In the non-canonical pathway, cytosolic LPS activates caspase-4/5 and caspase-11, triggering pyroptosis by cleaving GSDMD. In addition, the cleavage of GSDMD results in efflux of K^+, ultimately mediating the assembly of NLRP3 inflammasome, resulting in the cleavage of pro-IL-1β and pro-IL-18. In the caspase-3-mediated pathway, active caspase-3 cleaves GSDME to form GSDME-N, inducing pyroptosis. In the caspase-8-mediated pathway, inhibiting TAK1 induces the activation of caspase-8, which cleaves GSDMD, resulting in pyroptosis. In the granzyme-mediated pathway, CAR-T cells rapidly activate caspase-3 in target cells by releasing GzmB, and then GSDME was activated, causing extensive pyroptosis. In addition, GzmA and GzmB in cytotoxic lymphocytes enter target cells through perforin and induce pyroptosis. GzmA hydrolyzes GSDMB, and GzmB directly activates GSDME. Source: [17] / Springer Nature / CC BY 4.0.

GSDME is normally expressed in normal cells and suppressed in tumor cells. These findings suggested that the GSDME and caspase-3 provide novel inspiration of chemotherapy-mediated pyroptosis of tumor. Further research had demonstrated that along with caspase-1/11, caspase-8 also carried out a cleavage procedure on GSDMD, ultimately resulting in pyroptosis [15, 16] (Figure 11.1).

The analysis results of GSDMs and their upstream proteins make controlled regulation of pyroptosis feasible. Huang and coworkers found that pyroptosis occurred in tumor cells during chimeric antigen receptor T (CAR-T) cell therapy and was accompanied by cytokine release syndrome (CRS) [18]. CAR-T-induced CRS may impede the efficacy of CAR-T therapy. Thus, understanding the mechanisms of CRS becomes indispensable. Huang et al. discovered CAR-T cells activated caspase-3 through release of granzyme B (GzmB), and caspase-3 in turn triggered GSDME to give rise to pyroptosis. The inflammatory cytokines released by tumor cells further triggered macrophages to activate caspase-1 and GSDMD, which were the key step of CRS. Thus, the inhibition of GSDME or caspase-1 expression will effectively alleviate CRS. Moreover, Wu et al. uncovered disulfiram, a molecule inhibitor with GSDMD-target properties. The disulfiram prevented pyroptosis and the release of cytokines. Detailly, while disulfiram did not prevent formation of IL-1β and cleaved GSDMD, it prohibited pore formation and the release of IL-1β. This inhibitory effect of disulfiram on downstream of GSDMD might assist in the treatment of various inflammatory diseases [19]. Next, the pathways to induce noncaspase-dependent pyroptosis were disclosed. Shao et al. have reported that natural killer (NK) cells and cytotoxic T cells clear gasdermin B (GSDMB)-high-expression cells through pyroptosis. The NK cells and cytotoxic T cells participated in pyroptosis by releasing granzyme A (GzmA) to cleave GSDMB. In addition, interferon-γ (IFN-γ) upregulated the expression of GSDMB to promote pyroptosis. This study established a noncaspase-dependent pyroptosis mechanism and redefined the path of pyroptosis that can only be activated by caspases [20].

11.3 Pyroptosis and Tumor Immunotherapy

The removal of dead or senescent cells is vital in preserving normal organism function. Once a cell has aged or been irreversibly damaged, it undergoes PCD, primarily through apoptosis, and alerts the immune system for clearance. Preservation of the organism's homeostasis is ensured by the phagocytosis or removal of a substantial number of cells during regular metabolism [21]. In abnormal situations, a substantial number of cells may die anomalously, thus releasing cellular contents. These molecules are referred to as damage-associated molecular patterns (DAMPs) and act as triggers for the immune system, provoking immune response to recruit cells and repair the threat to restore normal physiological conditions. The immune system has a critical function in protecting the organism through its ability to monitor and eliminate abnormal cells to maintain homeostasis. The recognition and removal of mutant cells by the immune system is crucial to prevent tumor formation. Nonetheless, if this surveillance and removal system becomes dysfunctional or impaired, tumors may occur. In these instances, the immune system is unable to identify and remove anomalous cells, which can grow and potentially result in tumors. Therefore, the immune system's effective operation is crucial for impeding the advancement and propagation of tumors by competently detecting and eradicating mutated cells [22].

Malignant tumors grow rapidly and present challenges for effective suppression. They maneuver cleverly to evade the immune system's clearance mechanisms, inducing a "cold" mode of the immune response. The tumor microenvironment (TME) features hypoxia, as a result of tumor cells' vigorous proliferation and intense anaerobic respiration, leading to substantial nutrient consumption and the generation of high levels of lactic acid [23]. The hypoxic and acidic TME creates an environment that contributes to immunosuppression. Within TME, tumor cells produce various immunosuppressive metabolites and metabolic enzymes including adenosine, kynurenine, indoleamine 2,3-dioxygenase (IDO), arginase 1 (ARG1), and glutaminase (GLS). These substances act to suppress effector T cells, NK cells, and M1 macrophages, while simultaneously enhancing the function of immunosuppressive regulatory T cells and myeloid-derived suppressor cells (MDSCs) [24]. The surface antigens expressed by tumor cells also contribute to immune escape and further enhance the immunosuppressive nature of the TME. First, the dendritic cells (DCs) uptake and present tumor antigens to $CD8^+$ T cells and activate them. The recognition of tumor surface antigens is necessary for activated $CD8^+$ T cells to eliminate tumor cells. Notably, most tumor cells exhibit downregulated major histocompatibility complex (MHC) I antigen expression, increasing the difficulty of $CD8^+$ T cells to recognize tumor cells [25–27]. The unique characteristics of the TME, including hypoxia, acidity, and the generation of immunosuppressive metabolites and enzymes, collectively hinder immune responses within the TME. These immunosuppressive mechanisms enable tumors to avoid the immune system's surveillance and clearance mechanisms, ultimately promoting tumor growth and survival [28–30].

Certainly, conventional tumor treatments such as radiotherapy and chemotherapy have been extensively utilized and treat tumors through various mechanisms, including apoptosis and necrosis [31]. Apoptosis and necrosis have been studied for many years, and their impact on the immune response is well understood. Necrosis is a proinflammatory form of cell death in the context of the immune response. It is distinguished by the rupture of cell membranes, release of intracellular contents, and consequent activation of the immune system, culminating in inflammation. The released cellular components act as danger signals, stimulating immune cells and promoting immune response against tumor [32, 33]. Apoptosis, however, is typically viewed as a form of PCD that sustains, rather than strengthens, the immune response. The dying cells experience regulated fragmentation and are eliminated by immune cells, including macrophages [34]. Apoptosis is executed without causing notable inflammation or discharge of intracellular contents to potentially trigger an immune response. Consequently, it is widely regarded as a mechanism that enables the maintenance of immune homeostasis, rather than specifically stimulating immune activation against tumors [35, 36]. Another type of PCD known as pyroptosis has gained enormous attention in recent years. It is considered a programmed and active form of proinflammatory cell death, characterized by cell swelling, pore formation, and the release of proinflammatory cytokines such as IL-1β and IL-18, thus attracting immune cells and strengthening immune responses. In summary, the proinflammatory effect of pyroptosis and the release of danger signals have the potential to induce an immune response against tumors, thereby enhancing antitumor immunity [37, 38]. In summary, pyroptosis is considered a PCD that can elicit amplified immune responses. Additional exploration into comprehending and utilizing these distinct cell death modes may have therapeutic implications for innovative cancer treatments.

The process of pyroptosis entails the activation of the intracellular sensor NLRP3 and others in response to various stimuli, including DAMPs, PAMPs, cell membrane disruption, cellular osmotic pressure imbalance, ions overload, and so on. Upon NLRP3 activation, apoptosis-associated proteins are recruited, resulting in the formation of inflammasomes [6, 39]. Inflammasomes detect various microbial, stress, and injury signals and trigger caspase-1, resulting in the secretion of proinflammatory molecules and subsequent pyroptosis. Caspase-1 is present in a dimerized state and, after activation, undergoes protein hydrolysis, resulting in the production of fully active p20/p10 species. Activated caspase-1 performs two significant cleavage events. Initially, it cuts pro-IL-18, transforming it into its mature form, IL-18, which is a proinflammatory cytokine. Second, it cleaves pro-IL-1β, leading to the discharge of IL-1β, another potent proinflammatory cytokine [40, 41]. Eventually, GSDMD-C recruits caspase-1 again, creating a positive feedback loop during caspase-1 activation. GSDMD-N undergoes oligomerization and forms pores in the cell membrane, causing loss of intracellular contents and leading to cell lysis (Figure 11.2) [12, 31]. The formation of pores by GSDMD-N is crucial for the execution of pyroptosis. These pores allow for the release of intracellular contents, including cytokines and DAMPs, and also facilitate the infiltration of

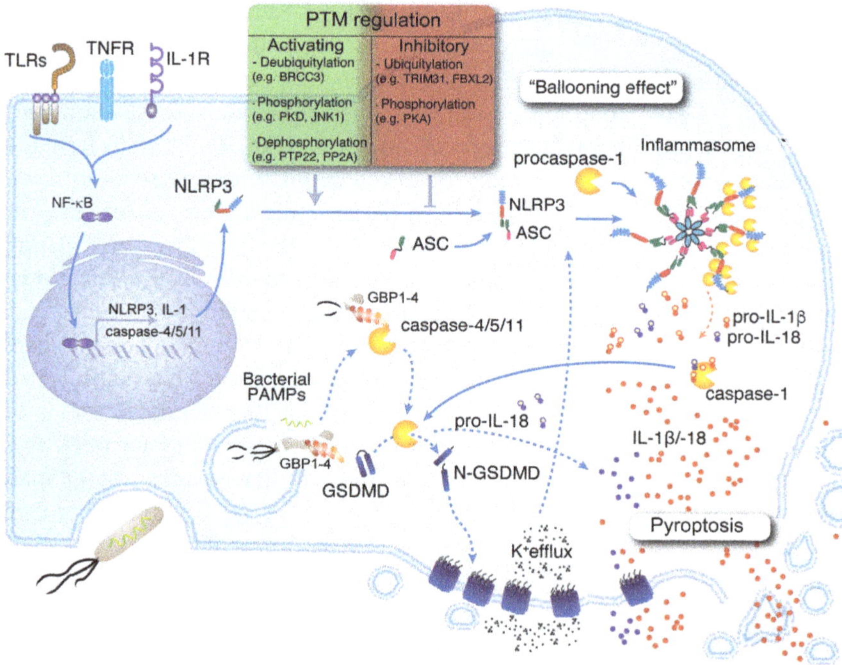

Figure 11.2 Pyroptosis is an inflammatory form of cell death triggered by intracellular sensors such as NLRP3 that detect DAMPs, PAMPs, membrane disturbances, osmotic imbalances, and ion efflux. Upon activation, these sensors recruit the adapter ASC, forming a micron-sized structure called the inflammasome. These oligomeric structures act as platforms for the activation of caspase-1. Active caspase-1 cleaves the interleukin-1 family cytokines IL-1β and IL-18. Caspase-1 also activates GSDMD, unmasking the N-terminal domain, which forms pores in the plasma membrane, in turn enabling the release of mature IL-1β and IL-18 and causing cell swelling (a "ballooning effect") and pyroptosis. Source: [31] / Springer Nature / CC BY 4.0.

immune cells into the affected area. The perforation of the cell membrane by GSDMD-N is a distinctive feature of pyroptosis and contributes to the proinflammatory [10, 11]. Overall, the activation of NLRP3 and the subsequent assembly of the inflammasome, in addition to the caspase-1-mediated cleavage of cytokines and GSDMD, perform imperative functions in pyroptosis. This process causes the secretion of proinflammatory cytokines and creates pores in the cell membrane, thus contributing to the active and proinflammatory components of pyroptosis.

Recently, inducing pyroptosis as a potential method of regulating PCD and strengthening autoimmunity has demonstrated promising capabilities in tumor therapy. In 2020, two papers published in *Nature* illustrate pyroptosis not only kills tumor cells but also induces a strong inflammatory response to stimulate antitumor immune response. Zhang et al. found and demonstrated that GSDME inhibits tumor growth by activating antitumor immunity. The tumors knocked out of the GSDME-expressing gene exhibited enhanced proliferation. Actually, the expression of GSDME is suppressed in most tumors. However, the ectopic expression in GSDME inhibits tumor growth through cytotoxic lymphocytes, and the absence of granzyme or cytotoxic lymphocytes will reduce this effect. In summary, GSDME enhances phagocytosis of macrophages and improves the quantity and efficacy of NK cells and $CD8^+$ T cells [42]. In addition, Shao et al. proposed a bioorthogonal system in which the tumor imaging probe phenylalanine trifluoroborate (Phe-BF_3) was combined with nanoparticle loaded with active GSDM protein. The nanoparticles selectively enter tumor cells and release active GSDM protein, leading to pyroptosis (Figure 11.3a). The antitumor immunity and eradication of the entire tumor were achieved, with only 15% of tumor cells undergoing pyroptosis. However, tumor regression did not occur when T cells were depleted or in the presence of immunodeficiency (Figure 11.3b). These findings suggested that GSDM-mediated tumor suppression is related to the immune response and that the release of DAMPs due to pyroptosis contributes to the reversal of the immunosuppressive environment. Together, these two studies provide compelling evidence that pyroptosis therapies and immunotherapies may have strong synergistic effects and that pyroptosis induction has a strong potential for enhancing immune response [43].

11.3.1 Ions Interference Therapy

The regulation and concentration of ions within cells are tightly controlled. Ion channels and transporters selectively transport ions across the cell membrane, ensuring precise management of intracellular ion concentrations. Any disturbance in ion balance can adversely affect cellular function including maintenance of osmotic pressure and acid balance, activation of signal pathway, and structuring of proteins [44, 45]. For example, calcium ions (Ca^{2+}) are vital for neuronal communication, muscle contraction, and intracellular signaling [46]. Sodium ions (Na^+) and potassium ions (K^+) are crucial for establishing and maintaining the resting membrane potential in excitable cells [47]. Chloride ions (Cl^-) are involved in regulating cell volume and maintaining the electroneutrality of cells [48].

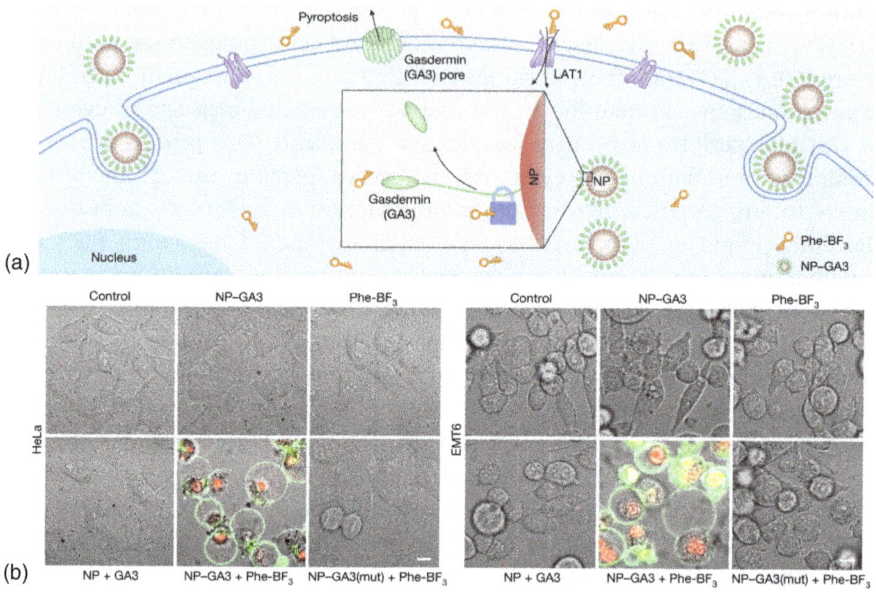

Figure 11.3 (a) Cartoon of the experimental design. Purified GSDMA3(N + C) was conjugated to a nanoparticle to obtain NP–GSDMA3. (b) Confocal images of the treated HeLa and EMT6 cells. Scale bars, 20 μm. Propidium iodide (PI) and annexin V–fluorescein isothiocyanate (FITC) were added to the cells 15 min before imaging. Source: Reproduced with permission from [43], 2020 / Springer Nature.

In conclusion, ions play a fundamental role in cellular activities, ranging from signal transduction to membrane potential regulation. Their presence and proper distribution within cells are essential for maintaining cellular homeostasis and ensuring the normal functioning of various physiological processes. Understanding the role of ions in cellular function is crucial for advancing our knowledge of cell biology and developing targeted therapies for diseases related to ion imbalances.

Due to the critical role of ions in cellular processes and disease pathology, there is growing interest in utilizing the manipulation of ion distribution as a promising approach for therapeutic interventions. Researchers are actively investigating various strategies aimed at regulating ion concentrations within cells and specific cellular compartments as potential targets for therapy [49]. This avenue of research offers promising prospects for developing effective treatments for a variety of conditions. To summarize, comprehending the intricate involvement of ions in cellular processes provides avenues for therapeutic progress. Through the precise regulation of ion distribution, we have the opportunity to develop innovative treatments for a wide array of diseases by harnessing the fundamental significance of ions in both cellular physiology and pathology.

In 2020, the research group led by Bu et al. proposed an innovative strategy for intervening in cellular processes by regulating the distribution of functional ions. This strategy has evolved into a highly effective antitumor technology known as ion interference therapy (IIT) (Figure 11.4). IIT for tumor treatment typically

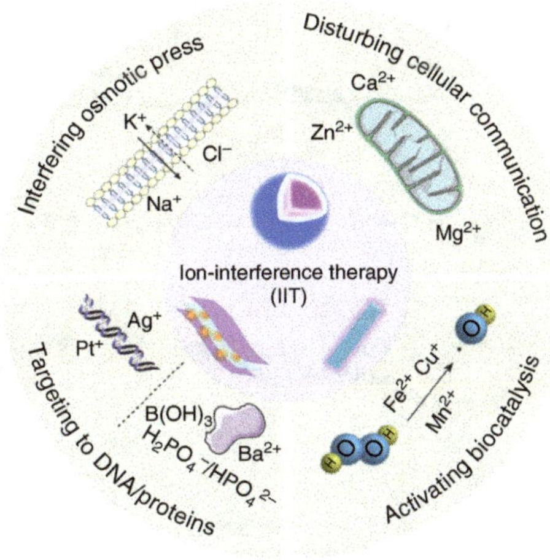

Figure 11.4 Bioactive nanomaterials for ion interference antitumor therapy. Source: [50] / John Wiley & Sons / CC BY 4.0.

involves mechanisms such as inducing osmotic pressure imbalance within cells, activating intracellular biocatalytic processes, and disrupting cell signal pathways [50]. However, the use of single ions or small molecules containing ions often encounters challenges such as rapid blood circulation, nonspecific biological distribution, and dose-limited systemic toxicity, which restrict their broader applications in the field of biomedicine. As a result, researchers have developed various nanomaterials that serve as repositories for ions utilized in IIT including $CaCO_3$, NaCl, CaO_2, Fe_3O_4, MnO_2, etc. [51–55].

In recent studies, researchers discovered that IIT further induces pyroptosis for enhanced immunity response. One notable study conducted by Zhang and coworkers involved the synthesis of phospholipid-coated $Na_2S_2O_8$ nanoparticles (Figure 11.5) [56]. These nanoparticles had the unique ability to gradually degrade and release Na^+ and $S_2O_8{}^{2-}$ ions in situ. The $S_2O_8{}^{2-}$ ions were subsequently converted into toxic $^{\cdot}SO_4{}^-$ and $^{\cdot}OH$ species. Moreover, the excessive presence of Na^+ ions led to a surge in cell osmotic pressure, triggering the activation of caspase-1 and, ultimately, pyroptosis. This process could further synergistically induce the immunogenic cell death, facilitating the activation of a systemic antitumor immune response. Consequently, this immune response effectively inhibited tumor recurrence and metastasis. Moreover, Engelke and coworkers reported for the first time that phospholipid-coated MIL-100(Fe) metal-organic framework (MOF) induces pyroptosis for tumor therapy [57]. As illustrated in Figure 11.6, when the material entered tumor cells via endocytosis, it underwent a process where a significant quantity of iron ions were released under the acidic conditions within the lysosomes. This release of iron ions subsequently triggered the activation of the caspase-1/GSDMD pyroptosis pathway. As a result, the affected tumor cells underwent a cellular process known as bubbling, along with the release of IL-1β, leading to the ultimate elimination of the tumor cells.

Lin and coworkers proposed biodegradable K_3ZrF_7:Yb/Er upconversion nanoparticles (ZrNPs) as pyroptosis inducers for cancer immunotherapy [58]. Tumor

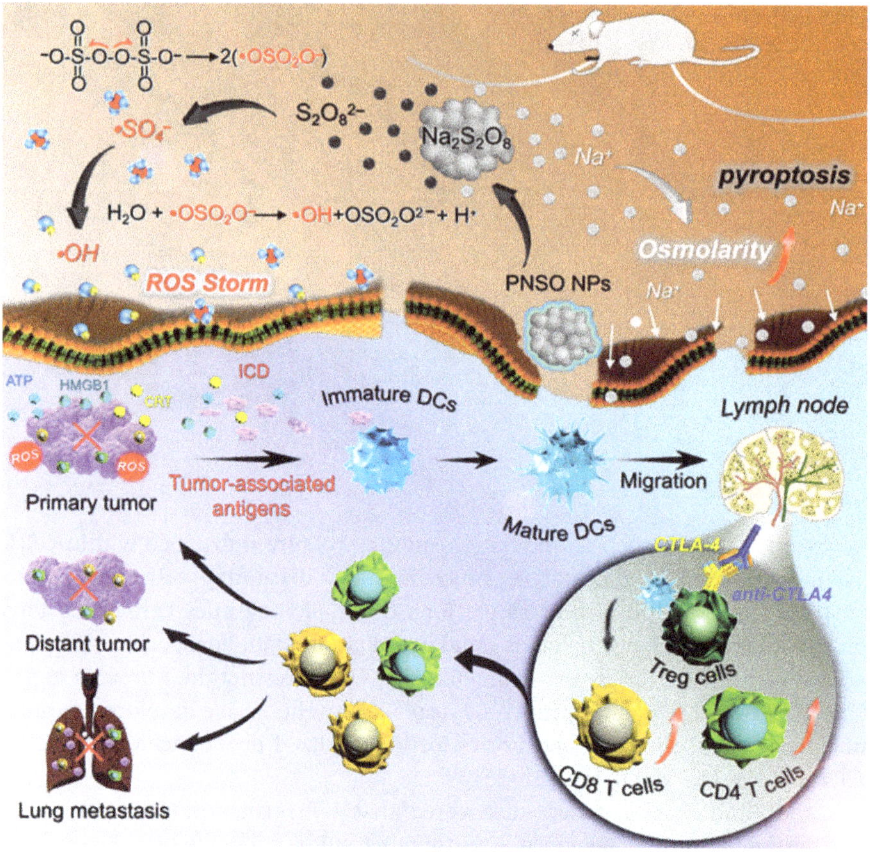

Figure 11.5 Illustration of therapeutic mechanism of $Na_2S_2O_8$ nanoparticles. Source: Reproduced with permission from [56], 2020 / American Chemical Society.

cells internalized ZrNPs which formed an ion pool within the cells and released significant amounts of K^+ and $[ZrF_7]^{3-}$ ions. The presence of these ions induced osmolarity changes, ultimately leading to increased reactive oxygen species (ROS) levels. These changes in osmolarity and ROS levels triggered pyroptosis mediated by caspase-1/GSDMD pathway. In addition, the ZrNPs performed excellent immune activation and tumor tracking in vivo. In further study, Lin and coworkers discovered that Ca^{2+} nanomodulators (CaNMs) could be used as pyroptosis inducers for cancer immunotherapy through mitochondrial Ca^{2+} overload-triggered pyroptosis (Figure 11.7) [59]. CaNMs were biodegradable materials that could rapidly increase mitochondrial Ca^{2+} concentration after entering cells. The sudden surge of Ca^{2+} ions in mitochondria quickly broke the dynamic balance of Ca^{2+}, causing mitochondrial Ca^{2+} overload, eventually leading to the increase of ROS and the release of cytochrome C. Cytochrome C further activated caspase-3 to cleave GSDME. The lysed GSDME-N-terminal domain bound to cell membrane phospholipids to form pores, leading to pyroptosis. Moreover, Lin and coworkers further demonstrated that the zeolitic imidazolate framework-8 nanoparticles (ZIF-8 NPs)-induced pyroptosis

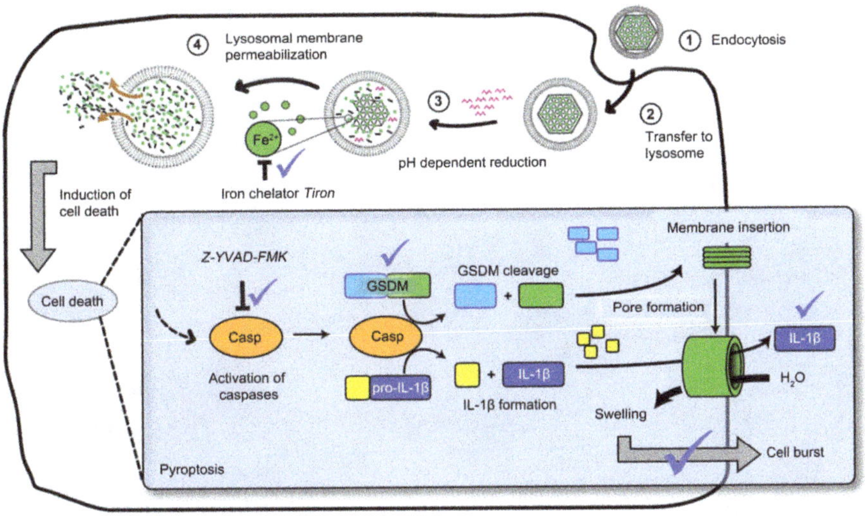

Figure 11.6 Uptake of Lip-MOF nanoparticles leads to pyroptosis of cells in acidic environments. Pyroptosis is mediated by caspase activation, which triggers gasdermin activation and interleukin-1β-release, resulting in cell swelling and lysis. Source: [57] / John Wiley & Sons /

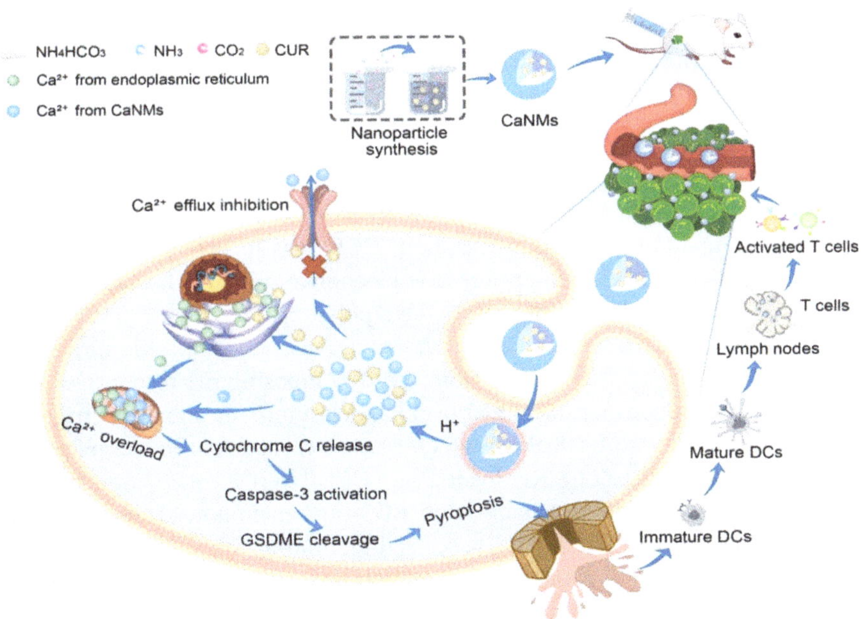

Figure 11.7 Biodegradable Ca^{2+} nanomodulators as the pyroptosis inducers to activate pyroptosis through mitochondrial Ca^{2+} overload for cancer immunotherapy. Source: Reproduced with permission from [59], 2022 / John Wiley & Sons.

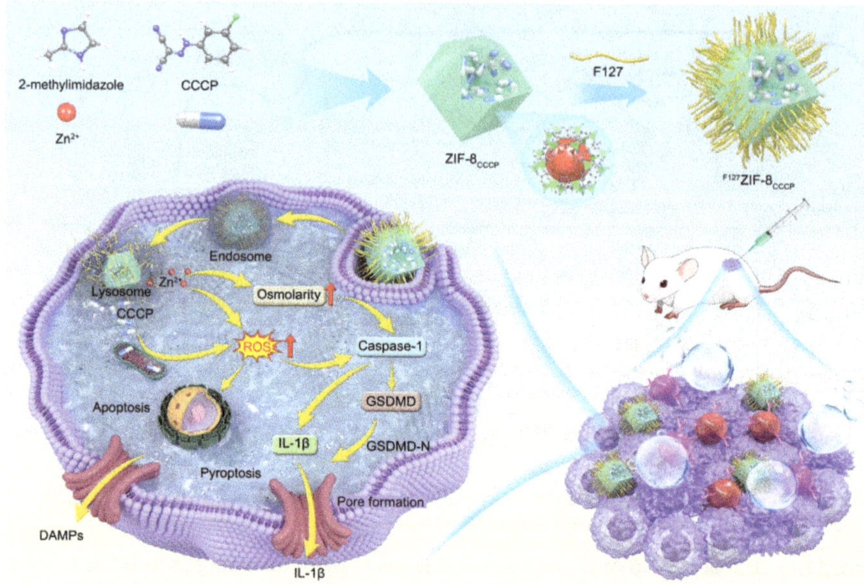

Figure 11.8 Schematic illustration of the fabrication and mechanism to induce pyroptosis of $^{F127}ZIF\text{-}8_{CCCP}$ nanoparticles for cancer immunotherapy. $^{F127}ZIF\text{-}8_{CCCP}$ NPs were prepared via assembling Zn^{2+} ions with 2-methylimidazole and CCCP, and Pluronic F127 was further modified. Once endocytosis by tumor cells, $^{F127}ZIF\text{-}8_{CCCP}$ NPs realize pH-sensitive Zn^{2+} ions and CCCP release to induce a sudden surge in ions, intracellular osmolarity, and ROS, activating caspase-1/GSDMD-dependent pyroptosis pathway. Source: Reproduced with permission from [60], 2023 / John Wiley & Sons.

and enhanced immune response via Zn^{2+} ions (Figure 11.8) [60]. To assemble ZIF-8 NPs for pyroptosis therapy, a methanol solution containing Zn^{2+} ions and 2-methylimidazole was utilized. Additionally, the mitochondrial depolarizing agent carbonyl cyanide 3-chlorophenylhydrazone (CCCP) was incorporated to enhance the efficacy of pyroptosis. Furthermore, the biocompatibility of the nanoparticles was improved by the addition of polyethylene-polypropylene glycol (F127). Once inside the cell, the pH-responsive ZIF-8 NPs underwent decomposition, releasing Zn^{2+} ions. The release of Zn^{2+} ions from the decomposing pH-responsive ZIF-8 NPs led to an imbalance in intracellular osmotic pressure. This imbalance triggered the activation of caspase-1/GSDMD, which plays a crucial role in the pyroptosis pathway. In addition, the ZIF-8 NPs loaded with CCCP were found to induce higher levels of ROS production. This increase in ROS further enhanced the pyroptosis damage and the release of DMAPs. These exciting findings highlight the advances made in ITT. They further expanded research on ions overload therapy, preparing alkaline sodium bicarbonate nanoparticles ($NaHCO_3$ NPs) by a simple and fast microemulsion method for amplified cancer immunotherapy (Figure 11.9) [61]. The use of weakly alkaline $NaHCO_3$ NPs was found to effectively deplete lactic acid in the TME, thereby regulating the acidic conditions of the tumor. This

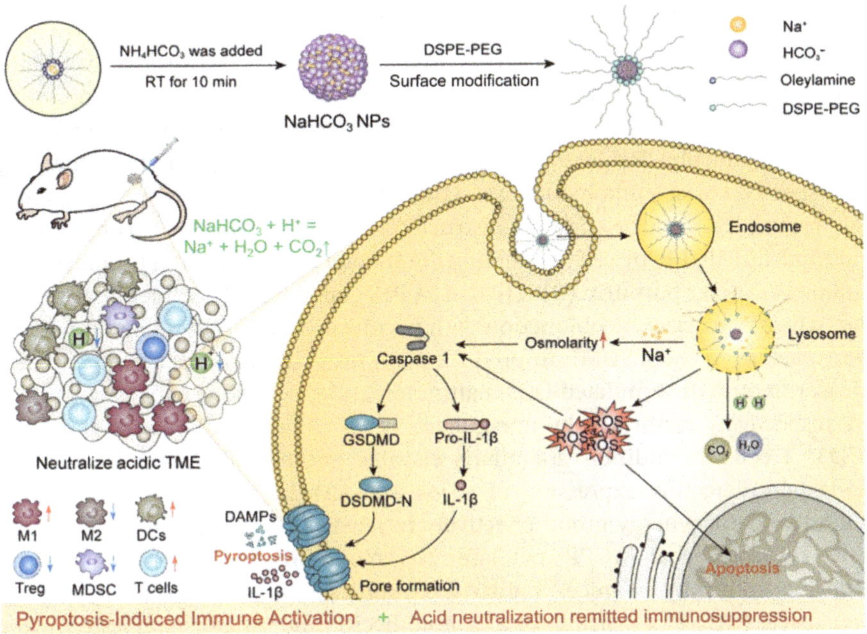

Figure 11.9 Schematic illustration of the fabrication and mechanism of sodium bicarbonate nanoparticles ($NaHCO_3$ NPs) for amplified cancer immunotherapy. $NaHCO_3$ NPs were prepared via a microemulsion method, and DSPE-PEG was further modified. The obtained $NaHCO_3$ NPs regulate lactic acid metabolism through acid-base neutralization and thus reverse the immunosuppressive tumor in a mildly acidic environment. Furthermore, it can release high amounts of Na^+ ions and induce a surge in intracellular osmolarity, so as to activate the pyroptosis pathway and improve immune responses. Source: Reproduced with permission from [61], 2023 / John Wiley & Sons.

regulatory effect on tumor acidity is crucial in alleviating immunosuppression. Simultaneously, the large production of Na^+ ions via the $NaHCO_3$ NPs caused an imbalance in cellular osmolarity, activating the pyroptosis pathway. Through acid-base neutralization and ion interference therapy, the $NaHCO_3$ NPs not only reversed immunosuppression but also activated the immune response. In vivo experiments further proved that the $NaHCO_3$ NPs exhibited significant inhibition of both distal and proximal tumors. Overall, the use of $NaHCO_3$ NPs represents a novel approach that combines amplified immunotherapy with lactic acid depletion therapy, providing a new paradigm in tumor treatment. Similarly, Lin and coworkers also synthesized phospholipid-coated sodium citrate nanoparticles (PSCT NPs) via a simple microemulsion method [62]. The PSCT NPs not only released amount of Na^+ ions to activate caspase-1/GSDMD, but also caused a unique caspase-8/GSDMC-mediated pyroptosis through $C_6H_5O_7^-$ ions release to interfere with the tricarboxylic acid cycle. As a result, effect of T cells, memory T cells, and DCs was significantly increased, proving that dual-pathway pyroptosis induction achieved synergistic IIT and immunotherapy.

11.3.2 TME-Responsive Pyroptosis Therapy

Researchers have made significant progress in developing TME-responsive nanoparticles that precisely control the induction of pyroptosis in tumors, optimize therapeutic outcomes, and stimulate the immune response. These nanomaterials are specifically designed to respond to unique TME properties, enabling tumor-specific therapy and reducing systemic toxicity [63].

For instance, Sun and coworkers constructed a ROS/glutathione (GSH)-responsive nano-prodrug (MCPP) by self-assembling the chemotherapeutic drug paclitaxel and the photosensitizer purpurin (P18) [64]. The P18 generated ROS under laser irradiation, which was used in conjunction with paclitaxel to administer chemotherapy, in turn, activated the GSDME to produce pyroptosis. In vivo experiments, MCPP and laser treatment stimulated DCs maturation and T-cell clone expansion, which were respectively confirmed by upregulated of CD103 expression and increasing of $CD3^+$ T cells. In addition, a similarly elevated number of memory T cells was observed by detecting expression of CD44 and CD62L on T cells. The MCPP, laser, and PD-L1 synergy group effectively resisted tumor recurrence. The above results suggested that MCPP and laser treatment initiated pyroptosis, leading to specific immunity in mice. Similarly, Xu and coworkers used a polyprodrug nanoparticle (CCNP) consisting of a camptothecin prodrug and a photosensitizer (Ce6) via hydrophobic interactions (Figure 11.10) [65]. The CCCP was GSH/ROS dual-responsive and able to release the chemotherapeutic drug camptothecin (CPT) under TME, which, in combination with photodynamic therapy (PDT), induced pyroptosis, enhanced immune effects and effectively suppressed primary/distal tumors. Figure 11.10 showed CCNP binding via π–π stacking and hydrophobic interaction. Under TME, the thioether bond of CCCP was broken by ROS and GSH, which then released CPT and Ce6. The CPT and PDT acted together to activate caspase-3/GSDME-mediated pyroptosis, elevated cellular ROS, and relieved hypoxia. Further, the reversal of TME resulted in a decrease in MDSC, leading to an increase in M1 tumor-associated macrophages (TAM), $CD4^+$/$CD8^+$ T cells, memory T cells, and DCs. The cellular contents released from pyroptosis led to systemic immune activation, and distal tumors thus underwent a reversal of immunosuppression consistent with that of the primary tumors.

In conclusion, the development of TME-responsive nanomaterials represents a promising approach to achieving accurate control of pyroptosis induction in tumors while maximizing therapeutic effects and stimulating immune response. These nanomaterials exploit the unique characteristics of the TME to precisely deliver therapeutic agents, induce pyroptosis, and reverse the TME toward an immunostimulatory state. Continued research in this field holds great potential for the advancement of tumor therapy and the realization of more effective and targeted treatments.

11.3.3 Demethylation-Activated Pyroptosis

Numerous past studies have proved that chemotherapy effects via a caspase-3-mediated apoptosis pathway [66, 67]. However, caspase-3 was found to acts as

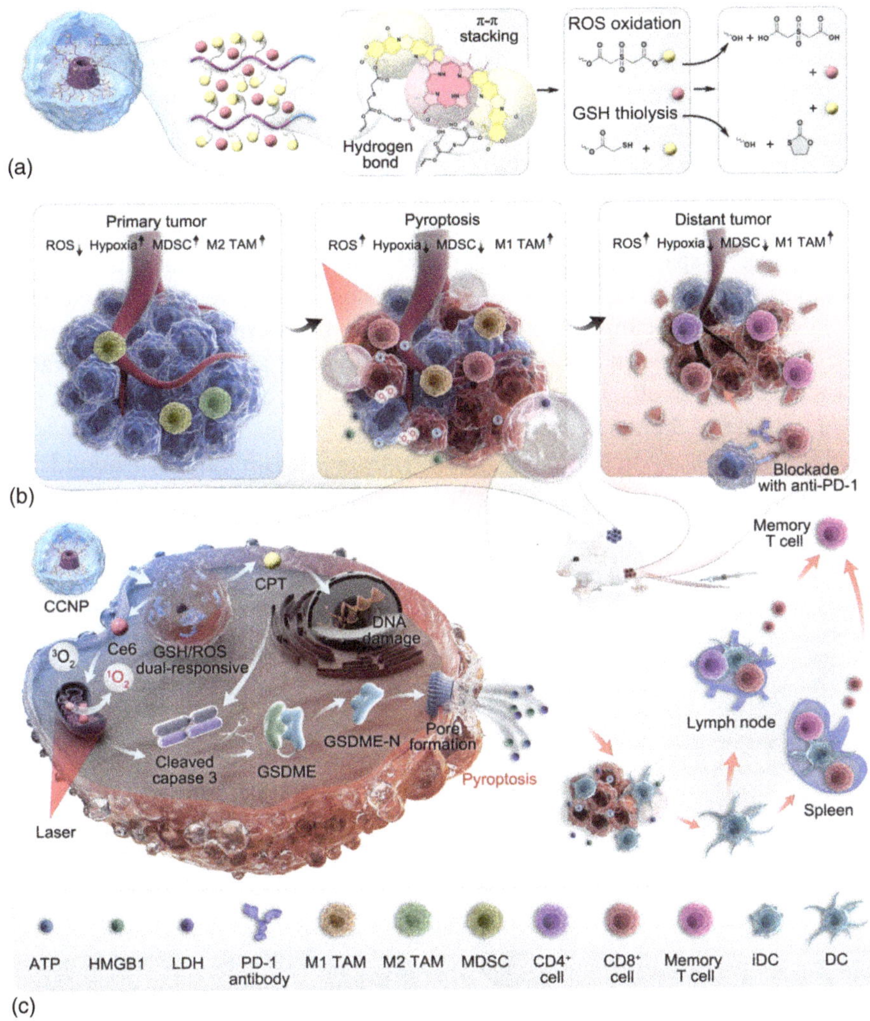

Figure 11.10 Schematic illustration of CCNP for antitumor immune response by precisely inducing pyroptosis. Source: Reproduced with permission from [61], 2022 / Elsevier.

upstream protein to cleave GSDME and induce pyroptosis either. Recent studies suggested that the GSDME is widely suppressed or hypermethylation in most tumor cells, which is the main reason chemotherapy generally induces apoptosis rather than pyroptosis. Apoptosis and pyroptosis are distinct modes of cell death with varying functional outcomes and immunogenicity. Apoptosis, characterized by DNA fragmentation, nuclear condensation, and the formation of apoptotic bodies, usually results in noninflammatory cell death. In contrast, pyroptosis elicits a markedly inflammatory response by releasing intracellular contents. This immune reaction serves as an alert to the immune system, drawing immune cells and fostering an antitumor immune response [68].

The decrease in GSDME expression in tumor cells hampers the transformation of caspase-3 activity into pyroptosis, restricting the advantages of pyroptosis and the linked immune response. Chemotherapeutic agents that instigate apoptosis mostly in GSDME-lacking tumor cells might not trigger commensurate immunogenicity and could be unsuccessful at efficiently activating the immune system. This may lead to diminished clearance of tumor cells and potential therapy resistance [14, 69, 70]. The contribution of DFNA5 gene hypermethylation to GSDME deletion prompted Zhang et al. to employ decitabine (DAC) and chemotherapeutic nanomedicines to spark pyroptosis and augment immune effects, as illustrated in the diagram (Figure 11.11). Tumor cells in hormonal mice were pretreated with DAC to demethylate, before delivering LipoDDP to the site of the tumor. As anticipated, GSDME expression in tumor tissue was elevated, and there was a notable increase in effector T cells, DCs, and memory T cells. Successful demethylation has enabled a range of therapies that were once employed for inducing apoptosis in tumors to be converted to pyroptosis, thereby promoting immune activation [71]. Additionally, Yang et al. utilized bivalent gold nanoparticles in conjunction with

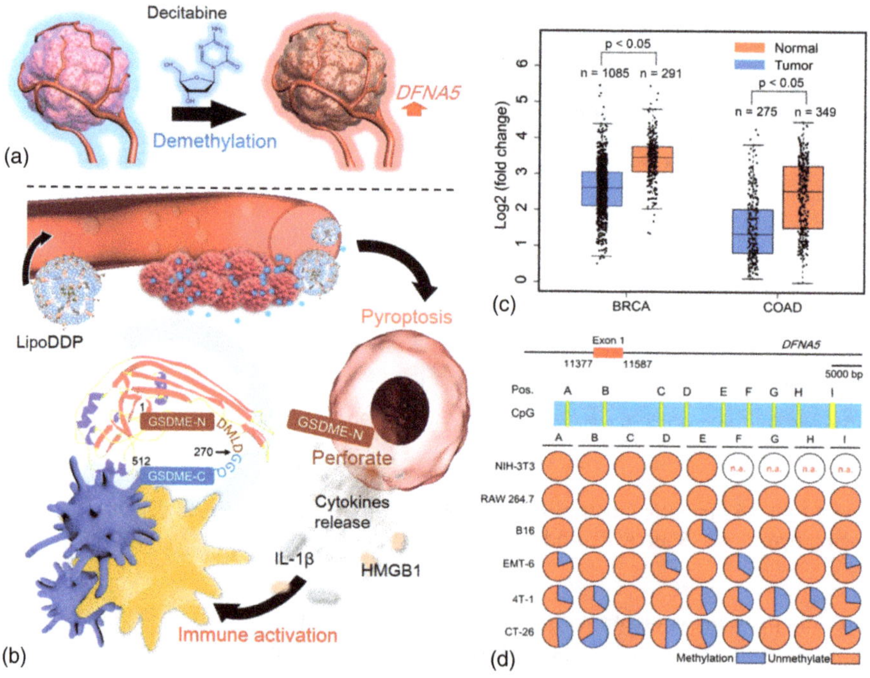

Figure 11.11 DFNA5 is downregulated in tumor cells of breast cancer (BRCA) and adenocarcinoma (COAD) patients ascribed to DNA methylation. (a) Schematic illustration of demethylation for tumor cells by DAC. (b) Therapeutic process of the immune activation triggered by DAC/chemotherapeutics through pyroptosis pathway. (c) The Cancer Genome Atlas (TCGA) database of DFNA5 expression in BRCA and COAD patients. The boxplot analysis shows log2 (TPM + 1) on a log scale. (d) Schematic diagram of the CpG islands of DFNA5. The red box shows the location of exon 1. The position of a dense CpG island with nine allelic methylation status (Pos.) is displayed with yellow stripes. Source: Reproduced with permission from [71], 2019 / American Chemical Society.

temperature-sensitive radiofrequency thermal ablation and DAC co-treatment to activate caspase-3 and upregulate GSDME. This process successfully converted apoptosis to pyroptosis and elicited a robust ICD effect to provoke a systemic antitumor immune response [72].

11.3.4 The Other Pyroptosis Therapies

Sonodynamic therapy (SDT) is an objective mode of noninvasive treatment that seeks to treat diseases by utilizing sound sensitizers to generate ROS, heat, or mechanical damage [73]. Not the same as PDT, which employs laser as input source, the SDT leverages ultrasonic waves, enabling deeper tissue penetration and safer energy input. By sensitizing cells to sound waves, SDT has the capability of triggering pyroptosis and presenting a precise and controlled approach for tumor therapy. Nonetheless, to fully realize the potential of SDT-induced pyroptosis in revolutionizing tumor treatment, the challenges such as the selection and optimization of sound sensitizers and the elucidation of the underlying mechanisms, must be addressed. Further research and clinical trials are needed to evaluate the efficacy and safety of this emerging modality [74, 75]. Chen and coworkers reported that the liposome nanoparticles loaded with protoporphyrin IX, CaO_2, and DAC for pyroptosis induction and tumor therapy (Figure 11.12) [76]. In this system, ROS produced by SDT and Ca^{2+} ions released by CaO_2 cooperatively

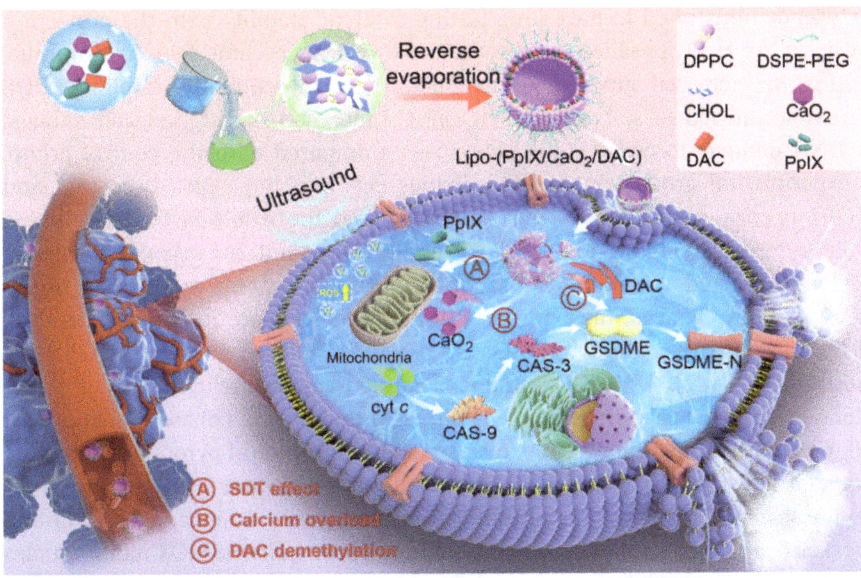

Figure 11.12 Schematic illustration for the construction of the lipo-(PpIX/CaO_2/DAC) nanoliposomes and the potential therapeutic mechanism of lipo-(PpIX/CaO_2/DAC)-activated SDT in damaging the mitochondrial Ca^{2+} buffering capacity to synergistically accomplish calcium overload for mitochondrial cyt *c* release, CAS-9/3 activation and GSDME cleavage into GSDME-N domains to induce the specific cancer-cell pyroptosis. Source: Reproduced with permission from [76], 2023 / Elsevier.

induced mitochondrial damage and promoted the release of cytochrome c (cyt c), thereby activating caspase-9 and caspase-3 proteins in turn. On the other hand, the expression of GSDME was upregulated by DNA methylation inhibitor DAC. The combination of caspase-3 and GSDME effectively activated the process of pyroptosis, thus realizing the pyroptosis caused by ultrasonic. Similarly, Chen et al. used PCN-224 nanomaterials loaded with LY-364947 and coated on erythrocyte membranes, with injected DAC molecules into the tail vein at the same time to achieve ultrasonic-induced pyroptosis and immunotherapy [77].

Li and coworkers selected ZnO_2 as the donor of H_2O_2, and induced pyroptosis by releasing high concentration of H_2O_2 in the TME [78]. The sudden surge of H_2O_2 after degradation of high concentration of ZnO_2 ($HZnO_2$) activated NLRP3 inflammasome and caspase-1 protein, leading to GSDMD cleavage and IL-1β maturation. The lysed GSDMD released the gasdermin-N domain, which caused cell expansion and produced large bubbles, leading to pyroptosis.

Glucose oxidase (GOx) consumes glucose to generate gluconic acid and hydrogen peroxide, which is widely used in tumor starvation therapy. Starvation therapy is a treatment that eliminates tumor cells by interrupting their energy supply [79, 80]. The Kazunori Kataoka's team in Japan demonstrated for the first time that GOx-induced starvation therapy also causes pyroptosis [81]. As shown in Figure 11.13a,b, in order to achieve tumor-responsive GOx consumption, the authors constructed a polyion complex vesicle with a thioketal structure, which effectively encapsulated GOx through a self-assembly process (GOD@PICsomes). GOx was demonstrated to have long-lasting catalytic stability with the protection of vesicles. As H_2O_2 production increased, the vesicles became more hydrophilic, resulting in increased glucose permeability. This subsequently amplified ROS production and led to starvation. GOD@PICsomes had the highest cell damage and DNA damage to cells (Figure 11.13c–e). Compared with the control group, the experimental group can induce obvious cell bubbling, CRT exposure, and HMGB1 release, which proves that starvation and the resulting oxidative stress induce pyroptosis (Figure 11.13f–h). In addition, Li and coworkers prepared a NIR light-triggered thermal response pyroptosis system for synergistic tumor immunotherapy. That is, mesoporous Pt nanoparticles were modified with GOx and thermosensitive material p(OEOMA-*co*-MEMA) (PCM) to obtain ^{PCM}Pt/GOx nanocomposites [82]. ^{PCM}Pt/GOx produced a mild temperature in the tumor region under near-infrared laser irradiation. Then, PCM underwent a phase transition at a mild temperature, resulting in GOx release. The occurrence of pyroptosis led to the leakage of DAMPs into TME, which triggered pyroptosis and systemic antitumor immune responses for synergistic tumor immunotherapy. In recent work, Li and coworkers further developed a self-destructing copper quinone GOx nanoparticles (CQG NPs) treatment platform, which could trigger synergistic immunotherapy with pyroptosis and cuproptosis through "Tai Chi mindset" to fight refractory dormant tumors (Figure 11.14) [83]. CQG NPs could destroy the antioxidant defense mechanism of tumor cells by inhibiting the nuclear factor-erythroid 2-related factor 2 (NRF2)-quinone oxidoreductase 1 (NQO1) signaling pathway. In addition, combined with its excellent multi-enzyme activity, it activated NLRP3-mediated

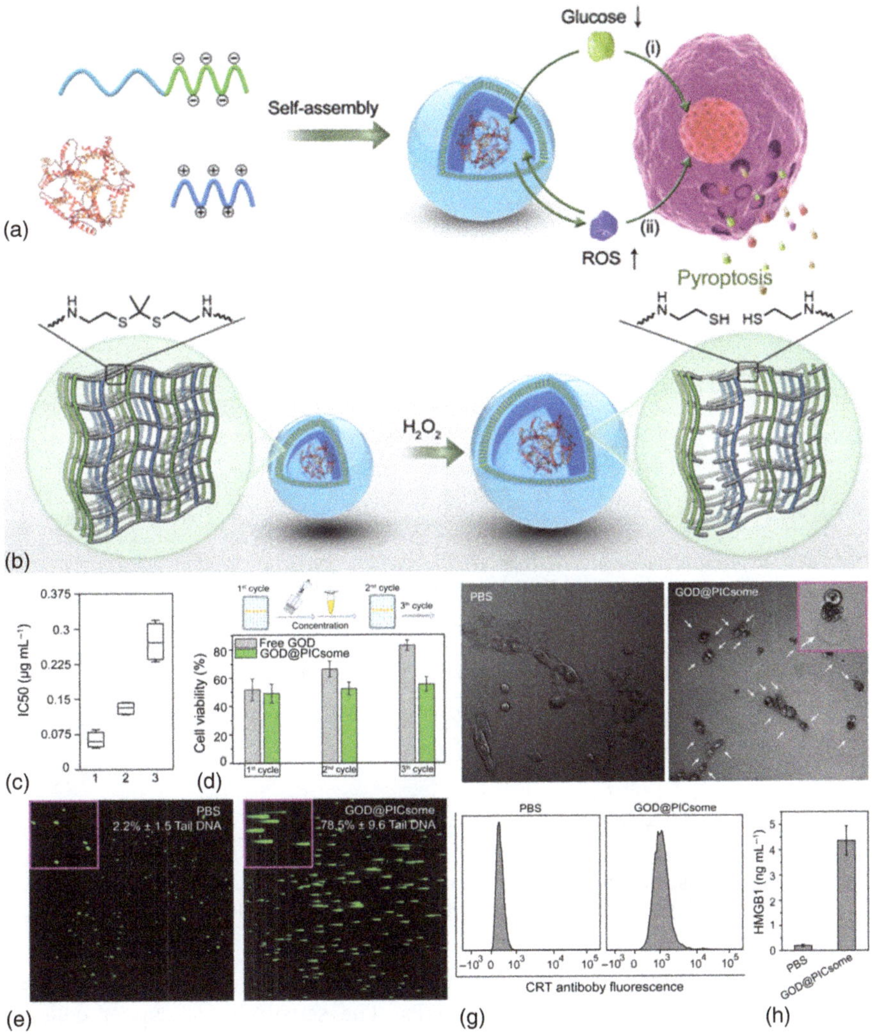

Figure 11.13 (a) Schematic illustration of the ROS-responsive GOD-loaded therapeutic nanoreactor. (b) Schematic illustration of vesicle swelling. (c) IC_{50} values of GOD (1), GOD@PICsomes (2), and GOD-loaded nonresponsive nanoreactors (3). (d) Cytotoxicity of recycled GOD and GOD@- PICsomes from old culture medium at their IC_{50} concentration. (e) DNA damage of 4T1 cell treated with GOD@PICsomes at GOD concentration of 0.2 μg ml^{-1}. (f) Bright field images of 4T1 cells at 12 hours post-incubation with GOD@PICsomes. The white arrows indicate the pyroptotic cell death characterized with large bubbles blowing from the plasma membrane. The level of immune simulants, (g) calreticulin and (h) HMGB1, for 4T1 cells at 12 hours post incubation with GOD@PICsomes. Source: Reproduced with permission from [81], 2020 / John Wiley & Sons.

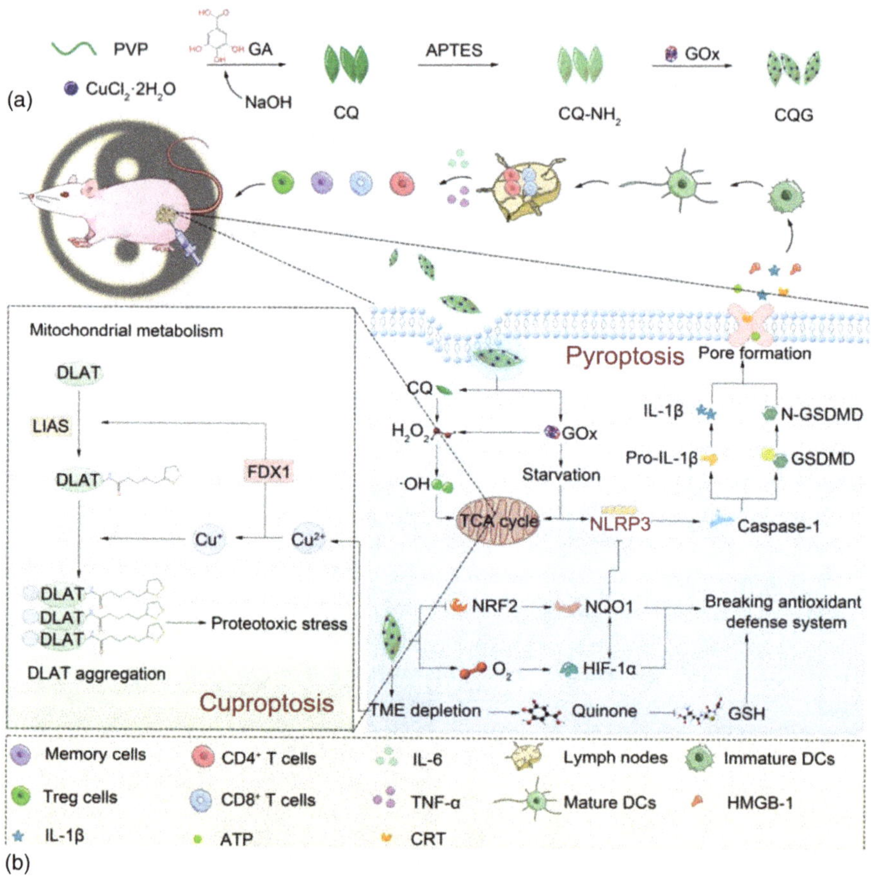

Figure 11.14 Illustration of the material synthesis and therapeutic process. (a) Schematic diagram of the synthesis of copper quinone GOx nanoparticles (CQG NPs). (b) Anticancer mechanism of synergistic activation of immunity by CQG-induced pyroptosis and cuproptosis. Source: Reproduced with permission from [83], 2023 / John Wiley & Sons.

pyroptosis. At the same time, cuproptosis could be triggered by copper ions released by self-induced decomposition of CQG NPs. The sensitivity of cancer cells to cuproptosis consumed endogenous copper chelators through the Michael addition reaction between GSH and quinone ligands and starvation-induced glucose deficiency. More importantly, CQG NPs-induced pyroptosis and cuproptosis could promote the remodeling of immunosuppressive TME, enhance the infiltration of immune cells into tumors, and activate strong systemic immunity.

11.4 Summary and Outlook

Pyroptosis has garnered significant attention lately owing to its multifaceted pathways and diverse inducers. What distinguishes pyroptosis apart from other cell death modes is its potential to release various DAMPs, thus triggering localized inflammation and stimulating immune responses. This unique attribute makes pyroptosis an exceptional immunotherapeutic mode with potential for novel treatment approaches. More controlled methods to induce pyroptosis are currently needed to explore. The techniques involve using small molecules, genetic modifications, and nanomaterials-based delivery systems. These methods target specific components within the pyroptosis pathway to activate or enhance pyroptosis.

Due to the good designability, long-lasting in vivo circulation and specific recognition, the use of nanomaterials as pyroptosis adjuvants or drug platforms has a promising future in the field of pyroptosis and immunotherapy. There have been numerous studies that have opened up nanomaterial-based pyroptosis induction to new depths. It is worth noting that less toxic/nontoxicized induction of pyroptosis remains to be explored more to mitigate systemic toxicity and excessive inflammatory responses. Improving the immunosuppressive environment to enhance the immune response induced by pyroptosis is also a worthy field of attention.

In conclusion, the complex pathways and varied triggers of pyroptosis render it a fascinating field in the realm of immunotherapy. The inextricable relationship between pyroptosis and immune induction provides opportunities for formulating innovative treatment strategies. With further strides in comprehending the fundamental mechanisms and investigating inducers of pyroptosis, significant advances in the development of pyroptosis-derived treatments that amplify therapeutic efficacy and boost immune responses in forthcoming years are anticipated.

References

1 Zychlinsky, A., Prevost, M.C., and Sansonetti, P.J. (1992). *Shigella flexneri* induces apoptosis in infected macrophages. *Nature* 358: 167–169.

2 Brennan, M.A. and Cookson, B.T. (2000). Salmonella induces macrophage death by caspase-1-dependent necrosis. *Molecular Microbiology* 38: 31–40.

3 Cookson, B.T. and Brennan, M.A. (2001). Pro-inflammatory programmed cell death. *Trends in Microbiology* 9: 113–114.

4 Galluzzi, L., Vitale, I., Aaronson, S.A. et al. (2018). Molecular mechanisms of cell death: recommendations of the Nomenclature Committee on Cell Death 2018. *Cell Death Differentiation* 25: 486–541.

5 Frank, D. and Vince, J.E. (2019). Pyroptosis versus necroptosis: similarities, differences, and crosstalk. *Cell Death Differentiation* 26: 99–114.

6 Schroder, K. and Tschopp, J. (2010). The inflammasomes. *Cell* 140: 821–832.

7 Martinon, F., Burns, K., and Tschopp, J. (2002). The inflammasome: a molecular platform triggering activation of inflammatory caspases and processing of proIL-β. *Molecular Cell* 10: 417–426.

8 Kayagaki, N., Warming, S., Lamkanfi, M. et al. (2011). Non-canonical inflammasome activation targets caspase-11. *Nature* 479: 117–121.

9 Kayagaki, N., Wong, M.T., Stowe, I.B. et al. (2013). Noncanonical inflammasome activation by intracellular LPS independent of TLR4. *Science* 341: 1246–1249.

10 Shi, J., Zhao, Y., Wang, K. et al. (2015). Cleavage of GSDMD by inflammatory caspases determines pyroptotic cell death. *Nature* 526: 660–665.

11 Kayagaki, N., Stowe, I.B., Lee, B.L. et al. (2015). Caspase-11 cleaves gasdermin D for non-canonical inflammasome signalling. *Nature* 526: 666–671.

12 He, W.t., Wan, H., Hu, L. et al. (2015). Gasdermin D is an executor of pyroptosis and required for interleukin-1β secretion. *Cell Research.* 25: 1285–1298.

13 Ding, J., Wang, K., Liu, W. et al. (2016). Pore-forming activity and structural autoinhibition of the gasdermin family. *Nature* 535: 111–116.

14 Wang, Y., Gao, W., Shi, X. et al. (2017). Chemotherapy drugs induce pyroptosis through caspase-3 cleavage of a gasdermin. *Nature* 547: 99–103.

15 Orning, P., Weng, D., Starheim, K. et al. (2018). Pathogen blockade of TAK1 triggers caspase-8–dependent cleavage of gasdermin D and cell death. *Science* 362: 1064–1069.

16 Sarhan, J., Liu, B.C., Muendlein, H.I. et al. (2018). Caspase-8 induces cleavage of gasdermin D to elicit pyroptosis during Yersinia infection. *Proceedings of the National Academy of Sciences of the United States of America* 115: E10888–E10897.

17 Yu, P., Zhang, X., Liu, N. et al. (2021). Pyroptosis: mechanisms and diseases. *Signal Transduction and Targeted Therapy* 6: 128.

18 Liu, Y., Fang, Y., Chen, X. et al. (2020). Gasdermin E–mediated target cell pyroptosis by CAR T cells triggers cytokine release syndrome. *Science Immunology* 5: eaax7969.

19 Hu, J.J., Liu, X., Xia, S. et al. (2020). FDA-approved disulfiram inhibits pyroptosis by blocking gasdermin D pore formation. *Nature Immunology* 21: 736–745.

20 Zhou, Z., He, H., Wang, K. et al. (2020). Granzyme A from cytotoxic lymphocytes cleaves GSDMB to trigger pyroptosis in target cells. *Science* 368: eaaz7548.

21 Chi, X., Huang, M., Tu, H. et al. (2023). Innate and adaptive immune abnormalities underlying autoimmune diseases: the genetic connections. *Science China Life Sciences* 66: 1482–1517.

22 Hiam-Galvez, K.J., Allen, B.M., and Spitzer, M.H. (2021). Systemic immunity in cancer. *Nature Reviews Cancer* 21: 345–359.

23 Lambert, A.W., Pattabiraman, D.R., and Weinberg, R.A. (2017). Emerging biological principles of metastasis. *Cell* 168: 670–691.

24 Leone, R.D. and Powell, J.D. (2020). Metabolism of immune cells in cancer. *Nature Reviews Cancer* 20: 516–531.

25 Waldman, A.D., Fritz, J.M., and Lenardo, M.J. (2020). A guide to cancer immunotherapy: from T cell basic science to clinical practice. *Nature Reviews Immunology* 20: 651–668.

26 Jhunjhunwala, S., Hammer, C., and Delamarre, L. (2021). Antigen presentation in cancer: insights into tumour immunogenicity and immune evasion. *Nature Reviews Cancer* 21: 298–312.

27 Reeves, E. and James, E. (2017). Antigen processing and immune regulation in the response to tumours. *Immunology* 150: 16–24.

28 Verdegaal, E.M.E., de Miranda, N.F.C.C., Visser, M. et al. (2016). Neoantigen landscape dynamics during human melanoma–T cell interactions. *Nature* 536: 91–95.

29 Kong, K.-F., Fu, G., Zhang, Y. et al. (2014). Protein kinase C-η controls CTLA-4–mediated regulatory T cell function. *Nature Immunology* 15: 465–472.

30 Snyder, A., Makarov, V., Merghoub, T. et al. (2014). Genetic basis for clinical response to CTLA-4 blockade in melanoma. *The New England Journal of Medicine* 371: 2189–2199.

31 Bertheloot, D., Latz, E., and Franklin, B.S. (2021). Necroptosis, pyroptosis and apoptosis: an intricate game of cell death. *Cellular Molecular Immunology* 18: 1106–1121.

32 Galluzzi, L., Kepp, O., Chan, F.K., and Kroemer, G. (2017). Necroptosis: mechanisms and relevance to disease. *Annual Review of Pathology* 12: 103–130.

33 Aaes, T.L., Kaczmarek, A., Delvaeye, T. et al. (2016). Vaccination with necroptotic cancer cells induces efficient anti-tumor immunity. *Cell Reports* 15: 274–287.

34 Mehrotra, P. and Ravichandran, K.S. (2022). Drugging the efferocytosis process: concepts and opportunities. *Nature Reviews Drug Discovery* 21: 601–620.

35 Kroemer, G., Galluzzi, L., Kepp, O., and Zitvogel, L. (2013). Immunogenic cell death in cancer therapy. *Annual Review of Immunology* 31: 51–72.

36 Galluzzi, L., Buqué, A., Kepp, O. et al. (2017). Immunogenic cell death in cancer and infectious disease. *Nature Reviews Immunology* 17: 97–111.

37 Wu, D., Wang, S., Yu, G.C., and Chen, X.Y. (2021). Cell death mediated by the pyroptosis pathway with the aid of nanotechnology: prospects for cancer therapy. *Angewandte Chemie International Edition* 60: 8018–8034.

38 Liu, W., Peng, J., Xiao, M. et al. (2023). The implication of pyroptosis in cancer immunology: current advances and prospects. *Genes Diseases* 10: 2339–2350.

39 Strowig, T., Henao-Mejia, J., Elinav, E., and Flavell, R. (2012). Inflammasomes in health and disease. *Nature* 481: 278–286.

40 Chui, A.J., Okondo, M.C., Rao, S.D. et al. (2019). N-terminal degradation activates the NLRP1B inflammasome. *Science* 364: 82–85.

41 Sandstrom, A., Mitchell, P.S., Goers, L. et al. (2019). Functional degradation: a mechanism of NLRP1 inflammasome activation by diverse pathogen enzymes. *Science* 364: eaau1330.

42 Zhang, Z., Zhang, Y., Xia, S. et al. (2020). Gasdermin E suppresses tumour growth by activating anti-tumour immunity. *Nature* 579: 415–420.

43 Wang, Q., Wang, Y., Ding, J. et al. (2020). A bioorthogonal system reveals antitumour immune function of pyroptosis. *Nature* 579: 421–426.

44 Banerjee, I., Behl, B., Mendonca, M. et al. (2018). Gasdermin D restrains type I interferon response to cytosolic DNA by disrupting ionic homeostasis. *Immunity* 49: 413–26.e415.

45 López-Hernández, T., Puchkov, D., Krause, E. et al. (2020). Endocytic regulation of cellular ion homeostasis controls lysosome biogenesis. *Nature Cell Biology* 22: 815–827.

46 Giorgi, C., Marchi, S., and Pinton, P. (2018). The machineries, regulation and cellular functions of mitochondrial calcium. *Nature Reviews Molecular Cell Biology* 19: 713–730.

47 Li, Y., Lin, J.Y., Wang, P.Y. et al. (2022). Tumor microenvironment-responsive yolk-shell NaCl@virus-inspired tetrasulfide-organosilica for ion-interference therapy via osmolarity surge and oxidative stress amplification. *ACS Nano* 16: 7380–7397.

48 Ma, C., Zhang, Y.D., Zhang, Y. et al. (2022). A pH-targeted and NIR-responsive NaCl-nanocarrier for photothermal therapy and ion-interference therapy. *Nanomedicine* 39: 102460.

49 Wei, Y.B., Wang, J.M., Wu, S.X. et al. (2022). Nanomaterial-based zinc ion interference therapy to combat bacterial infections. *Frontiers in Immunology* 13: 899992.

50 Liu, Y., Zhang, M., and Bu, W. (2020). Bioactive nanomaterials for ion-interference therapy. *VIEW* 1: e18.

51 Dong, Z., Feng, L., Hao, Y. et al. (2018). Synthesis of hollow biomineralized $CaCO_3$-polydopamine nanoparticles for multimodal imaging-guided cancer photodynamic therapy with reduced skin photosensitivity. *Journal of the American Chemical Society* 140: 2165–2178.

52 Jiang, W., Yin, L., Chen, H. et al. (2019). NaCl nanoparticles as a cancer therapeutic. *Advanced Materials* 31: 1904058.

53 Zhang, M., Song, R., Liu, Y. et al. (2019). Calcium-overload-mediated tumor therapy by calcium peroxide nanoparticles. *Chem* 5: 2171–2182.

54 Zhao, Z., Wang, W., Li, C. et al. (2019). Reactive oxygen species–activatable liposomes regulating hypoxic tumor microenvironment for synergistic photo/chemodynamic therapies. *Advanced Functional Materials* 29: 1905013.

55 Lin, L.S., Song, J., Song, L. et al. (2018). Simultaneous fenton-like ion delivery and glutathione depletion by MnO_2-based nanoagent to enhance chemodynamic therapy. *Angewandte Chemie International Edition* 57: 4902–4906.

56 Liu, Y., Zhen, W., Wang, Y. et al. (2020). $Na_2S_2O_8$ nanoparticles trigger antitumor immunotherapy through reactive oxygen species storm and surge of tumor osmolarity. *Journal of the American Chemical Society* 142: 21751–21757.

57 Ploetz, E., Zimpel, A., Cauda, V. et al. (2020). Metal-organic framework nanoparticles induce pyroptosis in cells controlled by the extracellular pH. *Advanced Materials.* 32: 1907267.

58 Ding, B., Sheng, J., Zheng, P. et al. (2021). Biodegradable upconversion nanoparticles induce pyroptosis for cancer immunotherapy. *Nano Letters* 21: 8281–8289.

59 Zheng, P., Ding, B., Zhu, G. et al. (2022). Biodegradable Ca^{2+} Nanomodulators activate pyroptosis through mitochondrial Ca^{2+} overload for cancer immunotherapy. *Angewandte Chemie International Edition* 61: e202204904.

60 Ding, B., Chen, H., Tan, J. et al. (2023). ZIF-8 nanoparticles evoke pyroptosis for high-efficiency cancer immunotherapy. *Angewandte Chemie International Edition* 62: e202215307.

61 Ding, B., Zheng, P., Tan, J. et al. (2023). Sodium bicarbonate nanoparticles for amplified cancer immunotherapy by inducing pyroptosis and regulating lactic acid metabolism. *Angewandte Chemie International Edition* 62: e202307706.

62 Li, J., Ding, B., Tan, J. et al. (2023). Sodium citrate nanoparticles induce dual-path pyroptosis for enhanced antitumor immunotherapy through synergistic ion overload and metabolic disturbance. *Nano Letters* 23: 10034–10043.

63 Wan, S.C., Ye, M.J., Yang, Q.C. et al. (2023). Diselenide-based dual-responsive prodrug as pyroptosis inducer potentiates cancer immunotherapy. *Advanced Healthcare Materials* 12: 2202135.

64 Xiao, Y., Zhang, T., Ma, X. et al. (2021). Microenvironment-responsive prodrug-induced pyroptosis boosts cancer immunotherapy. *Advanced Science* 8: 2101840.

65 Ma, X., Su, W., Ye, M. et al. (2023). Endogenous/exogenous stimulies inspired polyprodrug nano-inducer switches pyroptosis path for promoting antitumor immunity. *Nano Today* 48: 101727.

66 Woo, M., Hakem, R., Soengas, M.S. et al. (1998). Essential contribution of caspase 3/CPP32 to apoptosis and its associated nuclear changes. *Genes Development* 12: 806–819.

67 Kitada, S., Andersen, J., Akar, S. et al. (1998). Expression of apoptosis-regulating proteins in chronic lymphocytic leukemia: correlations with in vitro and in vivo chemoresponses. *Blood* 91: 3379–3389.

68 Zhai, Z., Yang, F., Xu, W. et al. (2022). Attenuation of rheumatoid arthritis through the inhibition of tumor necrosis factor–induced Caspase 3/Gasdermin E–mediated pyroptosis. *Arthritis Rheumatology* 74: 427–440.

69 Rogers, C., Fernandes-Alnemri, T., Mayes, L. et al. (2017). Cleavage of DFNA5 by caspase-3 during apoptosis mediates progression to secondary necrotic/pyroptotic cell death. *Nature Communications* 8: 14128.

70 Wang, H., Gao, Z., Jiao, D. et al. (2023). A microenvironment dual-responsive nano-drug equipped with PD-L1 blocking peptide triggers immunogenic pyroptosis for prostate cancer self-synergistic immunotherapy. *Advanced Functional Materials* 33: 2214499.

71 Fan, J.X., Deng, R.H., Wang, H. et al. (2019). Epigenetics-based tumor cells pyroptosis for enhancing the immunological effect of chemotherapeutic nanocarriers. *Nano Letters* 19: 8049–8058.

72 Zhang, Q., Shi, D., Guo, M. et al. (2022). Radiofrequency-activated pyroptosis of Bi-valent gold nanocluster for cancer immunotherapy. *ACS Nano* 17: 515–529.

73 Wang, Z.F., Wang, M., Qian, Q.R. et al. (2023). Dual-targeted nanoformulation with Janus structure for synergistic enhancement of sonodynamic therapy and chemotherapy. *Chinese Chemical Letters* 34: 107853.

74 Liang, S., Deng, X., Ma, P. et al. (2020). Recent advances in nanomaterial-assisted combinational sonodynamic cancer therapy. *Advanced Materials* 32: 2003214.

75 Qian, X., Zheng, Y., and Chen, Y. (2016). Micro/nanoparticle-augmented sonodynamic therapy (SDT): breaking the depth shallow of photoactivation. *Advanced Materials* 28: 8097–8129.

76 Zhang, Z., Zhang, X., Zhao, S. et al. (2023). Nanosonodynamic effect-promoted mitochondrial dysfunction augments calcium overload for Gasdermin E-induced pyroptotic antitumor therapy. *Chemical Engineering Journal* 455: 140869.

77 Chen, Z., Liu, W., Yang, Z. et al. (2023). Sonodynamic-immunomodulatory nanostimulators activate pyroptosis and remodel tumor microenvironment for enhanced tumor immunotherapy. *Theranostics* 13: 1571–1583.

78 Zheng, P., Ding, B., Zhu, W. et al. (2022). Biodegradable hydrogen peroxide nanogenerator for controllable cancer immunotherapy via modulating cell death pathway from apoptosis to pyroptosis. *Chemical Engineering Journal* 450: 137967.

79 Yu, S., Chen, Z., Zeng, X. et al. (2019). Advances in nanomedicine for cancer starvation therapy. *Theranostics* 9: 8026–8047.

80 Wang, M., Wang, D.M., Chen, Q. et al. (2019). Recent advances in glucose-based nanocomposites for tumor therapy. *Small* 15: 1903895.

81 Li, J., Anraku, Y., and Kataoka, K. (2020). Self-boosting catalytic nanoreactors integrated with triggerable crosslinking membrane networks for initiation of immunogenic cell death by pyroptosis. *Angewandte Chemie International Edition* 59: 13526–13530.

82 Zhu, G., Zheng, P., Wang, M. et al. (2023). Near-infrared light-triggered thermoresponsive pyroptosis system for synergistic tumor immunotherapy. *Advanced Healthcare Materials* https://doi.org/10.1002/adhm.202302095.

83 Qiao, L.Y., Zhu, G.Q., Jiang, T.F. et al. (2023). Self-destructive copper carriers induce pyroptosis and cuproptosis for efficient tumor immunotherapy against dormant and recurrent tumors. *Advanced Materials* https://doi.org/10.1002/adma.202308241.

12

NIR Light-Activated Conversion Nanomaterials for Photothermal/Immunotherapy

Yaru Zhang and Zhiyao Hou

Guangzhou Medical University, School of Basic Medical Sciences, Xinzao Road, Panyu District, Guangzhou 511436, China

12.1 Introduction

According to the International Agency for Research on Cancer of the World Health Organization, the incidence of cancer may increase to 29.5 million by 2040 [1]. Even today, cancer is one of the leading causes of human deaths worldwide. Tumor metastases are the main reason for nearly 90% of all cancer-related deaths. For many patients, when they were diagnosed with cancer, metastasis had already occurred. It was noticed that over 80% of patients diagnosed with lung cancer present with metastatic disease [2–6]. Although chemotherapy, hormonal therapy, and radiotherapy serve palliative purposes in metastatic cancers and offer a modest extension of survival, the therapeutic effect is still unsatisfactory.

Cancer immunotherapy, which aims to activate the host immune system to fight cancer cells [7–10], has made great progress and has been applied to clinical studies, including monoclonal antibodies [11], dendritic cell (DC)-based vaccine [12, 13], chimeric antigen receptor (CAR) T cells [14], whole-cell vaccine [15, 16] and immune checkpoint inhibitors [17, 18]. However, several recent studies have revealed that these therapeutic treatments exhibited inconsistent therapeutic responses in different patients. In addition, the cascading effects of inflammatory media, organ toxicity, hematopoietic system dysfunction, and other side effects also limit the clinical optimization of these approaches [19–22]. A satisfactory tumor therapy strategy should not only eliminate the primary tumors but also activate the host immune system to eliminate the metastatic and residual tumor cells. Combining local therapy with immunotherapy is a good choice to improve the therapeutic effect while making full use of the benefits of immunotherapy [23–25].

In recent years, photothermal therapy (PTT) has emerged as a promising and minimally invasive treatment strategy for primary tumors. This approach utilizes tumor-targeted photothermal agents (PATs) that convert light energy into heat under near-infrared (NIR) light irradiation, effectively eradicating tumor cells [26, 27]. By local administration of PATs and minimally invasive NIR radiation,

Photofunctional Nanomaterials for Biomedical Applications, First Edition.
Edited by Chunxia Li and Jun Lin.

hyperthermia prompted by PTT can be controlled to minimize the damage to non-targeted tissues. Notably, recent findings have shown that hyperthermia induced by PTT can trigger dying tumor cells to release antigens, proinflammatory cytokines, and immunogenic intracellular substrates, thereby fostering immune activation. The efficacy of PTT relies heavily on the PATs. Primarily, ideal PATs for PTT should have strong optical absorption, especially in the NIR region (700–1300 nm) [28, 29], due to the fact that water and blood cells have little absorption to NIR light, allowing deep tissue penetration of light to stimulate the optical-sensitive nanoparticles and minimize the damage in the surrounding healthy tissue [30]. So far, a wide range of NIR-responsive PATs have been explored, covering noble metal nanostructures [31, 32], carbon-based nanomaterials [33], organic dyes and polymers [34, 35], as well as semiconductor nanomaterials [36, 37].

Researchers have recognized the potential benefits of PTT as a strategic complement to address certain inherent limitations of immunotherapy. First, PTT effectively mitigates tumor burden through direct tumor ablation, which covers the shortage of immunotherapy. Second, PTT plays a pivotal role in enhancing tumor immunogenicity by inducing immunogenic cell death (ICD) and tumor ablation, resulting in the release of damage-associated molecular patterns (DAMPs) and tumor-associated antigens (TAAs) [38–41]. Third, the elevation in temperature induced by PTT positively influences tumor physiology, augmenting tumor blood flow and impacting the distribution and accumulation of small molecules within tumors, such as immune checkpoint inhibitors [42]. Last but not least, immunotherapy makes up for the deficiency of PTT in preventing metastasis and tumor recurrence by promoting adaptive immunity and establishing long-term immune memories [43–45]. Therefore, in this chapter, focusing on the latest developments of inorganic nanomaterials in PTT, we first introduced the light-to-heat conversion mechanism of inorganic nanomaterials. Second, we summarized the application of common inorganic nanomaterials, such as noble metal nanoparticles, semiconductor nanomaterials, and carbon-based materials. In addition, the mechanism of nanomaterial-based PTT against cancer was also expounded. Finally, we reviewed recent advances in nanomaterial-based photothermal immunotherapy and discussed challenges and future outlook.

12.2 The Photothermal Conversion Mechanism

The photothermal conversion occurs when materials undergo lattice vibration or electronic oscillation upon absorbing electromagnetic radiation, leading to an increase in their temperature. The mechanism of photothermal conversion varies among different materials used for this purpose.

Certain noble metal nanoparticles, such as Au NPs and Ag NPs, exhibit a distinctive local surface plasmon resonance (LSPR) effect. In the presence of light, free electrons on the surface of these inorganic nanomaterials are excited, forming an electron cloud that asymmetrically spreads around the nanoparticle [46]. The coulomb restoring force produced by the displacement between the positively

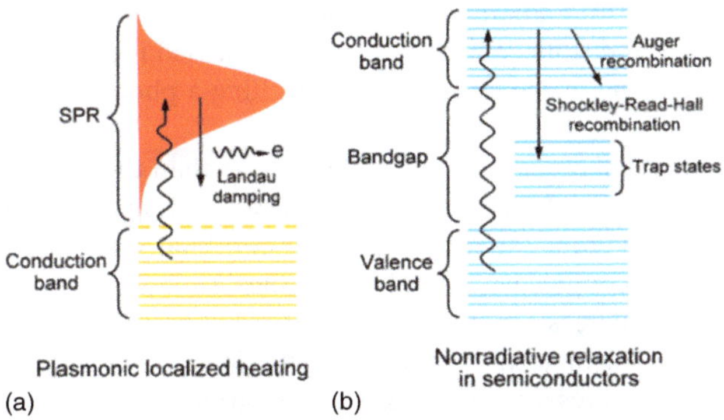

Figure 12.1 Different mechanisms of the photothermal effect. Source: Reproduced with permission from [50], 2018 / The Royal Society of Chemistry.

charged nucleus and the negatively charged electron cloud causes a series of continuous oscillations of the electron cloud on the surface of particles [47]. When the frequency of the light aligns well with the LSPR frequency of plasmonic metal nanoparticles, these nanoparticles absorb a significant amount of light, generating highly amplified and localized electric fields near the nanoparticle surface. Due to this process being resonant at a specific frequency of the light, it is termed LSPR [48, 49]. The surface plasmon oscillation decays by radiating its energy resulting in light scattering or decays nonradiative resulting in the conversion of absorbed light to heat (Figure 12.1a) [51–53]. Numerous researches have demonstrated that variations in shape, dimension, composition, dielectric properties, and external environment significantly impact the photothermal conversion efficiency (PCE) of these metal nanoparticles [54].

Semiconductor photothermal materials exhibit distinct photothermal conversion mechanisms owing to their unique structural characteristics, which can be divided into two categories according to the principle of light absorption: (i) Semiconductors with defect structure: the first category includes semiconductors with a defect structure, such as copper chalcogenide and certain transition metal oxides. The defect structure prompts carrier migration on the surface, leading to free carrier absorption and manifesting the LSPR effect akin to metal nanoparticles. Unlike precious metal nanoparticles, the LSPR level of semiconductors depends on surface carrier concentration, remaining largely unrelated to morphology. Consequently, semiconductors with a defect structure offer enhanced light stability [36, 55, 56]. (ii) Semiconductors with intrinsic band gap absorption: the second category encompasses semiconductors exhibiting intrinsic band gap absorption, including nitrides, carbides, borides, and sulfides of transition metals. Light absorption in these materials hinges on their intrinsic absorption band gap, situated between the valence band and conduction band. The band gap width is contingent on the semiconductor type, as illustrated in Figure 12.1b. Photon energy received by the electron in the valence band induces a direct transfer to the conduction band, generating holes in the valence band and forming electron–hole pairs. Subsequently, the generated energy is transferred to the material's surface holes through either radiative relaxation (photon) or nonradiative relaxation (phonon), and excited electrons then return to the ground state. When the energy is released in the form of phonon, it induces local heating of the crystal lattice, eliciting a photothermal effect [57].

For most carbon-based materials, the π bond is usually weaker than the σ bond due to the weak bonding strength of the electronic bond. The electrons involved in these bonds are easily excited from the π orbital to the π orbital with lower input energy. Consequently, the conjugated π bond can induce a red shift in the absorption spectrum. With an increase in the number of π bonds, the energy gap between the highest occupied molecular orbital and the lowest unoccupied molecular orbital decreases. In the presence of sunlight across various wavelengths, the extensive conjugated π bond in graphene and its derivatives facilitates electronic excitation and sunlight absorption, giving these materials a black appearance. When exposed to light, electronic transitions from low to high orbits may occur within the molecule. The excited electrons initiate polarization relaxation through electron–phonon coupling, transferring the absorbed light energy from the excited electron to the vibration of the entire crystal lattice, thereby causing the temperature of the material to rise (Figure 12.1c) [58].

12.3 Classification of Inorganic Photothermal Materials

12.3.1 Noble Metal Nanomaterials

Metal nanoparticles possess a multitude of appealing characteristics, including strong scattering, light absorption, and the LSPR, which collectively contribute to their outstanding photothermal capabilities. Interestingly, the LSPR attribute of metal nanoparticles can be finely tuned by modifying their nanostructure size, shape, and composition, aligning them with diverse biological applications. Additionally, metal nanoparticles demonstrate commendable photostability, enabling the utilization of higher light excitation energy and facilitating extended monitoring periods [59, 60]. These advantages of metal nanoparticles offer great potential for PTT applications.

Au nanomaterials, as the most representative noble metal nanomaterial, exhibit good biocompatibility and resistance to easy oxidation, enabling their stabilization in vivo. Various Au nanostructures, such as nanorods [61, 62], nanoshells [63, 64], and nanoclusters [65, 66], have been extensively investigated. Generally, Au nanomaterials are excited by the laser in the NIR-I region and exhibit a PCE of around 30% [62–64]. The photothermal conversion ability of metal plasmonics relies on the optical LSPR, a collective oscillation of free electrons in nanoparticles resulting in strong light absorption at specific wavelengths determined by the nanostructure's shape and size [67]. By using this property, Wang et al. demonstrated that the absorption wavelength of Au nanoparticles can be red-shifted to the NIR-II region through the incorporation of cavities into the nanostructure [68]. In a related study,

Chen et al. presented a range of hybrid structures to facilitate the growth of smaller Au nanoparticles on Au nanorods (Au-on-AuNR). This was achieved by finely tuning the interfacial energy and growth kinetics [69]. By adjusting the number and morphology of Au domains on AuNRs, their LSPRs could be effectively and continuously tuned in the NIR-visible range (Figure 12.2a). The multiple islands resulted in increased longitudinal absorption, diffusing across the visible, NIR-I, and even NIR-II window (1060 nm). The calculated PCE for different structures ranged from 26.1% to 67.2% at 1060 nm. The significant improvement in PCE was attributed to plasma coupling within particles of adjacent Au branches. In vitro and in vivo experiments further demonstrated the nanostructures' commendable photothermal performance and photoacoustic imaging capabilities. Although Au NPs have shown promise in clinical applications, challenges remain, including cytotoxicity, low photothermal efficiency, high cost, and time-intensive preparation. Notably, some modified materials like liposomes, copolymers, and peptides are frequently employed to enhance biocompatibility and photothermal efficiency. The resolution of these issues will shape the future role of noble metal nanomaterials as PTAs.

In contrast to Au NPs, Ag NPs exhibit a more pronounced and narrower plasmon resonance performance. The heightened absorption intensity, extinction intensity, and scattering cross-section collectively contribute to the elevated PCE of Ag NPs [71, 72]. Zhang et al. synthesized GO@Ag nanocomposite incorporating doxorubicin (DOX) and DSPE-PEG$_{2000}$-NGR for PTT. The LSPR effect of Ag NPs made this nanocomposite have a strong tumor photothermal ablation effect [73]. However, the chemical stability of Ag NPs is not as robust as that of Au NPs, and Ag NPs can induce significant toxicity in vivo [74, 75]. Therefore, the standalone use of Ag nanoparticles is seldom employed in biosensing or biomedicine. Ag NPs are typically utilized in conjunction with gold nanoparticles. For instance, Wilhelm et al. engineered hybrid nanoparticles by encapsulating Au shells around anisotropic Ag nanocores (Figure 12.2b). In vivo experiments demonstrated that the Au shells could shield the Ag cores from the harsh biological environment, preserving their excellent heating properties [70].

In addition, nanomaterials with multi-noble metal components also have satisfactory photothermal properties, which are widely used in the biological field. Wang et al. synthesized a multifunctional therapeutic nanoplatform Au_2Pt-PEG-Ce6 with a PCE of 31.5% at 808 nm (Figure 12.3a) [76]. Au_2Pt-PEG-Ce6 nanomaterials could be a promising candidate for PTT in the NIR bio-window due to their stable nanostructure and good photothermal capability. Based on this, their team synthesized AuPtAg-GOx nanomaterials, which exhibited mild photothermal properties under 1064 nm irradiation for tumor therapy (Figure 12.3b) [77]. The photothermal conversion efficiencies of AuPtAg at 808 and 1064 nm were calculated to be 47% and 41%, respectively. In conclusion, AuPtAg-GOx can achieve mild PTT augmented by starvation therapy and further promotes immunotherapy effects through mild PTT. Such cascade-promoting synergistic strategies may be generally applicable for highly efficient cancer therapy.

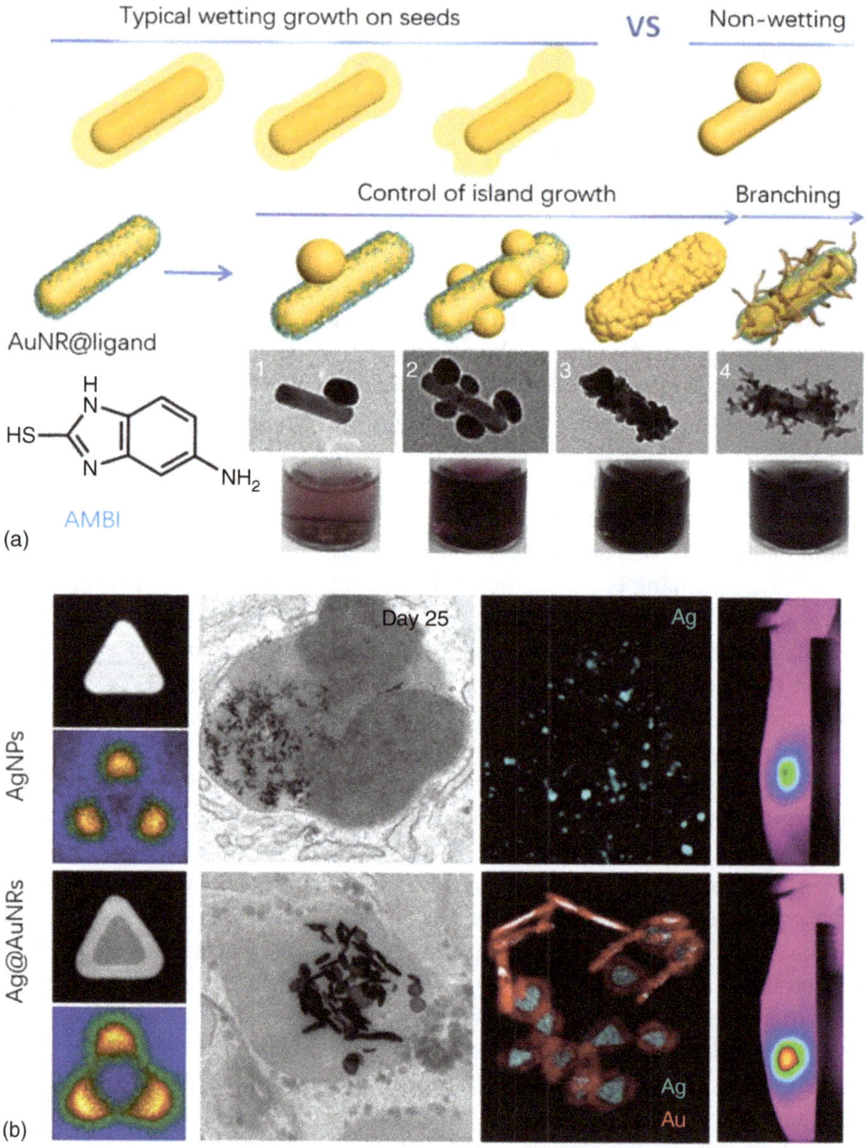

Figure 12.2 (a) Schematic illustration of the typical wetting and nonwetting Au-on-AuNR nanostructures. Source: Reproduced with permission from [69], 2020/ John Wiley & Sons. (b) Schematic illustration of the characterization of Ag NPs and Ag@Au NPs and their comparison of PTT effects. Source: Reprinted with permission from Ref. [70]; © 2018, American Chemical Society.

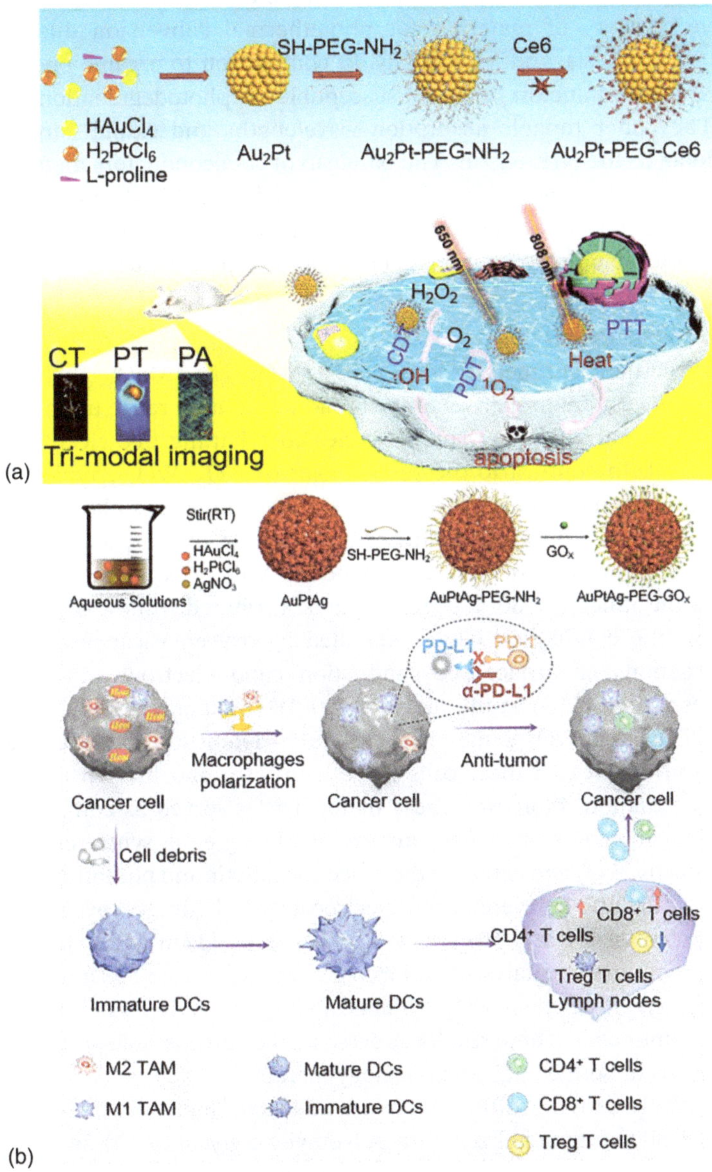

Figure 12.3 (a) Schematic illustrations for the fabrication of Au_2Pt-PEG-Ce6 nanoplatform and multimodal imaging-guided synergistic PTT/PDT/chemodynamic therapy (CDT) by employing tumor microenvironment. Source: Reproduced with permission from [76] 2020, / Elsevier (b) Schematic illustration of the formation of AuPtAg-GOx nanozyme and the mechanism for tumor immunotherapy induced by starvation therapy-augmented mild photothermal therapy. Source: [77] / John Wiley & Sons/ CC-BY4.0.

12.3.2 Semiconductor Nanomaterials

Semiconductor materials, such as metal oxides and chalcogenides, represent another distinctive category of materials for photothermal conversion due to their low cost, facile synthesis, and low toxicity. In comparison to organic photothermal materials, semiconductors are less susceptible to photodegradation or photobleaching. They offer tunable absorption wavelengths and exhibit strong extinction coefficients in the NIR region. The bandgap of semiconductors dictates light absorption based on the generation of free-charge carriers within the material. In this section, we summarize some commonly used inorganic semiconductor nanomaterials for PTT, including transition metal oxides and perovskites.

12.3.2.1 Transition Metal Oxides

There are some oxygen-deficient metal oxides, such as WO_{3-x} and MoO_{3-x}, which exhibit strong photo absorption properties in the broad wavelength range up to the NIR [78]. WO_{3-x}, characterized by oxygen vacancies contributing free electrons, displays a strong LSPR. In contrast to the yellow color of WO_3, WO_{3-x} typically appears blue and can take on various oxygen-deficient stoichiometries like $WO_{2.9}$, $WO_{2.83}$, and $WO_{2.72}$ [78, 79]. Besides the intrinsic absorption of WO_3 at 480 nm (with an indirect band gap of 2.6 eV), oxygen-deficient WO_{3-x} exhibits a broad absorption peak in the range of 480–1800 nm. This is attributed to new discrete energy bands below the conduction band, generated by oxygen vacancies, and the collective oscillations of surface-free conduction band electrons inducing surface plasmon resonance (SPR) in the range of 700–1800 nanometers [80]. Liu et al. developed PEGylated WO_{3-x} nanoparticles [81], demonstrating effective PTT against 4T1 murine breast cancer cells in both in vitro and in vivo under 980 nm NIR laser irradiation. Moreover, these nanoparticles served as computed tomography (CT) imaging contrast agents concurrently. Liang et al. synthesized a range of oxygen-deficient WO_3 and explored their electrocatalytic and photothermal properties [82]. The Vis-NIR light-enhanced electrocatalytic hydrogen evolution reaction (the overpotential [52 mV, 10 mV cm^{-1}], Tafel slope [37 mV dec^{-1}]) was comparable to the noble metal catalysts (Pt, Pt/C). When exposed to 808 nm NIR laser irradiation, the oxygen-deficient WO_3 exhibited an impressive 41.6% inhibition rate against MCF-7 tumor cells. These results underscore the considerable potential of WO_3 for applications in both energy and oncology therapies.

Yang et al. innovatively designed ultrasmall oxygen-deficient molybdenum oxide nanoparticles (MoO_x NPs) functionalized with polyethylene glycol (PEG) for PTT. These MoO_x NPs were incorporated into down-conversion nanoparticles with dendritic mesoporous silica through an electrostatic interaction with additional Nd ion doping (Figure 12.4) [83]. The introduction of oxygen defects reduced Mo^{6+} ions to Mo^{5+} or even lower state, causing a color change from white to blue and a substantial enhancement in NIR absorption. The PCE of this nanosystem reached an impressive 51.5% (at 1064 nm). Moreover, the induced hyperpyrexia substantially enhanced singlet oxygen ($^1O_2$) generation, enabling synergistic tumor treatment through combined PTT and photothermal-enhanced CDT. Tan et al.

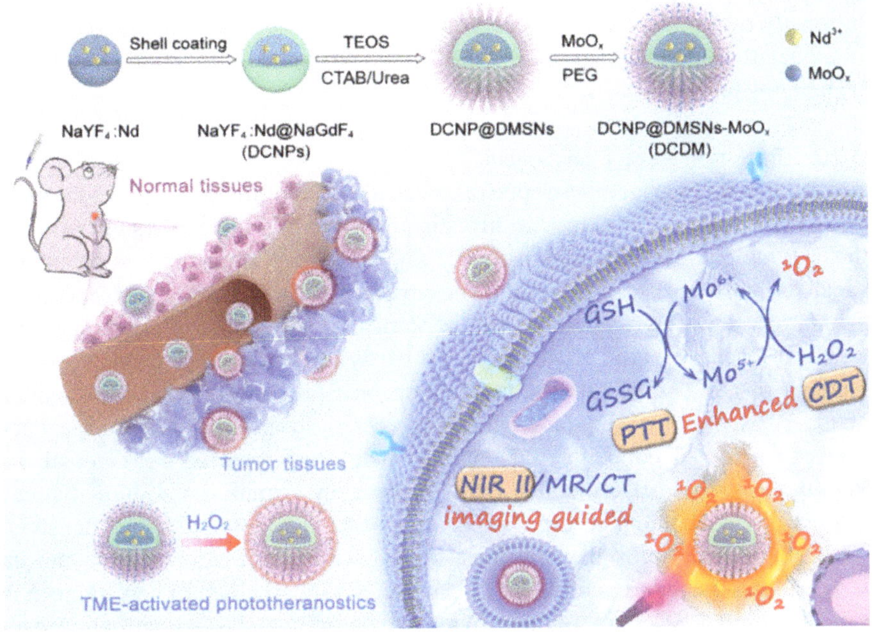

Figure 12.4 Schematic illustration of the preparation process and action mechanisms of the DCDM. Source: Reproduced with permission from [83], 2020 / American Chemical Society.

presented a MoO_{3-x} nanostructure with NIR-II absorption activated for efficient PTT in the NIR-II window (1064 nm) [84]. The MoO_{3-x}-based PTT agents not only exhibited excellent photothermal performance (PCE = 46.9% at 1064 nm) but also demonstrated satisfactory metabolism within the body, signifying their significant potential for clinical applications.

Additionally, certain ultrasmall vanadium oxide nanodots (VO_x NDs) with oxygen deficiency exhibited the ability to encapsulate DOX and convert 785 nm NIR into heat, allowing for a dual approach involving combined chemotherapy and PTT [85].

Among the transition metal oxides, iron oxide nanoparticles possess dual functionalities of NIR absorption and magnetism. Comparative studies between individual magnetic Fe_3O_4 nanoparticles and clustered Fe_3O_4 nanoparticles have demonstrated that the latter induces a higher temperature increase due to its robust absorption in the NIR region [86]. Likewise, Gazeau et al. investigated the photothermal properties and the dispersion of Fe_3O_4 with various configurations, such as isolated Fe_3O_4, colloidosome, and supraball [87]. In vivo experiments revealed that Fe_3O_4 was predominantly distributed in the extracellular matrix (ECM) and macrophages, while supraballs were partially located in the ECM and tumor cells. Despite the fact that the enhancement in PTT biological effects was not significant, the dispersed Fe_3O_4 exhibited a higher overall heating capacity than colloidosomes. Thus, by tuning the self-assembly process of the nanocrystals, researchers can control the distribution of inorganic nanoparticles in tissues and cells so that NIR light can more accurately radiate various biological responses. Additionally, owing to its unique superparamagnetic properties, Fe_3O_4 is frequently combined with other photosensitizers in PTT for simultaneous magnetic resonance (MR) imaging [88, 89].

12.3.2.2 Transition Metal Chalcogenides

Transition metal chalcogenides represent a crucial class of inorganic photothermal materials currently under extensive investigation owing to their robust absorption in the NIR region, favorable photostability, and additional advantages, such as low cytotoxicity, cost-effectiveness, and widespread availability. Particularly, copper sulfide (CuS), a widely recognized p-type semiconductor material, demonstrates inherent absorption in the NIR region, primarily attributed to d–d energy band transitions of Cu^{2+} ions, rather than relying on SPR. The pioneering research on copper-based chalcogenide compounds employed as photothermal agents (thioglycolate-coated copper sulfide nanoparticle) was reported by Li et al. in 2000 [90]. The nanoparticles, which are approximately 3 nm in size, exhibit distinct absorption in the NIR region. Upon exposure to laser irradiation at power densities of 24 and 64 W cm^{-2} for five minutes, approximately 50% and 88% of cancer cells were eradicated, respectively. Crucially, neither laser irradiation alone nor the mere presence of copper sulfide nanocrystals without laser treatment led to cell death, underscoring the specific cancer cell-killing capability arising from the laser-induced photothermal effect. To enhance the PCE of CuS, Tian et al. introduced a novel 3D flower-like CuS superstructure photothermal conversion material using a hydrothermal method [37]. These CuS superstructures could rapidly convert 980 nm laser energy into thermal energy, leading to the temperature of the aqueous CuS dispersion to increase in a concentration- and irradiation time-dependent manner. The study revealed that the faceted end planes of the crystalline CuS superstructures serve as laser-cavity mirrors for the 980 nm laser, enhancing reflection, absorption, and photothermal conversion. Besides the morphology, self-doped copper sulfide can also affect its absorption properties, which then affect its photothermal efficiency.

Tian et al. innovatively synthesized a hole-doped Cu_9S_5 photothermal conversion material [56], with an impressive extinction coefficient of 1.2×10^9 M^{-1} cm^{-1} at 980 nm. Remarkably, the PCE reached 25.7%, surpassing that of gold nanorods (22%) under comparable conditions. Li et al. developed another hole-doped copper sulfide, i.e. $Cu_{7.2}S_4$ nanocrystals [91]. Through self-doping, the plasmon resonance wavelength of these nanocrystals was tuned to match the illumination laser wavelength, resulting in an outstanding PCE of 56.7%. Furthermore, studies indicated that compounding copper sulfide with metals could enhance the absorption of the composite material, consequently boosting its photothermal efficiency.

Selenium, tellurium, and sulfur belong to the same chemical group and share similar properties. Copper selenide, copper telluride, and copper sulfide are all categorized as p-type semiconductors. Due to the SPR effect, copper selenide and copper telluride exhibit strong absorption in the NIR region, making them suitable for photothermal cancer treatment. Hessel et al. successfully synthesized oleylamine-coated $Cu_{2-x}Se$ nanocrystals through high-temperature liquid-phase synthesis [36]. To make these nanocrystals suitable for biological applications, amphoteric polymers were employed for hydrophilic modification. The polymer-coated $Cu_{2-x}Se$ with strong LSPR in the NIR range (600–1100 nm) and exhibited a good photothermal-induced cell-killing ability. Wen et al. successfully synthesized $Cu_{2-x}Te$ NPs through the reaction of copper(I) chloride with tri-octylphosphine telluride (TOPTe) [92]. The resulting $Cu_{2-x}Te$ NPs are hydrophobic, featuring oleic acid as surface ligands. These $Cu_{2-x}Te$ NPs exhibit dual enzyme-like activities, including glutathione oxidase (GSHOx) and peroxidase (POD). Under NIR-II light irradiation, the enzymatic activity of $Cu_{2-x}Te$ NPs is significantly enhanced, leading to the eradication of primary tumors. Remarkably, the oxidative stress induced by $Cu_{2-x}Te$ NPs contributes to the reversal of the immunosuppressive tumor microenvironment.

In addition to the copper chalcogenide nanocrystals, an emerging and noteworthy category is the 2D transition metal dichalcogenides (TMDs), which have garnered considerable interest in the last decade. These materials, akin to graphene-like structures, encompass over 40 nanocompounds with the MX_2 stoichiometry [93]. Exhibiting a layered structure, each layer is held together by strong covalent bonds, while the van der Waals forces between distinct sheets are relatively weak. Relying on the strong NIR absorbance, low toxicity, excellent dispersibility, high PCE as well as good photothermal stability, single-layer transition metal chalcogenides have been shown to be promising candidates as efficient PTAs [94]. MoS_2 nanosheets, as representative TMDs, are frequently employed in PTT. Photons with energy greater than the bandgap can be easily absorbed by monolayer MoS_2 to generate heat because of the existence of a direct bandgap. Typically, Chou et al. first prepared chemically exfoliated MoS_2 nanosheets as a new NIR PTA, and the results demonstrated that the as-made MoS_2 nanosheets had an excellent ability to cause the photothermal-triggered death of cancer cells in vitro [95]. Tan et al. synthesized ultrasmall single-layer MoS_2 nanodots with different phases (1T and 2H phase) and explored their phase-dependent performances as nanoagents for photoacoustic imaging (PAI)-guided PTT in the NIR-II window [96]. $MoSe_2$, as another representative 2D transition metal dichalcogenide, possesses a narrow bandgap (1.3–1.7 eV). Based on their high photothermal conversion capacity, $MoSe_2$ nanomaterials are explored as promising PTT agents with excellent biocompatibility [97, 98]. In the study by Qu et al., $MoSe_2/CoSe_2$@PEG nanosheets (PCE = 62.5%) were synthesized, exhibiting enhanced NIR absorption compared to $MoSe_2$@PEG (PCE = 58.3%) due to Co doping (Figure 12.5) [99]. Importantly, the presence of Co and O_2 induced the oxidation of Mo^{4+} to form soluble molybdate, allowing most $MoSe_2/CoSe_2$@PEG to be metabolized via urine and feces within two weeks, demonstrating outstanding biodegradability. With $MoSe_2$-based POD-mimic and catalase (CAT) activities, this system holds great potential for synergistic PTT/CDT. Additionally, Hu et al. synthesized Bi_2S_3 nanoflowers with a unique structure resembling 3D crumpled-paper-like nanosheets, significantly enhancing reflectivity, NIR absorption, and PCE (64.3%). In vivo experiments demonstrated that Bi_2S_3 nanoflowers effectively ablated cancer cells at low concentrations and

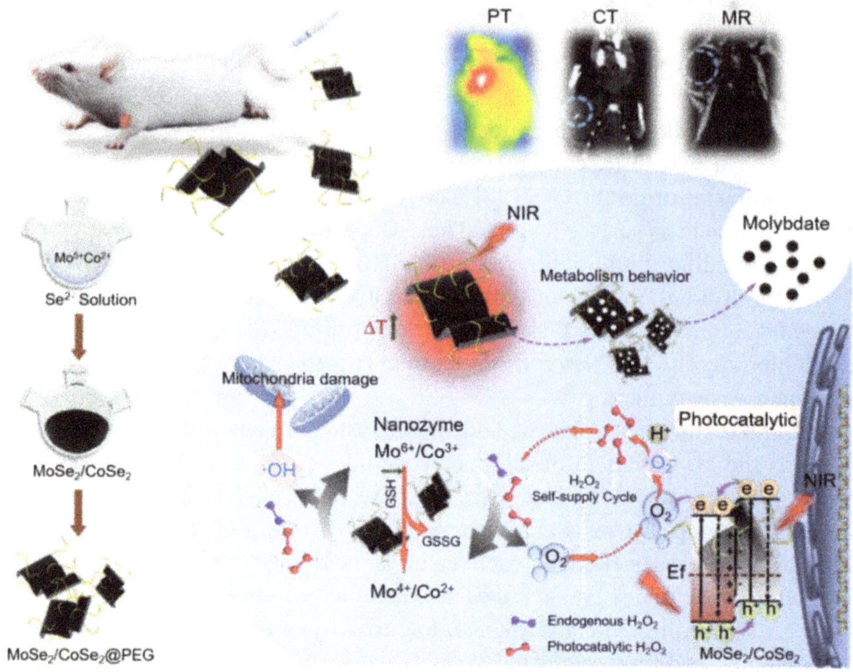

Figure 12.5 Schematic illustration of preparation of $MoSe_2/CoSe_2$ nanosheets and mechanism for synergistic therapy and imaging. Source: Reproduced with permission from [99] 2020, / John Wiley & Sons.

low laser power density [100]. Liu et al. developed a PTT agent using WS_2-PEG nanosheets, not only achieving efficient in vivo photothermal tumor ablation but also integrating X-ray CT and PAI [101]. In a subsequent study, Gd^{3+} doping into WS_2-PEG further enhanced the capability of MR imaging [102].

12.3.3 Carbon-Based Materials

Carbon-based nanomaterials have emerged as attractive PTAs with unique chemical and physical properties, including low toxicity, good stability, straightforward surface modification, low cost for large-scale production, broad optical absorptions, and reasonable photothermal properties. Graphene-based nanostructures, including graphene, graphene oxide (GO), carbon dots, reduced graphene oxide (rGO), carbon nanotubes (CNTs), and porous carbon NPs, have garnered extensive attention as PTAs in the realm of cancer therapy [103–106].

Graphene, a kind of single-layer sheet with a sp^2-hybridized carbon structure, has garnered significant attention since its discovery by Geim and Novoselov in 2004 [107, 108]. Beyond its outstanding physical properties, such as robust electrical conductivity, high thermal conductivity, and a large surface area [109], graphene and its derivatives boast desirable hydrophilicity, dispersity, and biocompatibility, broadening their applications in the biomedical field [110–113].

Due to the poor aqueous solubility of pristine graphene, graphene derivatives such as GO and rGO have been synthesized. GO is composed of single-atom carbon sheets with distinct functional groups in various locations, particularly carboxylates in the boundary area and hydroxyl and epoxy groups on the bottom surface. These groups serve as sites for subsequent surface modification. The first report in which PEGylated GO was shown to exhibit excellent biocompatibility and PCE in vivo was from Liu's group [33]. Subsequent studies have explored various GO-based PTAs. For example, Yang et al. took advantage of low toxicity and high surface area of GO to synthesize a GO/$BaHoF_5$/PEG nanocomposite for effective tumor PTT [114]. Chen et al. used polyacrylic acid-g-polylactic acid (PAA-g-PLLA) to modify GO to build a tumor microenvironment-responsive nanosystem for accurate PTT (Figure 12.6a) [115]. In vitro experiments revealed that this GO-based scaffold induced apoptosis in cancerous cells (with a death rate exceeding 95%) while simultaneously promoting the growth of adipose-derived stem cells. Reduced graphene oxide (rGO), a widely recognized derivative of graphene, can be produced through various methods, including chemical reduction [117], microwave reduction [118], thermal reduction [119], photocatalyst reduction [120], and solvothermal reduction [121]. Following the reduction process, the resulting rGO lacks oxygen-functional groups and can restore an aromatic lattice, thereby directly enhancing its NIR absorption properties. As a typical example, Robinson et al. developed biocompatible rGO sheets of approximately 20 nm through noncovalent functionalization with amphiphilic PEGylated polymer chains, resulting in a six-fold increase in NIR absorbance compared to GO [122].

Carbon nanotubes are composed of sp^2 carbon and are typical one-dimensional nanomaterials. They can be categorized into two types: single-walled (SWCNTs) and multi-walled (MWCNTs). SWCNTs exhibit a tubular structure rolled from a graphene sheet, whereas MWCNTs consist of a stack of graphene sheets rolled up into concentric cylinders [123, 124]. These distinct types possess varied properties, leading to specific photothermal effects. To address the potential toxicity and hydrophobic nature of bare CNTs, an effective approach involves chemical modification or functionalization to enhance aqueous solubility, particularly for biomedical applications. Shen et al. fabricated a PEG-2000@Fe_3O_4@carbon quantum dots@SWNT nanocomposite for PTT, PDT, chemotherapy, fluorescence imaging, and MR imaging [116]. Gu et al. constructed an artificial tobacco mosaic virus by coupling RGD peptide and capsid subunit-mimetic dendrons on the surface of SWNTs (Figure 12.6b) and realized precise chemotherapy and PTT in the tumor site [125]. Benefiting from the cylinder nested structure and higher surface average electron content, MWCNTs exhibit superior thermal performance compared to SWCNTs. Cai et al. synthesized a PEG-MnO-MWNT nanosystem, realizing dual-modality mapping-guided PTT in metastatic lymph nodes [126]. This PTA holds promising potential for treating metastatic lymph nodes in future clinical applications. Nevertheless, the limited water dispersion constrains their application in PTT. Therefore, surface modification with suitable materials is crucial when employing carbon-based nanosheets as PTT agents.

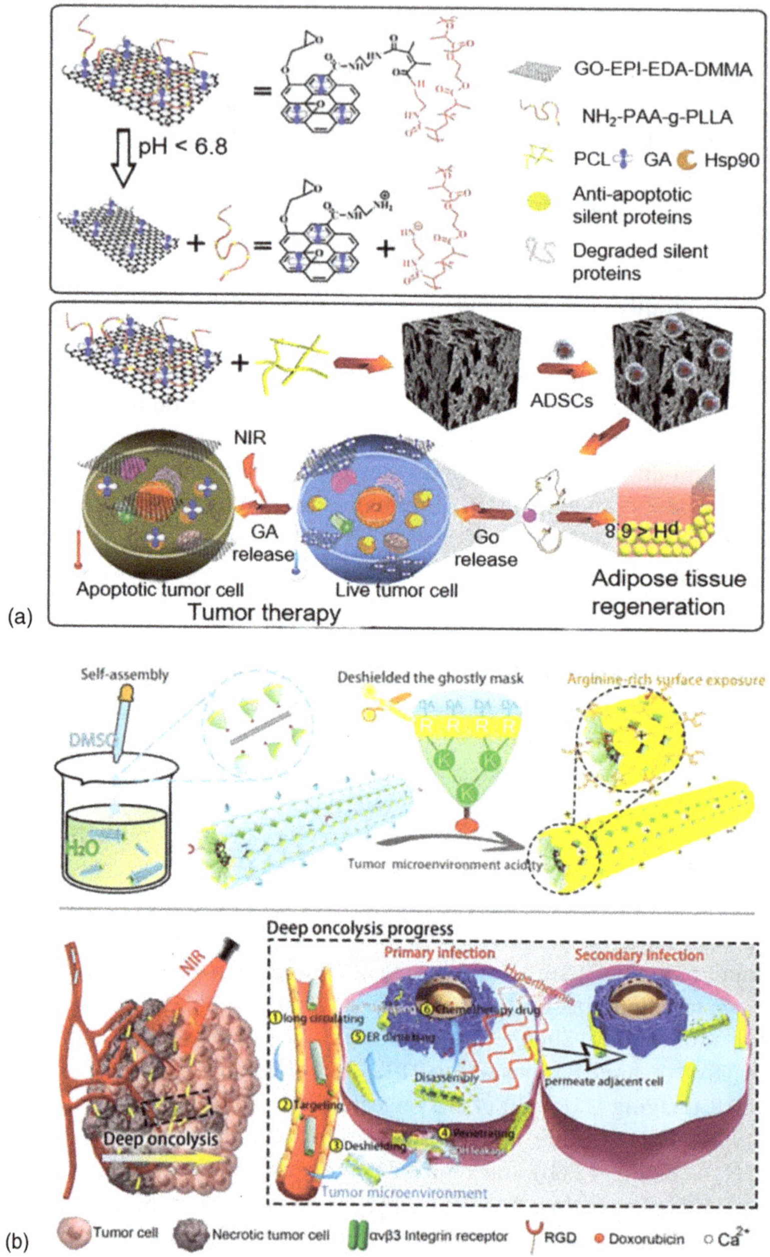

Figure 12.6 (a) Schematic illustration of synthesis and therapeutic procedure of the ADSC-loaded GO/GA-polymer scaffold. Source: Reproduced with permission from [115] 2019, / John Wiley & Sons. (b) Schematic illustration of synthesis and multidrug-resistant cancer therapy of ATMVs. Source: Reproduced with permission from [116] 2020, / John Wiley & Sons.

12.3.4 Other Types of PTAs

In addition to the aforementioned four major classes of inorganic PTAs, there are alternative light absorbers that exhibit efficient anticancer properties for PTT. Lanthanide ions, in particular, have been frequently employed in the synthesis of hybrid PTAs due to their unique transitions from intra-4f or 4f–5d. This process results in the generation of higher-energy photons [127]. A typical example is described in the work of Carrasco et al., where they established that Nd^{3+}-doped LaF_3 nanoparticles functioned as PTAs under 808 nm light irradiation. This effectiveness was attributed to the pronounced light absorption of Nd^{3+} ions [128]. Such a photothermal property is commonly observed in other PTAs based on Nd^{3+} ions [128–131]. In addition, various quantum dots (QDs) have been applied in PTT. For instance, Shao's group synthesized 3.8 nm Bi_2O_2Se QDs using a straightforward top-down method, displaying a high PCE (35.7%) and excellent photothermal stability. These QDs exhibited remarkable PTT efficiency both in vitro and in vivo [132]. Additionally, PbS/CdS/ZnS QDs can also serve as PTAs [133].

Apart from the aforementioned materials, another noteworthy category is black phosphorus (BP). One phosphorus atom can connect three phosphorus atoms through covalent bonds, and single-layers of BP are connected with each other by van der Waals forces [134]. The adjustable bandgap of BP layers caused strong absorption in the ultraviolet and NIR windows [135]. Moreover, BP displays moderate carrier mobility, excellent biocompatibility, and impressive biodegradability in vivo, making it suitable for various biomedical applications, including PTT, PDT, and drug delivery [136, 137]. Zhang et al. created a BP hydrogel loaded with emetine, a stress granule inhibitor, for precise tumor delivery with NIR light-controlled drug release [138]. This team further developed a minimally invasive therapeutic intravenous catheter, which was modified with anti-EpCAM antibody on the surface and internally filled with BP nanosheets. This therapeutic catheter demonstrated the capability to capture 2.1% of the total circulating tumor cells (CTCs) within five minutes and achieved 100% kill efficiency for CTCs through BP nanosheet-mediated PTT [139]. Importantly, BP nanosheets appear to undergo degradation into harmless phosphate in vivo [140]. Furthermore, these inorganic PTAs may serve as starting points for designing novel nanoplatforms with appealing functionalities [141, 142].

12.4 Mechanisms of PTT and Immunotherapy

12.4.1 Mechanism of PTT

PTT utilizes nanomaterials with NIR-absorbing properties to convert light energy into thermal energy, inducing tumor cell death and subsequent ablation. Specifically, PTAs, also known as photosensitizers, are localized within the tumor site and subsequently exposed to NIR laser irradiation within the range of 800–1350 nm. After that, these photosensitizers generate substantial heat, leading to tumor cell death without causing side effects to normal cells. Thermal-induced cell death can be classified as apoptosis, necrosis, and, importantly, immunogenic cell death (ICD). In addition to directly eliminating tumor cells, PTT also facilitates other therapies, such as immunotherapy and chemotherapy, to achieve a synergistic anticancer effect [143, 144]. The response of tumor cells to elevated temperatures is the underlying mechanism for the efficacy of PTT.

12.4.2 Response of Tumor Cells to Heat Stress

Temperature, a pivotal parameter in biological systems, governs the viability of cells, organs, and organisms. Sustained exposure to direct heat exceeding 42 °C for one hour can cause irreversible cellular damage, including cell membrane cleavage and cytoskeleton destruction [145]. Temperatures surpassing 48 °C for a few minutes lead to protein denaturation and nucleic acid damage [146, 147]. When tissue temperatures exceed 60 °C, as achieved in PTT, instantaneous cell death is induced through protein denaturation and plasma membrane destruction [148]. The unique sensitivity of tumor cells to heat forms the foundation of PTT. Tumor cells are more sensitive to heat compared to normal cells, owing to the mild acidic and hypoxic tumor microenvironment [149, 150]. Additionally, heat-shock proteins (HSPs) produced by tumor cells during heat stress can form complexes with tumor-specific peptides, potentially stimulating the body to generate tumor-specific cytotoxic T cells. Notably, HSPs produced by normal cells under heat stress lack this immunostimulatory function [149, 151]. Another study revealed that heat-induced apoptotic leukemia cells significantly increase the expression of costimulatory molecules (e.g. CD80, CD86, and CD40) on the surface of dendritic cells (DCs), triggering interleukin (IL)-12 secretion and the activation of adaptive immunity [152]. The tumor-specific immune response provoked by heat stress makes PTT an optimal complement to immunotherapy.

12.4.3 PTT-Induced Necrosis and Apoptosis

Possible modes of direct cell death induced by PTT include necrosis and apoptosis [153]. The main difference between apoptosis and necrosis is the integrity of the plasma membrane. Necrosis, characterized by the loss of cytoplasmic membrane integrity and the release of cellular contents, such as DAMPs, triggers abnormal inflammatory and immunogenic responses [154]. In contrast, apoptosis utilizes external cell membrane markers like phosphatidylserine to generate an “eat me” signal, initiating phagocytosis without eliciting inflammation [155]. The induction of these two death mechanisms depends on the intensity and duration of NIR irradiation. High-energy irradiation (e.g. 30 W cm^{-2}, two minutes) leads to necrosis, while low-energy irradiation (e.g. 5 W cm^{-2}, 10 minutes) promotes apoptosis. This explains why necrosis is the predominant PTT-induced cell death mode in in vitro cell experiments [153]. Apoptosis, mediated by PTT through the intrinsic mitochondrial pathway, is a more efficient and “cleaner” mode of tumor cell death compared to necrosis [156]. Researchers had managed to modulate light intensity and duration to prioritize apoptosis as the predominant cell death mode in

PTT, mitigating potential abnormal side effects and enhancing treatment success. Despite this, the distinct advantage of inducing instantaneous cell death through physical membrane disruption, unaffected by drug resistance, remains noteworthy. In subsequent preclinical studies, the efficacy and safety should be weighed, and appropriate NIR irradiation conditions should be selected.

12.4.4 PTT-Induced Immunogenic Cell Death

ICD represents a distinct form of apoptotic demise that would be noticed by the immune system [41]. It is characterized by the release of DAMPs, which are not detectable in living cells but show immunostimulatory effects upon exposure to dying cells [40, 157]. Most DAMPs have the capacity to engage pattern recognition receptors (PRRs), which are expressed in immune cells, leading to potent immunostimulatory effects [158]. PTT-generated DAMPs, such as high mobility group B1 (HMGB1), cell surface-exposed calreticulin (CRT), extracellular adenosine triphosphate (ATP), heat-shock protein (e.g. HSP90 and HSP70), are essential endogenous molecules that maintain cellular integrity and physiological function under normal conditions [159]. When cells undergo heat-induced damage, HMGB1 and ATP are released from cells, acting as "find me" signals to attract antigen-presenting cells (APCs). Additionally, CRT is displayed on the cell membrane surface, serving as an "eat me" signal to enhance phagocytosis by APCs. HSPs can form complexes with tumor antigens, effectively activating APCs. Subsequently, APCs migrate to lymphatic organs, where they present tumor antigens to T cells. Activated APCs (mainly DCs) can then prime $CD4^+$ and $CD8^+$ T cells and thereby trigger immunogenic T helper 1 (Th1) cell and cytotoxic T lymphocyte (CTL) responses, respectively (Figure 12.7) [160].

It is crucial to recognize that PTT-induced ICD requires an appropriate thermal dose, and the principle of "more is better" does not universally apply. Sweeney et al. investigated the thermal "window" of ICD using Prussian blue nanoparticle (PBNP) for PTT in a neuroblastoma animal model [161]. The criteria for identifying

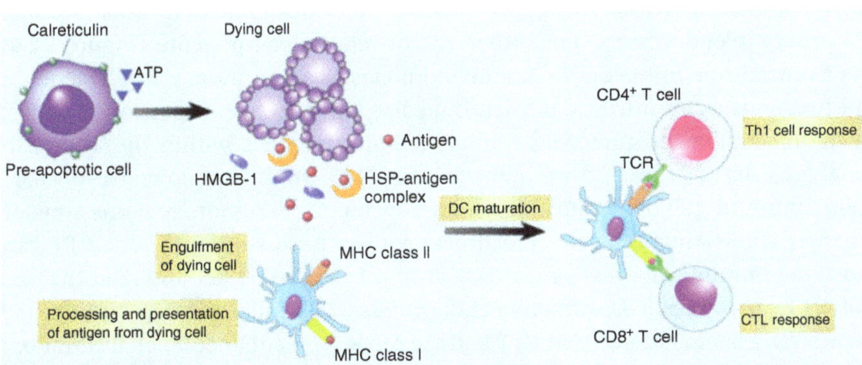

Figure 12.7 Characteristics of ICD. Source: Reproduced with permission from [160] 2019, / John Wiley & Sons.

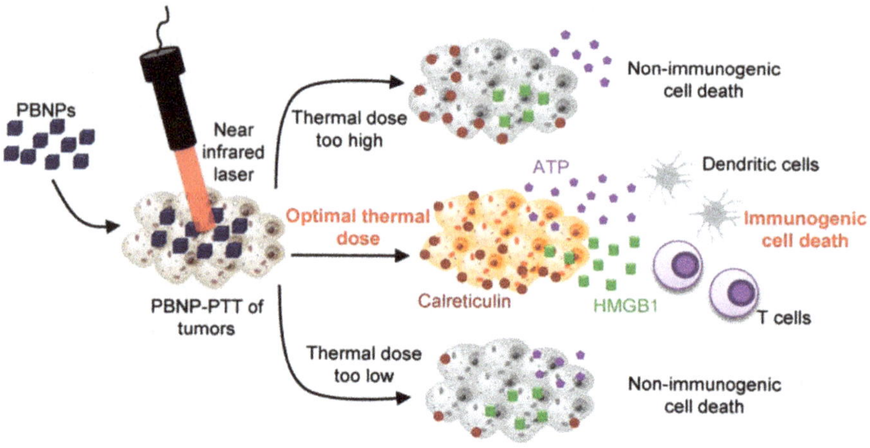

Figure 12.8 PTT generates ICD when an optimal thermal dose is administered to tumors. Source: Reproduced with permission from [161] 2018, / John Wiley & Sons.

cells undergoing ICD involve a decrease in intracellular ATP and HMGB1 levels, coupled with an increase in surface-expressed CRT. Their findings indicated that ICD markers were significantly expressed within an optimal temperature range (63.3–66.4 °C), while higher temperatures (e.g. 83.0–83.5 °C) or lower temperatures (e.g. 50.7–52.7 °C) resulted in lower expression (Figure 12.8). Subsequent vaccination studies in neuroblastoma models validated these observations. The flexibility observed in inducing ICD in this research may have wider relevance across various PTT strategies based on nanomaterials. It suggests that the "more is better" approach may not be suitable for these therapies, emphasizing the existence of an optimal treatment window.

12.4.5 The Impact of PTT on Tumor Microenvironment

The tumor microenvironment (TME), ECM, immunosuppressive cells, stromal fluid, tumor blood vessels, and other tumor-related components (Figure 12.9) plays a crucial role in the development and metastasis of tumors [162]. The dense ECM functions as an intrinsic barrier, impeding the effective dispersion of drugs, and the infiltration of numerous immunosuppressive cells within the TME can strongly restrict the attack from immune cells [163, 164]. How to convert "cold" tumors (immune cell depletion) into immunogenic "hot" tumors remains a major challenge. Interestingly, several preclinical studies have suggested that PTT can reverse the immunosuppressive characteristics of the TME [165–169]. On the one hand, PTT can induce ICD, attracting endogenous CTLs and activating the immune system. Wang demonstrated that PTT had the capacity to enhance T cell infiltration, reduce regulatory T cell (Treg) numbers, and potentiate checkpoint blockade therapy [170]. In another study, Sun highlighted that mild PTT (below 45 °C) not only sensitized immune cold tumors to enhanced anti-PD1 immunotherapy

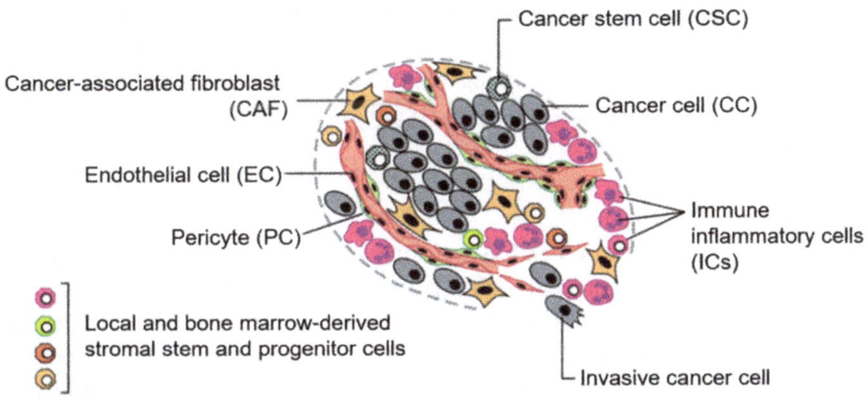

Figure 12.9 The Cells in the TME. Source: Reproduced with permission from [162] 2011, / Elsevier.

but also stimulated a systemic immune response, effectively suppressing tumor metastasis and reconstructing the cold TME in mouse models (4T1 and B16F10) [171]. On the other hand, photothermia increases blood perfusion, facilitating the infiltration of exogenous immune cells and enhancing the retention of nanodrugs at the tumor site [42, 172–174]. Numerous preclinical studies have demonstrated that this feature significantly enhances the efficacy of adoptive cell therapy [173–176]. Despite the promising potential of PTT on the TME, the inflammatory responses induced by PTT in tumors are seldom addressed, potentially leading to severe side effects and attenuating the antitumor immune responses [177, 178].

12.5 Nanomaterial-Based Photothermal/Immunotherapy

12.5.1 PTT-Synergized ICB Therapy

Immune checkpoints are a group of protein molecules that generate either costimulatory or inhibitory signals in the immune response. Their normal function is to prevent autoimmune reactions against healthy tissues [179]. However, during malignant progression, cancer cells can exploit this mechanism to evade recognition and elimination by the immune system, inducing T-cell anergy and establishing an immunosuppressive TME. Immune checkpoint blocking (ICB) therapy, which blocks the inhibitory immune signaling pathway with antibody-mediated immune checkpoint inhibitors, can impede tumor immune escape and activate the immune system to attack cancer cells. Recently, several monoclonal antibodies, such as cytotoxic T lymphocyte antigen 4 (CTLA 4), programmed cell death protein 1/programmed death-ligand 1 (PD-1/PD-L1), have been extensively studied in clinical settings and achieved clinical success in treating certain types of cancer, including melanoma, hepatoma, colorectal cancer, renal cell carcinoma, bladder urothelial carcinoma, nonsmall-cell lung cancer, and Hodgkin lymphoma [180–183].

However, ICB therapy still has some limitations, such as relatively low overall objective response rates and elevated risk of serious autoimmune adverse effects [184, 185]. The combination of immunogenic PTT and ICB therapy has been widely reported and shown to have great potential for enhancing therapeutic effects against noninflamed tumors.

12.5.1.1 CTLA-4 Checkpoint

In the human immune system, T cells play a crucial role, and their activation is finely regulated by two distinct signals: the stimulatory signal mediated by the cell surface molecule CD28 and the inhibitory signal facilitated by CTLA-4 [186]. Acting as a signal blocker in lymph nodes, CTLA-4 impedes the early activation of both naive and memory T cells by binding to CD80/86, thereby hindering the functionality of host immunity [187]. Overexpression of CTLA-4 is a primary reason why tumor cells manage to evade immune surveillance [188–190]. The first CTLA-4 inhibitor, ipilimumab, received approval from the US Food and Drug Administration (FDA) in 2011 for the treatment of melanoma [191]. Wang et al. took the first step to combine SWNTs-based PTT with anti-CTLA-4(αCTLA-4) antibody [144]. As illustrated in Figure 12.10, the thermal ablation induced by SWNTs led to the release of TAAs and danger signals, promoting DC maturation. Subsequently, the antigens were presented, initiating the recruitment of tumor-specific CD8$^+$ T cells. Simultaneously, aCTLA-4 effectively attenuated the immunosuppressive activity of regulatory T cells, increasing the CD4$^+$ and CD8$^+$ T cell to regulatory T cell ratios. Consequently, in synergy with aCTLA-4 therapy, SWNT-based PTT exhibited inhibitory effects on cancer cell growth in both subcutaneous tumor and lung metastasis models.

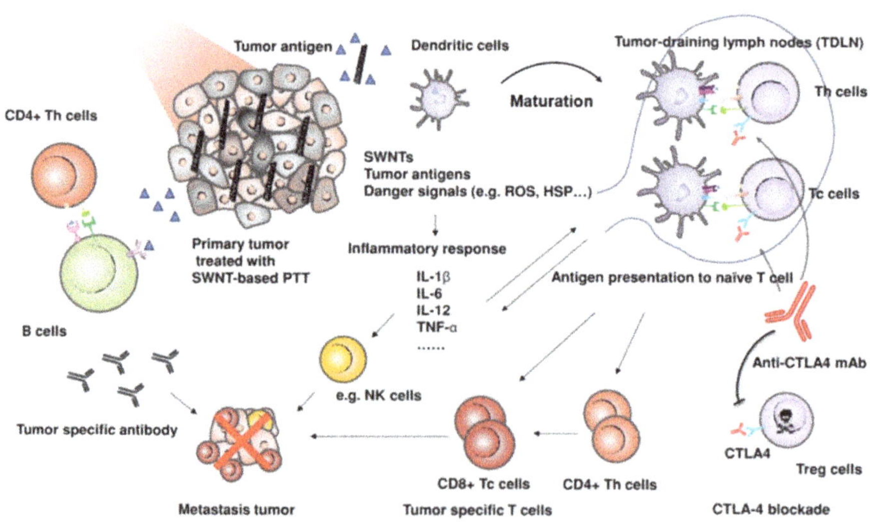

Figure 12.10 The hypothesized mechanism of antitumor immune responses induced by SWNT-based PTT in combination with the anti-CTLA-4 therapy. Source: Reproduced with permission from [144] 2014, / John Wiley & Sons.

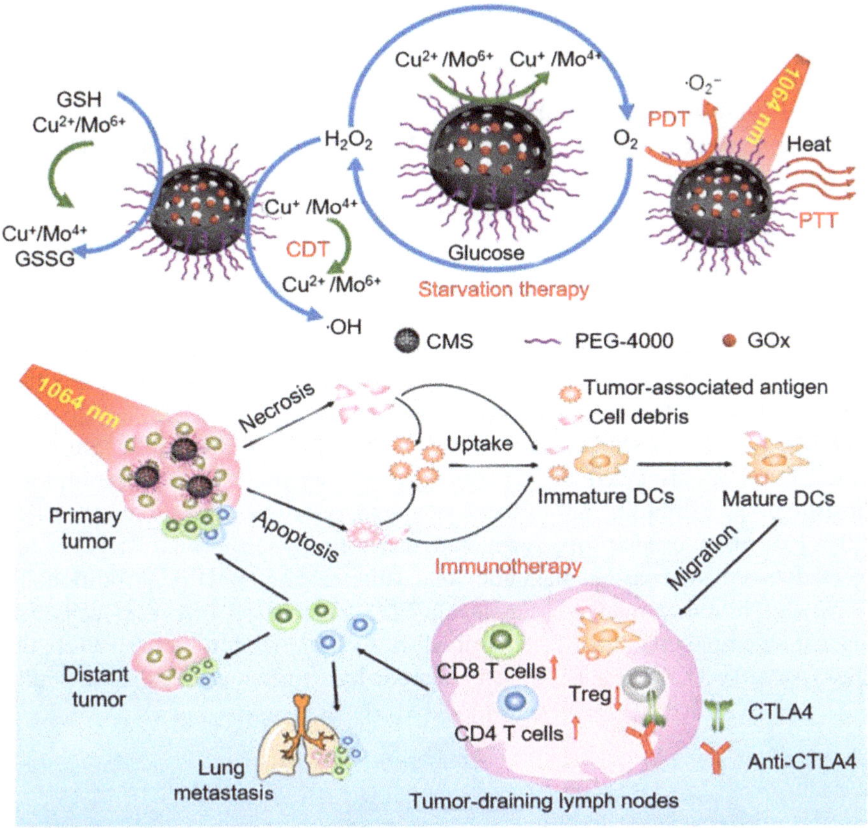

Figure 12.11 Schematic illustration of fabrication and mechanism of PEGylated CMS@GOx for chemodynamic therapy/starvation therapy/phototherapy/immunotherapy. Source: Reproduced with permission from [192] 2019, / John Wiley & Sons.

Additionally, Chang et al. devised a multifunctional cascade bioreactor based on PEGylated hollow mesoporous Cu_2MoS_4 (CMS) loaded with glucose oxidase (GOx) to combine CDT/starvation therapy/phototherapy/immunotherapy for the treatment of metastatic mammary tumors (Figure 12.11) [192]. CMS harboring multivalent elements ($Cu^{1+/2+}$, $Mo^{4+/6+}$) show excellent ·OH generation ability by Fenton-like reaction to realize CDT, as well as consume overexpressed glutathione (GSH) in TME for alleviating antioxidant capability of the tumor by glutathione peroxidase-like activity. In addition, CMS with catalase-like activity could react with endogenous H_2O_2 to generate O_2 for accelerating glucose depletion to obtain starvation therapy. After 1064 nm laser irradiation, CMS shows a significant phototherapeutic effect (PCE 63.3%) and $\cdot O_2^-$-generation capability. When coupled with aCTLA-4, the CMS@GOx-based synergistic therapy not only effectively eliminates primary tumors but also exerts inhibitory effects on lung metastasis by activating the body's antitumor immunity. These findings demonstrate that the combination of PTT and immune checkpoint inhibitors can effectively suppress both primary tumors and metastatic tumors and offer a novel treatment approach for metastatic cancers.

12.5.1.2 PD-1/PD-L1 Checkpoint

The interaction between programmed cell death protein-1 (PD-1) on T cells and programmed death-ligand 1 (PD-L1) hinders T cell proliferation and triggers the activation of T cell apoptosis [193]. Many inhibitors targeting the PD-1/PD-L1 pathway, such as pembrolizumab, nivolumab, atezolizumab, avelumab, durvalumab, and cemiplimab have been approved by the FDA. Moreover, numerous other studies investigating PD-1/PD-L1 are currently in clinical trials. However, the efficacy of most PD-1 inhibitors is limited by low immune response rates, with only 12.5% of eligible patients exhibiting a response to PD-1 inhibition monotherapy, as reported in 2018 [194]. The potential of PTT to enhance immune responses may lead to more favorable outcomes in studies involving PD-1 inhibitors. For example, Shi et al. synthesized monodispersed superparamagnetic core-shell $CoFe_2O_4@MnFe_2O_4$ nanostructures, exhibiting an optimal magnetocrystalline anisotropy (K value of $1.5 \times 10^4\,J\,m^{-3}$), crucial for exceptional magnetic hyperthermia (MHT) when subjected to an alternating magnetic field (Figure 12.12a) [195]. Additionally, the negative immune regulatory molecule PD-L1 expressed in cancer cells was mitigated by employing anti-PD-L1 antibodies. As illustrated in Figure 12.12b, the $CoFe_2O_4@MnFe_2O_4$ exhibited a robust magnetic hyperthermia effect in situ within

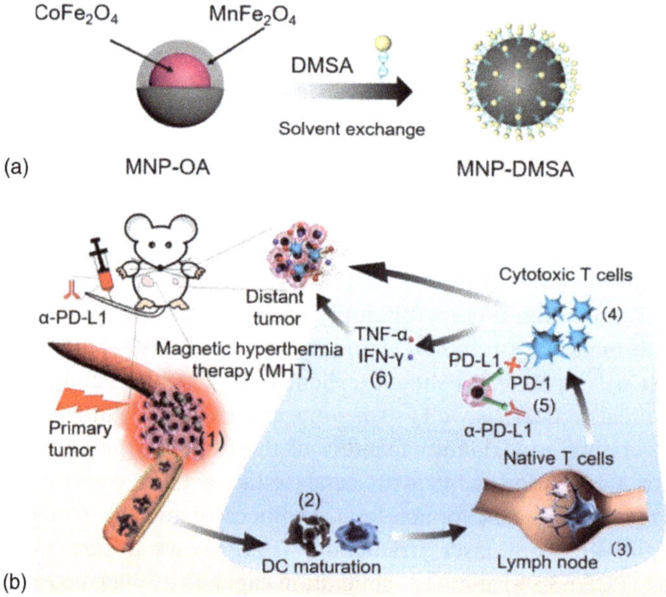

Figure 12.12 (a) Schematic diagram of $CoFe_2O_4@MnFe_2O_4$ nanoparticles. (b) The mechanism of antitumor immune responses to metastatic tumors, which are induced by MHT in combination with α-PD-L1 treatment. Source: Reproduced with permission from [195] 2020, / American Chemical Society.

the tumor, leading to tumor cell ablation and TAAs generation. Upon recognition and uptake by DCs, these TAAs migrated to lymph nodes, where they were presented to T cells, initiating a T-cell immune response. The subsequent combination with the immune checkpoint inhibitor anti-PD-L1 (αPD-L1) antibody significantly enhanced T-cell infiltration and the secretion of antitumor cytokines. This synergistic approach of MHT and immunotherapy effectively suppressed the growth of both primary and metastatic tumors. In another study, Zhang et al. developed a hollow structured Mo-doped Cu_9S_5 (CuMo) nanozyme loaded with GSH-responsive nitric oxide (NO) donors (sodium nitroprusside, SNP) for boosting immune response by mild PTT (Figure 12.13) [196]. The CuMo-SNP possesses photothermal and glutathione oxidase/peroxidase-like enzymatic activities, which lead to reactive oxygen species (ROS) generation and enhance intracellular oxidative stress toward mitochondria. Meanwhile, the continuously generated NO by GSH-responsive SNP impairs mitochondria protective autophagy to recover the ROS-induced oxidative damage for mitochondria, leading to effective block of ATP production. The limited ATP level can restrain the synthesis of intracellular heat stress-induced HSPs, thus achieving enhanced therapeutic effect of mild PTT upon 1064 nm laser irradiation.

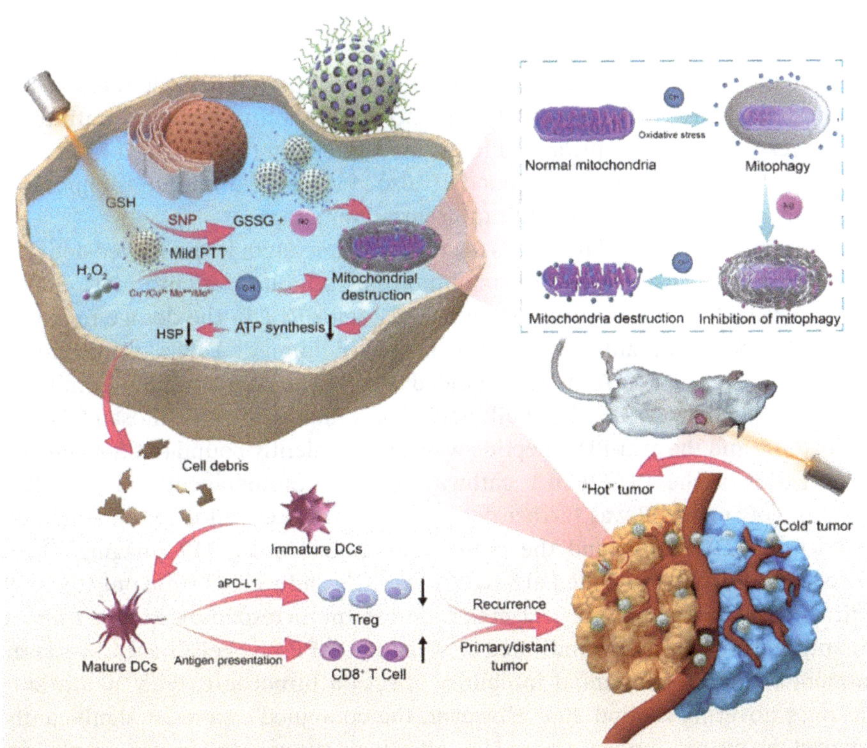

Figure 12.13 Schematic presentation of synergistic cancer therapy mechanism of the NIR-triggered CuMo-SNP. Source: Reproduced with permission from [196] 2023, / Elsevier.

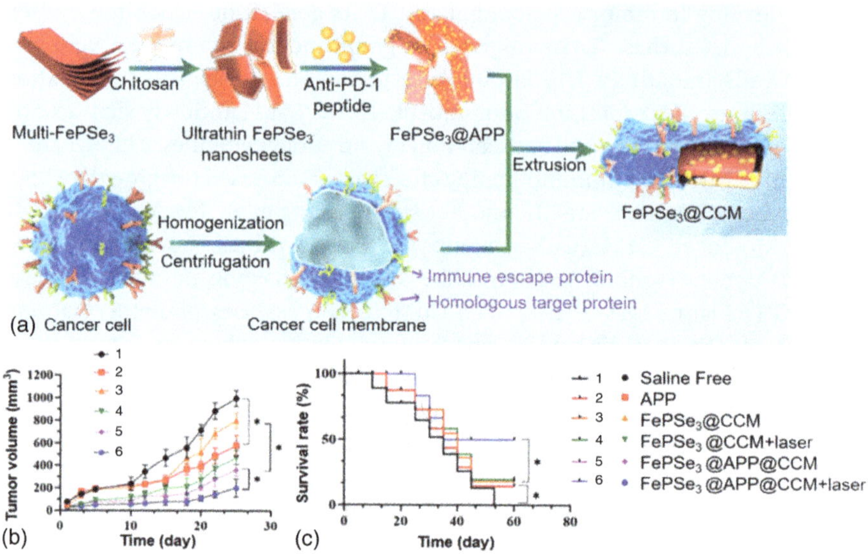

Figure 12.14 (a) Schematic illustration of $FePSe_3$@APP@CCM. (b) Tumor growth curve from the different treatment groups in 30 days. (c) Overall survival rates from different treatment groups in 60 days. $^*p < 0.05$. Source: Reproduced with permission from [197] 2020, / John Wiley & Sons.

Subsequently, when combined with anti-PD-L1 checkpoint blockade therapy, the immunostimulatory TME is further enhanced, promoting systemic antitumor immune responses against primary and distant tumors. In this study, PTT not only efficiently ablates the primary tumor but also reverses the immunosuppressive TME, reinforcing the efficacy of antitumor immunotherapy.

Harnessing photothermal materials to create a nanosystem with targeted delivery capability for loading antibodies or small molecule inhibitors offers a more convenient and reliable alternative to integrating ICB with PTT in the dosage regimen. Fang et al. reported an anti-PD-1 peptide-loaded biomimetic $FePSe_3$ semiconductor nanosheets for MR/PA imaging-guided synergistic PTT-immune combined therapy [197] (Figure 12.14). The bulk $FePSe_3$ was modified with chitosan (CS) for drug loading, and the anti-PD-1 peptide was then covalently bound to CS-stabilized $FePSe_3$ to block the PD-1/PDL1 pathway for immunotherapeutic effects. The surface of $FePSe_3$@APP was further decorated with cancer cell membrane (CCM) derived from CT26 cells, and the presence of iron enabled MR imaging. Once exposed to NIR irradiation, the $FePSe_3$@APP@CCM-induced PTT not only directly eradicated cancer cells but also triggered robust immune responses. Results from in vivo antitumor experiments indicated that neither APP nor $FePSe_3$@CCM + laser treatment exerted a substantial inhibitory effect on tumor size, with 60-day survival rates hovering around 20%. However, the combined treatment significantly enhanced the overall survival rate to 50%, effectively suppressing tumor sizes below 200 mm^3.

12.5.1.3 Other Immune Checkpoints

TAMs, possessing distinctive phagocytic capabilities, emerge as pivotal targets for immunotherapy [198]. Most tumor cells express an increased level of transmembrane protein named CD47, which can inhibit phagocytosis by binding with macrophage signal-regulated protein a (SIRPa) [199]. The administration of monoclonal antibodies targeting CD47 has been explored in various preclinical human cancer models [200–205], demonstrating the potential to elicit $CD8^+$ T cell-mediated antitumor responses [206]. To improve phagocytosis of macrophages, Guo et al. developed a CD47-targeting bismuth selenide nanoparticle (Ab-PEG-Bi_2Se_3), strategically combining PTT with immunotherapy centered on macrophage phagocytosis [207]. The CD47 antibody is linked to the nanoparticle, binding to the CD47 molecule on the tumor cell surface and inhibiting the "don't eat me" signal, thereby activating macrophages to phagocytize tumor cells. The therapeutic outcomes in vivo demonstrated no observed tumor recurrence throughout the entire experimental duration. In a parallel strategy, Xie et al. utilized black phosphorus-based PTT along with anti-CD47 immunotherapy for lymphoma treatment in BALB/c mice. This combined approach demonstrated significant inhibition of tumor growth, leading to a macroscopic tumor-free status in five out of six mice by day 19 [208]. Besides, several other immunotherapy targets have been used in combination with PTT preclinically, such as indoleamine 2,3-dioxygenase (IDO) [209, 210] and the mitogen-activated protein kinases pathway. Furthermore, Li's team prepared an endoplasmic reticulum (ER) targeted PdPtCu nanozyme ($PNBCT^{ER}$) to boost immunotherapy [211]. $PNBCT^{ER}$ not only realizes the combination therapy of PDT, PTT, and CDT but also damages the ER of tumor cells and activates antitumor immune response, which breaks through the immune blockade of TME. Finally, the NLG919 blocks the tryptophan/kynurenine immune escape pathway and reverses the immunosuppressive TME (Figure 12.15a). Their team also built a nanozyme therapeutic agent (PNDPL), which mainly consists of PtBi nanozymes, lactate oxidase (LOX), and the IDO inhibitor NLG919 [212]. The combination of lactate depletion-induced starvation therapy and PTT, along with the blocking of IDO-mediated immune escape, effectively inhibits tumor growth and reverses immunosuppressive microenvironment, thus preventing tumor metastasis (Figure 12.15b). This study represents the first investigation into the synergistic antitumor effects of lactate metabolism regulation and IDO-related immunotherapy.

The integration of ICB therapy with PTT, as demonstrated in the aforementioned studies, significantly improves the response rates of immune checkpoint inhibitors and enhances overall cancer prognosis. The immune mechanisms underlying both ICB therapy and PTT aim to restore immunity to natural levels within the human body. What remains to be explored is the comprehensive assessment of the internal immune interactions within each component and modality in combined PTT and ICB therapy approaches. This evaluation is essential for optimizing the overall treatment effect by adjusting drug doses, therapeutic frequency, and duration.

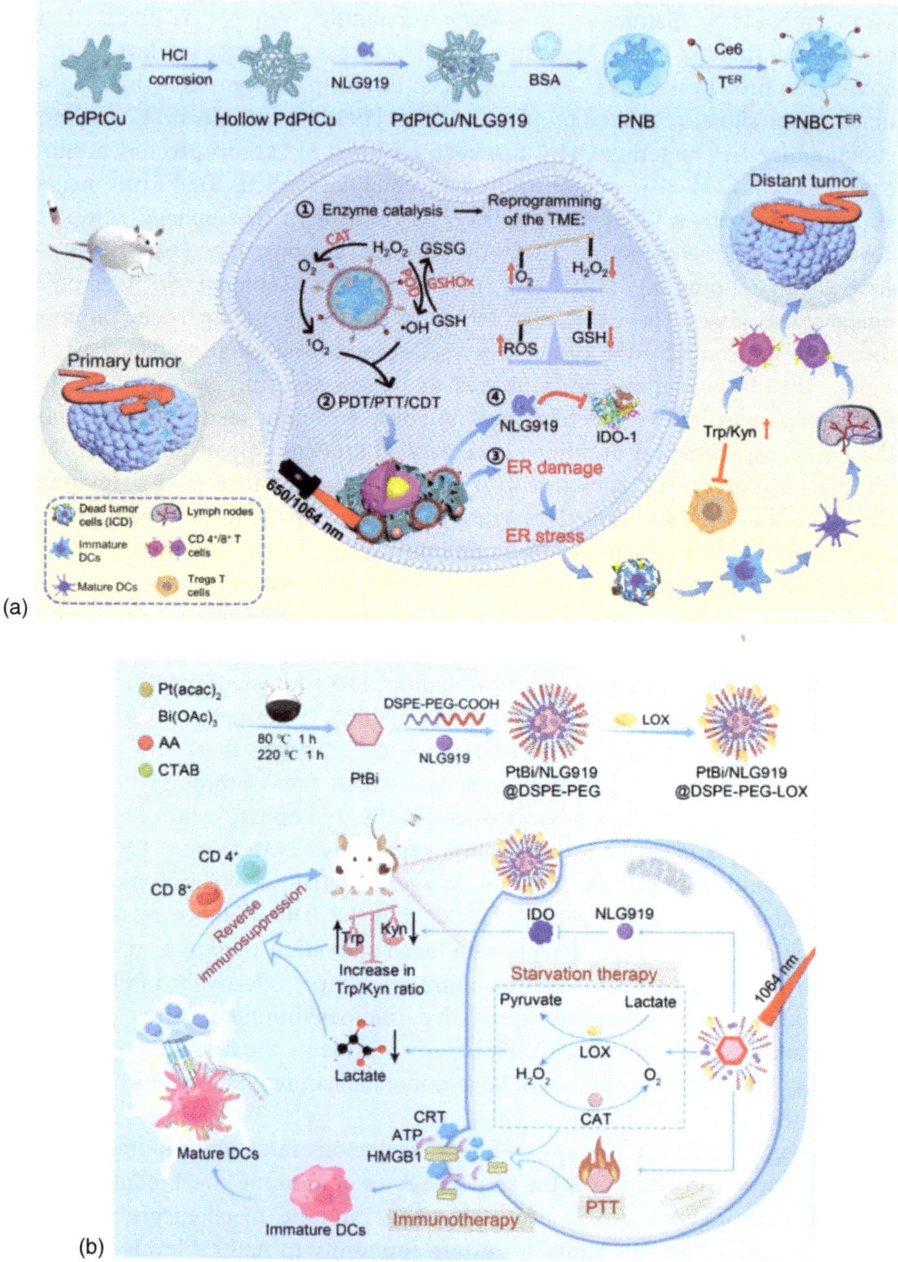

Figure 12.15 (a) Schematic diagram of PdPtCu nanozymes-mediated combination therapy through amplifying ER stress and blocking IDO-mediated immune escape. Source: Reproduced with permission from [211] 2023, / John Wiley & Sons. (b) Schematic illustrations for the construction of PtBi/NLG919@DSPE-PEG-LOX nanozymes and the synergistic effect of starvation therapy, PTT, and immunotherapy. Source: Reproduced with permission from [212] 2023, / John Wiley & Sons.

12.5.2 PTT-Synergized Immunoadjuvant Therapy

Immune adjuvants, characterized by structural similarities to bacterial or viral components, function as agents that stimulate the immune system through pattern recognition receptors (PRRs) [213]. Activation of PRRs results in the release of cytokines, inflammasome maturation, and immune cell activation [214]. Toll-like receptors (TLRs) are the most extensively studied PRRs, serving as innate immune sensors on cell membranes capable of detecting both internal DAMPs and external pathogen-associated molecular patterns. These signals are then converted into downstream signals, initiating subsequent immune cell responses [215]. So far, researchers have identified 10 human TLR protein subfamilies denoted as TLR-1 to -10. Each of these subfamilies can recognize distinct ligands (e.g. peptides, proteins, microRNA, and specific small molecules) [216, 217]. However, some limitations, such as the limited drug accumulation during systemic administration and severe immune-related adverse events, like anemia, lymphopenia, and flu-like symptoms associated with circulating drugs, have significantly constrained their widespread application [218, 219]. Fortunately, recent research has demonstrated that nanotechnology offers a promising approach to load immune adjuvants into different nanocarriers for synergistically enhanced PTT and immunostimulatory effects.

CpG ODNs, a synthetic DNA fragment, could interact with TLR-9 in the endosomal vesicles of B cells and DCs to cause a subsequent cytokine response [220]. Lu et al. designed HCuSNP-CpG conjugates using chitosan-coated hollow CuS nanoparticles assembled with CpG (Figure 12.16) [221]. Once exposed to NIR irradiation, HCuSNP-CpG can be decomposed into chitosan-CpG nanocomplexes (Chi-CpG-NPs) and small CuS nanoparticles (SCuSNPs). Simultaneously, HCuSNPs-based PTT effectively ablates primary tumors and releases TAAs. Both PTT-generated TAAs and the immunostimulatory chitosan-CpG could be taken up by DCs, subsequently resulting in the activation of NK cells and the maturation of DCs. Activated DCs subsequently migrate to the lymph nodes, facilitating T-cell activation and proliferation, ultimately achieving systemic antitumor immunity against untreated distant tumors. In a separate investigation, Ran et al. utilized clinically approved components, including superparamagnetic iron oxide, CpG, and monomethoxypoly (ethylene glycol)-poly(lactic-*co*-glycolic acid)-poly-L-lysine triblock copolymers to develop a magnetic-responsive immunostimulatory nanoagent (MINP) [222]. After PTT treatment, TAAs were released from tumor debris and recognized by TLR9 on DCs in the presence of CpG. Activated DCs then secreted TNF-α, IL-6, and IFN-γ, initiating T cell-mediated immune responses for sustained tumor inhibition.

Another small molecule, R837, which has been successfully approved by the FDA as a broad-spectrum antiviral drug, is being studied as a TLR7/8 agonist in synthesis strategies as well. Jiang et al. combined PTT, CDT, and immunotherapy in a multifunctional MoS_2-CuO heteronanocomposites by co-loading bovine serum albumin (BSA) and R837 (Figure 12.17) [223].The obtained MoS_2-CuO@BSA/R837 (MCBR) nanoplatforms realize the excellent computed tomography/infrared

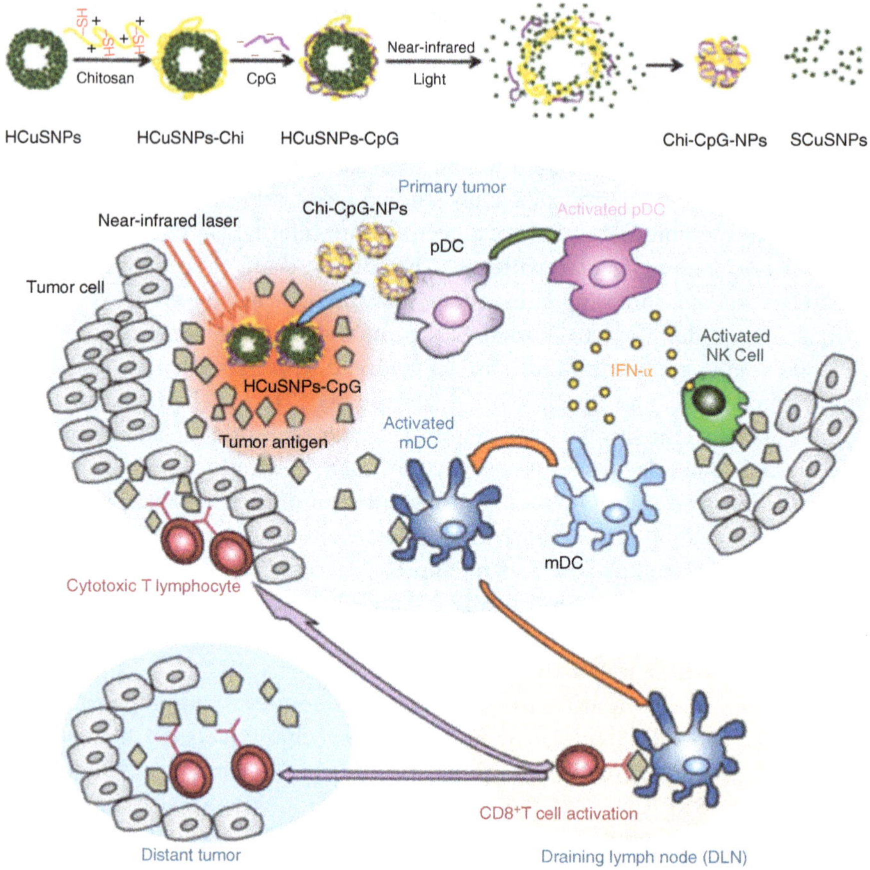

Figure 12.16 Schematic diagram of combined photothermal and immunotherapy using chitosan-coated hollow copper sulfide nanoparticles. Source: Banerjee et al. [217]/American Chemical Society/CC-BY 4.0.

thermal/MR multi-mode bioimaging as well as significantly enhanced antitumor efficacy of synergetic PTT/CDT/immunotherapy. The MCBR nanoplatforms effectively eliminate primary tumors and release TAAs via MCBR-based PTT and CDT. These TAAs, when combined with R837 as an immune-stimulating adjuvant, provoke robust immunological responses by promoting DC maturation, cytokine secretion, and increasing the numbers of $CD4^+$ T cells and $CD8^+$ T cells. The multifunctional nanoplatform MCBR demonstrates remarkable efficacy in inhibiting both primary and metastatic tumors through the integrated cancer therapy.

Other immune adjuvants, such as TLR-3 agonist (polyinosinic-polycytidylic acid [Poly I: C]) and TLR-4 agonist (monophosphoryl lipid A [MPLA]), have shown encouraging preclinical results when utilized in combination therapy with PTT [224–226]. The extensive study of immune adjuvants is evident in vaccine development, such as FDA-approved hepatitis B virus vaccine containing CpG, small molecule drugs, like R837 and R848, have exhibited significant side effects when administered orally or intravenously, limiting their application to topical

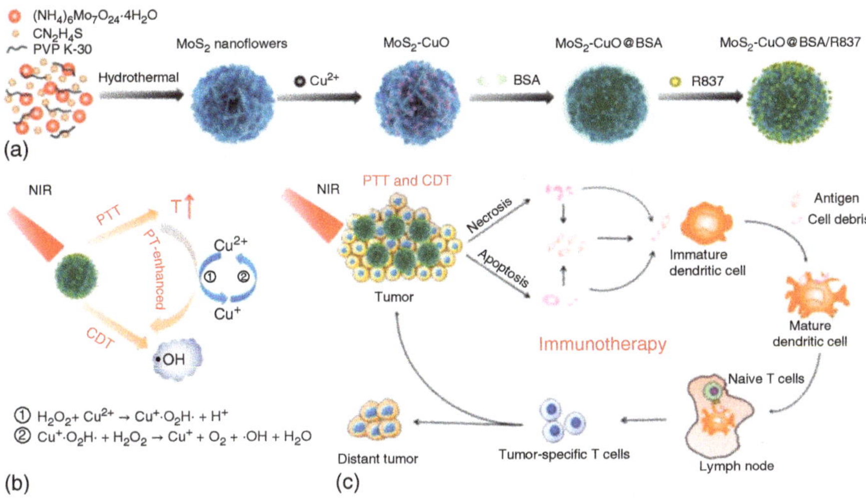

Figure 12.17 Schematic illustration of (a) fabrication of the MoS_2-CuO@BSA/R837 nanoplatforms; (b) the mechanism of MoS_2-CuO@BSA/R837 nanoplatforms for synergistic PTT and CDT. (c) the mechanism of antitumor immune responses triggered by the MoS_2-CuO@BSA/R837 nanoplatforms based synergistic PTT/CDT/immunotherapy. Source: Reproduced with permission from [223] 2021, / Elsevier.

use only [227, 228]. Exploring the use of thermal-responsive controlled-release nanovectors could potentially mitigate toxicity concerns, and safety assessments should be emphasized in future studies.

In summary, incorporating an immunoadjuvant presents a pragmatic strategy to elevate immunogenicity and enhance immune response rates in the context of in situ autologous cancer vaccines released through hyperthermia-ablated tumors. Nevertheless, numerous challenges endure in the clinical implementation of hyperthermia-triggered immunoadjuvant therapy. Further research is essential to fine-tune the immunoadjuvant load capacity in hyperthermia-triggered therapy agents, enhance the delivery and degradability of these agents containing immunoadjuvants, and elucidate the synergistic effects between immunoadjuvants and hyperthermia agents. Additionally, a deeper exploration of the mechanisms through which immunoadjuvants amplify antitumor immunity via diverse immune regulation pathways is necessary to ensure their safety and effectiveness and to establish a foundation for the development of combination therapies.

12.5.3 PTT-Synergized Adoptive Cellular Immunotherapy

Adoptive cell therapy (ACT) aims to enhance the antitumor immune response by infusing T cells that have been manipulated ex vivo. These T cells are typically genetically engineered to express either chimeric antigen receptors (CARs) or T cell receptors (TCR) [229]. CARs have the capability to activate immune cells, particularly NK cells, while recognizing antigens on the cell surface without the constraints of major histocompatibility complex molecules, thereby eliciting an efficient immune response targeted at tumors [230]. CAR-T therapy has shown remarkable success in hematologic tumors but faces challenges in treating solid tumors. The ability of CARs to migrate and invade tumor cells is diminished, due to the presence of dense fibrotic matrix in solid tumors and the lack of expression of chemokines involved in the infiltration of T cells into tumor tissues [38].

PTT has been reported to improve the tumor-infiltrating ability of CAR-T cells and create an immune-favorable TME, which provides a promising approach for the improvement of the curative effect of CAR-T therapy. Li et al. developed a multifunctional "turn-on" fluorescent nanocatalyst (APHA@CM) operating in the NIR-II range to enhance CAR-T cell therapy for solid tumors. The construction of APHA@CM involved encapsulating horseradish peroxidase (HRP)-loaded Au/PDA NPs and Ag_2S QDs with CAR-T cell membranes. This APHA@CM, equipped with excellent multimodal imaging capabilities, serves to precisely guide the nanocatalyst-induced regulation of the TME and CAR-T cell therapy within specific spatial and temporal windows. On the one hand, the oxidase-like activity of Au NPs inhibited the glycolytic metabolism of tumor cells, reducing lactate efflux and ultimately enhancing the activation of CAR-T cells within the tumor.

On the other hand, the HRP could alleviate the hypoxic environment of the tumor, enhance the synergistic sonodynamic/photothermal therapy induced by Au/PDA NPs, promote ICD of NALM 6 cells, and further augment CAR-T cell-mediated immune microenvironment reprogramming. The application of APHA@CM to treat NALM 6 solid tumors not only resulted in complete tumor elimination but also established a lasting immune memory effect, effectively inhibiting tumor metastasis and recurrence [231].

In ACT, NK cells offer greater off-the-shelf utility and enhanced safety compared to T cells. However, the effectiveness of NK cell therapy is significantly hampered by the formidable physical barriers within the tumor tissue and the lack of NK cell recognition signals on tumor cells. Lin et al. developed a nano-immunomodulator, AuNSP@αCD16 (CD16 antibody encoding plasmid), to reshape the TME and enhance the antitumor effects of adoptive NK cells (Figure 12.18). The AuNSP characterized by a sea urchin-like gold core and a cationic polymer shell, exhibited high gene transfection efficiency and a stable NIR-II photothermal capacity. Through mild photothermal intervention, AuNSP partially dismantled tumors and disrupted dense physical barriers, establishing a more permeable TME conducive to NK cell infiltration. Furthermore, AuNSP facilitated αCD16 gene transfection and enhanced NK cell recognition and activation through CD16-mediated antibody-dependent cellular cytotoxicity. The designed AuNSP@αCD16 induced an immune-favorable TME for NK cells performing killing functions against solid tumors, increasing the release of cytolytic granules and proinflammatory cytokines, which ultimately achieved a robustly boosted NK cell-based immunotherapy [232].

Compared to conventional treatment approaches, ACT offers the advantages of reduced toxicity, minimal side effects, enhanced targeting, and immune memory function. While ACT as adjuvant therapy for malignant tumors can demonstrate clinical efficacy due to the immunosuppressive TME, the direct transfer of in vitro amplified cultures into patients is not ideal. Interestingly, the incorporation of PTT proves effective in reversing the immunosuppressive TME, resulting in a synergistic

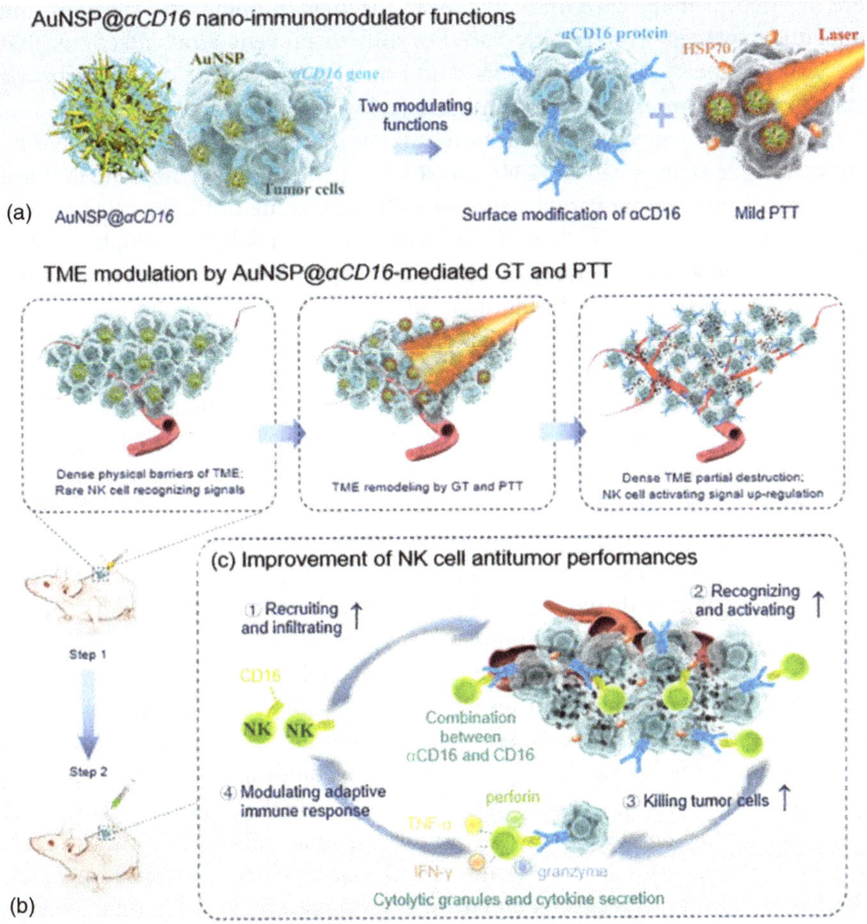

Figure 12.18 Schematic diagram of AuNSP@αCD16-mediated TME modulation for robustly boosting NK cell antitumor performances. (a) The intended functions of the AuNSP@αCD16 nano-immunomodulator: modification of the tumor surface by αCD16 and induction of mild PTT by laser irradiation. (b) AuNSP@αCD16-mediated TME modulation: destruction of dense physical barriers and up-regulation of NK cell recognizing and activating signals. (c) The improved NK cell antitumor performances after AuNSP@αCD16 TME remodeling based on the NK-cancer immunity cycle. Source: Reproduced with permission from [232] 2022, / Elsevier.

therapeutic impact. As an immune enhancement strategy, ACT elevates the immune response within the tolerable range for the human body. Addressing the short- and long-term immune risks, along with the high cost associated with ACT, remains crucial for the clinical advancement of PTT-synergized ACT.

12.5.4 PTT-Synergized Therapeutic Cancer Vaccine

Cancer vaccines protect body health by priming the innate and adaptive immune responses against cancers and play a pivotal role in cancer immunotherapy.

There are two primary categories of cancer vaccines: prophylactic vaccines and therapeutic vaccines. Prophylactic vaccines aim to prevent viral infections that could lead to cancer, such as hepatitis B and human papillomavirus. Therapeutic vaccines, on the other hand, utilize antigens and adjuvants to activate the immune system and target cancer cells. Currently, many therapeutic vaccines, including breast cancer [233], liver cancer [234], lung cancer [235], and melanoma [236] have been explored for treating various cancers. Although some of these vaccines have progressed to clinical trials, their therapeutic efficacy has been suboptimal. For example, peptide vaccines relying on glucolipids like GD3 and GM2 demonstrated weak immune responses, hindering their effectiveness [237]. The herpes simplex virus type 1 oncolytic vaccine (Talimogene laherparepvec) induces excessive toxicity and side effects, limiting its clinical application [238, 239]. Additionally, DC-based vaccines demonstrated promising effects in melanoma mouse models but fell short in achieving long-term immune responses in clinical trials [240–242]. The selection of suitable combination therapy and delivery strategies is crucial for optimizing the efficacy of cancer vaccines.

Therapeutic cancer vaccines combined with PTT have been extensively studied. Liu et al. developed an amphiphilic nanoplatform using Au nanorods (AuNRs) conjugated with the polyTLR-7/8 agonist imidazoquinoline to enhance immunotherapy against distant tumors [44]. Through intricate Au-S bonds, imidazoquinoline and the antigen-binding R9-GPLGLAG-PEG are linked to AuNRs, forming a versatile nanoplatform termed AuNRs-IMDQ-R9-PEG. This nanoplatform not only induces tumor cell death via PTT, but also facilitates the formation of a sustained immune vaccine by noncovalently binding released antigens to R9. In mice with bilateral tumor models, the Gel-BPQD-CCNVs + NIR + aPD-1 group exhibited the smallest recurrent tumor size and highest survival rate (70% within 60 days), as well as the fewest lung metastatic foci. Another approach by Mei et al. involved the development of a personalized photothermal vaccine (Gel-BPQD-CCNVs) combined with aPD-1 to prevent tumor relapse and metastasis [45]. Surgically removed tumor cell membrane-coated black phosphorus quantum dot nanovesicles (BPQD-CCNVs) were prepared and loaded into a thermosensitive hydrogel containing GM-CSF and lipopolysaccharide (LPS) (Figure 12.19). Local heat generated by BPQDs induced the sustained release of GM-CSF and LPS, recruiting and activating DCs to further activate T cells. Furthermore, aPD-1 antibody reinvigorated the exhausted $CD8^+$ T cells. In this strategy, BPQDs did not directly eliminate cancer cells but aimed to modify the TME and regulate the release of GM-CSF and LPS. For the primary tumor, the Gel-BPQD-CCNVs + NIR + aPD-1 group demonstrated the smallest recurrence tumor size and a 70% survival rate within 60 days. More importantly, mice treated with Gel-BPQD-CCNVs + NIR + aPD-1 exhibited the fewest lung metastatic foci, with a 40% survival rate on day 60, surpassing all other groups, which did not survive beyond day 52.

In conclusion, cancer vaccines demonstrate potent antitumor effects on distant tumors, effectively suppressing tumor relapse or metastasis. Numerous studies have documented a sustained immune memory response, underscoring the

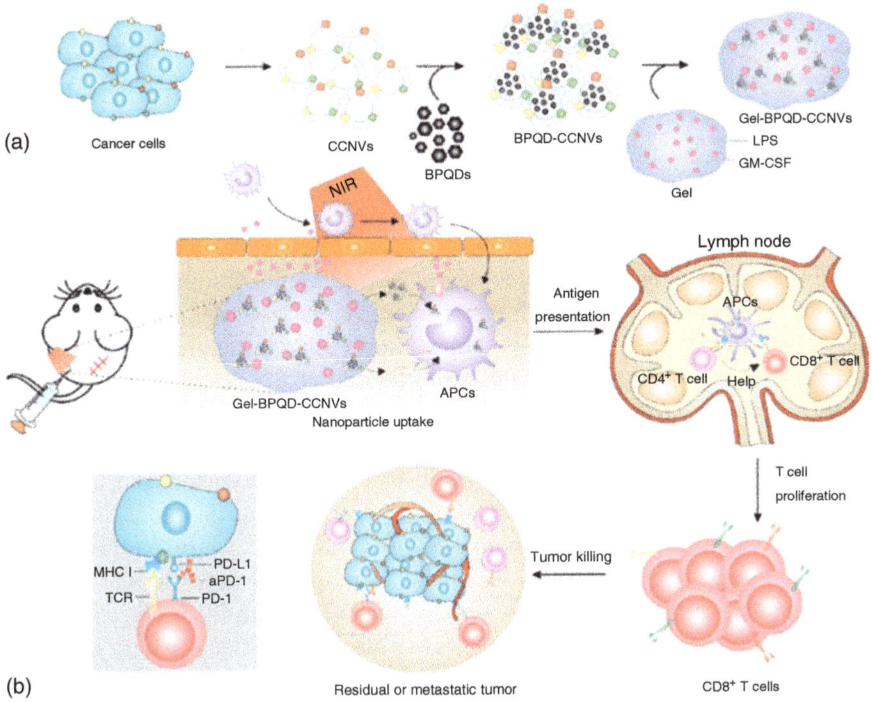

Figure 12.19 Schematic illustration and characterization of BPQD-CCNVs combined with aPD-1 antibody for postsurgical cancer immunotherapy. Source: Reproduced with permission from [45] 2019, / American Chemical Society.

potential of PTT-combined cancer vaccine-based immunotherapy as a promising therapeutic strategy for advanced, disseminated cancers or the prevention of post-operative recurrence.

12.6 Summary and Outlook

In conclusion, the latest research progress of PTT combined with immunotherapy based on nanomedicine is discussed in this chapter, including the mechanism of PTT and four types of PTAs, the mechanism of immunotherapy, and the research progress of different types of immunotherapies combined with PTT. PTT can eliminate tumor in situ, but its effect on tumor metastasis and recurrence is rather poor. Immunotherapy, including immune checkpoint inhibition, immune vaccine, and immune adjuvants, can have a certain clinical effect, but the low immune response rate, large individual differences, and immunosuppressive TME seriously limit its clinical efficacy. Many preclinical studies, as reported in this article, have highlighted the outstanding benefits of combination therapies. Until now, a few strategies can be summarized in PTT-immunotherapy studies: (i) Sequential injection of immunomodulators and PTAs: in this strategy, immunomodulators and PTAs are administered separately. PTAs primarily focus on eradicating the primary tumor, while subsequent immune responses are augmented by immunomodulators. (ii) Nanocarrier-mediated delivery of PTAs and immunomodulators: this method involves encapsulating both the PTAs and immunomodulators within a nanocarrier. Sometimes, the PTAs itself can serve as the carrier. (iii) TME modification: PTT is applied not for tumor elimination but to modify the TME. This approach is commonly observed in combination with therapies constrained by immunosuppressive TME, such as ACT. (iv) Controlled release of immunomodulators using PTA-generated heat: Heat generated by PTAs is employed to precisely regulate the release of immunomodulators. Encapsulating PTAs and immunomodulators with nanovectors favors the targeting and accumulation of both agents at the tumor site, protecting the cargo from rapid degradation and reducing side effects as well, but with a significant preparation cost. Separate injection of immunomodulators and PTAs may lead to severe systemic immune side effects.

Nevertheless, despite the promising achievements in PTT-synergized immunotherapy, the application of nanomedicines in PTT remains in the proof-of-concept phase and encounters several limitations in advancing toward clinical transformation. First, it is established that PTT can induce necrosis and elicit a proinflammatory response. However, inflammation has been identified as a factor that promotes tumor growth and metastasis in numerous studies and is significantly correlated with cancer incidence. Moreover, inflammation can dampen the immune response by suppressing tumor immunosurveillance and immunoediting. Although inflammation is a critical complication of PTT, it is seldom addressed in most studies within this field. Understanding the intricate relationship between cancer and PTT-induced inflammation is imperative. Secondly, the optimal temperature for PTT remains ambiguous. Varied studies present conflicting conclusions, and a standardized benchmark is lacking. While many studies confirm that PTT induces an immune response, there is a gap in analyzing the least harmful temperature to animals that yields the most effective therapeutic outcomes. Additionally, the power density of the NIR laser used in several studies has surpassed recommended limits ($0.33\,W\,cm^{-2}$ for 808 nm, $0.73\,W\,cm^{-2}$ for 980 nm, American National Standard for Safe Use of Lasers, ANSI Z136.1-2007). Future investigations should address these ignored factors. Thirdly, the potential risk of adverse immune responses in PTT-combined immunotherapy remains unknown. Research efforts have predominantly focused on immune enhancement and antitumor efficacy, neglecting the possible impact on the peripheral immune system. Subsequent studies should incorporate experiments to assess this aspect. Last but not least, it is crucial to note that all the studies discussed in this review are at the preclinical stage. The safety and feasibility of innovative nanoplatforms must be rigorously evaluated before progressing to clinical trials. Continued exploration of the synergized therapy on a multidisciplinary, multi-approach basis is essential. It is anticipated that clinical research on PTT-combined immunotherapy will yield valuable results in the future, opening up more possibilities for personalized cancer treatments.

References

1 Wen, H., Tamarov, K., Happonen, E. et al. (2021). Inorganic nanomaterials for photothermal-based cancer theranostics. *Advanced Therapeutics* 4: 2000207.

2 Lambert, A.W., Pattabiraman, D.R., and Weinberg, R.A. (2017). Emerging biological principles of metastasis. *Cell* 168: 670–691.

3 Steeg, P.S. (2006). Tumor metastasis: mechanistic insights and clinical challenges. *Nature Medicine* 12: 895–904.

4 Kraljevic, P.S., Sedic, M., Bosnjak, H. et al. (2011). Metastasis: new perspectives on an old problem. *Molecular Cancer* 10: 1–14.

5 Schroeder, A., Heller, D.A., Winslow, M.M. et al. (2012). Treating metastatic cancer with nanotechnology. *Nature Reviews Cancer* 12: 39–50.

6 Zhang, P., Zhai, Y., Cai, Y. et al. (2019). Nanomedicine-based immunotherapy for the treatment of cancer metastasis. *Advanced Materials* 31: 1904156.

7 Blattman, J.N. and Greenberg, P.D. (2004). Cancer immunotherapy: a treatment for the masses. *Science* 305: 200–205.

8 Rosenberg, S.A. (2001). Progress in human tumour immunology and immunotherapy. *Nature* 411: 380–384.

9 Couzin-Frankel, J. (2015). Cancer immunotherapy. *American Association for the Advancement of Science* 28: 71–74.

10 Mellman, I., Coukos, G., and Dranoff, G. (2011). Cancer immunotherapy comes of age. *Nature* 480: 480–489.

11 Maggon, K. (2007). Monoclonal antibody "gold rush". *Current Medicinal Chemistry* 14: 1978–1987.

12 Zhang, X., Gordon, J.R., and Xiang, J. (2002). Advances in dendritic cell-based vaccine of cancer. *Cancer Biotherapy and Radiopharmaceuticals* 17: 601–619.

13 Burgdorf, S.K., Fischer, A., Myschetzky, P.S. et al. (2008). Clinical responses in patients with advanced colorectal cancer to a dendritic cell based vaccine. *Oncology Reports* 20: 1305–1311.

14 Curran, K.J. and Brentjens, R.J. (2015). Chimeric antigen receptor T cells for cancer immunotherapy. *Journal of Clinical Oncology: Official Journal of the American Society of Clinical Oncology* 33: 1703–1706.

15 Yadav, D., Lu, C., and Yadav, R. (2014). Vaccine therapy for pancreatic cancer: a battle against deadly cancer. *Journal of Cancer Science and Therapy* 6: 268–277.

16 Aurisicchio, L., Pallocca, M., Ciliberto, G., and Palombo, F. (2018). The perfect personalized cancer therapy: cancer vaccines against neoantigens. *Journal of Experimental and Clinical Cancer Research* 37: 1–10.

17 Buchbinder, E.I. and Desai, A. (2016). CTLA-4 and PD-1 pathways: similarities, differences, and implications of their inhibition. *American Journal of Clinical Oncology* 39: 98–106.

18 Gatalica, Z., Vanderwalde, A.M., Rose, I. et al. (2016). Distribution of PD-L1 expression in diverse cancer types: experience with over 10,000 cases. *Journal of Clinical Oncology* 34: 11548.

19 Kaufman, H.L., Russell, J., Hamid, O. et al. (2016). Avelumab in patients with chemotherapy-refractory metastatic merkel cell carcinoma: a multicentre, single-group, open-label, phase 2 trial. *The Lancet Oncology* 17: 1374–1385.

20 Sade-Feldman, M., Jiao, Y.J., Chen, J.H. et al. (2017). Resistance to checkpoint blockade therapy through inactivation of antigen presentation. *Nature Communications* 8: 1136.

21 Dearden, H., Long, G., Menzies, A.M. et al. (2018). Safety of resuming anti-PD-1 in patients with immune-related adverse events (irAEs) during combined anti-CTLA-4 and anti-PD1 in metastatic melanoma. *Annals of Oncology* 29: 250–255.

22 Costa, R., Costa, R.B., Talamantes, S.M. et al. (2017). Analyses of selected safety endpoints in phase 1 and late-phase clinical trials of anti-PD-1 and PD-L1 inhibitors: prediction of immune-related toxicities. *Oncotarget* 8: 67782–67789.

23 Tang, C., Wang, X., Soh, H. et al. (2014). Combining radiation and immunotherapy: a new systemic therapy for solid tumors? *Cancer Immunology Research* 2: 831–838.

24 Formenti, S.C. and Demaria, S. (2013). Combining radiotherapy and cancer immunotherapy: a paradigm shift. *JNCI: Journal of the National Cancer Institute* 105: 256–265.

25 Niu, L., Chen, J., He, L. et al. (2013). Combination treatment with comprehensive cryoablation and immunotherapy in metastatic pancreatic cancer. *Pancreas* 42: 1143–1149.

26 Vankayala, R. and Hwang, K.C. (2018). Near-infrared-light-activatable nanomaterial-mediated phototheranostic nanomedicines: an emerging paradigm for cancer treatment. *Advanced Materials* 30: 1706320.

27 Xu, L., Mou, F., Gong, H. et al. (2017). Light-driven micro/nanomotors: from fundamentals to applications. *Chemical Society Reviews* 46: 6905–6926.

28 Weissleder, R. (2001). A clearer vision for in vivo imaging. *Nature Biotechnology* 19: 316–317.

29 Jain, P.K., Huang, X., El-Sayed, I.H., and El-Sayed, M.A. (2008). Noble metals on the nanoscale: optical and photothermal properties and some applications in imaging, sensing, biology, and medicine. *Accounts of Chemical Research* 41: 1578–1586.

30 Vogel, A. and Venugopalan, V. (2003). Mechanisms of pulsed laser ablation of biological tissues. *Chemical Reviews* 103: 577–644.

31 Skrabalak, S.E., Chen, J., Au, L. et al. (2007). Gold nanocages for biomedical applications. *Advanced Materials* 19: 3177–3184.

32 Chen, H., Shao, L., Ming, T. et al. (2010). Understanding the photothermal conversion efficiency of gold nanocrystals. *Small* 6: 2272–2280.

33 Yang, K., Zhang, S., Zhang, G. et al. (2010). Graphene in mice: ultrahigh in vivo tumor uptake and efficient photothermal therapy. *Nano Letters* 10: 3318–3323.

34 Sheng, Z., Hu, D., Zheng, M. et al. (2014). Smart human serum albumin-indocyanine green nanoparticles generated by programmed assembly for dual-modal imaging-guided cancer synergistic phototherapy. *ACS Nano* 8: 12310–12322.

35 Yang, K., Xu, H., Cheng, L. et al. (2012). In vitro and in vivo near-infrared photothermal therapy of cancer using polypyrrole organic nanoparticles. *Advanced Materials* 24: 5586–5592.

36 Hessel, C.M., Pattani, V.P., Rasch, M. et al. (2011). Copper selenide nanocrystals for photothermal therapy. *Nano Letters* 11: 2560–2566.

37 Tian, Q., Tang, M., Sun, Y. et al. (2011). Hydrophilic flower-like CuS superstructures as an efficient 980 nm laser-driven photothermal agent for ablation of cancer cells. *Advanced Materials* 23: 3542–3547.

38 Wells, A.D. and Malkovsky, M. (2000). Heat shock proteins, tumor immunogenicity and antigen presentation: an integrated view. *Immunology Today* 21: 129–132.

39 Zou, L., Wang, H., He, B. et al. (2016). Current approaches of photothermal therapy in treating cancer metastasis with nanotherapeutics. *Theranostics* 6: 762–772.

40 Casares, N., Pequignot, M.O., Tesniere, A. et al. (2005). Caspase-dependent immunogenicity of doxorubicin-induced tumor cell death. *The Journal of Experimental Medicine* 202: 1691–1701.

41 Kepp, O., Senovilla, L., Vitale, I. et al. (2014). Consensus guidelines for the detection of immunogenic cell death. *Oncoimmunology* 3: e955691.

42 Dunne, M., Regenold, M., and Allen, C. (2020). Hyperthermia can alter tumor physiology and improve chemo-and radio-therapy efficacy. *Advanced Drug Delivery Reviews* 163: 98–124.

43 Chen, D.S. and Mellman, I. (2013). Oncology meets immunology: the cancer-immunity cycle. *Immunity* 39: 1–10.

44 Liu, X., Su, Q., Song, H. et al. (2021). PolyTLR7/8a-conjugated, antigen-trapping gold nanorods elicit anticancer immunity against abscopal tumors by photothermal therapy-induced in situ vaccination. *Biomaterials* 275: 120921.

45 Ye, X., Liang, X., Chen, Q. et al. (2019). Surgical tumor-derived personalized photothermal vaccine formulation for cancer immunotherapy. *ACS Nano* 13: 2956–2968.

46 Kelly, K.L., Coronado, E., Zhao, L.L., and Schatz, G.C. (2003). The optical properties of metal nanoparticles: the influence of size, shape, and dielectric environment. *ACS Publications* 107: 668–677.

47 Huang, X., Neretina, S., and El-Sayed, M.A. (2009). Gold nanorods: from synthesis and properties to biological and biomedical applications. *Advanced Materials* 21: 4880–4910.

48 Mayer, K.M. and Hafner, J.H. (2011). Localized surface plasmon resonance sensors. *Chemical Reviews* 111: 3828–3857.

49 Willets, K.A. and Van Duyne, R.P. (2007). Localized surface plasmon resonance spectroscopy and sensing. *Annual Review of Physical Chemistry* 58: 267–297.

50 Gao, M., Zhu, L., Peh, C.K., and Ho, G.W. (2019). Solar absorber material and system designs for photothermal water vaporization towards clean water and energy production. *Energy and Environmental Science* 12: 841–864.

51 Kim, M., Lin, M., Son, J. et al. (2017). Hot-electron-mediated photochemical reactions: principles, recent advances, and challenges. *Advanced Optical Materials* 5: 1700004.

52 Hu, M., Chen, J., Li, Z.Y. et al. (2006). Gold nanostructures: engineering their plasmonic properties for biomedical applications. *Chemical Society Reviews* 35: 1084–1094.

53 Brongersma, M.L., Halas, N.J., and Nordlander, P. (2015). Plasmon-induced hot carrier science and technology. *Nature Nanotechnology* 10: 25–34.

54 Wang, M., Liang, Y., Zhang, Z. et al. (2019). Ag@ Fe_3O_4@ C nanoparticles for multi-modal imaging-guided chemo-photothermal synergistic targeting for cancer therapy. *Analytica Chimica Acta* 1086: 122–132.

55 Chen, R., Wu, Z., Zhang, T. et al. (2017). Magnetically recyclable self-assembled thin films for highly efficient water evaporation by interfacial solar heating. *RSC Advances* 7: 19849–19855.

56 Tian, Q., Jiang, F., Zou, R. et al. (2011). Hydrophilic Cu_9S_5 nanocrystals: a photothermal agent with a 25.7% heat conversion efficiency for photothermal ablation of cancer cells in vivo. *ACS Nano* 5: 9761–9771.

57 Cheng, Y., Chang, Y., Feng, Y. et al. (2018). Deep-level defect enhanced photothermal performance of bismuth sulfide-gold heterojunction nanorods for photothermal therapy of cancer guided by computed tomography imaging. *Angewandte Chemie International Edition* 57: 246–251.

58 Vélez-Cordero, J.R. and Hernandez-Cordero, J. (2015). Heat generation and conduction in PDMS-carbon nanoparticle membranes irradiated with optical fibers. *International Journal of Thermal Sciences* 96: 12–22.

59 Sönnichsen, C., Reinhard, B.M., Liphardt, J., and Alivisatos, A.P. (2005). A molecular ruler based on plasmon coupling of single gold and silver nanoparticles. *Nature Biotechnology* 23: 741–745.

60 Jain, P.K., Lee, K.S., El-Sayed, I.H., and El-Sayed, M.A. (2006). Calculated absorption and scattering properties of gold nanoparticles of different size, shape, and composition: applications in biological imaging and biomedicine. *The Journal of Physical Chemistry B* 110: 7238–7248.

61 Zhou, B., Song, J., Wang, M. et al. (2018). BSA-bioinspired gold nanorods loaded with immunoadjuvant for the treatment of melanoma by combined photothermal therapy and immunotherapy. *Nanoscale* 10: 21640–21647.

62 Wang, W.N., Pei, P., Chu, Z.Y. et al. (2020). Bi_2S_3 coated Au nanorods for enhanced photodynamic and photothermal antibacterial activities under NIR light. *Chemical Engineering Journal* 397: 125488.

63 Lee, S.Y. and Shieh, M.J. (2020). Platinum (II) drug-loaded gold nanoshells for chemo-photothermal therapy in colorectal cancer. *ACS Applied Materials & Interfaces* 12: 4254–4264.

64 Wang, Z., Li, S., Zhang, M. et al. (2017). Laser-triggered small interfering RNA releasing gold nanoshells against heat shock protein for sensitized photothermal therapy. *Advanced Science* 4: 1600327.

65 Duan, Q., Yang, M., Zhang, B. et al. (2021). Gold nanoclusters modified mesoporous silica coated gold nanorods: enhanced photothermal properties and fluorescence imaging. *Journal of Photochemistry and Photobiology B: Biology* 215: 112111.

66 Liu, P., Yang, W., Shi, L. et al. (2019). Concurrent photothermal therapy and photodynamic therapy for cutaneous squamous cell carcinoma by gold nanoclusters under a single NIR laser irradiation. *Journal of Materials Chemistry B* 7: 6924–6933.

67 Ohulchanskyy, T.Y., Kopwitthaya, A., Jeon, M. et al. (2013). Phospholipid micelle-based magneto-plasmonic nanoformulation for magnetic field-directed, imaging-guided photo-induced cancer therapy. *Nanomedicine: Nanotechnology Biology and Medicine* 9: 1192–1202.

68 Wang, J., Sun, J., Wang, Y. et al. (2020). Gold nanoframeworks with mesopores for Raman-photoacoustic imaging and photo-chemo tumor therapy in the second near-infrared biowindow. *Advanced Functional Materials* 30: 1908825.

69 Jia, J., Liu, G., Xu, W. et al. (2020). Fine-tuning the homometallic interface of Au-on-Au nanorods and their photothermal therapy in the NIR-II window. *Angewandte Chemie International Edition* 132: 14551–14556.

70 Espinosa, A., Curcio, A., Cabana, S. et al. (2018). Intracellular biodegradation of Ag nanoparticles, storage in ferritin, and protection by a Au shell for enhanced photothermal therapy. *ACS Nano* 12: 6523–6535.

71 Dos Santos, C.A., Seckler, M.M., Ingle, A.P. et al. (2014). Silver nanoparticles: therapeutical uses, toxicity, and safety issues. *Journal of Pharmaceutical Sciences* 103: 1931–1944.

72 Wei, L., Lu, J., Xu, H. et al. (2015). Silver nanoparticles: synthesis, properties, and therapeutic applications. *Drug Discovery Today* 20: 595–601.

73 Shi, J., Wang, L., Zhang, J. et al. (2014). A tumor-targeting near-infrared laser-triggered drug delivery system based on GO@Ag nanoparticles for chemo-photothermal therapy and X-ray imaging. *Biomaterials* 35: 5847–5861.

74 Akter, M., Sikder, M.T., Rahman, M.M. et al. (2018). A systematic review on silver nanoparticles-induced cytotoxicity: physicochemical properties and perspectives. *Journal of Advanced Research* 9: 1–16.

75 Liao, C., Li, Y., and Tjong, S.C. (2019). Bactericidal and cytotoxic properties of silver nanoparticles. *International Journal of Molecular Sciences* 20: 449.

76 Wang, M., Chang, M.Y., Chen, Q. et al. (2020). Au_2Pt-PEG-Ce6 nanoformulation with dual nanozyme activities for synergistic chemodynamic therapy/phototherapy. *Biomaterials* 252: 120093.

77 Wang, M., Chang, M.Y., Zheng, P. et al. (2023). A noble AuPtAg-GOx nanozyme for synergistic tumor immunotherapy induced by starvation therapy-augmented mild photothermal therapy. *Advanced Science* 9: 2202332.

78 Wu, C.M., Naseem, S., Chou, M.H. et al. (2019). Recent advances in tungsten-oxide-based materials and their applications. *Frontiers in Materials* 6: 49.

79 Manthiram, K. and Alivisatos, A.P. (2012). Tunable localized surface plasmon resonances in tungsten oxide nanocrystals. *Journal of the American Chemical Society* 134: 3995–3998.

80 Yan, J., Wang, T., Wu, G. et al. (2015). Tungsten oxide single crystal nanosheets for enhanced multichannel solar light harvesting. *Advanced Materials* 27: 1580–1586.

81 Liu, J., Han, J., Kang, Z. et al. (2014). In vivo near-infrared photothermal therapy and computed tomography imaging of cancer cells using novel tungsten-based theranostic probe. *Nanoscale* 6: 5770–5776.

82 Liang, H., Xi, H., Liu, S. et al. (2019). Modulation of oxygen vacancy in tungsten oxide nanosheets for vis-NIR light-enhanced electrocatalytic hydrogen production and anticancer photothermal therapy. *Nanoscale* 11: 18183–18190.

83 Dong, Y., Dong, S., Wang, Z. et al. (2020). Multimode imaging-guided photothermal/chemodynamic synergistic therapy nanoagent with a tumor microenvironment responded effect. *ACS Applied Materials & Interfaces* 12: 52479–52491.

84 Zhou, Z., Wang, X., Zhang, H. et al. (2021). Activating layered metal oxide nanomaterials via structural engineering as biodegradable nanoagents for photothermal cancer therapy. *Small* 17: 2007486.

85 Mao, F., Liu, Y., Ma, L. et al. (2019). Green synthesis of ultra-small VO_x nanodots for acidic-activated HSP60 inhibition and therapeutic enhancement. *Biomaterials* 194: 94–104.

86 Shen, S., Wang, S., Zheng, R. et al. (2015). Magnetic nanoparticle clusters for photothermal therapy with near-infrared irradiation. *Biomaterials* 39: 67–74.

87 Nicolas-Boluda, A., Yang, Z., Guilbert, T. et al. (2021). Self-assemblies of Fe_3O_4 nanocrystals: toward nanoscale precision of photothermal effects in the tumor microenvironment. *Advanced Functional Materials* 31: 2006824.

88 Zhou, Z., Tian, R., Wang, Z. et al. (2017). Artificial local magnetic field inhomogeneity enhances T2 relaxivity. *Nature Communications* 8: 15468.

89 Zhou, Z., Song, J., Tian, R. et al. (2017). Activatable singlet oxygen generation from lipid hydroperoxide nanoparticles for cancer therapy. *Angewandte Chemie International Edition* 129: 6592–6596.

90 Li, Y., Lu, W., Huang, Q. et al. (2010). Copper sulfide nanoparticles for photothermal ablation of tumor cells. *Nanomedicine* 5: 1161–1171.

91 Li, B., Wang, Q., Zou, R. et al. (2014). $Cu_{7.2}S_4$ nanocrystals: a novel photothermal agent with a 56.7% photothermal conversion efficiency for photothermal therapy of cancer cells. *Nanoscale* 6: 3274–3282.

92 Wen, M., Ouyang, J., Wei, C. et al. (2019). Artificial enzyme catalyzed cascade reactions: antitumor immunotherapy reinforced by NIR-II light. *Angewandte Chemie International Edition* 58: 17425–17432.

93 Chen, Y., Tan, C., Zhang, H., and Wang, L. (2015). Two-dimensional graphene analogues for biomedical applications. *Chemical Society Reviews* 44: 2681–2701.

94 Zhu, S., Gong, L., Xie, J. et al. (2017). Design, synthesis, and surface modification of materials based on transition-metal dichalcogenides for biomedical applications. *Small Methods* 1: 1700220.

95 Chou, S.S., Kaehr, B., Kim, J. et al. (2013). Chemically exfoliated MoS_2 as near-infrared photothermal agents. *Angewandte Chemie International Edition* 125: 4254–4258.

96 Zhou, Z., Li, B., Shen, C. et al. (2020). Metallic 1T phase enabling MoS_2 nanodots as an efficient agent for photoacoustic imaging guided photothermal therapy in the near-infrared-II window. *Small* 16: 2004173.

97 Gao, H., Bi, Y., Wang, X. et al. (2017). Near-infrared guided thermal-responsive nanomedicine against orthotopic superficial bladder cancer. *ACS Biomaterials Science and Engineering* 3: 3628–3634.

98 Wang, Y., Zhao, J., Chen, Z. et al. (2019). Construct of $MoSe_2/Bi_2Se_3$ nanoheterostructure: multimodal CT/PT imaging-guided PTT/PDT/chemotherapy for cancer treating. *Biomaterials* 217: 119282.

99 Li, Y., Jia, R., Lin, H. et al. (2021). Synthesis of $MoSe_2/CoSe_2$ nanosheets for NIR-enhanced chemodynamic therapy via synergistic in-situ H_2O_2 production and activation. *Advanced Functional Materials* 31: 2008420.

100 Xiao, Z., Xu, C., Jiang, X. et al. (2016). Hydrophilic bismuth sulfur nanoflower superstructures with an improved photothermal efficiency for ablation of cancer cells. *Nano Research* 9: 1934–1947.

101 Cheng, L., Liu, J., Gu, X. et al. (2014). PEGylated WS_2 nanosheets as a multifunctional theranostic agent for in vivo dual-modal CT/photoacoustic imaging guided photothermal therapy. *Advanced Materials* 26: 1886–1893.

102 Cheng, L., Yuan, C., Shen, S. et al. (2015). Bottom-up synthesis of metal-ion-doped WS_2 nanoflakes for cancer theranostics. *ACS Nano* 9: 11090–11101.

103 Augustine, S., Singh, J., Srivastava, M. et al. (2017). Recent advances in carbon based nanosystems for cancer theranostics. *Biomaterials Science* 5: 901–952.

104 Lan, M., Guo, L., Zhao, S. et al. (2018). Carbon dots as multifunctional phototheranostic agents for photoacoustic/fluorescence imaging and photothermal/photodynamic synergistic cancer therapy. *Advanced Therapeutics* 1: 1800077.

105 Mendes, R.G., Bachmatiuk, A., Büchner, B. et al. (2013). Carbon nanostructures as multi-functional drug delivery platforms. *Journal of Materials Chemistry B* 1: 401–428.

106 Liu, Z., Robinson, J.T., Tabakman, S.M. et al. (2011). Carbon materials for drug delivery and cancer therapy. *Materials Today* 14: 316–323.

107 Novoselov, K.S., Geim, A.K., Morozov, S.V. et al. (2004). Electric field effect in atomically thin carbon films. *Science* 306: 666–669.

108 Novoselov, K.S., Geim, A.K., Morozov, S.V. et al. (2005). Two-dimensional gas of massless dirac fermions in graphene. *Nature* 438: 197–200.

109 Park, J.U., Nam, S., Lee, M.S., and Lieber, C.M. (2012). Synthesis of monolithic graphene-graphite integrated electronics. *Nature Materials* 11: 120–125.

110 Yang, K., Feng, L., and Liu, Z. (2016). Stimuli responsive drug delivery systems based on nano-graphene for cancer therapy. *Advanced Drug Delivery Reviews* 105: 228–241.

111 Cheng, S.J., Chiu, H.Y., Kumar, P.V. et al. (2018). Simultaneous drug delivery and cellular imaging using graphene oxide. *Biomaterials Science* 6: 813–819.

112 Shen, H., Zhang, L., Liu, M., and Zhang, Z. (2012). Biomedical applications of graphene. *Theranostics* 2: 283–294.

113 Shin, S.R., Li, Y.C., Jang, H.L. et al. (2016). Graphene-based materials for tissue engineering. *Advanced Drug Delivery Reviews* 105: 255–274.

114 Chang, X., Zhang, M., Wang, C. et al. (2020). Graphene oxide/$BaHoF_5$/PEG nanocomposite for dual-modal imaging and heat shock protein inhibitor-sensitized tumor photothermal therapy. *Carbon* 158: 372–385.

115 Bai, G., Yuan, P., Cai, B. et al. (2019). Stimuli-responsive scaffold for breast cancer treatment combining accurate photothermal therapy and adipose tissue regeneration. *Advanced Functional Materials* 29: 1904401.

116 Zhang, M., Wang, W., Cui, Y. et al. (2018). Magnetofluorescent Fe_3O_4/carbon quantum dots coated single-walled carbon nanotubes as dual-modal targeted imaging and chemo/photodynamic/photothermal triple-modal therapeutic agents. *Chemical Engineering Journal* 338: 526–538.

117 Bai, H., Li, C., and Shi, G. (2011). Functional composite materials based on chemically converted graphene. *Advanced Materials* 23: 1089–1115.

118 Schwenke, A.M., Hoeppener, S., and Schubert, U.S. (2015). Microwave synthesis of carbon nanofibers-the influence of MW irradiation power, time, and the amount of catalyst. *Journal of Materials Chemistry A* 3: 23778–23787.

119 Li, X., Wang, H., Robinson, J.T. et al. (2009). Simultaneous nitrogen doping and reduction of graphene oxide. *Journal of the American Chemical Society* 131: 15939–15944.

120 Pei, S. and Cheng, H.M. (2012). The reduction of graphene oxide. *Carbon* 50: 3210–3228.

121 Dubin, S., Gilje, S., Wang, K. et al. (2010). A one-step, solvothermal reduction method for producing reduced graphene oxide dispersions in organic solvents. *ACS Nano* 4: 3845–3852.

122 Robinson, J.T., Tabakman, S.M., Liang, Y. et al. (2011). Ultrasmall reduced graphene oxide with high near-infrared absorbance for photothermal therapy. *Journal of the American Chemical Society* 133: 6825–6831.

123 Saifuddin, N., Raziah, A., and Junizah, A. (2013). Carbon nanotubes: a review on structure and their interaction with proteins. *Journal of Chemistry* 676815.

124 Rao, R., Pint, C.L., Islam, A.E. et al. (2018). Carbon nanotubes and related nanomaterials: critical advances and challenges for synthesis toward mainstream commercial applications. *ACS Nano* 12: 11756–11784.

125 Wu, H., Zhong, D., Zhang, Z. et al. (2020). Bioinspired artificial tobacco mosaic virus with combined oncolytic properties to completely destroy multidrug-resistant cancer. *Advanced Materials* 32: 1904958.

126 Wang, S., Zhang, Q., Yang, P. et al. (2016). Manganese oxide-coated carbon nanotubes as dual-modality lymph mapping agents for photothermal therapy of tumor metastasis. *ACS Applied Materials & Interfaces* 8: 3736–3743.

127 Xie, X., Li, Z., Zhang, Y. et al. (2017). Emerging ≈ 800 nm excited lanthanide-doped upconversion nanoparticles. *Small* 13: 1602843.

128 Rosal, B., Pérez-Delgado, A., Carrasco, E. et al. (2016). Neodymium-based stoichiometric ultrasmall nanoparticles for multifunctional deep-tissue photothermal therapy. *Advanced Optical Materials* 4: 782–789.

129 Carrasco, E., Del Rosal, B., Sanz-Rodríguez, F. et al. (2015). Intratumoral thermal reading during photo-thermal therapy by multifunctional fluorescent nanoparticles. *Advanced Functional Materials* 25: 615–626.

130 Bednarkiewicz, A., Wawrzynczyk, D., Nyk, M., and Strek, W. (2011). Optically stimulated heating using Nd^{3+} doped $NaYF_4$ colloidal near infrared nanophosphors. *Applied Physics B* 103: 847–852.

131 Lv, R., Yang, G., He, F. et al. (2014). LaF_3: Ln mesoporous spheres: controllable synthesis, tunable luminescence and application for dual-modal chemo-/photo-thermal therapy. *Nanoscale* 6: 14799–14809.

132 Xie, H., Liu, M., You, B. et al. (2020). Biodegradable Bi_2O_2Se quantum dots for photoacoustic imaging-guided cancer photothermal therapy. *Small* 16: 1905208.

133 Rosal, B., Carrasco, E., Ren, F. et al. (2016). Infrared-emitting QDs for thermal therapy with real-time subcutaneous temperature feedback. *Advanced Functional Materials* 26: 6060–6068.

134 Kou, L., Chen, C., and Smith, S.C. (2015). Phosphorene: fabrication, properties, and applications. *The Journal of Physical Chemistry Letters* 6: 2794–2805.

135 Xu, Y., Wang, Z., Guo, Z. et al. (2016). Solvothermal synthesis and ultrafast photonics of black phosphorus quantum dots. *Advanced Optical Materials* 4: 1223–1229.

136 Huang, J., He, B., Zhang, Z. et al. (2020). Aggregation-induced emission luminogens married to 2D black phosphorus nanosheets for highly efficient multimodal theranostics. *Advanced Materials* 32: 2003382.

137 Gui, R., Jin, H., Wang, Z., and Li, J. (2018). Black phosphorus quantum dots: synthesis, properties, functionalized modification and applications. *Chemical Society Reviews* 47: 6795–6823.

138 Xie, J., Fan, T., Kim, J.H. et al. (2020). Emetine-loaded black phosphorus hydrogel sensitizes tumor to photothermal therapy through inhibition of stress granule formation. *Advanced Functional Materials* 30: 2003891.

139 Wang, D., Ge, C., Liang, W. et al. (2020). In vivo enrichment and elimination of circulating tumor cells by using a black phosphorus and antibody functionalized intravenous catheter. *Advanced Science* 7: 2000940.

140 Anju, S., Ashtami, J., and Mohanan, P. (2019). Black phosphorus, a prospective graphene substitute for biomedical applications. *Materials Science and Engineering: C* 97: 978–993.

141 Lei, W., Liu, G., Zhang, J., and Liu, M. (2017). Black phosphorus nanostructures: recent advances in hybridization, doping and functionalization. *Chemical Society Reviews* 46: 3492–3509.

142 Ziletti, A., Carvalho, A., Campbell, D.K. et al. (2015). Oxygen defects in phosphorene. *Physical Review Letters* 114: 046801.

143 Nam, J., Son, S., Ochyl, L.J. et al. (2018). Chemo-photothermal therapy combination elicits anti-tumor immunity against advanced metastatic cancer. *Nature Communications* 9: 1074.

144 Wang, C., Xu, L., Liang, C. et al. (2014). Immunological responses triggered by photothermal therapy with carbon nanotubes in combination with anti-CTLA-4 therapy to inhibit cancer metastasis. *Advanced Materials* 26: 8154–8162.

145 Harmon, B., Corder, A., Collins, R. et al. (1990). Cell death induced in a murine mastocytoma by 42-47 C heating in vitro: evidence that the form of death changes from apoptosis to necrosis above a critical heat load. *International Journal of Radiation Biology* 58: 845–858.

146 Galán, A., García-Bermejo, L., Troyano, A. et al. (2001). The role of intracellular oxidation in death induction (apoptosis and necrosis) in human promonocytic cells treated with stress inducers (cadmium, heat, X-rays). *European Journal of Cell Biology* 80: 312–320.

147 Roti, J.L. (2008). Cellular responses to hyperthermia (40-46 C): cell killing and molecular events. *International Journal of Hyperthermia* 24: 3–15.

148 Knavel, E.M. and Brace, C.L. (2013). Tumor ablation: common modalities and general practices. *Techniques in Vascular and Interventional Radiology* 16: 192–200.

149 Hildebrandt, B., Wust, P., Ahlers, O. et al. (2002). The cellular and molecular basis of hyperthermia. *Critical Reviews in Oncology/Hematology* 43: 33–56.

150 Wang, S., Zhang, Q., Luo, X.F. et al. (2014). Magnetic graphene-based nanotheranostic agent for dual-modality mapping guided photothermal therapy in regional lymph nodal metastasis of pancreatic cancer. *Biomaterials* 35: 9473–9483.

151 Pockley, A.G. (2003). Heat shock proteins as regulators of the immune response. *The Lancet* 362: 469–476.

152 Feng, H., Zeng, Y., Graner, M.W., and Katsanis, E. (2002). Stressed apoptotic tumor cells stimulate dendritic cells and induce specific cytotoxic T cells. *Blood* 100: 4108–4115.

153 Melamed, J.R., Edelstein, R.S., and Day, E.S. (2015). Elucidating the fundamental mechanisms of cell death triggered by photothermal therapy. *ACS Nano* 9: 6–11.

154 Kono, H. and Rock, K.L. (2008). How dying cells alert the immune system to danger. *Nature Reviews Immunology* 8: 279–289.

155 Riedl, S.J. and Salvesen, G.S. (2007). The apoptosome: signalling platform of cell death. *Nature Reviews Molecular Cell Biology* 8: 405–413.

156 Pérez-Hernández, M., Del Pino, P., Mitchell, S.G. et al. (2015). Dissecting the molecular mechanism of apoptosis during photothermal therapy using gold nanoprisms. *ACS Nano* 9: 52–61.

157 Asadzadeh, Z., Safarzadeh, E., Safaei, S. et al. (2020). Current approaches for combination therapy of cancer: the role of immunogenic cell death. *Cancers* 12: 1047.

158 Zitvogel, L., Kepp, O., and Kroemer, G. (2010). Decoding cell death signals in inflammation and immunity. *Cell* 140: 798–804.

159 Gao, J., Wang, W., Pei, Q. et al. (2020). Engineering nanomedicines through boosting immunogenic cell death for improved cancer immunotherapy. *Acta Pharmacologica Sinica* 41: 986–994.

160 Duan, X., Chan, C., and Lin, W. (2019). Nanoparticle-mediated immunogenic cell death enables and potentiates cancer immunotherapy. *Angewandte Chemie International Edition* 58: 670–680.

161 Sweeney, E.E., Cano-Mejia, J., and Fernandes, R. (2018). Photothermal therapy generates a thermal window of immunogenic cell death in neuroblastoma. *Small* 14: 1800678.

162 Hanahan, D. and Weinberg, R.A. (2011). Hallmarks of cancer: the next generation. *Cell* 144: 646–674.

163 Dart, A. (2019). Immunosuppressive lymphatics. *Nature Reviews Cancer* 19: 6–7.

164 Cassim, S. and Pouyssegur, J. (2019). Tumor microenvironment: a metabolic player that shapes the immune response. *International Journal of Molecular Sciences* 21: 157.

165 Shi, C., Li, M., Zhang, Z. et al. (2020). Catalase-based liposomal for reversing immunosuppressive tumor microenvironment and enhanced cancer chemo-photodynamic therapy. *Biomaterials* 233: 119755.

166 Phuengkham, H., Ren, L., Shin, I.W., and Lim, Y.T. (2019). Nanoengineered immune niches for reprogramming the immunosuppressive tumor microenvironment and enhancing cancer immunotherapy. *Advanced Materials* 31: 1803322.

167 Chen, Q., Wang, C., Zhang, X. et al. (2019). In situ sprayed bioresponsive immunotherapeutic gel for post-surgical cancer treatment. *Nature Nanotechnology* 14: 89–97.

168 Song, M., Liu, T., Shi, C. et al. (2016). Bioconjugated manganese dioxide nanoparticles enhance chemotherapy response by priming tumor-associated macrophages toward M1-like phenotype and attenuating tumor hypoxia. *ACS Nano* 10: 633–647.

169 Nie, W., Wu, G., Zhang, J. et al. (2020). Responsive exosome nano-bioconjugates for synergistic cancer therapy. *Angewandte Chemie International Edition* 59: 2018–2022.

170 Ma, Y., Zhang, Y., Li, X. et al. (2019). Near-infrared II phototherapy induces deep tissue immunogenic cell death and potentiates cancer immunotherapy. *ACS Nano* 13: 11967–11980.

171 Huang, L., Li, Y., Du, Y. et al. (2019). Mild photothermal therapy potentiates anti-PD-L1 treatment for immunologically cold tumors via an all-in-one and all-in-control strategy. *Nature Communications* 10: 4871.

172 Stapleton, S., Dunne, M., Milosevic, M. et al. (2018). Radiation and heat improve the delivery and efficacy of nanotherapeutics by modulating intratumoral fluid dynamics. *ACS Nano* 12: 7583–7600.

173 Chen, Q., Hu, Q., Dukhovlinova, E. et al. (2019). Photothermal therapy promotes tumor infiltration and antitumor activity of CAR T cells. *Advanced Materials* 31: 1900192.

174 Chen, Z., Pan, H., Luo, Y. et al. (2021). Nanoengineered CAR-T biohybrids for solid tumor immunotherapy with microenvironment photothermal-remodeling strategy. *Small* 17: 2007494.

175 Miller, I.C., Zamat, A., Sun, L.K. et al. (2021). Enhanced intratumoural activity of CAR T cells engineered to produce immunomodulators under photothermal control. *Nature Biomedical Engineering* 5: 1348–1359.

176 Ma, W., Zhu, D., Li, J. et al. (2020). Coating biomimetic nanoparticles with chimeric antigen receptor T cell-membrane provides high specificity for hepatocellular carcinoma photothermal therapy treatment. *Theranostics* 10: 1281–1295.

177 Balkwill, F., Charles, K.A., and Mantovani, A. (2005). Smoldering and polarized inflammation in the initiation and promotion of malignant disease. *Cancer Cell* 7: 211–217.

178 Wu, Q., Yu, J., Li, M. et al. (2018). Nanoengineering of nanorattles for tumor treatment by CT imaging-guided simultaneous enhanced microwave thermal therapy and managing inflammation. *Biomaterials* 179: 122–133.

179 Koyama, S., Akbay, E.A., Li, Y.Y. et al. (2016). Adaptive resistance to therapeutic PD-1 blockade is associated with upregulation of alternative immune checkpoints. *Nature Communications* 7: 10501.

180 Garon, E.B., Rizvi, N.A., Hui, R. et al. (2015). Pembrolizumab for the treatment of non-small-cell lung cancer. *New England Journal of Medicine* 372: 2018–2028.

181 Ferris, R.L., Blumenschein, G. Jr., Fayette, J. et al. (2016). Nivolumab for recurrent squamous-cell carcinoma of the head and neck. *New England Journal of Medicine* 375: 1856–1867.

182 Korman, A., Chen, B., Wang, C. et al. (2007). Activity of anti-PD-1 in murine tumor models: role of "host" PD-L1 and synergistic effect of anti-PD-1 and anti-CTLA-4 (48.37). *The Journal of Immunology* 178: 82.

183 Curran, M.A., Montalvo, W., Yagita, H., and Allison, J.P. (2010). PD-1 and CTLA-4 combination blockade expands infiltrating T cells and reduces regulatory T and myeloid cells within B16 melanoma tumors. *Proceedings of the National Academy of Sciences* 107: 4275–4280.

184 Kalbasi, A. and Ribas, A. (2020). Tumour-intrinsic resistance to immune checkpoint blockade. *Nature Reviews Immunology* 20: 25–39.

185 Pitt, J.M., Vétizou, M., Daillère, R. et al. (2016). Resistance mechanisms to immune-checkpoint blockade in cancer: tumor-intrinsic and-extrinsic factors. *Immunity* 44: 1255–1269.

186 Wolchok, J.D. and Saenger, Y. (2008). The mechanism of anti-CTLA-4 activity and the negative regulation of T-cell activation. *The Oncologist* 13: 2–9.

187 Pardoll, D.M. (2012). The blockade of immune checkpoints in cancer immunotherapy. *Nature Reviews Cancer* 12: 252–264.

188 Paulsen, E.E., Kilvaer, T.K., Rakaee, M. et al. (2017). CTLA-4 expression in the non-small cell lung cancer patient tumor microenvironment: diverging prognostic impact in primary tumors and lymph node metastases. *Cancer Immunology, Immunotherapy* 66: 1449–1461.

189 Chae, Y.K., Arya, A., Iams, W. et al. (2018). Current landscape and future of dual anti-CTLA4 and PD-1/PD-L1 blockade immunotherapy in cancer; lessons learned from clinical trials with melanoma and non-small cell lung cancer (NSCLC). *Journal for Immunotherapy of Cancer* 6: 1–27.

190 Bajor, D.L., Mick, R., Riese, M.J. et al. (2018). Long-term outcomes of a phase I study of agonist CD40 antibody and CTLA-4 blockade in patients with metastatic melanoma. *Oncoimmunology* 7: e1468956.

191 Lipson, E.J. and Drake, C.G. (2011). Lpilimumab: an anti-CTLA-4 antibody for metastatic melanoma. *Clinical Cancer Research* 17: 6958–6962.

192 Chang, M., Wang, M., Wang, M. et al. (2019). A multifunctional cascade bioreactor based on hollowstructured Cu_2MoS_4 for synergetic cancer chemo-dynamic therapy/starvation therapy/phototherapy/immunotherapy with remarkably enhanced efficacy. *Advanced Materials* 31: 1905271.

193 Salmaninejad, A., Valilou, S.F., Shabgah, A.G. et al. (2019). PD-1/PD-L1 pathway: basic biology and role in cancer immunotherapy. *Journal of Cellular Physiology* 234: 16824–16837.

194 Haslam, A. and Prasad, V. (2019). Estimation of the percentage of US patients with cancer who are eligible for and respond to checkpoint inhibitor immunotherapy drugs. *JAMA Network Open* 2: e192535.

195 Pan, J., Hu, P., Guo, Y. et al. (2020). Combined magnetic hyperthermia and immune therapy for primary and metastatic tumor treatments. *ACS Nano* 14: 1033–1044.

196 Zhang, Y., Zhou, Z., Gao, Z. et al. (2023). Mild photothermal treatment sensitized immune checkpoint blockade therapy based on ATP-exhausted nanoenzymes. *Chemical Engineering Journal* 474: 145677.

197 Fang, X., Wu, X., Li, Z. et al. (2021). Biomimetic anti-PD-1 peptide-loaded 2D $FePSe_3$ nanosheets for efficient photothermal and enhanced immune therapy with multimodal MR/PA/thermal imaging. *Advanced Science* 8: 2003041.

198 DeNardo, D.G. and Ruffell, B. (2019). Macrophages as regulators of tumour immunity and immunotherapy. *Nature Reviews Immunology* 19: 369–382.

199 Barclay, A.N. and Van den Berg, T.K. (2014). The interaction between signal regulatory protein alpha (SIRPα) and CD47: structure, function, and therapeutic target. *Annual Review of Immunology* 32: 25–50.

200 Chao, M.P., Alizadeh, A.A., Tang, C. et al. (2011). Therapeutic antibody targeting of CD47 eliminates human acute lymphoblastic leukemia. *Cancer Research* 71: 1374–1384.

201 Gholamin, S., Mitra, S.S., Feroze, A.H. et al. (2017). Disrupting the CD47-SIRPα anti-phagocytic axis by a humanized anti-CD47 antibody is an efficacious treatment for malignant pediatric brain tumors. *Science Translational Medicine* 9: eaaf2968.

202 Willingham, S.B., Volkmer, J.P., Gentles, A.J. et al. (2012). The CD47-signal regulatory protein alpha (SIRPa) interaction is a therapeutic target for human solid tumors. *Proceedings of the National Academy of Sciences* 109: 6662–6667.

203 Chao, M.P., Alizadeh, A.A., Tang, C. et al. (2010). Anti-CD47 antibody synergizes with rituximab to promote phagocytosis and eradicate non-Hodgkin lymphoma. *Cell* 142: 699–712.

204 Jaiswal, S., Jamieson, C.H., Pang, W.W. et al. (2009). CD47 is upregulated on circulating hematopoietic stem cells and leukemia cells to avoid phagocytosis. *Cell* 138: 271–285.

205 Majeti, R., Chao, M.P., Alizadeh, A.A. et al. (2009). CD47 is an adverse prognostic factor and therapeutic antibody target on human acute myeloid leukemia stem cells. *Cell* 138: 286–299.

206 Tseng, D., Volkmer, J.P., Willingham, S.B. et al. (2013). Anti-CD47 antibody-mediated phagocytosis of cancer by macrophages primes an effective antitumor T-cell response. *Proceedings of the National Academy of Sciences* 110: 11103–11108.

207 Zhaoming, G.Y.L., Zhou, H., Zheng, K. et al. (2019). CD47-targeted bismuth selenide nanoparticles actualize improved photothermal therapy by increasing macrophage phagocytosis of cancer cells. *Colloids and Surfaces B: Biointerfaces* 184: 110546.

208 Xie, Z., Peng, M., Lu, R. et al. (2020). Black phosphorus-based photothermal therapy with aCD47-mediated immune checkpoint blockade for enhanced cancer immunotherapy. *Light: Science and Applications* 9: 161.

209 Liu, Y., Lu, Y., Zhu, X. et al. (2020). Tumor microenvironment-responsive prodrug nanoplatform via co-self-assembly of photothermal agent and IDO inhibitor for enhanced tumor penetration and cancer immunotherapy. *Biomaterials* 242: 119933.

210 Li, Y., Liu, X., Pan, W. et al. (2020). Photothermal therapy-induced immunogenic cell death based on natural melanin nanoparticles against breast cancer. *Chemical Communications* 56: 1389–1392.

211 Xie, Y.L., Wang, M., Qian, Y.R. et al. (2023). Novel PdPtCu nanozymes for reprogramming tumor microenvironment to boost immunotherapy through endoplasmic reticulum stress and blocking IDO-mediated Immune escape. *Small* 19: 2303596.

212 Xie, Y.L., Wang, M., Qiao, L.Y. et al. (2023). Photothermal-enhanced dual inhibition of lactate/kynurenine metabolism for promoting tumor immunotherapy. *Small Methods* 7: 2300945.

213 Kawai, T. and Akira, S. (2010). The role of pattern-recognition receptors in innate immunity: update on toll-like receptors. *Nature immunology* 11: 373–384.

214 Liston, A. and Masters, S.L. (2017). Homeostasis-altering molecular processes as mechanisms of inflammasome activation. *Nature Reviews Immunology* 17: 208–214.

215 Kawai, T. and Akira, S. (2011). Toll-like receptors and their crosstalk with other innate receptors in infection and immunity. *Immunity* 34: 637–650.

216 Wang, Y., Song, E., Bai, B., and Vanhoutte, P.M. (2016). Toll-like receptors mediating vascular malfunction: lessons from receptor subtypes. *Pharmacology and Therapeutics* 158: 91–100.

217 Banerjee, S., Thompson, W.E., and Chowdhury, I. (2021). Emerging roles of microRNAs in the regulation of toll-like receptor (TLR)-signaling. *Frontiers in Bioscience (Landmark edition)* 26: 771.

218 Lu, R., Groer, C., Kleindl, P.A. et al. (2019). Formulation and preclinical evaluation of a toll-like receptor 7/8 agonist as an anti-tumoral immunomodulator. *Journal of Controlled Release* 306: 165–176.

219 Dovedi, S.J., Melis, M.H., Wilkinson, R.W. et al. (2013). Systemic delivery of a TLR7 agonist in combination with radiation primes durable antitumor immune responses in mouse models of lymphoma. *Blood* 121: 251–259.

220 Klinman, D.M. (2004). Immunotherapeutic uses of CpG oligodeoxynucleotides. *Nature Reviews Immunology* 4: 249–259.

221 Guo, L., Yan, D.D., Yang, D. et al. (2014). Combinatorial photothermal and immuno cancer therapy using chitosan-coated hollow copper sulfide nanoparticles. *ACS Nano* 8: 5670–5681.

222 Guo, Y., Ran, Y., Wang, Z. et al. (2019). Magnetic-responsive and targeted cancer nanotheranostics by PA/MR bimodal imaging-guided photothermally triggered immunotherapy. *Biomaterials* 219: 119370.

223 Jiang, F., Ding, B., Liang, S. et al. (2021). Intelligent MoS_2-CuO heterostructures with multiplexed imaging and remarkably enhanced antitumor efficacy via synergetic photothermal therapy/chemodynamic therapy/immunotherapy. *Biomaterials* 268: 120545.

224 Xu, L., Zhang, W., Park, H.B. et al. (2019). Indocyanine green and poly I: C containing thermo-responsive liposomes used in immune-photothermal therapy prevent cancer growth and metastasis. *Journal for Immunotherapy of Cancer* 7: 1–14.

225 Nam, J., Son, S., Park, K.S., and Moon, J.J. (2021). Photothermal therapy combined with neoantigen cancer vaccination for effective immunotherapy against large established tumors and distant metastasis. *Advanced Therapeutics* 4: 2100093.

226 Jia, Y.P., Shi, K., Yang, F. et al. (2020). Multifunctional nanoparticle loaded injectable thermoresponsive hydrogel as NIR controlled release platform for local photothermal immunotherapy to prevent breast cancer postoperative recurrence and metastases. *Advanced Functional Materials* 30: 2001059.

227 Gunzer, M., Riemann, H., Basoglu, Y. et al. (2005). Systemic administration of a TLR7 ligand leads to transient immune incompetence due to peripheral-blood leukocyte depletion. *Blood* 106: 2424–2432.

228 Goldstein, D., Hertzog, P., Tomkinson, E. et al. (1998). Administration of imiquimod, an interferon inducer, in asymptomatic human immunodeficiency virus-infected persons to determine safety and biologic response modification. *Journal of Infectious Diseases* 178: 858–861.

229 Wang, Z. and Cao, Y.J. (2020). Adoptive cell therapy targeting neoantigens: a frontier for cancer research. *Frontiers in Immunology* 11: 176.

230 Ingegnere, T., Mariotti, F.R., Pelosi, A. et al. (2019). Human CAR NK cells: a new non-viral method allowing high efficient transfection and strong tumor cell killing. *Frontiers in Immunology* 10: 957.

231 Li, H., Yang, X., Wang, Z. et al. (2023). A near-infrared-II fluorescent nanocatalyst for enhanced CAR T cell therapy against solid tumor by immune reprogramming. *ACS Nano* 17: 11749–11763.

232 Lin, X., Li, F., Gu, Q. et al. (2022). Gold-seaurchin based immunomodulator enabling photothermal intervention and αCD16 transfection to boost NK cell adoptive immunotherapy. *Acta Biomaterialia* 146: 406–420.

233 Solinas, C., Aiello, M., Migliori, E. et al. (2020). Breast cancer vaccines: heeding the lessons of the past to guide a path forward. *Cancer Treatment Reviews* 84: 101947.

234 Buonaguro, L. and Consortium, H (2017). New vaccination strategies in liver cancer. *Cytokine and Growth Factor Reviews* 36: 125–129.

235 Tagliamento, M., Rijavec, E., Barletta, G. et al. (2018). CIMAvax-EGF, a therapeutic non-small cell lung cancer vaccine. *Expert Opinion on Biological Therapy* 18: 829–835.

236 Dany, M., Nganga, R., Chidiac, A. et al. (2016). Advances in immunotherapy for melanoma management. *Human Vaccines and Immunotherapeutics* 12: 2501–2511.

237 Klein, O., Schmidt, C., Knights, A. et al. (2011). Melanoma vaccines: developments over the past 10 years. *Expert Review of Vaccines* 10: 853–873.

238 Andtbacka, R.H.I., Collichio, F.A., Amatruda, T. et al. (2013). OPTiM: a randomized phase III trial of talimogene laherparepvec (T-VEC) versus subcutaneous (SC) granulocyte-macrophage colony-stimulating factor (GM-CSF) for the treatment (tx) of unresected stage IIIB/C and IV melanoma. *Journal of Clinical Oncology* 31: LBA9008.

239 Puzanov, I., Milhem, M.M., Andtbacka, R.H.I. et al. (2014). Primary analysis of a phase 1b multicenter trial to evaluate safety and efficacy of talimogene laherparepvec (T-VEC) and ipilimumab (ipi) in previously untreated, unresected stage IIIB-IV melanoma. *American Society of Clinical Oncology* 32: 9029.

240 Nestle, F.O., Alijagic, S., Gilliet, M. et al. (1998). Vaccination of melanoma patients with peptide-or tumorlysate-pulsed dendritic cells. *Nature Medicine* 4: 328–332.

241 Lesterhuis, W.J., Schreibelt, G., Scharenborg, N.M. et al. (2011). Wild-type and modified gp100 peptide-pulsed dendritic cell vaccination of advanced melanoma patients can lead to long-term clinical responses independent of the peptide used. *Cancer Immunology, Immunotherapy* 60: 249–260.

242 Greene, J.M., Schneble, E.J., Jackson, D.O. et al. (2016). A phase I/IIa clinical trial in stage IV melanoma of an autologous tumor-dendritic cell fusion (dendritoma) vaccine with low dose interleukin-2. *Cancer Immunology, Immunotherapy* 65: 383–392.

13

Near-Infrared Region-Responsive Antimicrobial Nanomaterials for the Treatment of Multidrug-Resistant Bacteria

Manlin Qi[1], Shangyan Shan[1], Biao Dong[2], and Lin Wang[1]

[1]Jilin University, School of Stomatology, Jilin Provincial Key Laboratory of Tooth Development and Bone Remodeling, Department of Oral Implantology, Qinghua Road 1500, 130021, Changchun, P. R. China
[2]Jilin University, College of Electronic Science and Engineering, State Key Laboratory on Integrated Optoelectronics, Department of Biomedical Engineering, Qianjin Str.2699, 130012, Changchun, P. R. China

Drugs will be found before microbes overwhelm existing ones.
Prof. Laura Piddock, 2020

13.1 Introduction

Bacterial infections pose a persistent threat to global health, affecting various organs and potentially leading to critical conditions such as endocarditis, meningitis, and systemic sepsis. These infections place a significant economic strain on individuals, healthcare systems, and societies at large. The advent of penicillin in the 1940s marked the beginning of the antibiotic era. Nonetheless, antibiotic resistance has escalated to a major public health issue, complicating the treatment of bacterial infections, prolonging illnesses, inflating healthcare expenditures, and increasing the risk of severe complications. Antibiotic resistance is now a leading cause of mortality worldwide, trailing only ischemic heart disease and stroke. A 2019 analysis across 204 countries estimated 1.27 million direct fatalities due to antibiotic resistance, with an associated 4.95 million deaths [1]. In the United States alone, multidrug-resistant (MDR) bacteria account for over 600,000 hospitalizations annually [2]. The emergence of MDR and pan-drug-resistant (PDR) bacteria, or "superbugs," has led to infections that defy current antimicrobial treatments. Consequently, the fight against MDR and PDR pathogens is an urgent global health challenge. Although numerous recommendations and guidelines have been proposed for managing MDR bacterial infections, their effectiveness is often short-lived [3, 4]. The pipeline for new antimicrobials is dwindling, with rapid resistance development anticipated even for new drugs. A WHO report highlights that only a handful of the 32 antibiotics in clinical development are considered

Photofunctional Nanomaterials for Biomedical Applications, First Edition.
Edited by Chunxia Li and Jun Lin.

innovative against priority pathogens, and there is a notable scarcity of high-quality antimicrobials.

The rise of nanotechnology has opened innovative avenues to address MDR and biofilm-associated infections [5]. Nanomaterials disrupt traditional bacterial resistance mechanisms, offering several advantages in the fight against MDR bacteria [6–8]: (i) Enhanced antibacterial efficacy due to a larger surface-to-volume ratio, facilitating increased bacterial interaction and improved antimicrobial action. The modifiable size and shape of nanomaterials further optimize their antibacterial properties. (ii) Broad-spectrum functionality, with nanomaterials inherently lacking antibiotic resistance, unlike conventional antibiotics that target specific bacteria. (iii) Multifunctional capabilities, enabling the concurrent delivery of antimicrobial action and reduced collateral damage. This includes targeting resistant bacteria while preserving beneficial microorganisms to maintain microbiome balance and precise localization of therapeutic effects. (iv) Synergistic approaches, where combined antimicrobial strategies like photodynamic therapy (PDT), photothermal therapy (PTT), sonodynamic therapy (SDT), chemodynamic therapy (CDT), antibiotic therapy, and metal ion antimicrobials can be used to synergistically enhance antimicrobial and antibiofilm effects. (v) Improved drug delivery, with nanomaterials serving as carriers for antimicrobial agents, thereby minimizing drug loss, mitigating side effects, and boosting therapeutic effectiveness.

Biofilms consist of bacterial communities that form dense, sessile colonies, often with antibiotic concentrations up to 1000 times higher than those found in planktonic bacteria [9]. Moreover, biofilms commonly reside in deep tissues, presenting significant challenges to their eradication. In the near-infrared (NIR) region, two biological optical windows are recognized: the NIR-I window, ranging from 700 to 950 nm, and the NIR-II window, spanning 1000–1700 nm [10]. The notable benefits of NIR light, such as reduced optical absorption and scattering, deep tissue penetration, and minimal tissue damage, have propelled the use of NIR-triggered nanomaterials to the forefront of biomedical research, particularly in oncology and antibacterial treatments. NIR-responsive antimicrobial strategies, mainly PTT and PDT, leverage NIR light to suppress the growth of both planktonic bacteria and biofilms. However, the bactericidal efficacy of either PDT or PTT alone, when applied at biosafe levels and light intensities, falls short in tackling MDR bacteria. Consequently, researchers are increasingly turning to a variety of synergistic treatments to effectively target pathogenic bacteria.

In this chapter, we initiate our discourse with an overview of the antibacterial mechanisms intrinsic to photofunctional antibacterial nanomaterials. The second part contrasts the structural and functional nuances of these nanomaterials, shedding light on the strategies employed to surmount the resistance of MDR pathogens. Concluding this treatise, we address the salient challenges that persist in the field of photofunctional antibacterial nanomaterials including common PDT or PTT resistance mechanisms and MDR bacteria drug resistance mechanisms. In synthesizing these perspectives, our objective is to chart a prospective course for inquiry and innovation in the therapeutic modalities targeting MDR bacterial infections.

13.2 The Antibacterial Mechanisms of Photofunctional Antibacterial Nanomaterials

Photofunctional antibacterial nanomaterials, engineered to combat bacterial infections through light-activated mechanisms, have gained increasing interest due to their potential applications in various medical domains, including medicine, wound dressings, and medical device coatings [11]. These nanomaterials present a promising strategy for addressing bacterial infections, particularly those caused by antibiotic-resistant strains. However, the excitation wavelength of some nanomaterials is predominantly in the ultraviolet (UV) or visible spectrum, which restricts their use. To circumvent this limitation, researchers have employed upconversion nanoparticles (UCNPs) to effectively extend the excitation wavelength into the NIR region [12]. Typically, photofunctional antibacterial nanomaterials are involved in the antimicrobial mechanisms of either PDT or PTT. The appeal of both PDT and PTT lies in their non-invasiveness, low toxicity, and potential to reduce antibiotic resistance. Consequently, these therapies have garnered significant scholarly interest in recent years. The antimicrobial mechanisms and key molecules or nanomaterials for PDT and PTT are summarized below.

PDT entails the synergistic use of a photosensitizer (PS), light of a specific wavelength, and molecular oxygen [12]. The interaction of these elements leads to the production of reactive oxygen species (ROS), such as singlet oxygen ($^1O_2$), hydroxyl radicals (·OH), and superoxide anions ($O_2{\cdot}^-$), which possess high cytotoxic potential and are effective in bacterial eradication. PDT can be categorized into two types: type I and type II. In type I PDT, ROS generation is initiated by electron transfer reactions that occur independently of direct O_2 involvement, leading to the formation of radical species that react with O_2 to yield ROS, including ·OH and $O_2{\cdot}^-$. Conversely, type II PDT primarily generates ROS through the direct interaction of the excited state of PS with O_2, resulting in the production of $^1O_2$ [13]. The bactericidal effects of ROS are multifaceted, including [14]: (i) compromising bacterial cell membrane integrity, resulting in permeabilization and leakage of cellular contents; (ii) causing protein denaturation or dysfunction, which disrupts critical cellular processes and enzyme activities; (iii) inflicting oxidative damage to DNA, which can cause mutations, strand breaks, and interfere with DNA replication and repair mechanisms; (iv) disrupting bacterial metabolism, leading to inhibition of metabolic pathways or energy depletion; and (v) contributing to the reduction of antibiotic resistance.

PTT is a therapeutic modality that employs lasers or other light sources to produce heat within targeted tissues or cells, thereby causing their destruction [15]. In PTT, nanoparticles (NPs), typically composed of gold, silver, or carbon, are engineered to specifically target and accumulate on or within bacterial cells [16]. Upon NIR light exposure, these NPs efficiently transform the absorbed light into thermal energy via plasmonic resonance. This results in localized heating, which elevates the temperature in the immediate vicinity, affecting bacterial cells [17]. The primary outcomes of thermal damage include: (i) compromised cell membrane integrity of bacteria;

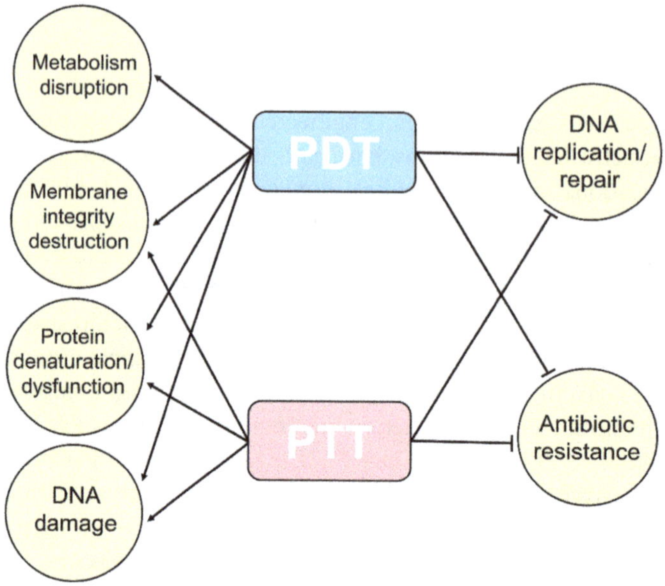

Figure 13.1 The antibacterial mechanisms of photofunctional antibacterial nanomaterials.

(ii) protein denaturation or impairment; (iii) oxidative damage to DNA, encompassing strand breaks and replication issues; and (iv) reduction in antibiotic resistance [18] (Figure 13.1).

It is crucial to meticulously calibrate the dose and parameters for both PDT and PTT to ensure therapeutic efficacy. The success of these treatments hinges on generating sufficient ROS or achieving a temperature high enough to eradicate bacterial cells while preserving surrounding healthy tissue. Thus, dosage control, along with the careful adjustment of NIR light intensity and exposure time, is key to maintaining this delicate balance. Furthermore, the varying susceptibilities of different bacterial strains to PDT or PTT must be considered. Certain pathogens may exhibit heightened resistance to oxidative or thermal stress from PTT, necessitating tailored treatment parameters or alternative therapeutic strategies.

13.3 Photofunctional Nanomaterials and Antibacterial Activity Against MDR Bacteria

Current research predominantly addresses the challenge of Methicillin-resistant *Staphylococcus aureus* (MRSA), with a focus on PDT, PTT, and its integration with multimodal treatments. However, literature from the past five years indicates a scarcity of studies exploring the use of PDT as a standalone treatment for MDR bacteria. This section examines NIR-responsive treatment strategies for MDR bacterial strains and provides a preliminary assessment of their sterilization efficacy.

13.3.1 Representative NIR PDT Photosensitizers

13.3.1.1 NIR-Responsive Porphyrins

Porphyrins, encompassing various derivatives and analogs, are extensively utilized as photosensitizers in antimicrobial PDT (aPDT) [19]. Known for their potent light-absorbing capabilities within the UV and visible spectrums, porphyrins generally have excitation wavelengths between 400 and 700 nm. Despite the efficacy of aPDT, the utility of porphyrins is often restricted by their excitation wavelength limitations [20]. To address this, there is a concerted effort to engineer new porphyrin derivatives with excitation wavelengths extending into the NIR spectrum. For instance, Xu et al. developed a cationic porphyrin, 5,10,15,20-tetrakis-(4-*N*-methylpyridyl)-porphyrin (TMPyP), capable of targeting specific bacteria and initiating adaptive PDT/PTT processes. TMPyP, when reduced to phlorin in hypoxic conditions by facultative anaerobes such as *Escherichia coli* and *Salmonella Typhimurium* exhibits strong NIR absorption and exceptional photothermal conversion capability, achieving over 99% antimicrobial efficacy through PTT. In the presence of oxygen, TMPyP serves as a conventional photosensitizer, effectively killing aerobic bacteria (>99.9%) via PDT (Figure 13.2a) [21]. Moreover, a cationic porphyrin integrated with a graphene nanoribbon (GNR) nanocomposite (Pp4N/GNR), through simple supramolecular self-assembly, demonstrates bimodal functionality, producing significant levels of ROS for PDT and generating heat for PTT under light irradiation at 660 and 808 nm, respectively. This dual-modality approach synergistically eradicates a broad spectrum of Gram-positive, Gram-negative, and MRSA strains, both in vitro and in vivo (Figure 13.2b) [22].

Porphyrin-based porous organic polymers (POPs or PPOPs), synthesized through covalent bonding of porphyrins or metalloporphyrins, have attracted significant attention for their roles as photocatalysts, electrocatalysts, and biomimetic catalysts [25]. The extended coupling on their frameworks endows PPOPs with a large, narrow bandgap and robust light absorption in the therapeutic window of 700–1000 nm, positioning them as promising agents for antibacterial material innovation. When integrated with conventional wound dressings, phototherapy using POPs enhances anti-infective efficacy in wound management. These polymers are responsive across the full light spectrum and can be activated by white light. MRSA growth was substantially impeded following a five-minute exposure to white light at an intensity of 60 mW cm^{-2} using a concentration of POPs at 50 μg ml^{-1}. Furthermore, MRSA biofilms experienced significant disruption when treated with 200 μg ml^{-1} of POPs (Figure 13.2c) [23]. In another study, a PPOP-based antibacterial agent with a monitoring function was developed. $FePPOP_{BFPB}$ was synthesized by integrating BFPB (4-{2,2-bis[(4-formylphenoxy)methyl]-3-(4-formylphenoxy) propoxy} benzaldehyde) with pyrrole through a facile aromatic substitution reaction. This reaction utilized BFPB's C-centric tetrahedral structure to yield materials with a three-dimensional interconnected porous architecture, exhibiting remarkable chemical and thermal stability, high-density catalytic sites, and superior NIR absorption. Thanks to these properties, $FePPOP_{BFPB}$ demonstrated potent dual

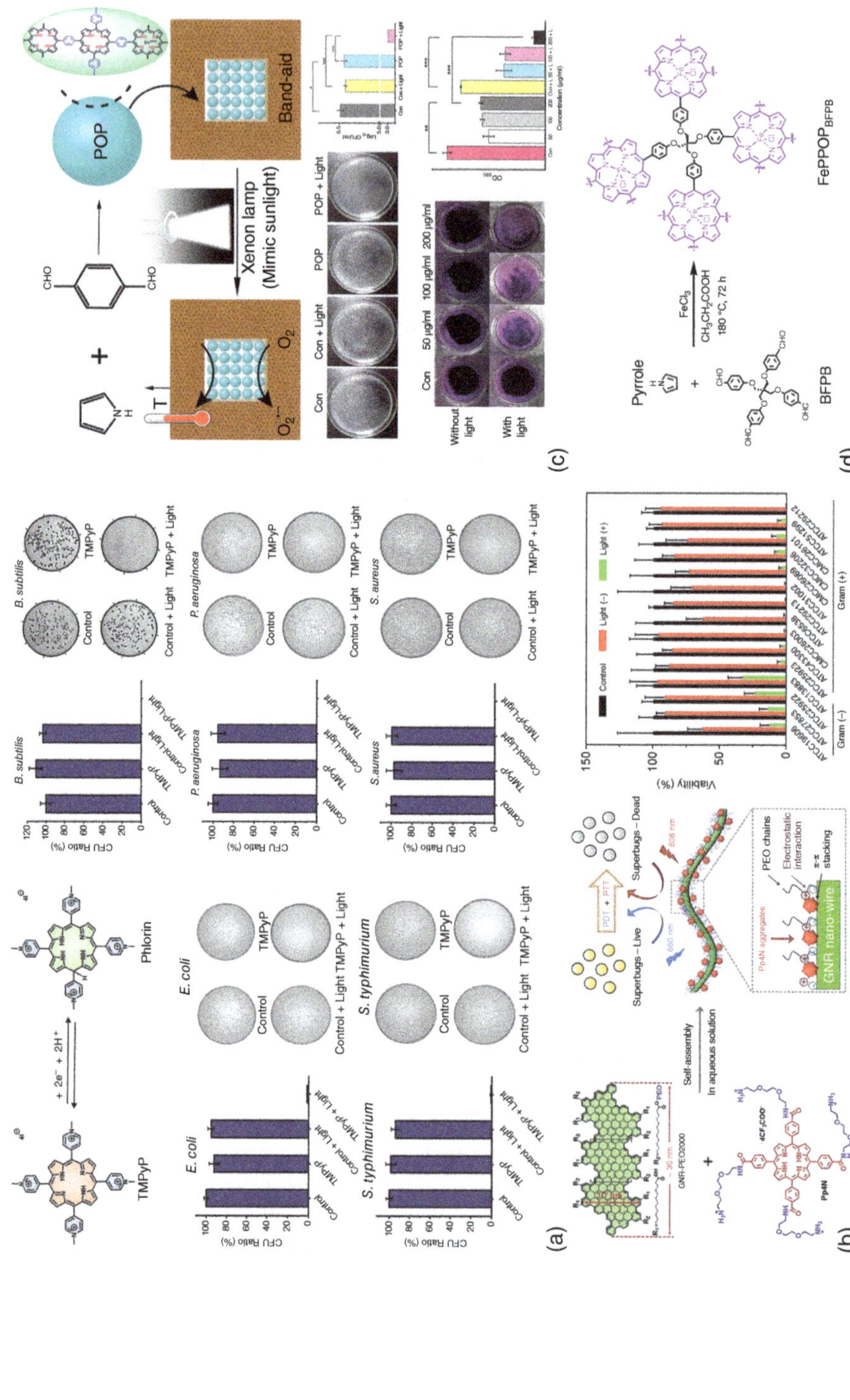
TMPyP
Phlorin
E. coli
S. typhimurium
B. subtilis
P. aeruginosa
S. aureus
Control
Control + Light
TMPyP + Light
CFU Ratio (%)
Superbugs – Live
Superbugs – Dead
PDT + PTT
Self-assembly
In aqueous solution
GNR nano-wire
PEO chains
Electrostatic interaction
π–π stacking
Viability (%)
Gram (−)
Gram (+)
POP
Band-aid
Xenon lamp
(Mimic sunlight)
Con
Con + Light
POP + Light
Without light
With light
50 μg/ml
100 μg/ml
200 μg/ml
Pyrrole
BFPB
FePPOP$_{BFPB}$
180 °C, 72 h
(a)
(b)
(c)
(d)

Figure 13.2 The structure of porphyrin derivatives and antibacterial effects. (a) Colony-forming unit ratio of *E. coli*, *S. typhimurium*, *B. subtilis*, *P. aeruginosa*, and *S. aureus* incubated with/without TMPyP (0.10 mM) after 15 minutes NIR irradiation. Source: Reproduced with permission from Hu et al. [21], 2022 / John Wiley & Sons. (b) Relative bacterial viability of *A. baumannii*, *P. aeruginosa*, *E. coli*, *K. pneumonia*, MSSA, MRSA, *B. subtilis*, *S. epidermidis*, γ-Streptococcus, *S. lentus*, VRE, and VSE in the absence and presence of the composite (Pp4N/GNR = 40 μm/30 μg ml^{-1}) treated with and without 15 minutes of 808 nm light irradiation followed by 15 min of 660 nm light irradiation. Source: Reproduced with permission from Yu et al. [22], 2019 / John Wiley & Sons. (c) Antibacterial and antibiofilm activity of POP against MRSA biofilms at different concentrations with or without white light irradiation. Source: Reproduced with permission from Li et al.[23], 2019 / American Chemical Society. (d) Schematic illustration of the synthesis of $FePPOP_{BFPB}$ by condensation reaction between pyrrole and BFPB. Source: Reproduced with permission from Li et al. [24], 2020 / American Chemical Society.

functionality, enabling rapid, precise, and selective bacterial detection alongside enhanced antibacterial efficacy (Figure 13.2d) [24].

13.3.1.2 NIR-Responsive Phthalocyanines

Phthalocyanines (Pcs) constitute a group of compounds with extensive conjugated systems, structurally akin to naturally occurring porphyrins. However, due to their inherent hydrophobicity, Pcs are typically amalgamated with other substrates for use as proficient PSs [26]. For example, boronic acid-functionalized phthalocyanine (PcN4-BA) exhibits aggregation-enhanced photodynamic effects. Its capacity to form uniform nanostructured self-assemblies in aqueous environments, coupled with effective ROS production and boronic acid-mediated targeting, renders PcN4-BA highly effective in aPDT. Remarkably, in Figure 13.3a, when subjected to a 655 nm laser at 0.4 W cm^{-2} for 10 minutes at a 50 nM concentration, PcN4-BA nearly eradicated Extended-spectrum beta-lactamase (ESBL) *E. coli* and MRSA [27]. Furthering the field, Sun et al. have extended the excitation wavelength of Pcs from the red to the NIR spectrum. They synthesized two nickel-coordinated phthalocyanines (NiPc) to assess their antibacterial efficacy, leveraging the solar-Fenton reaction pathway. The achiral variants, aNiPc crystal, and aNiPc@SNC, surpassed their chiral counterparts, sNiPc crystal and sNiPc@SNC, in antibacterial performance (Figure 13.3b) [28]. Additionally, Liu et al. engineered a photothermal agent by conjugating silicon phthalocyanine with fluorescein (SiPc-FITC) through covalent bonding, yielding a dual-function agent with both fluorescence and photothermal imaging capabilities. Antimicrobial assays revealed that this agent could significantly inhibit and eliminate *S. aureus* and *E. coli* within minutes of exposure to a low-intensity 808 nm NIR laser [29].

13.3.2 NIR-Responsive PTT Agents

13.3.2.1 Gold Nanoparticles and Derived Nanostructures

The field of antimicrobial therapy has seen the advent of various gold-based nanomaterials. Gold nanorods (GNRs), possessing elongated and anisotropic structures, facilitate the tuning of NIR absorption peaks to desired therapeutic

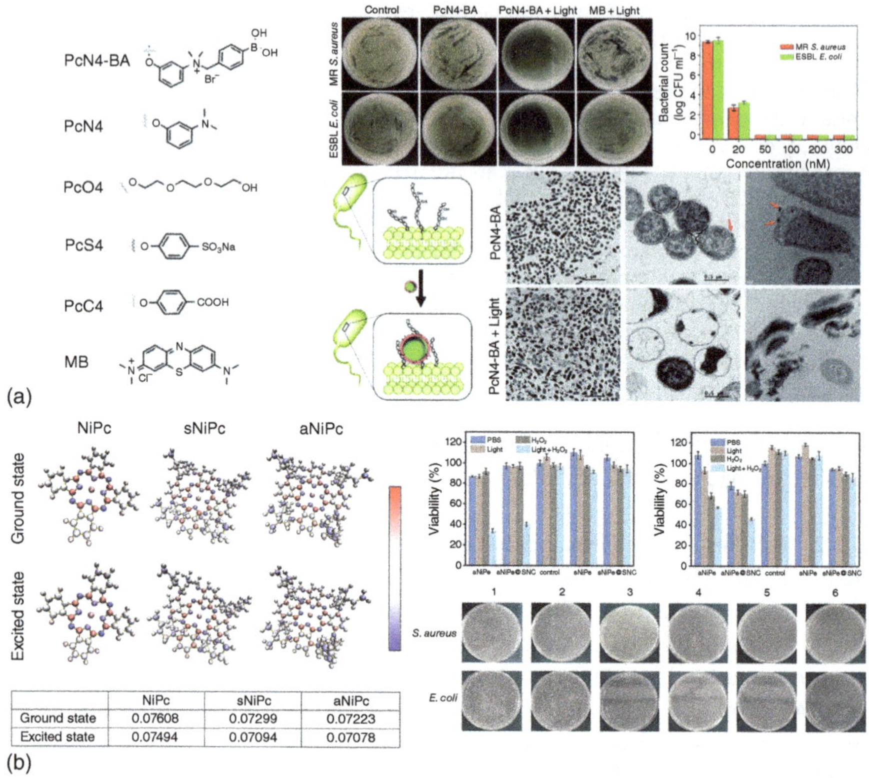

	NiPc	sNiPc	aNiPc
Ground state	0.07608	0.07299	0.07223
Excited state	0.07494	0.07094	0.07078

Figure 13.3 The structure of phthalocyanines derivatives and antibacterial effects. (a) Molecular structures of different PSs and corresponding photodynamic antibacterial effect. Concentration: PcN4-BA and MB 50 μM, 655 nm, 0.4 W cm^{-2}, 10 minutes. Source: Lee et al. [27]/Royal Society of Chemistry/CC BY 3.0. (b) Hirshfeld charges of the optimized configuration of NiPcs in the ground state and in excited state and antibacterial effects. Concentration: NiPcs 10 μM, H_2O_2 100 μM, and NIR light (>600 nm, 60 mW cm^{-2}) for 30 minutes. Source: Reproduced with permission from Zhao et al. [28] / John Wiley & Sons.

wavelengths [30]. A novel theranostic nanoprobe, comprising GNRs coated with polyethylene glycol-modified thiol, curcumin, and a cell-targeting DNA aptamer, has been demonstrated to effectively target and combat MRSA USA300, with GNRs providing both the PTT and diagnostic capabilities. The treatment's effectiveness enhanced when the superbug underwent a 10-hour incubation with nanocapsules, achieving a 98% eradication rate within 30 minutes [31]. Gold nanoshells, consisting of a dielectric core usually made of silica and enveloped by a thin gold layer, exhibit plasmonic resonance within the NIR spectrum, rendering them suitable for PTT applications. A notable development is a NIR-II responsive nanohybrid, integrating a lipopolysaccharide (LPS)-targeting aptamer and an imipenem-encapsulated liposome with a gold shell, designed specifically to eradicate MDR *Pseudomonas aeruginosa* (MDR-*PA*). Exposure to the NIR laser allowed the aptamer to effectively eliminate MDR-PA, resulting in a complete absence of bacterial colonies in the culture plate [32]. Gold nanoclusters are characterized by their diminutive size and unique optical properties, including NIR fluorescence. There are reports of NIR-assisted in situ reduction of antimicrobial peptide-protected gold nanoclusters for the sequential eradication of bacteria and cancer cells [33]. Another structure, the gold nanocage, features a hollow configuration with porous walls and a prominent NIR absorption peak conducive to efficient light absorption. Such nanocages can facilitate PTT and targeted drug release upon NIR illumination. Gold silver nanocages (GSNCs) have emerged as versatile carriers for drug delivery and NIR PTT. Their intrinsic antibacterial efficacy is often compromised by the substantial gold coating and the hollow silver core, which contains minimal silver species [34]. To circumvent this, a strategy exploiting the aggregation-induced photothermal (AIP) effect was proposed. This approach involves the use of high silver-loading GSNCs with thiolated cysteamine (Cyh), which induces the aggregation of GSNCs, enhancing NIR photothermal effects and promoting silver release. The GSNC-Cyh complex synergistically eradicates biofilms from MRSA and MDR *E. coli*, with an efficacy of up to 98% and 95%, respectively [35]. Furthermore, the facile synthesis of AuAg yolk–shell cubic nanoframes (YSCNFs) with adjustable sizes offers a new dimension to PTT. Even at a low-power density of $0.27\,\text{W cm}^{-2}$, YSCNFs have demonstrated a photothermal conversion efficiency of 65.6%. Specifically, AuAg YSCNFs-60 significantly improved antibacterial effectiveness, increasing MRSA eradication rates from approximately 50% to 90% [36]. Table 13.1 shows part of gold nanomaterials against MDR bacteria.

13.3.2.2 Carbon Nanotubes

Carbon nanotubes (CNTs) possess remarkable capabilities for efficient absorption of NIR light and subsequent heat generation [37, 38]. Moreover, through the process of functionalization, carbon nanotubes can be precisely tailored for specific bacterial targeting, making them a promising tool in the realm of nanomedicine. NIR laser irradiation of CNTs encased in a polypyrrole coating within a core-shell architecture (CNTs@PPy) resulted in a notable temperature rise, which validates

Table 13.1 The application of representative NIR-responsive gold nanomaterials against MDR bacteria.

Morphology	NIR	Condition	Species	Efficiency	References
GNRs@PEG@Cur@Apt	808 nm	0.92 μm $1\,\text{W cm}^{-2}$	MRSA USA300	98%	[31]
Gold nanoshells	1064 nm	$150\,\mu\text{g ml}^{-1}$ $1\,\text{W cm}^{-2}$	MDR-*PA*	100%	[32]
GSNC-Cyh	1064 nm	$64\,\mu\text{g ml}^{-1}$ $0.5\,\text{W cm}^{-2}$	MRSA MDR *E. coli*	98% 95%	[35]
AuAg YSCNFs-60	808 nm	1.66×10^{10} particles ml^{-1} $0.27\,\text{W cm}^{-2}$	MRSA	90%	[36]

the substantial photothermal properties of the material for the eradication of *P. aeruginosa*. Moreover, laser treatment with CNTs@PPy at 250 and 500 μg ml^{-1} achieved approximately a 70% reduction in *P. aeruginosa* viability [39].

13.3.2.3 Graphene Oxide

When subjected to NIR irradiation, GO generates heat, which compromises bacterial cell integrity [40]. Demonstrating its potential, few-layer graphene (FLG) and few-layer GO (FLGO) films, even under low-power NIR illumination (0.150 W cm^{-2}), have shown promise as light-activated antibacterial platforms. These films have been reported to effectively eliminate *S. aureus* and *Staphylococcus epidermidis*, achieving eradication rates of approximately 99% for planktonic cells and 85% for biofilms [41]. Additionally, antibody-modified GO-coated gold nanoparticles have been developed for rapid bacterial detection and subsequent ablation [42]. In a separate study, two-dimensional (2D) reduced graphene oxide (r-GO) demonstrated a 25% reduction in MDR superbug populations through photothermal effects after just 30 minutes of light exposure. This r-GO was integrated into a system with WO_3 nanowires to suppress light-induced electron-hole pair recombination, enhancing the photocatalytic eradication of MDR pathogens [43]. Furthermore, Poly(ethylene sulfide)-nitrosylated nanoparticles (PES-RSNO NPs) were encapsulated within a GO-doped hydrogel composed of poly(vinyl alcohol) and poly(acrylic acid) interpenetrating networks (designated as GO@PVA–PAA). In vivo assays indicated that the dual-mode antibacterial hydrogel significantly curtailed bacterial proliferation, eradicating $99.84 \pm 0.13\%$ of MRSA in the absence of light via a phage-mimicking mechanism. Moreover, when exposed to 808 nm NIR light for 20 minutes, the hydrogel executed a swift sterilization process, achieving a $97.13 \pm 0.41\%$ antibacterial efficacy through combined photothermal and NO therapy [44].

13.3.2.4 Semiconductor Nanoparticles

Semiconductor NPs, including metal sulfides such as copper sulfide (CuS) and silver sulfide (Ag_2S), have garnered interest in NIR PTT applications [45]. These NPs are proficient at absorbing NIR light and converting it to heat, targeting and neutralizing bacterial threats. Beyond their photothermal effects, the ions released from these materials, specifically Cu^{2+} and Ag^+, exhibit inherent bactericidal properties by disrupting bacterial membranes and interfering with enzymes, proteins, and DNA functions [46]. Further investigations have revealed that metal sulfides like MoS_2, Bi_2S_3, and $AgBiS_2$ also possess antibacterial capabilities. For instance, LA-MoS_2-IR780-blended poly(vinyl alcohol) nanofibers (MoS_2-IR780), fabricated via electrospinning, endow nanofibrous membranes with dual-mode PDT/PTT antibacterial activity. This synergistic antibacterial system, even at lower temperatures, has demonstrated broad-spectrum sterilization, with negligible colony formation of *S. aureus*, *E. coli*, and MRSA at a 50 μg ml^{-1} concentration in the MoS_2-IR780 NIR group [47]. Thin-layer MoS_2-based multivalent glycosheets for the targeted, light-driven therapy of MDR *P. aeruginosa* on wounds [48]. Zinc atom–doped g-C_3N_4 and Bi_2S_3 nanorod heterojunctions (CN–Zn/ Bi_2S_3) showed

99.2% and 99.6% antibacterial efficiency against MRSA and *E. coli*, respectively, under 10 minutes of NIR irradiation [49]. Bovine serum albumin templated bismuth-copper sulfide nanocomposites (BSA-BiZ/CuxS NCs) exhibited substantial 90% biofilm inhibition efficiency and >75% eradication efficiency of MRSA under NIR irradiation [50]. UCNPs@$AgBiS_2$, employing a combination of PDT and PTT, have been able to reduce MRSA populations by nearly 99.9% at a concentration of 40 $\mu g\,ml^{-1}$. Transcriptomic analyses suggest that UCNPs@$AgBiS_2$ disrupt MRSA's galactose metabolism, compromise its transport systems, and interfere with quorum sensing (QS), contributing to its high antibacterial effectiveness [51]. Cu_2MoS_4 nanoplates (CMS NPs) exhibited dual catalytic properties akin to enzymes, with their oxidase- and peroxidase-like activities being notably enhanced by NIR-II light, thereby improving ROS generation. Nanoenzyme like activity and photothermal effect are beneficial for killing MDR *E. coli* and MDR *S. aureus*. CMS NPs at a concentration with 40 $\mu g\,ml^{-1}$ can achieve an 8-log inactivation efficiency against MDR *E. coli* and a 6-log inactivation efficiency of MDR *S. aureus* with NIR-II light irradiation [52] (Figure 13.4).

13.3.3 NIR-Responsive PDT/PTT Agents

13.3.3.1 NIR Cyanine Dyes

NIR cyanine (Cy) dyes, particularly those with PDT functionality, are organic compounds that have captured the interest of the medical community. These dyes can absorb light in the NIR spectrum, which is optimal for deep tissue penetration [53]. Indocyanine Green (ICG), a water-soluble tricarbocyanine dye, stands out as a prominent NIR cyanine dye in PDT [54]. It is FDA-approved and has been deemed safe for human use in controlled conditions, with applications spanning several decades in dental, ocular, and wound infections [55]. ICG's application in PDT shows considerable promise for antimicrobial purposes, but research and clinical trials are ongoing to further ascertain its efficacy and safety across various conditions. Therapeutic parameters such as ICG concentration, light dosage, and exposure duration are tailored to the specific bacteria and infection type. Despite ICG's effectiveness, researchers are actively enhancing its photostability and specificity for PDT through modifications, such as altering its absorption and emission characteristics, and improving its solubility in aqueous or lipid environments for diverse applications and drug delivery systems. A novel composite nanoplatform comprising calcium peroxide (CaO_2) and ICG encapsulated in polydopamine NPs (CaO_2/ICG@PDA NPs) was engineered, providing O_2 self-supplying and heat-sensitizing properties to boost ICG's PDT performance against MRSA biofilms. Under NIR light, there was a significant reduction in MRSA biofilm biomass by 67.6% ± 6.8% in the CaO_2/ICG@PDA group [56]. Besides, ICG-functionalized hexagonal Mn_3O_4 nanosheets (Mn_3O_4HNSs@ICG) were applied in synergetic combating MRSA infections [57]. Except for MRSA, the antibacterial efficiencies of Mn_3O_4HNSs@ICG showed 99.63%, 99.36%, and 96.74% toward *E. faecalis*, *E. coli*, and *P. aeruginosa*, respectively. Another notable NIR Cy dye, IR780, offers a more intense fluorescence and greater stability than ICG, commonly employed

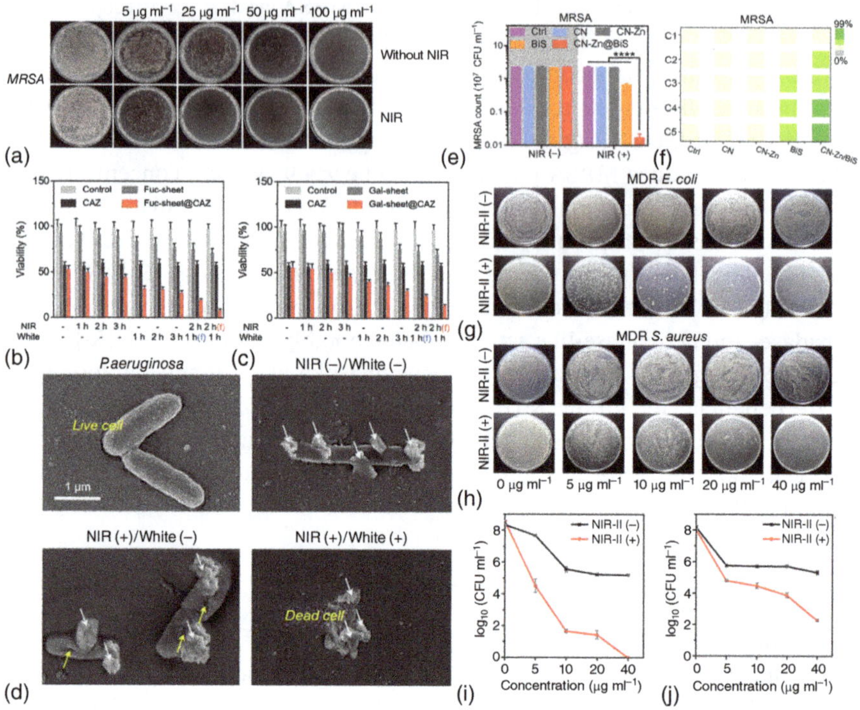

Figure 13.4 The semiconductor NPs and antibacterial effects. (a) antibacterial results of MoS_2-IR780 without or with NIR (power density: 1.5 W cm^{-2}). Source: Zhang et al. [47]. © 2023/Elsevier/CC BY 4.0. (b) Relative bacterial viability of *P. aeruginosa* in the absence and presence of CAZ, Fuc-sheet, or Fuc-sheet@CAZ with and without light irradiation for the indicated time scales (c) Relative bacterial viability of *P. aeruginosa* in the absence and presence of CAZ, Gal-sheet, or Gal-sheet@CAZ with and without light irradiation for the indicated time scales. (d) TEM images of *P. aeruginosa* in the absence and presence of Fuc-sheet@CAZ without and with single NIR irradiation (two hours) or double-light irradiation (two hours of NIR followed by one hour of white light). The wavelength of NIR and white light is 808 nm (1 W cm^{-2}) and 400–700 nm (1 W cm^{-2}), respectively. Source: Reproduced with permission from Hu et al. [48] 2019, / John Wiley & Sons. (e) MRSA counts calculated from spread-plate assays after treatment with CN, CN–Zn, BiS, or CN–Zn/BiS (200 µg ml^{-1}) under 808 nm light irradiation. (f) Orthogonal experiment on antibacterial activities of CN, CN–CN, CN–Zn, BiS, or CN–Zn/BiS and gradient concentrations toward MRSA. Source: Reproduced with permission from Li et al. [49] 2019, / John Wiley & Sons. Antibacterial activity of CMS NPs in vitro. Photographs of bacterial colonies of (g) MDR *E. coli* and (h) MDR *S. aureus* treated by CMS NPs without and with NIR-II light irradiation (1064 nm, 1 W cm^{-2}). Bacterial inactivation efficiency of (i) MDR *E. coli* and (j) MDR *S. aureus* exposed to CMS NPs without and with NIR-II light irradiation (1 W cm^{-2}) for 10 minutes. Source: Reproduced with permission from Shan et al. [52] 2020, / John Wiley & Sons.

in PDT. It also serves as an ultrasound sensitizer to augment ROS production under low-intensity ultrasound [58]. However, IR780's clinical utility is hindered by its poor solubility, rapid clearance, and potential toxicity in biological fluids. Nanoencapsulation of IR780 enhances its solubility, extends its bioavailability, and mitigates toxicity.

It is noteworthy that while ICG and IR780 also act as photothermal agents, their photothermal effects are generally less pronounced compared to gold NPs or carbon-based materials [59]. Yet, the combination of IR780 with nitric oxide (NO) gas therapy has proven to be highly effective in MRSA eradication and biofilm removal, with the highest antibacterial ratios reaching up to $99.61 \pm 0.16\%$ [60].

13.3.3.2 NIR QDs

NIR quantum dots (NIR QDs) are semiconductor NPs with exceptional optical properties within the NIR region of the electromagnetic spectrum, typically between 700 and 900 nm. Their ability to operate effectively within the NIR window has made them highly sought after for diverse applications, such as imaging, sensing, phototherapy, and notably, antibacterial treatment [61]. Commonly, NIR QDs possess dual PDT/PTT functionalities, contributing to their potent antibacterial properties. Recent studies have highlighted the efficacy of graphene QDs (GQDs), black phosphorus QDs (BP QDs), and Ag_2S QDs in bacterial disinfection [62].

GQDs, 20 nm-wide carbon nanosheets, have sparked considerable scientific interest due to their unique spin structure. Renowned for their superior optical performance and biocompatibility, GQDs are extensively utilized in various biomedical applications, including biosensors, drug delivery, PDT, and PTT [63]. GQDs display excitation-dependent emission within the 320–480 nm range, a factor that may constrain certain biological uses. To overcome this limitation, nitrogen-doped GQDs (N-GQDs) have been developed to harness the more favorable NIR-II bio-window for photothermal antibacterial therapy [64]. These N-GQDs, featuring graphitic nitrogen sites that act as paramagnetic F centers, demonstrate efficient NIR absorption and photothermal conversion. With a high doping level of graphitic nitrogen (3.6 at%), N-GQDs exhibit extended absorption into the NIR-II region and achieve photothermal conversion efficiencies of 77.8% at 808 nm and 50.4% at 1064 nm. When subjected to NIR laser irradiation, MRSA cultures treated with N-GQDs at a concentration of 200 $\mu g\ ml^{-1}$ show a survival rate below 3%, emphasizing the N-GQDs potent photothermal antibacterial properties of N-GQDs (Figure 13.5a,b) [65]. Additional research has led to the creation of amino-functionalized N-GQDs, which serve as effective PSs with pronounced PDT effects. These N-GQDs are capable of eradicating MDR pathogens using a low-energy ultrashort pulse laser. They also display enhanced two-photon properties, making them excellent contrast agents for two-photon imaging (TPI) in three-dimensional studies. Notably, the amino-functionalized N-GQDs outperform their non-modified counterparts in both PDT effectiveness and two-photon functionality [66].

BP with its remarkable charge carrier mobility and extensive optical absorption spectrum, is a compelling material in phototherapy. Notably, BPQDs showcase extraordinary $^1O_2$ generation and substantial heat production upon NIR irradiation [69]. These BPQDs are esteemed as photothermal agents given their excellent NIR extinction coefficients, high photothermal conversion efficiency, and low cytotoxicity. Innovatively, hydrogels encapsulating epigallocatechin gallate (EGCG)-modified BPQDs have been synthesized, which are expected to serve multiple therapeutic

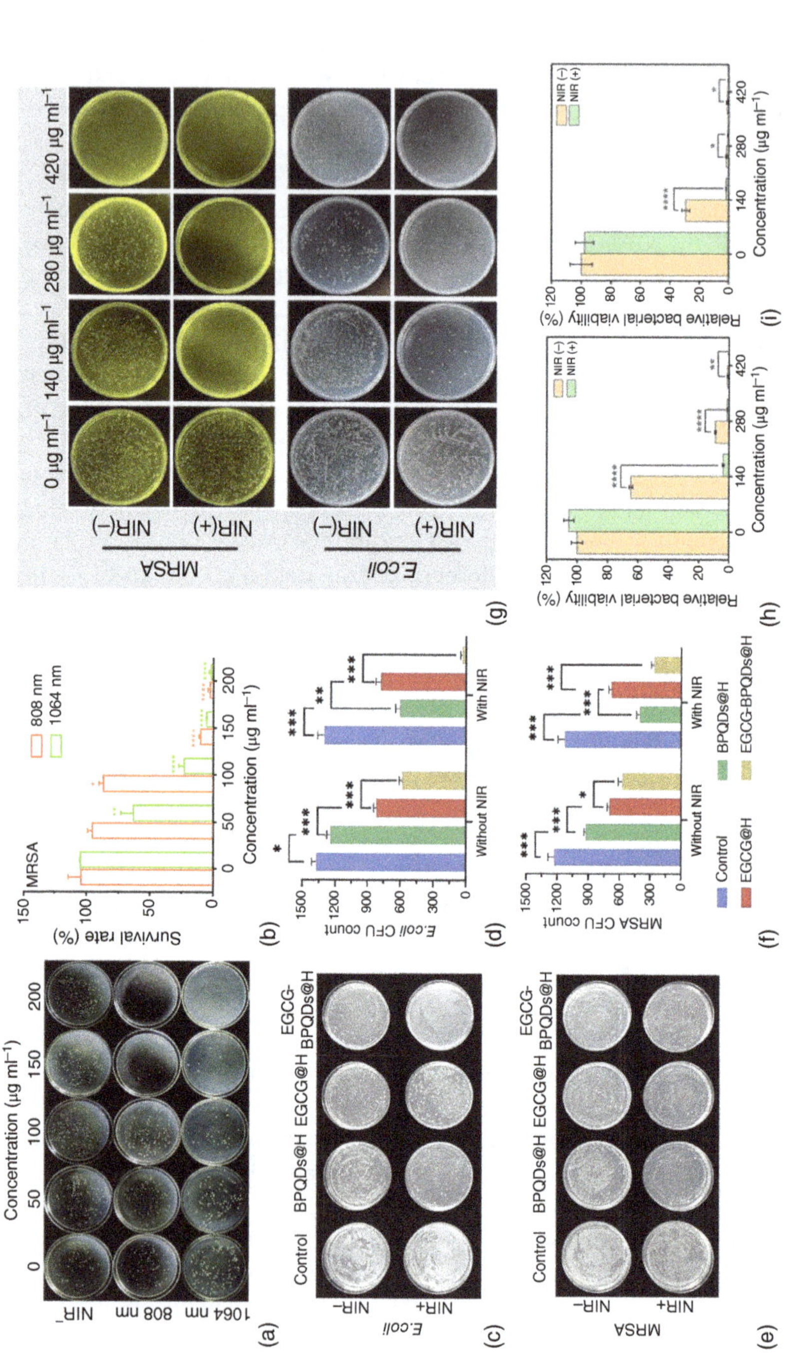
(a)
Concentration (μg ml^{-1})
0
50
100
150
200
NIR$^-$
808 nm
1064 nm
(b)
MRSA
808 nm
1064 nm
Survival rate (%)
Concentration (μg ml^{-1})
(c)
E.coli
NIR−
NIR+
Control
BPQDs@H
EGCG@H
EGCG-BPQDs@H
(d)
E.coli CFU count
Without NIR
With NIR
(e)
MRSA
NIR−
NIR+
Control
BPQDs@H
EGCG@H
EGCG-BPQDs@H
(f)
MRSA CFU count
Without NIR
With NIR
Control
EGCG@H
BPQDs@H
EGCG-BPQDs@H
(g)
0 μg ml^{-1}
140 μg ml^{-1}
280 μg ml^{-1}
420 μg ml^{-1}
MRSA
NIR(−)
NIR(+)
E.coli
NIR(−)
NIR(+)
(h)
Relative bacterial viability (%)
Concentration (μg ml^{-1})
NIR (−)
NIR (+)
(i)
Relative bacterial viability (%)
Concentration (μg ml^{-1})
NIR (−)
NIR (+)

Figure 13.5 Antibacterial effects of QDs related materials. (a) and the survival rate (b) of MRSA after receiving treatments with various concentrations of N-GQD aqueous solution without or with 808 ($0.4\,W\,cm^{-2}$) or 1064 nm laser ($1.0\,W\,cm^{-2}$) irradiation. Source: Reproduced from Geng et al. [65]/ Royal Society of Chemistry. (c–f) Digital photos of bacterial colonies on agar plates of *E. coil* and MRSA with different treatments and corresponding bacterial colonies on agar plates were calculated. Source: Xu et al. [67]. 2021/Springer Nature/CC BY 4.0. (g) Photographs of bacterial colonies formed after treatment with different concentrations of NP hydrogel accompanied by exposure or unexposure to NIR laser at the power density of $1.8\,W\,cm^{-2}$ for four minutes. (h and i) The corresponding survival rates of MRSA and *E. coli*. Source: Reproduced with permission from Du et al. [68] 2022, / Elsevier.

functions, including antibacterial activity and promotion of epithelial and vascular regeneration. The peak photothermal conversion efficiency of these hydrogels reaches 46.7%, with a demonstrated superior capacity for $^1O_2$ production. When applied, the EGCG-BPQD hydrogel system substantially reduced bacterial viability to merely 3.2% for *E. coli* and 11.3% for MRSA (Figure 13.5c–f) [67]. Furthermore, the integration of BPQDs with PEGylated chitosan NPs (CS NPs) facilitates the delivery of the antibiotic amikacin (PEG@CS/BPQDs-AM NPs), offering an effective treatment modality for chronic obstructive pulmonary disease-related lung infections. Remarkably, the synthesized NPs (at $30\,\mu g\,ml^{-1}$ concentration) demonstrated an antibacterial efficiency exceeding 90% against a spectrum of tested bacteria, including Gram-negative *E. coli*, *P. aeruginosa*, and Gram-positive *S. aureus* [70]. Furthermore, a multifaceted approach combining gas therapy, peptides, and BPQDs was devised for the synergistic treatment of bacterial skin abscesses. A cationic amphiphilic copolymer, mPEG-PLL-g-IP (PELI), was synthesized by copolymerizing poly-L-lysine (PLL) with isopentenyl grafts (PLL-g-IP) with methoxy poly(ethylene glycol) (mPEG). *S*-nitrosocysteamine (SNO), serving as a NO donor, was electrostatically attached to black BPQDs to form BPQD-SNO complexes. These complexes, along with mPEG-PLL-g-IP, were amalgamated through electrostatic interactions to produce PELI@BPQD-SNO NPs. Upon NIR irradiation, the BPQDs converted light into heat, enabling the controlled release of NO, while the degradation of BPQDs facilitated the release of the cationic copolymer mPEG-PLL-g-IP. The ensuing oxidative damage from NO to the bacterial cell membrane enhanced permeability and susceptibility to heat and the cationic polymers, culminating in a synergistic antibacterial effect. Consequently, the PELI@BPQD-SNO/NIR treatment achieved bactericidal rates of up to 99.6% for *S. aureus* and 96.7% for *P. aeruginosa* [71].

Ag_2S QDs, with their pronounced NIR fluorescence properties and biocompatibility, have also demonstrated significant promise for biomedical applications [72]. Ag_2S QDs capped with a silver-binding peptide (AgBP2) exhibited exceptional fluorescence emission within the NIR-II window, and maintained physical and photostability in aqueous solutions. The photothermal conversion efficiency of these QDs was measured at 63%. Upon exposure to NIR light for 25 minutes, AgBP2-Ag_2S QDs displayed a remarkably high disinfection rate of 99.06% against *E. coli* [73]. In a separate study by Du et al., a NIR-responsive hydrogel encapsulating functional inorganic Ag_2S QDs was engineered for use as a wound dressing. Subjected to an 808 nm NIR laser, the NP hydrogel showcased a high photothermal conversion efficiency (57.3%) and formidable antibacterial activity, eliminating 99.7% and 99.8% of *E. coli* and MRSA, respectively, within just four minutes. In vitro studies indicated that the hydrogel's volumetric response to NIR irradiation could regulate the release of silver ions (Figure 13.5g–i) [68].

13.3.3.3 Aggregation-Induced Emission Luminogens

Aggregation-induced emission (AIE) photosensitizers represent an innovative class of compounds that defy the typical photophysical behaviors of conventional photosensitizers [74]. Traditional photosensitizers often experience quenching or diminished luminescence upon aggregation, while AIE photosensitizers become more luminescent and exhibit augmented photodynamic capabilities, leading to substantial ROS production in aggregated states [75]. These properties render AIE photosensitizers particularly effective against a spectrum of microbial pathogens, including bacteria, fungi, and viruses.

The concept of luminogens with aggregation-induced emission (AIEgens) was pioneered by Tang's research group in 2001 [76]. Utilizing the inherent negative surface charge of bacteria, the incorporation of cationic groups into AIEgens fosters a strong bacterial affinity via electrostatic interactions. Building on this principle, Tang's group developed a triphenylamine thiophen pyridinium (TTPy)-based cationic AIEgen, unveiling the antibacterial mechanisms unique to this type of AIEgens. TTPy's narrow singlet–triplet energy gap aids the intersystem crossing from the excited singlet state (S1) to the triplet state (T1), enhancing the generation of the triplet excited state and, consequently, ROS production [77]. Furthermore, the cationic nature of TTPy fosters electrostatic adherence to the teichoic acids on Gram-positive bacterial surfaces, while its hydrophobicity allows penetration through lipid membranes, thus enabling intracellular access. Leveraging these characteristics, Tang's group synthesized four anion–π AIEgens (TTCPy-1, TTCPy-2, TTCPy-3, and TTCPy-4) designed for bacterial imaging and the eradication of MDR bacteria [78]. These AIEgens, benefiting from the pronounced donor–acceptor effect and optimal energy level gaps, facilitate the intersystem crossing process conducive to type I ROS generation. Consequently, even at low concentrations, these novel AIEgens exhibit potent activity against MDR bacterial strains. Under white light irradiation, over 99% of MRSA were killed at a concentration of 0.25×10^{-6} m of TTCPy series except for TTCPy-2, suggesting a high photodynamic killing efficiency of TTCPy-1, TTCPy-3, and TTCPy-4 toward MRSA.

AIEgens have extended their utility to PTT applications. Recent advancements include the development of NIR-emitting AIEgens that harness nonradiative excited state energy for photothermal and photoacoustic effects, converting it efficiently into heat [79]. For instance, a NIR-II emissive polymer AIEgen, poly(dithieno[3,2-b:2′,3′-d]pyrrole-alt-benzo[1,2-c:4,5-c′]bis([1,2,5]thiadiazole)) (PDTPTBT), was synthesized specifically for the treatment of diabetic skin infections. L-PDTPTBT exhibited a photothermal conversion efficiency of 52% in an aqueous solution [80]. With just eight minutes of NIR irradiation at a modest power of 0.8 W cm^{-2}, the solution temperature reached up to 55 °C, demonstrating substantial bactericidal effects in vitro. Increasing the irradiation intensity to $1.0\,W\,cm^{-2}$ resulted in the near-total annihilation of MRSA bacteria, evidenced by an 8 log10 reduction [80]. Tang's group has also engineered an AIE-based material that concurrently exhibits photodynamic and photothermal activities, effectively targeting Gram-positive bacteria such as *S. aureus*. Under low-intensity NIR laser exposure (808 nm at 0.1 and $0.2\,W\,cm^{-2}$), these AIE NPs exhibit potent antibacterial action, successfully suppressing nearly 100% of *E. coli* and *S. aureus* colony-forming units [81]. The structure of phthalocyanines derivatives and antibacterial effects are summarized in Figure 13.6.

13.4 Limitations and Challenges

So far, most of the literature focuses mainly on the efficacy of PDT/PTT, but ignores the potential resistance role of bacteria. Although PDT or PTT do not cause bacterial resistance, they may trigger bacterial resistance to oxidative stress or thermal damage. In this chapter, we analyze the potential resistance mechanisms of bacteria so that researchers can better optimize NIR photofunctional antibacterial materials.

13.4.1 Common PDT or PTT Resistance Mechanism

13.4.1.1 Oxidative Stress Defense

Bacteria have enzymatic antioxidant systems and produce enzymes that scavenge ROS [82]. Some of the key enzymes include superoxide dismutase (SOD), catalase, and peroxidases. Bacteria also can synthesize and accumulate non-enzymatic antioxidants like glutathione, thioredoxin, and ascorbic acid. which could directly scavenge ROS and regenerate other antioxidants. Bacterial regulatory proteins detect oxidative stress and turn on the production of protective genes. In response to oxidative stress, the OxyR and SoxRS systems in *E. coli*, for example, regulate the expression of antioxidant enzymes [83].

Bacteria control the intracellular concentrations of metal ions such as iron, copper, and manganese, which can accelerate the production of ROS via Fenton chemistry. Bacteria can reduce ROS generation by strictly managing metal ion concentrations [84]. Some bacteria include efflux pumps that can transport hazardous chemicals out of the cell, including ROS [85]. This reduces the intracellular concentration of ROS and protects cells from oxidative damage. Bacteria can alter their metabolic pathways to reduce ROS production. For example, they may switch to less aerobic or anaerobic metabolism under oxidative stress conditions to minimize the generation of ROS.

In addition, by working together, bacteria are able to withstand oxidative damage. a matrix of extracellular polymeric substances in biofilms can offer defense against oxidative stress by physically securing the bacteria inside and making bacteria less vulnerable to ROS attack. QS is a technique of bacterial communication that enables bacteria to coordinate their behavior in response to population density. When some

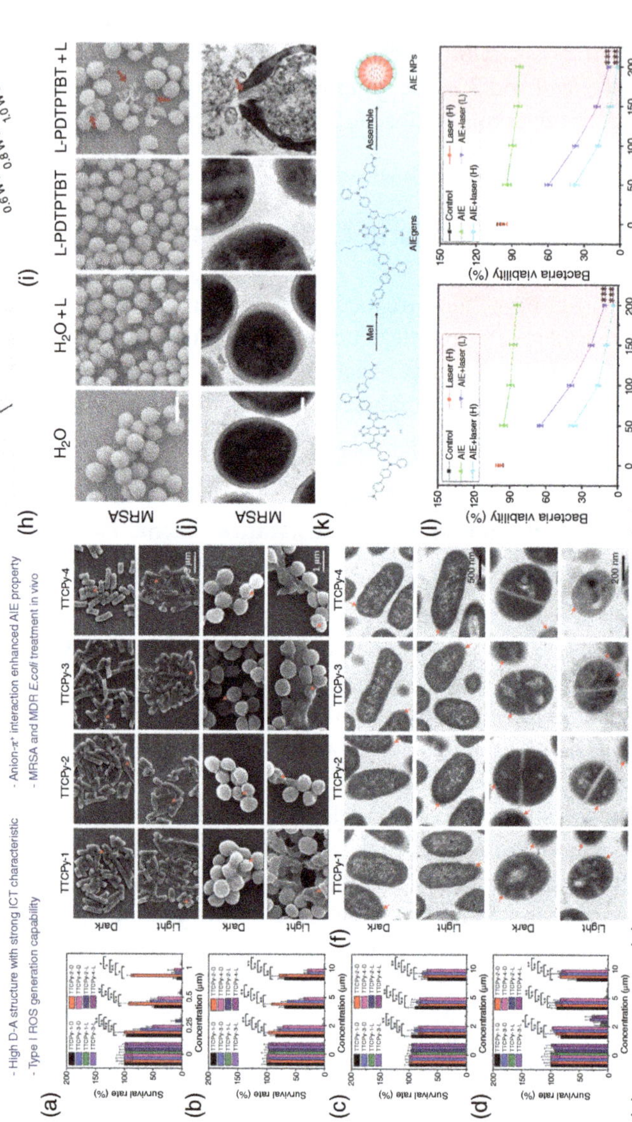

(a)
(b)
(c)
(d)
(e)
(f)
(g)
(h)
(i)
(j)
(k)
(l)
(m)
TTCPy-1
TTCPy-2
TTCPy-3
TTCPy-4
- High D-A structure with strong ICT characteristic
- Type I ROS generation capability
- Anion-π+ interaction enhanced AIE property
- MRSA and MDR E.coli treatment in vivo
Survival rate (%)
Concentration (μm)
Dark
Light
MRSA CFU (ml⁻¹)
H2O + L
L-PDTPTBT
L-PDTPTBT + L
Origin
0.6 W cm⁻²
0.8 W cm⁻²
1.0 W cm⁻²
H2O
MRSA
MeI
Assemble
AIEgens
AIE NPs
Control
Laser (H)
AIE
AIE+laser (L)
AIE+laser (H)
Bacteria viability (%)
AIE concentration (μg ml⁻¹)

Figure 13.6 The structure of phthalocyanines derivatives and antibacterial effects. (a) the structure of TTCPy series. Survival rate of (b) MRSA, (c) *E. coli*, (d) *E. coli* top 10, and (e) MDR *E. coli* at different concentrations in darkness or upon white light irradiation. FE-SEM images of (f) *E. coli* and MRSA in darkness or upon white light irradiation. (g) TEM images of *E. coli* and MRSA in darkness or upon white light irradiation. Concentration: $16\,mW\,cm^{-2}$ for 10 minutes, $n = 3$. Source: Xiao et al. [78]/John Wiley & Sons/CC BY 4.0. (h) The structure of L-PDTPTBT (i) In vitro MRSA killing property of L-PDTPTBT upon different powers ("+L" stands for under 808 nm NIR irradiation). (j) Morphology changes of MRSA after different treatments characterized by SEM. Scale bar: 1 μm. (k) Leakage of cytoplasm from MRSA after different treatments characterized by TEM. Scale bar: 100 nm. Concentration of L-PDTPTBT: $200\,\mu g\,ml^{-1}$, PDTPTBT equivalent. NIR irradiating condition: $0.8\,W\,cm^{-2}$, eight minutes. Source: Reproduced with permission from Huang et al. [80], 2022 / Elsevier. (l) Synthetic route of AIEgens (molecule 12) and AIE NPs. (m) *S. aureus* and *E. coli* incubated with various concentrations of AIE NPs after different treatments. Source: Reproduced with permission from Wang et al. [81], 2022 / American Chemical Society.

bacteria detect a hazard, such as oxidative stress, they use QS to boost antioxidant defenses [86].

13.4.1.2 Thermal Stress Defense

Bacteria could produce some specialized proteins and regulate gene networks to initiate the transcription of these proteins. When exposed to heat stress, heat shock proteins (HSPs) prevent the aggregation and denaturation of proteins, ensuring that they maintain their functional structures [87]. Small HSPs stabilize proteins prevent their aggregation under heat stress conditions and help protect proteins from irreversible damage. When the temperature rises, the transcription of coding for chaperones and HSPs is initiated by the sigma factor RpoH (σ32) in *E. coli* [88].

Bacteria can repair DNA, protein, and membrane lipids [89]. The process of nucleotide excision repair and homologous recombination correct DNA lesions caused by thermal stress. Bacteria utilize protease to help degrade misfolded or denatured proteins, and repair and refold damaged proteins [90]. Besides, bacteria may adjust the lipid composition of their membranes, incorporating more heat-resistant lipids to maintain membrane fluidity and stability. Similarly, biofilms can defend against thermal stress. Bacteria collectively adjust their behavior and gene expression patterns in response to changing conditions by QS [91].

13.4.2 MDR Bacteria Drug Resistance Mechanism

MDR pathogens are defined as non-susceptibility to more than one agent in over three antimicrobial categories [92]. They have developed resistance to multiple classes of antibiotics and are thus difficult to treat with standard antibiotic therapies. This resistance can make infections caused by these bacteria more challenging to manage and potentially life-threatening. The WHO has identified all six major pathogens *E. coli*, *S. aureus*, *Klebsiella pneumoniae*, *Streptococcus pneumoniae*, *Chlamydia baumannii*, and *P. aeruginosa* that contribute to the burden of antimicrobial resistance (AMR) in 2019 as key pathogens [1]. MRSA is one of the most common MDR bacteria. MRSA is resistant to many antibiotics, including methicillin

Table 13.2 MDR antibiotics.

Type of antibiotics	Representative antibiotics	Bacterial resistance mechanisms
Beta-Lactam Antibiotics	Penicillins Cephalosporins Carbapenems	Bacteria produce beta-lactamase enzymes to break down antibiotics.
Fluoroquinolones	Ciprofloxacin Levofloxacin Moxifloxacin	Bacteria mutations in their DNA gyrase or topoisomerase genes.
Aminoglycosides	Gentamicin Amikacin Tobramycin	Bacteria modify bacterial ribosomal components or through the acquisition of resistance genes.
Tetracyclines	Doxycycline Minocycline	Bacteria may have efflux pumps or ribosomal protection proteins.
Macrolides	Erythromycin Clarithromycin	Bacteria modified target sites on ribosomes, preventing antibiotics from binding effectively.
Glycopeptides	Vancomycin	Bacteria alter the structure of their cell wall to avoid the action of antibiotics.

and other beta-lactam antibiotics [93]. It can cause infections in various parts of the body, such as the skin, bloodstream, and lungs. ESBL-producing bacteria such as ESBL-*E. coli* and ESBL-*K. pneumoniae* can break down a wide range of beta-lactam antibiotics, including penicillins and cephalosporins, by ESBLs enzymes [94]. Carbapenem-resistant Enterobacteriaceae (CRE) bacteria including CRE-*E. coli* and CRE-*K. pneumoniae* are resistant to carbapenem antibiotics, which are often considered the last resort for treating severe infections [95]. Vancomycin-resistant *Enterococci* (VRE) were found in the gastrointestinal tract developing resistance to vancomycin. Common classes of antibiotics to which MDR bacteria can be resistant include such as beta-lactam antibiotics, fluoroquinolones, and aminoglycosides [96]. The details of these antibiotics are summarized in Table 13.2. It is important to note that the development and spread of MDR bacteria is a global concern. Proper antibiotic stewardship, infection control measures, and the development of new antibiotics are crucial in combating these resistant strains.

13.5 Conclusions

In conclusion, the exploration and optimization of photofunctional antibacterial materials, particularly those responsive to NIR light, represent a significant advance in the fight against MDR bacterial infections. NIR phototherapy, encompassing both photodynamic and photothermal modalities, offers a promising alternative to traditional antibiotics, with the potential to circumvent the growing issue of AMR.

This chapter has highlighted the efficacy of various NIR-responsive materials, which have been engineered to maximize photothermal and photodynamic effects. The unique properties of these materials allow for deep tissue penetration, targeted bacterial eradication, and the integration of multiple therapeutic strategies. However, the struggle against MDR bacteria is a pressing global concern, necessitating not only innovative treatments but also responsible antibiotic stewardship and the continual development of novel antimicrobial agents. The potential for bacteria to develop resistance mechanisms to these therapies – through oxidative or thermal stress defenses – remains a challenge. Understanding and preempting these mechanisms is crucial for the continued success of NIR photofunctional materials. As this field progresses, it is imperative that we approach it with a balanced view, considering both the potential and the limitations of NIR-based therapies in a clinical setting. The insights gained from this research can pave the way for more effective, safer, and more sustainable antibacterial strategies in the future.

References

1 Murray, C.J.L., Ikuta, K.S., Sharara, F. et al. (2022). Global burden of bacterial antimicrobial resistance in 2019: a systematic analysis. *Lancet* 399: 629–655.

2 Jernigan, J.A., Hatfield, K.M., Wolford, H. et al. (2020). Multidrug-resistant bacterial infections in U.S. hospitalized patients, 2012–2017. *New England Journal of Medicine* 382: 1309–1319.

3 Bharadwaj, A., Rastogi, A., Pandey, S. et al. (2022). Multidrug-resistant bacteria: their mechanism of action and prophylaxis. *BioMed Research International* 2022: 5419874.

4 Sy, C.L., Chen, P.Y., Cheng, C.W. et al. (2022). Recommendations and guidelines for the treatment of infections due to multidrug resistant organisms. *Journal of Microbiology, Immunology, and Infection* 55: 359–386.

5 Maria Diez-Pascual, A. (2018). Antibacterial activity of nanomaterials. *Nanomaterials* 8: 359.

6 Wu, W., Duan, M., Shao, S. et al. (2023). Recent advances in nanomaterial-mediated bacterial molecular action and their applications in wound therapy. *Biomaterials Science* 11: 6748–6769.

7 Cai, T., Fang, G., Tian, X. et al. (2019). Optimization of antibacterial efficacy of noble-metal-based core-shell nanostructures and effect of natural organic matter. *ACS Nano* 13: 12694–12702.

8 Gao, W., Chen, Y., Zhang, Y. et al. (2018). Nanoparticle-based local antimicrobial drug delivery. *Advanced Drug Delivery Reviews* 127: 46–57.

9 Trubenova, B., Roizman, D., Moter, A. et al. (2022). Population genetics, biofilm recalcitrance, and antibiotic resistance evolution. *Trends in Microbiology* 30: 841–852.

10 Kenry, D. and Y., Liu, B. (2018). Recent advances of optical imaging in the second near-infrared window. *Advanced Materials* 30: e1802394.

11 Overchuk, M., Weersink, R.A., Wilson, B.C., and Zheng, G. (2023). Photodynamic and photothermal therapies: synergy opportunities for nanomedicine. *ACS Nano* 17: 7979–8003.

12 Sun, J., Fan, Y., Ye, W. et al. (2021). Near-infrared light triggered photodynamic and nitric oxide synergistic antibacterial nanocomposite membrane. *Chemical Engineering Journal* 417: 128049.

13 Chen, D., Xu, Q., Wang, W. et al. (2021). Type I photosensitizers revitalizing photodynamic oncotherapy. *Small* 17: e2006742.

14 Van Acker, H. and Coenye, T. (2017). The role of reactive oxygen species in antibiotic-mediated killing of bacteria. *Trends in Microbiology* 25: 456–466.

15 Chen, J., Ning, C., Zhou, Z. et al. (2019). Nanomaterials as photothermal therapeutic agents. *Progress in Materials Science* 99: 1–26.

16 Xu, M., Li, L., and Hu, Q. (2021). The recent progress in photothermal-triggered bacterial eradication. *Biomaterials Science* 9: 1995–2008.

17 Hu, X., Zhang, H., Wang, Y. et al. (2022). Synergistic antibacterial strategy based on photodynamic therapy: progress and perspectives. *Chemical Engineering Journal* 450: 138129.

18 Qi, X., Xiang, Y., Cai, E. et al. (2023). Inorganic-organic hybrid nanomaterials for photothermal antibacterial therapy. *Coordination Chemistry Reviews* 496: 215426.

19 Jin, T., Cheng, D., Jiang, G. et al. (2022). Engineering naphthalimide-cyanine integrated near-infrared dye into ROS-responsive nanohybrids for tumor PDT/PTT/chemotherapy. *Bioactive Materials* 14: 42–51.

20 Jin, L., Lv, S.B., Miao, Y.Y. et al. (2021). Recent development of porous porphyrin-based nanomaterials for photocatalysis. *ChemCatChem* 13: 140–152.

21 Hu, H., Wang, H., Yang, Y.C. et al. (2022). A bacteria-responsive porphyrin for adaptable photodynamic/photothermal therapy. *Angewandte Chemie, International Edition* 61: e202200799.

22 Yu, Z.H., Li, X.S., Xu, F.G. et al. (2020). A supramolecular-based dual-wavelength phototherapeutic agent with broad-spectrum antimicrobial activity against drug-resistant bacteria. *Angewandte Chemie, International Edition* 59: 3658–3664.

23 Li, Z., Feng, X., Gao, S. et al. (2019). Porous organic polymer-coated band-aids for phototherapy of bacteria-induced wound infection. *ACS Applied Bio Materials* 2: 613–618.

24 Li, D., Fang, Y., and Zhang, X. (2020). Bacterial detection and elimination using a dual-functional porphyrin-based porous organic polymer with peroxidase-like and high near-infrared-light-enhanced antibacterial activity. *ACS Applied Materials & Interfaces* 12: 8989–8999.

25 Yao, X.Y., Chen, K.H., Qiu, L.Q. et al. (2021). Ferric porphyrin-based porous organic polymers for CO photocatalytic reduction to syngas with selectivity control. *Chemistry of Materials* 33: 8863–8872.

26 Salih Agirtas, M., Karatas, C., and Ozdemir, S. (2015). Synthesis of some metallophthalocyanines with dimethyl 5-(phenoxy)-isophthalate substituents and

evaluation of their antioxidant-antibacterial activities. *Spectrochimica Acta. Part A, Molecular and Biomolecular Spectroscopy* 135: 20–24.

27 Lee, E., Li, X., Oh, J. et al. (2020). A boronic acid-functionalized phthalocyanine with an aggregation-enhanced photodynamic effect for combating antibiotic-resistant bacteria. *Chemical Science* 11: 5735–5739.

28 Zhao, Y., An, F.H., Meng, J. et al. (2022). Nickel phthalocyanines as potential aggregation-induced antibacterial agents: fenton-like pathways driven by near-infrared light. *Chemphotochem* 6.

29 Zhang, W., Wang, X., Ma, G. et al. (2023). Fluorescence and photothermal dual-readout phthalocyanine-fluorescein conjugate for detection and photothermal sterilization of anaerobic bacteria. *Sensors and Actuators B: Chemical* 134042.

30 Zhou, J., Cao, Z.L., Panwar, N. et al. (2017). Functionalized gold nanorods for nanomedicine: past, present and future. *Coordination Chemistry Reviews* 352: 15–66.

31 Thorat, N.D., Dworniczek, E., Brennan, G. et al. (2021). Photo-responsive functional gold nanocapsules for inactivation of community-acquired, highly virulent, multidrug-resistant MRSA. *Journal of Materials Chemistry B* 9: 846–856.

32 Pan, W., Wu, B., Nie, C. et al. (2023). NIR-II responsive nanohybrids incorporating thermosensitive hydrogel as sprayable dressing for multidrug-resistant-bacteria infected wound management. *ACS Nano* 11253–11267.

33 Zhu, S.X., Wang, X.Y., Li, S.L. et al. (2020). Near-infrared-light-assisted in situ reduction of antimicrobial peptide-protected gold nanoclusters for stepwise killing of bacteria and cancer cells. *ACS Applied Materials & Interfaces* 12: 11063–11071.

34 Ling, C., Wang, X., and Shen, Y. (2021). Advances in hollow inorganic nanomedicines for photothermal-based therapies. *International Journal of Nanomedicine* 16: 493–513.

35 Qin, Z., Zheng, Y., Du, T. et al. (2021). Cysteamine: a key to trigger aggregation-induced NIR-II photothermal effect and silver release booming of gold-silver nanocages for synergetic treatment of multidrug-resistant bacteria infection. *Chemical Engineering Journal* 414: 128779.

36 Zhang, H.F., Tan, J., Yang, X.Q. et al. (2022). Size-tunable yolk-shell gold-silver nanostructures for photothermal treatment of multidrug-resistant bacteria. *ACS Applied Nano Materials* 5: 10818–10828.

37 Behnam, M.A., Emami, F., Sobhani, Z. et al. (2018). Novel combination of silver nanoparticles and carbon nanotubes for plasmonic photo thermal therapy in melanoma cancer model. *Advanced Pharmaceutical Bulletin* 8: 49–55.

38 Zhao, Y., Zhao, T., Cao, Y. et al. (2021). Temperature-sensitive lipid-coated carbon nanotubes for synergistic photothermal therapy and gene therapy. *ACS Nano* 15: 6517–6529.

39 Tondro, G.H., Behzadpour, N., Keykhaee, Z. et al. (2019). Carbon@polypyrrole nanotubes as a photosensitizer in laser phototherapy of *Pseudomonas aeruginosa*. *Colloids and Surfaces. B, Biointerfaces* 180: 481–486.

40 Ji, H.W., Sun, H.J., and Qu, X.G. (2016). Antibacterial applications of graphene-based nanomaterials: recent achievements and challenges. *Advanced Drug Delivery Reviews* 105: 176–189.

41 Henriques, P.C., Pereira, A.T., Bogas, D. et al. (2021). Graphene films irradiated with safe low-power NIR-emitting diodes kill multidrug resistant bacteria. *Carbon* 180: 10–21.

42 Tegou, E., Magana, M., Katsogridaki, A.E. et al. (2016). Terms of endearment: bacteria meet graphene nanosurfaces. *Biomaterials* 89: 38–55.

43 Pramanik, A., Dhar, J.A., Banerjee, R. et al. (2023). WO(3) nanowire-attached reduced graphene oxide-based 1D-2D heterostructures for near-infrared light-driven synergistic photocatalytic and photothermal inactivation of multidrug-resistant superbugs. *ACS Applied Bio Materials* 6: 919–931.

44 Gong, C., Guan, W., Liu, X. et al. (2022). Biomimetic bacteriophage-like particles formed from probiotic extracts and NO donors for eradicating multidrug-resistant *Staphylococcus aureus*. *Advanced Materials* 34: e2206134.

45 Chai, S., Xie, Y., and Yang, L. (2022). Antibacterial applications of elemental nanomaterials. *Current Opinion in Solid State & Materials Science* 26: 101043.

46 Bezza, F.A., Tichapondwa, S.M., and Chirwa, E.M.N. (2020). Fabrication of monodispersed copper oxide nanoparticles with potential application as antimicrobial agents. *Scientific Reports* 10: 16680.

47 Zhang, L., Yang, J., Zhu, X.Y. et al. (2023). Electrospun nanofibrous membranes loaded with IR780 conjugated MoS_2-nanosheet for dual-mode photodynamic/photothermal inactivation of drug-resistant bacteria. *Environmental Technology & Innovation* 30: 103098.

48 Hu, X.L., Chu, L.Y., Dong, X.J. et al. (2019). Multivalent glycosheets for double light driven therapy of multidrug-resistant bacteria on wounds. *Advanced Functional Materials* 29: 1806986.

49 Li, Y., Liu, X.M., Tan, L. et al. (2019). Eradicating multidrug-resistant bacteria rapidly using a multi functional g-C_3N_4@ Bi_2S_3 nanorod heterojunction with or without antibiotics. *Advanced Functional Materials* 29: 1900946.

50 Nain, A., Huang, H.H., Chevrier, D.M. et al. (2021). Catalytic and photoresponsive BiZ/Cu(x)S heterojunctions with surface vacancies for the treatment of multidrug-resistant clinical biofilm-associated infections. *Nanoscale* 13: 18632–18646.

51 Chen, S., Chu, Z., Cao, L. et al. (2022). Antibacterial mechanism and transcriptomic analysis of a near-infrared triggered upconversion nanoparticles@ $AgBiS_2$ for synergetic bacteria-infected therapy. *Nano Research* 15: 9298–9308.

52 Shan, J., Yang, K., Xiu, W. et al. (2020). Cu(2) MoS(4) nanozyme with NIR-II light enhanced catalytic activity for efficient eradication of multidrug-resistant bacteria. *Small* 16: e2001099.

53 Cao, J., Chi, J., Xia, J. et al. (2019). Iodinated cyanine dyes for fast near-infrared-guided deep tissue synergistic phototherapy. *ACS Applied Materials & Interfaces* 11: 25720–25729.

54 Li, C.N., Gan, Y.Y., Li, Z.S. et al. (2023). Neutrophil-inspired photothermo-responsive drug delivery system for targeted treatment of bacterial infection and endotoxins neutralization. *Biomaterials Research* 27.

55 Carr, J.A., Franke, D., Caram, J.R. et al. (2018). Shortwave infrared fluorescence imaging with the clinically approved near-infrared dye indocyanine green. *Proceedings of the National Academy of Sciences of the United States of America* 115: 4465–4470.

56 Zhang, H.X., Zou, Y., Lu, K.Y. et al. (2024). A nanoplatform with oxygen self-supplying and heat-sensitizing capabilities enhances the efficacy of photodynamic therapy in eradicating multidrug-resistant biofilms. *Journal of Materials Science and Technology* 169: 209–219.

57 Zhang, X.W., Min, Y.H., Zhang, Q. et al. (2022). Functionalized MnO nanosheets with photothermal, photodynamic, and oxidase-like activities triggered by low-powered near-infrared light for synergetic combating multidrug-resistant bacterial infections. *Advanced Healthcare Materials* 11.

58 Liu, S.L., Zhang, W., Chen, Q.Q. et al. (2021). Multifunctional nanozyme for multimodal imaging-guided enhanced sonodynamic therapy by regulating the tumor microenvironment. *Nanoscale* 13: 14049–14066.

59 Zhao, X., Zhao, H., Wang, S. et al. (2021). A tumor-targeting near-infrared heptamethine cyanine photosensitizer with twisted molecular structure for enhanced imaging-guided cancer phototherapy. *Journal of the American Chemical Society* 143: 20828–20836.

60 Jiang, D., Pan, L., Yang, X. et al. (2023). Photo-controllable burst generation of peroxynitrite based on synergistic interactions of polymeric nitric oxide donors and IR780 for enhancing broad-spectrum antibacterial therapy. *Acta Biomaterialia* 159: 259–274.

61 Tao, W., Ji, X.Y., Xu, X.D. et al. (2017). Antimonene quantum dots: synthesis and application as near-infrared photothermal agents for effective cancer therapy. *Angewandte Chemie, International Edition* 56: 11896–11900.

62 Sun, Z.B., Xie, H.H., Tang, S.Y. et al. (2015). Ultrasmall black phosphorus quantum dots: synthesis and use as photothermal agents. *Angewandte Chemie, International Edition* 54: 11526–11530.

63 Chung, S., Revia, R.A., and Zhang, M. (2021). Graphene quantum dots and their applications in bioimaging, biosensing, and therapy. *Advanced Materials* 33: e1904362.

64 Sun, T., Tang, M., Shi, Y., and Li, B. (2022). MXenes quantum dots for biomedical applications: recent advances and challenges. *Chemical Record* 22: 876–902.

65 Geng, B., Li, Y., Hu, J. et al. (2022). Graphitic-N-doped graphene quantum dots for photothermal eradication of multidrug-resistant bacteria in the second near-infrared window. *Journal of Materials Chemistry B* 10: 3357–3365.

66 Kuo, W.S., Shao, Y.T., Huang, K.S. et al. (2018). Antimicrobial amino-functionalized nitrogen-doped graphene quantum dots for eliminating multidrug-resistant species in dual-modality photodynamic therapy and bioimaging under two-photon excitation. *ACS Applied Materials & Interfaces* 10: 14438–14446.

67 Xu, S., Chang, L., Hu, Y. et al. (2021). Tea polyphenol modified, photothermal responsive and ROS generative black phosphorus quantum dots as nanoplatforms for promoting MRSA infected wounds healing in diabetic rats. *Journal of Nanobiotechnology* 19: 362.

68 Du, T., Xiao, Z., Cao, J. et al. (2022). NIR-activated multi-hit therapeutic Ag(2)S quantum dot-based hydrogel for healing of bacteria-infected wounds. *Acta Biomaterialia* 145: 88–105.

69 Li, Y., Liu, Z., Hou, Y. et al. (2017). Multifunctional nanoplatform based on black phosphorus quantum dots for bioimaging and photodynamic/photothermal synergistic cancer therapy. *ACS Applied Materials & Interfaces* 9: 25098–25106.

70 Li, Z.B., Luo, G.H., Hu, W.P. et al. (2020). Mediated drug release from nanovehicles by black phosphorus quantum dots for efficient therapy of chronic obstructive pulmonary disease. *Angewandte Chemie, International Edition* 59: 20568–20576.

71 Chen, M., Zhou, J., Ran, P. et al. (2022). Photoactivated release of nitric oxide and antimicrobial peptide derivatives for synergistic therapy of bacterial skin abscesses. *Advanced Healthcare Materials* 11: e2200199.

72 Cui, D., Han, L., Jiang, W. et al. (2023). $Ag_{2-3X}Bi_XS$ quantum dots as single-component theranostic agents for second near-infrared fluorescence imaging-guided photothermal therapy. *ACS Applied Nano Materials* 6: 1303–1314.

73 Sun, P., Li, K., Liu, X. et al. (2023). Peptide-mediated aqueous synthesis of NIR-II emitting Ag_2S quantum dots for rapid photocatalytic bacteria disinfection. *Angewandte Chemie* 135: e202300085.

74 Hu, F., Xu, S., and Liu, B. (2018). Photosensitizers with aggregation-induced emission: materials and biomedical applications. *Advanced Materials* 30: 1801350.

75 Yu, H., Chen, B., Huang, H. et al. (2022). AIE-active photosensitizers: manipulation of reactive oxygen species generation and applications in photodynamic therapy. *Biosensors-Basel* 12: 348.

76 Sun, J., Li, H., Gu, X., and Tang, B.Z. (2021). Photoactivatable biomedical materials based on luminogens with aggregation-induced emission (AIE) characteristics. *Advanced Healthcare Materials* 10: 2101177.

77 Li, T., Wu, Y., Cai, W. et al. (2022). Vision defense: efficient antibacterial AIEgens induced early immune response for bacterial endophthalmitis. *Advanced Science* 9: e2202485.

78 Xiao, P., Shen, Z., Wang, D. et al. (2022). Precise molecular engineering of type I photosensitizers with near-infrared aggregation-induced emission for image-guided photodynamic killing of multidrug-resistant bacteria. *Advanced Science* 9: 2104079.

79 He, W., Zhang, T., Bai, H. et al. (2021). Recent advances in aggregation-induced emission materials and their biomedical and healthcare applications. *Advanced Healthcare Materials* 10: 2101055.

80 Huang, Y., Li, D., Wang, D. et al. (2022). A NIR-II emissive polymer AIEgen for imaging-guided photothermal elimination of bacterial infection. *Biomaterials* 286: 121579.

81 Wang, W., Wu, F., Zhang, Q. et al. (2022). Aggregation-induced emission nanoparticles for single near-infrared light-triggered photodynamic and photothermal antibacterial therapy. *ACS Nano* 16: 7961–7970.

82 Borisov, V.B., Siletsky, S.A., Nastasi, M.R., and Forte, E. (2021). ROS defense systems and terminal oxidases in bacteria. *Antioxidants* 10: 839.

83 Mendez, V., Rodriguez-Castro, L., Duran, R.E. et al. (2022). The OxyR and SoxR transcriptional regulators are involved in a broad oxidative stress response in *Paraburkholderia xenovorans* LB400. *Biological Research* 55: 7.

84 Li, X.-Z., Plesiat, P., and Nikaido, H. (2015). The challenge of efflux-mediated antibiotic resistance in gram-negative bacteria. *Clinical Microbiology Reviews* 28: 337–418.

85 Yamasaki, S., Zwama, M., Yoneda, T. et al. (2023). Drug resistance and physiological roles of RND multidrug efflux pumps in *Salmonella enterica*, *Escherichia coli* and *Pseudomonas aeruginosa*. *Microbiology-Sgm* 169: 001322.

86 Lee, J. and Zhang, L. (2015). The hierarchy quorum sensing network in *Pseudomonas aeruginosa*. *Protein & Cell* 6: 26–41.

87 Dabbaghizadeh, A. and Tanguay, R.M. (2020). Structural and functional properties of proteins interacting with small heat shock proteins. *Cell Stress & Chaperones* 25: 629–637.

88 Williamson, K.S., Dlakic, M., Akiyama, T., and Franklin, M.J. (2023). The *Pseudomonas aeruginosa* RpoH (σ32) regulon and its role in essential cellular functions, starvation survival, and antibiotic tolerance. *International Journal of Molecular Sciences* 24: 1513.

89 Zhang, Y.-M. and Rock, C.O. (2008). Membrane lipid homeostasis in bacteria. *Nature Reviews Microbiology* 6: 222–233.

90 Schramm, F.D., Schroeder, K., and Jonas, K. (2020). Protein aggregation in bacteria. *FEMS Microbiology Reviews* 44: 54–72.

91 Yin, W., Wang, Y., Liu, L., and He, J. (2019). Biofilms: the microbial “protective clothing” in extreme environments. *International Journal of Molecular Sciences* 20: 3423.

92 Wolfensberger, A., Kuster, S.P., Marchesi, M. et al. (2019). The effect of varying multidrug-resistence (MDR) definitions on rates of MDR gram-negative rods. *Antimicrobial Resistance and Infection Control* 8: 193.

93 Liu, W.-T., Chen, E.-Z., Yang, L. et al. (2021). Emerging resistance mechanisms for 4 types of common anti-MRSA antibiotics in *Staphylococcus aureus*: a comprehensive review. *Microbial Pathogenesis* 156: 104915.

94 Khadka, C., Shyaula, M., Syangtan, G. et al. (2023). Extended-spectrum beta-lactamases producing *Enterobacteriaceae* (ESBL-PE) prevalence in Nepal: a systematic review and meta-analysis. *The Science of the Total Environment* 901: 166164.

95 Suay-Garcia, B. and Teresa Perez-Gracia, M. (2019). Present and future of carbapenem-resistant *Enterobacteriaceae* (CRE) infections. *Antibiotics-Basel* 8: 122.

96 Shaikh, S., Fatima, J., Shakil, S. et al. (2015). Antibiotic resistance and extended spectrum beta-lactamases: types, epidemiology and treatment. *Saudi Journal of Biological Sciences* 22: 90–101.

14

Photoelectrochemical Nanomaterials for Biosensing Applications

Qianqian Sun[1] *and Piaoping Yang*[2]

[1] *Shandong University, School of Chemistry and Chemical Engineering, Institute of Frontier Chemistry, Binhai Road, Qingdao 266237, P. R. China*
[2] *Harbin Engineering University, College of Material Sciences and Chemical Engineering, Ministry of Education, Key Laboratory of Superlight Materials and Surface Technology, Nantong Street, Harbin 150001, P. R. China*

14.1 Introduction

Photoelectrochemical nanomaterials are those materials that have excellent optical and electrical properties and dimensions on the nanoscale. They typically exhibit properties such as strong light absorption, high fluorescence quantum yield, and tunable electronic energy levels [1, 2]. In recent years, photoelectrochemical nanomaterials have become a research hotspot in the biomedical field, with a wide range of potential applications. The special optical and electrical properties of photoelectrochemical nanomaterials give them unique advantages in bioimaging, drug delivery, photothermal therapy, and biosensing. This review focuses on the applications of photoelectrochemical nanomaterials in biosensing. Biosensing is an important technology that enables highly sensitive and selective detection of biomolecules [3–7]. The special surface properties and tunable optical properties of photoelectrochemical nanomaterials make them ideal for biosensors. This biosensing technology based on photoelectrochemical nanomaterials can be used in the detection of biomolecules, the study of protein interactions, and the analysis of cell signaling. Finally, an outlook on its challenges and potential clinical applications is given.

14.2 Classification of Photoelectrochemical Materials

Photoelectrochemical analysis originated in 1839 when A.E. Becquerel found that light makes the potential difference between different parts of semiconductor materials (later called the "photoelectric effect") [8, 9], and in 1887 when Hertz, while studying the metal-vacuum interface, found the phenomenon of photoelectron emission [10]. Photoelectrochemical is the study of the effect of light on optical

Photofunctional Nanomaterials for Biomedical Applications, First Edition.
Edited by Chunxia Li and Jun Lin.

electrodes or interface materials and the conversion of light energy into electrical energy. The essence of photoelectrochemical analysis refers to the photochemical phenomenon induced by the change in photocurrent due to the transfer of electrons between the excited state of the photoelectrochemical material and the electron donor or acceptor, and therefore, the process is accompanied by the conversion of light energy into electrical and chemical energy [11].

Photoelectrochemical biosensing combines photoelectrochemistry with sensor-based bioanalysis. As described in Chapter 1, a typical photoelectrochemical biosensor consists of an excitation light source, a photoactive material, a biorecognition element, a signal converter, and a signal output device [1, 12]. The detection principle of the sensor can be briefly summarized as follows: light can excite a specific photoactive material, causing it to generate a photocurrent or photovoltage. When the target analyte is added to the system, the output signal from this photoactive material changes accordingly. Under certain conditions, a function is formed between the change in the electrical output signal and the concentration of the target analyte. Through this functional relationship and the measured change in the electrical signal, detection, and analysis of the target analyte can be achieved (Figure 14.1) [13–16]. Combining the advantages of optical and electrochemical detection, photoelectrochemical biosensors have developed rapidly due to their simplicity and high efficiency and have found wide applications, especially in the fields of disease diagnosis, pathogen detection, drug analysis, and biological research [2, 17].

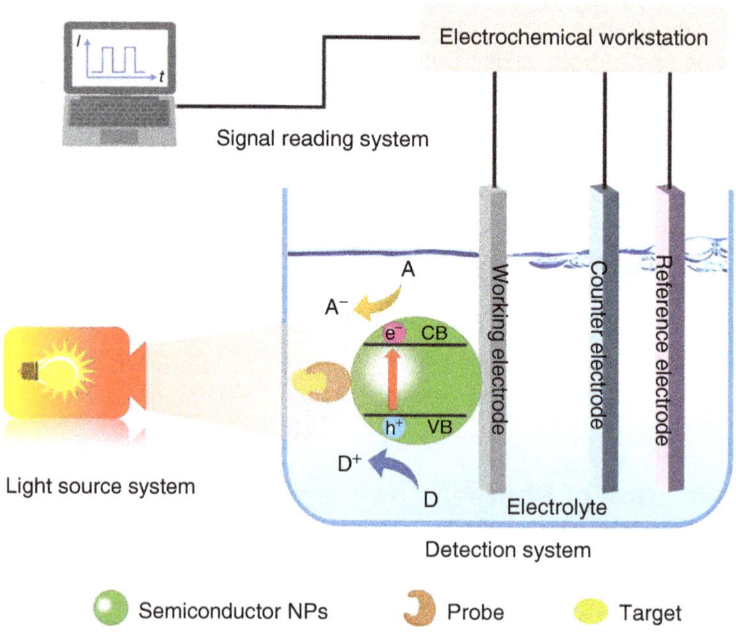

Figure 14.1 Diagram of typical photoelectrochemical biosensing system. Source: Zhang et al. [1]/MDPI/CC-BY 4.0.

Photoactive material is one of the core components of photoelectrochemical biosensors, which directly affects the analytical performance of the sensor. Excellent photoelectrochemical materials can significantly improve the detection sensitivity of biosensors [18]. At present, photoelectric materials commonly used in photoelectrochemical biosensors can be divided into inorganic materials, organic materials, composite materials (inorganic–inorganic composite, inorganic–organic composite), and so on, according to their different properties.

14.2.1 Inorganic Photoelectrochemical Materials

Inorganic photoelectrochemical semiconductors are divided into two main types: monomers and compounds. Among them, the mono-type semiconductor includes semiconductor materials composed of a single element, including but not limited to silicon, germanium, selenium, and so on. Compound semiconductors are composed of two or more different elements and can be further divided into metal selenides, sulfides, oxides, arsenides, etc. They have diverse electrical and optical properties and occupy an important position in the field of photoelectrochemical chemistry. The performance of compound semiconductors depends on their energy band structure, which determines their ability to absorb and emit light at different wavelengths. By changing the composition of the compounds, the band gap width of the material can be adjusted, which in turn regulates its optical and electrical properties to suit different application scenarios.

Inorganic optoelectronic materials can be divided into two categories according to the nature of the main carriers: N-type semiconductors and P-type semiconductors. The main carriers in N-type semiconductors are electrons [19, 20]. Common N-type semiconductor materials include:

I. Traditional semiconductor materials with N-type characteristics obtained by doping (e.g. silicon (doped with elements such as phosphorus, arsenic, or antimony), germanium (the same can be doped with elements such as phosphorus, arsenic or antimony), gallium arsenide (doped with elements such as selenium, tellurium or silicon), and gallium nitride (doped with elements such as silicon)). These semiconductors are usually made by doping the pure semiconductor with elements (impurities) that provide additional electrons. These impurities usually have one more valence electron than the semiconductor. The choice of dopant depends on the lattice and electronic structure of the host semiconductor and the desired electron concentration.
II. Two-dimensional materials such as transition metal sulfides (e.g. MoS_2) have a layered structure and can exhibit N-type properties through doping or natural defects.
III. Metal oxides (e.g. TiO_2 and ZnO) are wide bandgap metal oxides with N-type properties and are often used as photocatalysts or in optoelectronic devices.
IV. Graphitic carbon nitride (g-C_3N_4) and sulfide semiconductors (e.g. Bi_2S_3) also commonly exhibit N-type properties.

The main carriers in P-type semiconductors are photogenerated holes with a low concentration of free electrons, and the current generation mechanism depends mainly on positively charged holes [21, 22]. Common P-type semiconductor materials include:

(I) Traditional semiconductor materials with P-type characteristics obtained by doping (e.g. silicon (doped with elements such as boron, aluminum, gallium, or indium), germanium (which can also be doped with elements of boron, aluminum or gallium), gallium arsenide (doped with elements of zinc or magnesium), gallium nitride (doped with elements of magnesium or cobalt), etc.). In other words, P-type semiconductors are made by doping the material with electron-accepting elements that usually have one less valence electron than the semiconductor.

(II) Metal oxides (e.g. Cu_2O and CuO) are important P-type semiconductor materials used in the preparation of various optoelectronic devices.

(III) Sulfide semiconductors (e.g. PbS) have narrow band gap and high carrier mobility properties and usually exhibit P-type characteristics.

14.2.2 Organic Photoelectrochemical Materials

Organic photoelectrochemical materials are a class of organic molecules or polymers containing carbon that are capable of generating an electric charge in response to light or light emission in response to an applied electric field. These materials have special light absorption and luminescence properties that make them very useful in optoelectronics (e.g. photoelectrochemical biosensors) [23]. Organic photoelectrochemical materials can be divided into two main categories: small-molecule organic materials and polymeric organic materials.

I. Small-molecule organic materials: Small-molecule organic materials are usually prepared as thin films for sensors by vapor deposition or solution treatment. Small-molecule organic materials include clinolines, phthalocyanines, bipyridines, chlorophylls, azoles, fullerenes, and their derivatives (e.g. C60 and [6]-phenyl-C61-butyric acid methyl ester) and small-molecule transmissive materials (e.g. orange phosphorus and copper phthalocyanine).

II. Polymeric organic materials: Polymeric organic materials are long-chain polymers consisting of repeating units that can be prepared by solution processing techniques such as spin coating, inkjet printing, or doctor blade coating. They include conductive polymers (e.g. polythiophene, polyaniline, polypyrrole, and polyfluorene) and electroluminescent polymers (e.g. polyethylene terephthalate and polyfluorene-styrene analogs). These organic photoelectrochemical materials have great potential in sensing and biomedical applications due to their tunable electronic energy levels, broad spectral absorption range, excellent mechanical flexibility, and ability to be structurally modified by chemical synthesis.

14.2.3 Composite Photoelectrochemical Materials

Although a single inorganic or organic photoelectrochemical material can achieve a certain level of photoelectrochemical conversion efficiency on its own, its performance is often limited by its respective limitations in light absorption, carrier separation, and electron migration. To overcome these limitations, a strategy has been developed to construct highly efficient composite photoelectrochemical materials. This strategy involves combining two or more photoelectrochemical materials with complementary energy levels to extend the visible light absorption range of the materials and to facilitate efficient separation and fast migration of photogenerated carriers by creating a stepped energy level pattern. Currently, researchers have constructed various types of composite photoelectrochemical material structures, such as P–N heterojunctions [24, 25], Z-type heterojunctions [26, 27], and type I heterojunctions [28], through careful design. The heterostructures of these composite materials can significantly improve the photoelectrochemical performance compared to a single material, which has been promoted and used in many applications. For example, organic–inorganic heterojunctions based on polypyrrole (PPy) and Bi_2WO_6 have been reported to be used in the construction of photoelectrochemical biosensors for the highly sensitive immunoassay of creatine kinase-MB (CK-MB) [29]. Owing to the advantages of higher absorption coefficient and carrier mobility in the visible region of PPy, the response signal of Bi_2WO_6 can be successfully sensitized, opening a new perspective for the use of organic–inorganic heterojunctions in biosensing.

14.3 Introduction to Biorecognition Elements

The biorecognition element is one of the other core components of a photoelectrochemical biosensor, modified or immobilized on the surface of the sensor or dispersed in solution, which modifies the photocurrent by in situ generated electron donors or acceptors. Enrichment and localization of the target biomolecule on the sensor is achieved by specific recognition and binding to the target biomolecule [30]. Common biorecognition elements include antibodies, enzymes, nucleic acids, molecularly imprinted polymers, and bioreceptors [31]. These biorecognition elements are highly specific and selective and can specifically bind to antigenic regions, substrates, or specific sequences of target biomolecules to achieve recognition and detection of target biomolecules. Below is a brief introduction to some common biometric elements:

(I) *Antibodies*: Antibodies are highly specific proteins produced by organisms that bind to antigenic regions of target molecules. Modification of antibodies on the surface of sensors allows highly selective recognition of specific antigens. Due to the high specificity and affinity between antibodies and antigens, antibodies are widely used as biorecognition elements in biosensors, such as for the detection of disease markers and pathogens.

(II) *Enzymes*: Enzymes are a class of biocatalysts that catalyze specific chemical reactions and can convert a substrate into a product through a specific enzyme-substrate reaction. Modification of enzymes on the surface of sensors can enable the detection of specific substrates. Enzymes are widely used as biorecognition elements in biosensors, such as glucoamylase for glucose sensors and peroxidase for bio-oxidant sensors.

(III) *Nucleic acids*: Nucleic acids (e.g. DNA and RNA) have specific sequence and structural features that allow for specific pairing with their complementary nucleic acid sequences or structures. By synthesizing DNA or RNA molecules with specific sequences, highly selective recognition of specific target sequences can be achieved. Nucleic acids are widely used as biorecognition elements in DNA sensors and RNA sensors.

(IV) *Molecularly imprinted polymers*: Molecularly imprinted polymers are formed by copolymerizing a target molecule with a functional monomer and then, after removal of the target molecule, forming a polymer with recognition sites for the target molecule. Molecularly imprinted polymers have specific molecular recognition properties and can achieve highly selective recognition of specific target molecules. Molecularly imprinted polymers are used as biorecognition elements in biosensors, such as for drug detection.

(V) *Bioreceptors*: Bioreceptors are proteins found in living organisms that specifically recognize and bind to certain molecules (such as hormones or neurotransmitters), usually found on cell membranes. Cell membrane receptors are proteins on the surface of cells that bind to specific ligands. Modification of cell membrane receptors on the surface of sensors enables highly selective recognition of specific ligands. Cell membrane receptors are used as biorecognition elements in cellular sensors and drug screening.

These biorecognition elements enable the recognition and detection of different types of biomolecules according to the specificity of the target molecule and the recognition mechanism. When constructing photoelectrochemical biosensors, the selection of suitable biorecognition elements is the key to achieving high sensitivity, high selectivity, and fast response.

14.4 Factors Affecting the Photocurrent Signal

In simple terms, a photoelectrochemical sensor is a sensor that uses the photoelectrochemical effect to detect the presence and concentration of an analyte and obtains information about the target substance through changes in photocurrent [32, 33]. Photocurrent is the electric current generated by a material under light irradiation. It is formed by photogenerated excited electrons (or holes) driven by an internal electric field. The strength of the photocurrent is directly related to the photoelectrochemical conversion efficiency of the photoelectrochemical biosensor, that is, the efficiency with which the incident light is converted into electrical energy. Overall, the strength of the photocurrent is affected by a number of factors, including the pH of the buffer solution, the type of photoelectrochemical material, the cleanliness of the working electrode, the incubation time of the photoelectrochemical material, the type of reference electrode, the intensity of the excitation light, the magnitude of the bias applied, the degree of energy level matching between the photoelectrochemical material and the sensitizer, the type and concentration of the electron donor or acceptor, and the photostability of the photoelectrochemical material. All of these factors can affect the generation and intensity of the photocurrent. To be more precise:

(I) *pH of the buffer solution*: changes in pH can affect the ionic concentration of the electrolyte and the rate of charge transfer, thus affecting the strength of the photocurrent.
(II) *Photoelectrochemical material properties*: material properties such as energy band structure, absorption coefficient, carrier lifetime, and mobility can affect photocurrent generation and transport efficiency.
(III) *Cleanliness of the working electrode*: impurities or contaminants on the surface of the electrode can impede the injection or extraction of photogenerated carriers, thus affecting photocurrent generation and transport.
(IV) *Incubation time of the photoelectrochemical material*: the incubation time of the material can affect its lattice structure and surface morphology, which in turn affects the strength of the photocurrent.
(V) *The type of reference electrode*: the choice of reference electrode can affect the electrode potential and reaction kinetics, which is crucial to ensure the reproducibility of the experiment.
(VI) *The intensity of the excitation light*: Stronger light will usually produce more photogenerated carriers and increase the photocurrent.
(VII) *Amount of bias applied*: correct application of bias to the electrode will promote the separation of photogenerated carriers and increase the photocurrent.
(VIII) *Degree of energy level matching between the photoelectrochemical material and the sensitizer*: a good energy level matching between the two can effectively promote the transfer of electrons from the sensitizer to the conductive substrate, thus influencing the strength of the photocurrent.
(IX) *The type and concentration of electron donor or acceptor*: Adding the right amount of electron donor or acceptor can adjust the energy level structure of the photoelectrochemical material, thus affecting the magnitude of the photocurrent.
(X) *Photostability of photoelectrochemical materials*: Materials with good photostability can maintain efficient photoelectric conversion performance under prolonged light exposure.

In summary, the precise control and measurement of the photocurrent is very important in photoelectrochemical sensor applications. By measuring the change in photocurrent, the concentration of an analyte can be inferred. For example, if an analyte increases photogenerated carrier production or transport efficiency, then the photocurrent will increase in the presence of that substance. Conversely, if the analyte decreases the number of carriers or the transport efficiency, the photocurrent will decrease. Detection of the photocurrent allows quantitative or qualitative analysis of specific chemical substances. The generation of a photocurrent is a complex process that is co-regulated by several physical and chemical factors. Therefore, understanding the relationship between the strength of the photocurrent and the performance of photoelectrochemical sensors by optimizing the above factors can help in the design and fabrication of more efficient and sensitive sensors for use in biomedical detection and other fields.

14.5 Signal Amplification and Bursting Strategies

In the design and application of photoelectrochemical biosensors, the amplification strategy and the bursting strategy of the photocurrent signal are two key aspects to be considered [34]. These strategies directly affect the sensitivity and accuracy of the biosensors.

14.5.1 Photocurrent Signal Amplification Strategies

In many bio-detection applications, the concentration of the target biomolecules may be very low, such as pathogen detection, disease marker detection. Therefore, in addition to some influencing factors of the photocurrent signal mentioned in the previous subsection, the sensor may also fail to detect enough signal to accurately indicate the presence of the target molecule if the signal amplification strategy is not used properly. By using signal amplification strategies, the sensor's response to the target molecule can be significantly improved, thereby increasing its sensitivity [35, 36]. At the same time, signal amplification can increase the signal generated by the target elements, thereby lowering the detection limit. This is important for applications that require the detection of target molecules at very low concentrations (e.g. early disease diagnosis and environmental monitoring). Signal amplification strategies can also increase the signal-to-noise ratio, thereby improving the accuracy and reliability of detection. This is particularly important in situations where high-precision detection is required (e.g. drug concentration monitoring and biochemical analysis). In addition, signal amplification is expected to significantly increase the dynamic detection range of the sensor, enabling it to accurately measure target elements over a wider concentration range.

Currently, researchers have developed a variety of signal amplification strategies. The following are some common enhancement techniques:

I. *Use of special nanomaterials*: The use of nanomaterials with special optoelectronic properties is a common strategy for signal enhancement. For example, gold nanoparticles and quantum dots are widely used for signal amplification [37, 38]. These materials have high surface-to-volume ratios and excellent optoelectronic properties that can significantly enhance optoelectronic signals.
II. *Enzyme-labeled amplification strategy*: Enzyme-labeled amplification is another common strategy. Enzymes are used as markers to catalyze the conversion of a large number of substrates into products that amplify the detection signal. For example, alkaline phosphatase and horseradish peroxidase are both widely utilized in this method [39–44].
III. *DNA amplification strategies*: Rolling circle amplification (RCA) and hybridization chain reaction (HCR) are two frequently employed methods for signal

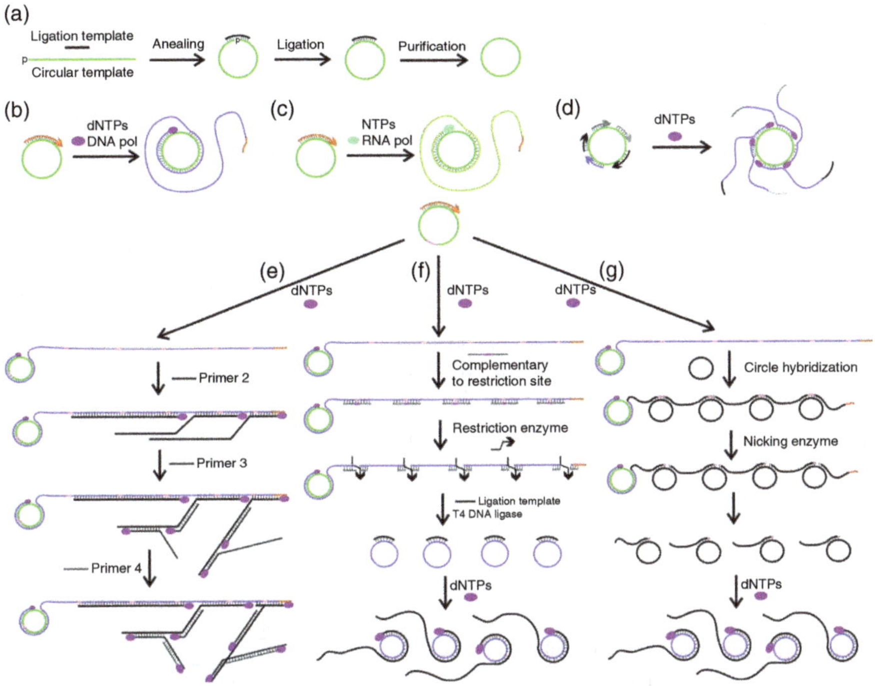

Figure 14.2 Schematic illustration of the basic principles of RCA. (a) Making a DNA circle by template-mediated enzymatic ligation. (b) Linear RCA reaction to generate long ssDNA with a premade circle and target nucleic acids. (c) Linear RCA reaction to generate long RNA with a premade DNA circular template. (d) Multiprimed RCA to generate multiple copies of the RCA product from a single circle. (e) Exponential DNA amplification by hyper-branched RCA. (f) Exponential DNA amplification generates multiple circles from the RCA products. (g) Exponential DNA amplification by primer generation using nicking enzymes. Source: Reproduced with permission from Ali et al. [45]/The Royal Society of Chemistry.

amplification in photoelectrochemical biosensors. RCA uses the shearing action of DNA enzymes and the linking action of DNA ligase to amplify the target DNA sequence into a long chain form by cyclic amplification, which improves the intensity of the photoelectrochemical signal (Figure 14.2) [45, 46]. The HCR signal amplification strategy, on the other hand, exploits the hybridization and separation properties of DNA strands to amplify the signal through multiple hybridization and separation processes (Figure 14.3) [47, 48]. In addition to the single signal amplification strategies described above, there are signal cycle amplification strategies (including strand substitution amplification [49], DNA walker-mediated signal cycle amplification [50], and DNAzyme-mediated signal cycle amplification [51]) and signal cascade amplification strategies implemented by coupling multiple signal amplification strategies.

IV. *Surface-enhanced Raman scattering (SERS) strategy*: SERS is an optically based signal enhancement strategy that exploits the surface plasmon resonance effect of metal nanoparticles [52], such as gold or silver, which can significantly enhance the Raman signal, enabling the detection of single molecules.

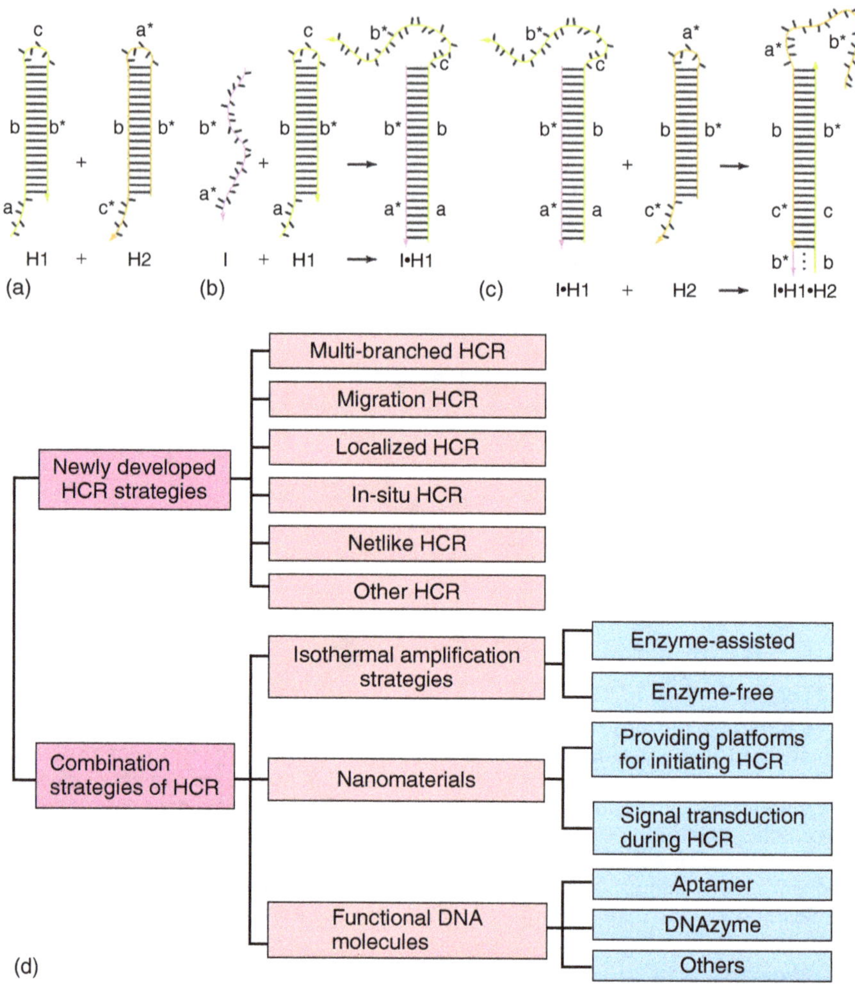

Figure 14.3 (a–c) Fundamentals of HCR. The initiator (I) triggers a chain hybridization reaction between alternating H1 and H2 hairpins to form a nicked double helix. Source: Reproduced with permission from Bi et al. [47]/The Royal Society of Chemistry. (d) Recent development of HCR strategies in biosensors. Source: Reproduced with permission from Zhang et al. [48]/American Chemical Society.

These are just some of the more commonly applied signal amplification strategies for photoelectrochemical biosensors, which can be used individually or in combination to achieve higher sensitivity, selectivity, and response. Depending on the specific application requirements and sensor design, selecting the appropriate signal amplification strategy is key to improving the performance of photoelectrochemical biosensors. The following section focuses on the enhancement of photocurrent signals by quantum dot sensitization and plasmon resonance effects on noble metal surfaces.

Quantum dot sensitization is a technique that uses quantum dots in semiconductor nanomaterials to enhance the photocurrent signal [53, 54]. In this process,

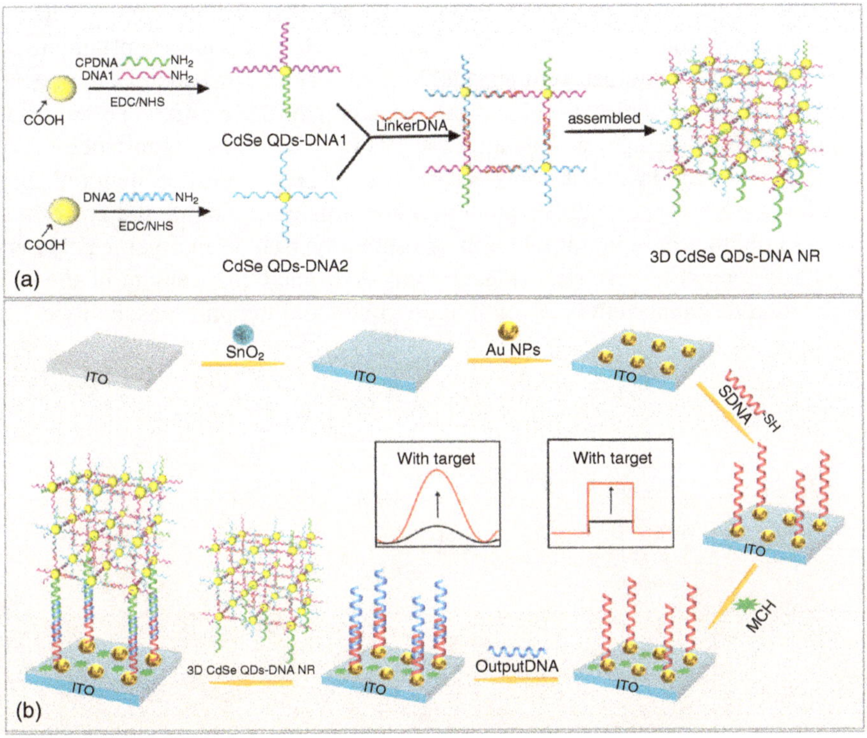

Figure 14.4 (a) Schematic diagram of the formation of 3D CdSe QDs-DNA nanonetworks and (b) incorporation of the DNA walker multiple amplification for HIV detection. Source: Reproduced with permission from Cai et al. [55]/Elsevier.

quantum dots act as a photosensitive material that absorbs photons and converts them into electron–hole pairs. Specifically, quantum dot sensitization can enhance the photocurrent signal by adjusting the size and band structure of the quantum dots, exploiting the multiple electron excitation effect, improving the electron transport properties, and increasing the photoelectrochemical conversion efficiency. As shown in Figure 14.4, Zhou's team designed a photoelectrochemical biosensor based on CdS quantum dots and DNA nanostructures [55]. This sensor takes advantage of the narrow band gap and excellent visible light absorption properties of CdS quantum dots and the high loading capacity of DNA nanostructures to efficiently sensitize SnO_2 electrodes. Through this combination, the sensor achieves efficient recognition and detection of human immunodeficiency virus.

Precious metal surface plasmon resonance is a phenomenon of collective oscillations between the free electrons of nanoparticles of precious metals (e.g. gold, silver, and platinum) under light irradiation and incident light waves [56–59]. This oscillation can excite plasma excitations at the interface between the metal and the medium (usually air or water). Resonance occurs when the frequency of the plasma oscillations matches the frequency of the incident light, at which point the light absorption and scattering of the system is greatly enhanced [60–64]. In the design of photoelectrochemical biosensors, the use of noble metal surface plasmon resonance can amplify the photocurrent signal. Specifically, noble metal surface plasmon resonance can amplify photocurrent signals through mechanisms such as interfacial enhancement effects, optical field localization effects, and SERS effects. These effects take advantage of the special optical properties of noble metal nanostructures to enhance the interaction strength between light and the photosensitive material, thus increasing the photoelectrochemical conversion efficiency. Zhang and coworkers designed a photoelectrochemical biosensor using P-type covalent organic polymers (COPs) as photoactive materials (Figure 14.5) [65]. The enhancement of the surface plasmon resonance effect of gold nanoparticles and the photosensitivity effect

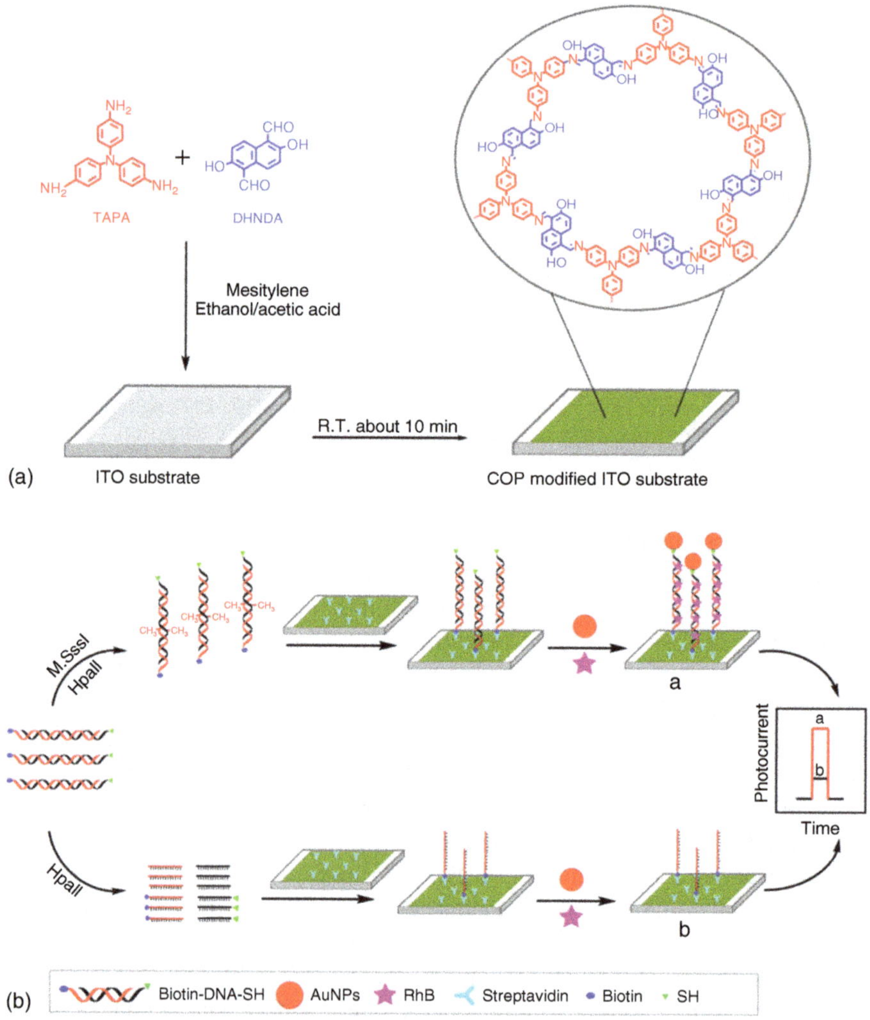

Figure 14.5 (a) In situ synthesis of a porous COP onto an ITO glass substrate and (b) development of a PEC biosensor for M.SssI MTase assay. Source: Reproduced with permission from Cui et al. [65]/American Chemical Society.

of rhodamine B could significantly amplify the photocurrent signal, and a desirable detection of bacterial SssI methyltransferase was obtained.

14.5.2 Photocurrent Signal Bursting Strategies

Photocurrent signal bursting is typically used in sensors based on a bursting mechanism that works on the principle that the interaction of the analyte with the sensor causes the signal to decrease [66]. The following are some of the bursting strategies:

(I) *Energy transfer*: using techniques such as fluorescence resonance energy transfer or SERS, energy transfer occurs when the target analyte binds to the probe, resulting in the bursting of the photocurrent signal. For example, fluorescence resonance energy transfer involves two fluorescent substances, and when one is excited, energy can be transferred nonradiatively to the other, resulting in a signal burst.

(II) *Competitive binding*: the biorecognition element in the designed sensor competes with the target analyte for binding, and as the concentration of the target analyte increases, more of the biorecognition element binds to the target analyte, which can reduce binding to the electrode surface, resulting in a decrease in photocurrent.

(III) *Introduction of light-absorbing materials*: Introduction of materials that can absorb light energy, such as specific metal-organic frameworks that can absorb light energy, which can lead to a burst of photocurrent when they bind to the analyte.

(IV) *Harnessing intermolecular interactions*: Noncovalent forces such as electrostatic interactions between molecules, hydrogen bonding, or van der Waals forces alter the spatial conformation or activity of the biometric element, causing the photocurrent to burst.

(V) *Depletion of electron donors*: usually involves a chemical substance or process that can react with and deplete electrons from a light-activated substance. In photoelectrochemical sensors, bursting can be achieved by depleting the electron donor to reduce the chances of light-induced recombination of electron–hole pairs, resulting in a reduction of the photocurrent. The following are several possible mechanisms by which the photocurrent signal can be burst by depletion of the electron donor.

   A *Introduction of chemical oxidants*: Strong oxidants can act as electron acceptors and react with electron donors on the surface of the photoactivated electrode, thereby reducing the number of electrons on the electrode surface. Such oxidants are ferricyanide, manganese dioxide, etc., which trap electrons and reduce the photocurrent.

   B *Use of enzyme-catalyzed reactions*: specific enzymes can catalyze the oxidation of electron donors (e.g. NADH), thereby depleting the electron donor and reducing the number of electrons available for photoconversion.

C *Using biomolecule binding*: certain biomolecules (e.g. DNA and proteins) can bind to the electrode surface and consume electrons through their own electron adsorption properties, leading to a signal burst.

D *Disruption of the electron transport chain*: In some biosensors, the binding of the target analyte can interfere with the electron transport chain in the cell, consuming the electron donor and leading to a decrease in photocurrent.

E *Preparation of chemically modified electrodes*: specific groups or molecules are introduced to the surface of the electrode by chemical modification, which can improve the ability of the electrode to adsorb electrons, thus consuming the electron donor and reducing the photocurrent.

(VI) *Physical masking*: Physically masking the photoactive region on the electrode, for example by using membranes or colloidal particles, can reduce the light irradiation to the electrode and thus reduce the photocurrent.

When designing photoelectrochemical sensors, amplification or bursting strategies are often rationally selected based on the needs of a particular application and the nature of the target analyte being detected, and multiple strategies may be combined to achieve optimal sensing performance. It is also important to ensure that these strategies do not degrade the specificity and accuracy of the sensor. The following section focuses on the resonance energy transfer effect, competitive binding effect, and depletion of electron donor strategies that have been more widely used in photocurrent signal bursting applications to date.

Resonance energy transfer was first discovered in the field of fluorescence, and the process involves two molecules: an energy donor that absorbs light energy and is in an excited state and an energy acceptor that receives energy from the donor. This energy transfer does not involve the emission and re-absorption of photons but is transferred directly through dipole-dipole interactions. For this energy transfer to occur, there must be sufficient spectral overlap between the emission spectrum of the donor and the absorption spectrum of the acceptor. Resonance energy transfer plays an important role in photocurrent signal burst applications. By introducing a molecule or nanoparticle capable of accepting resonance energy transfer as a donor, the photocurrent signal can be effectively burst. This strategy can be used to modulate the intensity and response of the photocurrent signal. As shown in Figure 14.6, Lin and coworkers fabricated a resonance energy transfer photoelectrochemical biosensor (R-CDs NS/HA/dsDNA/AuNCs) for the detection of small nucleic acid molecules based on gold nanocages (AuNCs) and carbon dotted self-assembled nanosheets (R-CDs NS) [67]. The absorption by the AuNCs overlapped well with the emission by the R-CDs NS, which can be used as a photocurrent bursting agent for the photosensitive system consisting of TiO_2 and R-CDs NS. Through the resonance energy transfer between R-CDs NS and AuNCs, the complexation of photogenerated electron–hole pairs generated in R-CDs NS can be accelerated, which can effectively burst the photocurrent signal in the system.

Competitive binding is the phenomenon of two or more molecules competing to bind to the same receptor molecule. This is commonly seen in biochemical processes

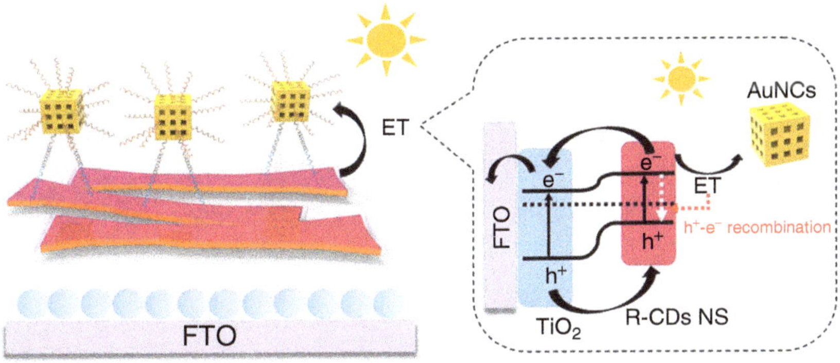

Figure 14.6 Principle of the energy transfer-based photoelectrochemical sensing platform. Source: Reproduced with permission from Yang et al. [67]/American Chemical Society.

such as the binding of enzymes to substrates, antibodies to antigens and drugs to their targets. In photoelectrochemical sensors, the competitive binding effect can be used to detect the presence and concentration of a target molecule (analyte). A common application is to use competitive binding to burst the photocurrent signal, allowing sensitive detection of analytes. The sensitivity of a photoelectrochemical enzyme sensor for the detection of glucose was improved by Wu et al (Figure 14.7) [68]. By introducing a competitive reaction with ascorbic acid (AA), which, in addition to acting as an electron donor that can trap the photogenerated holes of ZnO, is also able to scavenge the H_2O_2 produced during the oxidation of glucose. Therefore, the detection range, detection stability, detection limit, and sensitivity of this photoelectrochemical enzyme sensor for glucose detection were significantly improved in the presence of AA.

The electron-consuming donor strategy refers to the use of substances with the ability to consume electrons as acceptors for accepting electrons, which is used to regulate the degree of bursting of photocurrent signals and achieve selective detection. This strategy can achieve selective bursting, intensity modulation, and dynamic response control of photocurrent signals by selecting appropriate consuming electron donors and optimizing their properties.

As shown in Figure 14.8, a split immunoassay strategy for synergistic modulation of the photocurrent signal with cation exchange and electron donor concentration modulation has been proposed by Wei et al [69]. $MoSe_2$/CdSe, as a Z-type heterojunction, has a good visible light absorption capacity and a low carrier complexation rate, which can provide a strong initial photocurrent signal. The mesoporous silica nanomicrospheres uniformly loaded with silver nanoparticles will generate a large amount of silver ions as the initiator of the cation exchange reaction after the acidolysis reaction. It was also found that Ag^+ could replace Cd^{2+} in CdSe and maintain the crystal structure of $MoSe_2$, which would disrupt the heterojunction structure and introduce defects, resulting in a reduced photocurrent response. In addition, AA oxidase loaded on silica microspheres can act as an electron donor depletion agent, further increasing the degree of bursting of the photocurrent signal and thus improving the sensitivity and reliability of the sensor.

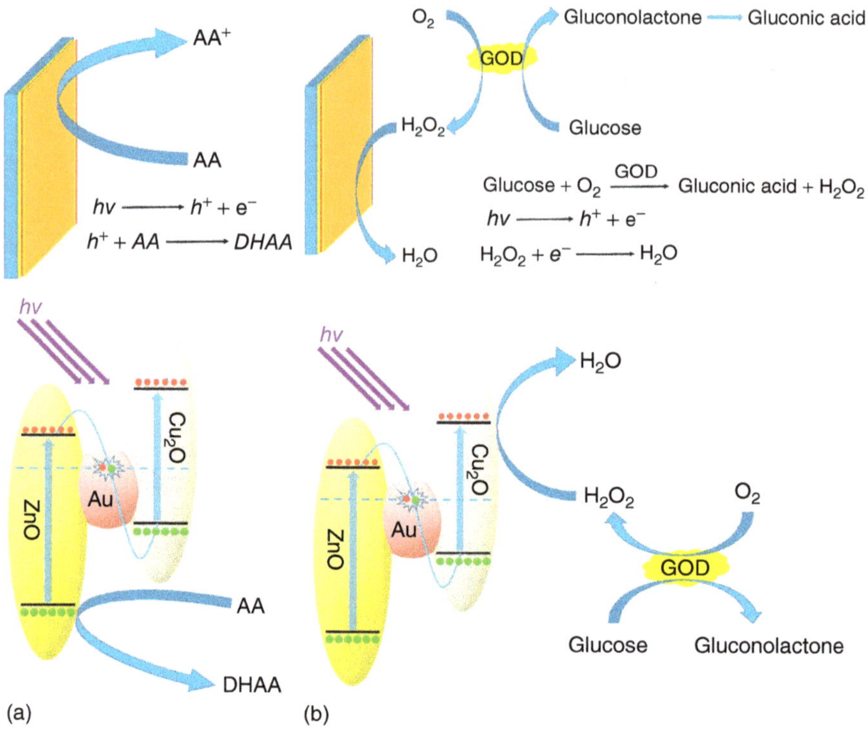

Figure 14.7 Scheme of electron transfer in different status: in the presence of AA (a) and glucose (b). Source: Reproduced with permission from Chen et al. [68]/Elsevier.

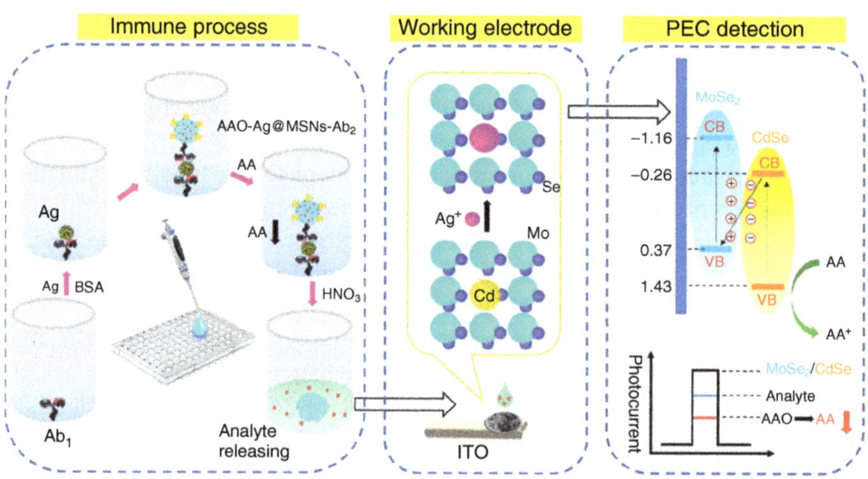

Figure 14.8 Fabrication process of the photoelectrochemical immunosensor. Source: Reproduced with permission from Leng et al. [69]/American Chemical Society.

14.6 Applications of Photoelectrochemical Biosensors

Photoelectrochemical biosensors do have a wide range of applications in the field of bioanalysis, which can be classified into the following four categories according to different detection targets and methods: direct photoelectrochemical detection, photoelectrochemical enzyme detection, photoelectrochemical nucleic acid detection, and photoelectrochemical immunoassay. Each of these four applications has its own characteristics and advantages in photoelectrochemical biosensors, providing important tools and methods for biomedicine, environmental monitoring, food safety, and other fields. These four applications are discussed in detail below.

14.6.1 Direct Photoelectrochemical Detection

In direct photoelectrochemical detection, a target biomolecule is immobilized directly on the surface of the sensor, and the presence and concentration of the target biomolecule are detected by a signal generated by a photoelectrochemical reaction [70–73]. This method does not require the use of additional biometric elements and uses the direct interaction of the photoelectrochemically active material with the target molecule to achieve detection. Direct photoelectrochemical detection is commonly used to detect small molecules, drugs, etc. This method is simple and fast but may not be able to detect biomolecules that are not photoelectrochemically active. Song's group used the specific cleavage ability of glutathione (GSH) at disulfide bonds to convert cystine to cysteine, which acts as an electron donor for the Pt NCs-cystine-TMNTs sensors, leading to an increase in photocurrent (Figure 14.9) [74]. Shao's research team constructed a novel-visible-light-driven photoelectrochemical sensor based on an iron hydroxyl oxide-modified bismuth vanadate composite ($BiVO_4$/FeOOH) and used it for the indirect detection of

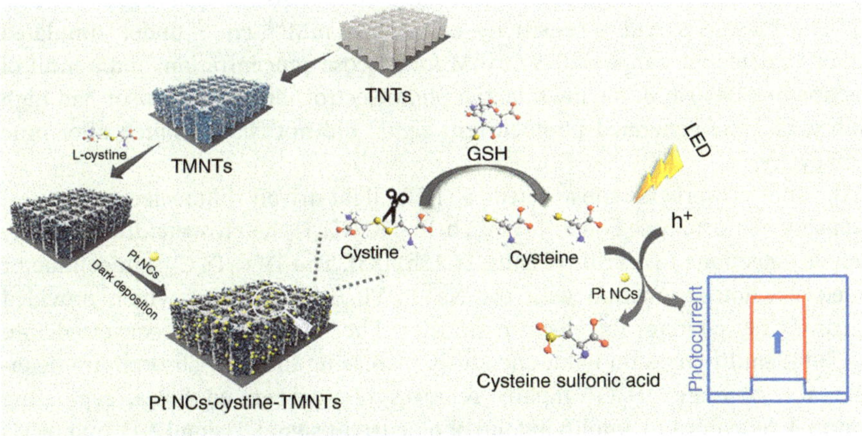

Figure 14.9 Schematic illustration of the fabrication of Pt NCs-cystine-TMNTs and their mechanism of photo-to-current conversion for GSH sensing. Source: Reproduced with permission from Wang et al. [74]/American Chemical Society.

dopamine [75]. The linear response of the photoelectrochemical sensor was in the ranges of 0.2–40 μM and 40–1400 μM, with a detection limit of 0.09 μM. Hao's group prepared a nonenzymatic photoelectrochemical glucose sensor with a $BiVO_4$ nanoporous film as the electrode for accurate blood glucose monitoring in diabetic patients [76]. Under visible light irradiation, glucose molecules adsorbed on the surface of the $BiVO_4$ electrode were oxidized by photogenerated holes, leading to an increase in photocurrent. The sensor has a linear detection range of 0–5 mM for glucose, with a minimum detection limit of 0.13 μmol L^{-1}. By constructing a p–n heterojunction cuprous oxide/copper@micropore carbon sensitized g-C_3N_4 substrate, Fei and coworkers achieved sensitive detection of L-cysteine based on the interaction between L-cysteine and cuprous oxide/copper@micropore carbon, which can form Cu—S bonds [77]. The detection limit was 1.26×10^{-9} M, and the linear range was 1.0×10^{-8} to 4.0×10^{-6} M.

14.6.2 Photoelectrochemical Enzyme Detection

Photoelectrochemical enzyme detection refers to the modification of an enzyme as a biorecognition element on the surface of a sensor [78–81]. The enzyme, as a biorecognition element, can specifically recognize and catalyze the reaction of a particular substrate, and the resulting product or consumed substrate can be detected by photoelectrochemical methods. This method is very effective for the detection of some enzyme substrates or enzyme inhibitors. Common photoelectrochemical enzyme detection includes glucosidase sensors, peroxidase sensors, and so on. Photoelectrochemical enzyme detection has high selectivity and sensitivity and can achieve quantitative detection of substrates, which is widely used in biochemical analysis, clinical diagnosis, etc. Wu and coworkers fabricated a visible-light-driven photoelectrochemical enzyme biosensor using carbon dots/branched titanium dioxide as the photoactive material and glucose oxidase as the recognition element, and this sensor can detect glucose concentration under visible light (Figure 14.10a) [82]. The biosensor had a sensitivity of 1.648 μA mM^{-1} cm^{-2} under simulated daylight, a detection range of 0–9.00 mM for glucose concentration, and a limit of detection of 0.0430 mM. Meanwhile, the photoelectrochemical biosensor had high stability, and the generated photocurrent signal did not change much after nine on/off cycles.

Han and coworkers constructed a NIR light-driven photoelectrochemical biosensing platform based on a $NaYF_4$:Yb/Er-TiO_2-Ti_3C_2 heterostructure for highly sensitive detection of D-serine (Figure 14.10b) [83]. The TiO_2-Ti_3C_2 semiconductor formed a Schottky junction with the $NaYF_4$:Yb/Er optosensor, which provided an irreversible channel for electron transfer. The NIR light was converted into absorbable multi-emission light, effectively increasing the light absorption. Meanwhile, the generated H_2O_2 rapidly removes photogenerated holes, separating electron–hole pairs and amplifying the photocurrent signal. Using Fe_2O_3 nanorods as a photoactive material, Wang and coworkers achieved a significant enhancement of the anodic photocurrent by spontaneously binding 2,3-dihydroxynaphthalene (generated by the hydrolysis of their respective characteristic substrates by

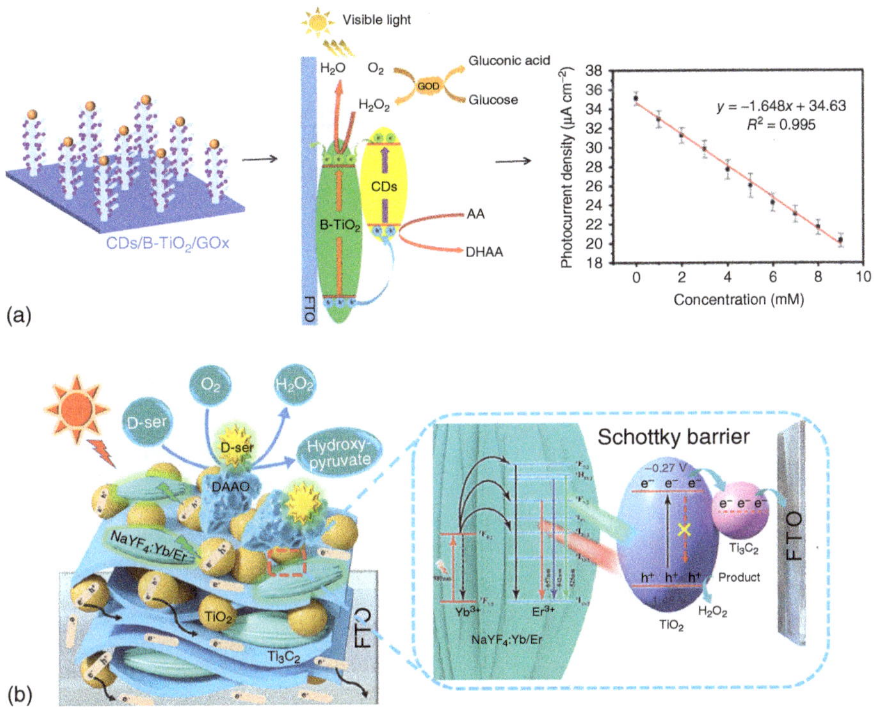

Figure 14.10 (a) Photoelectrochemical detection mechanism of glucose by CDs/BTiO$_2$/GO$_x$. Source: Yuan et al. [82]/American Chemical Society. (b) NaYF$_4$:Yb/Er-TiO$_2$-Ti$_3$C$_2$-based near-infrared light-driven biosensing platform for D-serine detection. Source: Reproduced with permission from Huang et al. [83]/American Chemical Society.

β-glucosidase and lipase, respectively) to the surface of the material in situ to generate oxygen vacancies, which in turn introduced a new defect state energy level to trap photogenerated holes and promote charge separation [84]. The assay has good sensitivity, selectivity, and reproducibility for the detection of disease markers such as β-glucosidase and lipase.

14.6.3 Photoelectrochemical Nucleic Acid Detection

Photoelectrochemical nucleic acid detection refers to the use of nucleic acids as biorecognition elements modified on the surface of a sensor that uses a nucleic acid probe to specifically recognize a target DNA or RNA sequence [85–87]. When the target sequence is present, the probe can form a complementary double-stranded structure with it, altering the electrochemical properties of the electrode surface to generate an electrical signal. Common nucleic acid photoelectrochemical detectors include DNA sensors and RNA sensors. Photoelectrochemical nucleic acid detection is highly specific and selective and can achieve quantitative or qualitative detection of specific target sequences, which is widely used in genetic analysis, mutation analysis, pathogen detection, and other fields. Li's team used TiO$_2$@Au as the

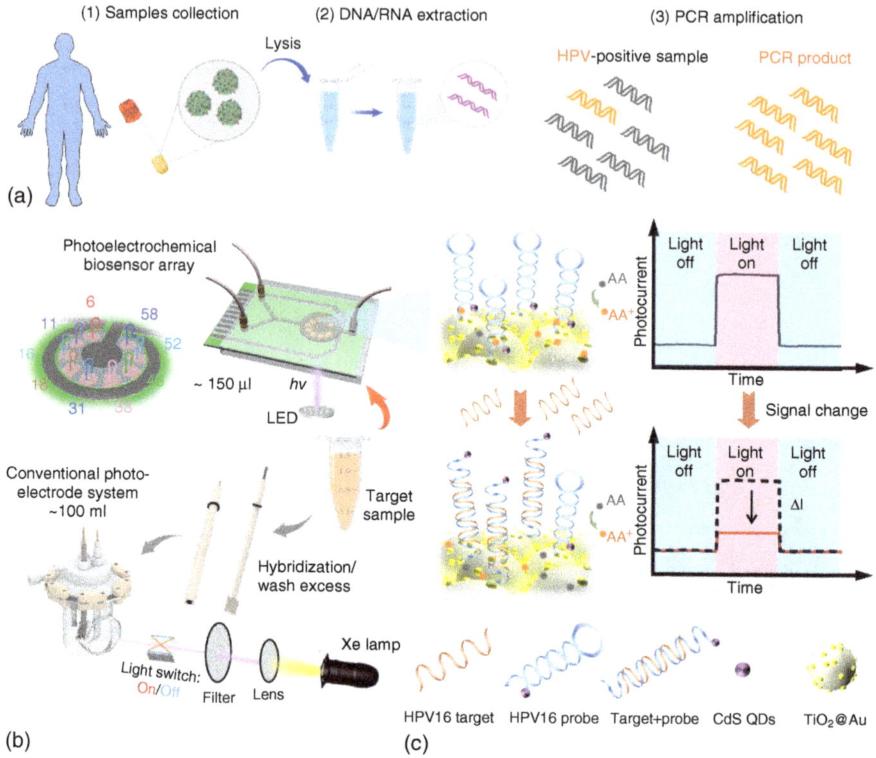

Figure 14.11 Overview of photoelectrochemical biosensor array platform. (a) Pretreatment workflow of HPV genes in clinical samples. (b) Schematic illustration of the photoelectrochemical biosensor array setup for HPV genotyping and a comparison between the proposed photoelectrochemical biosensor array and conventional photoelectrochemical testing system. (c) Schematic depiction of the fabricated PEBA for detecting HPV-related genes. Source: Reproduced with permission from Sun et al. [88]/Elsevier.

photoelectrochemically active material and CdS quantum dot-labeled DNA hairpin probes anchored on TiO_2@Au as recognition elements for the detection of high-risk human papillomavirus (HPV) genotyping (Figure 14.11) [88]. When the HPV target is added, the probe undergoes a conformational change that causes the CdS quantum dots to move away from the TiO_2@Au, resulting in a decrease in the photoelectrochemical signal. When combined with nucleic acid amplification technology, the photoelectrochemical biosensor was found to be sensitive to nine HPV genotypes down to 0.1 copies μL^{-1}, with a linear range of 0.6–600 copies μL^{-1}. The research team evaluated 40 clinical samples using the sensor, with a 97.5% concordance rate with commercial kits.

Jie and coworkers developed a novel multifunctional photoelectrochemical biosensor based on $AgInS_2$/ZnS quantum dots sensitized WSe_2 nanoflowers and DNA nanostructured signaling probes to achieve ultrasensitive "on-off" detection of human tumor necrosis factor-α and methylase Dam MTase [89]. When tumor necrosis factor-α is present, the $AgInS_2$/ZnS signal probe is attached to the

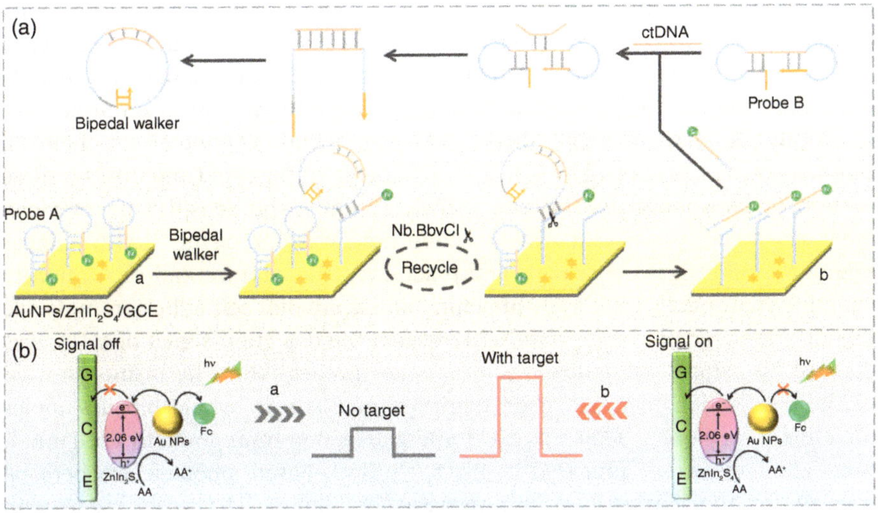

Figure 14.12 Illustration of the proposed photoelectrochemical biosensor for the detection of ctDNA: (a) Assembly of annular bipedal DNA walker, construction of the photoelectrochemical biosensor, and (b) the detection mechanism of the designed photoelectrochemical biosensor. Source: Reproduced with permission from Cui et al. [90]/Elsevier.

WSe_2 nanoflower modified electrode by binding to the aptamer and generates an amplified signal for the "on" detection of tumor necrosis factor-α. In addition, Dam MTase as a second target induced methylation of the hairpin HDam, causing it to be cleaved by the endonuclease Dpn I, resulting in the release of the $AgInS_2$/ZnS signal probe for "off" detection of Dam MTase signals. This work develops a photoelectrochemical biosensor with dual-target analysis, which is of great value for practical applications. Li and coworkers constructed a photoelectrochemical sensor for the detection of circulating tumor DNA signals based on $ZnIn_2S_4$@AuNPs and a bipedal DNA walker (Figure 14.12) [90]. Probe A, modified with ferrocene, was immobilized on the surface of $ZnIn_2S_4$@AuNPs through Au-S bonds, and ferrocene competed with $ZnIn_2S_4$@AuNPs for energy. Upon triggering the reaction with the target, the dumbbell wheel-like structure transforms into a bipedal DNA walker, promoting signal amplification and achieving sensitive detection of circulating tumor DNA in the range of 1 fM–100 nM. Lin's group used the AuNPs/GaN Schottky junction with a CA125 aptamer attached to its surface as a photoelectrochemical biosensor to specifically detect CA125 [91]. The photocurrent signal generated by AuNPs/GaN was found to be related to the concentration of CA125, and the detection limit of the protein concentration was 0.3 U mL^{-1}.

14.6.4 Photoelectrochemical Immunoassay

Photoelectrochemical immunoassay refers to the use of antibodies as biorecognition elements modified on the surface of the sensor through the specific binding of antigen-antibody to achieve the detection of target molecules with high selectivity and sensitivity [92–95]. Photoelectrochemical immunoassays are widely used in biomedical fields. Common photoelectrochemical immunoassays include photoelectrochemical immunosensors and photoelectrochemical immunochromatography. A new cathodic dual-signal amplification photoelectrochemical immunoassay was developed by Guo and coworkers for the rapid and ultrasensitive detection of phosphorylated histone H2AX, an early and sensitive genotoxicity biomarker, in cell lysates [38]. The detection limit was 6.87 pg mL^{-1} with a linear range of 0.01–500 ng mL^{-1}. Yang's group achieved the capture and separation of circulating tumor cells (CTCs) by attaching anti-epithelial cell adhesion molecule antibodies to magnetic Fe_3O_4 nanospheres, and the detection signal of CTCs was enhanced by using Cu_2O-nucleic acid aptamer probes [96]. The results showed that the sensor demonstrated good detection of CTCs in whole blood samples with a linear range of 3–3000 cells mL^{-1} and a detection limit down to 1 cell mL^{-1}. Zhang and coworkers prepared a $Bi_2O_3/CuBi_2O_4$-based photoelectrochemical biosensor for the competitive immunoassay of aflatoxin B1 with a detection limit as low as 297.4 fg mL^{-1} and a linear range of 1.4 pg mL^{-1}–280 ng mL^{-1} [97]. The excellent stability, specificity, repeatability, and reproducibility of the sensor were verified by spiking experiments. A new enzyme-free signal-off photoelectrochemical immunoassay based on an up-conversion composite material $NaYF_4$:Yb,Er was developed by Tang's team [98]. $NaYF_4$:Yb,Er was used as a secondary light source absorbing NIR excitation light and loaded with mercaptopropionic acid-modified CdTe quantum dots as a matching semiconductor photoelectrochemical signal-generating material. At the same time, CuO nanoparticles were labeled on the detection antibody, and the immunocomplexes were obtained by sandwich-type immunoreaction, and acid treatment was used to obtain Cu^{2+}, which played a dual synergistic role in attenuating the photoelectrochemical signals (on the surfaces of the CdTe quantum dots and the $NaYF_4$:Yb,Er, respectively) to achieve the highly sensitive detection of cancer markers such as α-fetoprotein.

14.7 Challenges and Potential Clinical Applications

Photoelectrochemical biosensors are biosensors that combine optical and electrochemical detection methods to convert the recognition and detection of biomolecules into photoelectrochemical signals. Such sensors typically have the advantages of high sensitivity, high specificity, low cost, and rapid detection and have the potential for a wide range of applications in a variety of fields such as diagnostics, biomarker detection, and environmental monitoring. However, photoelectrochemical biosensors still face a number of challenges in clinical applications:

(1) *Stability and reproducibility*: the performance of many sensors is affected by temperature, pH, humidity, and storage conditions, which can affect their stability and reproducibility in real clinical environments.

(2) *Selectivity*: In complex biological samples (e.g. blood and urine), the presence of different substances can interfere with the sensor response. Developing sensors with high selectivity that can specialize in identifying and quantifying specific biomarkers is a major challenge.
(3) *Sensitivity*: *Sensors* need to be extremely sensitive when detecting low-abundance biomarkers. This requires the sensor to be able to detect molecules at very low concentrations in complex biological samples.
(4) *Anti-interference*: Interfering substances in biological samples can affect the performance of the sensor, and the *development* of sensors that can work accurately in complex biological systems is an important issue.
(5) *Rapid detection*: *Clinical* applications often require rapid results for timely diagnosis and treatment decisions. Improving the response speed of sensors is therefore another challenge.
(6) *Integration and miniaturization*: To facilitate the integration of sensors into portable or wearable devices, they need to be further miniaturized and integrated to facilitate real-time monitoring at the point of care or in the field.
(7) *Versatility*: Some *clinical* applications may require the simultaneous detection of multiple biomarkers. Therefore, there is a trend to develop multifunctional sensors capable of detecting multiple analytes simultaneously.

Despite these challenges, photoelectrochemical biosensors have great potential for clinical applications, and their potential clinical applications include

(1) *Early diagnosis of diseases*: Early detection and diagnosis of diseases can be achieved by using photoelectrochemical biosensors to detect specific biomarkers (e.g. proteins, nucleic acids, or cells). For example, sensors can be developed for the early diagnosis of cancer or for monitoring blood glucose levels in people with diabetes.
(2) *Drug screening and treatment monitoring*: Sensors can be used for drug screening and treatment monitoring. They can detect changes in the concentration of drugs and biomarkers, assess the effects of drugs on cells or biomolecules, and help doctors adjust treatment plans to achieve personalized medicine.
(3) *Detection of infectious diseases*: In the event of an epidemic outbreak, photoelectrochemical biosensors can rapidly detect pathogens such as viruses and bacteria and help control the spread of disease.

It is important to note that overcoming these challenges and realizing potential clinical applications will require interdisciplinary collaboration, including continued research and innovation in fields such as materials science, biochemistry, medicine, and engineering. Meanwhile, despite the potential clinical applications of photoelectrochemical biosensors, their performance and feasibility need to be fully validated in clinical laboratories and clinical trials to ensure accuracy and reliability in diagnosis and treatment. However, we are always confident that with the advancement of technology, the application of photoelectrochemical biosensors in clinical diagnosis and personalized medicine will become more and more promising.

References

1 Zhang, L., Zhu, Y., and Zhao, W. (2021). Recent advances of nanostructured materials for photoelectrochemical bioanalysis. *Chemosensors* 10: 14.

2 Ma, X., Kang, J., Wu, Y. et al. (2022). Recent advances in metal/covalent organic framework-based materials for photoelectrochemical sensing applications. *TrAC Trends in Analytical Chemistry* 157: 116793.

3 Cardoso, A.R., Frasco, M.F., Serrano, V. et al. (2021). Molecular imprinting on nanozymes for sensing applications. *Biosensors* 11: 152.

4 Qin, J., Wang, W., Gao, L., and Yao, S.Q. (2022). Emerging biosensing and transducing techniques for potential applications in point-of-care diagnostics. *Chemical Science* 13: 2857–2876.

5 Luo, J., Shang, J., Li, D. et al. (2023). Overview of biosensors and its application in animal disease detection. *China Animal Health Inspection* 40: 72–79.

6 Yu, J., Yao, Z., He, K. et al. (2023). Nanomaterials-based optical biosensors for detection of mycotoxins in traditional Chinese medicine. *Chinese Journal of Analytical Chemistry* 4: 472–483.

7 Flynn, C.D., Chang, D., Mahmud, A. et al. (2023). Biomolecular sensors for advanced physiological monitoring. *Nature Reviews Bioengineering* 1: 560–575.

8 Yin, K., Li, D., and Meng, Q. (2021). Copper-zinc-tin-sulfur-selenium thin film solar cells. *Physics* 6: 405–412.

9 Meng, L., Wang, X., Ding, L., and Chen, Y. (2022). Organic solar cells. *Science Focus* 17: 13–19.

10 Zhang, Y. (2019). Experimental techniques used in topological insulators and realization of quantum spin hall effect. *IOP Conference Series: Materials Science and Engineering* 563: 022033.

11 Kong, W., Zhu, D., Zhang, Y. et al. (2023). Electron donor coordinated metal-organic framework to enhance photoelectrochemical performance. *Angewandte Chemie International Edition* 62: 202308514.

12 Zhao, W., Xu, J., and Chen, H. (2015). Photoelectrochemical bioanalysis: the state of the art. *Chemical Society Reviews* 44: 729–741.

13 Shu, J. and Tang, D. (2019). Recent advances in photoelectrochemical sensing: from engineered photoactive materials to sensing devices and detection modes. *Analytical Chemistry* 92: 363–377.

14 Zhuang, J., Tang, D., Lai, W. et al. (2015). Target-induced nano-enzyme reactor mediated hole-trapping for high-throughput immunoassay based on a split-type photoelectrochemical detection strategy. *Analytical Chemistry* 87: 9473–9480.

15 Gu, N. and Liu, S. (2020). Introduction to biosensors. *Journal of Materials Chemistry B* 8: 3168–3170.

16 Devadoss, A., Sudhagar, P., Terashima, C. et al. (2015). Photoelectrochemical biosensors: new insights into promising photoelectrodes and signal amplification strategies. *Journal of Photochemistry and Photobiology C: Photochemistry Reviews* 24: 43–63.

17 Shang, M., Gao, Y., Zhang, J. et al. (2020). Signal-on cathodic photoelectrochemical aptasensing of insulin: plasmonic Au activated amorphous MoS_x

photocathode coupled with target-induced sensitization effect. *Biosensors and Bioelectronics* 165: 112359.

18 Sui, C., Li, F., Wu, H. et al. (2019). Photoelectrochemical biosensor for 5hmC detection based on the photocurrent inhibition effect of ZnO on MoS_2/C_3N_4 heterojunction. *Biosensors and Bioelectronics* 142: 111516.

19 Sun, Z., Sun, K., Gao, M. et al. (2022). Optimizing pt electronic states through formation of a schottky junction on non-reducible metal-organic frameworks for enhanced photocatalysis. *Angewandte Chemie International Edition* 61: 202206108.

20 Bai, X., Yu, Z., Li, Y. et al. (2023). N-type semiconductive polymers based on pyrene-1,5,6,10-tetracarboxyl diimide. *Chinese Journal of Polymer Science* 41: 1584–1590.

21 Wang, X., Guan, D., Li, F. et al. (2021). Magnetic and optical field multi-assisted Li-O_2 batteries with ultrahigh energy efficiency and cycle stability. *Advanced Materials* 34: 2104792.

22 Deng, X., Li, R., Wu, S. et al. (2019). Metal-organic framework coating enhances the performance of Cu_2O in photoelectrochemical CO_2 reduction. *Journal of the American Chemical Society* 141: 10924–10929.

23 Lee, K., Wan, Y., Huang, Z. et al. (2023). Organic optoelectronic materials: a rising star of bioimaging and phototherapy. *Advanced Materials* 2306492.

24 Liu, M., Wang, Y., Lu, C. et al. (2024). Localized oxidation embellishing strategy enables high-performance perovskite solar cells. *Angewandte Chemie International Edition* 202318621.

25 Zhang, Y., Wu, T., Liu, D. et al. (2022). Photoelectrochemical immunosensor for the sensitive detection of neuron-specific enolase based on the effect of Z-scheme $WO_3/NiCo_2O_4$ nanoarrays p-n heterojunction. *Biosensors and Bioelectronics* 213: 114452.

26 Gao, Y., Li, M., Zeng, Y. et al. (2022). Tunable competitive absorption-induced signal-on photoelectrochemical immunoassay for cardiac troponin Z-scheme metal–organic framework heterojunctions. *Analytical Chemistry* 94: 13582–13589.

27 Pei, F., Feng, S., Zhang, Y. et al. (2022). A photoelectrochemical immunosensor based on Z-scheme CdS composite heterojunction for aflatoxin B1. *Biosensors and Bioelectronics* 214: 114500.

28 Gao, Y., Zeng, Y., Liu, X., and Tang, D. (2022). Liposome-mediated in situ formation of type-I heterojunction for amplified photoelectrochemical immunoassay. *Analytical Chemistry* 94: 4859–4865.

29 Zhu, L., Lu, L., Wang, H. et al. (2019). Enhanced organic-inorganic heterojunction of polypyrrole@Bi_2WO_6: fabrication and application for sensitive photoelectrochemical immunoassay of creatine kinase-MB. *Biosensors and Bioelectronics* 140: 111349.

30 Tian, Y., Xu, G., Cai, K. et al. (2023). Emerging biotransduction strategies on soft interfaces for biosensing. *Nanoscale* 15: 80–91.

31 Wu, J., Liu, H., Chen, W. et al. (2023). Device integration of electrochemical biosensors. *Nature Reviews Bioengineering* 1: 346–360.

32 Peng, J., Zu, L., Ren, Z. et al. (2023). Synergistic integration of moox co-catalyst with oxygen vacancies-engineered on $BiVO_4$ for enhancing the sensitivity of photoelectrochemical DNA biosensing. *Chemical Engineering Journal* 477: 146989.

33 Wang, H., Zhang, S., Zhang, Y. et al. (2023). Magnetically controlled and addressable photoelectrochemical sensor array with self-calibration for the label-free detection of amyloid β-proteins. *Analytical Chemistry* 95: 16169–16175.

34 Hou, Y., Jia, B., Sheng, P. et al. (2021). Aptasensors for mycotoxins in foods: recent advances and future trends. *Comprehensive Reviews in Food Science and Food Safety* 21: 2032–2073.

35 Wang, Y., Rong, Y., Ma, T. et al. (2023). Photoelectrochemical sensors based on paper and their emerging applications in point-of-care testing. *Biosensors and Bioelectronics* 236: 115400.

36 Zhao, C. and Ding, S. (2019). Perspective on signal amplification strategies and sensing protocols in photoelectrochemical immunoassay. *Coordination Chemistry Reviews* 391: 1–14.

37 Zhang, J., Zhang, S., Zhang, Y. et al. (2022). Colloidal quantum dots: synthesis, composition, structure, and emerging optoelectronic applications. *Laser and Photonics Reviews* 17: 2200551.

38 Zhang, B., Li, F., Shen, L. et al. (2023). A cathodic photoelectrochemical immunoassay with dual signal amplification for the ultrasensitive detection of DNA damage biomarkers. *Biosensors and Bioelectronics* 224: 115052.

39 Su, L., Tong, P., Zhang, L. et al. (2019). Photoelectrochemical immunoassay of aflatoxin B1 in foodstuff based on amorphous TiO_2 and $CsPbBr_3$ perovskite nanocrystals. *The Analyst* 144: 4880–4886.

40 Zhu, Y., Zhang, N., Ruan, Y. et al. (2016). Alkaline phosphatase tagged antibodies on gold nanoparticles/TiO_2 nanotubes electrode: a plasmonic strategy for label-free and amplified photoelectrochemical immunoassay. *Analytical Chemistry* 88: 5626–5630.

41 Zhao, C., Zhou, J., Wu, K. et al. (2020). Plasmonic enhanced gold nanoclusters-based photoelectrochemical biosensor for sensitive alkaline phosphatase activity analysis. *Analytical Chemistry* 92: 6886–6892.

42 Liu, L., Fan, G., Zhang, J., and Zhu, J. (2018). Ultrasensitive cathode photoelectrochemical immunoassay based on TiO_2 photoanode-enhanced 3D Cu_2O nanowire array photocathode and signal amplification by biocatalytic precipitation. *Analytica Chimica Acta* 1027: 33–40.

43 Yu, L., Liu, Y., Zhu, L. et al. (2020). Boosting the biocatalytic precipitation with enzyme-loaded liposomes: toward a general platform for amplified photoelectrochemical immunoassay. *Analytica Chimica Acta* 1115: 1–6.

44 Lv, J., Xie, J., Mohamed, A.G.A. et al. (2022). Photoelectrochemical energy storage materials: design principles and functional devices towards direct solar to electrochemical energy storage. *Chemical Society Reviews* 51: 1511–1528.

45 Ali, M.M., Li, F., Zhang, Z. et al. (2014). Rolling circle amplification: a versatile tool for chemical biology, materials science and medicine. *Chemical Society Reviews* 43: 3324–3341.

46 Gao, Y., Huang, K., Wang, F. et al. (2022). Recent advances in biological detection with rolling circle amplification: design strategy, biosensing mechanism, and practical applications. *The Analyst* 147: 3396–3414.

47 Bi, S., Yue, S., and Zhang, S. (2017). Hybridization chain reaction: a versatile molecular tool for biosensing, bioimaging, and biomedicine. *Chemical Society Reviews* 46: 4281–4298.

48 Zhang, C., Chen, J., Sun, R. et al. (2020). The recent development of hybridization chain reaction strategies in biosensors. *ACS Sensors* 5: 2977–3000.

49 Li, H., Cao, Y., Wu, T. et al. (2021). Programmable DNA circuits for flexible and robust exciton-plasmon interaction-based photoelectrochemical biosensing. *Analytical Chemistry* 93: 11043–11051.

50 Liu, F., Zhao, J., Liu, X. et al. (2023). PEC-SERS dual-mode detection of foodborne pathogens based on binding-induced DNA walker and C_3N_4/MXene-Au NPs accelerator. *Analytical Chemistry* 95: 14297–14307.

51 Chen, C., Zhou, X., Wang, Z. et al. (2022). Core-shell Au@PtAg modified TiO_2-Ti_3C_2 heterostructure and target-triggered dnazyme cascade amplification for photoelectrochemical detection of ochratoxin a. *Analytica Chimica Acta* 1216: 339943.

52 Xu, Y., Zhang, Y., Li, C. et al. (2023). SERS as a probe of surface chemistry enabled by surface-accessible plasmonic nanomaterials. *Accounts of Chemical Research* 56: 2072–2083.

53 Song, H., Lin, Y., Zhang, Z. et al. (2021). Improving the efficiency of quantum dot sensitized solar cells beyond 15% via secondary deposition. *Journal of the American Chemical Society* 143: 4790–4800.

54 Yu, Y., Ma, T., and Huang, H. (2023). Semiconducting quantum dots for energy conversion and storage. *Advanced Functional Materials* 33: 2213770.

55 Cai, Q., Wu, D., Li, H. et al. (2021). Versatile photoelectrochemical and electrochemiluminescence biosensor based on 3D CdSe QDs-DNA nanonetwork-SnO_2 nanoflower coupled with DNA walker amplification for HIV detection. *Biosensors and Bioelectronics* 191: 113455.

56 Karimi-Maleh, H., Kumar, B.G., Rajendran, S. et al. (2020). Tuning of metal oxides photocatalytic performance using Ag nanoparticles integration. *Journal of Molecular Liquids* 314: 113588.

57 Li, Y., Wei, Y., He, W. et al. (2023). Ordered macroporous structured TiO_2-based photocatalysts for CO_2 reduction: a review. *Chinese Chemical Letters* 34: 108417.

58 Tsao, C., Narra, S., Kao, J. et al. (2024). Dual-plasmonic Au@Cu_7S_4 yolk@shell nanocrystals for photocatalytic hydrogen production across visible to near infrared spectral region. *Nature Communications* 15: 413.

59 Zhou, J., He, W., Liu, H., and Huang, C. (2021). Energy flow during the plasmon resonance-driven photocatalytic reactions on single nanoparticles. *ACS Catalysis* 12: 847–853.

60 Guo, R., Xia, C., Bi, Z. et al. (2023). Recent progress of photothermal effect on photocatalytic reduction of CO_2. *Fuel Processing Technology* 241: 107617.

61 Liu, D. and Xue, C. (2021). Plasmonic coupling architectures for enhanced photocatalysis. *Advanced Materials* 33: 2005738.

62 Liu, G., Xu, J., Chen, T., and Wang, K. (2022). Progress in thermoplasmonics for solar energy applications. *Physics Reports* 981: 1–50.

63 Wang, C., Shi, Y., Yang, D., and Xia, X. (2018). Combining plasmonics and electrochemistry at the nanoscale. *Current Opinion in Electrochemistry* 7: 95–102.

64 Wu, F., Xia, S., Wei, J. et al. (2023). Metallic heterostructures for plasmon-enhanced electrocatalysis. *ChemPhysChem* 24: 202200881.

65 Cui, L., Shen, J., Li, C. et al. (2021). Construction of a dye-sensitized and gold plasmon-enhanced cathodic photoelectrochemical biosensor for methyltransferase activity assay. *Analytical Chemistry* 93: 10310–10316.

66 Wu, C., Deng, H., Ding, Q. et al. (2023). Au nano-flower/organic polymer heterojunction-based cathode photochemical biosensor with reduction-accelerated quenching effect of porphyrin manganese. *Journal of Hazardous Materials* 445: 130510.

67 Yang, J., Luo, F., Wang, J. et al. (2021). Ultrasensitive photoelectrochemical biosensor for microrna-155 based on energy transfer between Au nanocages and red emission carbon dot-assembled nanosheets coupled with the duplex-specific nuclease enzyme-assisted target recycling strategy. *Analytical Chemistry* 94: 1482–1490.

68 Chen, D.L., Wang, X.H., Zhang, K.X. et al. (2020). Glucose photoelectrochemical enzyme sensor based on competitive reaction of ascorbic acid. *Biosensors and Bioelectronics* 166: 112466.

69 Leng, D., Zhao, J., Ren, X. et al. (2021). $MoSe_2$/CdSe heterojunction destruction by cation exchange for photoelectrochemical immunoassays with a controlled-release strategy. *Analytical Chemistry* 93: 10712–10718.

70 Zhao, X., Zhou, S., Shen, Q. et al. (2012). Fabrication of glutathione photoelectrochemical biosensor using graphene-CdS nanocomposites. *The Analyst* 137: 3697.

71 Zhao, K., Yan, X., Gu, Y. et al. (2016). Self-powered photoelectrochemical biosensor based on CdS/RGO/ZnO nanowire array heterostructure. *Small* 12: 245–251.

72 Yu, L., Zhu, Y., Liu, Y. et al. (2018). Ferroelectric perovskite oxide@TiO_2 nanorod heterostructures: preparation, characterization, and application as a platform for photoelectrochemical bioanalysis. *Analytical Chemistry* 90: 10803–10811.

73 Tang, J., Kong, B., Wang, Y. et al. (2013). Photoelectrochemical detection of glutathione by IrO_2-hemin-TiO_2 nanowire arrays. *Nano Letters* 13: 5350–5354.

74 Wang, T., Wang, Z., Hao, J. et al. (2023). Improved sensitivity and selectivity of glutathione detection through target-driven electron donor generation in photoelectrochemical electrodes. *Analytical Chemistry* 95: 13242–13249.

75 Wang, Y., Wang, D., Dong, S. et al. (2022). A visible-light-driven photoelectrochemical sensing platform based on the $BiVO_4$/FeOOH photoanode for dopamine detection. *Electrochimica Acta* 414: 140207.

76 Wang, S., Li, S., Wang, W. et al. (2019). A non-enzymatic photoelectrochemical glucose sensor based on $BiVO_4$ electrode under visible light. *Sensors and Actuators B: Chemical* 291: 34–41.

77 Huang, L., Li, J., Wang, Y. et al. (2023). A light-driven enzyme-free photoelectrochemical sensor based on HKUST-1 derived Cu_2O/Cu@microporous carbon with

g-C_3N_4 p-n heterojunction for ultra-sensitive detection of L-cysteine. *Carbon* 215: 118466.

78 Zhu, J., Huo, X., Liu, X., and Ju, H. (2015). Gold nanoparticles deposited polyaniline-TiO_2 nanotube for surface plasmon resonance enhanced photoelectrochemical biosensing. *ACS Applied Materials & Interfaces* 8: 341–349.

79 Zhao, W., Ma, Z., Xu, J., and Chen, H. (2013). In situ modification of a semiconductor surface by an enzymatic process: a general strategy for photoelectrochemical bioanalysis. *Analytical Chemistry* 85: 8503–8506.

80 Wang, Y., Tang, J., Zhou, T. et al. (2014). Reversible chemical tuning of charge carriers for enhanced photoelectrochemical conversion and probing of living cells. *Small* 10: 4967–4974.

81 Pardo-Yissar, V., Katz, E., Wasserman, J., and Willner, I. (2003). Acetylcholine esterase-labeled CdS nanoparticles on electrodes: photoelectrochemical sensing of the enzyme inhibitors. *Journal of the American Chemical Society* 125: 622–623.

82 Yuan, R., Yan, B., Lai, C. et al. (2023). Carbon dot-modified branched TiO_2 photoelectrochemical glucose sensors with visible light response. *ACS Omega* 8: 22099–22107.

83 Huang, L., Liang, Z., Zhang, F. et al. (2022). Upconversion $NaYF_4$:Yb/Er-TiO_2-Ti_3C_2 heterostructure-based near-infrared light-driven photoelectrochemical biosensor for highly sensitive and selective D-serine detection. *Analytical Chemistry* 94: 16246–16253.

84 Yan, M., Gu, M., Yan, Z. et al. (2023). 2,3-Dihydroxynaphthalene invoked surface oxygen vacancy effect on Fe_2O_3 nanorods for photoanodic signal transduction tactic. *Biosensors and Bioelectronics* 232: 115286.

85 Ma, Z., Ruan, Y., Xu, F. et al. (2016). Protein binding bends the gold nanoparticle capped DNA sequence: toward novel energy-transfer-based photoelectrochemical protein detection. *Analytical Chemistry* 88: 3864–3871.

86 Li, C., Wang, H., Shen, J., and Tang, B. (2015). Cyclometalated iridium complex-based label-free photoelectrochemical biosensor for DNA detection by hybridization chain reaction amplification. *Analytical Chemistry* 87: 4283–4291.

87 Chao, J., Zhu, D., Zhang, Y. et al. (2016). DNA nanotechnology-enabled biosensors. *Biosensors and Bioelectronics* 76: 68–79.

88 Sun, Y., Liu, J., Peng, X. et al. (2023). A novel photoelectrochemical array platform for ultrasensitive multiplex detection and subtype identification of HPV genes. *Biosensors and Bioelectronics* 224: 115059.

89 Cai, Q., Li, H., Dong, W., and Jie, G. (2023). Versatile photoelectrochemical biosensor based on AIS/ZnS QDs sensitized-WSe_2 nanoflowers coupled with DNA nanostructure probe for "on-off" assays of TNF-α and MTase. *Biosensors and Bioelectronics* 241: 115704.

90 Cui, Y., Yan, H., Sun, Z. et al. (2023). A photoelectrochemical biosensor based on $ZnIn_2S_4$@AuNPs coupled with circular bipedal DNA walker for signal-on detection of circulating tumor DNA. *Biosensors and Bioelectronics* 231: 115295.

91 Hu, D., Liang, H., Wang, X. et al. (2020). Highly sensitive and selective photoelectrochemical aptasensor for cancer biomarker CA125 based on AuNPs/GaN schottky junction. *Analytical Chemistry* 92: 10114–10120.

92 Shu, J. and Tang, D. (2017). Current advances in quantum-dots-based photoelectrochemical immunoassays. *Chemistry-An Asian Journal* 12: 2780–2789.

93 Zhao, W., Han, Y., Zhu, Y. et al. (2015). DNA labeling generates a unique amplification probe for sensitive photoelectrochemical immunoassay of HIV-1 p24 antigen. *Analytical Chemistry* 87: 5496–5499.

94 Zhao, W., Chen, R., Dai, P. et al. (2014). A general strategy for photoelectrochemical immunoassay using an enzyme label combined with a CdS quantum dot/TiO_2 nanoparticle composite electrode. *Analytical Chemistry* 86: 11513–11516.

95 Zhang, N., Ma, Z., Ruan, Y. et al. (2016). Simultaneous photoelectrochemical immunoassay of dual cardiac markers using specific enzyme tags: a proof of principle for multiplexed bioanalysis. *Analytical Chemistry* 88: 1990–1994.

96 Luo, J., Liang, D., Zhao, D., and Yang, M. (2020). Photoelectrochemical detection of circulating tumor cells based on aptamer conjugated Cu_2O as signal probe. *Biosensors and Bioelectronics* 151: 111976.

97 Chen, X., Wu, W., Zhang, Q. et al. (2022). Z-scheme $Bi_2O_3/CuBi_2O_4$ heterojunction enabled sensitive photoelectrochemical detection of aflatoxin B1 for health care, the environment, and food. *Biosensors and Bioelectronics* 214: 114523.

98 Luo, Z., Qi, Q., Zhang, L. et al. (2019). Branched polyethylenimine-modified upconversion nanohybrid-mediated photoelectrochemical immunoassay with synergistic effect of dual-purpose copper ions. *Analytical Chemistry* 91: 4149–4156.

15

X-Ray-Induced Photodynamic Therapy for Deep-Seated Tumors

Jinliang Liu

Shanghai University, School of Environmental and Chemical Engineering, 99 Shangda Road, Shanghai, 200444, China

15.1 Introduction

Cancer poses a significant global health threat, claiming millions of lives annually [1]. Currently, the primary clinical treatments employed are chemotherapy and radiotherapy. However, these methods often lead to severe systemic toxicity and adverse reactions. Photodynamic therapy (PDT) stands out as a noninvasive and highly selective approach to cancer treatment, offering specific spatiotemporal selectivity and minimal invasiveness, thereby reducing damage to healthy tissues [2–4]. In the cancer treatment process, PDT utilizes a combination of a photosensitizer (PS), light (electromagnetic radiation), and oxygen to destroy tumor cells. This highly localized and nonmutagenic process minimizes the adverse impacts of the treatment. The energy transmitted by light can be transferred from the photoactive PS to molecular oxygen, subsequently producing singlet oxygen ($^1O_2$), a highly active and toxic substance. This reacts rapidly with cellular organelles, causing oxidative damage, eventually leading to tumor cell death and the closure of peripheral blood vessels. In general, the therapeutic action of PDT on tumors involves three systems: tumor cell death induced by reactive oxygen species (ROS), damage to tumor-associated vasculature, and initiation of a specific immune response against tumor cells. Since PS causes cytotoxicity only under specific types of light irradiation, PDT can selectively destroy malignant tumors while protecting healthy organs like the liver and the spleen. Therefore, the development of PDT could bring new opportunities for future cancer treatments.

PDT unfolds through a tripartite synergy of light, a photosensitizing agent, and oxygen, adhering to a three-stage paradigm [5, 6]. Initially, the PS, upon excitation by light of an appropriate wavelength, transitions electrons from a ground state to a singlet-excited state, subsequently shifting to a triplet-excited state via intersystem crossing [7]. This activation initiates two critical pathways: type I, where the excited PS transfers electrons to substrates like H_2O or $^3O_2$, generating ROS such as $O_2^{\cdot -}$ or $^{\cdot}OH$, and type II, which includes energy transfer to triplet-state oxygen

Photofunctional Nanomaterials for Biomedical Applications, First Edition.
Edited by Chunxia Li and Jun Lin.

($^3O_2$) to produce singlet oxygen ($^1O_2$). The prevalence of either pathway depends on the PS type and oxygen levels [8]. The generated ROS, including superoxide anions ($O_2^{\bullet-}$), hydrogen peroxide (H_2O_2), hydroxyl radicals ($^{\bullet}OH$), and singlet oxygen ($^1O_2$), interact with essential biological molecules such as cysteine, methionine, tyrosine, histidine, tryptophan, DNA, and others, causing damage on tumor cells and, consequently, the tumor tissue.

The effectiveness of PDT heavily relies on the penetration of light into biological tissue, which poses challenges in treating deeper malignancies due to its wavelength dependency. Traditional photosensitizers, activated by ultraviolet (UV), visible (Vis), or near-infrared (NIR) light, falter in depth due to substantial light absorption and scattering by skin, limiting their therapeutic reach to a mere few millimeters to a centimeter [9–11]. X-rays, however, present a promising solution. Their minimal absorption and scattering in tissues break the depth barrier imposed by conventional PDT, marking them as a promising modality for treating deep-seated tumors. Since the discovery by Wilhelm Röntgen in 1895, X-rays have found extensive application in diagnostic imaging and radiation therapy. Approximately 60% of cancer patients benefit from radiation therapy (RT), where X-rays ionize DNA directly or generate cytotoxic free radicals, inducing apoptosis. However, traditional RT is hampered by poor X-ray attenuation in living tissues and hypoxic tumor resistance, leading to insufficient energy deposition at the target site.

In 2006, a revolutionary concept known as X-ray-induced PDT (X-PDT), utilizing nanoparticles to address these challenges, was emerged [12]. X-PDT is divided into two main categories: one involves nanoscintillator-PS combinations, where X-rays stimulate nanoscintillators to emit X-ray-excited optical luminescence (XEOL), activating the PS to produce $^1O_2$. The other, radiodynamic therapy (RDT), utilizes metal–complex or metal-based hybrid structures, which do not require fluorescence for PS activation but instead rely on energy transfer within nanomaterials to generate ROS like $^1O_2$, $O_2^{\bullet-}$, and $^{\bullet}OH$. Indeed, the term X-PDT, initially synonymous with the first strategy, is increasingly used interchangeably with RDT.

Typically, X-PDT involves three components [13] (Figure 15.1): X-rays as the excitation source, nanoscintillators as the X-ray absorbing antennas and energy conversion mediums, and the PS as the agent producing therapeutic ROS. Nanoscintillators, upon X-ray exposure, generate abundant electron–hole pairs via the photoelectric or Compton effect [14, 15]. These pairs migrate within the crystal lattice, recombining at emission centers to release photons. Through the absorption, release photons. Through the absorption of photons, organic photosensitizers trigger the generation of ROS via electron or energy transfer processes, leading to significant cytotoxic effects and the elimination of tumors. Consequently, when exposed to X-ray radiation, nanoscintillators play a crucial role in converting energy to activate photosensitizers, resulting in a notable ROS production within tumor tissues and presenting a promising approach for treating deep-seated tumors [16]. Considering the nearly limitless penetration of X-rays and the systemic delivery of nanomaterials through intravenous injection or other methods, X-PDT holds great promise for advancing PDT in treating deep tumors.

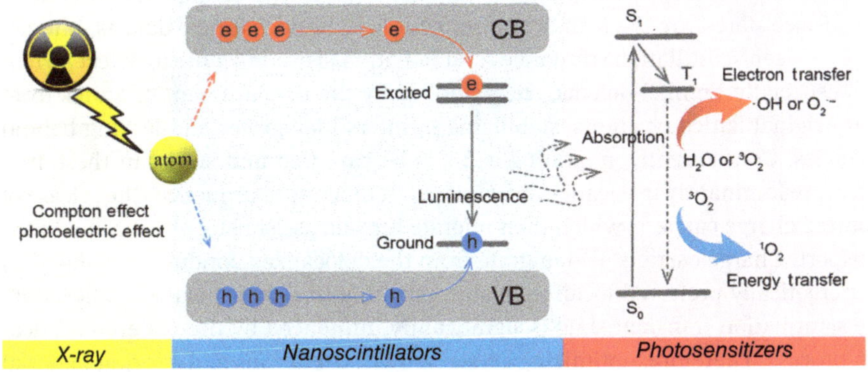

Figure 15.1 General mechanism of indirect excitation of X-PDT. Source: Reproduced with permission from He et al. [13]/American Chemical Society.

This chapter delineates the interaction mechanisms between X-rays and materials, highlights the common and promising X-ray-sensitive materials for therapeutic purposes, and discusses the latest advancements in the therapeutic applications of X-ray-activated nanosystems. It concludes by envisioning the current trends and prospects of X-ray-activated nanosystem designs alongside the technical challenges and difficulties in their clinical translation.

15.2 Mechanisms of Interaction Between X-Rays and Scintillation Materials

The luminescence effectiveness of nanocrystals upon X-ray excitation plays a crucial role in the practical efficacy of X-PDT for cancer treatment. This interaction primarily unfolds over three stages: (i) conversion, (ii) energy transfer (ET) and charge carrier migration, and (iii) scintillation. Initially, the conversion process occurs due to the interaction between substances and photons/ionizing particles, where the scintillation material absorbs the irradiation energy, generating high-energy charge carriers. The corresponding mechanism of energy absorption varies with the specific type of radiation energy involved. For X-rays, it can be articulated as [17]:

$$\frac{I}{I_0} = e^{-\mu x} \tag{15.1}$$

where I_0 and I represent the incident and transmitted radiation intensities of X-rays, respectively. μ denotes the linear absorption coefficient, and x signifies the material thickness. Materials with high atomic numbers typically exhibit substantial mass absorption coefficients, enhancing X-ray absorption and conversion, thereby increasing the probability of X-ray-related luminescence.

The second stage primarily involves the migration process of electrons and holes to the luminescent centers, significantly impacting energy loss, electron–hole (e–h) capture, and timing performance. Defects in the scintillator lattice, mainly caused by vacancies and interstitials, grain boundaries, impurities or dopants, interfaces, and surface states, are the primary culprits for loss. Besides innate defects, interactions between scintillation substances and radiation also contribute to defect formation, especially Frenkel defects, characterized by the displacement of atoms from their original lattice positions, stabilizing in interstitial spaces and leaving behind vacancies. Consequently, migrating *e–h* can be captured and stored in these trap states, predominantly in vacancy defect states. Temperature influences the release of captured charge carriers, which then migrate further, assisted by phonon-mediated transport. Charge carriers also migrate from the delocalized conduction band (CB) to energetically preferred localized states, resulting in slower polarons' migration. The scintillation transient state is significantly influenced by the material fabrication process. Therefore, optimizing crystal growth conditions and altering material morphology can mitigate lattice defect formation, enhancing *e–h* migration rates and thus curtailing luminescence delay and nonradiative losses.

In the final stage, charge carriers are captured through potential transition pathways at emission centers (typically impurity/dopant ions within the lattice) and are elevated to excited states. According to selection rules, *e–h* pairs with radiative recombination emit fluorescence in the UV/Vis region. In rapid emission scenarios, it is crucial to choose luminescent centers that exhibit electric dipole-allowed transitions, such as the widely employed $4f^{n-1}5d$–$4f^{n}$ transitions in lanthanide ions. If electric dipole transitions are forbidden, electrons and holes must undergo additional processes, resulting in emission delays and significantly reducing efficiency in rapid-timing scintillation applications. Therefore, besides considering conversion and ET efficiencies, scintillator materials are expected to select ions/molecules with allowed transitions to ensure efficient radiative recombination in applications requiring rapid decay times.

The parameter quantifying the overall conversion efficiency of X-rays to UV/Vis is determined by the material's intrinsic and extrinsic properties and can be computed using the equations below [18–20]:

$$N_{ph} = \frac{E \cdot S \cdot Q}{\beta \cdot E_g} \tag{15.2}$$

$$\eta = \frac{E_{vis} \cdot N_{ph}}{E} \tag{15.3}$$

The X-ray beam's energy per photon is denoted by E, with S and Q representing the quantum efficiencies of ET for electron-hole pairs and subsequent luminescence. The bandgap of the scintillator is referred to as E_g, and a parameter β is used to symbolize the ratio of the average energy needed to create electron-hole pairs to the E_g value. Considering the presence of matrix defects and impurities, which prevent complete capture of generated *e–h* pairs by luminescent centers, β is not the ideal value of 1, and researchers estimate it to be between 2 and 3 for most materials [18–20].

When the luminescence band of X-ray-excited nanocrystals aligns with the absorption band of the PS or active nitrogen species precursors, these X-rays can serve as an internal light source in vivo, activating functional molecules to produce

ROS through Fluorescence Resonance Energy Transfer (FRET). Additionally, electrons generated in the nanocrystal valence band can be captured by electron acceptors like oxygen molecules, leading to the formation of superoxide radicals ($O_2^{\cdot-}$). Meanwhile, holes in the CB can be trapped by electron donors like water molecules, resulting in the creation of hydroxyl radicals ($HO^{\cdot}$). This process enhances RT effects by generating a variety of radicals. Generally, nanocrystals used for X-PDT should meet the following criteria [16, 21]. (i) Their emission spectrum should closely match the absorption spectrum of the PS to ensure efficient activation and ROS production; (ii) they should exhibit high luminescent efficiency, facilitating ET to the PS; (iii) they should possess superior ionizing radiation absorption capability than surrounding tissues to minimize radiation damage to healthy tissue; and (iv) they should demonstrate high stability and low toxicity, minimizing toxic effects on the human body.

15.3 X-Ray-Sensitive Materials

The fundamental principle of X-PDT is to utilize X-rays to activate the PS. The interaction between X-rays and matter follows either the photoelectric effect or the Compton effect. Essentially, elements with high atomic numbers (high-Z) exhibit strong absorption or scattering of X-ray photons due to their large atomic absorption cross-sections. High-Z elements are commonly employed as dopants or host ions in nanoscintillator matrices. This section reviews recent advancements in new X-ray-sensitive materials, particularly those with promising applications in biomedicine.

15.3.1 Metallic Materials

Due to the inefficiency of organic PSs in absorbing high-energy radiation photons, heavy metal inorganic nanomaterials play a crucial role in nanoscintillator materials.

15.3.1.1 Lanthanide-based Nanophosphors

Rare-earth scintillating nanoparticles (rare-earth SCNPs) are a type of rare-earth compound nanoparticles that exhibit the XEOL characteristic of rare-earth elements. They are considered a promising category of SCNPs with potential applications in biological imaging and therapy [22–27]. The first rare-earth doped scintillators were reported in the late 1980s. After decades of development, these scintillator systems have demonstrated satisfactory performance in scintillation thanks to their exceptional chemical and optical properties. These properties include low toxicity, narrow emission bands, long luminescence life, and strong photo-stability. To date, a variety of nanocrystals have been synthesized by doping rare-earth elements into matrices, such as oxides, fluorides, oxysulfides, phosphates, hafnates, and tungstates, which typically exhibit characteristic emissions of lanthanide activators [28–32]. They are the most extensively studied class of nanophosphors to date.

The sharp linear spectral lines of f–f transitions in rare-earth elements result from the minimal energy difference in their 4f electron layers. These transitions are unaffected by the spectral position within the embedding matrix, offering flexibility in adjusting both their intensity and fine structure. Rich in 4f electron shells, the X-ray-excited emission region of rare-earth elements includes UV, green, yellow, and orange-red areas. Trivalent cerium (Ce^{3+}) based nanophosphors are commonly used in the UV area for X-PDT research [9, 33–36]. For instance, Yang and colleagues developed a multifunctional core–shell radiosensitizer (SCNPs@DMSN@CeOx-PEG, referred to as SSCP), which was meticulously designed and constructed by coating adjustable mesoporous silica on the nanocrystals, followed by the in situ integration of cerium oxide nanoparticles to boost X-PDT capabilities (Figure 15.2) [37]. The resulting SSCP radiosensitizer exhibited a noticeable blue shift in the UV region, which allowed it to effectively capture the UV light converted by the scintillating nanoparticles (SCNPs) core, leading to the

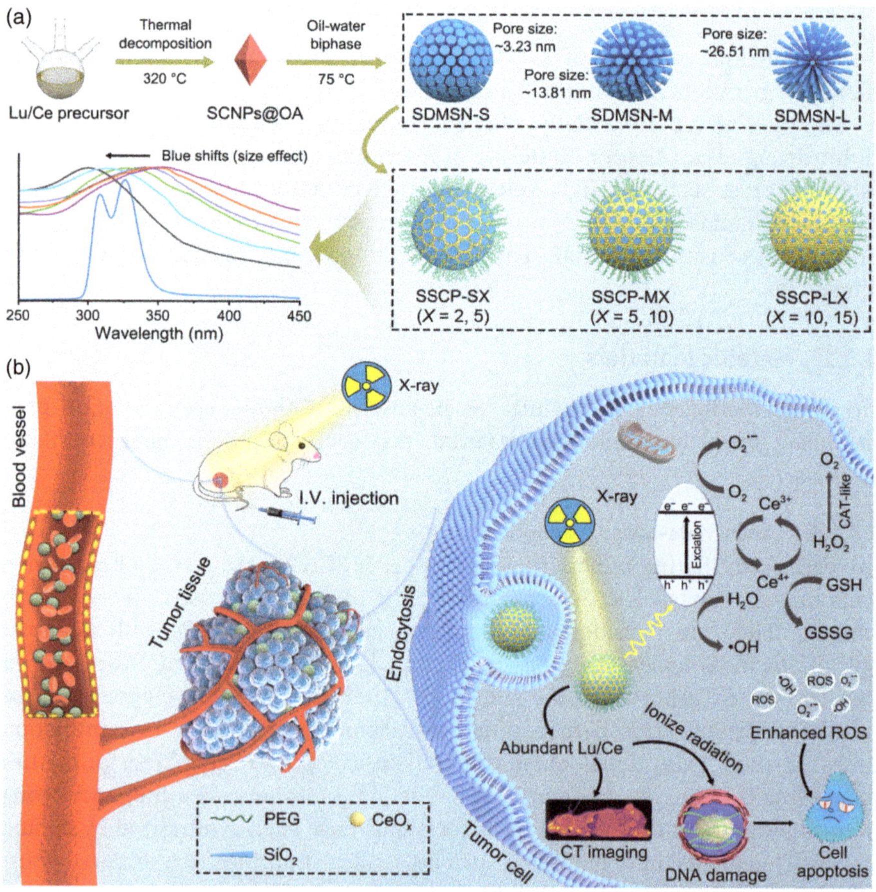

Figure 15.2 Schematic Illustration of SSCP with (a) tunable structural and (b) the mechanisms of SSCP-mediated X-PDT. Source: Reproduced with permission from Liu et al. [37]/American Chemical Society.

separation of light-induced electron-hole (e^--h^+) pairs and the production of ROS. Additionally, cerium oxide demonstrated a notable ability to consume glutathione, thereby enhancing ROS accumulation, along with peroxidase-like activity to alleviate hypoxia, ultimately boosting the effectiveness of radiotherapy. With a rich content of Lu and Ce elements, the CT imaging performance of SSCP surpasses that of conventional contrast agents (iodixanol) by approximately 3.79 times, indicating significant promise for both preclinical imaging and clinical applications. Dinakaran et al. developed a novel nanoparticle (NP) for radiation-activated PDT (radio PDT) by using poly(ethylene glycol) methyl ether-block-poly(lactide-co-glycolide) (PEG-PLGA) nanospheres to encapsulate the nanoscintillator LaF_3:Ce^{3+} and the PS protoporphyrin IX (PpIX) [35]. This radio PDT NP enabled the excitation of PpIX for the generation of $^1O_2$, as the emission spectrum of LaF_3:Ce^{3+} (peaking at 305 nm) overlapped with the absorption spectrum of PpIX. When exposed to X-ray irradiation (8 Gy), this radio PDT NP displayed enhanced therapeutic efficacy against prostate cancer (PC3) cells compared to radiation treatment alone.

Trivalent terbium (Tb^{3+}) exhibits a characteristic green emission band, with a peak near 544 nm attributed to the $^5D_4 \rightarrow {}^7F_5$ transition. Terbium-doped nanophosphors have higher X-ray absorption efficiency and conversion efficiency of X-rays to visible light, making them a widely studied class of nanophosphors [31, 38–43]. Yang and coworkers [31] studied the optimization of Tb^{3+} doping levels (5–25 mol%) in core nanoparticles to enhance X-ray-excited luminescence efficiency. Emission spectra analysis under X-ray irradiation revealed characteristic bands at 489, 546, 584, and 612 nm attributed to Tb^{3+} transitions. Optimal luminescence was observed at 15 mol% Tb^{3+} doping in $NaGdF_4$, with higher concentrations causing luminescence quenching due to cross-relaxation effects. This study highlights the importance of host selection, shell layer design, and pH control in optimizing X-ray-excited luminescence for potential applications in nanoparticle-based imaging and sensing technologies.

In X-ray-activated PDT studies, a typical combination is Tb^{3+} with the PS Rose Bengal (RB) [44–46]. Ren et al. reported the Tb-doped core–shell–shell (CSS) nanophosphors with the PS RB, further functionalized by the targeting agent cRGD peptide to form RGD–CSS–RB nanocomplexes for X-PDT applications [40]. The green emission of CSS nanoparticles under X-ray excitation aligns with RB absorption, facilitating $^1O_2$ generation. Under 4 Gy X-ray exposure, U87 MG glioblastoma cell viability significantly decreased to $30.3 \pm 1.8\%$, indicating the application potential of RGD–CSS–RB nanocomplexes in X-PDT. Feng et al. developed another CSS-structured nanoparticle via a dry solvothermal method [41]. Through surface ligand exchange, the nanoparticle $NaErF_4$:0.5%Tm@$NaYF_4$@$NaGdF_4$:15%Tb was covalently coupled with the RB and targeting molecule FA, further modified with polyethylene glycol succinyl amide carbonate to yield UIRPPs with a diameter of about 77 nm. Under the excitation of X-rays, these UIRPPs exhibit characteristic emissions of terbium, matching the absorption of RB. The efficient FRET between Tb^{3+} and RB leads to the activation of RB for the generation of $^1O_2$. In vivo antitumor experiments showed that the combination therapy involving UIRPPs and X-ray irradiation (ultra-low X-ray dose, 1 Gy) exhibited the most pronounced tumor growth inhibition compared to the control group. These findings underscore the promising therapeutic efficacy of UIRPPs for the treatment of deep-seated tumors.

The transition between the $^4f_0\ ^5d_1 \rightarrow ^4f_1\ ^5d_0$ configurations of Ce^{3+} can produce an emission peak at 585 nm, achieving yellow emission [47, 48]. Jain and colleagues reported gadolinium aluminum garnet ($Gd_{2.98}Ce_{0.02}Al_5O_{12}$, GAG) magnetic luminescent SCNPs using a sol–gel technique. Subsequently, these nanoparticles were coated with mesoporous silica and loaded with the PS RB to produce nanocomposite materials denoted as GAG@mSiO_2@RB [47]. Upon X-ray exposure (55 kV, 40 mA), the $^1O_2$ yield of GAG@mSiO_2@RB nanocomposite materials is four times that of RB alone. Additionally, GAG@mSiO_2 nanoparticles have a higher magnetic mass susceptibility (χ_m) (9.75×10^{-5} emu/gOe). The remarkable paramagnetic and PDT properties of GAG@mSiO_2@RB nanocomposites make them suitable for deep cancer treatment.

Europium (Eu^{3+}/Eu^{2+}) can produce a notable orange-red emission, with its maximum peak around 613 nm (peak position may vary slightly with the matrix) [22, 49–51]. Xie and coworkers presented a novel X-PDT approach mediated by $SrAl_2O_4$:Eu^{2+} (denoted as SAO). The system is composed of a core made of SAO and a silica coating loaded with the photosensitizer Merocyanine 540 (MC540), in which SAO converts X-ray photons into visible photons that match the excitation wavelength of MC540 (Figure 15.3) [50]. This ET from SAO effectively activates MC540 to generate cytotoxic singlet oxygen. Notably, both in vitro and in vivo studies revealed that a low X-ray dose (0.5 Gy, single dose) was adequate to induce damage to cancer cells. Specifically, in a mouse U87MG xenograft model, SAO-mediated X-PDT was observed to induce significant tumor growth inhibition and even shrinkage, while sparing normal tissues from adverse effects.

Furthermore, luminescence of two or more colors can be achieved by co-doping with multiple rare-earth ions. Chang and coauthors [52] synthesized a new type of SCNP with a CSS structure, ScNP-$NaLuF_4$:Gd (35%), Eu (15%)@$NaLuF_4$:Gd (40%)@$NaLuF_4$:Gd (35%), Tb (15%). These nanoparticles can be excited by X-rays to generate red light at 614 and 695 nm (from Eu^{3+}) for luminescent imaging, as well as green light at 543 nm (from Tb^{3+}) to activate the loaded PS RB through FRET to generate $^1O_2$ for PDT. The introduction of the intermediate shell layer is to prevent the energy back-transfer between Eu^{3+} in the core and Tb^{3+} in the outer shell. The incorporation of Gd^{3+} ions in the core, intermediate shell, and outer shell serves as antennas to transfer the absorbed 313 nm ($^6P_{7/2} \rightarrow ^8S_{7/2}$) to the Eu^{3+} and the Tb^{3+} ions. The utilization of $NaLuF_4$ as the matrix is to help synthesize monodisperse nanoparticles, when doping Gd^{3+}, Eu^{3+}, and Tb^{3+} ions, and to prevent concentration quenching.

15.3.1.2 Metal Cluster Nanomaterials

Metal cluster nanomaterials have garnered growing attention among researchers in therapeutic applications for their remarkable chemical stability, adjustable optical characteristics, and compatibility with biological systems [53–55]. Osakada et al. conducted a comprehensive investigation on biomolecule-stabilized metal clusters,

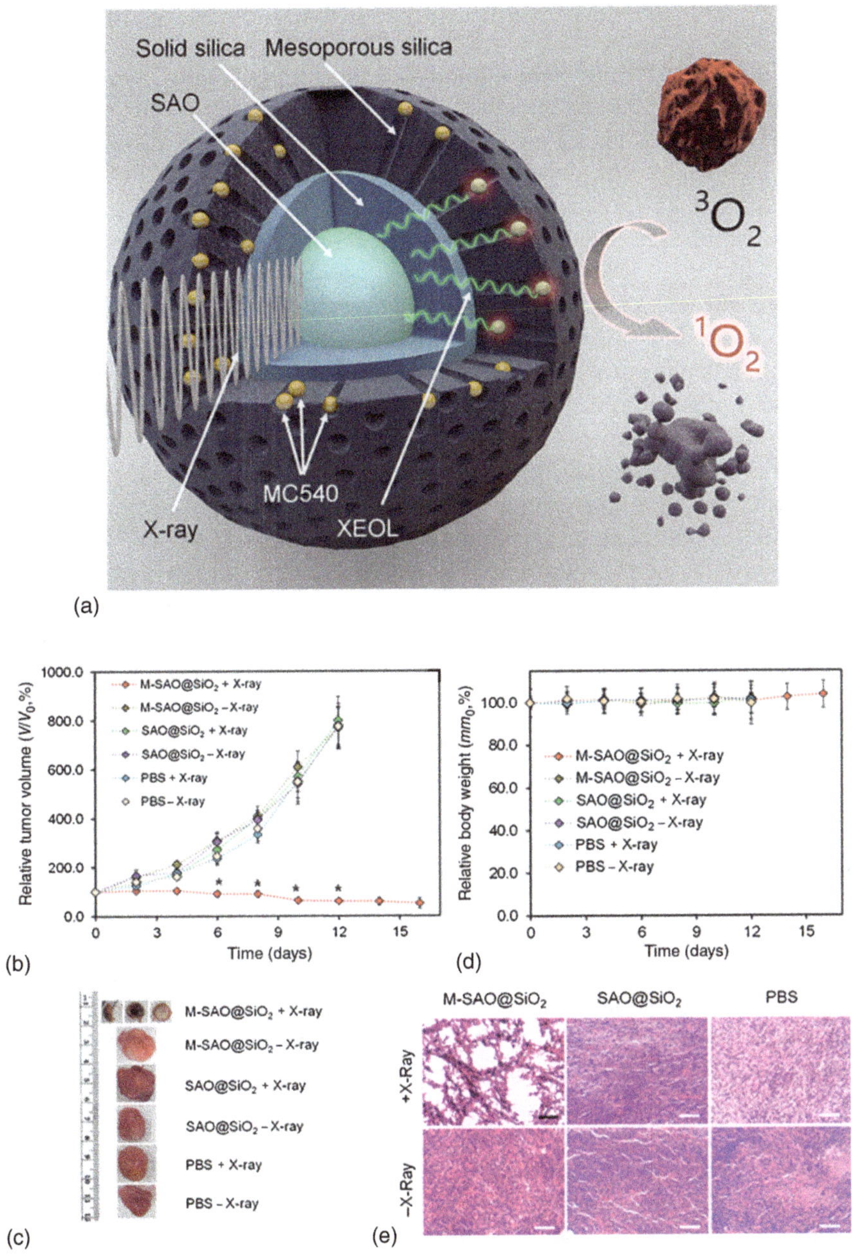

Figure 15.3 Schematic illustration of the working mechanism of X-PDT (a) and X-PDT for in vivo tumor therapy (b–e). (b) Tumor growth curves. (c) Photographs of representative tumors taken from Groups 1–6. (d) Body weight curves. (e) H&E staining on tumor tissues taken from Groups 1–6. Source: Reproduced with permission from Chen et al. [50]/American Chemical Society.

including single-stranded DNA-stabilized Ag clusters (Ag_x-ssDNA), lysozyme VI or bovine serum albumin-stabilized Au clusters (Au_8-Lyz or Au_{25}-BSA) [53]. Surprisingly, among these clusters, Au_{25}-BSA exhibited the lowest absolute quantum yield when excited by visible light but showed the strongest X-ray luminescence upon X-ray exposure. Au nanomaterials have garnered significant interest as X-ray-responsive agents for CT and RT applications. Chang and coworkers systematically studied the size-dependent effects of gold nanoparticles (AuNPs) in enhancing CT and RT therapy in images. In vivo studies showed that AuNPs of ~13 nm size exhibited excellent CT contrast and significant radiation sensitization effect [56].

Bismuth polyoxometalate and multimetal oxide clusters, such as nanoclusters, have shown good application prospects as a new type of radiosensitizer in tumor therapy [57, 58]. For instance, Gu and coworkers investigated bismuth polytungstate (BiP_5W_{30}) nanoclusters as radiosensitizers to modulate the tumor microenvironment and enhance RT efficacy [59]. Upon X-ray irradiation, owing to the presence of heavy Bi and W atoms, the BiP_5W_{30} nanoclusters could enhance the radiation dose deposition in the tumor region, and because of their unique electronic structure and multi-electron nature, they can also consume glutathione (GSH) through redox reactions and facilitate the conversion of H_2O_2 into $^{\bullet}OH$ to boost ROS generation. Moreover, by coupling with reduced graphene oxide (0.5 nm) to promote electron–hole separation, further enhanced radiocatalytic activity can be achieved, showing significant anticancer effects against HeLa human cervical tumor-bearing mice upon X-ray exposure.

Zhao and coworkers developed another radiosensitization system using Gd-containing polyoxometalates-conjugated chitosan (GdW_{10}@CS), which aims to enhance radiosensitivity through both extrinsic and intrinsic mechanisms. By generating cytotoxic ROS with high-energy X-ray stimulation and delivering hypoxia-inducible factor-1α (HIF-1α) siRNA to down-regulate HIF-1α expression, the nanospheres help overcome radioresistance by depleting intracellular glutathione and inhibiting DNA repair (Figure 15.4) [60]. Through a combination of external and internal radiosensitization strategies, these nanospheres enhance the therapeutic efficiency of radiotherapy. By synergistically leveraging ROS generation and GSH depletion, as well as suppressing DNA repair, the GdW_{10}@CS nanospheres provide a promising approach to improve radiosensitivity in hypoxic tumors. This innovative radiosensitizer strategy shows the potential to enhance the efficacy of radiotherapy for treating hypoxic tumors.

15.3.1.3 Long-Afterglow Luminescent Nanomaterials

Long-afterglow luminescent materials are a classic type of luminescent material that has the ability to retain excitation energy within the material via electronic traps and then slowly release photons over a long period after the removal of the excitation source [61–63]. This distinct behavior has been extensively researched in enhancing signal-to-noise ratio and continuous tracking of biological imaging without the need for excitation sources. However, traditional long-afterglow luminescent materials are usually excited by ultraviolet or visible light. Due to the short observation

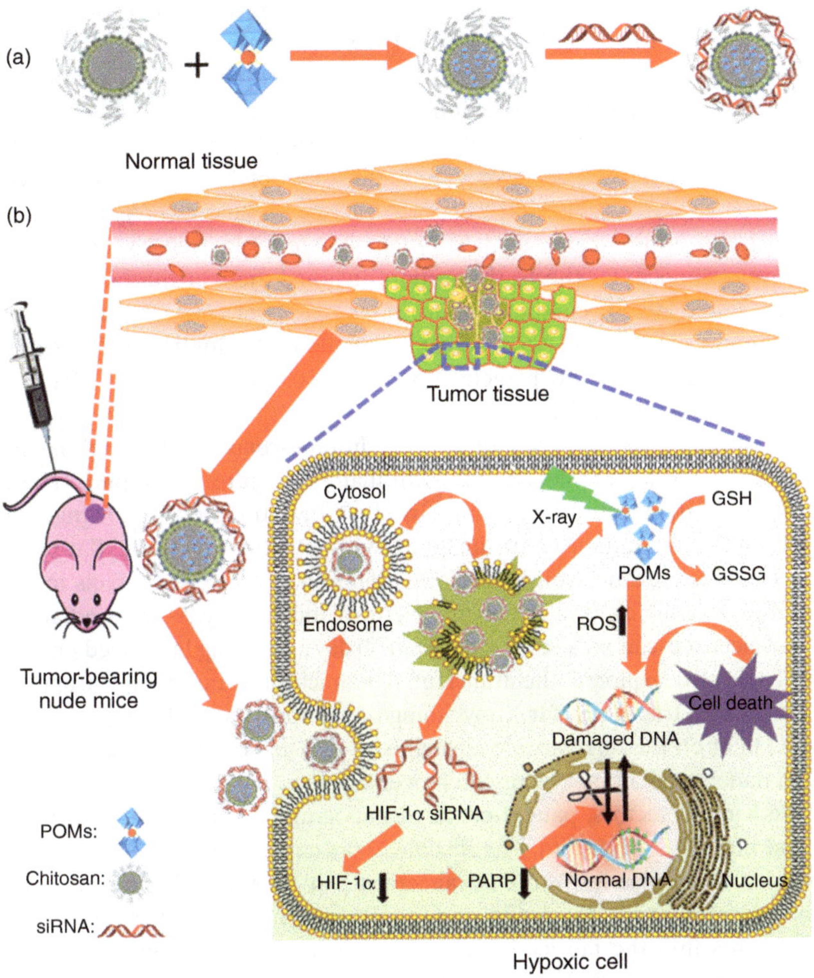

Figure 15.4 Schematic illustration of the Gd-containing polyoxometalates-conjugated chitosan (GdW_{10}@CS nanosphere) as a siRNA delivery platform for efficient radiosensitization efficacy of radiotherapy against hypoxic tumor cells. Source: Reproduced with permission from Yono et al. [60]/American Chemical Society.

time and low tissue penetration depth when excited in vivo, their applications in living systems are limited. In recent years, research on X-ray-activated and reactivated long-afterglow luminescent nanomaterials and their biological applications have been conducted.

In a study by Hao et al., $ZnGa_2O_4$:Cr PLNPs with a hydrodynamic diameter of 141 nm were synthesized. These nanoparticles emit light around 700 nm, originating from the energy level transition of Cr^{3+} ($^2E \rightarrow {}^4A_2$), and display persistent afterglow luminescence for up to six hours post-X-ray source removal. Interestingly, these PLNPs could be efficiently reactivated by X-rays (45 kVp, 0.5 mA) even at depths of up to 20 mm. Moreover, after removing the X-ray source, renewable whole-body imaging in vivo was successfully achieved with oral or intravenous injection of the polyethylene glycol (PEG) modified PLNPs [64]. In their study, Shen and coworkers observed that nanophosphors composed of bulk $LiGa_5O_8{:}Cr^{3+}$ (LGO:Cr) scintillation material exhibited persistent NIR luminescence under X-ray irradiation [65]. Using nanophosphors with an average diameter of 100.9 ± 31.7 nm as LGO:Cr cores and wrapping a silicon dioxide layer with a thickness of 25.6 ± 2.5 nm, intense NIR luminescence was displayed at 720 nm attributed to the $^2E \rightarrow {}^4A_2$ transition of Cr^{3+}. Persistent luminescence signals remained detectable even 24 hours after the X-rays were stopped. Furtherly, the authors enclosed the PS naphthalocyanine in mesopores and coupled cetuximab to the outer surface of LGO:Cr. This system enabled continuous NIR luminescence imaging on the H1299 lung tumor model in situ within one minute at a low preactivation dose of 0.02 Gy, and the X-PDT at a relatively higher therapeutic dose of 6 Gy.

Liu and coworkers developed a new X-ray luminescence extension imaging technique using lanthanide-doped nanoscintillators (Figure 15.5) [66]. These nanoscintillators demonstrate ultralong-lived X-ray trapping, allowing for high-resolution, three-dimensional imaging without the need for flat-panel detectors. The mechanism involves the formation of anion Frenkel defects in the nanocrystal lattice, which trap energetic electrons induced by X-ray irradiation. The trapped electrons can persist for more than 30 days, enabling long-lived photon trapping. They also explored the transient dynamics of X-ray energy trapping and the creation of trap states for X-ray storage. Experimental and computational results shed light on the microscopic mechanisms of X-ray energy conversion and trapping in nanoscintillators. The research opens new possibilities for applications in wearable X-ray detectors, patient-centered radiography, imaging-guided therapeutics, and high-energy physics. The flexible X-ray detector fabricated using this technique shows promise for future advancements in X-ray imaging technologies with unprecedented spatial resolution and special imaging capabilities. This study provides insights into the fabrication of persistent luminescent nanocrystals for optogenetics, low-dose radiotherapy, and expansion microscopy. These findings contribute to the advancement of X-ray imaging technologies and offer new opportunities for the development of next-generation X-ray detectors with enhanced performance and versatility.

15.3.1.4 Quantum Dots

Colloidal semiconductor nanocrystals known as quantum dots have diameters smaller than the exciton Bohr radius, exhibiting size-dependent optical and electronic properties due to their discrete electronic energy states, distinguishing them from traditional organic dyes. This characteristic has spurred extensive research in the fields of biological imaging and biosensing. The luminescence intensity of CdSe@ZnS quantum dots is proportional to the X-ray dose. Utilizing the FRET effect, photosensitizers coupled with CdSe@ZnS quantum dots for X-ray-activated PDT treatment showed remarkable cytotoxic effects on H460 human lung cancer cells [67].

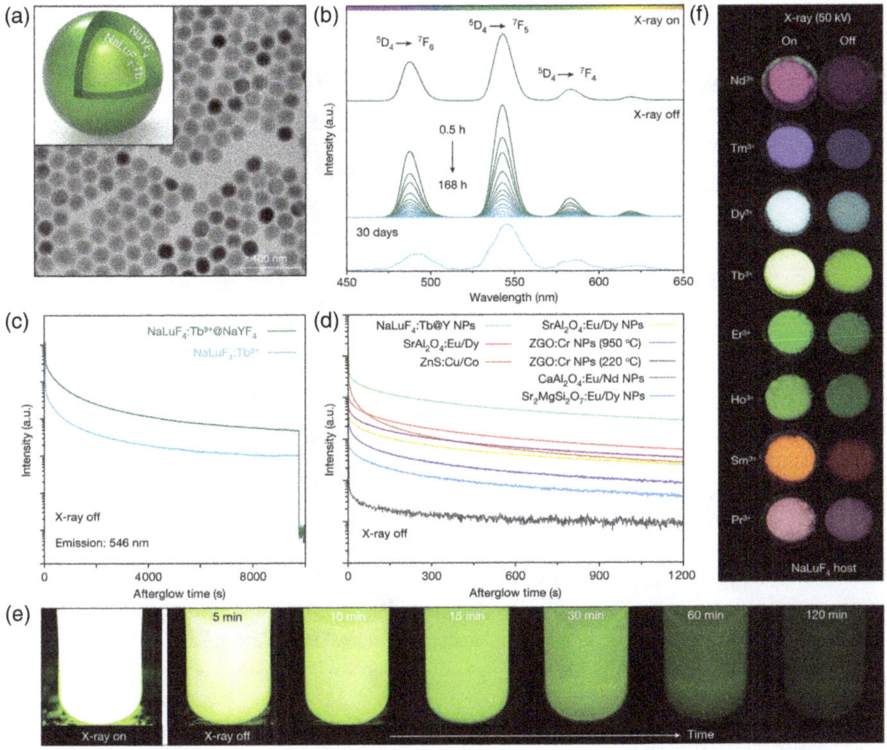

Figure 15.5 (a) Transmission electron microscope (TEM) micrograph of $NaLuF_4$:Tb (15 mol%)@$NaYF_4$ nanocrystals. (b) Radioluminescent emission spectra of the core–shell nanocrystals. (c) Radioluminescent intensity of $NaLuF_4$:Tb (15 mol%) and $NaLuF_4$:Tb (15 mol%)@$NaYF_4$ nanocrystals. (d) Comparison of afterglow decay profiles of various phosphors after cessation of X-ray excitation. (e) Afterglow photographs of $NaLuF_4$:Tb (15 mol%)@$NaYF_4$ nanocrystals. (f) Radioluminescence and afterglow of $NaLuF_4$ nanocrystals doped with various activators. Source: Reproduced with permission from Ou et al. [66]/Springer Nature.

The production of trap-free, high-quality quantum dots is essential for efficient X-ray-induced luminescence under high-energy irradiation. Jensen and coworkers detailed the controlled fabrication of ZnSe quantum dots within a continuous-flow microfluidic reactor and investigated their X-ray luminescence properties [68]. The emission of X-ray luminescence shifts with the change of quantum dot size, indicating that the luminescence is related to the bound exciton. Moreover, the intensity of X-ray luminescence is directly proportional to the radiation dose.

In recent years, halide lead perovskite, as a new class of X-ray-sensitive material, has garnered considerable attention due to its exceptional performance in X-ray detection and radiography. Liu and co-workers have demonstrated the efficacy of all-inorganic halide lead perovskite nanocrystal scintillators (9.6 nm, $Z_{max} = 82$) in converting high-energy X-ray photons into abundant visible light photons (Figure 15.6) [69]. These scintillators exhibit strong X-ray luminescence in the visible light spectrum and the emitted light wavelength can be readily tailored by

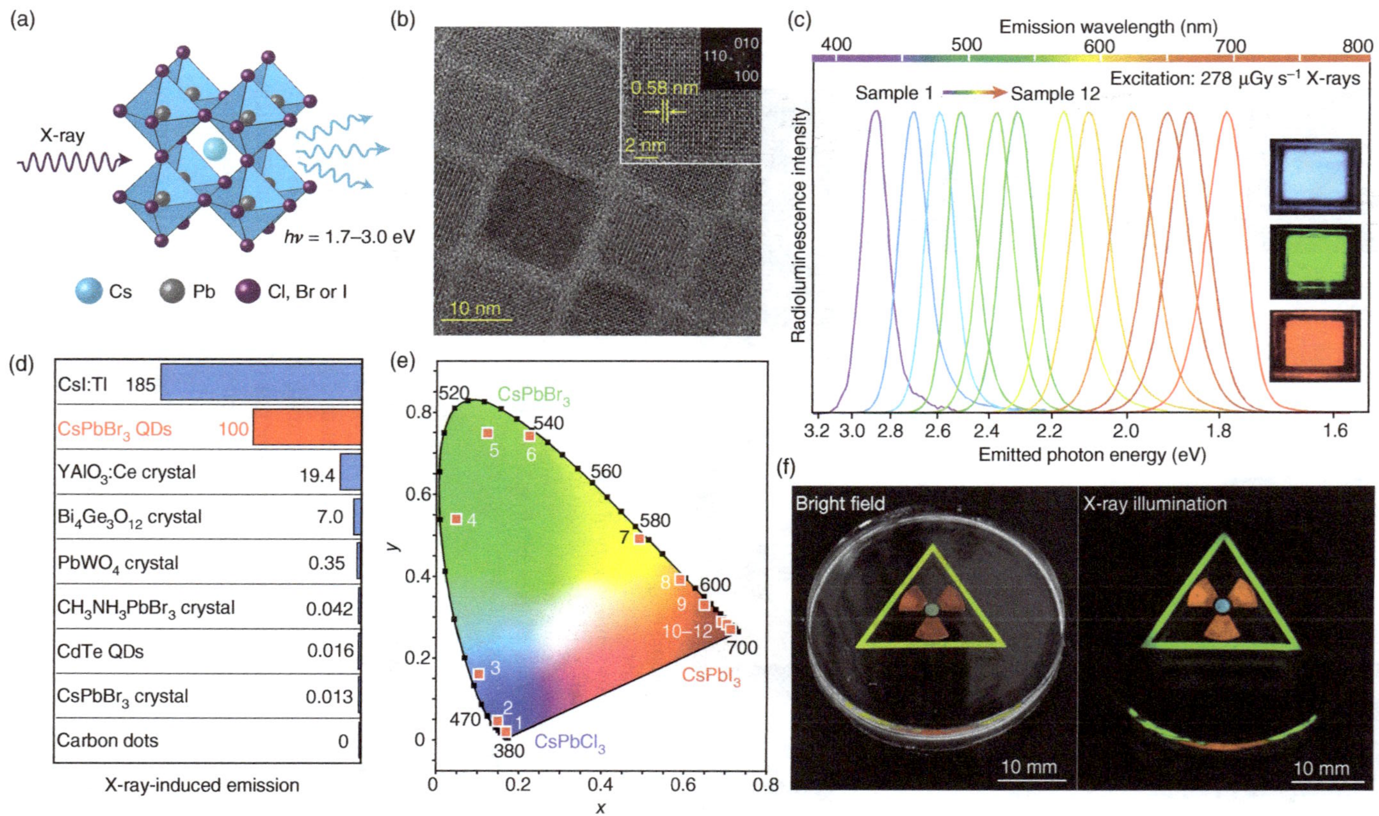

Figure 15.6 (a) Schematic representation of X-ray-induced luminescence of energy $h\nu$, generated in an all-inorganic perovskite lattice with a cubic crystal structure. (b) Low-resolution TEM image of the as-synthesized $CsPbBr_3$ nanocrystals. (c) Tunable luminescence spectra of the perovskite QDs. (d) Comparison of the optical sensitivity of various scintillator materials. (e) CIE (Commission Internationale de l'Eclairage) chromaticity coordinates of the X-ray-induced visible emissions measured for samples 1–12. (f) Multicolour X-ray scintillation from three types of perovskite nanocrystal scintillator. Source: Reproduced with permission from Chen et al. [69]/Springer Nature.

substituting the anionic components of the precursor during synthesis. Moreover, the X-ray luminescence intensity of halide lead perovskite surpasses that of traditional quantum dots, such as CdTe quantum dots (Z_{max} = 52), with a relatively low X-ray attenuation coefficient. Mechanism studies have shown that in the early stages of energy state transition, high-energy X-rays mainly interact with the lattice atoms of the perovskite through the photoelectric effect. Subsequently, a substantial number of high-energy electrons and holes are rapidly heated to the edges of the valence band and CB, respectively, leading to the rapid emission of X-ray luminescence from the highly emissive triplet-excited state. Integration of these innovative nanoscintillators with commercial flat-panel imagers enables electronic circuit board inspection under low-dose X-ray irradiation, marking a significant advancement in X-ray imaging device fabrication technology. Recent strides have also been made in utilizing halide lead perovskite encapsulated in a polymer matrix (such as polystyrene and polymethyl methacrylate) as a luminescent probe for cell imaging. This approach overcomes the chemical instability caused by polar molecules and toxic substances, underscoring the potential of halide lead perovskite in therapeutic applications.

15.3.1.5 Metal–Organic Complexes

Metal–organic complexes, such as iridium, ruthenium, and hafnium complexes, exhibit considerable promise in therapeutic applications. For example, Liu et al. reported a new type of coordination polymer nanoparticles based on hafnium ions (Hf^{4+}) and aggregation-induced emission (AIE) PS, which achieved synchronous radiotherapy and radiation dynamic therapy under X-ray irradiation through coordination with the high-Z element Hf (Figure 15.7) [70]. In this system, Hf^{4+} acts not only as a radiosensitizer to capture X-rays and deposit radiation energy to enhance RT but also facilitates ET from X-rays to TPEDC-DAC for RDT. Following two-step lipid modification, Hf-AIE-PEG exhibited good biocompatibility. Compared to TPEDC-DAC for RDT and Hf-CAIE-PEG (CAIE: AIE molecule without photodynamic effect) for RT, Hf-AIE-PEG exhibited a more significant inhibitory effect on tumor growth under X-ray irradiation. Under X-ray exposure, Hf-AIE-PEG-DBCO (DBCO: Dibenzocyclooctyne) can concurrently generate effective ·OH and SO through the radiosensitization of Hf and Hf-mediated ET from X-rays to TPEDC-DAC, demonstrating a stronger growth inhibition effect than conventional PDT under white light exposure.

Recently, multinuclear complexes of transition metals with high heavy metal content, adjustable ligand environments, and favorable biocompatibility have become potential candidate materials for X-ray-sensitive materials in therapeutic applications [71–73]. Kirakci and co-workers reported two tetranuclear copper iodide complexes involving triphenylphosphine and pyridine ($[Cu_4I_4(PPh_3)_4]$ and $[Cu_4I_4Py_4]$) and investigated their X-ray luminescence properties [71]. The X-ray luminescence emission spectra of $[Cu_4I_4(PPh_3)_4]$ and $[Cu_4I_4Py_4]$ exhibit emission peaks at 547 nm and 580–585 nm, respectively, with luminescence intensities surpassing those of the conventional scintillator $Bi_4Ge_3O_{12}$ by 4 times and 2.5 times, respectively. Analysis of the temperature-dependent X-ray luminescence and ultraviolet-excited

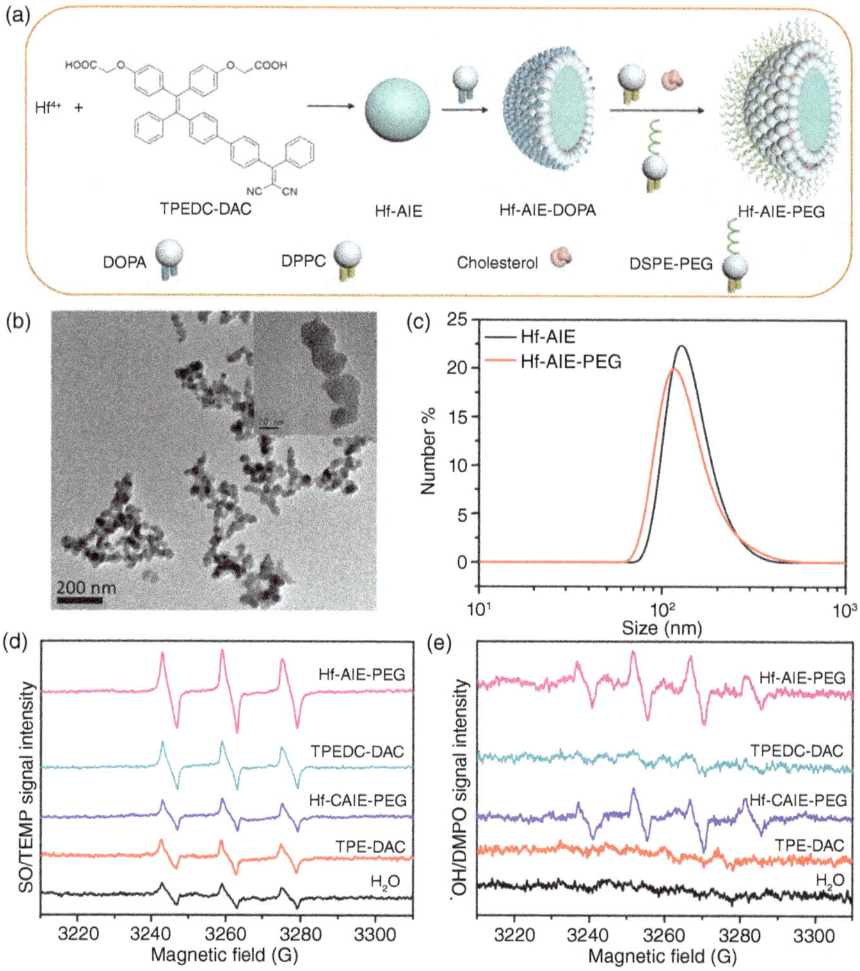

Figure 15.7 (a) Schematic illustration of fabrication and modification of Hf-AIE. (b) TEM images of Hf-AIE. (c) Hydrodynamic diameters of Hf-AIE and Hf-AIE-PEG nanoparticles. (d) The electron spin resonance (ESR) spectra to detect SO generated by Hf-AIE-PEG, TPEDC-DAC, Hf-CAIE-PEG, TPE-DAC, and H_2O under X-ray irradiation. (e) The ESR spectra to detect $^{\bullet}OH$ produced by Hf-AIE-PEG, TPEDC-DAC, Hf-CAIE-PEG, TPE-DAC, and H_2O under X-ray irradiation. Source: Reproduced with permission from Liu et al. [70]/John Wiley & Sons.

luminescence indicated the same origin of the luminescence from the triplet-state emission.

Octahedral molybdenum and tungsten cluster complexes have exhibited promising potential in biomedical applications, with a detailed investigation into their X-ray luminescence properties. A new generation of X-ray-activated nanoparticles composed of octahedral molybdenum cluster compounds $(n\text{-}Bu_4N)_2[Mo_6I_8(OCOCF_3)_6)]$ was synthesized by Lang and co-workers, using a nanoprecipitation method [72]. Due to the presence of heavy element components, the nanoparticles (180 ± 80 nm) effectively absorbed X-rays. Upon X-ray irradiation, the transmission of primary and secondary Auger electrons led to the formation of an excited triplet state capable of generating $^1O_2$ upon interaction with oxygen. In fact, the increased level of ROS in HeLa human cervical cancer cells was shown to be proportional to the radiation dose.

Shestopalov and coworkers conducted a systematic exploration of the X-ray luminescence characteristics of eight octahedral molybdenum and tungsten cluster complexes $[(M_6I_8)L_6]_n$ (M = Mo or W, L = I^-, NO^{3-}, Ots^- or OH^-/H_2O) [74]. The W_6-based complexes exhibited superior X-ray luminescence intensity compared to their Mo_6-based counterparts, attributed to their favorable X-ray absorption capacity and photoluminescence performance. Given this advancement, it is anticipated that other metal cluster complexes, such as rhenium-based materials ($Z = 75$), may demonstrate exceptional X-ray luminescence properties.

15.3.1.6 Metal-Organic Frameworks (MOFs)

MOFs are composed of secondary building units (SBUs) of metal ions/clusters and organic linkers, known for their crystal structure, expansive surface area, and significant porosity [75, 76]. The heavy atom effect of metal connection nodes and the flexibility of adjustable functional organic ligands make MOFs a promising class of X-ray-responsive materials with potential therapeutic applications [77, 78].

In 2014, Lin and coworkers first reported the design of well-structured MOFs with X-ray-induced luminescence. In their study, MOFs (size of ~1 μm) were constructed by $M_6(\mu_3\text{-}O)_4(\mu_3\text{-}O)_4\text{-(carboxylate)}_{12}$ (M = Zr or Hf) SBUs and anthracene-based dicarboxylate organic scintillator connecting ligands as emitters [79]. Upon X-ray exposure, similar emission peaks were observed in both Zr-MOFs and Hf-MOFs, ranging from 400 to 600 nm. Moreover, the X-ray luminescence intensities of Hf-MOFs and Zr-MOFs were about 24 times and 11 times greater than that of the individual anthracene ligands, respectively. Investigation into the mechanism revealed that X-ray photons were absorbed by Zr-SBUs and Hf-SBUs, leading to the generation of fast electrons that interacted with the bridging anthracene organic molecules, resulting in the emission of visible photons from their excited states.

Expanding on prior work, further studies have been conducted on hafnium-based metal-organic layers (MOLs) as PSs for X-ray-triggered colon cancer PDT [80]. Two-dimensional Hf-MOLs, with a thickness of 1.2 nm and a diameter of around 500 nm, were synthesized using $[Hf_6O_4(OH)_4(HCO_2)_6]$ SBUs and tricarboxylic ligands derived from $Ir[bpy(ppy)_2]^+$ or $[Ru(bpy)_3]^{2+}$. These MOLs exhibited an exceptionally high $^1O_2$ quantum yield of 0.97 (studied in O_2-saturated CH_2Cl_2–MeOH (9 : 1) solution for $Ir[bpy(ppy)_2]^+$) [81], and a $^1O_2$ quantum yield of 0.73 (studied in air-saturated CD_3OD for $[Ru(bpy)_3]^{2+}$) (bpy = 2,20-bipyridine, ppy = 2-phenylpyridine) [82], making them highly efficient X-ray PSs. High loading efficiencies were achieved for Hf-BPY-Ir and Hf-BPY-Ru, reaching 67% and 59% respectively. When exposed to Xe lamp illumination filtered at 400 nm, Zr-BPY-Ir and Hf-BPY-Ir exhibited $^1O_2$ generation capabilities approximately 3.7 and ~2.7 times higher than those of Zr-BPY-Ru and Hf-BPY-Ru, respectively, consistent with the notable $^1O_2$ quantum yield of $Ir[BPY(ppy)_2]^+$ PS. Interestingly, Hf-BPY-Ir and Hf-BPY-Ru demonstrated $^1O_2$ generation capabilities about 3.1 and ~5.3 times greater than those of Zr-BPY-Ir and Zr-BPY-Ru, respectively, suggesting a crucial role of Hf-SBU in absorbing X-ray energy and efficiently transferring it to the photosensitizer bridging ligand. Additionally, significant tumor shrinkage was observed in CT26 and MC38 murine colorectal tumor-bearing mice treated with Hf-BPY-Ir/Hf-BPY-Ru and subjected to X-ray irradiation, underscoring the potential of X-ray-activated cancer PDT.

Furthermore, Lin and colleagues introduced two novel types of MOLs fabricated using Hf_{12} and Hf_6-SBUs in conjunction with photosensitizer ligands Ir(bpy)[(dF(CF$_3$)ppy]$^{2+}$ derived from [dF(CF$_3$)ppy = 2-(2,4-difluorophenyl)-5-(trifluoromethyl)pyridine] for synchronous radiotherapy and X-PDT (Figure 15.8) [78]. When irradiated with X-rays, HO$^{\cdot}$ produced by Hf-SBUs leads to RT enhancement, while $^1O_2$ and O_2^- can be synchronously produced through ET from Hf-SBU to PS ligands. Importantly, the planar structure of Hf-MOLs significantly promotes the

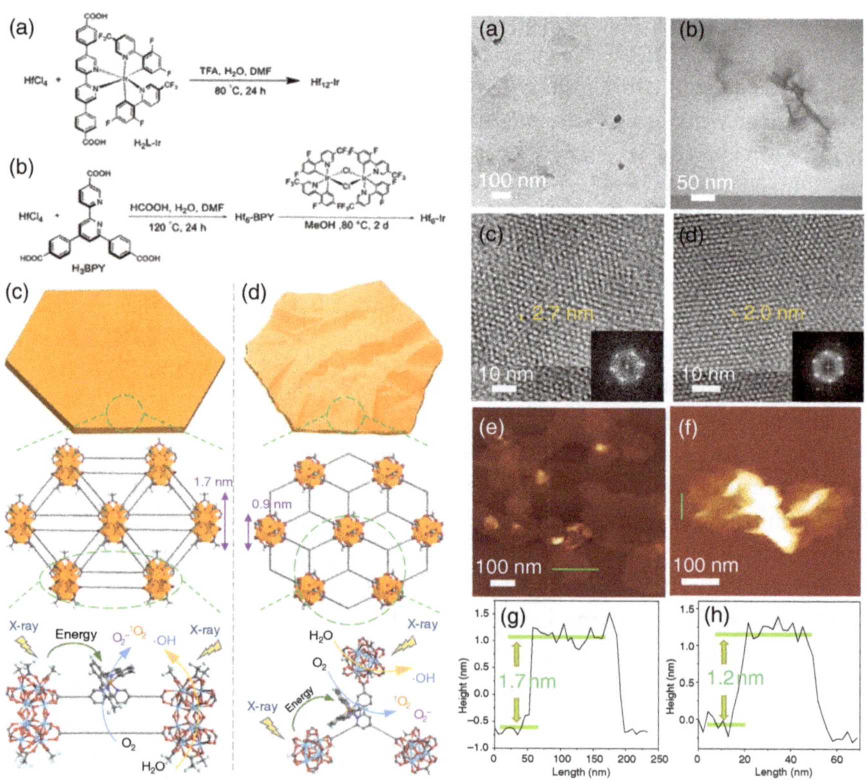

Figure 15.8 Left: Schematic showing the synthesis, morphologies, and topological structures of an X-ray triggered ROS generation by Hf_{12}-Ir (a,c) and Hf_6-Ir (b,d) and right: TEM image (a), HRTEM image and fast Fourier transform (FFT) pattern (c), AFM topography (e), and height profile (g) of Hf_{12}-Ir. TEM image (b), HRTEM image and FFT pattern (d), AFM topography (f), and height profile (h) of Hf_6-Ir. Source: Reproduced with permission from Lan et al. [78]/American Chemical Society.

diffusion of cytotoxic ROS into the MC38 murine colorectal cancer cell environment. Through anticancer efficacy evaluation using MC38 murine colorectal tumor mice, significant tumor regression was observed, indicating the great potential of reducing X-ray doses for cancer treatment.

The synthesis of uranium-MOFs as a novel scintillator with bright X-ray luminescence was reported. Wang and colleagues described the creation of a uranium organic framework (SCU-9) exhibiting vibrant green X-ray luminescence attributed to the HOMO–LUMO transition of UO_2^{2+} ions [83]. Compared to the conventional CsI:Tl scintillator, SCU-9 maintained luminescence intensity at 65% of the original level after exposure to an X-ray dose of 53 Gy.

15.3.2 Nonmetallic Materials

15.3.2.1 Organic Materials

Most previously reported X-ray scintillators have been confined to either inorganic phosphors or organic metal complexes containing heavy metals. While organic phosphors offer inherent advantages such as abundant resources and high mechanical flexibility, their weak X-ray absorption and inefficient exciton utilization have impeded the advancement of scintillators based on organic molecules. These organic phosphors primarily consist of light elements like carbon, hydrogen, and nitrogen, resulting in limited X-ray absorption capabilities. In addition, the lack of strong spin-orbit coupling in traditional organic phosphors leads to fluorescence emission only from singlet excitons postirradiation. Consequently, roughly 75% of the excitons remain underutilized due to the dark state characteristics of triplet excitons in nonmetallic organic phosphors [84, 85]. In a study by Huang and colleagues [86], a range of organic materials incorporating halogen atoms, along with oxygen and nitrogen atoms integrated into the molecular structure, were designed and synthesized to enhance n–π* transitions and facilitate the intersystem crossing (ISC) process. By modifying the iodine atom at various positions, three isomers (*ortho*-ITC, *meta*-ITC, and *para*-ITC) of 9,9′-(6-iodophenoxy-1,3,5-triazine-2,4-diyl)bis(9*H*-carbazole) were synthesized. It was observed that these phosphors exhibited strong luminescence under X-ray excitation (Figure 15.9). Subsequently, they reported a low-dose X-ray PDT approach based on pure organic phosphor nanoscintillators [87]. These nanoscintillators were capable of generating significant amounts of singlet oxygen upon X-ray exposure, thereby enhancing the efficacy of X-PDT. Notably, under low-dose X-ray irradiation (0.4 Gy), the nanoscintillator demonstrated good therapeutic effects on deep-seated tumors in vivo, with minimal adverse effects on healthy organs. Tang and coworkers [88] reported the first pure organic AIE nanoscintillator (TBDCRNPs) capable of producing $^1O_2$ and free radicals ($HO^{\bullet-}$) upon direct X-ray irradiation to enable effective X-PDT. Using the distinctive AIE group triphenylamine (TPA) as the electron donor, a pure organic AIE nanoscintillator (TBDCR NPs) exhibited aggregation-induced emission, formation of triplet excitons, and generation of ROS.

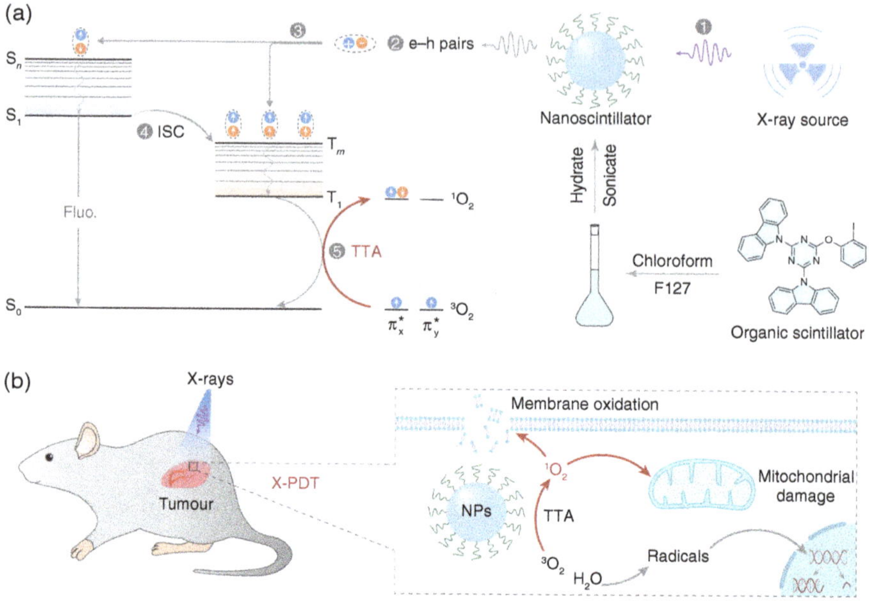

Figure 15.9 (a) Process to prepare organic nanoparticles in aqueous solution and subsequent $^1O_2$ generation under X-ray irradiation. (b) Illustration of the mechanism in the treatment of deep-seated tumors in vivo. Source: Reproduced with permission from Wang et al. [86]/Springer Nature.

TBDCR was further encapsulated into lipid-PEG NPs with a size of about 3.8 nm, and TBDCR NPs produced a large amount of $^1O_2$ and more toxic $HO^{\bullet-}$ upon both light excitation and direct X-ray exposure, without inorganic converters. Due to the PDT induced by $^1O_2$ and $HO^{\bullet-}$ and the deeper tissue penetration capability of X-rays, TBDCR NPs exhibited excellent tumor eradication effects on both HeLa cancer cells and radioresistant HelaR tumors in vitro and in vivo.

15.3.2.2 Nonmetallic Inorganic Materials

In the realm of therapeutic applications, nonmetallic-based nanomaterials, such as black phosphorus and selenium nanomaterials, have garnered significant interest owing to their distinct optical characteristics, anticancer potential, and exceptional biocompatibility [89–91]. A study by Chen and coworkers detailed the fabrication of poly(lactic-*co*-glycolic acid) coated black phosphorus quantum dots (PLGA-SS-D@BPQDs), which can serve as an effective radiosensitizer for precise radiotherapy of tumors [90]. Upon exposure to X-ray radiation (4 Gy), these nanoparticles effectively generated cytotoxic $^1O_2$ to induce cell apoptosis. Furthermore, the impact of X-ray irradiation on A375 tumor-bearing mice treated with PLGA-SS-D@BPQDs was assessed, revealing a notable reduction in tumor size (Figure 15.10). In a separate study, Yu and co-workers [89] introduced PEG-modified selenium nanoparticles (PSNP) as radiosensitizers for A549 human lung cancer cell radiotherapy. Upon X-ray exposure, PSNP instigated ROS production through secondary electron generation, leading to the demise of A549 lung cancer cells.

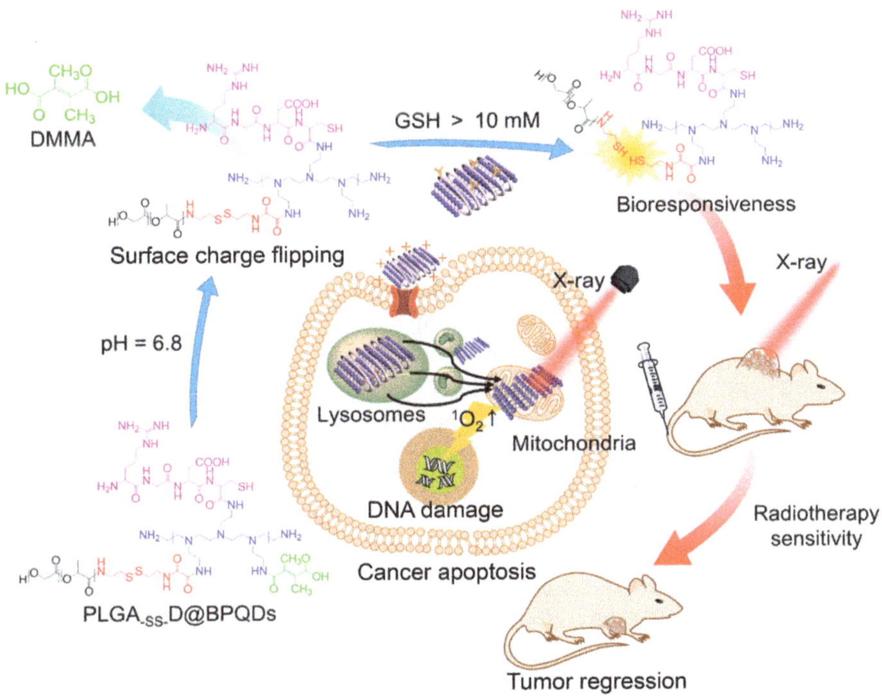

Figure 15.10 Rational design and application of PLGA-SS-D@BPQDs to tumor radiotherapy. Source: Reproduced with permission from Chan et al. [90]/American Chemical Society.

Subsequent investigations revealed an augmentation in Caspase-3 activity within both the PSNP treatment and the X-ray irradiation group, suggesting that apoptosis serves as the primary mechanism of cellular demise.

Recent advancements have showcased the potential of tellurium-based nanomaterials, encompassing nanoneedles, quantum dots, nanorods, and nanoplates, for therapeutic purposes due to their unique optical attributes and minimal toxicity [92–94]. Despite this, the utilization of tellurium-based nanomaterials as radiosensitizers or imaging contrast agents remains unexplored. It is hypothesized that tellurium-based nanomaterials could serve as X-ray-sensitive therapeutic agents, presenting a promising avenue for future research and development.

15.4 X-Ray-Activated Therapy

15.4.1 Type I X-Ray-Excited PDT

The classification of PDT includes type I and type II, which are based on the different photosensitizers and substrates required to generate ROS [95, 96], type II PDT relies on oxygen to produce ROS, while type I PDT utilizes water instead of oxygen as a substrate. Despite this difference, type I PDT exhibits significant tumoricidal effects on hypoxic tumors, gaining popularity in X-PDT applications. A myriad of type I PDT PSs have been investigated for cancer therapy, ranging from conventional metal oxides like TiO_2 and ZnO to novel emerging nanomaterials such as gC_3N_4 and MOFs (DBBC-UiO). It should be pointed out that X-ray-stimulated type I PDT is more commonly referred to as RDT, as mentioned in the introduction section. Here, we will not overly emphasize the subtle differences in definitions.

In a study by Shi and coworkers in 2015, the generation of type I ROS in X-PDT was explored using core–shell $LiYF_4:Ce^{3+}@SiO_2@ZnO$ nanoparticles (SZNPs) [33]. The diameter of the synthesized SZNP is 33.8 nm, with a uniform size distribution. Its surface is coated with two layers of SiO_2 (3.5 nm) and ZnO (3.6 nm). The scintillator $LiYF_4:Ce^{3+}$ served as a mediator between X-rays and the inorganic PS ZnO. Upon X-ray exposure, Ce^{3+} doped $LiYF_4$ emitted UV light, activating ZnO to arise electron-hole pairs, leading to the production of ·OH. This study proposed a therapeutic modality akin to type I PDT for tumors. Compared with traditional PDT, X-ray-triggered PDT in conjunction with RT offers enhanced tissue penetration and reduced risk of high radiation exposure. Bu and coworkers synthesized the $LiLuF_4:Ce^{3+}@SiO_2@Ag_3PO_4@pt\text{-}(IV)$ nanosystem using a layer-by-layer assembly method, achieving high-efficiency X-PDT [36]. The sequential synthesis involved $LiLuF_4:Ce^{3+}$ nanoparticle preparation via thermal decomposition, SiO_2 coating, amino group attachment, and subsequent Ag_3PO_4 and Pt(IV) conjugation. XRD spectra confirmed the presence of $LiLuF_4$ and Ag_3PO_4 (Figure 15.11). Under

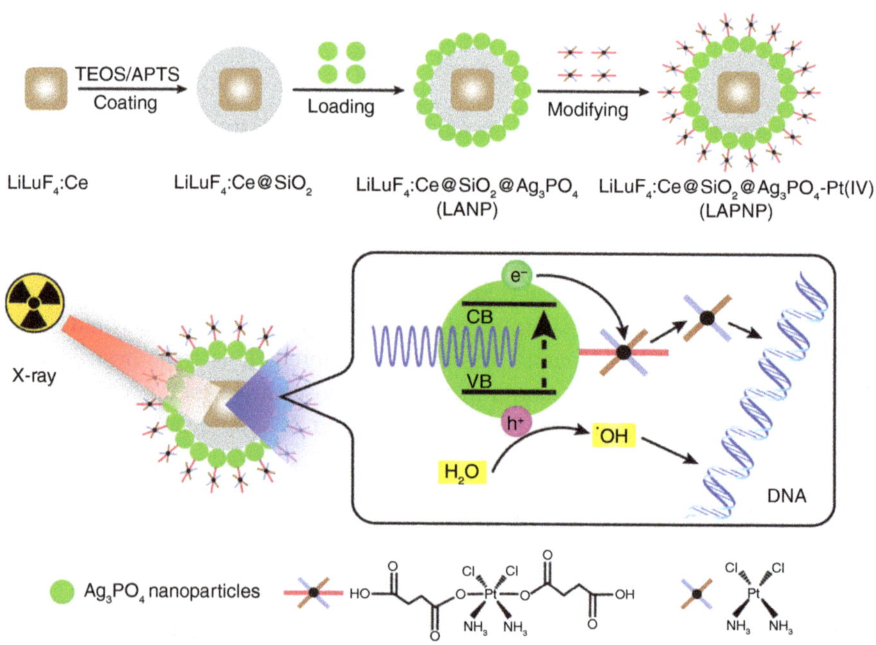

Figure 15.11 Diagram illustrating the preparation and mechanisms of X-PDT using $LiLuF_4:Ce@SiO_2@Ag_3PO_4@Pt(IV)$ Nanoparticles (LAPNP). Source: Reproduced with permission from Chan et al. [90]/American Chemical Society.

80 kV X-ray irradiation, the emission bands of $LiLuF_4:Ce^{3+}$ at 305 and 325 nm, corresponding to the transitions from 5d to 4f of the doped Ce^{3+}, triggered electron-hole pair generation in Ag_3PO_4 upon light absorption, leading to ·OH production for tumor cell destruction. Notably, the use of Pt(IV) further increased ·OH production, as it greatly facilitated hole separation upon X-ray exposure. Through these strategic interventions, the researchers achieved remarkable X-PDT performance, demonstrating ROS-induced DNA damage in hypoxic cancer cells comparable to normoxic conditions, suggesting that the nanosystem could be used for X-ray-excited type I PDT in hypoxic tumors. The efficient production of ROS under low-dose X-ray irradiation is a challenge for X-PDT. In this work, the radiation required for type I PDT is reduced to 4 Gy, which is sufficient to achieve effective X-PDT in vivo.

Antosh and coworkers developed a Cu–Cy nanosystem that also employed typical X-ray-excited type I PDT in mice tumors, showing significant tumor size reduction. The conjugation of nanoparticles with pH-low insertion peptide enhances the therapeutic effect, demonstrating the importance of targeted delivery in cancer treatment. The study highlights the promising role of Cu-Cy nanoparticles in overcoming the challenges of traditional PDT and offers a new approach to treating mammalian cancer [97]. Notably, Chen and coworkers described in vivo X-PDT based on R-AIE-Au using an extremely low dose of 1Gy, suggesting greater impetus for X-ray-excited type I PDT compared to the accompanying type II PDT. They also evaluated the uptake efficiency, cytotoxicity, and anticancer efficacy of R-AIE-Au nanosensitizers in various cancer cell lines, including U87MG, HepG2, and PC3. The results showed promising outcomes in terms of tumor regression, minimal toxicity, and efficient cancer cell killing [98]. Overall, these studies, whether applied independently or in conjunction with conventional type II PDT, have advanced the progression of low-dose X-PDT based on type I mechanisms.

15.4.2 Type II X-Ray-Excited PDT

PDT involves the utilization of PSs that produce ROS upon activation by light. Due to its less invasive nature and higher efficiency compared to X-rays, PDT has garnered considerable attention in cancer treatment. When photons are absorbed, the PS transitions from the ground state to the excited state, transferring energy to oxygen molecules ($^3O_2$), producing cytotoxic $^1O_2$ or other ROS. However, the efficacy of photosensitizers is hindered by light scattering and reflection in tissues, which limits their effectiveness in treating deep tumors. To address this challenge, a nanosystem responsive to NIR light was developed by combining upconversion nanoparticles (UCNPs) with photosensitizers, but it is still difficult to achieve sufficient penetration. To overcome this obstacle, X-rays have been used to initiate PDT due to their unparalleled tissue penetration ability, providing a solution to this critical limitation.

Zhang et al. designed β-$NaLuF_4:Tb^{3+}$ spherical nanoparticles (NPs) with ultra-strong green XEOL, possessing smaller nonradiative relaxation probability and enhanced X-ray absorption mass coefficient. These NPs were well-tailored to match the absorption spectrum of the photosensitizer RB, serving as an efficient energy converter for X-PDT [99]. By covalently linking the NPs with RB, an impressive FRET efficiency of up to 94.29% was achieved, resulting in a high singlet oxygen yield. The in vivo X-PDT efficacy was evaluated with nude mice bearing $HepG_2$ tumor xenografts. The synthesized NPs–RB nanocomposite demonstrated

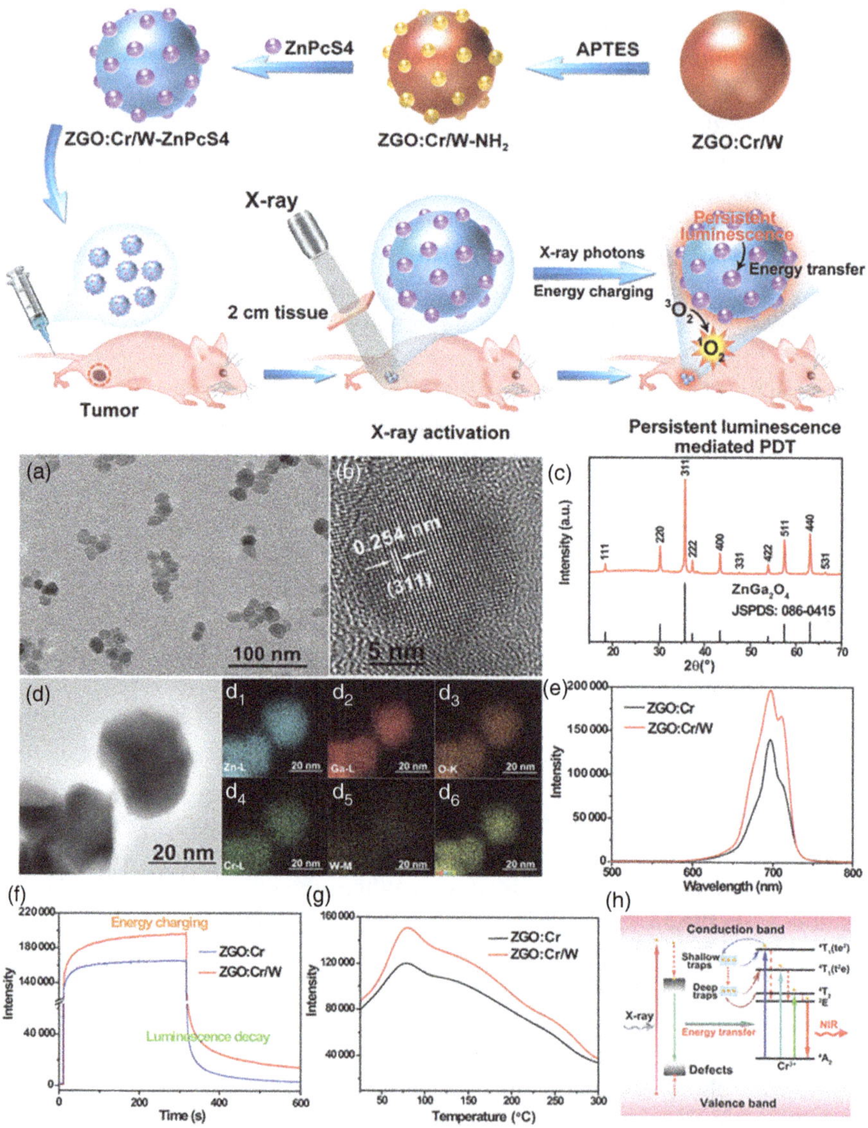

Figure 15.12 Schematic illustration of X-ray-activated PLNP-mediated PDT nanoplatform and ZGO:Cr/W PLNP characteristics: (a) TEM image, (b) High-resolution TEM image, (c) X-ray diffraction analysis, (d) STEM-EDS elemental mapping, (e) Persistent luminescence spectra, (f) energy charging and luminescence decay process, (g) thermoluminescence glow curve, and (h) schematic of persistent luminescence mechanism in ZGO:Cr/W PLNPs.
Source: Reproduced with permission from Song et al. [100]/John Wiley & Sons.

good biocompatibility and exhibited a significant antitumor efficiency of up to $80 \pm 12.3\%$ at a total X-ray dose of only 0.19 Gy, proving the viability of low-dose X-PDT in vivo for the first time. This study offers a promising platform for X-PDT of deep-seated tumors.

While X-ray-triggered PDT can penetrate deep tissues, the elevated radiation dose can also harm healthy tissues. Yang and co-workers prepared $ZnGa_2O_4$:Cr/W (ZGO:Cr/W) persistent luminescent nanoparticles (PLNPs) with a size of approximately 70 nm (Figure 15.12) [100]. Subsequently, they developed an X-ray-triggered nanosystem by combining ZGO:Cr/W PLNPs with the photosensitizer Zn(II) phthalocyanine tetrasulfonate ($C_{32}H_{16}N_8O_{12}$S4Zn, $ZnPcS_4$). The efficacy of ZGO:Cr/W-$ZnPcS_4$ in producing $^1O_2$ was appraised using the fluorescence probe 2,7-dichlorofluorescein (DCFH). After X-ray irradiation ceased, the fluorescence intensity continued to increase, producing afterglow luminescence. Furthermore, significant anticancer effects were achieved both in vitro and in vivo, even with a 0.18 Gy radiation.

15.4.3 Combined Type I and Type II X-Ray-Excited PDT

The combination of type I and type II in X-ray-mediated PDT offers several advantages. Type I PDT involves the generation of free radicals like hydroxyl radicals ($HO^{\bullet -}$) and superoxide radicals ($O_2^{\bullet -}$) through an electron transfer mechanism, while type II PDT generates $^1O_2$ through energy exchange with surrounding oxygens. By combining both types of PDT, X-ray-mediated PDT can achieve a more comprehensive and effective treatment approach. Type I PDT, which is less oxygen-dependent, is particularly beneficial for hypoxia-tolerant tumor treatment. Additionally, the use of both type I and type II ROS in X-PDT allows for a synergistic effect, enhancing the overall therapeutic efficacy. Tang and coworkers [88] developed a novel approach using an organic AIE nanocrystal for X-ray-induced type I and type II PDT (Figure 15.13). Unlike conventional X-PDT methods that rely on inorganic scintillators, the AIE nanoscintillator, TBDCR NPs, can directly generate both type I and type II ROS under X-ray irradiation. The introduction of heteroatoms (oxygen, nitrogen, and sulfur atoms) enhances X-ray harvesting and ROS generation, with TBDCR demonstrating aggregation-enhanced ROS production, particularly in the generation of oxygen-independent hydroxyl radicals ($HO^{\bullet -}$, type I). The TBDCR NPs are encapsulated in a PEG crystalline shell, providing a rigid intraparticle microenvironment that further boosts ROS generation. Notably, TBDCR NPs exhibit bright near-infrared fluorescence and significant $^1O_2$ and $HO^{\bullet -}$ production. Experimental results from both in vitro and in vivo studies have confirmed the remarkable antitumor efficacy of TBDCR NPs against HeLa tumors and radiation-resistant HeLaR tumors. The innovative use of a pure organic AIE nanoscintillator for X-PDT represents a significant advancement in cancer therapy, offering a promising alternative to traditional inorganic scintillators. This research opens new possibilities for the development of organic scintillators with superior X-ray harvesting and ROS generation capabilities, potentially revolutionizing the clinical treatment of solid tumors through X-PDT.

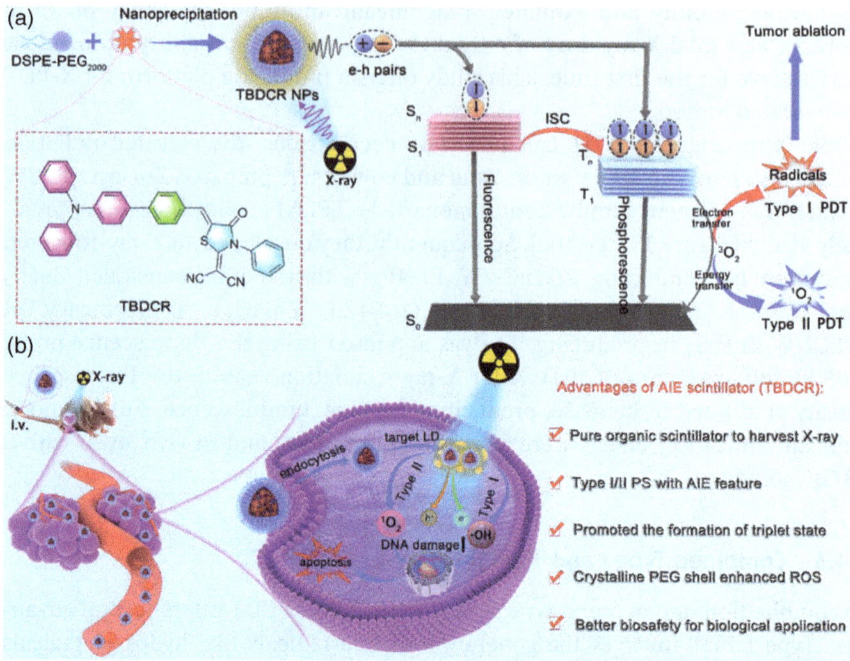

Figure 15.13 Schematic illustration of TBDCR NPs structure and preparation for X-PDT. (a) Preparation of TBDCR NPs followed by $^1O_2$ and radical ROS generation. (b) Application of TBDCR NPs on imaging-guided X-PDT therapy. Source: Yu et al. [88]/ John Wiley & Sons / CC BY 4.0.

15.4.4 X-Ray-Induced Generation of RNS for Dynamic Therapy

Reactive nitrogen species (RNS) are a category of nitrogen-containing reactive molecules crucial for redox reactions, particularly in the context of dynamic tumor therapy. Nitric oxide (NO) and peroxynitrite (ONOO–) are key components of RNS, with NO being a primary player generated in the body through the oxidation of L-arginine by nitric oxide synthase. It serves as a signaling molecule in various physiological processes such as neurotransmission, vasodilation, and macrophage immunity. ONOO–, on the other hand, acts as a secondary RNS formed through the interaction of NO and $O_2^{-\bullet}$, regulating signal transduction pathways and disrupting cellular redox balance. Its effects include protein structure alteration via tyrosine nitration, impairment of mitochondrial function, promotion of lipid peroxidation, and induction of DNA damage through modifications or breaks in the DNA strands.

NO can enhance oxygen levels in the tumor microenvironment by promoting vasodilation. It also functions as a radiosensitizer by inducing DNA damage. Recent research has focused on stimuli-responsive RNS donors and their potential applications in cancer therapy. Shi and colleagues engineered an X-ray-responsive nanosystem for on-demand NO generation, achieving depth-independent hypoxic radiosensitization [101]. X-ray exposure triggered the cleavage of the S—N bond

(150 kJ mol^{-1}) in S-nitrosothiol-SNO groups to release NO. Due to the inherent instability of the S—N bond, NO release could occur even in the absence of X-ray stimulation.

Hao and coworkers developed a soft X-ray-induced scintillator $NaYF_4$: Gd/Tb nanorod, which was further integrated with NO donor (RBS) to prepare a nanoplatform for NO generation [102]. Based on the ET process between $NaYF_4$:Gd/Tb nanorods and RBS, NO release triggered by ultra-low-dose soft X-ray (0.85 mGy) can be achieved in tissues as deep as 3 cm, surpassing the limitations of traditional UV/visible and NIR light sources. The modulation of X-ray tube voltage enabled the controlled release of NO, which is crucial for concentration-dependent NO gas therapy. Furthermore, the $NaYF_4$:Gd/Tb–RBS agents activated by X-rays demonstrated effective inhibition of tumor growth, suggesting a novel soft X-ray-activated gas emitter for in-depth tissue gas-sensitized tumor therapy. Dou et al. synthesized $NaGdF_4$: Eu^{3+} scintillating nanocrystals (NSC) by hydrothermal method, followed by $mSiO_2$ coating and loading of the NO donor s-nitrosoisoproterenol (SNO) and the PS indocyanine green (ICG) to obtain NSC@$mSiO_2$-SNO/ICG nanoparticles (Figure 15.14) [49]. The emission peaks of NSC@$mSiO_2$-SNO nanoparticles are located at 591, 616, 698, and 780 nm, corresponding to the different transitions of Eu^{3+} ($^5D_0 \rightarrow {}^7F_1$, $^5D_0 \rightarrow {}^7F_2$, $^5D_0 \rightarrow {}^7F_4$, and $^5D_0 \rightarrow {}^7F_6$). Specifically, the emission peak at 780 nm matches the absorption peak of ICG at 784 nm. When exposed to X-ray irradiation, these nanoparticles demonstrate the optimal generation of $^{\bullet}OH$ and $^1O_2$, thus can induce X-PDT. Additionally, they release significant levels of NO, enhancing the efficacy of X-PDT and improving hypoxia through NO-mediated vasodilation.

Hao and coworkers developed a novel soft X-ray-activated persistent luminescent nanotransducer by integrating $ZnGa_2O_4$:Mn (ZGO:Mn) nanoparticles with photosensitive NO donor (RBS) for precise and sustained NO release in deep tissue anticancer therapy [103]. Leveraging the advantage of ultra-low dose (as low as 0.9 mGy) soft X-ray-activated persistent luminescence of small ZGO:Mn, continuous NO release was achieved for approximately 40 minutes postirradiation cessation. Additionally, the green persistent luminescence could be regenerated through in situ soft X-ray exposure, enabling repeatable NO release in deep tissues (up to 24 mm). This NO-releasing platform exhibited effective anticancer effects both in vitro and in vivo, offering a novel strategy for depth-independent gas-sensitized therapeutic applications.

Gu and coworkers introduced an X-ray responsive nanosystem composed of TPGS-modified nanoscintillator particles (SCNPs) and a UV-responsive NO donor RBS for controllable ONOO– release (Figure 15.15) [104]. These SCNPs could convert X-rays into UV light, stimulating RBS to simultaneously produce NO and $O_2^{-\bullet}$, leading to excess ONOO- generation to enhance RT efficacy. RBS-T-SCNPs induced DNA damage directly through ONOO– and inhibited DNA repair, overcoming radiation resistance induced by hypoxia. Treatment of A549 human lung tumor-bearing mice with RBS-T-SCNPs showed significant tumor volume reduction after X-ray irradiation.

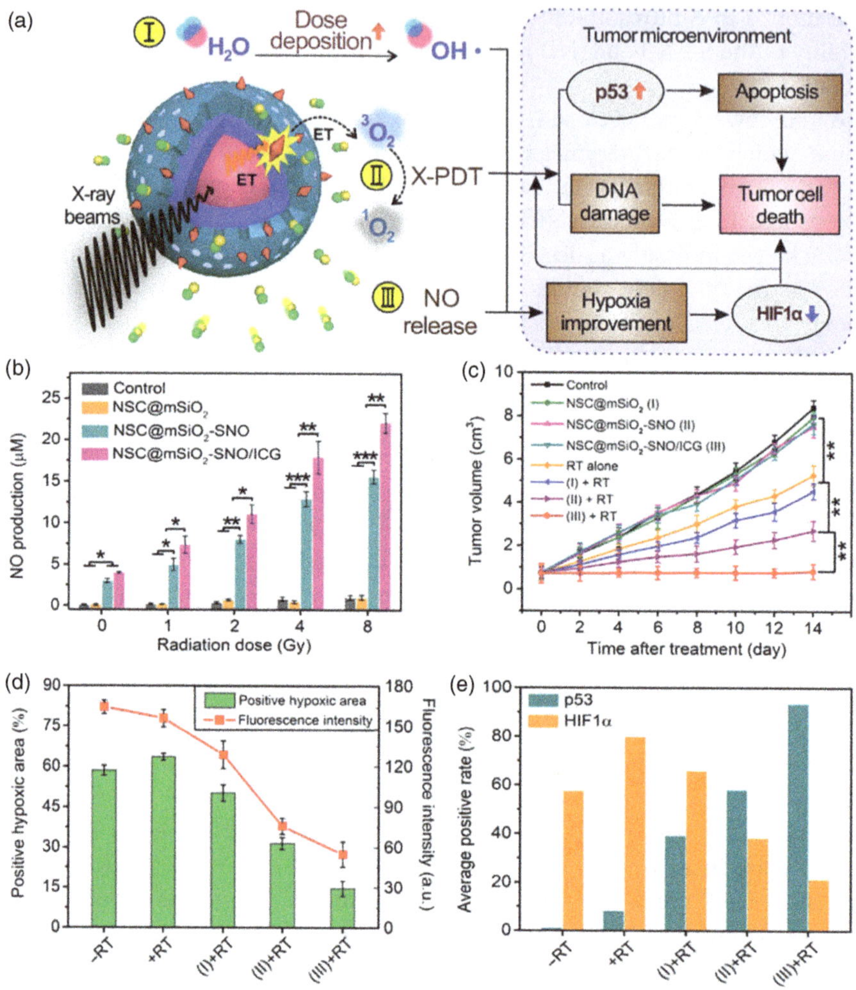

Figure 15.14 (a) Mechanism of NaGdF$_4$:Eu^{3+}@SiO$_2$-S-nitrosothiol (SNO)/indocyanine green (ICG) (NSC@mSiO$_2$-SNO/ICG) nanosystems for enhanced cancer treatment in hypoxic tumors. (b) NO production levels post varying X-ray doses. (c) Tumor growth inhibition in 4T1 tumor-bearing mice under different treatments. (d) Semiquantitative analysis of hypoxic regions and fluorescence intensities in ex vivo immunofluorescence images across treatments. (e) Average rates of p53 and HIF1α expression in immunohistochemical images. Source: He et al. [13]; © 2022 American Chemical Society. Source: Dou et al. [49]; ©2018 Ivyspring International Publisher. Terms of use.

Similarly, Krayschi and co-workers prepared Au-Fe_3O_4 nanoheterodimers functionalized with nitrosonium tetrafluoroborate to generate NO on the Fe_3O_4 surface and $O_2^{-\cdot}$ on the AuNPs surface upon X-ray exposure, subsequently generating ONOO– [105]. The controlled generation of RNS via X-ray activation presents a promising therapeutic approach, particularly in combating hypoxia-induced radiation resistance due to the high reactivity and cytotoxicity of RNS.

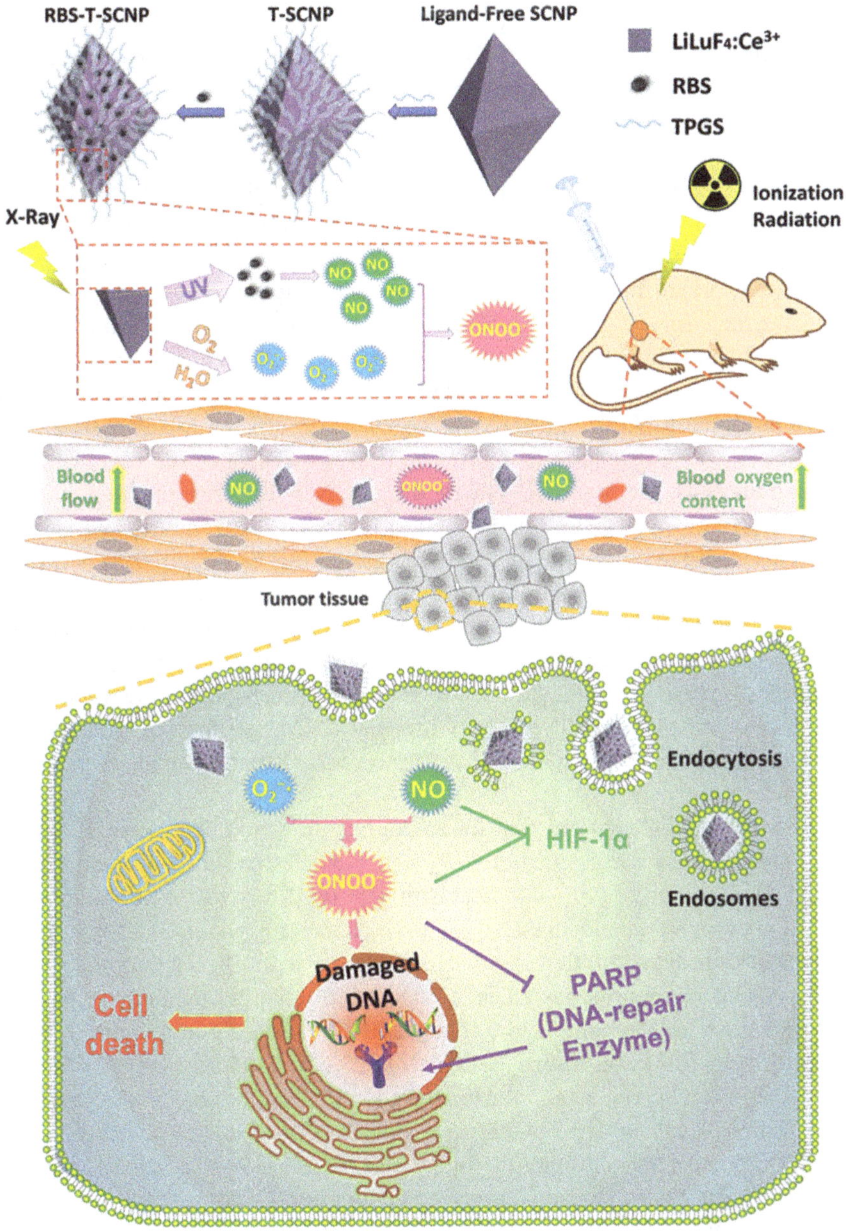

Figure 15.15 Schematic illustration of the synthetic route to RBS-T-SCNPs and the X-ray-controlled $ONOO^-$ generation for on-demand cancer radiotherapy. Source: Reproduced with permission from Du et al. [104]/John Wiley & Sons.

15.4.5 Synergistic Therapy

Monotherapy often fails to fully inhibit tumor growth completely, prompting the need for combining different treatment modalities to improve the three- to five-year survival rates of patients. The field of nanomaterials has seen significant progress, leading to the development of numerous nanomedicines that serve as the foundation for various nano-therapeutic platforms designed to enhance the effectiveness of tumor treatments. X-PDT can be synergized with chemotherapy, photothermal therapy (PTT), immunotherapy, and bacteriotherapy. This integration forms an enhanced X-PDT precision treatment platform for tumors, holding promising prospects in the realm of cancer therapy.

Su and coworkers introduced a multifunctional supramolecular cage named MOC-43, which serves as a multifunctional platform for X-PDT, radiotherapy, and chemotherapy, integrating both in vitro and in vivo diagnostic imaging with synergistic anticancer treatment [106]. This cage, incorporating multiple osmium atoms in its structure, functions as a dual X-PDT agent, efficiently photosensitizing the generation of $^1O_2$ under NIR light. The intricate design of the cage, with its cavities and portals, endows MOC-43 with drug-delivery carrier properties.

It capitalizes on host–guest interactions to transport, concentrate, and release poorly water-soluble coumarin molecules, highlighting its potential as a carrier for therapeutic agents. As a multifunctional therapeutic nanoplatform, MOC-43 achieves a comprehensive approach to breast cancer treatment, combining the forces of radiotherapy, chemotherapy, and X-PDT. It facilitates cancer cell tracking and tumor-targeted NIR imaging, enhancing the precision of the treatment. Furthermore, the platform leverages a host-guest strategy for the transport of anticancer drugs, streamlining the delivery process and maximizing the therapeutic impact.

Li and colleagues illustrated that by co-doping Gd^{3+} and Tb^{3+} ions, the limitations in hole-capture efficiency of CeF_3 nanoparticles could be overcome. The resulting Tb^{3+} co-doped CeF_3 nanoplatform was used for synergistic radio-PDT, complemented by metabolomics analysis [43]. Nanoplatforms measuring approximately 80 nm in diameter were synthesized using a hydrothermal approach and then enveloped in mesoporous silica. Notably, the Ce^{3+} ion exhibited a greater absorption cross section compared to Gd^{3+} and Tb^{3+}, leading to enhanced green XEOL emission from the co-doped CeF_3:Gd_{3+}, Tb^{3+} nanoparticles. Within the CeF_3 matrix, Gd^{3+} served as an intermediary for ET, facilitating the transmission of energy from Ce^{3+} to Tb^{3+} owing to the intermediate energy levels of Gd^{3+} between Ce^{3+} and Tb^{3+}. Through the incorporation of RB as a PS, the researchers employed singlet oxygen sensor green (SOSG) to quantify $^1O_2$ generation in this X-ray-induced nanoplatform. Notably, the fluorescence intensity of SOSG in the CeF_3:Gd^{3+}, Tb^{3+}@SiO_2@RB (CGTS-RB) group exhibited a significant rise with increasing X-ray irradiation doses, whereas no fluorescence alterations were observed in the other three groups, demonstrating the excellent ability of CGTS-RB to produce ample $^1O_2$ for effective X-PDT. In vivo therapy of 4T1 tumors affirmed the effectiveness of combined X-PDT and RT treatment. Remarkably, the therapeutic effectiveness of the CGTS-RB group at a modest dose of 3 Gy was comparable to that of the RT group under double the dose of radiation. Metabolomic analyses revealed notable alterations in metabolic products following both high (6 Gy) and low (3 Gy) radiation doses, particularly in the urea cycle. After the combined X-PDT and RT treatment with a single 3 Gy X-ray dose, amino acids involved in protein and DNA synthesis, such as alanine and aspartic acid, were downregulated, leading to tumor regression.

Chen and coworkers proposed a universal rapid synthesis method aimed at the large-scale production of monodisperse and uniform zinc-manganese-doped silicate (ZSM) nanoscintillators for the X-PDT treatment of tumors at low doses (Figure 15.16) [107]. Using a silica template method, biocompatible ZSM nanoscintillators measuring 101.2 ± 2.7 nm in diameter were synthesized. These nanoscintillators were further functionalized with biocompatible polyether compounds (PEG), cyclic arginine-glycine-aspartic acid (RGD) peptides for targeting, and the PS RB. X-ray diffraction analysis confirmed the rhombic α-Zn_2SiO_4 phase of the synthesized ZSM nanoparticles, validating the success of the approach in fabrication of crystalline scintillator materials. Upon X-ray exposure, ZSM nanoparticles exhibited intense XEOL peaking at 570 nm, closely overlapping with the absorption spectrum of RB. The incorporation of RB into ZSM led to a significant reduction in XEOL intensity, indicating efficient ET between ZSM and RB. To address the limitations of PDT and RT, the researchers tailored the doping of metals and endowed the nanoparticles with active targeting capabilities. Consequently, the combined RT and X-PDT treatment induced high-frequency DNA strand breaks,

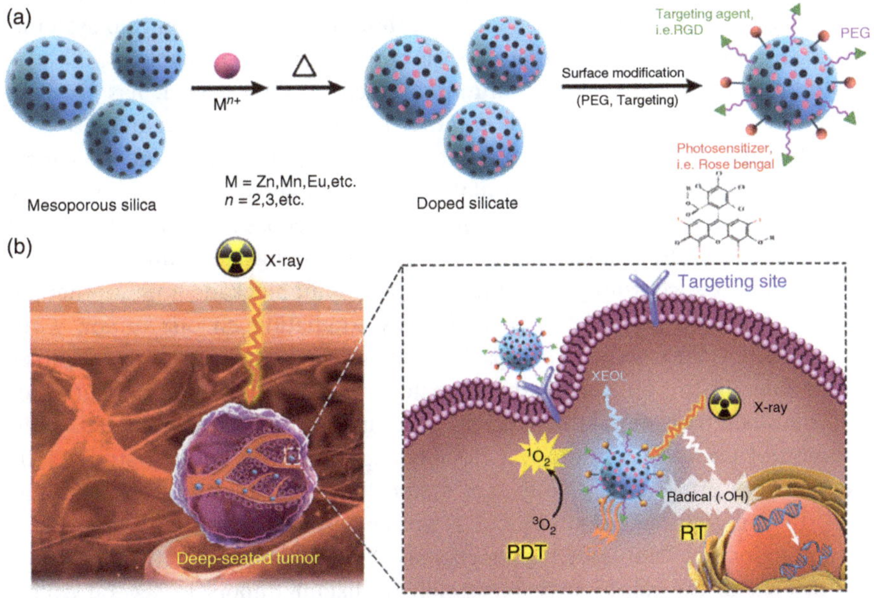

Figure 15.16 Schematic illustrations of the process of the preparation of RGD-ZSM-RB nanosensitizers (a) and of the mechanisms of RGD-ZSM-RB-mediated X-PDT (b). Source: Reproduced with permission from Sun et al. [107]/John Wiley & Sons.

enhancing DNA damage, and notably, X-ray irradiation resulted in substantial tumor size reduction, even when covered with a layer of pork.

Liu et al. introduced a novel heterojunction architecture comprising Lu-based upconversion phosphor nanocrystals ($NaLuF_4$:Yb, Tm) and Bi-based mesoporous materials ($Na_{0.2}Bi_{0.8}O_{0.35}F_{1.91}$:Yb) for loading FePc [108]. To enable dual-modal (upconversion luminescence [UCL]/CT) tumor imaging and trimodal (PTT/PDT/RT) tumor therapy, they utilized amphiphilic PEG functionalized with folic acid to encapsulate the compound, forming a nanocomposite named UCNP@NBOF-FePc-PFA. The unique structure of UCNP@NBOF-FePc-PFA resembles a landmine with a size of 50 nm. The optimal loading capacity for FePc was determined to be approximately 47%. Subsequent PEG coating enhanced water dispersibility. Evaluation of photothermal conversion efficiency yielded a value of 42.5% for both UCNP@NBOF and UCNP@NBOF-FePc-PFA. ROS generation studies showed that UCNP@NBOF-FePc-PFA produced ROS upon exposure to X-ray light and a 730 nm laser. In vitro experiments revealed potent phototoxic effects when UCNP@NBOF-FePc-PFA was subjected to combined irradiation with X-ray light and a 730 nm laser. UCL and CT imaging data suggested an optimal irradiation time of six hours postinjection. In vivo synergistic therapy studies indicated that UCNP@NBOF-FePc-PFA exhibited significant tumor suppression effects compared to single-treatment controls under irradiation with X-ray light and 730 nm laser. These findings suggest that combined trimodal therapy is a unique method for tumor therapy, with a therapeutic platform with a heterojunction structure for manufacturing novel functional nanoagents.

In contrast to most previous studies, which concentrated solely on simplistic experiments with localized tumor models, a thorough investigation into systemic immune responses to overcome metastasis is both crucial and valuable.

Immunotherapy, harnessing the immune system to fight cancer, has emerged as a proven and potent treatment approach. The synergistic combination of immunotherapy and X-PDT is poised to be a clinically significant anticancer strategy. Lin and coworkers encapsulated the small molecule inhibitor of the immune regulatory enzyme indoleamine 2,3-dioxygenase (IDO), IDOi, into DBP-Hf nMOFs [109]. By utilizing IDOi@DBP-Hf alongside a minimal 1.5 Gy X-PDT dose, they successfully implemented checkpoint blockade immunotherapy. Following treatment administration and X-ray irradiation on one tumor, while leaving another untreated for comparison, significant outcomes were observed 12 hours later in the CT26 immunocompetent mouse model. Tumors subjected to the combined treatment either regressed completely or were eliminated, while the growth of untreated tumors was notably suppressed. In another study [110], the researchers introduced a conjugate system based on ultrathin Hf-MOL and anti-PD-L1 (programmed death-ligand 1) antibodies, demonstrating the superior ROS production capability of the nanocrystals. This enhanced systemic antitumor immunity led to the rejection or regression of distant tumors in several cancer types, suggesting the potential of X-ray-based local therapy as a promising approach for managing primary and metastatic tumors in clinical settings with minimal radiation exposure.

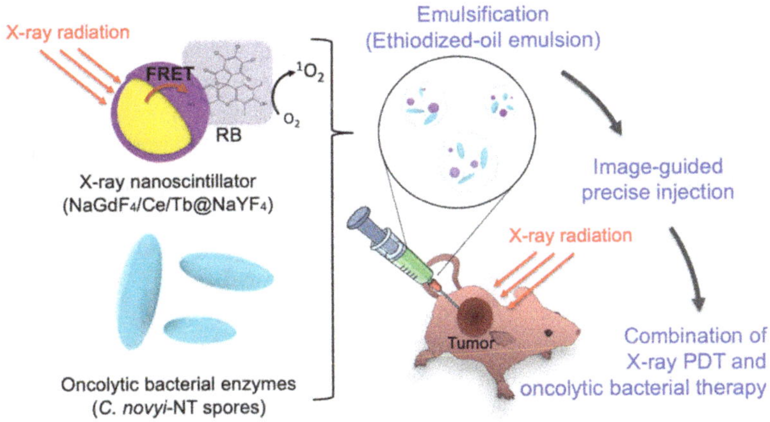

Figure 15.17 Schematic illustration of nano-bio therapeutic emulsion formulated with nanoscintillators, anaerobic *C. novyi*-NT spores, and ethiodized oil for the synergistic combination of X-ray photodynamic and anaerobic bacteriolytic therapy. Source: Reproduced with permission from Park et al. [42]/John Wiley & Sons.

Bacterial therapy has emerged as a novel treatment modality in cancer therapy, especially when combined with X-ray-based cancer treatments [111]. Facultative anaerobic bacteria can effectively target or deeply penetrate endogenous or radiation-induced hypoxic zones within solid tumors. Through the secretion of bacterial cytotoxins like cytolysin A, cell cycle arrest in the radiation-sensitive G2/M phase can be induced, leading to apoptosis and tumor breakdown. Kim and colleagues showcased the synergistic advantages of bacterial therapy in X-PDT for hypoxic tumors by utilizing novel anaerobic *Clostridium novyi*-NT spores and a therapeutic emulsion containing $NaGdF_4$:Tb, Ce@$NaGdF_4$ core/shell-structured nanoscintillators (Figure 15.17) [42]. Under X-ray irradiation, $NaGdF_4$:Tb,Ce@$NaGdF_4$ could emit luminescence to activate conjugated RB for ROS production, while the delivered *Clostridium novyi*-NT spores could target hypoxic tumor regions for further cancer cell eradication. In vivo tests combining X-PDT and bacterial therapy in a mouse tumor model revealed that at a 6 Gy radiation dose, significant cancer cell destruction occurred in both the normoxic periphery and the hypoxic core, underscoring the potentiation of X-PDT's antitumor efficacy in overcoming hypoxia through bacterial presence.

15.5 Conclusions and Perspectives

In the past 17 years, significant progress has been made in the field of X-ray-induced PDT. However, this method still faces several critical challenges. These include the simple coupling of nanoscintillators and photosensitizers, the hypoxic conditions within the tumor microenvironment (TME), and the intricate interplay between radiation and related cellular pathways or immune system regulation, among others. Effective strategies aim to enhance the efficacy of X-ray activation materials through methods as follows: (i) Designing and synthesizing innovative materials with high X-ray attenuation coefficients to bolster X-ray absorption capabilities. (ii) Improving intrinsic luminescence efficiency by choosing appropriate matrices, controlling size and morphology, altering the type and concentration of doping luminescent activators, and optimizing the ET process. (iii) Enhancing X-ray photocatalytic efficiency by coupling with noble metals (like gold) or reduced graphene oxide, promoting efficient separation of X-ray-induced electron-hole triplets.

Moreover, the uniformity of nanoscintillators and the relatively elevated radiation doses utilized in most published studies remain significant challenges. For example, the majority of nanoscintillators employed in X-PDT are predominantly comprised of inorganic nanoparticles, raising concerns regarding the biotoxicity of such materials. To overcome this impasse, it is imperative to develop alternative types of nanoscintillators and further investigate existing inorganic species, such as metal complexes, supramolecular structures, and cluster-molecule-based compounds. Concurrently, recent advancements in low-dose X-PDT represent a significant stride toward obtaining clinical approval for reduced radiation levels, with its potential applications exceeding those of high-dose X-PDT, traditional PDT, and radiotherapy. The ongoing progress in this field continues to drive related research efforts, leading us to believe that X-PDT will evolve from a mere proof-of-concept to a genuine clinical technology.

References

1 Siegel, R.L., Miller, K.D., Wagle, N.S., and Jemal, A. (2023). Cancer statistics. *Ca-a Cancer Journal for Clinicians* 73: 17–48.

2 Yanovsky, R.L., Bartenstein, D.W., Rogers, G.S. et al. (2019). Photodynamic therapy for solid tumors: a review of the literature. *Photodermatology Photoimmunology & Photomedicine* 35: 295–303.

3 Li, W.P., Yen, C.J., Wu, B.S., and Wong, T.W. (2021). Recent advances in photodynamic therapy for deep-seated tumors with the aid of nanomedicine. *Biomedicines* 9: 69.

4 Sun, B., Bte Rahmat, J.N., and Zhang, Y. (2022). Advanced techniques for performing photodynamic therapy in deep-seated tissues. *Biomaterials* 291: 121875.

5 Zhang, T., Zheng, H., Zhang, F. et al. (2023). Rare-earth scintillating nanoparticles for X-ray induced photodynamic therapy. *Journal of Luminescence* 261: 119862.

6 Niculescu, A.G. and Grumezescu, A.M. (2021). Photodynamic therapy-an up-to-date review. *Applied Sciences-Basel* 11: 3626.

7 Kwiatkowski, S., Knap, B., Przystupski, D. et al. (2018). Photodynamic therapy – mechanisms, photosensitizers and combinations. *Biomed Pharmacother* 106: 1098–1107.

8 Robertson, C.A., Evans, D.H., and Abraharnse, H. (2009). Photodynamic therapy (PDT): a short review on cellular mechanisms and cancer research applications for PDT. *Journal of Photochemistry and Photobiology* 96: 1–8.

9 Fan, W.P., Huang, P., and Chen, X.Y. (2016). Overcoming the Achilles' heel of photodynamic therapy. *Chemical Society Reviews* 45: 6488–6519.

10 Chen, C., Wu, C.S., Yu, J.M. et al. (2022). Photodynamic-based combinatorial cancer therapy strategies: tuning the properties of nanoplatform according to oncotherapy needs. *Coordination Chemistry Reviews* 461: 214495.

11 Chen, X.F., Song, J.B., Chen, X.Y., and Yang, H.H. (2019). X-ray-activated nanosystems for theranostic applications. *Chemical Society Reviews* 48: 3073–3101.

12 Chen, W. and Zhang, J. (2006). Using nanoparticles to enable simultaneous radiation and photodynamic therapies for cancer treatment. *Journal of Nanoscience and Nanotechnology* 6: 1159–1166.

13 He, L., Yu, X., and Li, W. (2022). Recent progress and trends in X-ray-induced photodynamic therapy with low radiation doses. *ACS Nano* 16: 19691–19721.

14 Stanton, I.N., Belley, M.D., and Nguyen, G. (2014). Europium- and lithium-doped yttrium oxide nanocrystals that provide a linear emissive response with X-ray radiation exposure. *Nanoscale* 6: 5284–5288.

15 Fränzel, W. and Gerlach, R. (2009). The irradiation action on human dental tissue by X-rays and electrons – a nanoindenter study. *Zeitschrift für Medizinische Physik* 19: 5–10.

16 Karnkaew, A., Chen, F., Zhan, Y.H. et al. (2016). Scintillating nanoparticles as energy mediators for enhanced photodynamic therapy. *Acs Nano* 10: 3918–3935.

17 Lee, P.A., Citrin, P.H., Eisenberger, P., and Kincaid, B.M. (1981). Extended x-ray absorption fine structure – its strengths and limitations as a structural tool. *Reviews of Modern Physics* 53: 769–806.

18 Robbins, D.J. (1980). On predicting the maximum efficiency of phosphor systems excited by ionizing radiation. *Journal of The Electrochemical Society* 127: 2694.

19 Zhou, F.G., Li, Z.Z., Lan, W. et al. (2020). Halide perovskite, a potential scintillator for X-Ray detection. *Small Methods* 4: 2000506.

20 Nikl, M. (2006). Scintillation detectors for x-rays. *Measurement Science and Technology* 17: R37–R54.

21 Ren, X.D., Hao, X.Y., Li, H.C. et al. (2018). Progress in the development of nanosensitizers for X-ray-induced photodynamic therapy. *Drug Discov Today* 23: 1791–1800.

22 Gai, S.L., Li, C.X., Yang, P.P., and Lin, J. (2014). recent progress in rare earth micro/nanocrystals: soft chemical synthesis, Luminescent Properties, and Biomedical Applications. *Chemical Reviews* 114: 2343–2389.

23 Fan, W., Tang, W., Lau, J. et al. (2019). Breaking the depth dependence by nanotechnology-enhanced X-ray-excited deep cancer theranostics. *Adv Mater* 31 (12): e1806381.

24 Wei, Z., Liu, Y.W., Li, B. et al. (2022). Rare-earth based materials: an effective toolbox for brain imaging, therapy, monitoring and neuromodulation. *Light: Science & Applications* 11: 224.

25 Chen, H., Moore, T., Qi, B. et al. (2013). Monitoring pH-triggered drug release from radioluminescent nanocapsules with X-ray excited optical luminescence. *Acs Nano* 7: 1178–1187.

26 Kirsanova, D., Polyakov, V., Butova, V. et al. (2021). The rare-earth elements doping of BaGdF(5) nanophosphors for X-ray photodynamic therapy. *Nanomaterials* 11: 3212.

27 Zhou, D., Gao, Y., Yang, Z. et al. (2023). Biomimetic upconversion nanoplatform synergizes photodynamic therapy and enhanced radiotherapy against tumor metastasis. *ACS Applied Materials & Interfaces* 15: 26431–26441.

28 Shi, S.K., Hossu, M., Hall, R., and Chen, W. (2012). Solution combustion synthesis, photoluminescence and X-ray luminescence of Eu-doped nanoceria CeO:Eu. *Journal of Materials Chemistry A* 22: 23461–23467.

29 Trojan-Piegza, J., Gierlotka, S., Zych, E., and Lojkowski, W. (2014). Spectroscopic studies of nanopowder and nanoceramics LaHfO:Pr scintillator. *Journal of the American Ceramic Society* 97: 1595–1601.

30 Espinoza, S., Juestel, T., and Haase, M. (2018). Colloidal LaPO:Gd nanocrystals: X-ray induced single line UV emission. *Nanoscale* 10: 22533–22540.

31 Ou, X.Y., Guo, T., Li, J.Y. et al. (2018). Autofluorescence-free immunoassay using X-ray scintillating nanotags. *Analytical Chemistry* 90: 6992–6997.

32 Sudheendra, L., Das, G.K., Cherry, S., and Kennedy, I.M. (2014). NaGdF:Eu nanoparticles for enhanced X-ray excited optical imaging. *Chemistry of Materials* 26: 1881–1888.

33 Zhang, C., Zhao, K.L., Bu, W.B. et al. (2015). Marriage of scintillator and semiconductor for synchronous radiotherapy and deep photodynamic therapy with diminished oxygen dependence. *Angewandte Chemie International Edition* 54: 1770–1774.

34 Orsi, D., Rimoldi, T., Pinelli, S. et al. (2018). New CeF-ZnO nanocomposites for self-lighted photodynamic therapy that block adenocarcinoma cell life cycle. *Nanomedicine-Uk* 13: 2311–2326.

35 Dinakaran, D., Sengupta, J., Pink, D. et al. (2020). PEG-PLGA nanospheres loaded with nanoscintillators and photosensitizers for radiation-activated photodynamic therapy. *Acta Biomaterialia* 117: 335–348.

36 Wang, H., Lv, B., Tang, Z.M. et al. (2018). Scintillator-based nanohybrids with sacrificial electron prodrug for enhanced X-ray-induced photodynamic therapy. *Nano Letters* 18: 5768–5774.

37 Liu, S., Fang, L., Ding, H. et al. (2022). Alternative strategy to optimize cerium oxide for enhanced X-ray-induced photodynamic therapy. *ACS Nano* 16: 20805–20819.

38 Abliz, E., Collins, J.E., Bell, H., and Tata, D.B. (2011). Novel applications of diagnostic X-rays in activating a clinical photodynamic drug: photofrin II through X-ray induced visible luminescence from "rare-earth" formulated particles. *Journal of X-Ray Science and Technology* 19: 521–530.

39 Chen, M.H., Jenh, Y.J., Wu, S.K. et al. (2017). Non-invasive photodynamic therapy in brain cancer by use of Tb-doped LaF nanoparticles in combination with photosensitizer through X-ray irradiation: a proof-of-concept study. *Nanoscale Research Letters* 12: 62.

40 Ren, Y.F., Rosch, J.G., Landry, M.R. et al. (2021). Tb-Doped core-shell-shell nanophosphors for enhanced X-ray induced luminescence and sensitization of radiodynamic therapy. *Biomaterials Science* 9: 496–505.

41 Feng, Y.S., Liu, X.M., Li, Q.Q. et al. (2022). A scintillating nanoplatform with upconversion function for the synergy of radiation and photodynamic therapies for deep tumors. *Journal of Materials Chemistry C* 10: 688–695.

42 Park, W., Cho, S., Kang, D. et al. (2020). Tumor microenvironment targeting nano-bio emulsion for synergistic combinational X-Ray PDT with oncolytic bacteria therapy. *Advanced Healthcare Materials* 9: 1901812.

43 Ahmad, F., Wang, X.Y., Jiang, Z. et al. (2019). Codoping enhanced radioluminescence of nanoscintillators for X-ray-activated synergistic cancer therapy and prognosis using metabolomics. *ACS Nano* 13: 10419–10433.

44 Tang, Y.G., Hu, J., Elmenoufy, A.H., and Yang, X.L. (2015). Highly efficient FRET system capable of deep photodynamic therapy established on X-ray excited mesoporous LaF_3:Tb scintillating nanoparticles. *ACS Applied Materials & Interfaces* 7: 12261–12269.

45 Zhang, W.L., Zhang, X.F., Shen, Y.L. et al. (2018). Ultra-high FRET efficiency NaGdF: Tb-Rose Bengal biocompatible nanocomposite for X-ray excited photodynamic therapy application. *Biomaterials* 184: 31–40.

46 Elmenoufy, A.H., Tang, Y.A., Hu, J. et al. (2015). A novel deep photodynamic therapy modality combined with CT imaging established X-ray stimulated silica-modified lanthanide scintillating nanoparticles. *Chemical Communications* 51: 12247–12250.

47 Jain, A., Koyani, R., Muñoz, C. et al. (2018). Magnetic-luminescent cerium-doped gadolinium aluminum garnet nanoparticles for simultaneous imaging and photodynamic therapy of cancer cells. *Journal of Colloid and Interface Science* 526: 220–229.

48 Jain, A., Koyani, R., Munoz, C. et al. (2018). Blue light triggered generation of reactive oxygen species from silica coated $Gd_3Al_5O_{12}$:Ce^{3+} nanoparticles loaded with rose Bengal. *Data Brief* 20: 1023–1028.

49 Dou, Y., Liu, Y., Zhao, F. et al. (2018). Radiation-responsive scintillating nanotheranostics for reduced hypoxic radioresistance under ROS/NO-mediated tumor microenvironment regulation. *Theranostics* 8: 5870–5889.

50 Chen, H.M., Wang, G.D., Chuang, Y.J. et al. (2015). Nanoscintillator-mediated X-ray inducible photodynamic therapy for in vivo cancer treatment. *Nano Letters* 15: 2249–2256.

51 Isikawa, M. and Guidelli, E. (2022). Microfluidic synthesis of theranostic nanoparticles with near-infrared scintillation: toward next-generation dosimetry in X-ray-induced photodynamic therapy. *ACS Applied Materials & Interfaces* 14: 324–336.

52 Hsu, C.C., Lin, S.L., and Chang, C.A. (2018). Lanthanide-doped core-shell-shell nanocomposite for dual photodynamic therapy and luminescence imaging by a single X-ray excitation source. *ACS Applied Materials & Interfaces* 10: 7859–7870.

53 Osakada, Y., Pratx, G., Sun, C. et al. (2014). Hard X-ray-induced optical luminescence via biomolecule-directed metal clusters. *Chemical Communications* 50: 3549–3551.

54 Lin, L.S., Yang, X.Y., Zhou, Z.J. et al. (2017). Yolk-shell nanostructure: an ideal architecture to achieve harmonious integration of magnetic-plasmonic hybrid theranostic platform. *Advanced Materials* 29: 1606681.
55 Ngwa, W., Kumar, R., Sridhar, S. et al. (2014). Targeted radiotherapy with gold nanoparticles: current status and future perspectives. *Nanomedicine-Uk* 9: 1063–1082.
56 Dou, Y., Guo, Y.Y., Li, X.D. et al. (2016). Size-tuning ionization to optimize gold nanoparticles for simultaneous enhanced CT imaging and radiotherapy. *ACS Nano* 10: 2536–2548.
57 Bi, H.T., He, F., Dong, Y.S. et al. (2018). Bismuth nanoparticles with "light" property served as a multifunctional probe for X-ray computed tomography and fluorescence imaging. *Chemistry of Materials* 30: 3301–3307.
58 Yu, N., Wang, Z.J., Zhang, J.L. et al. (2018). Thiol-capped Bi nanoparticles as stable and all-in-one type theranostic nanoagents for tumor imaging and thermoradiotherapy. *Biomaterials* 161: 279–291.
59 Zhou, R.Y., Wang, H.M., Yang, Y.F. et al. (2019). Tumor microenvironment-manipulated radiocatalytic sensitizer based on bismuth heteropolytungstate for radiotherapy enhancement. *Biomaterials* 189: 11–22.
60 Yong, Y., Zhang, C.F., Gu, Z.J. et al. (2017). Polyoxometalate-based radiosensitization platform for treating hypoxic tumors by attenuating radioresistance and enhancing radiation response. *ACS Nano* 11: 7164–7176.
61 Jiang, H., Wang, R., Zhang, Q. et al. (2022). A dual-functional nanoplatform based on NIR and green dual-emissive persistent luminescence nanoparticles for X-ray excited persistent luminescence imaging and photodynamic therapy. *Nanoscale* 14: 15451–15461.
62 Jiang, X., Gao, X., Li, L. et al. (2023). Enhancement of light and X-ray charging in persistent luminescence nanoparticle scintillators $Zn_2SiO_4:Mn^{2+}$, Yb^{3+}, Li^{+}. *ACS Applied Materials & Interfaces* 15: 21228–21238.
63 Dong, Y.Z., Du, T.Y., Wang, M.Y., and Han, X.X. (2023). X-ray-excited persistent luminescence of Tb^{3+}-doped fluoride nanoparticles for photodynamic therapy. *Acs Applied Nano Materials* 6: 23047–23052.
64 Xue, Z.L., Li, X.L., Li, Y.B. et al. (2017). X-ray-activated near-infrared persistent luminescent probe for deep-tissue and renewable in vivo bioimaging. *ACS Applied Materials & Interfaces* 9: 22132–22142.
65 Chen, H.M., Sun, X.L., Wang, G.D. et al. (2017). LiGaO:Cr-based theranostic nanoparticles for imaging-guided X-ray induced photodynamic therapy of deep-seated tumors. *Materials Horizons* 4: 1092–1101.
66 Ou, X.Y., Qin, X., Huang, B.L. et al. (2021). High-resolution X-ray luminescence extension imaging. *Nature* 590: 410–415.
67 Yang, W.S., Read, P.W., Mi, J. et al. (2008). Semiconductor nanoparticles as energy mediators for photosensitizer-enhanced radiotherapy. *International Journal of Radiation Oncology Biology Physics* 72: 633–635.
68 Guidelli, E.J., Lignos, I., Yoo, J.J. et al. (2018). Mechanistic insights and controlled synthesis of radioluminescent ZnSe quantum dots using a microfluidic reactor. *Chemistry of Materials* 30: 8562–8570.

69 Chen, Q.S., Wu, J., Ou, X.Y. et al. (2018). All-inorganic perovskite nanocrystal scintillators. *Nature* 561: 88–93.

70 Liu, J., Hu, F., Wu, M. et al. (2021). Bioorthogonal coordination polymer nanoparticles with aggregation-induced emission for deep tumor-penetrating radio- and radiodynamic therapy. *Advanced Materials* 33: e2007888.

71 Kirakci, K., Kubát, P., Fejfarová, K. et al. (2016). X-ray inducible luminescence and singlet oxygen sensitization by an octahedral molybdenum cluster compound: a new class of nanoscintillators. *Inorganic Chemistry* 55: 803–809.

72 Kirakci, K., Zelenka, J., Rumlová, M. et al. (2018). Octahedral molybdenum clusters as radiosensitizers for X-ray induced photodynamic therapy. *Journal of Materials Chemistry B* 6: 4301–4307.

73 Nam, J.S., Kang, M.G., Kang, J. et al. (2016). Endoplasmic reticulum-localized iridium(III) complexes as efficient photodynamic therapy agents via protein modifications. *Journal of the American Chemical Society* 138: 10968–10977.

74 Evtushok, D.V., Melnikov, A.R., Vorotnikova, N.A. et al. (2017). A comparative study of optical properties and X-ray induced luminescence of octahedral molybdenum and tungsten cluster complexes. *Dalton Transactions* 46: 11738–11747.

75 Lu, K.D., He, C.B., and Lin, W.B. (2014). Nanoscale metal-organic framework for highly effective photodynamic therapy of resistant head and neck cancer. *Journal of the American Chemical Society* 136: 16712–16715.

76 Lu, K.D., Aung, T., Guo, N.N. et al. (2018). Nanoscale metal-organic frameworks for therapeutic, imaging, and sensing applications. *Advanced Materials* 30: 1707634.

77 He, C.B., Liu, D.M., and Lin, W.B. (2015). Nanomedicine applications of hybrid nanomaterials built from metal-ligand coordination bonds: nanoscale metal-organic frameworks and nanoscale coordination polymers. *Chemical Reviews* 115 (19): 11079–11108.

78 Lan, G.X., Ni, K.Y., Veroneau, S.S. et al. (2018). Nanoscale metal-organic layers for radiotherapy-radiodynamic therapy. *Journal of the American Chemical Society* 140 (49): 16971–16975.

79 Wang, C., Volotskova, O., Lu, K.D. et al. (2014). Synergistic assembly of heavy metal clusters and luminescent organic bridging ligands in metal-organic frameworks for highly efficient X-ray scintillation. *Journal of the American Chemical Society* 136: 6171–6174.

80 Lan, G.X., Ni, K.Y., Xu, R.Y. et al. (2017). Nanoscale metal-organic layers for deeply penetrating X-ray-induced photodynamic therapy. *Angewandte Chemie International Edition* 56: 12102–12106.

81 Takizawa, S.Y., Aboshi, R., and Murata, S. (2011). Photooxidation of 1,5-dihydroxynaphthalene with iridium complexes as singlet oxygen sensitizers. *Photochemical & Photobiological Sciences* 10: 895–903.

82 GarciaFresnadillo, D., Georgiadou, Y., Orellana, G. et al. (1996). Singlet-oxygen ((1)Delta(g)) production by ruthenium(II) complexes containing polyazaheterocyclic ligands in methanol and in water. *Helvetica Chimica Acta* 79: 1222–1238.

83 Wang, Y.X., Yin, X.M., Liu, W. et al. (2018). Emergence of uranium as a distinct metal center for building intrinsic X-ray scintillators. *Angewandte Chemie International Edition* 57: 7883–7887.

84 Rao, A., Chow, P.C.Y., Gélinas, S. et al. (2013). The role of spin in the kinetic control of recombination in organic photovoltaics. *Nature* 500: 435.

85 Thompson, N.J., Wilson, M.W.B., Congreve, D.N. et al. (2014). Energy harvesting of non-emissive triplet excitons in tetracene by emissive PbS nanocrystals. *Nature Materials* 13: 1039–1043.

86 Wang, X., Shi, H., Ma, H. et al. (2021). Organic phosphors with bright triplet excitons for efficient X-ray-excited luminescence. *Nature Photonics* 15: 187–192.

87 Wang, X., Sun, W., Shi, H. et al. (2022). Organic phosphorescent nanoscintillator for low-dose X-ray-induced photodynamic therapy. *Nature Communications* 13: 5091.

88 Yu, Y., Xiang, L., Zhang, X. et al. (2023). Pure organic AIE nanoscintillator for X-ray mediated type I and type II photodynamic therapy. *Advanced Science* 10: 2302395.

89 Zhang, H.J., Sun, Q.J., Tong, L.L. et al. (2018). Synergistic combination of PEGylated selenium nanoparticles and X-ray-induced radiotherapy for enhanced anticancer effect in human lung carcinoma. *Biomedicine & Pharmacotherapy* 107: 1135–1141.

90 Chan, L., Gao, P., Zhou, W.H. et al. (2018). Sequentially triggered delivery system of black phosphorus quantum dots with surface charge-switching ability for precise tumor radiosensitization. *Acs Nano* 12: 12401–12415.

91 Chang, Y.Z., He, L.Z., Li, Z.B. et al. (2017). Designing core-shell gold and selenium nanocomposites for cancer radiochemotherapy. *ACS Nano* 11: 4848–4858.

92 Lin, Y., Wu, Y., Wang, R. et al. (2018). Two-dimensional tellurium nanosheets for photoacoustic imaging-guided photodynamic therapy. *Chemical Communications* 54: 8579–8582.

93 Huang, W., Huang, Y.Y., You, Y.Y. et al. (2017). High-yield synthesis of multifunctional tellurium nanorods to achieve simultaneous chemo-photothermal combination cancer therapy. *Advanced Functional Materials* 27: 1701388.

94 Yang, T., Ke, H.T., Wang, Q.L. et al. (2017). Bifunctional tellurium nanodots for photo-induced synergistic cancer therapy. *Acs Nano* 11: 10012–10024.

95 Foote, C.S. (1991). Definition of type I and type II photosensitized oxidation. *Photochemistry and Photobiology* 54: 659.

96 Baptista, M.S., Cadet, J., Ghogare, A.A. et al. (2017). Type I and type II photosensitized oxidation reactions: guidelines and mechanistic pathways. *Photochemistry and Photobiology* 93: 912–919.

97 Shrestha, S., Wu, J., Sah, B. et al. (2019). X-ray induced photodynamic therapy with copper-cysteamine nanoparticles in mice tumors. *Proceedings of the National Academy of Sciences of the United States of America* 116: 16823–16828.

98 Sun, W.J., Luo, L., Feng, Y.S. et al. (2020). Aggregation-induced emission gold clustoluminogens for enhanced low-dose X-ray-induced photodynamic therapy. *Angewandte Chemie International Edition* 59: 9914–9921.

99 Zhang, X.F., Lan, B., Wang, S.C. et al. (2019). Low-dose X-ray excited photodynamic therapy based on NaLuF:Tb-rose bengal nanocomposite. *Bioconjugate Chemistry* 30: 2191–2200.

100 Song, L., Li, P.P., Yang, W. et al. (2018). Low-dose X-ray activation of W(VI)-doped persistent luminescence nanoparticles for deep-tissue photodynamic therapy. *Advanced Functional Materials* 28: 1707496.

101 Fan, W.P., Bu, W.B., Zhang, Z. et al. (2015). X-ray radiation-controlled NO-release for on-demand depth-independent hypoxic radiosensitization. *Angewandte Chemie International Edition* 14026–14030.

102 Jiang, M.Y., Xue, Z.L., Li, Y.B. et al. (2020). A soft X-ray activated lanthanide scintillator for controllable NO release and gas-sensitized cancer therapy. *Nanoscale Horizons* 5: 268–273.

103 Xue, Z.L., Jiang, M.Y., Liu, H.R. et al. (2020). Low dose soft X-ray-controlled deep-tissue long-lasting NO release of persistent luminescence nanoplatform for gas-sensitized anticancer therapy. *Biomaterials* 263: 120384.

104 Du, Z., Zhang, X., Guo, Z. et al. (2018). X-ray-controlled generation of peroxynitrite based on nanosized LiLuF:Ce scintillators and their applications for radiosensitization. *Advanced Materials* 30: 1804046.

105 Klein, S., Harreiss, C., Menter, C. et al. (2018). NOBF-functionalized Au-FeO nanoheterodimers for radiation therapy: synergy effect due to simultaneous reactive oxygen and nitrogen species formation. *ACS Applied Materials & Interfaces* 10: 17071–17080.

106 Wang, Y.P., Duan, X.H., Huang, Y.H. et al. (2023). Radio- and photosensitizing Os(II)-based nanocage for combined radio-/chemo-/X-ray-induced photodynamic therapies, NIR imaging, and drug delivery. *ACS Applied Materials & Interfaces* 15: 43479–43491.

107 Sun, W.J., Shi, T.H., Luo, L. et al. (2019). Monodisperse and uniform mesoporous silicate nanosensitizers achieve low-dose X-ray-induced deep-penetrating photodynamic therapy. *Advanced Materials* 31: 1808024.

108 Liu, J., Zhang, J., Huang, F. et al. (2020). X-ray and NIR light dual-triggered mesoporous upconversion nanophosphor/Bi heterojunction radiosensitizer for highly efficient tumor ablation. *Acta Biomaterialia* 113: 570–583.

109 Lu, K.D., He, C.B., Guo, N.N. et al. (2018). Low-dose X-ray radiotherapy-radiodynamic therapy via nanoscale metal-organic frameworks enhances checkpoint blockade immunotherapy. *Nature Biomedical Engineering* 2: 600–610.

110 Ni, K.Y., Lan, G.X., Chan, C. et al. (2019). Ultrathin metal-organic-layer mediated radiotherapy-radiodynamic therapy. *Matter* 1: 1331–1353.

111 Pan, P., Dong, X., Chen, Y. et al. (2022). Engineered bacteria for enhanced radiotherapy against breast carcinoma. *ACS Nano* 16: 801–812.

16

Conclusions and Perspectives

Chunxia Li[1] *and Jun Lin*[2]

[1] *Shandong University, School of Chemistry and Chemical Engineering, Institute of Frontier Chemistry, Binhai Road, Qingdao 266237, P. R. China*
[2] *Changchun Institute of Applied Chemistry, Chinese Academy of Sciences, State Key Laboratory of Rare Earth Resource and Utilization, Changchun 130022, P. R. China*

Photofunctional nanomaterials have displayed the intriguing potentials and made substantial advances in the diverse biomedical fields. There is ample reason to believe that the applications of photofunctional nanomaterials in biomedicine will continue to expand and deepen in the future. The following are some directions that will needed to be developed and researched deeply.

(1) *Improving targeting and biocompatibility*: Researchers will continue to improve the surface modification and functionalization of photofunctional nanomaterials and optimize the biodegradability and immunoreactivity of existing materials to improve their targeting and biocompatibility in organisms. This will help to reduce damage to normal tissues and improve therapeutic efficacy.
(2) *Design for integrated multifunctionality*: Future research will continue to work on the design of multifunctional photofunctional nanomaterials to integrate multiple therapeutic and imaging functions. For example, combining imaging, therapeutic, and sensing functions to achieve one-stop treatment of diseases.
(3) *Integration of nanomaterials with gene therapy*: Nanomaterials can be used as gene carriers to combine gene therapy with photofunctional nanomaterials. This can be achieved by combining gene carriers with fluorescent nanomaterials or photothermal nanomaterials for the combined application of gene delivery and photofunctional therapy. This approach is expected to provide more effective means of gene therapy and light-controlled gene expression and regulation.
(4) *Biosensing and diagnostics*: Photofunctional nanomaterials also have great potential for biosensing and diagnostic applications. By combining these materials with specific biomolecules or cells, highly sensitive and selective biosensing can be achieved. This will contribute to early disease diagnosis, efficacy monitoring, and personalized medicine.

Photofunctional Nanomaterials for Biomedical Applications, First Edition.
Edited by Chunxia Li and Jun Lin.

(5) *Intelligence and control*: Future research will aim to achieve intelligence and controllability of photofunctional nanomaterials. By designing responsive material structures (e.g. capable of sensing changes in the microenvironment of an organism) or introducing response mechanisms to external stimuli, precise regulation of the behavior of photofunctional nanomaterials can be achieved. This will provide further opportunities for precision medicine and personalized treatment.
(6) *Personalized medicine*: With the development of precision medicine, the role of photofunctional nanomaterials in individualized treatment plans will become increasingly important, such as personalized drug release through targeted delivery of specific biomarkers.
(7) *Promoting clinical applications*: As the technology develops and clinical validation progresses, it is expected that photofunctional nanomaterials will be more widely used in clinical medicine. This will require more clinical trials and standardized production and preparation processes to ensure the safety and efficacy of the materials.
(8) *Interdisciplinary collaboration*: Close collaboration between chemistry, physics, biology, materials science, and medicine will accelerate the innovative application of photofunctional nanomaterials in biomedicine.

In conclusion, the application of photofunctional nanomaterials in biomedicine is promising. With in-depth research on these materials and continuous technological advancement, we believe they will bring more innovations and breakthroughs in the biomedical field, provide more effective solutions for the treatment and diagnosis of diseases, and ultimately make significant contributions to human health.

Index

Photofunctional Nanomaterials for Biomedical Applications, First Edition.
Edited by Chunxia Li and Jun Lin.

u

www.ingramcontent.com/pod-product-compliance
Lightning Source LLC
Chambersburg PA
CBHW080930240926
51849CB00003BA/428
* 9 7 8 3 5 2 7 3 5 3 3 3 0 *